The Solid Earth

AN INTRODUCTION TO GLOBAL GEOPHYSICS

C. M. R. FOWLER
University of Saskatchewan

Published by the Press Syndicate of the University of Cambridge
The Pitt Building, Trumpington Street, Cambridge CB2 1RP
40 West 20th Street, New York, NY 10011-4211, USA
10 Stamford Road, Oakleigh, Victoria 3166, Australia

First published 1990
Reprinted 1992 (twice), 1993

Printed in Canada

Library of Congress Cataloging-in-Publication Data

Fowler, C. M. R.
The solid earth: an introduction to global geophysics/C. M. R. Fowler
p. cm
Includes bibliographical references.
ISBN 0-521-37025-6. ISBN 0-521-38590-3 (pbk.)
1. Geophysics. 2. Earth. I. Title.
QC806.F625 1990
551 - dc20 89-17443
CIP

British Library Cataloguing in Publication Data

Fowler, C. M. R.
The solid earth.
1. Geophysics
I. Title
551

ISBN 0-521-37025-6 hardback
ISBN 0-521-38590-3 paperback

TO MY FAMILY

Magna opera domini exoquisita in omnes voluntates ejus.

The works of the Lord are great, sought out of all them that have pleasure therein.

Psalm 111 v. 2: at the entrances to the Cavendish Laboratories, Cambridge.

Contents

Preface

Geophysics is a diverse science. At its best it has the rigour of physics and the vigour of geology. Its subject is the earth. How does the earth work? What is its composition? How has it changed? Thirty years ago many of the answers to these questions were uncertain. We knew the gross structure of our planet and that earthquakes occurred, volcanoes erupted and high mountains existed, but we did not understand why. Today we have a general knowledge of the workings of the planet, although there is still much to be discovered.

My aim in writing this book was to convey in a fairly elementary way what we know of the structure and dynamics of the solid earth. The fabric of geophysics has changed dramatically in the decades since the discovery of plate tectonics. The book places a strong emphasis on geophysical research since the initial formulation of plate theory, and the discussion centres on the crust and upper mantle. It also outlines the recent increases in our knowledge of the planet's deeper interior.

To whom is this book addressed? It is designed to serve as an introduction to geophysics for senior undergraduates in geology or physics, and for graduate students in either subject who need to learn the elements of geophysics. My hope is that the book will give them a fairly comprehensive basis on which to build an understanding of the solid earth.

Part of the challenge in writing a geophysics text is to make the book accessible to both types of student. For instance, some students enter the study of geophysics from a background in the earth sciences, others from physics or mathematics: only a few enroll directly in geophysics programs. Geology students tend to know about rocks and volcanoes, but possess only the basics of calculus. In contrast, students of physics have good mathematical skills, but do not know the difference between a basalt and a granite. I have attempted throughout the book to explain for the geologists the mathematical methods and derivations and to include worked examples as well as questions. I hope that this will make the book useful to students who have only introductory calculus. For the nongeologists, I have tried to limit or explain the abundant geological terminology. There is a glossary of terms, to rescue physics students lost in the undergrowth of nomenclature.

For more advanced students of either geological or physical training I have in places included more mathematical detail than is necessary for a basic introductory course. This detail can easily be by-passed without either interrupting the continuity of the text or weakening the understanding of less mathematical students. Throughout the book I have attempted to give every step of logic so that students can understand why every equation and each conclusion is valid.

In general, I have tried to avoid the conventional order of textbooks, in which geophysical theory comes first, developed historically, followed in later sections by interesting and concrete examples. For instance, because the book focuses to a large extent on plate tectonic theory, which is basic to the study of the crust and mantle, this theory is introduced in its proper geophysical sense, with a discussion of rotation, motions on plate boundaries and absolute plate motions. Most geological texts avoid discussing this, relying instead on two-dimensional cartoons of ridges and subduction zones. I have met many graduate students who have no idea what a rotation pole is. Their instructors thought the knowledge irrelevant. Yet understanding tectonics on a sphere is crucial to geophysics because one cannot fully comprehend plate motion without it.

The next chapters of the book are concerned with past plate motions, magnetics, seismology and gravity. These are the tools with which plate tectonics was discovered. The exposition is not historical, although historical details are given. The present generation of geophysicists learned by error and discovery, but the next generation will begin with a complete structure on which to build their own inventions.

These chapters are followed by discussions of radioactivity and heat. The earth is a heat engine, and the discovery of radioactivity radically changed our appreciation of the physical aspects of the planet's history, thermal evolution and dynamics. The study of isotopes in the earth is now, perhaps unfairly, regarded as an area of geochemistry rather than of geophysics; nevertheless, the basic tools of dating, at least, should be part of any geophysics course. Understanding heat, on the other hand, is central to geophysics and fundamental to our appreciation of the living planet. All geology and geophysics, indeed the existence of life itself, depend on the earth's thermal behaviour. Heat is accordingly discussed in some detail.

The final chapters use the knowledge built up in the earlier ones to create an integrated picture of the complex operation of the oceanic and continental lithosphere, its growth and deformation. The workshops of geology – ridges and subduction zones – are described from both geophysical and petrological viewpoints. Sedimentary basins and continental margins employ most of the world's geophysicists. It is important that those who explore the wealth or perils of these regions know the broader background of their habitat.

SI units have been used except in cases where other units are clearly more appropriate. Relative plate motions are quoted in centimetres per year ($cm\,yr^{-1}$), not in metres per second. Geological time and ages are quoted in millions or billions of years (Ma or Ga) instead of seconds. Temperatures are quoted in degrees Centigrade (°C), not in Kelvin. Seismic velocities are in kilometres per second.

Most geophysicists look for oil. Some worry about earthquakes or landslips, or advise governments. Some are research workers or teach at universities. Uniting this diversity is a deep interest in the earth. Geophysics is a rigorous scientific discipline, but it is also interesting and fun. The student reader to whom this book is addressed will need rigour and discipline and often hard work, but the reward is an understanding of our planet. It is worth it.

Acknowledgements

Textbooks are not easy to write. They need to have an author, of course, but much of the work is done by an array of encouragers, teachers and critics. Lady Jeffreys introduced me to Sir Edward Bullard, who made geophysics sound fun. To Drummond Matthews, Dan McKenzie and Brian Kennett, as well as to those who were students with me, I owe much. Ships and rocks are much more exciting than a life crouched over computers and equations.

In writing this book I have been helped by many people. P. H. Fowler, D. P. McKenzie, E. G. Nisbet, A. Prugger, C. Sammis and P. J. Smith, together with several anonymous reviewers, read most or all of the manuscript, making many useful suggestions and pointing out errors. Since the manuscript was long the task was large. I am very grateful. For detailed critiques of individual chapters I am indebted to M. J. Bickle, P. van Calsteren, S. R. Fowler, D. Gubbins, E. Hegner, J. A. Jacobs, T. K. Kyser, J. B. Merriam, B. I. Pandit, J. A. Pearce, G. M. Purdy, G. Quinlan, R. S. White and P. J. Wyllie. The expertise and solid criticism of these kind people helped greatly in my quest for accuracy in so many fields in which I am not a specialist. Many people, including T. Atwater, D. R. Barraclough, R. M. Clowes, R. G. Coleman, A. M. Dziewonski, Sir Charles Frank, R. G. Gordon, A. G. Green, N. B. W. Harris, W. Haxby, E. Irving, J. A. Jackson, C. E. Keen, S. Klemperer, K. D. Klitgord, R. A. Langel, R. D. Lindwall, H. Nevanlinna, N. W. Peddie, J. A. Percival, G. M. Purdy, M. P. Ryan, J. G. Sclater, C. R. Scotese, A. G. Smith, J. F. Sweeney, J. Verrall and M. L. Zoback, have given advice on specific details, and sent prints of their figures or preprints and reprints of their articles. I am very appreciative of their help. The geodynamics lectures given in Cambridge by Geoff King provided the basis for my presentation of relative plate motions in Chapter 2. Most of all, I wish to acknowledge the generous and painstaking help of Walter Pilant, who read the text not once but twice, and pointed out much error and infelicity. Despite all this help there are bound to be mistakes. For these I apologize; they are entirely my fault, either as errors of understanding or as errors of printing. My hope is that they will be few.

I should like to thank W. G. E. Caldwell and H. E. Hendry, who as respective chairmen of the Department of Geological Sciences at the University of Saskatchewan provided the facilities and encouragement that enabled me to undertake this project. A. Heppner showed immense tolerance and patience in typing many versions of the enormous manuscript, while A. C. Williamson made it possible for me to write. Lindsay Embree drew many of the figures. Peter-John Leone's enthusiasm and the diligent, careful work of Mary Nevader and Glenn Cochran created a book from a pile of paper. Thank you.

I thank my children for their forbearance over the past few years. In fact, without a family I would probably not have had the opportunity and time to undertake the writing of this book; I would have been too busy doing other things such as writing papers and sitting on committees. Finally, I thank Euan for constant encouragement, help and advice.

Sources

I thank the following for granting permission to reprint material:

Academic Press, Figures 4.28, 4.29, 4.30, 8.1, 9.2 and 9.14
Adam Hilger, Problems 5.7, 5.9, 5.10 and 7.21
American Association for the Advancement of Science, Figures 3.8, 3.10 and 3.13
American Association of Petroleum Geologists, Figures 4.56 and 9.27
American Geophysical Union, Figures 2.9, 2.10, 3.9, 3.12, 3.22, 3.23, 4.20, 4.22, 4.32, 4.33, 6.6, 7.6, 7.7, 7.8, 7.10, 7.11, 7.12, 7.13, 7.14, 7.15, 7.16, 7.19, 8.5, 8.6, 8.7, 8.9, 8.11, 8.12, 8.16, 8.26, 8.30, 8.33, 8.36, 8.39, 8.41, 8.45, 8.47, 8.48, 8.53, 9.4, 9.13, 9.17, 9.18, 9.37, 9.47, 9.49, 9.50, 9.51 and 9.53.
American Institute of Physics, Figure 7.16
American Scientist, Figure 2.19
Annual Reviews Inc., Figures 5.11, 5.15, 7.26 and 9.29
Birkhauser Verlag AG, Figures 4.17 and 4.42
M. H. P. Bott, Figures 5.6 and 7.23
Canadian Society of Petroleum Geologists, Figure 9.21
Cornell University, Figure 8.37
Elsevier Science Publishers, Figures, 3.15, 3.21, 4.36, 5.14, 7.18, 8.10, 8.27, 8.32, 8.46, 9.1, 9.2, 9.6, 9.34 and 9.36
W. H. Freeman, Figure 4.4
Geological Association of Canada, Figures 7.3, 7.4, 7.29, 7.30, 7.31, 7.32, 7.33, 7.34, 7.35, 7.36 and 7.37
Geological Society of America, Figures 3.7, 3.24, 3.25, 3.26, 3.29, 8.4, 9.22, 9.48, 9.51 and 9.56
Geological Society of London, Figures 8.5 and 8.29
Geological Survey of Canada (reprinted with permission of the Minister of Supply and Services, Canada, 1989), Figure 8.48
C. J. Hawkesworth, Figure 9.8
W. Haxby, Figure 5.10 and cover
Jones and Bartlett Publishers, Boston, Figure 4.1
Copyright © 1981 by William Kaufmann Inc., all rights reserved, Figure 4.14
D. C. King-Hele, Figure 5.3
Kluwer Academic Publishers, Figures 3.30, 4.45, 8.27, 8.31, 8.42 and 9.6
Lunar and Planetary Institute Houston, Figure 9.54
McGraw-Hill Book Company, Figure 8.1

National Research Council of Canada, Problem 7.1, Figures 9.30, 9.33, 9.37 and 9.55

Nature, MacMillan Journals Ltd, Figures 2.4, 2.15, 2.16, 2.20, 4.57, 5.9, 8.14, 8.21, 8.22, 8.23, 8.28, 9.12, 9.16, 9.23, 9.39, 9.42, 9.43, 9.45 and 9.57

H. Nevanlinna, Figure 3.1

The Open University, Figure 7.24

Pergamon Press PLC, Figure 9.9

Princeton University Press, Figures 5.18 and 5.19

Royal Astronomical Society, Figures 3.17, 3.18, 4.10, 4.15, 7.20, 7.27, 8.7, 8.11, 8.12, 8.15, 8.17, 8.18, 8.24, 8.30, 8.38, 8.40, 8.43, 9.10, 9.11, 9.18, 9.24, 9.30, 9.31, 9.32, 9.45, 9.46 and 9.58

The Royal Society, Figures 3.3 and 8.47

Seismological Society of America, Figures 2.1 and 4.11

Society of Economic Palaeontologists and Mineralogists Pacific Section, Figure 3.27

Society of Exploration Geophysicists, Figures 4.44, 4.48 and 4.49

Springer-Verlag, Figures 4.2, 4.5, 4.6, 6.1, 7.28 and 8.20

Swiss Geological Society, Figures 9.20 and 9.21

United States Geological Survey, Figures 8.8, 8.51, 8.52 and 9.52

University of Chicago Press, © University of Chicago Press, Figures 7.28 and 9.7

John Wiley and Sons Inc., copyright © John Wiley, Figures 4.2, 5.17, 6.4, 6.7, 6.8 and 8.3

1

Introduction

Geophysics, the physics of the earth, is a huge subject which includes the physics of space and the atmosphere, of the oceans and of the interior of the planet. The heart of geophysics, though, is the theory of the solid earth. We now understand in broad terms how the earth's surface operates, and we have some notion of the workings of the deep interior. These processes and the tools by which they have been understood form the theme of this book. To the layperson, geophysics means many practical things. For Californians, it is earthquakes and volcanoes; for Texans and Albertans, it is oil exploration; for Africans, it is groundwater hydrology. The methods and practices of applied geophysics are not dealt with at length here because they are covered in many specialized textbooks. This book is about the earth, its structure and function from surface to centre.

Our search for an understanding of the planet goes back millennia to the ancient Hebrew writer of the Book of Job and to the Egyptians, Babylonians and Chinese. The Greeks first measured the earth, Galileo and Newton put it in its place, but the Victorians began the modern discipline of geophysics. They and their successors were concerned chiefly with understanding the *structure* of the earth, and they were remarkably successful. The results are summarized in the magnificent book *The Earth* by Sir Harold Jeffreys, which was first published in 1924. Since the Second World War the *function* of the earth's surface has been the focus of attention, especially since 1967 when geophysics was revolutionized by the discovery of *plate tectonics*, the theory that explains the function of the uppermost layers of the planet.

The rocks exposed at the surface of the earth are part of the *crust* (Fig. 1.1). This crustal layer, which is rich in silica, was identified by John Milne, Lord Rayleigh and Lord Rutherford. It is on average 35 km thick beneath continents and 7–8 km thick beneath oceans. Beneath this thin crust lies the *mantle*, which extends down some 2900 km to the earth's central *core*. The mantle (originally termed *Mantel* or 'coat' in German by Emil Wiechert in 1897, perhaps by analogy with Psalm 104) is both physically and chemically distinct from the crust, being rich in magnesium silicates. The crust has been derived from the mantle over the aeons by a series of melting and reworking processes. The boundary between the crust and mantle, which was delineated by Andrya Mohorovičić in 1909, is termed the Mohorovičić discontinuity, or *Moho* for short. The core of the earth was discovered by R. D. Oldham in 1906 and correctly delineated by

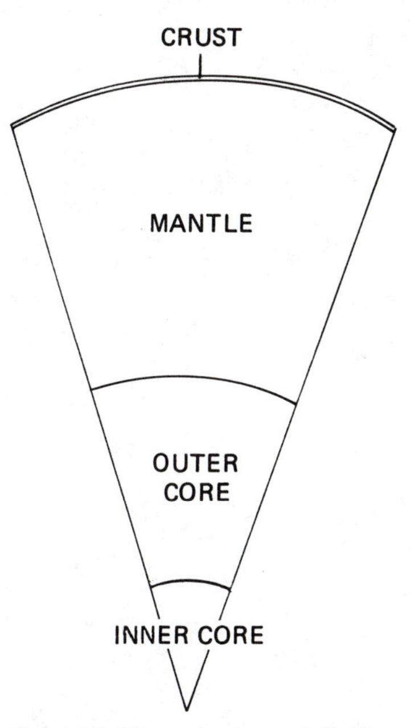

Figure 1.1. The major internal divisions of the earth. The crust and mantle are silicates, the outer core is an iron–oxygen or iron–sulphur alloy and the inner core is iron.

Beno Gutenberg in 1912 from studies of earthquake data. The core is totally different, both physically and chemically, from the crust and mantle. It is predominantly iron with lesser amounts of other elements. The core was established as being fluid in 1926 as the result of work on tides by Sir Harold Jeffreys. In 1929 a large earthquake conveniently occurred near Buller in the South Island of New Zealand. This, being on the other side of the earth from Europe, enabled Inge Lehmann, a Danish seismologist, to study the energy which had passed through the core. In 1936, on the basis of data from this earthquake she was able to show that the earth has an *inner core* within the liquid *outer core*. The inner core is solid.

The presence of ancient beaches and fossils of sea creatures in mountains thousands of feet above sea level was a puzzle and a stimulation to geologists from Pliny's time to the days of Leonardo and Hutton. On February 20, 1835, the young Charles Darwin was on shore resting in a wood near Valdivia, Chile, when suddenly the ground shook. In his journal *The Voyage of the Beagle* Darwin wrote: 'The earth, the very emblem of solidity, has moved beneath our feet like a thin crust over a fluid.' This was the great Concepción earthquake. Several days later, near Concepción, Darwin reported that 'Captain Fitz Roy found beds of putrid mussel shells still adhering to the rocks, ten feet above high water level: the inhabitants had formerly dived at low-water spring-tides for these shells.' The volcanoes erupted. The solid earth was active.

By the early twentieth century scientific opinion was that the earth had cooled from its presumed original molten state and the contraction which resulted from this cooling caused surface topography: the mountain ranges and the ocean basins. The well-established fact that many fossils, animals and plants found on separated continents must have had a common source was explained by either the sinking of huge continental areas to form the oceans (which is, and was then recognized to be, impossible) or the sinking beneath the oceans of land bridges which would have enabled the animals and plants to move from continent to continent.

In 1915 the German meteorologist Alfred Wegener published a proposal that the continents had slowly moved about. This theory of *continental drift*, which accounted for the complementarity of the shapes of coastlines on opposite sides of oceans and for the palaeontological, zoological and botanical evidence, was accepted by some geologists, particularly those from the southern hemisphere such as Alex Du Toit, but was generally not well received. Geophysicists quite correctly pointed out that it was physically impossible to move the continents through the solid rock which comprised the ocean floor. By the 1950s, however, work on the magnetism of continental rocks indicated that in the past the continents must have moved relative to each other; the *midocean ridges*, the earth's longest system of mountains, had been discovered, and continental drift was again under discussion. In 1962 the American geologist Harry H. Hess published an important paper on the workings of the earth. He proposed that continental drift had occurred by the process of *sea-floor spreading*. The midocean ridges marked the limbs of rising convection cells in the mantle. Thus, as the continents moved apart, new sea-floor material rose from the mantle along the midocean ridges to fill the vacant space. In the following decade the theory of plate tectonics, which was able successfully to account for the physical, geological and biological observations, was developed. This theory has become the unifying factor in the study of geology and geophysics. The main difference between plate tectonics and the early proposals of continental drift is that the continents are no longer thought of as ploughing through the oceanic rocks; instead, the oceanic rocks and the continents are together moving over the interior of the earth.

BIBLIOGRAPHY

Brush, S. J. 1980. Discovery of the earth's core. *Am. J. Phys.*, *48*, 705–24.

Darwin, C. R. 1845. *Journal of researches into the natural history and geology of the countries visited during the voyage of H.M.S. Beagle round the world, under the command of Capt. Fitz Roy R. N.*, 2nd ed. John Murray, London.

Du Toit, A. 1937. *Our wandering continents.* Oliver and Boyd, Edinburgh.

Gutenberg, B. 1913. Uber die Konstitution der Erdinnern, erschlossen aus Erdbebenbeobachtungen. *Phys. Zeit.*, *14*, 1217.

1914. Uber Erdbebenwellen, VIIA. Beobachtungen an Registrierungen von Fernbeben in Göttingen und Folgerungen über die Konstitution des Erdkörpers. *Nachr. Ges. Wiss. Göttingen, Math. Phys., Kl.1*, 1–52.

Hess, H. H. 1962. History of ocean basins. *In* A. E. J. Engel, H. L. James and B. F. Leonard, eds., Petrologic studies: A volume in honor of A. F. Buddington. *Geol. Soc. Am.*, 599–620.

Holmes, A. 1965. *Principles of physical geology.* Ronald Press, New York.

Jeffreys, H. 1926. The rigidity of the Earth's central core. *Mon. Not. R. Astron. Soc. Geophys. Suppl.*, *1*, 371–83. (Reprinted in H. Jeffreys, 1971. *Collected Papers, Vol. 1.* Gordon and Breach, New York.)

1976. *The earth.* 6th ed., Cambridge University Press, Cambridge.

Lehmann, I. 1936. *P′. Trav. Sci., Sect. Seis. U.G.G.I. (Toulouse)*, *14*, 3–31.

Milne, J. 1906. Bakerian Lecture – recent advances in seismology. *Proc. Roy. Soc. A*, *77*, 365–76.

Mohorovičić, A. 1909. Das Beben vom 8.X.1909. *Jahrbuch met. Obs. Zagreb*, 9, 1–63.

Oldham, R. D. 1906. The constitution of the earth as revealed by earthquakes. *Quart. J. Geol. Soc.*, *62*, 456–75.

Rutherford, E. 1907. Some cosmical aspects of radioactivity. *J. R. Astr. Soc.* Canada, May–June, 145–65.

Wegener, A. 1915. Die Entstehung der Kontinente und Ozeane.

1924. *The origin of continents and oceans.* Dutton, New York.

Wiechert, E. 1987. Uber die Massenvertheilungim Innern der Erde. *Nachr. Ges. Wiss. Göttingen*, 221–43.

General Books

Anderson, R. N. 1986. *Marine geology: A planet earth perspective.* Wiley, New York.

Brown, G. C., and Mussett, A. E. 1981. *The inaccessible earth.* Allen and Unwin, London.

Cattermole, P., and Moore, P. 1985. *The story of the earth.* Cambridge Univ. Press, Cambridge.

Clark, S. P. J. 1971. *Structure of the earth.* Prentice-Hall, Englewood-Cliffs, N.J.

Cloud, P. 1988. *Oasis in space: Earth history from the beginning.* Norton, New York.

Cole, G. H. A. 1986. *Inside a planet.* Hull Univ. Press, Hull, U.K.

van Andel, T. H. 1985. *New views on an old planet, continental drift and the history of earth.* Cambridge Univ. Press, Cambridge.

Wyllie, P. J. 1976. *The way the earth works.* Wiley, New York.

2

Tectonics on a Sphere

The Geometry of Plate Tectonics

2.1 Plate Tectonics

The earth has a cool and therefore mechanically strong outermost shell called the *lithosphere* (Greek *lithos*, 'rock'). The lithosphere is of the order of 100 km thick and comprises the crust and uppermost mantle. It is thinnest in the oceanic regions and thicker in continental regions, where its base is poorly understood. The *asthenosphere* (Greek *asthenia*, 'weak' or 'sick') is beneath the lithosphere. The high temperature and pressure which exist at the depth of the asthenosphere cause its viscosity to be low enough to allow viscous flow, on a geological time scale (millions of years, not seconds!). If the earth is viewed in purely mechanical terms, the mechanically strong lithosphere floats on the mechanically weak asthenosphere. Alternatively, if the earth is viewed as a heat engine, the lithosphere is an outer skin, through which heat is lost by conduction, and the asthenosphere is an interior shell through which heat is transferred by convection (Sect. 7.1).

The basic concept of *plate tectonics* is that the lithosphere is divided into a small number of nearly rigid *plates* (like curved caps on a sphere), which are moving over the asthenosphere. Most of the deformation which results from the plate motion – such as stretching, folding or shearing – takes place at the edge, or *boundary*, of a plate. Deformation inside the boundary is not significant.

A map of the *seismicity* (earthquake activity) of the earth (Fig. 2.1) outlines the plates very clearly because nearly all earthquakes as well as most of the earth's volcanism occur along the plate boundaries. These seismic belts are the zones in which differential movements between the nearly rigid plates occur. There are seven main plates, of which the largest is the Pacific plate, and numerous smaller plates such as Nazca, Cocos, and Scotia plates (Fig. 2.2).

The theory of plate tectonics, which describes the interactions of the plates and the consequences of these interactions, is based on several important assumptions:

1. The generation of new plate material occurs by seafloor spreading; that is, new oceanic lithosphere is generated along the active midocean ridges (see Chapters 3 and 8).
2. The new oceanic lithosphere, once created, forms part of a rigid plate; this plate may or may not include continental material.

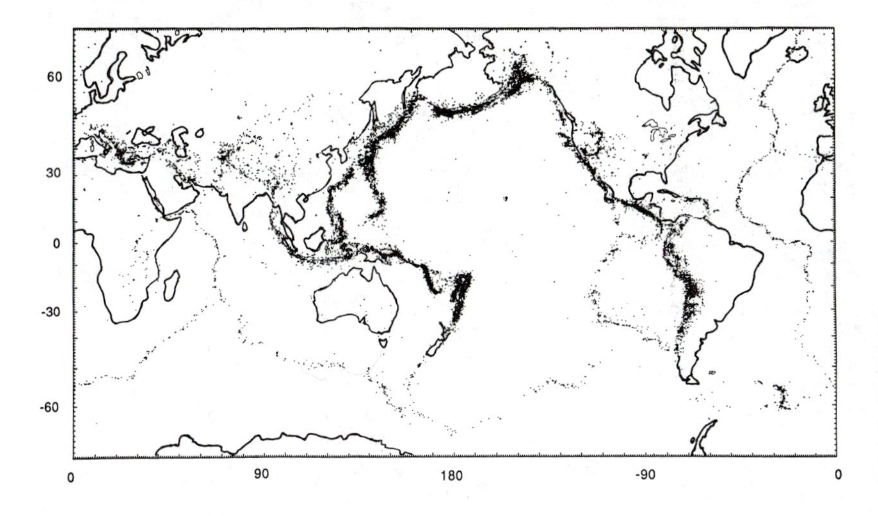

Figure 2.1. A computer plot of some 30,000 earthquakes which occurred between 1961 and 1967 at depths from 0 to 700 km. These earthquakes delineate the boundaries of the plates very well indeed. (From Barazangi and Dorman 1969.)

3. The earth's surface area remains constant; therefore, seafloor spreading must be balanced by consumption of plate elsewhere.
4. The lithospheric plates are capable of transmitting stresses over great horizontal distances without buckling; in other words, the relative motion between plates is taken up only along plate boundaries.

Plate boundaries are of three types:

1. Along *divergent* boundaries, also called accreting or constructive, plates are moving away from each other. At such boundaries new plate material, derived from the mantle, is added to the lithosphere. The divergent plate boundary is represented by the *midocean ridge system*, along the axis of which new plate material is produced (Fig. 2.3a).
2. Along *convergent* boundaries, also called consuming or destructive, plates are approaching each other. Most such boundaries are represented by the oceanic *trench, island arc* systems of *subduction zones* where one of the colliding plates descends into the mantle and is destroyed (Fig. 2.3c). The downgoing plate often penetrates the mantle to depths of about 700 km. Some convergent boundaries occur on land. Japan, the Aleutians and the Himalayas are the surface expression of convergent plate boundaries.
3. Along *conservative* boundaries, lithosphere is neither created nor destroyed. The plates move laterally relative to each other (Fig. 2.3e). These plate boundaries are represented by *transform faults*, of which the San Andreas fault in California, U.S.A. is a famous example. Transform faults can be grouped into six basic classes (Fig. 2.4). By far the most common type of transform fault is the ridge–ridge fault, which ranges from a few kilometres to hundreds of kilometres in length. Some very long ridge–ridge faults occur in the Pacific, equatorial Atlantic and southern oceans (Fig. 2.2, in which the modern distribution of types of plate boundary is also shown). Adjacent plates move relative to each other at rates up to about 15 cm yr^{-1}. The present-day rates for all the main plates are discussed in Section 2.4.

Although the plates are made up of both oceanic and continental

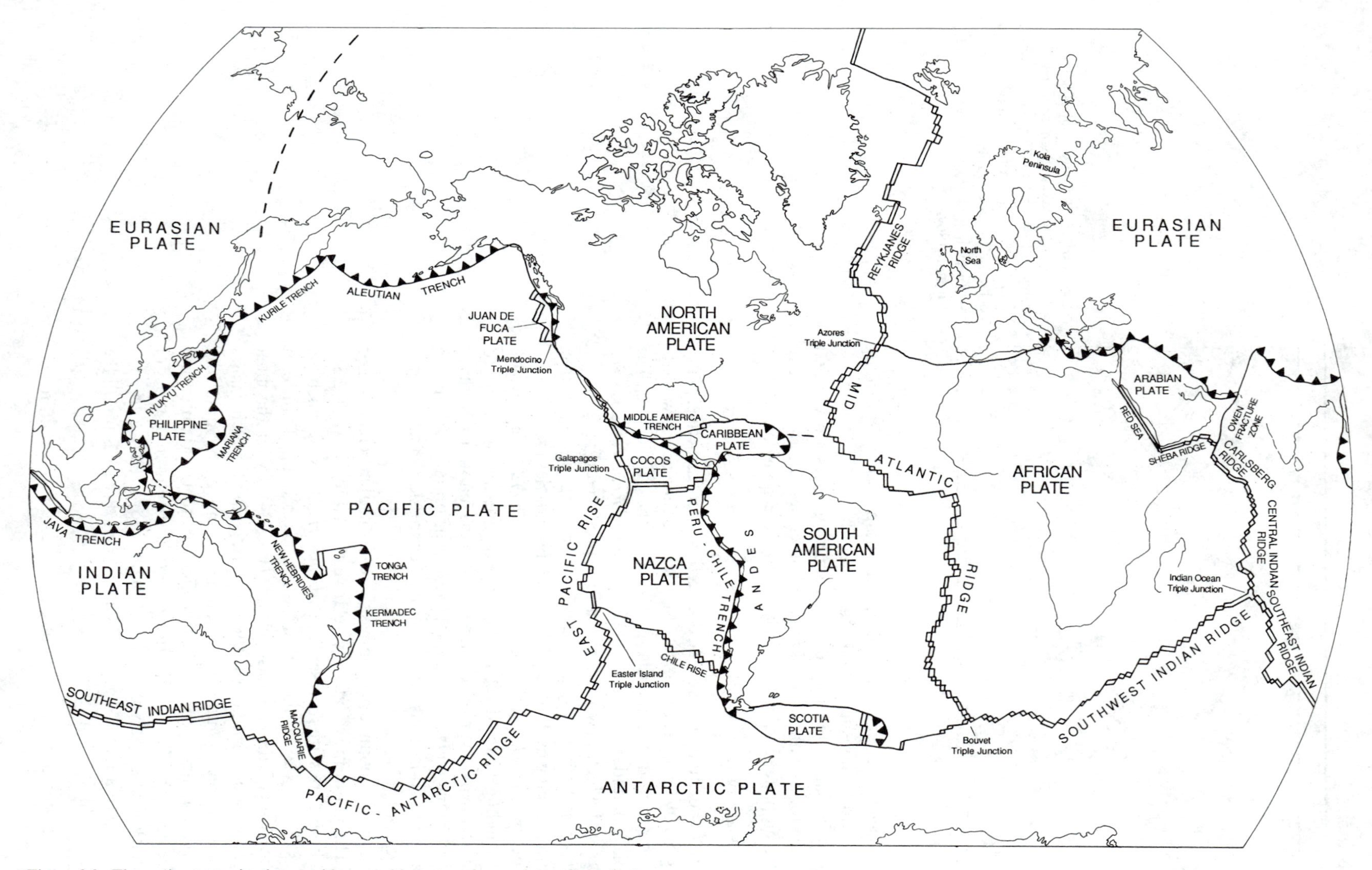

Figure 2.2. The major tectonic plates, midocean ridges, trenches and transform faults.

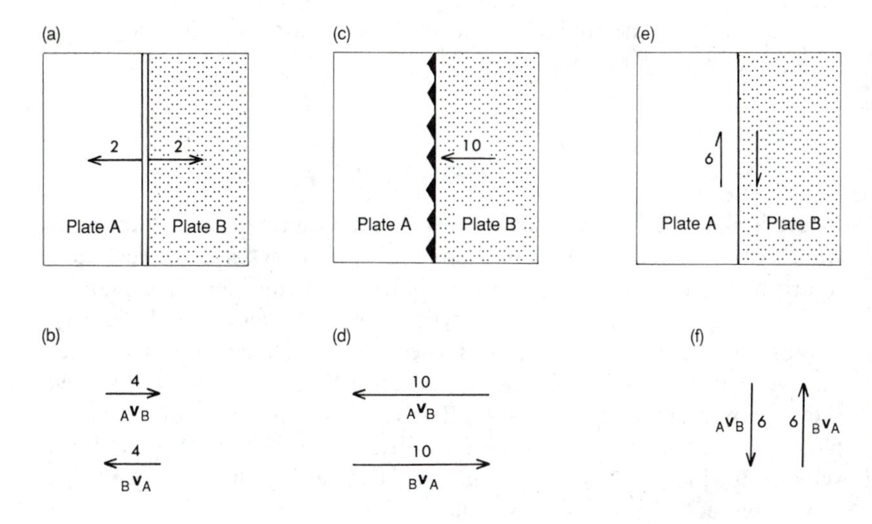

Figure 2.3. Three possible boundaries between plates A and B. (a) A *constructive* boundary (midocean ridge). The double line is the symbol for the ridge axis, and the arrows and numbers indicate the direction of spreading and relative rate of movement of the plates away from the ridge. In this example the half-spreading rate of the ridge (half-rate) is 2 cm yr^{-1}; that is, plates A and B are moving apart at 4 cm yr^{-1}, and each plate is growing at 2 cm yr^{-1}. (b) The relative velocities $_A\mathbf{v}_B$ and $_B\mathbf{v}_A$ for the ridge shown in (a). (c) A *destructive* boundary (subduction zone). The barbed line is the symbol for a subduction zone; the barbs are on the side of the overriding plate, pointing away from the subducting or downgoing plate. The arrow and number indicate the direction and rate of relative motion between the two plates. In this example, plate B is being subducted at 10 cm yr^{-1}. (d) The relative velocities $_A\mathbf{v}_B$ and $_B\mathbf{v}_A$ for the subduction zone shown in (c). (e) A *conservative* boundary (transform fault). The single line is a symbol for a transform fault. The half-arrows and number indicate the direction and rate of relative motion between the plates: in this example, 6 cm yr^{-1}. (f) The relative velocities $_A\mathbf{v}_B$ and $_B\mathbf{v}_A$ for the transform fault shown in (e).

material, usually only the oceanic part of any plate is created or destroyed. Obviously, seafloor spreading at a midocean ridge produces only oceanic lithosphere, but it is hard to understand why continental material usually is not destroyed at convergent plate boundaries. At subduction zones, where continental and oceanic materials meet, it is the oceanic plate which is *subducted* (and thereby destroyed). It is probable that if the thick, relatively low-density continental material (approximate continental crustal density, 2.8×10^3 kg m^{-3}) reaches a subduction zone, it may descend a short way, but because the mantle density is so much greater (approximate mantle density, 3.3×10^3 kg m^{-3}), the downwards motion does not continue. Instead, the subduction zone ceases to operate at that place and moves to a more favourable location. Mountains are built (orogeny) above subduction zones as a result of continental collisions. In other words, the continents are rafts of lighter material which remain on the surface while the denser oceanic lithosphere is subducted beneath either oceanic or continental lithosphere. The discovery that plates can include both continental and oceanic parts, but that only the oceanic parts are created or destroyed, removed the main objection to the theory of continental

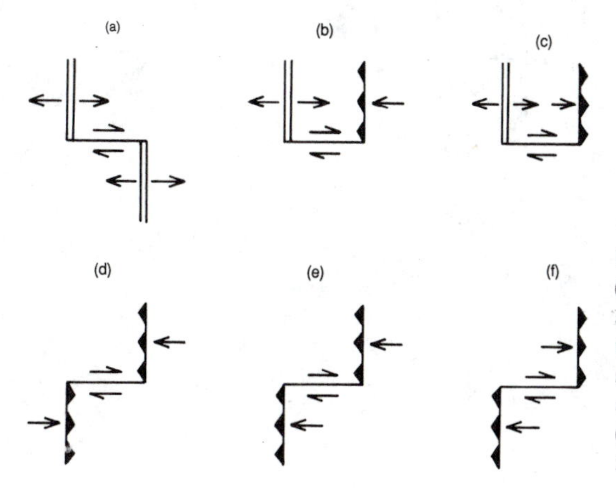

Figure 2.4. The six types of dextral (right-handed) transform faults. There also are six sinistral (left-handed) transform faults, mirror images of those shown here. (a) Ridge–ridge fault, (b) and (c) ridge–subduction-zone fault, (d), (e) and (f) subduction-zone–subduction-zone fault. (After Wilson 1965.)

drift, which was the unlikely concept that continents somehow were ploughing through oceanic rocks.

2.2 A Flat Earth

Before looking in detail at the motions of plates on the surface of the earth (which of necessity involves some spherical geometry), it is instructive to return briefly to the Middle Ages so that we can consider a flat planet.

Figures 2.3a, c, e show the three types of plate boundary and the ways they are usually depicted on maps. To describe the relative motion between the two plates A and B, we must use a vector that expresses their relative rate of movement (*relative velocity*). The velocity of plate A with respect to plate B is written ${}_{B}\mathbf{v}_{A}$ (i.e., if you are an observer on plate B, then ${}_{B}\mathbf{v}_{A}$ is the velocity at which you see plate A moving). Conversely, the velocity of plate B with respect to plate A is ${}_{A}\mathbf{v}_{B}$, and

$$ {}_{A}\mathbf{v}_{B} = -{}_{B}\mathbf{v}_{A} \tag{2.1} $$

Figures 2.3b, d, f illustrate these vectors for the three types of plate boundary.

To make our models more realistic, let us set up a two-plate system (Fig. 2.5a) and try to determine the more complex motions. The western boundary of plate B is a ridge which is spreading with a half-rate of $2\,\mathrm{cm\,yr^{-1}}$. This information enables us to draw ${}_{A}\mathbf{v}_{B}$ and ${}_{B}\mathbf{v}_{A}$ (Fig. 2.5b). Since we know the shape of plate B, we can see that its northern and southern boundaries must be transform faults. The northern boundary is *sinistral*, or left-handed; rocks are offset to the left as you cross the fault. The southern boundary is *dextral*, or right-handed; rocks are offset to the right as you cross the fault. The eastern boundary is ambiguous: ${}_{A}\mathbf{v}_{B}$ indicates that plate B is approaching plate A at $4\,\mathrm{cm\,yr^{-1}}$ along this boundary, which means that a subduction zone is operating there; but there is no indication as to which plate is being subducted. The two possible solutions for this

(a)

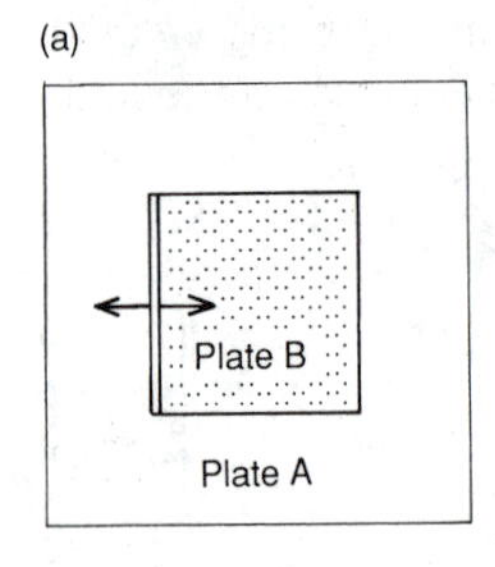

(b)

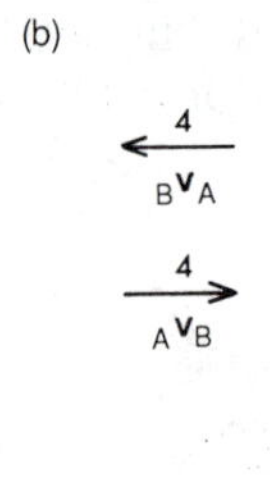

(c)

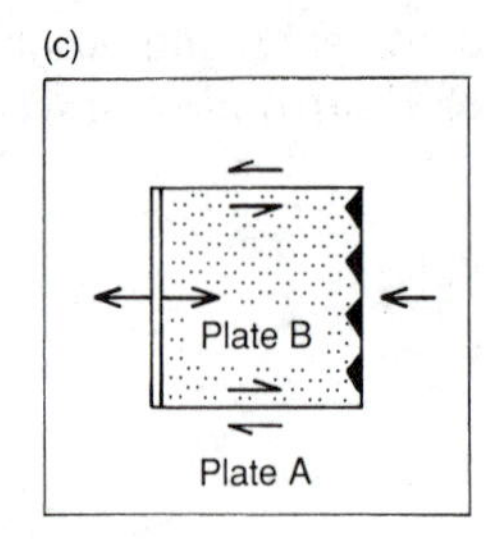

(d)

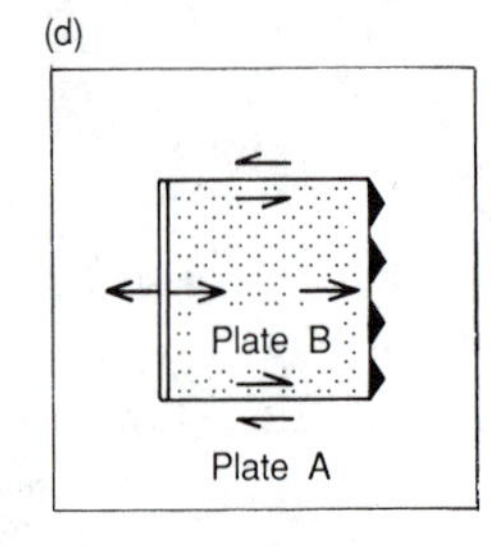

Figure 2.5. (a) A two-plate model on a flat planet. Plate B is shaded. The western boundary of plate B is a ridge from which sea floor spreads at a half-rate of $2\,\mathrm{cm\,yr^{-1}}$. (b) Relative velocity vectors ${}_{A}\mathbf{v}_{B}$ and ${}_{B}\mathbf{v}_{A}$ for the plates in (a). (c) One solution to the model shown in (a): The northern and southern boundaries of plate B are transform faults, the eastern boundary is a subduction zone with plate B overriding plate A. (d) Alternative solution for the model in (a): The northern and southern boundaries of plate B are transform faults, the eastern boundary is a subduction zone with plate A overriding plate B.

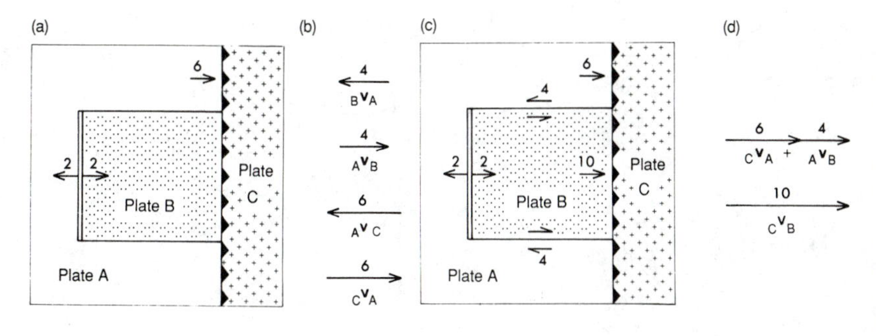

Figure 2.6. (a) A three-plate model on a flat planet. Plate A is unshaded. The western boundary of plate B is a ridge spreading at a half-rate of 2 cm yr^{-1}. The boundary between plates A and C is a subduction zone with plate C overriding plate A at 6 cm yr^{-1}. (b) Relative velocity vectors for the plates shown in (a). (c) The solution to the model in (a): Both the northern and southern boundaries of plate B are transform faults, and the eastern boundary is a subduction zone with plate C overriding plate B at 10 cm yr^{-1}. (d) Vector addition to determine the velocity of plate B with respect to plate C, $_C\mathbf{v}_B$.

model are shown in Figure 2.5c, d. Figure 2.5c shows plate A being subducted beneath plate B at 4 cm yr^{-1}. This means that plate B is increasing in length by 2 cm yr^{-1}, this being the rate at which new plate is formed at the ridge axis. Figure 2.5d shows plate B being subducted beneath plate A at 4 cm yr^{-1}, faster than new plate is being created at its western boundary (2 cm yr^{-1}); so eventually plate B will cease to exist on the surface of the planet.

If we introduce a third plate into the model, the motions become more complex still (Fig. 2.6a). In this example, plates A and B are spreading away from the ridge at a half-rate of 2 cm yr^{-1}, just as in Fig. 2.5a. The eastern boundary of plates A and C is a subduction zone, with plate A being subducted beneath plate C at 6 cm yr^{-1}. The presence of plate C does not alter the relative motions across the northern and southern boundaries of plate B; these boundaries are transform faults just as in the previous example. To determine the relative rate of plate motion at the boundary between plates B and C, we must use vector addition:

$$_C\mathbf{v}_B = {_C\mathbf{v}_A} + {_A\mathbf{v}_B} \tag{2.2}$$

This is demonstrated in Figure 2.6d: Plate B is being subducted beneath plate C at 10 cm yr^{-1}. This means that the net rate of destruction of plate B is $10 - 2 = 8$ cm yr^{-1}; eventually, plate B will be totally subducted, and a simple two-plate subduction model will be in operation.

So far the examples have been straightforward in that all relative motions have been in an east–west direction. (Vector addition was not really necessary; common sense would work equally well.) But now let us include motion in the north–south direction also. Figure 2.7a shows the model of three plates A, B and C: The western boundary of plate B is a ridge which is spreading at a half-rate of 2 cm yr^{-1}, the northern boundary of plate B is a transform fault (just as in the other examples) and the boundary between plates A and C is a transform fault with relative motion of 3 cm yr^{-1}. The motion at the boundary between plates B and C is unknown and must be determined by using Eq. 2.2. For this example it is necessary to draw a vector triangle to determine $_C\mathbf{v}_B$ (Fig. 2.7d). A solution to the problem is shown in Figure 2.7c: Plate B undergoes oblique subduction beneath plate C at 5 cm yr^{-1}. The other possible solution is for plate C to be subducted beneath plate B at 5 cm yr^{-1}. In that case, the boundary between plates B and C would not remain collinear with the boundary between plates A and C but would move to the east. (This is an example of the instability of a triple junction; see Sect. 2.6.)

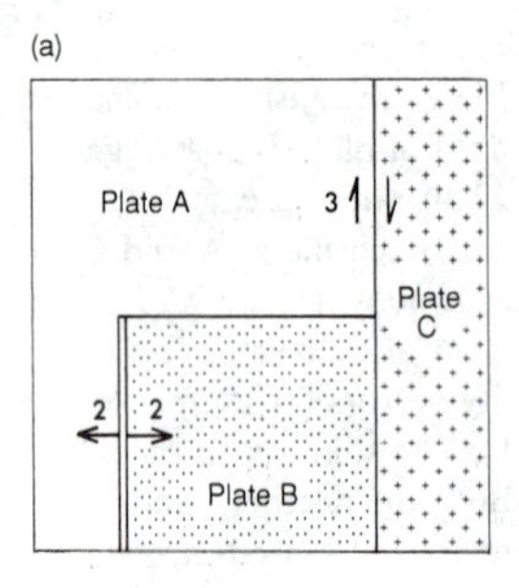

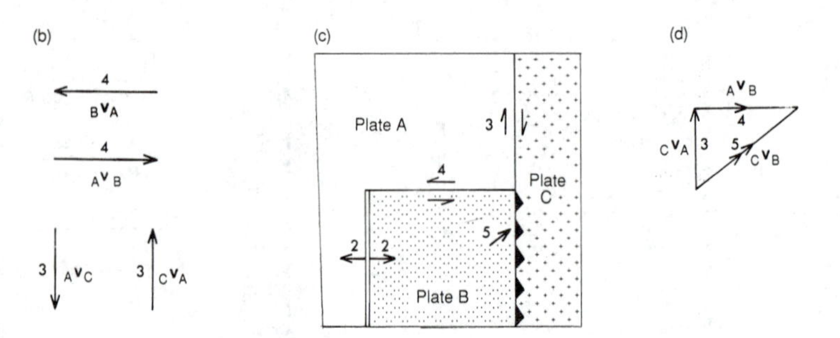

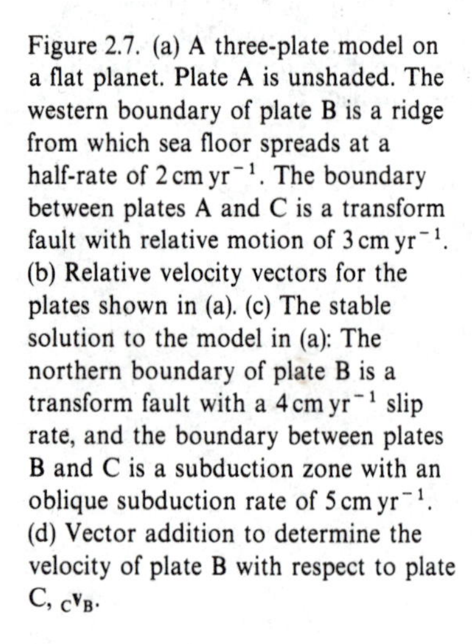

Figure 2.7. (a) A three-plate model on a flat planet. Plate A is unshaded. The western boundary of plate B is a ridge from which sea floor spreads at a half-rate of 2 cm yr^{-1}. The boundary between plates A and C is a transform fault with relative motion of 3 cm yr^{-1}. (b) Relative velocity vectors for the plates shown in (a). (c) The stable solution to the model in (a): The northern boundary of plate B is a transform fault with a 4 cm yr^{-1} slip rate, and the boundary between plates B and C is a subduction zone with an oblique subduction rate of 5 cm yr^{-1}. (d) Vector addition to determine the velocity of plate B with respect to plate C, $_C\mathbf{v}_B$.

These examples should give some idea of what can happen when plates move relative to each other and of the types of plate boundaries that occur in various situations. Some of the problems at the end of this chapter refer to a flat earth, such as we have assumed for these examples. The real earth, however, is spherical, so we need to understand some spherical geometry.

2.3 Rotation Vectors and Rotation Poles

To describe motions on the surface of a sphere we use *Euler's 'fixed point' theorem*, which states: 'The most general displacement of a rigid body with a fixed point is equivalent to a rotation about an axis through that fixed point'.

Taking a plate as a rigid body and the centre of the earth as a fixed point, we can restate this theorem: 'Every displacement from one position to another on the surface of the earth can be regarded as a rotation about a suitably chosen axis passing through the centre of the earth'.

This restated theorem was first applied by Bullard et al. (1965) in their paper on continental drift, in which they describe the fitting of the coastlines of South America and Africa. The 'suitably chosen axis' which passes through the centre of the earth is called the *rotation axis*, and it cuts the surface of the earth at two points called the *poles of rotation* (Fig. 2.8a). These are purely mathematical points and have no physical reality, but their positions describe the directions of motion of all points along the plate boundary. The magnitude of the angular velocity about the axis then defines the magnitude of the relative motion between the two plates. Because angular velocities behave as vectors, the relative motion between two plates can be written as $\boldsymbol{\omega} = \omega\mathbf{k}$, where $\mathbf{k}$ is a unit vector along the rotation axis and ω is the angular velocity. The sign convention used is that a rotation which is clockwise (or right-handed) when viewed from the centre of the earth along the rotation axis is positive. Viewed from outside the earth, a positive rotation is anticlockwise. Thus, one rotation pole is positive and the other is negative (Fig. 2.8b).

Consider a point X on the surface of the earth (Fig. 2.8c). At X the value of the relative velocity v between the two plates is

$$v = \omega R \sin\theta \tag{2.3}$$

where θ is the angular distance between the rotation pole P and the point X, R is the radius of the earth. Thus, the relative velocity is zero at the rotation

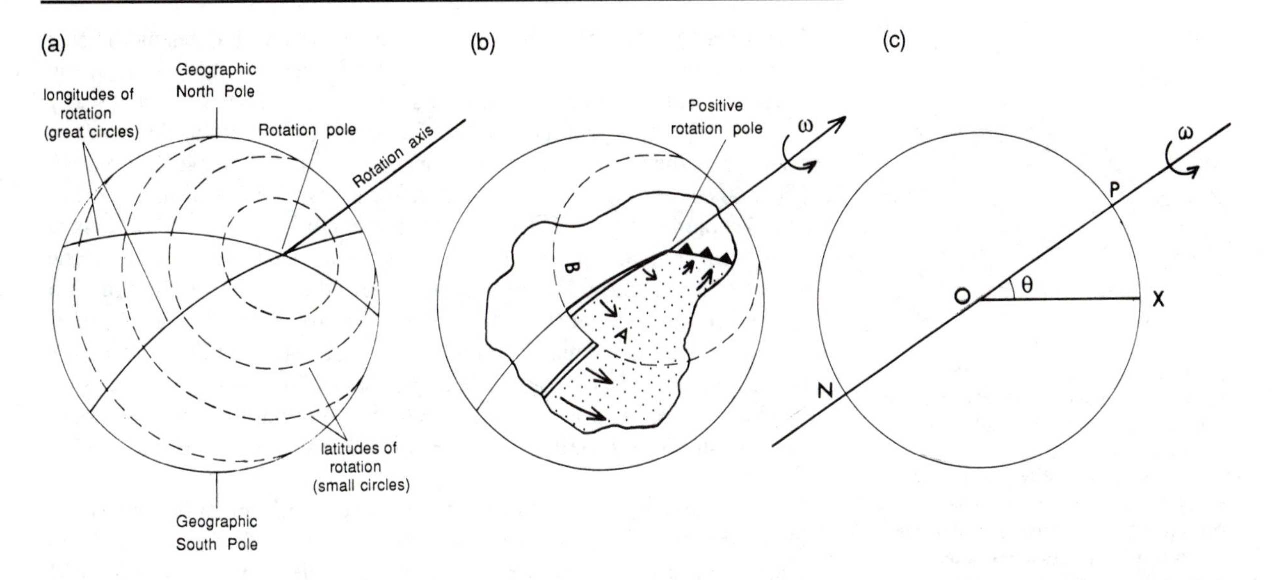

Figure 2.8. The movement of plates on the surface of the earth. (a) The lines of latitude of rotation around the rotation poles are small circles (shown dashed), whereas the lines of longitude of rotation are great circles (i.e., circles with the same diameter as the earth). Note that these lines of latitude and longitude of rotation are *not* the geographical lines of latitude and longitude because the poles for the geographical coordinate system are the North and South poles, not the rotation poles. (b) Constructive, destructive and conservative boundaries between plates A and B. Plate B is assumed to be fixed so that the motion of plate A is relative to plate B. The visible rotation pole is positive (motion is anticlockwise when viewed from outside the earth). Note that the spreading and subduction rates increase with distance from the rotation pole. The transform fault is an arc of small circle (shown dashed) and thus is perpendicular to the ridge axis. As the plate boundary passes the rotation pole, the boundary changes from ridge to subduction zone. (c) Cross section through the centre of the earth *O*. *P* and *N* are the positive and negative rotation poles, and *X* is a point on the plate boundary.

poles, where $\theta = 0°$ and 180°, and has a maximum value of ωR at 90° from the rotation poles. This factor of $\sin\theta$ means that the relative motion between two adjacent plates changes with position along the plate boundary, in contrast to the earlier examples for a flat earth. If by chance the plate boundary passes through the rotation pole, the nature of the boundary changes from divergent to convergent, or vice versa (Fig. 2.8b). Lines of constant velocity are small circles defined by $\theta =$ constant about the rotation poles.

2.4 Present-Day Plate Motions

2.4.1 Determination of Rotation Poles and Rotation Vectors

Several methods can be used to find the present-day *instantaneous poles of rotation* and *relative angular velocities* between pairs of plates. *Instantaneous* refers to a geological instant; it means a value averaged over a period of time ranging from a few years to a few million years, depending on the method used. These methods include the following:

1. A local determination of the direction of relative motion between two plates can be made from the strike of active transform faults. Methods of recognizing transform faults are discussed fully in Section 8.5. Since transform faults on ridges are much easier to recognize and more common than transform faults along destructive boundaries, this method is used primarily to find rotation poles for plates on either side of a midocean ridge. The relative motion at transform faults is parallel to the fault and is of constant value along the fault. This means that the faults themselves are arcs of small circles about the rotation pole. The rotation pole must therefore lie somewhere on the great circle which is perpendicular to that small circle. So, if two or more transform faults can be used, the intersection of the great circles is the position of the rotation pole (Fig. 2.9).

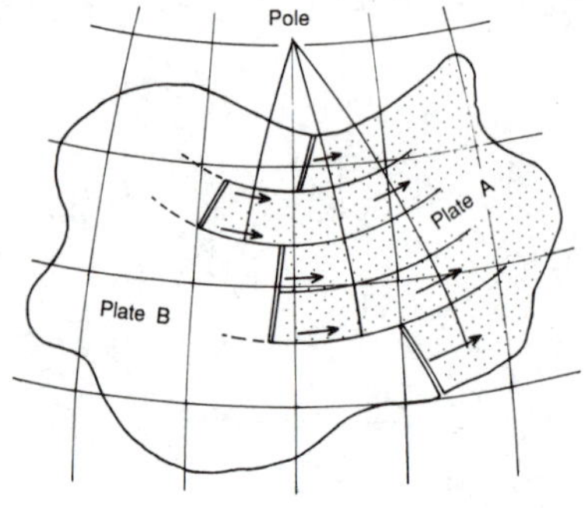

Figure 2.9. On a spherical earth the motion of plate A relative to plate B must be a rotation about some pole. All the transform faults on the boundary between plates A and B must be small circles concentric about the pole. Transform faults can be used to locate the pole: It lies at the intersection of the great circles which are perpendicular to the transfrom faults. Although ridges are generally perpendicular to the direction of spreading, this is not a geometrical requirement, so it is not possible to determine the relative motion or locate the pole from the ridge itself. (After Morgan 1968.)

2. The spreading rate of an accreting plate boundary changes as the sine of the angular distance θ from the rotation pole (Eq. 2.3). Thus, if the spreading rate at various locations along the ridge can be locally determined (from spacing of oceanic magnetic anomalies as discussed in Chapter 3), the rotation pole and angular velocity can then be estimated.
3. The analysis of data from an earthquake can give the direction of motion and the plane of the fault on which the earthquake occurred. This is known as a *fault plane solution* or a *focal mechanism.* Fault plane solutions for earthquakes along a plate boundary can give the direction of relative motion between the two plates. For example, earthquakes which occur on the transform fault between plates A and B, as illustrated in Figure 2.8b, indicate that there is right lateral motion across the fault. The location of the pole and the direction, though not the magnitude, of the motion can thus be estimated. (Fault plane solutions are discussed in Sect. 4.2.8.)
4. Where plate boundaries cross land, surveys of displacements can be used (over large distances and long periods of time) to determine the local relative motion. For example, stream channels and even roads, field boundaries and buildings may be displaced.
5. Satellites have made it possible to measure instantaneous plate motions with reasonable accuracy. One method uses a satellite laser-ranging system to determine differences in distance between two sites on the earth's surface over a period of years. The other method, known as *very-long-baseline interferometry* (VLBI), uses quasars for the signal source and terrestrial radio telescopes as the receivers. Again, the difference in distance between two telescope sites is measured over a period of years. Figure 2.10 compares the velocity of the Pacific Plate relative to the

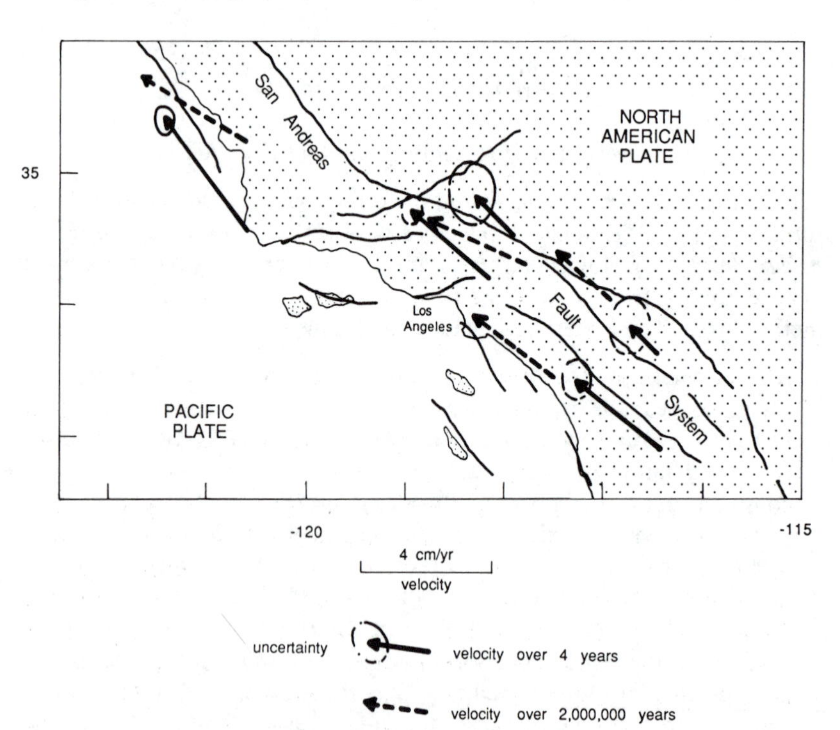

Figure 2.10. Velocity of the Pacific Plate relative to the North American Plate along the San Andreas Fault system in California. Solid arrows, velocities determined from VLBI data recorded over four years: dashed arrows, geologically determined velocities. (After Kroger et al. 1987.)

Table 2.1. *Rotation vectors for the present-day relative motion between some pairs of plates*

Plates	Positive pole position		Angular velocity
	Latitude	Longitude	(10^{-7} deg yr^{-1})
Africa–Antarctica	5.6°N	39.2°W	1.3
Africa–Eurasia	21.0°N	20.6°W	1.3
Africa–North America	78.8°N	38.3°E	2.5
Africa–South America	62.5°N	39.4°W	3.2
Australia–Antarctica	13.2°N	38.2°E	6.8
Pacific–Antarctica	64.3°S	96.0°E	9.1
South America–Antarctica	86.4°S	139.3°E	2.7
Arabia–Eurasia	24.6°N	13.7°E	5.2
India–Eurasia	24.4°N	17.7°E	5.3
Eurasia–North America	62.4°N	135.8°E	2.2
Eurasia–Pacific	61.1°N	85.8°W	9.0
Pacific–Australia	60.1°S	178.3°W	11.2
North America–Pacific	48.7°N	78.2°W	7.8
Cocos–North America	27.9°N	120.7°W	14.2
Nazca–Pacific	55.6°N	90.1°W	14.2
Nazca–South America	56.0°N	94.0°W	7.6

Note: The first plate moves anticlockwise with respect to the second plate as shown.
Source: After DeMets, Gordon, Argus and Stein (1990).

North American Plate along the San Andreas Fault system in California, as determined by VLBI and by sea-floor spreading and geological methods. The VLBI rates of motion averaged over four years are almost identical to the geologically determined rates. This is an impressive corroboration of relative plate motions.

An estimate of the present-day plate motions made by using all the ridge spreading rate estimates available in 1989 (277 estimates), trends of oceanic transform faults (121 estimates) and suitable earthquake slip vectors (724 estimates) is given in Table 2.1.

It is important to realize that a rotation with a high angular velocity ω does not necessarily mean that the relative motion along the plate boundary is also large (remember that Eq. 2.3 has a $\sin\theta$ factor multiplying the angular velocity ω). The distance between rotation pole and plate boundary is important. However, it is conventional to use the relative velocity at $\theta = 90°$ when quoting a relative velocity for two plates, even though neither plate may extend 90° from the pole.

2.4.2 Calculation of the Relative Motion at a Plate Boundary

After the instantaneous rotation pole and angular velocity have been determined for a pair of adjacent plates, they can be used to calculate the

Table 2.2. *Symbols used in calculations involving rotation poles*

Symbol	Meaning	Sign convention
λ_P	Latitude of rotation pole P	°N positive
λ_X	Latitude of point X on plate boundary	°S negative
ϕ_P	Longitude of rotation pole P	°W negative
ϕ_X	Longitude of point X on plate boundary	°E positive
$\mathbf{v}$	Velocity at point X on plate boundary	
v	Amplitude of velocity	
β	Azimuth of the velocity with respect to north N	Clockwise positive
R	Radius of the earth	
ω	Angular velocity about rotation pole P	

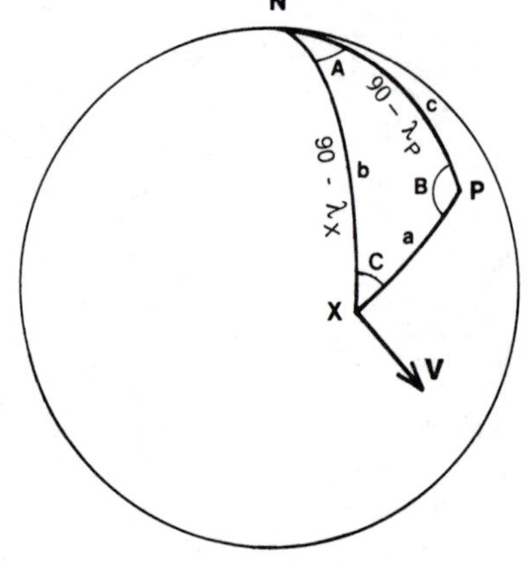

Figure 2.11. Diagram showing the relative positions of the positive rotation pole P and point X on the plate boundary. N is the North Pole. The sides of the spherical triangle NPX are all great circles, the sides NX and NP are lines of geographic longitude. The vector $\mathbf{v}$ is the relative velocity at point X on the plate boundary (note that it is perpendicular to PX). It is usual to quote the lengths of the sides of spherical triangles as angles (e.g., latitude and longitude when used as geographic coordinates).

direction and magnitude of the relative motion at any point along the plate boundary.

The notation and sign conventions used in the following pages are given in Table 2.2. Figure 2.11 shows the relative positions of the North Pole N, positive rotation pole P and point X on the plate boundary (compare with Fig. 2.8b). In the spherical triangle NPX, let the angles $\widehat{XNP} = A$, $\widehat{NPX} = B$ and $\widehat{PXN} = C$, and let the angular lengths of the sides of the triangle be $PX = a$, $XN = b$ and $NP = c$. Thus, the angular lengths b and c are known, but a is not:

$$b = 90 - \lambda_X \tag{2.4}$$

$$c = 90 - \lambda_P \tag{2.5}$$

Angle A is known, but B and C are not:

$$A = \phi_P - \phi_X \tag{2.6}$$

Equation 2.3 is used to obtain the magnitude of the velocity at point X:

$$v = \omega R \sin a \tag{2.7}$$

The azimuth of the velocity β is given by

$$\beta = 90 + C \tag{2.8}$$

To find the angles a and C to substitute into Eqs. 2.7 and 2.8, we use spherical geometry. Just as there are cosine and sine rules which relate the angles and sides of plane triangles, there are cosine and sine rules for spherical triangles:

$$\cos a = \cos b \cos c + \sin b \sin c \cos A \tag{2.9}$$

and

$$\frac{\sin a}{\sin A} = \frac{\sin c}{\sin C} \tag{2.10}$$

Substituting Eqs. 2.4–2.6 into Eq. 2.9 gives

$$\cos a = \cos(90 - \lambda_X)\cos(90 - \lambda_P) + \sin(90 - \lambda_X)\sin(90 - \lambda_P)\cos(\phi_P - \phi_X) \quad (2.11)$$

This can then be simplified to yield the angle a, which is needed to calculate the velocity from Eq. 2.7.

$$a = \cos^{-1}[\sin\lambda_X \sin\lambda_P + \cos\lambda_X \cos\lambda_P \cos(\phi_P - \phi_X)] \quad (2.12)$$

Substituting Eqs. 2.5 and 2.6 into Eq. 2.10 gives

$$\frac{\sin a}{\sin(\phi_P - \phi_X)} = \frac{\sin(90 - \lambda_P)}{\sin C} \quad (2.13)$$

Upon rearrangement this becomes

$$C = \sin^{-1}\left[\frac{\cos\lambda_P \sin(\phi_P - \phi_X)}{\sin a}\right] \quad (2.14)$$

Therefore, if the angle a is calculated from Eq. 2.12, angle C can then be calculated from Eq. 2.14, and finally, the relative velocity and its azimuth can be calculated from Eqs. 2.7 and 2.8. Note that the inverse sine function of Eq. 2.14 is double-valued.* Always check that you have the correct value for C.

Example: calculation of relative motion at a plate boundary

Calculate the present-day relative motion at 28°S 71°W on the Peru–Chile Trench using the Nazca–South America rotation pole given in Table 2.1. Assume the radius of the earth to be 6371 km.

$$\lambda_X = -28°, \qquad \phi_X = -71°$$

$$\lambda_P = 56°, \qquad \phi_P = -94°$$

$$\omega = 7.6 \times 10^{-7} \text{ deg yr}^{-1} = \frac{\pi}{180} \times 7.6 \times 10^{-7} \text{ radians yr}^{-1}$$

These values are substituted into Eqs. 2.12, 2.14, 2.7 and 2.8 in that order, giving:

$$a = \cos^{-1}[\sin(-28)\sin(56) + \cos(-28)\cos(56)\cos(-94 + 71)]$$

$$= 86.26° \quad (2.15)$$

$$C = \sin^{-1}\left[\frac{\cos(56)\sin(-94 + 71)}{\sin(86.26)}\right]$$

$$= -12.65° \quad (2.16)$$

$$v = \frac{\pi}{180} \times 7.6 \times 10^{-7} \times 6371 \times 10^5 \times \sin(86.26) \text{ cm yr}^{-1} \quad (2.17)$$

* An alternative way to calculate motion along a plate boundary and to avoid the sign ambiguities is to use vector algebra (see Altman 1986 or Cox and Hart 1986, p. 154).

$$= 8.43 \text{ cm yr}^{-1}$$

$$\beta = 90 - 12.65 \tag{2.18}$$

$$= 77.35°$$

Thus, the Nazca Plate is moving relative to the South American Plate at 8.4 cm yr^{-1} with azimuth 77°; the South American Plate is moving relative to the Nazca Plate at 8.4 cm yr^{-1}, azimuth 257° (Fig. 2.2).

2.4.3 Combination of Rotation Vectors

Suppose there are three rigid plates A, B and C and that the angular velocity of A relative to B, ${}_B\boldsymbol{\omega}_A$, and that of B relative to C, ${}_C\boldsymbol{\omega}_B$, are known. The motion of plate A relative to plate C, ${}_C\boldsymbol{\omega}_A$, can be determined by vector addition just as for the flat earth:

$$ {}_C\boldsymbol{\omega}_A = {}_C\boldsymbol{\omega}_B + {}_B\boldsymbol{\omega}_A \tag{2.19}$$

(Remember that in this notation the first subscript refers to the 'fixed' plate.)

Alternatively, Eq. 2.19 can be written as

$$ {}_A\boldsymbol{\omega}_B + {}_B\boldsymbol{\omega}_C + {}_C\boldsymbol{\omega}_A = 0 \tag{2.20}$$

since ${}_B\boldsymbol{\omega}_A = -{}_A\boldsymbol{\omega}_B$. The resultant vector ${}_C\boldsymbol{\omega}_A$ of Eq. 2.19 must be in the same plane as the two original vectors ${}_B\boldsymbol{\omega}_A$ and ${}_C\boldsymbol{\omega}_B$. Imagine the great circle on which these two poles lie; the resultant pole must also lie on that same great circle (Fig. 2.12). Note that this relationship (Eqs. 2.19 and 2.20) should be used only for infinitesimal movements or angular velocities and not for finite rotations. The theory of finite rotations is complex. (For a treatment of the whole theory of instantaneous and finite rotations, the reader is referred to Le Pichon et al. 1973.)

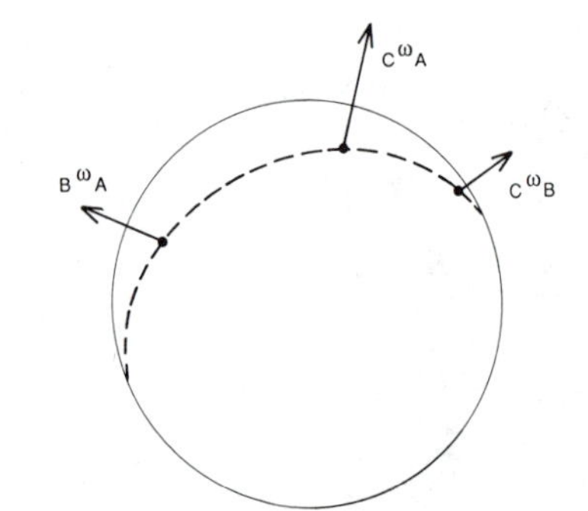

Figure 2.12. Relative rotation vectors ${}_B\boldsymbol{\omega}_A$ and ${}_C\boldsymbol{\omega}_B$ for the plates A, B and C. The dashed line is the great circle on which the two poles lie. The resultant rotation vector is ${}_C\boldsymbol{\omega}_A$ (Eq. 2.19). The resultant pole must also lie on the great circle because the resultant rotation vector has to lie in the plane of the two original rotation vectors.

Let the three vectors ${}_B\boldsymbol{\omega}_A$, ${}_C\boldsymbol{\omega}_B$, ${}_C\boldsymbol{\omega}_A$ be written as shown in Table 2.3. It is simplest to use a rectangular coordinate system through the centre of the earth, with the x–y plane being equatorial, the x axis passing through the Greenwich meridian and the z axis passing through the North Pole, as shown in Figure 2.13. The sign convention of Table 2.2 continues to apply. Then Eq. 2.19 can be written

$$x_{CA} = x_{CB} + x_{BA} \tag{2.21}$$

$$y_{CA} = y_{CB} + y_{BA} \tag{2.22}$$

and

$$z_{CA} = z_{CB} + z_{BA} \tag{2.23}$$

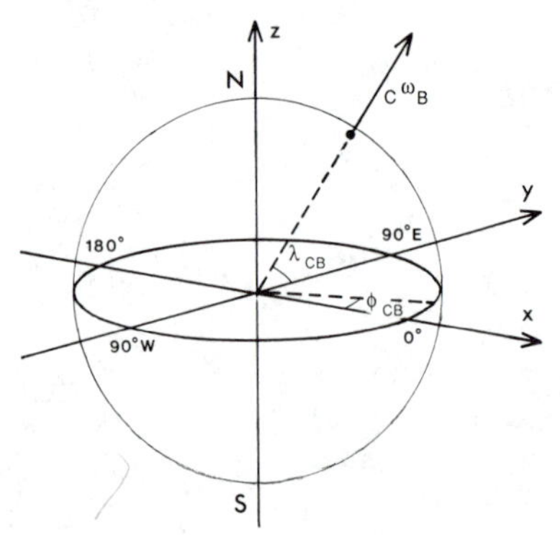

Figure 2.13. Rectangular coordinate system used in the addition of rotation vectors. The x–y plane is equatorial with the x-axis passing through 0° Greenwich and the z-axis through the North Pole. Notation and sign conventions are given in Table 2.3.

Table 2.3. *Notation used in addition of rotation vectors*

Rotation vector	Magnitude	Latitude of pole	Longitude of pole
${}_B\boldsymbol{\omega}_A$	${}_B\omega_A$	λ_{BA}	ϕ_{BA}
${}_C\boldsymbol{\omega}_B$	${}_C\omega_B$	λ_{CB}	ϕ_{CB}
${}_C\boldsymbol{\omega}_A$	${}_C\omega_A$	λ_{CA}	ϕ_{CA}

where x_{BA}, y_{BA} and z_{BA} are the x, y and z coordinates of the vector ${}_{B}\boldsymbol{\omega}_{A}$, and so on. Equations 2.21–2.23 become

$$x_{CA} = {}_{C}\omega_{B} \cos\lambda_{CB} \cos\phi_{CB} + {}_{B}\omega_{A} \cos\lambda_{BA} \cos\phi_{BA} \quad (2.24)$$

$$y_{CA} = {}_{C}\omega_{B} \cos\lambda_{CB} \sin\phi_{CB} + {}_{B}\omega_{A} \cos\lambda_{BA} \sin\phi_{BA} \quad (2.25)$$

$$z_{CA} = {}_{C}\omega_{B} \sin\lambda_{CB} + {}_{B}\omega_{A} \sin\lambda_{BA} \quad (2.26)$$

when the three rotation vectors are expressed in their x, y, z components. The magnitude of the resultant rotation vector ${}_{C}\boldsymbol{\omega}_{A}$ is

$${}_{C}\omega_{A} = \sqrt{x_{CA}^2 + y_{CA}^2 + z_{CA}^2} \quad (2.27)$$

and the pole position is given by

$$\lambda_{CA} = \sin^{-1}\left(\frac{z_{CA}}{{}_{C}\omega_{A}}\right) \quad (2.28)$$

and

$$\phi_{CA} = \tan^{-1}\left(\frac{y_{CA}}{x_{CA}}\right) \quad (2.29)$$

Note that this expression for ϕ_{CA} has an ambiguity of 180° (e.g., $\tan 30° = \tan 210° = 0.5774$, $\tan 110° = \tan 290° = -2.747$). This is resolved by adding or subtracting 180° so that

$$x_{CA} > 0 \quad \text{when} \quad -90° < \phi_{CA} < +90° \quad (2.30)$$

$$x_{CA} < 0 \quad \text{when} \quad |\phi_{CA}| > 90° \quad (2.31)$$

Example: addition of relative rotation vectors

Given the instantaneous rotation vectors in Table 2.1 for the Nazca Plate relative to the Pacific Plate and the Pacific Plate relative to the Antarctic Plate, calculate the instantaneous rotation vector for the Nazca Plate relative to the Antarctic Plate:

Plate	Rotation vector	Latitude of pole	Longitude of pole	Angular velocity (10^{-7} deg yr^{-1})
Nazca–Pacific	${}_{P}\boldsymbol{\omega}_{N}$	55.6°N	90.1°W	14.2
Pacific–Antarctica	${}_{A}\boldsymbol{\omega}_{P}$	64.3°S	96.0°E	9.1

To calculate the rotation vector for the Nazca Plate relative to the Antarctic Plate we apply Eq. 2.19:

$${}_{A}\boldsymbol{\omega}_{N} = {}_{A}\boldsymbol{\omega}_{P} + {}_{P}\boldsymbol{\omega}_{N} \quad (2.32)$$

Substituting the tabulated values into the equations for the x, y and z components of ${}_{A}\boldsymbol{\omega}_{N}$ (Eqs. 2.24–2.26) yields

$$x_{AN} = 9.1\cos(-64.3)\cos(96.0) + 14.2\cos(55.6)\cos(-90.1)$$
$$= -0.427 \quad (2.33)$$

$$y_{AN} = 9.1\cos(-64.3)\ \sin(96.0) + 14.2\cos(55.6)\ \sin(-90.1)$$
$$= -4.098 \qquad (2.34)$$

$$z_{AN} = 9.1\sin(-64.3) + 14.2\sin(55.6)$$
$$= 3.517 \qquad (2.35)$$

The magnitude of the rotation vector ${}_{A}\boldsymbol{\omega}_{N}$ can now be calculated from Eq. 2.27 and the pole position from Eqs. 2.28 and 2.29:

$${}_{A}\omega_{N} = \sqrt{0.427^2 + 4.098^2 + 3.517^2} = 5.417 \qquad (2.36)$$

$$\lambda_{AN} = \sin^{-1}\left(\frac{3.517}{5.417}\right) = 40.49^\circ \qquad (2.37)$$

$$\phi_{AN} = \tan^{-1}\left(\frac{-4.098}{-0.427}\right) = 180 + 84.05^\circ \qquad (2.38)$$

Therefore, the rotation for the Nazca Plate relative to the Antarctic Plate has a magnitude of 5.4×10^{-7} deg yr^{-1}, and the rotation pole is located at latitude 40.5°N, longitude 95.9°W.

The problems at the end of this chapter enable the reader to use these methods to determine motions along real and imagined plate boundaries.

2.5 Plate Boundaries Can Change with Time

The examples of plates moving upon a flat earth (Sect. 2.2) illustrated that plates and plate boundaries do not stay the same for all time. This observation remains true when we advance from plates moving on a flat model earth to plates moving on a spherical earth. The formation of new plates and destruction of existing plates are the most obvious global reasons why plate boundaries and relative motions change. For example, a plate may be lost down a subduction zone, such as happened when most of the Farallon and Kula plates were subducted under the North American Plate in the early Tertiary (discussed in Sect. 3.3.3). Alternatively, two continental plates may coalesce into one (with resultant mountain building). If the position of a rotation pole changes, all the relative motions also change. A drastic change in pole position of 90° would, of course, completely alter the status quo: Transform faults would become ridges and subduction zones, and vice versa! Changes in the trends of transform faults and magnetic anomalies on the Pacific Plate imply that the direction of seafloor spreading has changed and indicate that the Pacific–Farallon pole position changed slightly a number of times during the Tertiary.

Parts of plate boundaries can change locally, however, without any major 'plate' or 'pole' event occurring. Consider three plates A, B and C. Let there be a convergent boundary between plates A and B, and let there be strike slip faults between plates A and C and plates B and C, as illustrated in Figure 2.14a, b. From the point of view of an observer on plate C, the boundary of C (circled) will change with time because the plate to which it is adjacent will change from plate A to plate B. The boundary will remain a dextral (right-handed) fault, but the slip rate will change from 2 cm yr^{-1} to

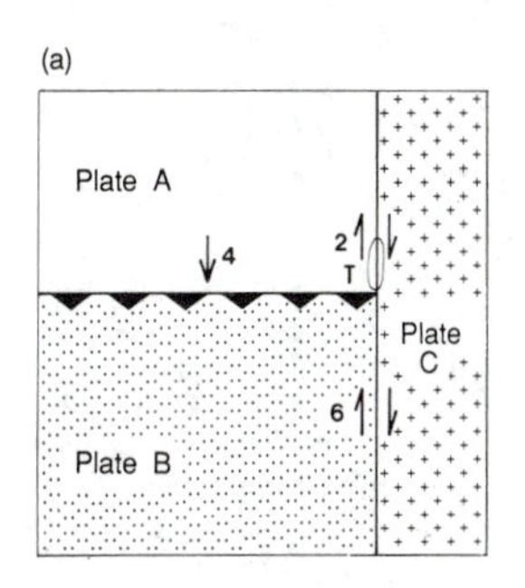

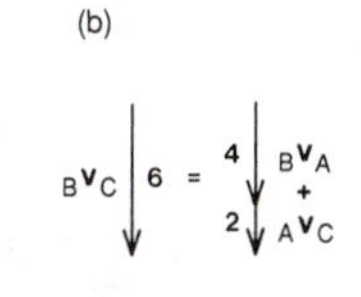

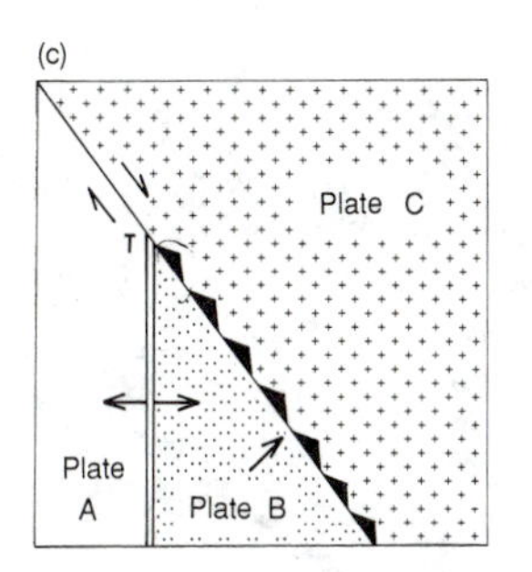

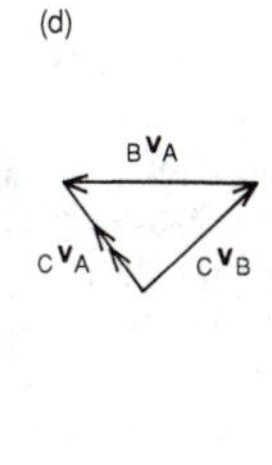

Figure 2.14. Two examples of a plate boundary locally changing with time. (a) A three-plate model. Point *T*, where plates A, B and C meet, is the triple junction. The western boundary of plate C consists of transform faults. Plate B is overriding plate A at 4 cm yr^{-1}. The circled part of the boundary of plate C changes with time. (b) Relative velocity vectors for the plates in (a). (c) A three-plate model. Point *T*, where plates A, B and C meet, is the triple junction. The boundary between plates A and B is a ridge, that between plates A and C is a transform fault and that between plates B and C is a subduction zone. The circled part of the boundary changes with time. (d) Relative velocity vectors for the plates in (c). In these examples, velocity vectors have been used rather than angular velocity vectors. This is justified, even for a spherical earth, because these examples are concerned only with small areas of the plates in the immediate vicinity of the triple junctions, over which the relative velocities are constant.

6 cm yr^{-1}. Relative to plate C, the subduction zone is moving northwards at 6 cm yr^{-1}. Another example of this type of plate boundary change is illustrated in Figure 2.14c, d. In this case, the relative velocities are such that the boundary between plates A and C is a strike slip fault, that between plates A and B is a ridge and that between plates B and C is a subduction zone. The motions are such that the ridge migrates slowly to the south relative to plate C, so the circled portion of plate boundary will change with time from subduction zone to transform fault.

These local changes in plate boundary are a geometric consequence of the motions of the three rigid plates and are not caused by any disturbing outside event. A complete study of all possible interactions of three plates is made in the next section. Such a study is very important because it enables us to apply the theory of rigid geometric plates to the earth and to deduce past plate motions from evidence in the local geological record. We can also predict details of future plate interactions.

2.6 Triple Junctions

2.6.1 *Stable and Unstable Triple Junctions*

A *triple junction* is the name given to a point at which three plates meet, such as the points *T* in Figure 2.14. A triple junction is said to be 'stable' when the relative motions of the three plates and the azimuth of their boundaries are such that the configuration of the junction does not change with time. The two examples shown in Figure 2.14 are in these terms stable. In both cases the triple junction moves along the boundary of plate C, locally changing this boundary. The relative motions of the plates and triple junction and the azimuths and types of plate boundaries of the whole system do not change with time. An 'unstable' triple junction exists only momentarily before evolving to a different geometry. If four or more plates meet at one point, the configuration is always unstable, and the system will evolve into two or more triple junctions.

As a further example, consider a triple junction where three subduction zones meet (Fig. 2.15): Plate A is overriding plates B and C, and plate C is overriding plate B. The relative velocity triangle for the three plates at the triple junction is shown in Figure 2.15b. Now consider how this triple junction evolves with time. Assume that plate A is fixed; then the positions of the plates at some later time are as shown in Figure 2.15c. The dashed

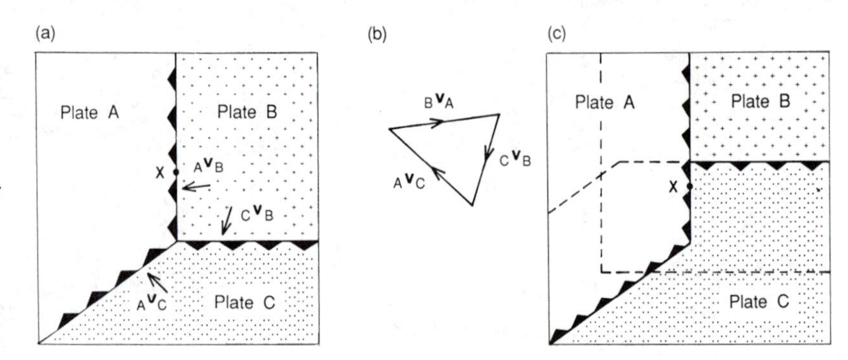

Figure 2.15. (a) A triple junction where three subduction zones intersect. Plate A overrides plates B and C, and plate C overrides plate B. ${}_A\mathbf{v}_B$, ${}_C\mathbf{v}_B$ and ${}_A\mathbf{v}_C$ are the relative velocities of the three plates in the immediate vicinity of the triple junction. (b) Relative velocity triangle for (a). (c) Geometry of the three subduction zones at some time later than in (a). The dashed lines show where plates B and C would have been had they not been subducted. The point X in (a) was originally on the boundary between plates A and B; now it is on the boundary between plates A and C. The original triple junction has changed its form. (After McKenzie and Morgan 1969.)

boundaries show where plates B and C would have been if they had not been overridden by plate A. The subduction zone between plates B and C has moved north along the north–south edge of plate A. Thus, the original triple junction (Fig. 2.15a) was unstable; however, the new triple junction (Fig. 2.15c) is stable (meaning that its geometry and the relative velocities of the plates are unchanging), though the triple junction itself continues to move northwards along the north–south edge of plate A. The point X is originally on the boundary of plates A and B. As the triple junction passes X, an observer there will see a sudden change in subduction rate and direction. Finally, X is a point on the boundary of plates A and C.

In a real situation, the history of the northward passage of the triple junction along the boundary of plates A and C could be determined by estimating the time at which the relative motion between the plates changed at a number of locations along the boundary. If such time estimates increase regularly with position along the plate boundary, it is probable that a triple junction migrated along the boundary and not that one of the plates had changed its relative motion (that would happen only once). It can be seen that although the original triple junction shown in Figure 2.15a is not stable, it would be stable if ${}_A\mathbf{v}_C$ were parallel to the boundary between plates B and C. In this special case, the boundary between B and C would not move in a north–south direction relative to A, and so the geometry of the triple junction would be unchanging in time. The other configuration in which the triple junction would be stable occurs when the edge of plate A, on both sides of the triple junction, is straight. This is, of course, the final configuration illustrated in Figure 2.15c.

Altogether there are sixteen possible types of triple junction, all shown in Figure 2.16. Of these sixteen triple junctions, one is always stable (the ridge–ridge–ridge junction) if oblique spreading is not allowed, and two are always unstable (the fault–fault–fault and fault–ridge–ridge junctions). The other thirteen junctions are stable under certain conditions. In the notation used to classify the types of triple junction, a ridge is written as R, transform fault as F and a subduction zone (or trench) as T. Thus, a ridge–ridge–ridge junction is RRR, a fault–fault–ridge junction is FFR, and so on. To examine the stability of any particular triple junction, it is easiest to draw a relative velocity triangle and include the azimuths of the plate boundaries. In Figure 2.16 the lengths of the lines AB, BC and AC are proportional and parallel to the relative velocities ${}_A\mathbf{v}_B$, ${}_B\mathbf{v}_C$ and ${}_A\mathbf{v}_C$. Thus,

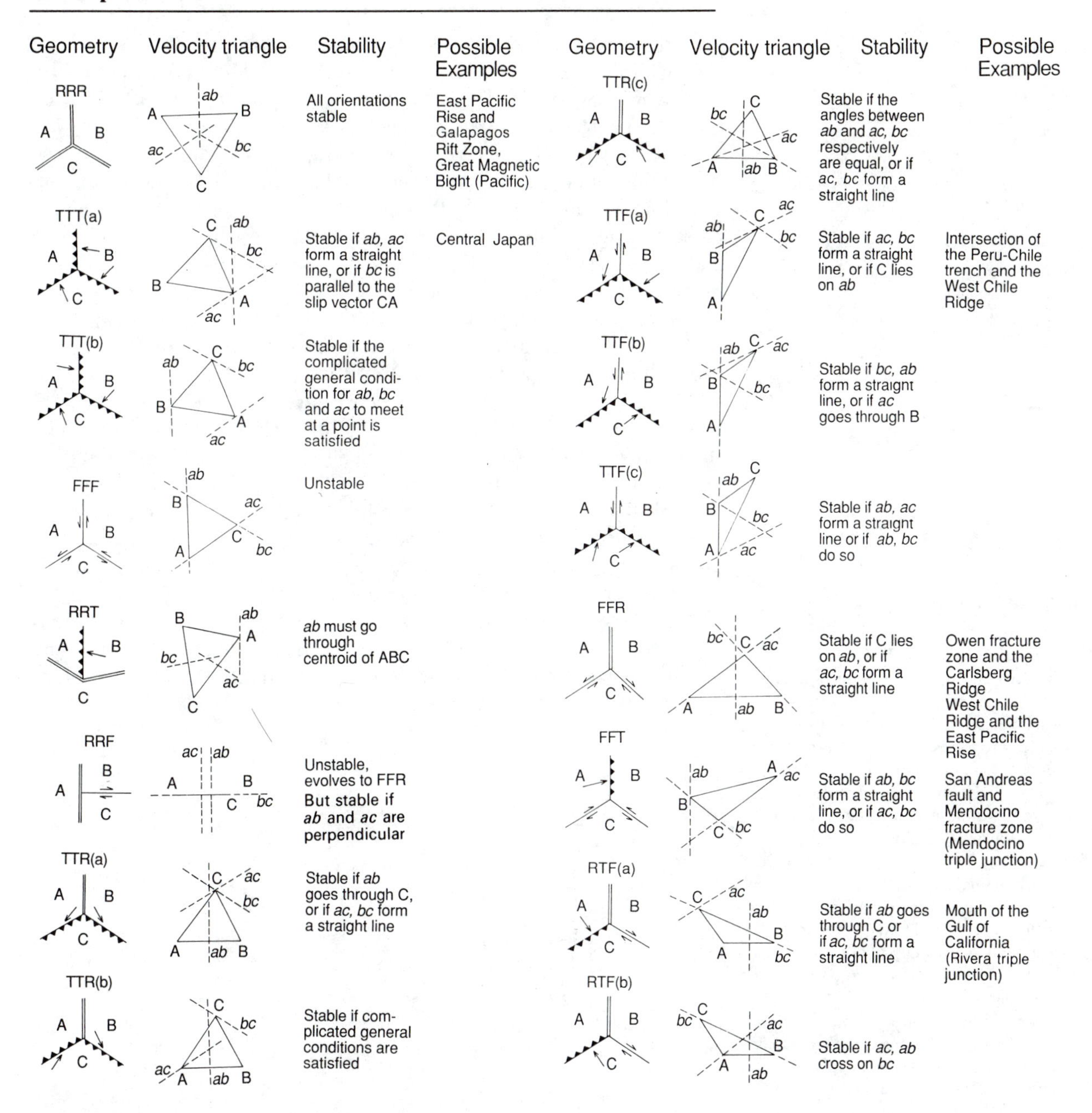

Figure 2.16. The geometry and stability of all possible triple junctions. In the categories represented by RRR, RTT, RTF and so on, R denotes ridge, T trench and F transform fault. The dashed lines *ab*, *bc* and *ac* in the velocity triangles represent velocities which leave the geometry of the boundary between plates A and B, B and C and A and C, respectively, unchanged. A triple junction is stable if *ab*, *bc* and *ac* meet at a point. Only an RRR triple junction (with ridges spreading symmetrically and perpendicular to their strikes) is always stable. (After McKenzie and Morgan 1969.)

the triangles are merely velocity triangles such as that shown in Figure 2.15b. The triple junction of Figure 2.15, type TTT(a) in Figure 2.16, is shown in Figure 2.17. The subduction zone between plates A and B does not move relative to plate A because plate A is overriding plate B. However, because all parts of the subduction zone look alike, any motion of the subduction zone parallel to itself would also satisfy this condition. Therefore, we can draw onto the velocity triangle a dashed line *ab*, which passes through point *A* and has the strike of the boundary between plates A and B (Fig. 2.17c). This line represents the possible velocities of the boundary between plates A and B which leave the geometry of these two plates unchanged. Similarly, we can draw a line *bc* which has the strike of the boundary between plates B and C and passes through point *C* (since the subduction zone is fixed on plate C) and a line *ac* which passes through *A* and has the strike of the boundary between plates A and C. The point at which the three dashed lines *ab*, *ac* and *bc* meet represents the velocity of a stable triple junction. Clearly, in Figure 2.17c, these three lines do not meet at a point; therefore, this particular plate boundary configuration is unstable. However, the three dashed lines would meet at a point (and the triple junction would be stable) if *bc* were parallel to *AC* or if *ab* and *ac* were parallel. This would mean that either the relative velocity between A and C, ${}_A\mathbf{v}_C$, was parallel to the boundary between plates B and C or the entire boundary of plate A was straight. These are the only two possible situations in which such a TTT triple junction is stable. By plotting the lines *ab*, *bc* and *ac* onto the relative velocity triangle, we can obtain the stability conditions more easily than we did in Figure 2.15.

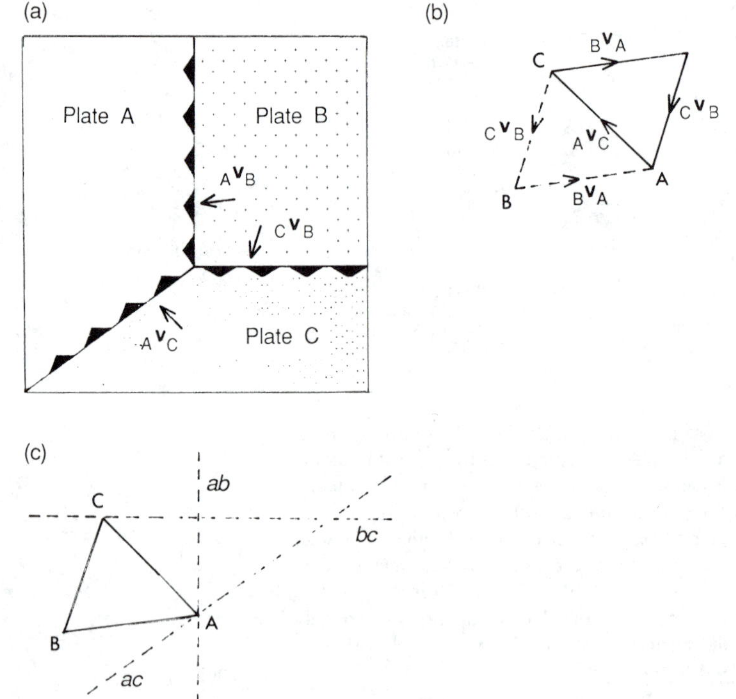

Figure 2.17. Determination of stability for a triple junction involving three subduction zones, TTT(a) of Fig. 2.16. (a) Geometry of the triple junction and relative velocities; this is the same example as Fig. 2.15a. (b) Relative velocity triangles. Sides *BA*, *CB* and *AC* represent ${}_B\mathbf{v}_A$, ${}_C\mathbf{v}_B$ and ${}_A\mathbf{v}_C$, respectively (i.e., the line from *A* to *C* is the velocity of plate C with respect to plate A, ${}_A\mathbf{v}_C$; and the line from *B* to *A* is the velocity of plate A with respect to plate B, ${}_B\mathbf{v}_A$). The upper triangle in the parallelogram is the relative velocity triangle of Figure 2.15. In the lower, dashed triangle the corner *A* represents the velocity of plate A; thus, for example, relative to plate A, plate C is moving with velocity of *AC*, ${}_A\mathbf{v}_C$. (c) The dashed lines *ab*, *bc* and *ac* drawn onto the velocity triangle *ABC* represent possible velocities of the boundary between plates A and B, plates B and C and plates A and C, respectively, which leave the geometry of that boundary unchanged. The triple junction is stable if these three dashed lines intersect at a point. In this case that would occur if *ab* were parallel to *ac* or if the velocity ${}_A\mathbf{v}_C$ were parallel to *bc*. If the geometry of the plate boundaries and relative velocities at the triple junction do not meet either of these conditions, then the triple junction is unstable and can only exist momentarily in geological time.

Figure 2.18 illustrates the procedure for a triple junction involving a ridge and two transform faults (type FFR). In this case, *ab*, the line representing motion along the ridge, must be the perpendicular bisector of *AB*, and *bc* and *ac*, representing motion along the faults, are collinear with *BC* and *AC*. This type of triple junction is stable only if line *ab* goes through *C* (both transform faults have the same slip rate) or if *ac* and *bc* are collinear (the boundary of plate C is straight). Choosing *ab* to be the perpendicular bisector of *AB* assumes that the ridge is spreading symmetrically and at right angles to its strike. This is usually the case. However, if the ridge does not spread symmetrically and/or at right angles to its strike, then *ab* must be drawn accordingly, and the stability conditions are different.

Figure 2.16 gives the conditions for stability of the various types of triple junction and also gives examples of some of the triple junctions occurring around the earth at present. Many research papers discuss the stability or instability of the Mendocino and Queen Charlotte triple junctions that lie off western North America. These are the junctions (at the south and north ends, respectively) of the Juan de Fuca Plate with the Pacific and North American plates and so are subjects of particular interest to North Americans because they involve the San Andreas Fault in California and the Queen Charlotte Fault in British Columbia. Another triple junction in that part of the Pacific is the Galapagos triple junction, where the Pacific, Cocos and Nazca plates meet; it is an RRR junction and thus is stable.

2.6.2 *Significance of Triple Junctions*

Work on the Mendocino triple junction, at which the Juan de Fuca, Pacific and North American plates meet at the northern end of the San Andreas Fault, shows why the stability of triple junctions is important for continental geology. The Mendocino triple junction is an FFT junction involving the San Andreas Fault, the Mendocino transform fault and the Cascade subduction zone. It is stable, as seen in Figure 2.16, provided that the San Andreas Fault and the Cascade subduction zone are collinear. It has, however, been suggested that the Cascade subduction zone is after all not exactly collinear with the San Andreas Fault and, thus, that the Mendocino triple junction is unstable. This instability would result in the northwards migration of the triple junction and the internal deformation of the continental crust of the western United States along preexisting zones of weakness. It would also explain many features such as the clockwise rotation of major blocks, such as the Sierra Nevada, and the regional extension and eastward stepping of the San Andreas transform. The details of the geometry of this triple junction are obviously of great importance to the regional evolution of the entire western United States. Much of the geological history of the area over approximately the past 30 million years may be related to the migration of the triple junction, and so a detailed knowledge of the plate motions is essential background for any explanation of the origin of Tertiary structures in this region. This subject is discussed further in Section 3.3.3. The motions of offshore plates can thus produce major structural changes even in the continent.

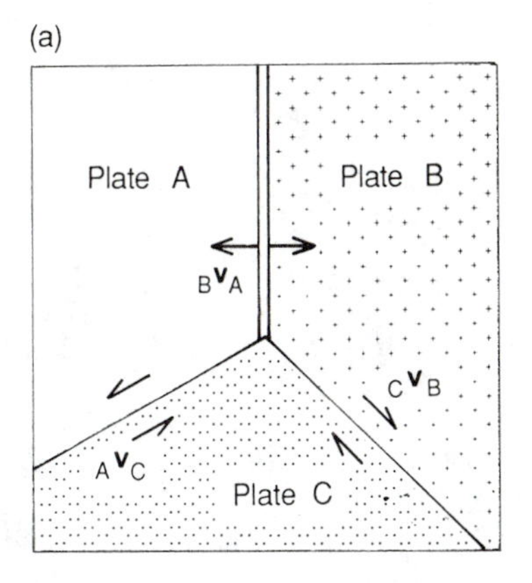

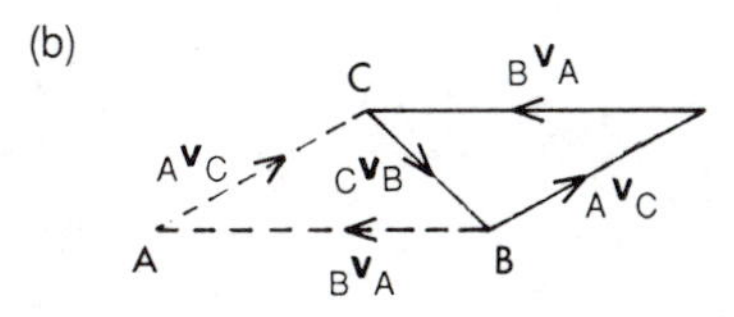

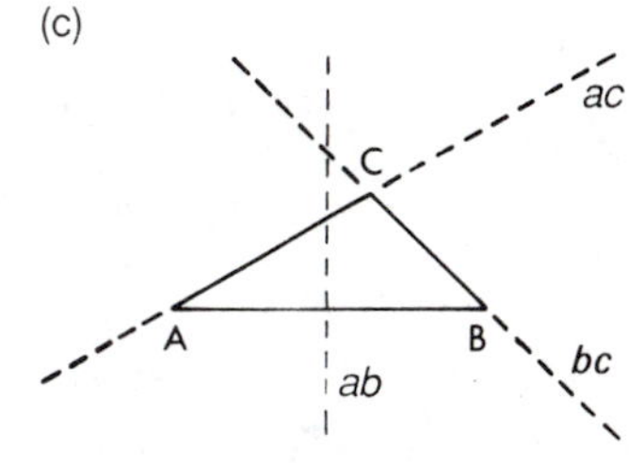

Figure 2.18. Determination of stability for a triple junction involving a ridge and two transform faults, FFR. (a) Geometry of the triple junction and relative velocities. (b) Relative velocity triangles (notation as for Fig. 2.17b). (c) Dashed lines drawn onto relative velocity triangle as for Figure 2.17c. This triple junction is stable if point *C* lies on the line *ab* or if *ac* and *bc* are collinear. The first condition is satisfied if the triangle *ABC* is isosceles (i.e., the two transform faults are mirror images of each other). The second condition is satisfied if the boundary of plate C with plates A and B is straight.

2.7 Absolute Plate Motions

Although most of the volcanism on the earth's surface is associated with the boundaries of plates, along the midocean ridges and subduction zones, some isolated volcanic island chains occur in the oceans. These chains of oceanic islands are unusual in several respects: They occur well away from the plate boundaries (i.e., they are *intraplate volcanoes*); the chemistry of the erupted lavas is significantly different from both midocean ridge and subduction zone lavas; the active volcano may be at one end of the island chain, with the islands aging with distance from that active volcano; the island chains appear to be arcs of small circles. These features, taken together, are consistent with the volcanic islands having formed as the plate moved over what is colloquially called a *hot spot*, a place where melt rises from deep in the mantle. Figure 2.19 shows four volcanic island and seamount chains in the Pacific Ocean. There is an active volcano at the southeastern end of each of the island chains. The Emperor–Hawaiian seamount chain is the best defined and most studied. The ages of the seamounts increase steadily from Loihi, the youngest and presently active, a submarine volcano which lies off the southeast coast of the main island of Hawaii, northwest through the Hawaiian chain. There is a pronounced change in strike where the volcanic rocks reach 43 million years old (Ma). The northern end of the Emperor seamount chain near the Kamchatka peninsula of the USSR is 78 Ma old. The change in strike at 43 Ma can most simply be explained by a change in the direction of movement of the Pacific Plate over the hot spot at that time. The chemistry of oceanic island lavas is discussed in Section 7.8.3 and the structure of the islands themselves in Section 8.7.

All the plate motions described so far in this chapter have been relative motions, that is, motions of the Pacific Plate relative to the North American Plate, the African Plate relative to the Eurasian Plate and so on. There is no fixed point on the earth's surface. *Absolute plate motions* are motions of the plates relative to some imaginary fixed point. One way of determining absolute motions is to suppose that the earth's mantle moves much more slowly than the plates so that it can be regarded as nearly fixed. Such absolute motions can be calculated from the traces of the oceanic island chains or the traces of continental volcanism, which are assumed to have formed as the plate passed over a hot spot with its source in the mantle. The absolute motion of a plate, the Pacific, for example, can be calculated from the traces of the oceanic island and seamount chains on it. The absolute motions of all the other plates can then be calculated from their motion relative to the Pacific Plate. Repeating the procedure, using hot spot traces from other plates, gives some idea of the validity of the assumption that hot spots are fixed. Figure 2.20 shows a determination of the present absolute plate motions. They are not fixed; the mantle is moving. Nevertheless, a framework of motion can be developed. The Pacific and Indian plates are moving fast and the North American Plate more slowly, while the Eurasian and African plates are hardly moving at all.

Figure 2.21 is a plot of the orientation of horizontal stress measured in the interior of the plates. The pattern of these intraplate stresses in the crust (stress is force per unit area) can be used to assess the forces acting on the

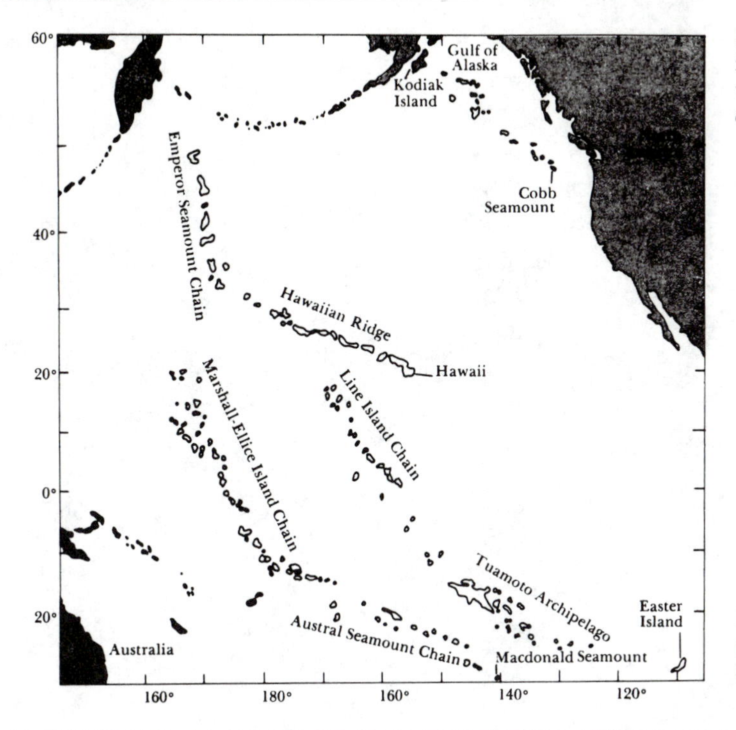

Figure 2.19. Four volcanic island chains in the Pacific Ocean. The youngest active volcano is at the southeast end of each chain. (From Dalrymple et al. 1973.)

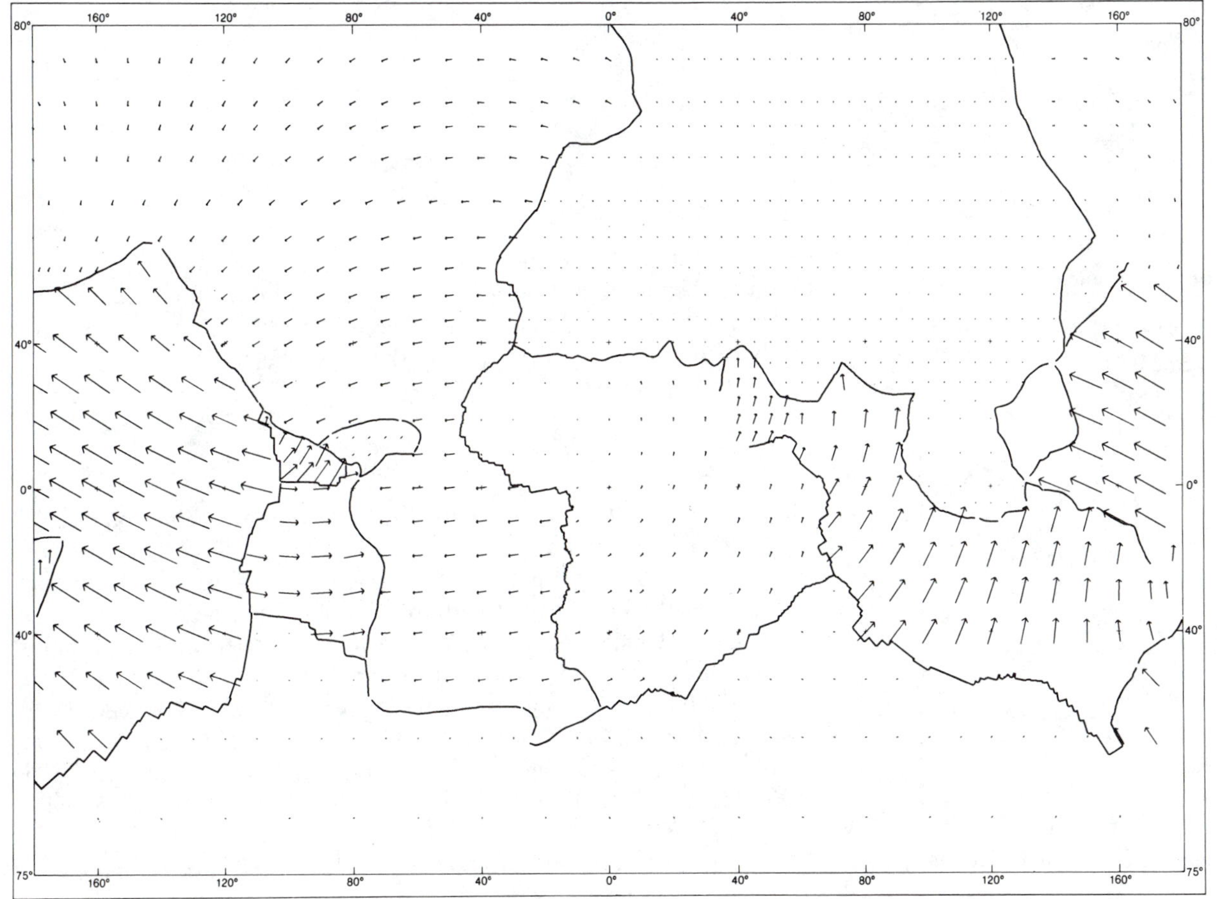

Figure 2.20. Absolute motions of the plates as determined from hot spot traces. (Courtesy M. L. Zoback, after Minster and Jordan 1978.)

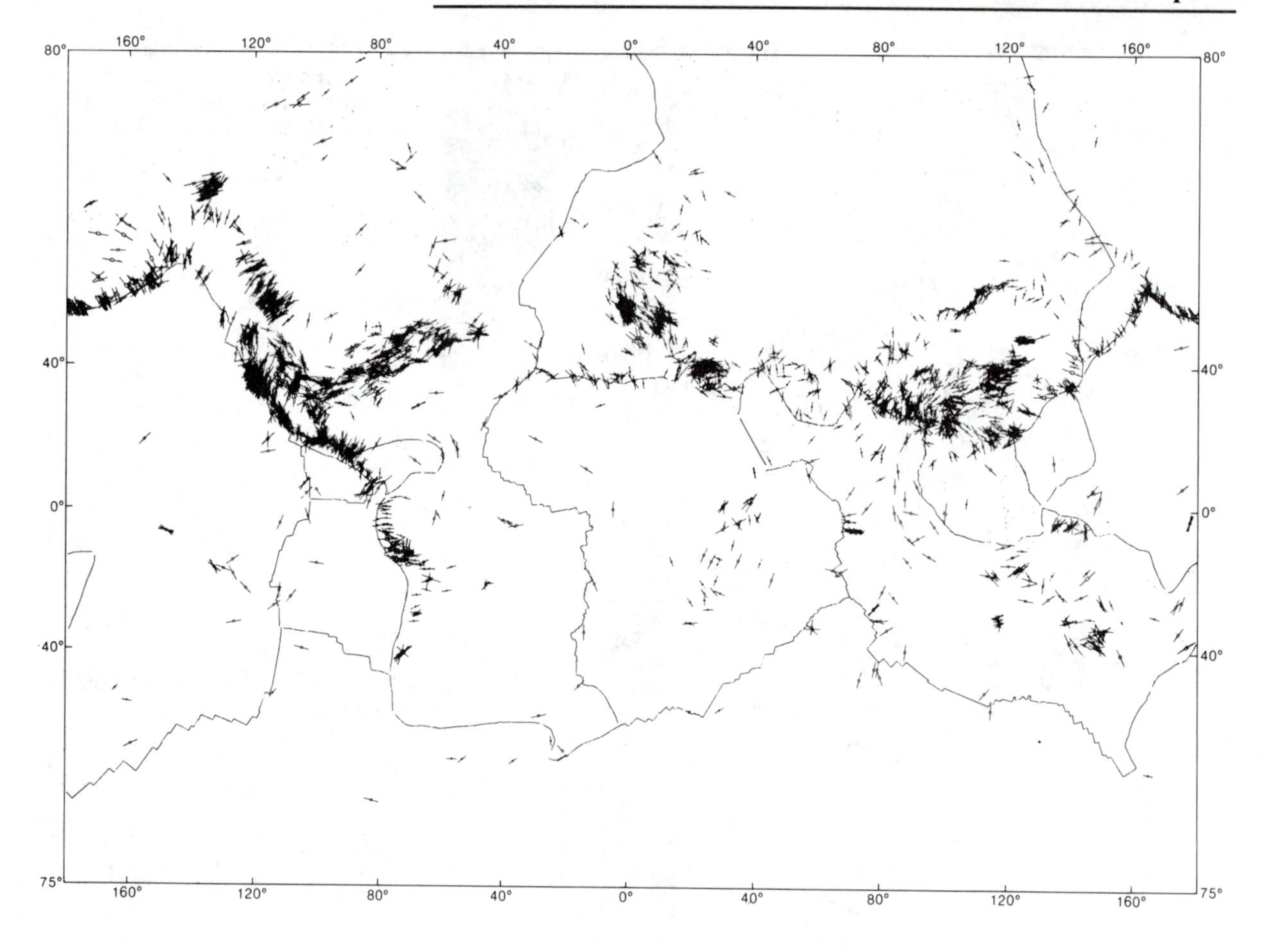

Figure 2.21. World stress map. Lines show orientation of the maximum horizontal compressive stress. (From Zoback et al. 1989)

plates. The direction of the stresses is correlated with the direction of the absolute plate motions (Fig. 2.20). This implies that the forces moving the plates around, which act along the edges of the plates (Sect. 7.10), are also partly responsible for the stresses in the lithosphere.

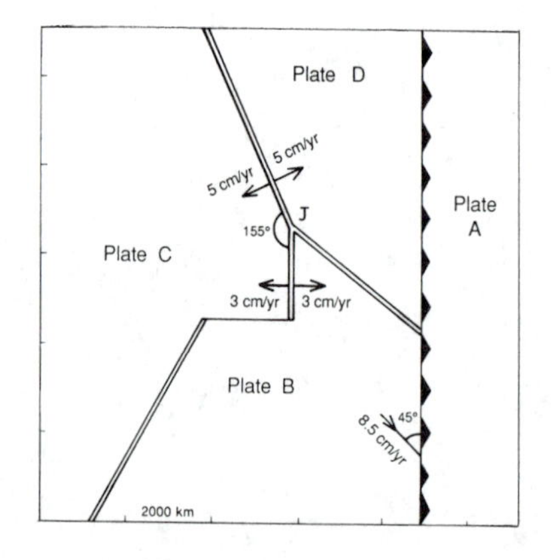

Figure 2.22. Map of part of a flat planet.

PROBLEMS

1. All plates A–D shown in Figure 2.22 move rigidly without rotation. All ridges add at equal rates to the plates on either side of them; the rates given on the diagram are half the plate separation rates. The trench forming the boundary of plate A does not consume A.

   Use the plate velocities and directions to determine by graphical means or otherwise (a) the relative motion between plates B and D and (b) the relative motion between the triple junction J and plate A. Where and when will J reach the trench? Draw a sketch of the geometry after this collision showing the relative velocity vectors and discuss the subsequent evolution. (From Cambridge University Natural Sciences Tripos 1B, 1982.)
2. Long ago and far away on a distant galaxy lived two tribes on a *flat*

planet known as Emit-on (Fig. 2.23). The tribe of the dark forces lived on the islands in the cold parts of the planet near *A*, and a happy light-hearted people lived on the sunny beaches near *B*. Both required a constant supply of fresh andesite to survive and collected the shrimp-like animals that prospered near young magnetic-anomaly patterns, for their food. Using the map and your knowledge of any other galaxy, predict the future of these tribes. (From Cambridge University Natural Sciences Tripos 1B, 1980.)

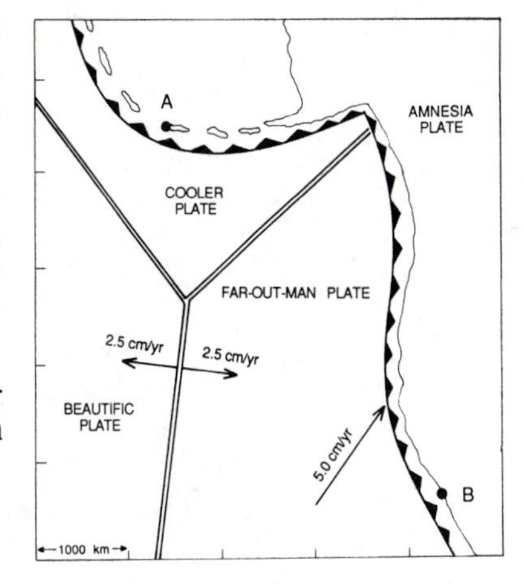

Figure 2.23. Map of part of Emit-on.

3. Use the map in Fig. 2.2 to do the following:
   (a) Calculate the relative present-day motions at the tabulated locations, using or calculating the appropriate poles from those given in Table 2.1.

Latitude	Longitude	Location
54°N	169°E	W. Aleutian Trench
52°N	169°W	E. Aleutian Trench
38°N	122°W	San Francisco–San Andreas Fault
26°N	110°W	Gulf of California
13°S	112°W	East Pacific Rise
36°S	110°W	East Pacific Rise
59°S	150°W	Antarctic–Pacific Ridge
45°S	169°E	S. New Zealand
55°S	159°E	Macquarrie Island
52°S	140°E	Southeast Indian Ridge
28°S	74°E	Southeast Indian Ridge
7°N	60°E	Carlsberg Ridge
22°N	38°E	Red Sea
55°S	5°E	Southwest Indian Ridge
52°S	5°E	Mid-Atlantic Ridge
9°N	40°W	Mid-Atlantic Ridge
35°N	35°W	Mid-Atlantic Ridge
66°N	18°W	Iceland
36°N	8°N	Gorringe Bank
35°N	25°E	E. Mediterranean
12°S	120°E	Java Trench
35°N	72°E	Himalayas
35°S	74°W	S. Chile Trench
4°S	82°W	N. Peru Trench
20°N	106°W	Middle America Trench

   (b) Plot the azimuth and magnitude of these relative motions on the map.
   (c) Plot the pole positions. Note how the relative motions change along plate boundaries as distance from the pole changes.
   (d) Discuss the nature of the plate boundary between the Indian/Australian and Pacific plates.

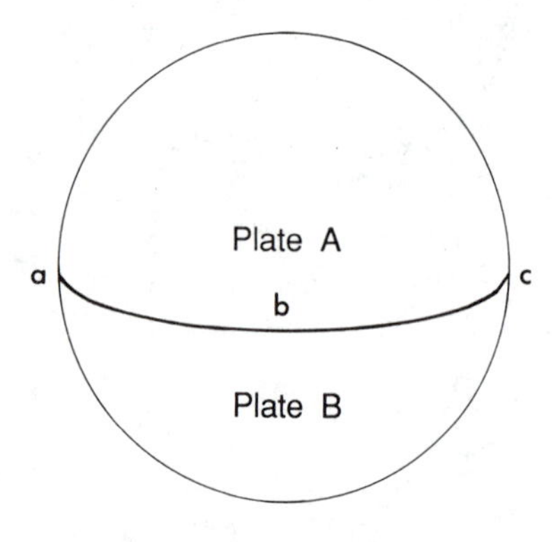

Figure 2.24. Planet Ares has just two plates.

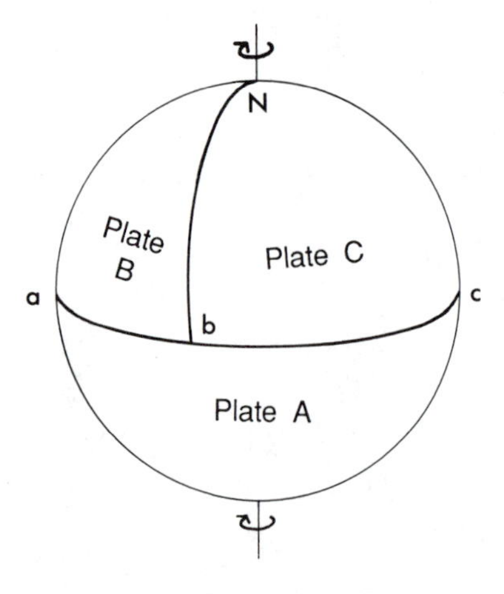

Figure 2.25. Tritekton has three plates.

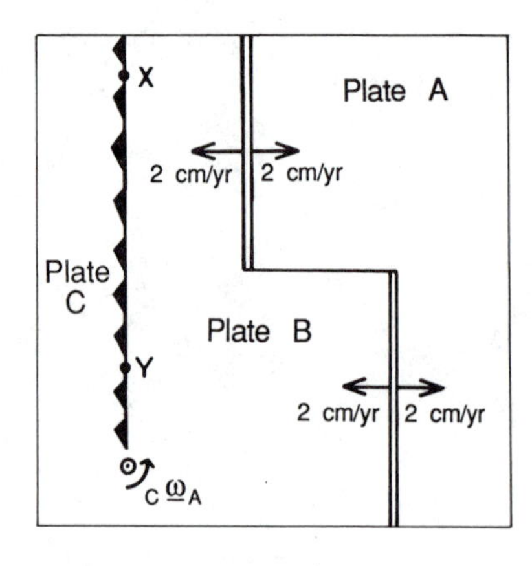

Figure 2.26. Map for problem 7

(e) Discuss the nature of the plate boundary between the Eurasian and African plates between the Azores and Gibraltar. How does this boundary differ in the Mediterranean?

(From G. C. P. King, personal communication.)

4. Define the pole of relative angular velocity for two plates and explain why it is a useful concept. Give two practical methods for finding poles of rotation from geophysical data. When these data are not available how would you try to locate the poles? (Cambridge University Natural Sciences Tripos IB, 1977.)
5. Ares is a planet with just two plates A and B (Fig. 2.24). Plate B comprises the lower hemisphere and plate A the upper, as shown. Points *a*, *b* and *c* lie on the equator, and point *d* is diametrically opposite *b*. The zero meridian passes through point *a*. The pole of rotation A relative to B is at 45°N 0°E. The amplitude of the angular velocity vector is 10^{-10} radians per terrestrial year. The radius of Ares is 3400 km.
   (a) What is the nature of the plate boundary between plates A and B?
   (b) State where magnetic lineations might be found and sketch the pattern that would be observed.
   (c) Calculate the relative velocity between plates A and B at locations *a*, *b*, *c* and *d*.
   (d) Sketch possible fault plane solutions for earthquakes occurring at locations *a*, *b*, *c* and *d*.
   (e) Discuss the possible existence of such a two-plate planet.
   (f) Discuss briefly how the stability or instability of a two-plate tectonic system depends upon the pole position and/or relative size of the two plates.
6. The lithosphere of Tritekton, a recently discovered spherical satellite of Jupiter, consists of three plates, as shown in Figure 2.25. Plate A is a hemisphere, and plates B and C are half a hemisphere each. Points *a*, *b* and *c* lie on its equator, and point *d* (not visible) lies diametrically opposite *b*. The zero meridian lies midway between *b* and *d* and passes through *a*. The pole of rotation B to A lies in plate B at latitude 30°N longitude 0°. The amplitude of the angular velocity vector through this pole is 3×10^{-9} radians per terrestrial year. The rotation pole C to B is at the north pole N, and the amplitude of the angular velocity vector is 6×10^{-9} rad yr^{-1}. The radius of Tritekton is 6000 km.
   (a) Find the coordinates of the angular velocity vector of C to A and its amplitude.
   (b) Draw velocity triangles for the triple junctions *b* and *d*.
   (c) Tritekton has a magnetic field that reverses every 10 million years. Draw the magnetic pattern created at the triple junction where extension is taking place. State whether it is at *b* or *d*.
   (d) The features at each plate boundary are trenches, transform faults or spreading ridges. Describe their distribution along each plate boundary.

(Cambridge University Natural Sciences Tripos IB, 1979.)

7. In Figure 2.26, the trench between B and C is consuming B only. The ridge between A and B is spreading symmetrically at right angles to its axis. The pole of rotation between A and C is fixed to C. The angular

velocity of A with respect to C is in the direction shown, and is 2×10^{-8} radians yr^{-1} in magnitude.

(a) Mark the poles of rotation and angular velocities of (i) plate B and (ii) the ridge axis with respect to plate C (remember that the pole of motion between two plates is that point which is stationary with respect to both of them).

(b) Show the direction and rate of consumption at X and Y.

(c) How long will it take for the ridge to reach (i) X and (ii) Y? Sketch the history of the triple junction between A, B and C with respect to C.

(Cambridge University Natural Sciences Tripos 1B, 1981.)

8. The three plates A, B and C meet at a ridge–ridge–ridge triple junction as shown in Figure 2.27. The ridge between plates A and B has a half-spreading rate of $2\,cm\,yr^{-1}$. Calculate
   (a) the half-spreading rates of the other two ridges and
   (b) the motion of the triple junction relative to plate C.
9. All the plates of problem 1 move rigidly without rotation. Discuss the difference between those motions and the motions that would take place were the plates on the surface of a spherical earth. (For example, by how much would the plate separation rates vary along the length of the ridge between plates B and C?)
10. Four flat plates are moving rigidly, without rotation, on a flat earth as illustrated in Figure 2.28.
    (a) Determine the relative motion vector of the Beautific–Jokers–Nasty (BJN) triple junction to the Albatross Plate.
    (b) Determine the relative motion vector of the Beautific–Nasty–Erratic (BNE) triple junction to the Albatross Plate.
    (c) Using these vectors, draw the locus of the future position of these junctions relative to the Albatross Plate. Mark points at 10 Ma intervals.
    (d) At 10 Ma intervals redraw the positions of the plate boundaries until both the Jokers Plate and the Nasty Plate have been consumed at the subduction zone boundary by the Albatross Plate.
    (e) Draw velocity triangles for triple junctions at the times when they pass through points A and B.
    (f) Draw the magnetic anomaly pattern formed by the BJN junction for 10 Ma, assuming that a recognizable anomaly appears at 2 Ma intervals.

    (From G. C. P. King, personal communication.)
11. Figure 2.29 depicts a mythical planet whose tectonic features comprise three plates A, B and C with boundaries as shown. The plate boundary A–B is a fault with sense of motion as shown and relative velocity of $4\,cm\,yr^{-1}$, and that between B and C is also a fault with velocity $3\,cm\,yr^{-1}$.
    (a) Using angular velocity vectors find the pole of relative motion between A and C, giving its geographic coordinates. Calculate the relative angular velocity in degrees per year (mythical planet radius $= 6000\,km$).
    (b) Describe the tectonic nature of the plate boundary between A and C, giving details of places where magnetic lineations might be found (mythical planet has a reversing magnetic field like the

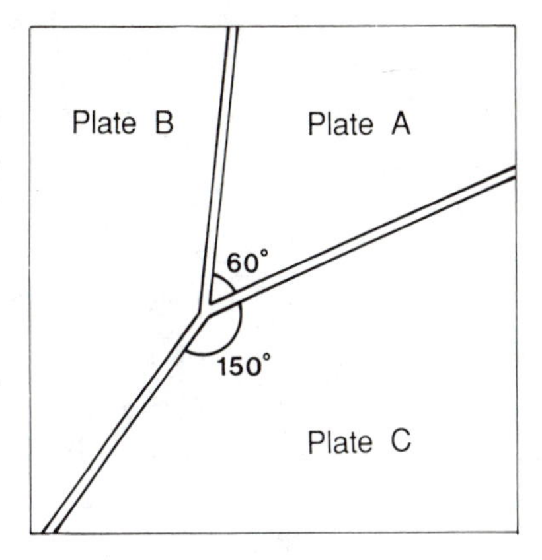

Figure 2.27. A ridge–ridge–ridge triple junction.

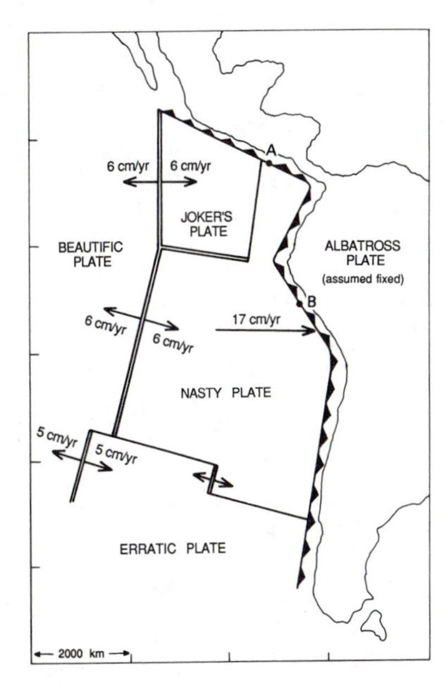

Figure 2.28. Plates on a flat earth.

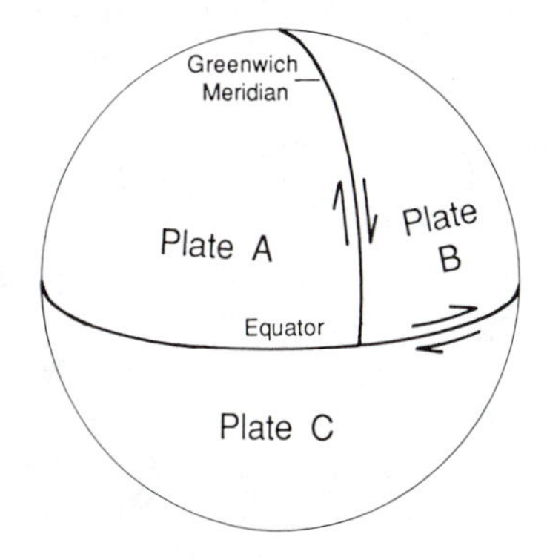

Figure 2.29. Mythical planet has three plates.

earth's), what the maximum spreading rate might be expected to be and any other geological and geophysical information that you can deduce.

(c) Consider the triple junction where A, B and C all meet on the visible side of mythical planet. What do you think will happen to it if the instantaneous velocity description we have given continues for several million years? In particular, think about the way ridges can only spread symmetrically and trenches consume on only one side. Is the junction likely to persist as we have drawn it? What about the junction on the other side of mythical planet? Is it similar?

BIBLIOGRAPHY

Altman, S. L. 1986. *Rotations, quaternians and double groups.* Oxford Univ. Press, New York.

Barazangi, M., and Dorman, J. 1969. World seismicity maps compiled from ESSA, Coast and Geodetic Survey, epicenter data, 1961–1967. *Bull. Seism. Soc. Am., 59*, 369–80.

Båth, M. 1979. *Introduction to seismology, 2d ed.* Birkhauser, Boston.

Bullard, E. C., Everett, J. E., and Smith, A. G. 1965. Fit of continents around the Atlantic. In P. M. S. Blackett, E. C. Bullard and S. K. Runcorn, eds., A Symposium on Continental Drift, *Roy. Soc. London, Phil. Trans. Ser. A, 258*, 41–75.

Carter, W. E., and Robertson, D. S. 1986. Studying the Earth by very-long-baseline interferometry. *Sci. Am., 255*, 46–54.

Chase, C. G. 1978. Plate kinematics: the Americas, East Africa and the rest of the world. *Earth Planet. Sci. Lett., 37*, 355–68.

Cox, A., ed. 1973. *Plate tectonics and geomagnetic reversals.* Freeman, San Francisco.

Cox, A., and Hart, R. B. 1986. *Plate tectonics: How it works.* Blackwell Scientific, Oxford.

Dalrymple, G. B., Silver, E. A., and Jackson, E. D. 1973. Origin of the Hawaiian islands. *Am. Scientist, 61*, 294–308.

DeMets, C., Gordon, R. G., Argus, D. F. and Stein, S. 1990. Current plate motions. *Geophys. J. Int, 101*, 425–78.

Dickinson, W. R., and Snyder, W. S. 1979. Geometry of triple junctions related to San Andreas transform. *J. Geophys. Res., 84*, 561–72.

Gubbins, D. 1990. *Seismology and plate tectonics.* Cambridge Univ. Press, Cambridge.

Irving, E. 1977. Drift of the major continental blocks since the Devonian. *Nature, 270*, 304–9.

Kroger, P. M., Lyzenga, G. A., Wallace, K. S., and Davidson, J. M. 1987. Tectonic motion in western United States inferred from very long baseline interferometry measurements, 1980–1986. *J. Geophys. Res., 92*, 14151–63.

Le Pichon, X., Francheteau, J., and Bonnin, J. 1973. *Plate tectonics*, Vol. 6 of *Developments in geotectonics.* Elsevier, Amsterdam.

McKenzie, D. P., and Morgan, W. J. 1969. Evolution of triple junctions. *Nature, 224*, 125–33.

McKenzie, D. P., and Parker, R. L. 1967. The north Pacific: An example of tectonics on a sphere. *Nature, 216*, 1276–80.

Minster, J. B., and Jordan, T. H. 1978. Present-day plate motions. *J. Geophys. Res., 83*, 5331–54.

Morgan, W. J. 1968. Rises, trenches, great faults and crustal blocks. *J. Geophys. Res.*, *73*, 1959–82.

1971. Convection plumes in the lower mantle. *Nature, 230*, 42–3.

Stein, R. S. 1987. Contemporary plate motion and crustal deformation. *Rev. Geophys.*, *25*, 855–63.

1988. Plate tectonic prediction fulfilled. *Physics Today* (Jan.), S42–4.

Sykes, L. R. 1967. Mechanism of earthquakes and nature of faulting on the mid-ocean ridges. *J. Geophys. Res.*, *72*, 2131–53.

Wilson, J. T. 1965. A new class of faults and their bearing on continental drift. *Nature, 207*, 343–7.

Ziegler, A. M., Scotese, C. R., McKerrow, W. S., Johnson, M. E., and Bambach, R. K. 1979. Paleozoic paleogeography. *Ann. Rev. Earth Planet. Sci.*, *7*, 473–502.

Zoback, M. L., and 28 others 1989. Global patterns of tectonic stress: A status report on the World Stress Map Project of the International Lithosphere Program. *Nature, 341*, 291–8.

3

Past Plate Motions

3.1 The Role of the Earth's Magnetic Field

3.1.1 Introduction

It should be clear from the preceding chapter that it is possible, without too much difficulty, to calculate the relative motions of pairs of plates at any location along their common boundary and to see what may occur in the future. This chapter deals with the past motions of the plates and shows how to reconstruct their previous interactions from evidence they have left.

Two important facts together make it possible to determine past plate motions. The first is that the earth's magnetic field has not always had its present (normal) polarity with the 'north' magnetic pole close to the north geographic pole* and the 'south' magnetic pole close to the south geographic pole. Over geological history the magnetic field has intermittently reversed. Thus, there have been times in the past when the north magnetic pole has been located close to the south geographic pole and the south magnetic pole has been located close to the north geographic pole; then the field is said to be reversed. The second fact is that, under certain circumstances (discussed in Sect. 3.2) rocks can record the earth's past (*palaeo*) *magnetic* field. Together, these facts enable us to estimate dates and past positions of the plates from magnetic measurements.

3.1.2 The Earth's Magnetic Field

To specify the geomagnetic field at any point on the earth's surface both a magnitude and a direction are required: The geomagnetic field is a vector quantity. It is far from being constant either in magnitude or in direction and varies spatially over the surface of the earth as well as in time. Systematic mapping of the magnetic field began some five hundred years ago with work by the early explorers of the Atlantic Ocean. The internationally agreed values of the geomagnetic field are updated and published every few years as the *International Geomagnetic Reference Field* (IGRF). Figure 3.1 shows the magnitude of the IGRF for 1980. To a first approximation the geomagnetic field is a *dipole field*. This means that the

* The geomagnetic pole presently situated in the Northern Hemisphere is in fact a south pole since it attracts the north poles of magnets (compass needles)! See Figure 3.2.

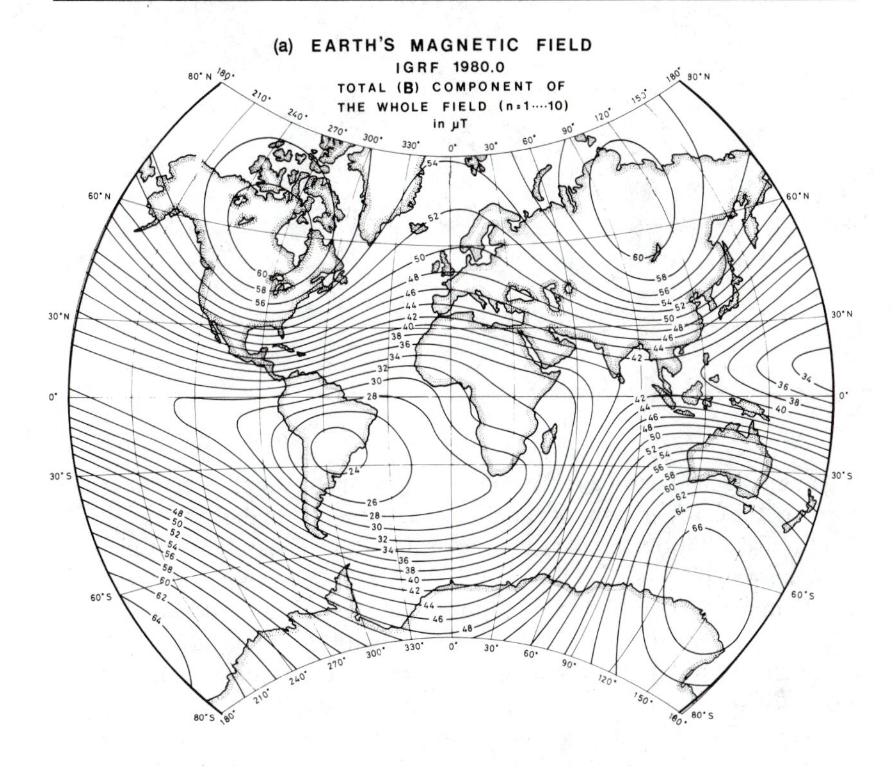

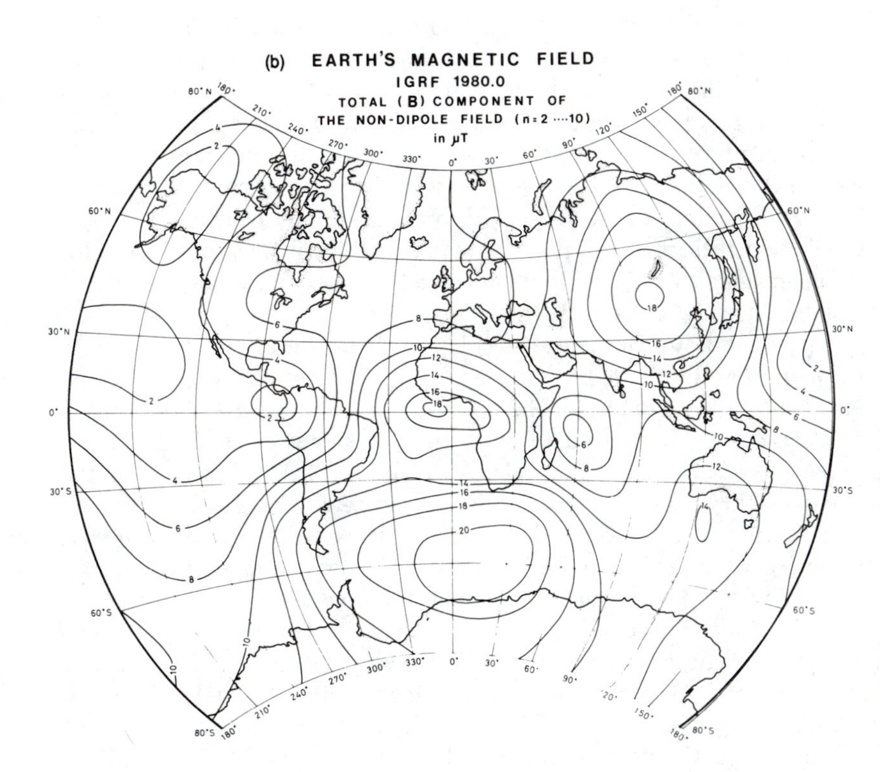

Figure 3.1. (a) Magnitude of the earth's magnetic field in microteslas (μT). (b) Magnitude of the nondipole component of the earth's magnetic field (magnetic field — best-fitting dipole field) in microteslas (μT). The field is the International Geomagnetic Reference Field for 1980. The nanotesla (nT $= 10^{-3}\,\mu$T), an SI magnetic field unit, is the same as gamma (γ), the magnetic field unit of the c.g.s. system. (From Nevanlinna et al. 1983.)

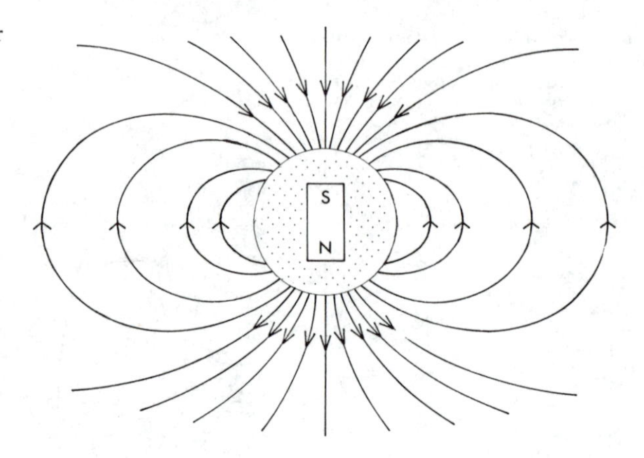

Figure 3.2. A magnetic dipole (e.g., a bar magnet) at the centre of the earth has a magnetic field which is a good first approximation to the earth's magnetic field. Note that the geomagnetic north pole is so called because the north end of the compass needle points towards it (it is therefore actually a south magnetic pole). Such a dipole field would also be produced by a uniformly magnetized earth, a uniformly magnetized core or particular current systems within the core (Fig. 7.23). The origin of the field and models of the core and possible reasons for the reversals of the earth's magnetic field are discussed further in Section 7.9.

geomagnetic field can be represented by a magnetic dipole situated at the centre of the earth (imagine a bar magnet at the centre of the earth, Fig. 3.2). This approximation was first pointed out by Sir William Gilbert in 1600 while he was physician to Queen Elizabeth I. The difference between the earth's magnetic field and the best dipole field is termed the *nondipole field*. Figure 3.1b shows the magnitude of the nondipole field for the IGRF 1980. At the earth's surface the nondipole field is small compared with the dipole field, though this is not the case for the field at the core–mantle boundary.

Today, the best-fitting dipole is aligned at about 11.5° to the earth's geographic north–south axis (spin axis). The *geomagnetic poles* are the two points at which the axis of this best-fitting dipole intersect the earth's surface. Now at 79°N, 71°W and 79°S, 109°E, they are called the geomagnetic north and geomagnetic south poles, respectively. The *geomagnetic equator* is the equator of the best-fitting dipole axis. The two points on the earth's surface at which the magnetic field is vertical and has no horizontal component are called the *magnetic poles*, or *dip poles*. The present north magnetic pole is at 76°N, 101°W, and the south magnetic pole is at 66°S, 141°E. The *magnetic equator* is the line along which the magnetic field is horizontal and has no vertical component. If the field were exactly a dipole field, these magnetic poles and this equator would be coincident with the geomagnetic north and south poles and equator. The various poles and equators are illustrated in Figure 3.3.

Figure 3.4 shows the magnetic field lines around the earth. The sun plays the major part in the shape of the field far from the earth and in the short-term variations of the field. The *solar wind*, a constant stream of ionized particles emitted by the sun, confines the earth's magnetic field to a region known as the *magnetosphere* and deforms the field lines so that the magnetosphere has a long 'tail', the *magnetotail*, which extends several million kilometres away from the sun. A shock wave, the *bow shock*, is produced where the solar wind is slowed by interaction with the earth's magnetic field. A turbulent zone within the bow shock is known as the *magnetosheath*, the inner boundary of which is called the *magnetopause*. The earth's magnetic field, therefore, shields the earth from most of the incident radiation, and the atmosphere absorbs most of the remainder. The *Van Allen radiation belts* are zones of charged particles trapped by the earth's magnetic field. Changes in the solar wind, and hence in the magnetic field,

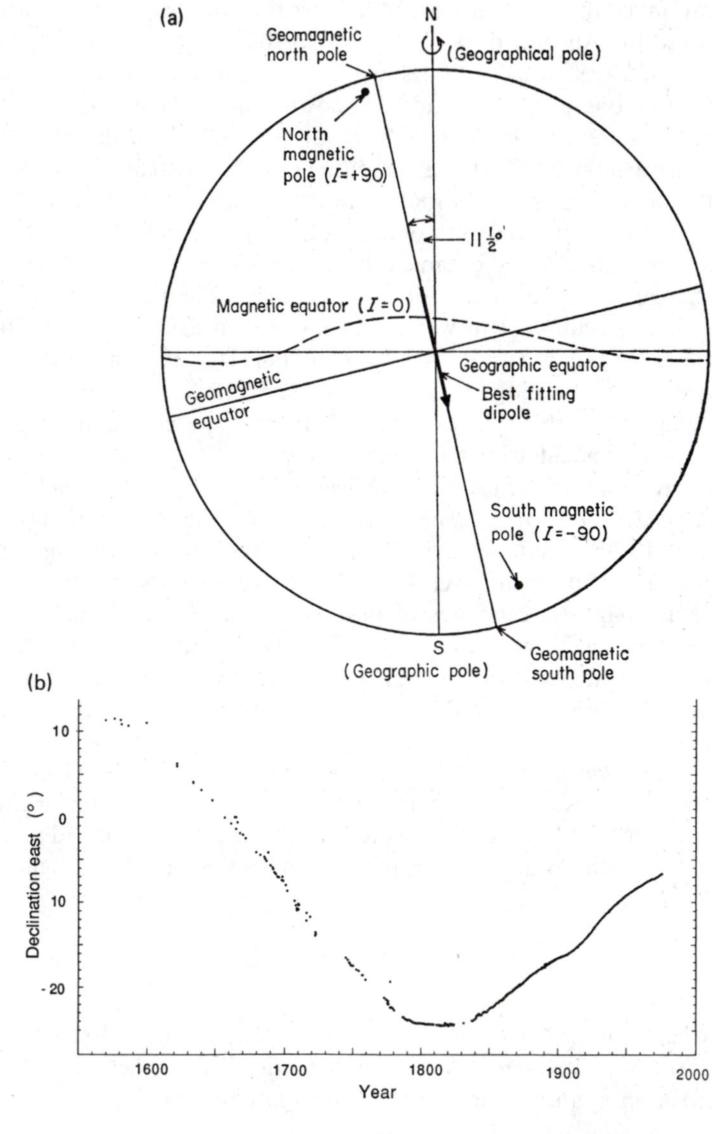

Figure 3.3. (a) Diagram illustrating the difference between the earth's geographic, geomagnetic and magnetic poles and equator. (From McElhinny 1973.) (b) The change in magnetic declination as observed from London over four centuries. (From Malin and Bullard 1981.)

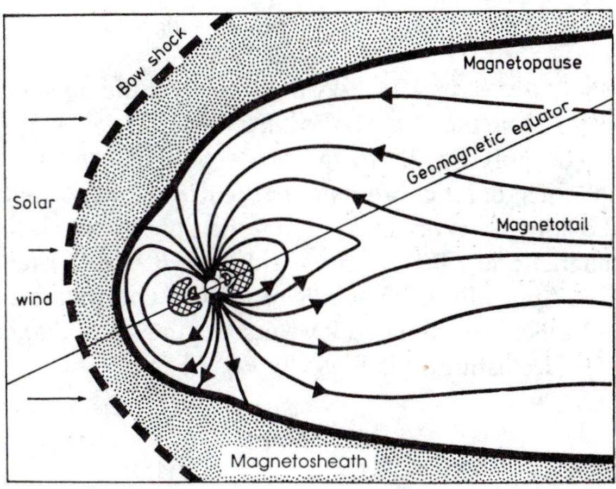

Figure 3.4. The earth's magnetic field far from the earth is controlled largely by the solar wind. (After Hutton 1976.)

can make these charged particles enter the upper atmosphere, where they cause the Auroral displays or the Northern and Southern lights. These irregularities, which are related to sunspot activity, cause short-term fluctuations in the magnetic field known as *magnetic storms.* The sun is also responsible for the *diurnal* (daily) variation in the earth's magnetic field. This variation, which has an amplitude of less than 0.5% of the total field, is the main short-period variation in the earth's magnetic field.

The long-term reversals of the magnetic field mentioned in Section 3.1.1 are used to date the oceanic lithosphere. However, the geomagnetic field changes on a much shorter time scale too: The pole wanders by a few degrees a century. This wandering has been measured throughout historical time and is termed *secular variation.* However, it appears that on average over geological time, the geomagnetic field axis has been aligned along the earth's spin axis (i.e., on average the geomagnetic poles have been coincident with the geographic poles). This means that, to a first approximation, *the geomagnetic field can be modelled as the field of a dipole aligned along the geographic north–south axis.* This assumption is critical to all palaeomagnetic work: If the dipole axis had wandered randomly in the past and had not on average been aligned along the geographic axis, all palaeomagnetic estimates of past positions of rock samples would be meaningless because they would be relative only to the position of the geomagnetic pole at the time each sample acquired its permanent magnetization and would have nothing at all to do with the geographical pole.

The *magnetic potential* from which the earth's magnetic field is derived can be expressed as an infinite series of spherical harmonic functions. The first term in this series is the potential due to a dipole situated at the centre of the earth. At any point **r**, from a dipole, the magnetic potential $V(\mathbf{r})$ is given by

$$V(\mathbf{r}) = \frac{1}{4\pi r^3}\mathbf{m}\cdot\mathbf{r} \tag{3.1}$$

where **m** is the *dipole moment,* a vector aligned along the dipole axis. For the earth **m** is 7.94×10^{22} A m^2 in magnitude. The magnetic field $\mathbf{B}(\mathbf{r})$ at any position **r** can be determined by differentiating the magnetic potential

$$\mathbf{B}(\mathbf{r}) = -\mu_0 \nabla V(\mathbf{r}) \tag{3.2}$$

where μ_0 $(= 4\pi \times 10^{-7}\ \mathrm{kg\,m\,A^{-2}\,s^{-2}})$ is the magnetic permeability of free space (A is the abbreviation for amp).

To apply Eq. 3.2 to the earth, we find it most convenient to work in spherical polar coordinates (r, θ, ϕ), (r is the radius, θ the colatitude and ϕ the longitude or azimuth on the sphere, as shown in Fig. A1.4). The magnetic field $\mathbf{B}(\mathbf{r})$ is written as $\mathbf{B}(\mathbf{r}) = (B_r, B_\theta, B_\phi)$ in this coordinate system. (See Appendix 1 for details of various coordinate systems.)

In these coordinates, if we assume that **m** is aligned along the *negative z* axis (see caption for Fig. 3.2), Eq. 3.1 is

$$V(\mathbf{r}) = -\frac{1}{4\pi r^3}\mathbf{m}\cdot\mathbf{r}$$

$$= -\frac{mr\cos\theta}{4\pi r^3}$$

$$= -\frac{m\cos\theta}{4\pi r^2} \tag{3.3}$$

Substitution of Eq. 3.3 into Eq. 3.2 gives the three components (B_r, B_θ, B_ϕ) of the magnetic field due to a dipole at the centre of the earth. The radial component of the field is B_r:

$$B_r(r,\theta,\phi) = -\mu_0\frac{\partial V}{\partial r}$$

$$= \frac{\mu_0 m\cos\theta}{4\pi}\frac{\partial}{\partial r}\left(\frac{1}{r^2}\right)$$

$$= -\frac{2\mu_0 m\cos\theta}{4\pi r^3} \tag{3.4}$$

The component of the field in the θ direction is B_θ:

$$B_\theta(r,\theta,\phi) = -\mu_0\frac{1}{r}\frac{\partial V}{\partial\theta}$$

$$= \frac{\mu_0 m}{4\pi r^3}\frac{\partial}{\partial\theta}(\cos\theta)$$

$$= -\frac{\mu_0 m\sin\theta}{4\pi r^3} \tag{3.5}$$

The third component is B_ϕ:

$$B_\phi(r,\theta,\phi) = -\mu_0\frac{1}{r\sin\theta}\frac{\partial V}{\partial\phi}$$

$$= 0 \tag{3.6}$$

Note that by symmetry there can obviously be no field in the B_ϕ direction. The total field strength at any point is

$$B(r,\theta,\phi) = \sqrt{B_r^2 + B_\theta^2 + B_\phi^2}$$

$$= \frac{\mu_0 m}{4\pi r^3}\sqrt{4\cos^2\theta + \sin^2\theta} \tag{3.7}$$

Along the north polar axis ($\theta = 0$) the field is

$$B_r(r,0,\phi) = -\frac{\mu_0 m}{2\pi r^3}$$

$$B_\theta(r,0,\phi) = 0 \tag{3.8}$$

On the equator ($\theta = 90°$) the field is

$$B_r(r,90,\phi) = 0$$

$$B_\theta(r,90,\phi) = -\frac{\mu_0 m}{4\pi r^3} \tag{3.9}$$

and along the south polar axis ($\theta = 180°$) the field is

$$B_r(r, 180, \phi) = \frac{\mu_0 m}{2\pi r^3}$$
$$B_\theta(r, 180, \phi) = 0 \tag{3.10}$$

If we define a constant B_0 as

$$B_0 = \frac{\mu_0 m}{4\pi R^3} \tag{3.11}$$

where R is the radius of the earth; then Eqs. 3.4 and 3.5 give the components of the magnetic field at the earth's surface.

$$B_r(R, \theta, \phi) = -2B_0 \cos\theta \tag{3.12}$$

$$B_\theta(R, \theta, \phi) = -B_0 \sin\theta \tag{3.13}$$

B_0 is most simply visualized in practice as being the equatorial field of the best-fitting dipole field (see Eq. 3.9). The strength of the field at the poles is about 6×10^{-5} teslas (T) or 6×10^4 nanoteslas (nT) (1 weber $m^{-2} = 1$ T), and at the equator it is about 3×10^{-5} T (Fig. 3.1a).

In geomagnetic work, the inward radial component of the earth's field is usually called Z (it is the downward vertical at the earth's surface) and is positive. The horizontal magnitude is called H.

Thus for a dipole,

$$Z(R, \theta, \phi) = -B_r(R, \theta, \phi) \tag{3.14}$$

$$H(R, \theta, \phi) = |B_\theta(R, \theta, \phi)| \tag{3.15}$$

At the surface of the earth and angle between the magnetic field and the horizontal is called the *inclination* I (Fig. 3.5).

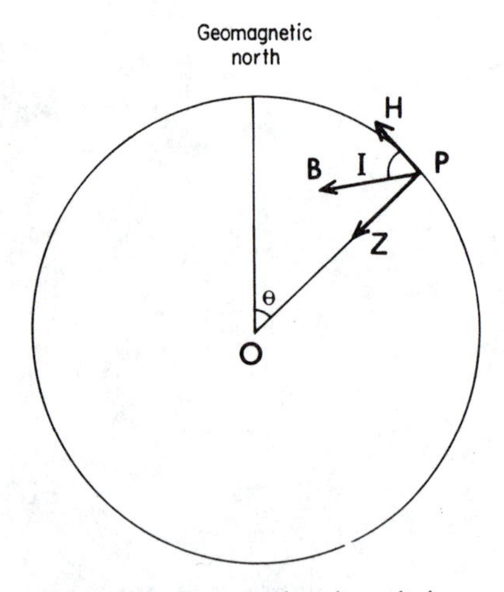

Figure 3.5. Cross section through the earth, illustrating the components of the dipole magnetic field at a location P on the surface. O is the centre of the earth, θ the colatitude of P, B the total field at P, Z and H the vertical and horizontal components of **B** and I the angle of inclination.

$$\tan I = \frac{Z}{H} \tag{3.16}$$

Substituting for B_r and B_θ from Eqs. 3.12 and 3.13 gives

$$\begin{aligned} \tan I &= \frac{2\cos\theta}{\sin\theta} \\ &= 2\cot\theta \\ &= 2\tan\lambda \end{aligned} \tag{3.17}$$

where λ is the magnetic latitude ($\lambda = 90 - \theta$). Equation 3.17 makes it a simple matter to calculate the magnetic latitude, given the angle of inclination, and vice versa. Mariners use the angle of inclination for navigational purposes.

The angle of *declination* is the azimuth of the horizontal component of the magnetic field. It is measured in degrees east or west of north. Mariners call it the variation or magnetic variation of the compass*. Figure 3.3b shows that the declination has varied over the last four centuries. In the case of a dipole field aligned along the earth's rotation axis, the declination would always be zero.

* The compass was a Chinese invention. Declination was described by Shen Kua in 1088.

3.1.3 Magnetization of Rocks

Rocks can become permanently magnetized by the earth's magnetic field. This fact has enabled geophysicists to track past movements of the plates.

As any volcanic rock cools, it passes through a series of critical temperatures at which the various grains of iron minerals acquire spontaneous magnetization. These critical temperatures, called the *Curie points*, or Curie temperatures, are different for each mineral [e.g., approximately 580°C for magnetite (Fe_3O_4), 680°C for haematite Fe_2O_3]. Below the *blocking temperature*, which is tens of degrees less than the Curie point for most minerals, the magnetized grains cannot be reoriented and on average remain with their magnetic moments parallel to the earth's magnetic field at that time. Both of these temperatures are much lower than the temperatures at which lavas crystallize (typically 1100–800°C), which means that magnetization becomes permanent some time after lavas solidify. How long afterwards depends on the physical size and other properties of the intrusion or flow and the rate of cooling, which depends in turn on its environment (see Sect. 7.11). This type of permanent residual magnetization is called *thermoremanent magnetization* (TRM) and is considerably larger in magnitude than the magnetism induced in the basalt by the earth's present field.

Sedimentary rocks can acquire remanent magnetization even though they have never been as hot as 500°C, but the remanent magnetization of sediments is generally very much less than that of igneous rocks. Sedimentary rocks can acquire magnetization in two ways: *depositional* or *detrital remanent magnetization* (DRM) and *chemical remanent magnetization* (CRM). DRM can be acquired, as indicated by its name, during the deposition of sedimentary rocks. If the sediments are deposited in still water, any previously magnetized grains will align themselves with their magnetic moments parallel to the earth's magnetic field. CRM is acquired in situ after deposition during the chemical growth of iron oxide grains, as in a sandstone. When the grains reach critical size, the magnetic field necessary to change their magnetization is stronger than the earth's field; so their magnetization is essentially permanent.

The degree to which a rock body can be magnetized by an external magnetic field is determined by the *magnetic susceptibility* of the rock. *Induced magnetization* is the magnetization of the rock **M** which is induced when the rock is put into the earth's magnetic field **B**; it is given by

$$\mu_0 \mathbf{M}(\mathbf{r}) = \chi \mathbf{B}(\mathbf{r}) \tag{3.18}$$

where χ is the magnetic susceptibility (a dimensionless quantity). Values of χ for basalts vary from about 10^{-4} to 10^{-1}, so the induced magnetization gives rise to a field which is very much weaker than the earth's field. Thermoremanent magnetization (TRM) is generally many times stronger than this induced magnetization. For any rock sample, the ratio of its remanent magnetization to the magnetization induced by the earth's present field is called the *Königsberger ratio* Q. Measured values of Q for oceanic basalts are in the range 1–160. Thus, an effective susceptibility for thermoremanent magnetization of basalt of about 10^{-3} to 10^{-1} appears to be reasonable. This permanent magnetization therefore produces a local field of perhaps 1% of the earth's magnetic field. Effective susceptibilities for sedimentary rocks are about two orders of magnitude less than for basalt.

The relationship between the angle of inclination and the magnetic latitude (Eq. 3.17) means that a measurement of the angle of inclination of the remanent magnetization of a suitable lava or sediment laid down on a continent immediately gives the magnetic palaeolatitude for the particular piece of continent. If the continent has not moved with respect to the pole since the rock cooled, then the magnetic latitude determined from the magnetization of the rock is the same as its present latitude. However, if the continent has moved, the magnetic latitude determined from the magnetization of the rock can be different from its present latitude. Thus, the angle of inclination provides a powerful method of determining the past latitudes (palaeolatitudes) of the continents. Unfortunately, it is not possible to use palaeomagnetic data to make a determination of palaeolongitude.

Example: calculation of palaeomagnetic latitude

Magnetic measurements have been made on a basalt flow presently at 47°N, 20°E. The angle of inclination of the remanent magnetization of this basalt is 30°. Calculate the magnetic latitude of this site at the time the basalt was magnetized.

The magnetic latitude λ is calculated by using Eq. 3.17:

$$\tan I = 2 \tan \lambda$$

I is given as 30°, so

$$\begin{aligned} \lambda &= \tan^{-1}\left(\frac{\tan 30^\circ}{2}\right) \\ &= \tan^{-1}(0.2887) \\ &= 16.1^\circ \end{aligned}$$

Therefore, at the time the sample was magnetized, it was at a magnetic latitude of 16°N, which indicates that between then and now the site has moved 31° northwards to its present position at 47°N.

If the angles of declination and of inclination of our rock sample are measured, the position of the palaeomagnetic pole can be calculated. To do this it is necessary to use spherical geometry, as in the calculations of Chapter 2. Figure 2.11 shows the appropriate spherical triangle if we assume N to be the present North Pole, P the palaeomagnetic North Pole and X the location of the rock sample. The cosine formula for a spherical triangle (e.g., Eq. 2.9) gives the geographic latitude of the palaeomagnetic pole P, λ_P, as

$$\cos(90 - \lambda_P) = \cos(90 - \lambda_X)\cos(90 - \lambda) + \sin(90 - \lambda_X)\sin(90 - \lambda)\cos D \tag{3.19}$$

where λ_X is the geographic latitude of the sample location, D the measured remanent declination and λ the palaeolatitude (given by Eq. 3.17). Simplifying Eq. 3.19 gives

$$\sin \lambda_P = \sin \lambda_X \sin \lambda + \cos \lambda_X \cos \lambda \cos D \tag{3.20}$$

After λ_P has been calculated, the sine formula for a spherical triangle (e.g.,

Eq. 2.10) can be used to give the difference between the longitudes of the palaeomagnetic pole and the sample location, $\phi_P - \phi_X$:

$$\sin(\phi_P - \phi_X) = \frac{\sin(90 - \lambda)\sin D}{\sin(90 - \lambda_P)}$$

$$= \frac{\cos\lambda \sin D}{\cos\lambda_P} \qquad \sin\lambda \geqslant \sin\lambda_P \sin\lambda_X \qquad (3.21)$$

$$\sin(180 + \phi_X - \phi_P) = \frac{\cos\lambda \sin D}{\cos\lambda_P} \qquad \sin\lambda < \sin\lambda_P \sin\lambda_X$$

Therefore, by using Eqs. 3.17 and 3.19–3.21, we can calculate past magnetic pole positions.

Example: calculation of latitude and longitude of palaeomagnetic pole

If the angle of declination for the basalt flow of the previous example is 80°, calculate the latitude and longitude of the palaeomagnetic pole.

The latitude of the palaeomagnetic pole is calculated by using Eq. 3.20 with $\lambda = 16°$ and $D = 80°$:

$$\sin\lambda_P = \sin 47 \sin 16 + \cos 47 \cos 16 \cos 80$$

Thus, $\lambda_P = 18°$.

The longitude of the palaeomagnetic pole can now be calculated from Eq. 3.21 after checking whether $\sin\lambda \geqslant \sin\lambda_P \sin\lambda_X$ or not:

$$\sin(\phi_P - \phi_X) = \frac{\cos 16 \sin 80}{\cos 18}$$

Thus, $\phi_P - \phi_X = 84°$, and so $\phi_P = 104°$. The position of the palaeomagnetic pole is 18°N, 104°E.

If palaeomagnetic pole positions can be obtained from rocks of different ages on the same continent, these poles can be plotted on a map. Such a plot is called a *polar wander path* (a name that is a 'fossil' of the older notion that it was the poles, not the continents, that drifted) and shows how the magnetic pole moved relative to that continent. If such polar wander paths from two continents coincide, then the two continents cannot have moved relative to each other during the times shown. However, if the paths differ, there must have been relative motion of the continents. Figure 3.6 shows polar wander paths for Europe and North America for the last 550 Ma. Although these two paths have almost the same shape, they are certainly not coincident. When the opening of the Atlantic Ocean is taken into account, however, the two paths can be rotated on top of each other, and they then do approximately coincide.

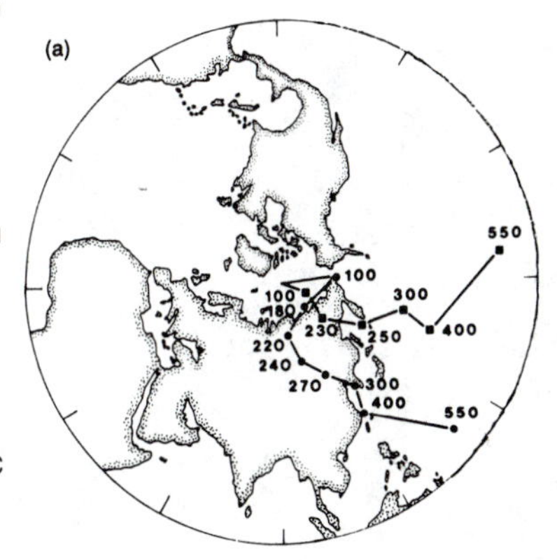

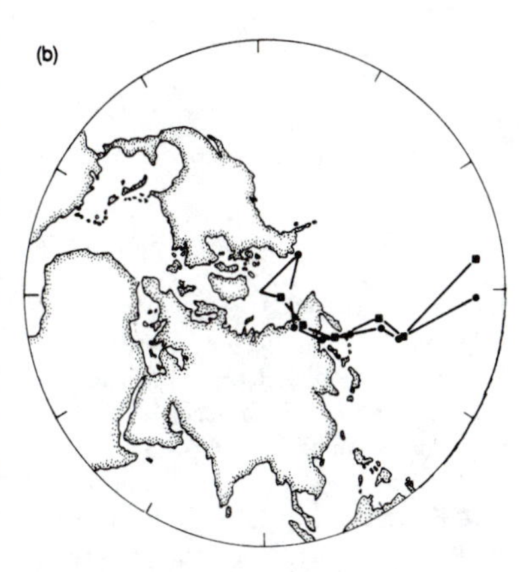

Figure 3.6. (a) Polar wander curves for North America (circles) and Europe (squares). (b) Polar wander curves for North America and Europe when allowance has been made for the opening of the Atlantic Ocean. The two curves are now almost coincident. (After McElhinney 1973.)

3.2 Dating the Oceanic Plates

3.2.1 Magnetic Stripes

To use measurements of the magnetic field to gain information about the magnetization of the crust, it is first necessary to subtract the regional value of the geomagnetic field (e.g., IGRF 1980). What remains is the magnetic anomaly. Over the oceans, magnetic field measurements are made by towing a magnetometer behind a ship. A magnetometer can be routinely towed while a research ship is on passage or doing other survey work that does not involve slow, tight manoeuvring. These marine instruments measure the magnitude B but not the direction of the total field $\mathbf{B}$. (Magnetometers that can measure both magnitude and direction are widely used on land as prospecting tools.) Marine magnetic anomalies are therefore anomalies in the magnitude (or total intensity) of the magnetic field.

The first detailed map of magnetic anomalies off the west coast of North America, published in 1961, showed what was then a surprising feature:

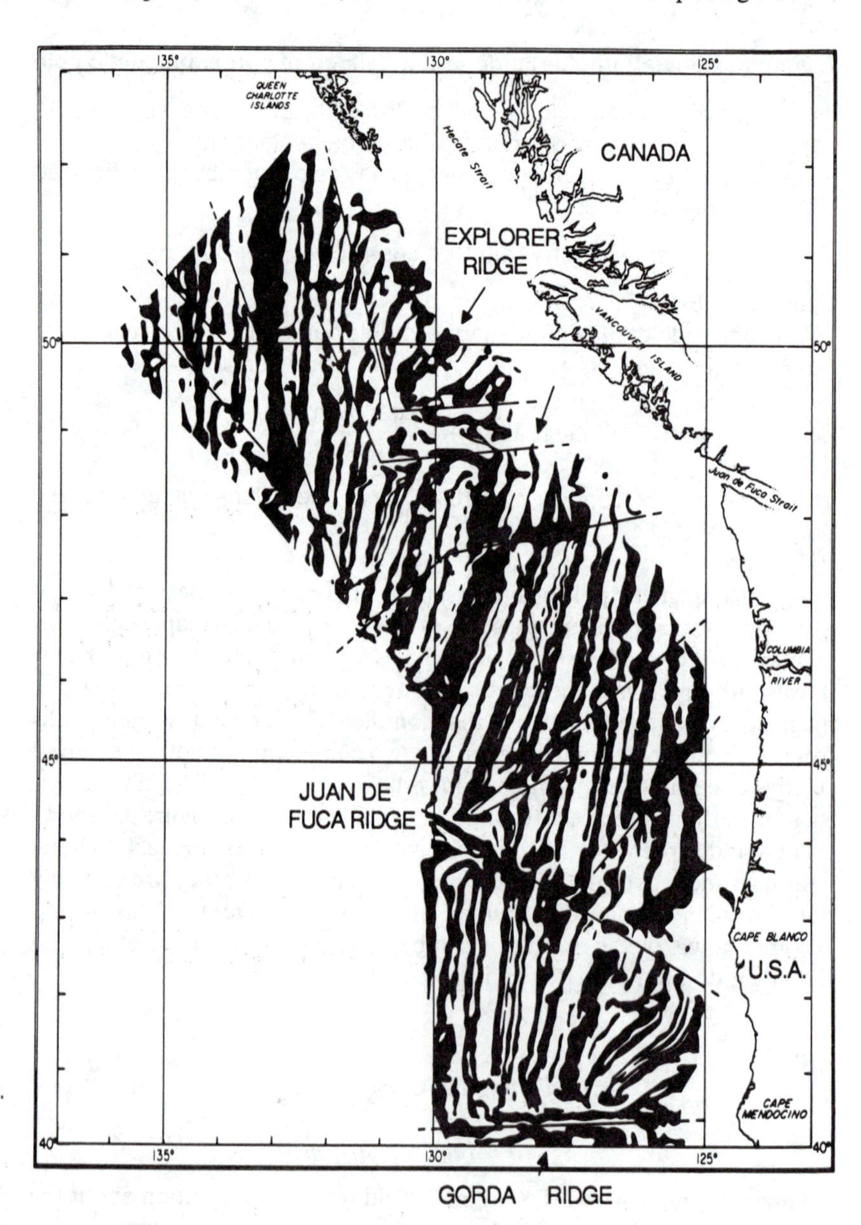

Figure 3.7. Anomalies of the total magnetic field southwest of Vancouver Island in the Pacific Ocean. Positive anomalies are shown in black and negative anomalies in white. Arrows indicate the axes of the three midocean ridges; straight lines indicate the major faults which offset the anomaly pattern (Fig. 3.22). This was the first large-scale map to show the details of the magnetic anomalies over an active midocean ridge. As such it was a vital piece of evidence in the development of the theories of sea-floor spreading and plate tectonics. (After Raff and Mason 1961.)

alternate stripes of anomalously high and low values of the magnetic field stretching over the entire region (Fig. 3.7). All subsequent magnetic anomaly maps show that these stripes are typical of oceanic regions. The stripes run parallel to and are generally symmetric about the axes of the midocean ridges. They are offset by fracture zones, are a few tens of kilometers in width and typically ± 500 nT in magnitude (Fig. 3.8a). This means that magnetic anomaly maps are very useful in delineating ridge axes and fracture zones.

The origin of these magnetic stripes was first correctly understood in 1963 by F. J. Vine and D. H. Matthews and independently by L. W. Morley. They realized that the idea of sea-floor spreading, which was then newly proposed by H. H. Hess, coupled with the then recently discovered evidence that intermittent reversals of the earth's magnetic field have taken place, provide the answer. The oceanic crust, which is formed along the axes of the midocean ridges as mafic material wells up, acts as a magnetic tape recorder that preserves past reversals of the magnetic field (Fig. 3.8b). The width of magnetic stripe is determined by the speed at which 'the tape' is moving (half-spreading rate) and the length of time between the magnetic

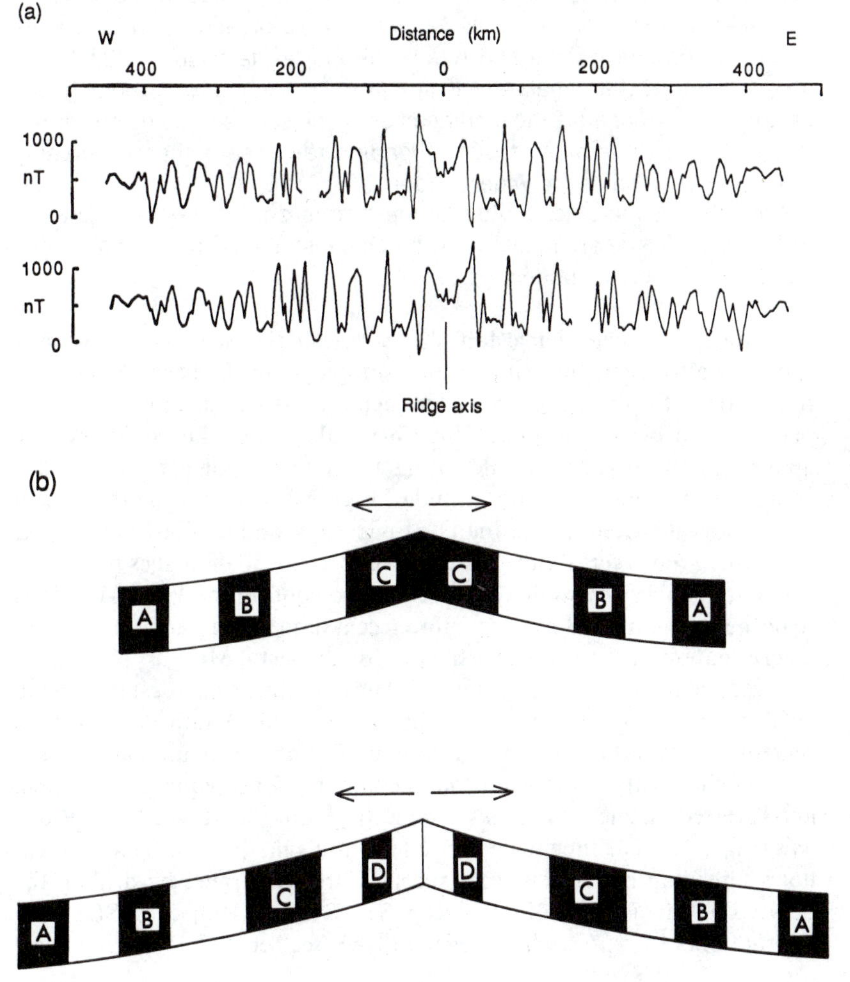

Figure 3.8. (a) A magnetic profile across the Pacific–Antarctic Ridge, plotted above its end-to-end reverse (mirror image), demonstrates the commonly observed symmetry of magnetic anomalies about the ridge axis. The half-spreading rate as determined from this profile is 4.5 cm yr^{-1}. Magnetic anomalies are generally about ± 500 nT in amplitude, about 1% of the earth's magnetic field. (After Pitman and Heirtzler 1966.) (b) Cross section through an idealized ridge illustrates the block model of normal and reverse magnetized material. In the upper drawing, the earth's magnetic field is positive (normal) and has been for some time; thus, the positively magnetized blocks C have been formed. Positive blocks A and B were intruded during previous times of normal polarity of the earth's magnetic field. The lower drawing shows the same ridge at a later time; another positive block D has been intruded, and the magnetic field is currently in a period of reverse polarity.

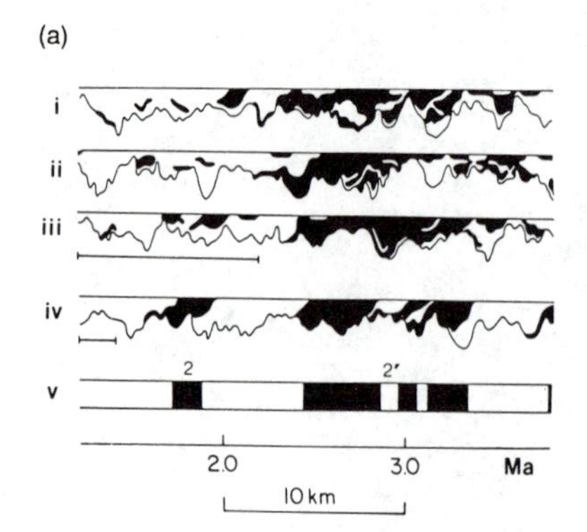

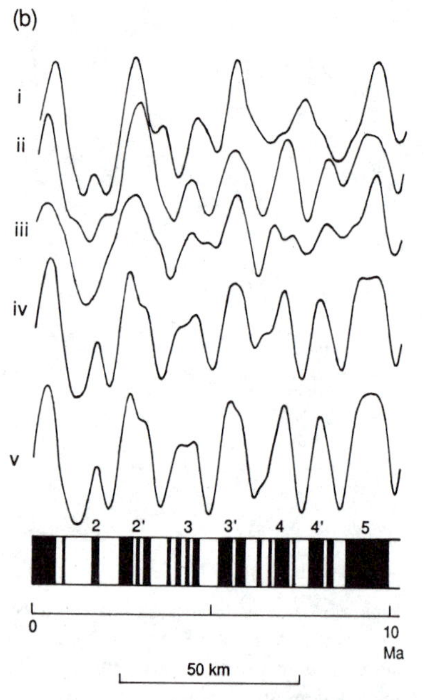

Figure 3.9. (a) Detail of magnetized layer generated by extrusion of lava randomly in time and spatially in an emplacement zone; half-spreading rate is $1\,\mathrm{cm\,yr^{-1}}$. (b) Theoretical magnetic anomalies generated by this random extrusion model. In both (a) and (b), (i), (ii) and (iii) are for an emplacement zone 10 km across, (iv) an emplacement zone 2 km across and (v) the block model, in which the emplacement zone has zero width. (From Schouten and Denham 1978.)

reversals. Thus, while the earth's magnetic field is in its normal polarity, a block of material is formed with a strong component of permanent magnetization parallel to the normal field. When the earth's field is reversed, new oceanic crust will have a strong component of permanent magnetization parallel to the reversed field. In this way a magnetically normal and reversed striped oceanic crust is formed, with the stripes parallel to the ridge axis. To 'decode' a magnetic anomaly pattern it is necessary to know either when the earth's field reversed or else the half-spreading rate of the ridge.

Example: variability of marine magnetic anomalies

Decoding magnetic anomalies is not as simple in practice as it sounds, partly because the magnetization of the oceanic crust does not conform to a perfect block model. The lava flows that make up the magnetized layer are not produced continuously along the ridge axis. Rather, they are extruded randomly in time and in space in an *emplacement zone* which is centred on the ridge axis. The effect of this can be seen in Figure 3.9. The block model gives way to a much more complex structure as the width of the emplacement zone is increased. The variability in magnetic anomalies increases accordingly. An emplacement zone 10 km across, which seems to be appropriate for the Mid-Atlantic Ridge, explains the variability of Atlantic magnetic anomalies. Pacific magnetic anomalies are much less variable, partly because the emplacement zone seems to be narrower but mainly because the much faster spreading rate means that the polarity reversals are much more widely spaced.

Nevertheless the block model for the magnetization of oceanic crust is widely used. All the anomalies shown in the rest of this chapter have been calculated for block models.

By 1966, researchers had established a *reversal time scale* extending back some 4 Ma by using potassium argon isotopic dating (see Sect. 6.7) to fit a time scale to the magnetic reversal sequence measured in continental lava piles and on oceanic islands. This time scale of reversal could then be applied to the oceans by calculating theoretical anomaly patterns for assumed spreading rates and latitudes. Figure 3.10 shows the theoretical anomalies calculated for the Juan de Fuca Ridge and the East Pacific Rise using this 1966 reversal time scale. These theoretical anomalies match the actual anomalies very well and were used to confirm the Vine–Matthews hypothesis. Figure 3.11 shows a more recent (and therefore more detailed) determination of the reversal sequence for the last 4 Ma.

A geomagnetic time scale extending back 80 Ma was first established in 1968 by assuming that the spreading rate in the South Atlantic had remained constant from 80 Ma until now. The observed anomaly pattern was matched with theoretical profiles computed for a sequence of normal and reversed magnetized blocks symmetric about the Mid-Atlantic Ridge axis (Fig. 3.12). The time scale derived from this model and the whole sea-floor spreading hypothesis were spectacularly confirmed by drilling the ocean bottom as part of the Deep Sea Drilling Project (DSDP), an international enterprise which began in 1968 (see Sect. 8.2.1). During Leg 3

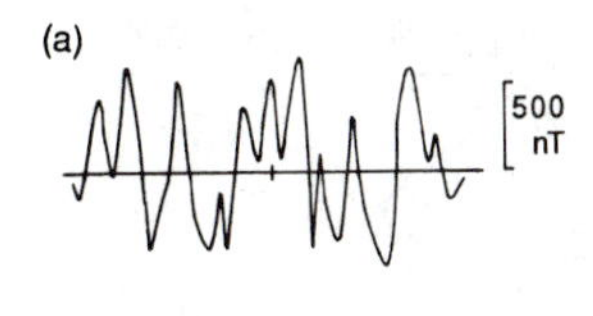

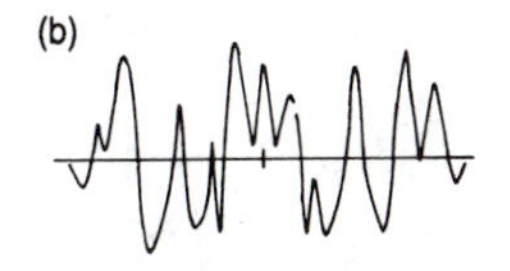

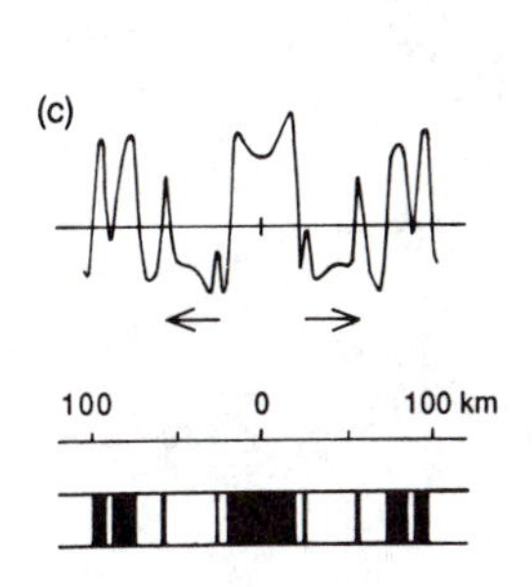

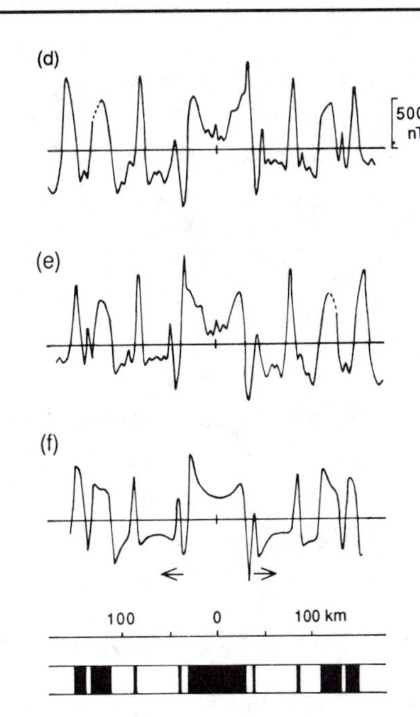

Figure 3.10. (a) Magnetic anomaly profile over the Juan de Fuca Ridge at 46°N, southwest of Vancouver Island, Canada. (b) Profile in (a) is reversed; note the symmetry. (c) Model magnetic anomaly profile calculated for this ridge assuming a half-spreading rate of 2.9 cm yr^{-1} and the magnetic reversal sequence shown below. Magnetic reversal time scale; black blocks represent periods of normal polarity. (d) Magnetic anomaly profile over the East Pacific Rise at 51°S. (e) Profile in (d) is reversed; note the symmetry. (f) Model magnetic anomaly profile calculated by assuming a half-spreading rate of 4.4 cm yr^{-1} and the magnetic reversal sequence shown below. Magnetic reversal time scale (same as on left but with a different horizontal scale); black blocks represent periods of normal polarity. (After Vine 1966.)

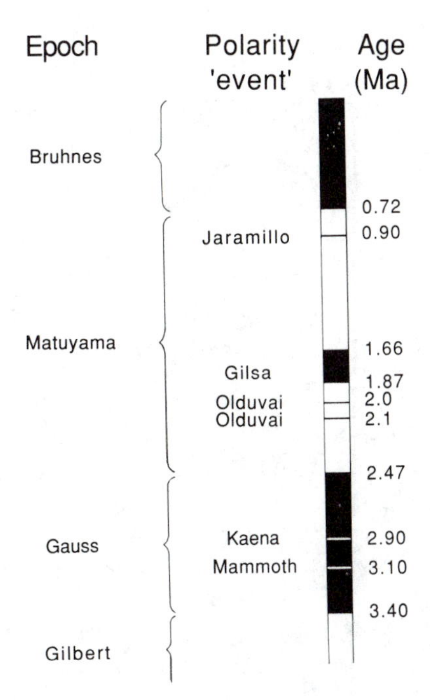

Figure 3.11. Details of the reversals of the earth's magnetic field that have occurred in the last four million years as determined from detailed radiometric dating of continental and oceanic island lavas and palaeomagnetism of marine sediments. The *epochs*, time intervals during which the earth's magnetic field was either predominantly normal or predominantly reversed in polarity, have been named after prominent scientists in the study of the earth's magnetic field. [William Gilbert was a sixteenth-century English physician, Karl Friedrich Gauss was a nineteenth-century German mathematician, Bernard Bruhnes (1906) was the first person to propose that the earth's magnetic field was reversed at the time the lavas were formed and Motonori Matuyama (1929) was the first person to attempt to date these reversals.] The *polarity events*, short fluctuations in magnetic polarity, are named after the geographic location where they were first recognized (e.g., Olduvai Gorge, Tanzania, the site of the early hominoid discoveries of Leakey; Mammoth, California, U.S.A.; Jaramillo Creek, New Mexico, U.S.A.) (Based on Ness et al. 1980.)

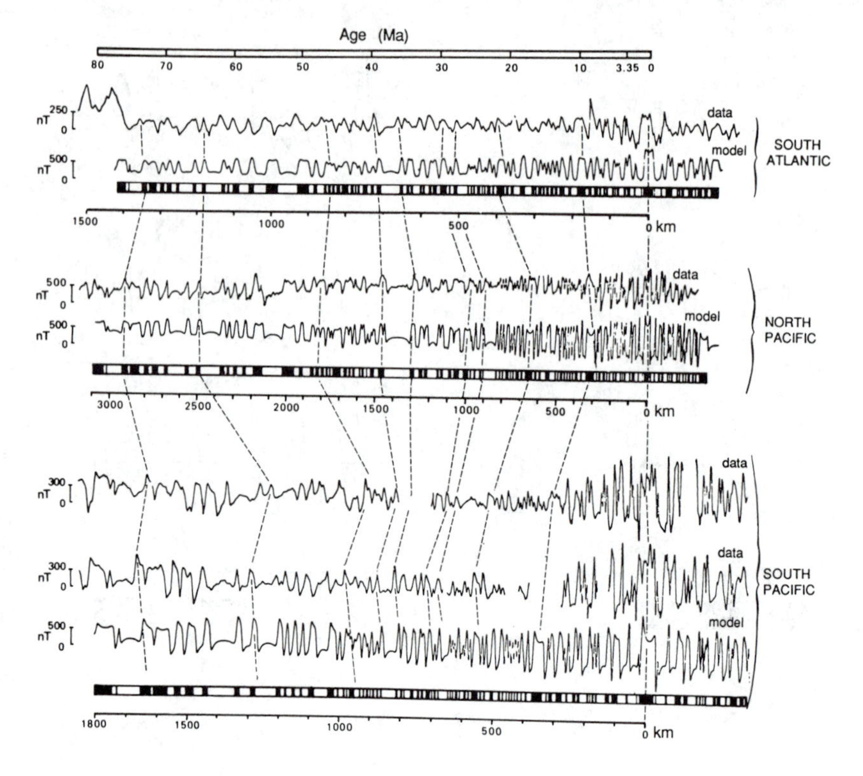

Figure 3.12. Geomagnetic reversal time scale for the last 80 Ma as proposed in 1968. The magnetic anomaly profiles with matching model profiles and model reversal sequences are shown for the South Atlantic, North Pacific and South Pacific. The South Atlantic time scale was made by assuming that the spreading rate there has been constant for the last 80 Ma. In comparison, spreading in the Pacific has clearly been both faster and more irregular (note the different distance scales for the Pacific data). Dashed lines connect specific magnetic anomalies numbered as in Figure 3.14. (After Heirtzler et al. 1968.)

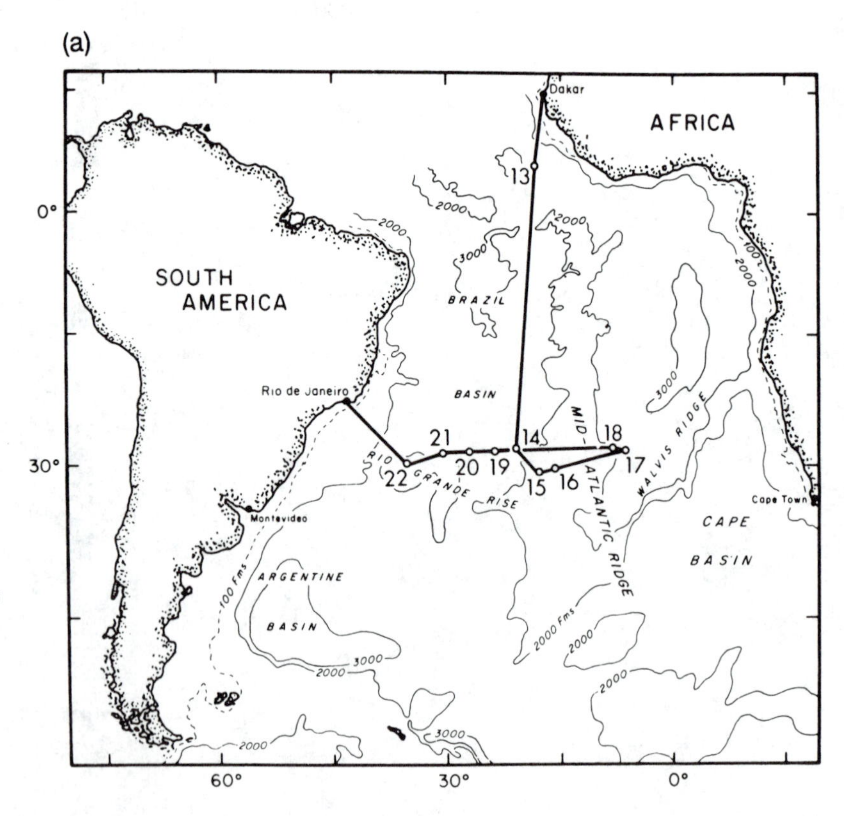

Figure 3.13. (a) Geographic location of drilling sites on leg 3 of the Deep Sea Drilling Project. Site numbers refer to the order in which the holes were drilled. (b) Age of the sediment immediately above the basalt basement versus the distance of the drill site, as shown in (a), from the axis of the Mid-Atlantic Ridge. (From Maxwell et al. 1970.)

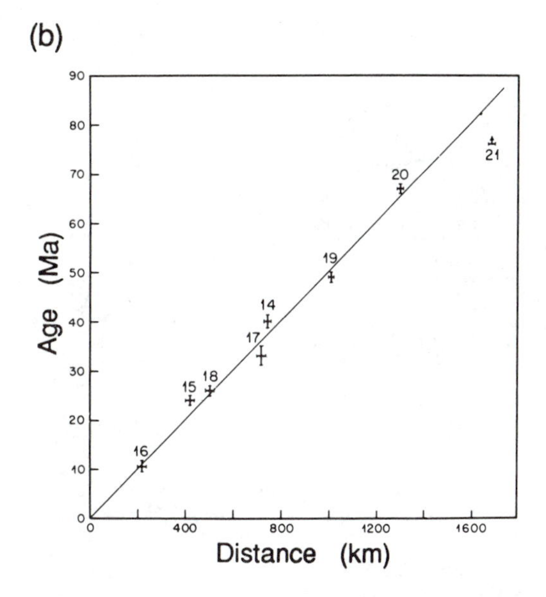

Figure 3.13 (*cont.*)

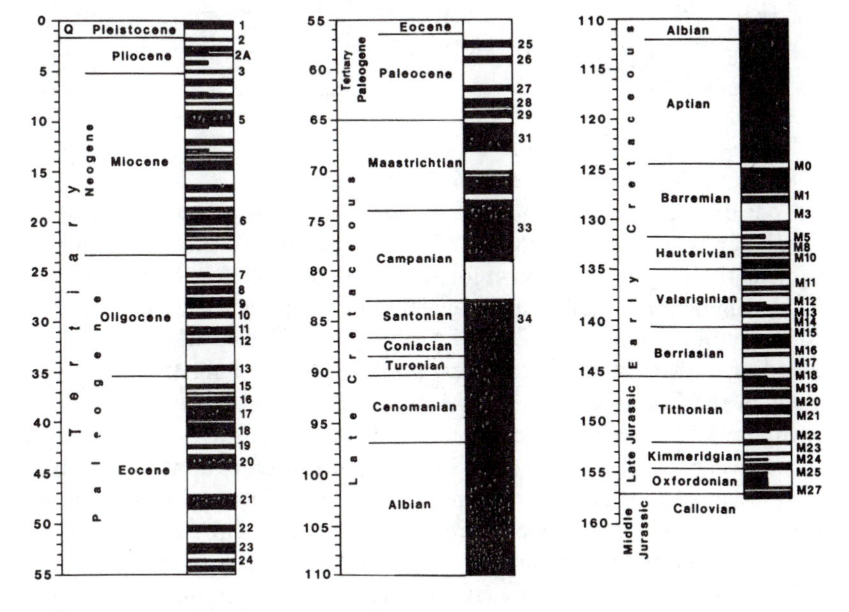

Figure 3.14. Geomagnetic reversal time scale. Black indicates periods of normal polarity for the earth's magnetic field; white, reversed polarity periods. The anomaly numbers, 1–34 and M0–M27, are on the right-hand side of the column; the age is on the left-hand side. Note that in the 0–90 Ma reversal sequence, the anomaly numbers refer to periods of normal polarity, whereas in the older sequence, anomaly numbers refer to periods of reversed polarity. (From Harland et al. 1990.)

of the DSDP a series of holes was drilled into the basalt at the top of the oceanic crust right across the Atlantic at 30°S (Fig. 3.13). It was not possible to date the lavas from the top of the crust by radiometric methods because they were too altered; instead, the basal sediments were dated by using fossils. The ages are therefore slightly younger than the lava ages would have been. Figure 3.13b shows these sediment ages plotted against the distance of their sites from the ridge axis. The straight line confirmed that spreading in the South Atlantic had been continuous and had been occurring at a fairly steady rate for the last 80 Ma, and that the geomagnetic time scale (Fig. 3.12) was reasonably accurate. Since then, considerable effort has been put into ensuring that geologic and magnetic time scales are

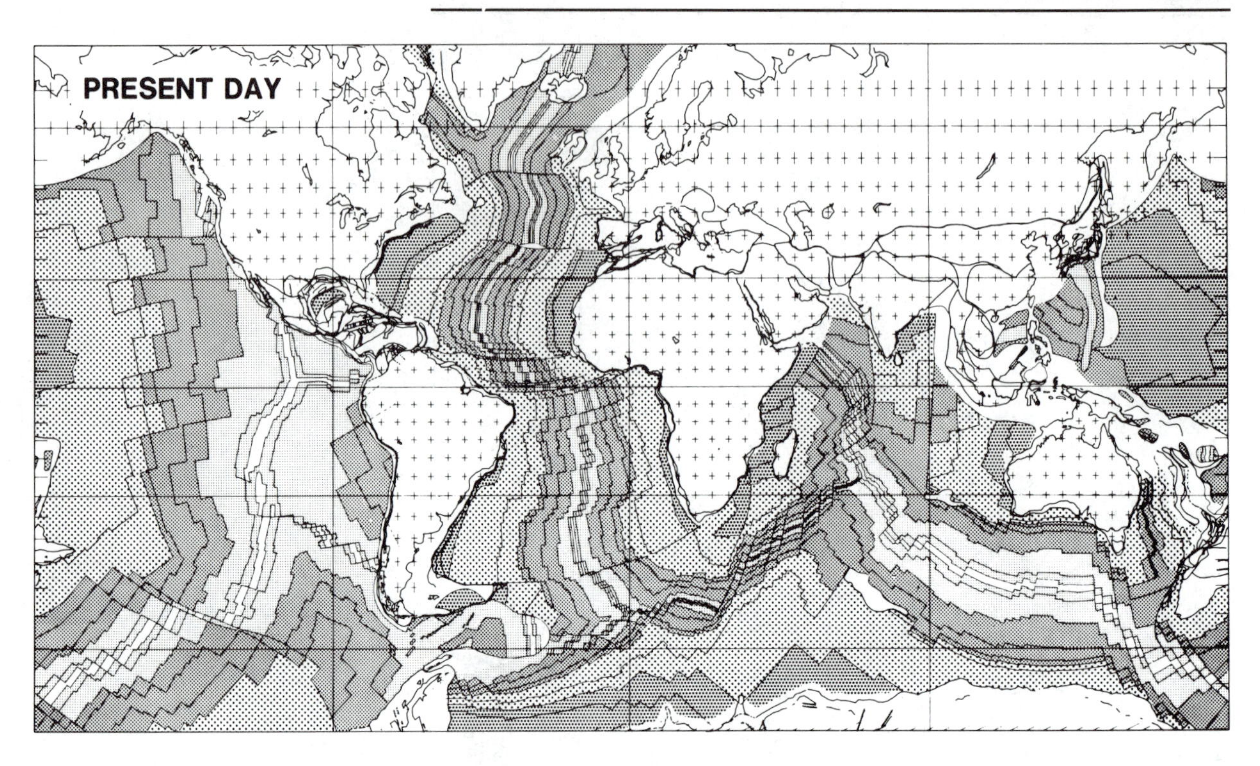

Figure 3.15. The age of the ocean floor determined by using magnetic anomaly data, basement ages from deep-sea drilling anomaly time scale and rotation poles and angles. White indicates the youngest ocean floor; fine light stipple, 2–23 Ma; fine dark stipple, 23–59 Ma; coarse light stipple, 59–84 Ma; coarse dark stipple, 84–144 Ma. (From Scotese et al. 1988.)

as precise as possible. Figure 3.14 shows a time scale, as determined in 1989, from the middle Jurassic to the present.

To use a geomagnetic time scale to date the oceanic plates it is necessary to recognize specific anomalies. Fortunately, the reversal sequence is sufficiently irregular (Fig. 3.14) for this to be possible for the trained eye. The prominent anomalies up to age 80 Ma have been numbered from one to thirty-three. For ages 125–157 Ma they are labelled MO to M27 (M standing for Mesozoic). Particularly prominent is the long *Magnetic Quiet Zone* in the Cretaceous (83–124 Ma) when no reversals occurred.

Figure 3.12 shows profiles from the Pacific and their corresponding theoretical profiles, as well as the South Atlantic profiles on which the time scale was based. It is clear that in contrast to the history of the Atlantic, spreading rates in the Pacific have changed markedly with time. This is also the case in the Indian Ocean. Figure 3.15 is an isochron map of the ocean floor. From this we can see that the ridges have undergone a number of changes in spreading rate and spreading direction in the past. The oldest parts of the ocean floor are Jurassic. This is an interesting fact in itself and the subject of conjecture about the density and stability of older oceanic lithosphere and causes of initiation of a subduction zone in any particular location.

3.2.2 Calculation of Marine Geomagnetic Anomalies

The marine magnetic anomaly patterns (e.g., Figs. 3.10 and 3.12) can give immediate values for the relative motion between two plates if specific anomalies can be identified and if the reversal time scale is known. The patterns can also be used to estimate the relative motion between the plate and the earth's magnetic pole. This follows in the same way that the remanent magnetism of continental lavas gives their magnetic latitude (Eq. 3.17). The magnitude and direction of the magnetization of oceanic crust depends on the latitude at which the crust was formed and is unaffected by later movements and position. This means that, although the magnetic anomaly resulting from this magnetization of the crust is dependent on the present location of that piece of crust, it can be used to determine the original latitude and orientation of the midocean ridge, though not the longitude.

Imagine a midocean ridge spreading symmetrically, producing infinitely long blocks of new crust (Fig. 3.16). The magnetic field measured above any block will include a contribution from the permanent magnetization of the block. Suppose first that at the time the block was formed, the ridge was at the equator, where the magnetic field is north–south and has no vertical component (Eq. 3.9), and was spreading east–west. In this case, the permanent magnetization of the block would be along the block, $\mathbf{M} = (0, M_y, 0)$, and so the lines of force cannot leave the block (the block being infinitely long). It is therefore impossible for the magnetic field of this block to affect the magnetic field outside the block. Such a block would not produce an anomaly!

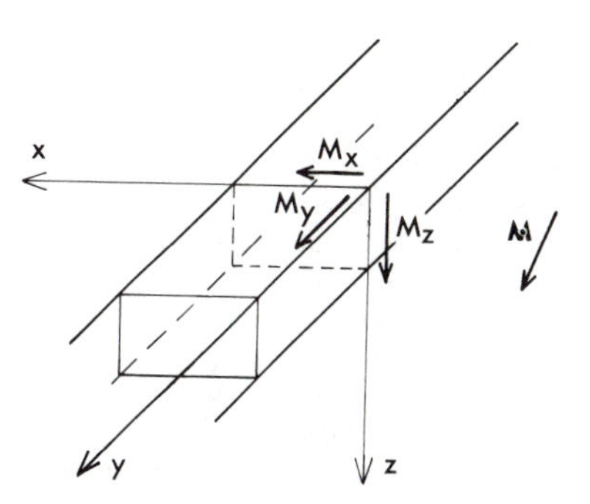

Figure 3.16. An infinitely long block of magnetized oceanic crust. The magnetization of this block $\mathbf{M}$ is resolved into three components (M_x, M_y, M_z). M_y is parallel to the block and cannot contribute to the magnetic field outside the block. M_z, the vertical component, and M_x, the horizontal component perpendicular to the block, both contribute to the magnetic field outside the block and hence to the resultant magnetic anomaly produced by the block.

Now imagine a ridge spreading east–west but not at the equator. The vertical component of the magnetic field as measured today by a ship above the block will be either reduced or increased, depending on whether the magnetic field at the time the block was formed was reversed or normal. Because, in this example, the ridge is assumed to be striking north–south, the horizontal magnetization is in the y direction and has no component in the x direction (because $B_\phi = 0$). Magnetization in the y direction cannot affect the magnetic field outside the block because the block is assumed to be infinitely long, and so it is only the vertical component of the magnetization that affects the magnetic field outside the block. The magnetic anomalies produced by a symmetrical pattern of such blocks are symmetrical.

For all spreading directions of the ridge that are not east–west, there is a component of the permanent magnetization in the x direction. The effect of this component M_x is complex, but it usually produces an asymmetry in the magnetic anomalies. In the vicinity of the magnetic poles, where the earth's magnetic field is almost vertical, the effects of the x component of magnetization are almost negligible; however, close to the magnetic equator the effect becomes important.

Calculation of synthetic magnetic anomalies for a block model (e.g., Fig. 3.8b) must be performed in the following order:

1. Calculate magnetization in the x and z directions for each block by assuming the orientation and latitude at which blocks were formed.
2. Calculate the field produced by these blocks along a line normal to the blocks and at a constant distance above them.
3. Add this field to the earth's magnetic field at the block's present-day latitude.
4. Calculate difference in magnitude between this resultant field and the earth's field. (For more details, see McKenzie and Sclater 1971.)

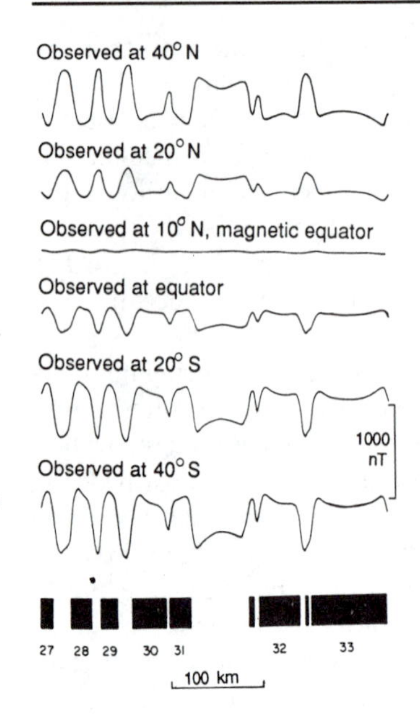

Figure 3.17. Synthetic magnetic anomaly profiles generated by a ridge at 40°S, spreading east–west with a half-rate of 3 cm yr^{-1}. The magnetized crust subsequently moved northwards to the latitudes shown, leaving the blocks striking north–south. Note that the shapes of the anomalies depend on the present latitude and that, if the magnetized crust moves across the equator, positively magnetized blocks give rise to negative anomalies. Black blocks denote periods of normal magnetic field; white blocks denote reversed field. Numbers are anomaly numbers (see Fig. 3.14). (From McKenzie and Sclater 1971.)

To visualize the effect of present-day latitude on the magnetic anomalies produced in a given block structure, consider hypothetical oceanic crust formed by east–west spreading of a midocean ridge at 40°S. Figure 3.17 shows the magnetic anomalies that this magnetized oceanic crust would produce if it later moved north from 40°S to a number of different latitudes. The present-day latitude is clearly an important factor in the shapes of the anomalies; note in particular that if the plate moves across the equator, positively magnetized blocks give rise to negative anomalies, and vice versa. (This is a result of B_r at today's latitude and M_z having opposite signs.)

Magnetic anomalies are dependent on the orientation of the ridge at which the crust was formed as well as the latitude. The orientation determines the relative values of M_x and M_y. Figure 3.18 shows anomalies that would be observed at 15°N. The magnetized blocks were produced by ridges at 40°S, striking in the three directions shown. The anomalies produced by the ridge striking N45°E and the ridge striking N45°W are identical. This ambiguity in the strike of the ridge is always there: Anomalies from a ridge striking $-\alpha$ could have come from a ridge striking $+\alpha$.

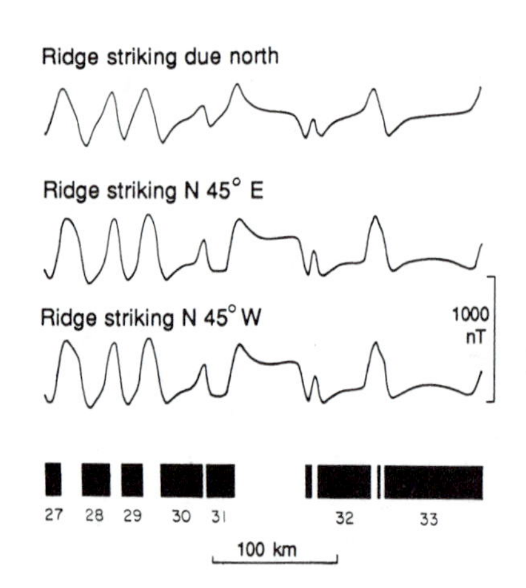

Figure 3.18. Synthetic magnetic anomaly profiles produced by the magnetized blocks now striking east–west at 15°N. The magnetized oceanic crust was produced at 40°S by a midocean ridge spreading at 3 cm yr^{-1} half-rate and striking due north, N45°E and N45°W. Note that the anomalies for the N45°E ridge and the N45°W ridge are exactly the same. Numbers are anomaly numbers (see Fig. 3.14). (From McKenzie and Sclater 1971.)

The thickness of the magnetized blocks that give rise to the oceanic magnetic anomalies is not well determined. In general, the anomalies depend primarily on the product of the magnetic susceptibility and layer thickness (i.e., 400 m of material with a susceptibility of 0.05 gives rise to almost the same magnetic anomaly as 2 km of material with a susceptibility of 0.01). At one time it was thought that the magnetized layer was only the very top of the oceanic crust (then termed layer 2A, discussed in Sect. 8.2.1) but this is not now thought necessarily to be the case. The minimum width of a block that can be detected from a profile observed at the sea surface is of the order of 0.5 km. Seabed magnetic anomaly measurements can give more resolution and detail, but such data are much more difficult and expensive to acquire (and to interpret) than seasurface measurements.

3.3 Reconstruction of Past Plate Motions

3.3.1 Introduction

A magnetic anomaly profile can be used to construct a magnetized block model of the ocean crust and to estimate the latitude and orientation of the midocean ridge which produced it. When several profiles are available and the magnetic anomalies are plotted on a map, as in Figure 3.7, such a block model is easily visualized. To determine the past movements of plates, a substantial amount of palaeomagnetic data is required. This data collection began with the development of magnetometers to detect submarines. The instruments were required to measure magnetic fields to about 1 nT, which was ideal for measuring marine magnetic anomalies.

Because the oldest oceanic lithosphere is Jurassic ($\sim$ 160 Ma old), magnetic anomaly data can only be used to trace the past motions of the plates back to that time. Continental magnetic data and other geological data provide evidence for motions of the plates prior to the Jurassic, but the data are necessarily sparser and more difficult to interpret. The remainder of this chapter provides sections on the geological histories of the Pacific, Indian and Atlantic oceans as established by deciphering magnetic anomalies. The full reconstruction of the geological history of an ocean is only possible if the ocean contains only ridges. For oceans such as the Indian and Pacific, in which subduction has taken place, there has been a loss of information which may prevent a full reconstruction. Deciphering the geological histories of the oceans, especially the early work by McKenzie and Sclater on the Indian Ocean, was one of the triumphs of plate tectonics.

3.3.2 The Atlantic Ocean

The magnetic anomaly map of the Atlantic Ocean is by far the simplest of the three major oceans (Fig. 3.19). The history of continental splitting and sea-floor spreading is almost completely preserved because, apart from the short lengths of the Puerto Rico Trench and the South Sandwich Trench, there are no subduction zones. The Mid-Atlantic Ridge is the plate boundary between the Eurasian, African and North and South American plates. Despite the changes in pairs of plates, the poles and instantaneous rotation rates are such that the spreading rate of the Mid-Atlantic Ridge does not vary greatly along its length.

The oldest identified anomaly in the South Atlantic is M11, which occurs just off the west coast of South Africa. Thus, Africa and South America must have started separating shortly before this time (135 Ma). The oldest anomaly in the central Atlantic is M25, identified off the east coast of North America and the northwest coast of Africa. Africa and North America therefore started to separate in the later Jurassic shortly before 160 Ma. This motion resulted in considerable faulting and folding in the Mediter-

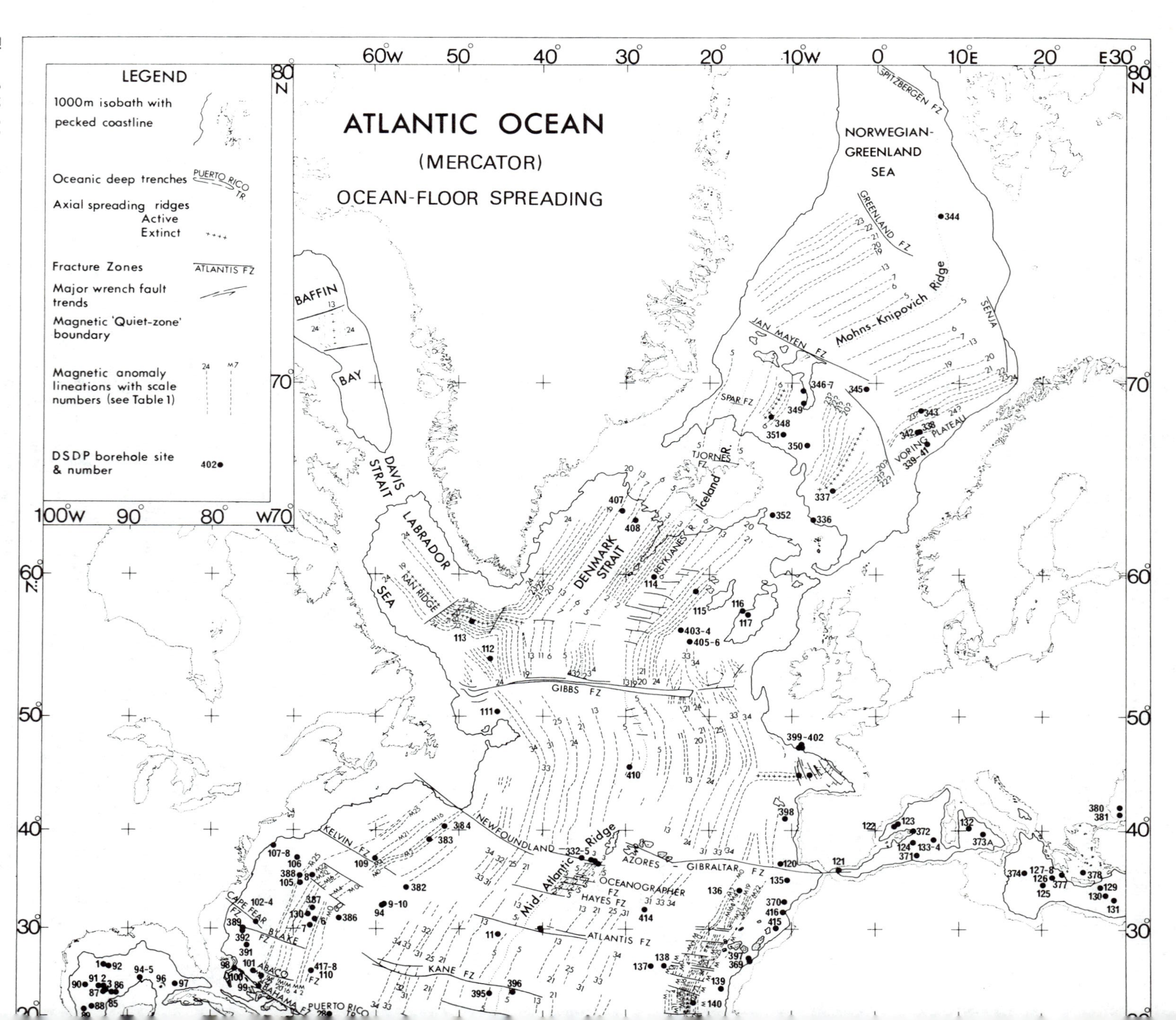

Figure 3.19. Magnetic anomalies over the Atlantic Ocean. (From Owen 1983.)

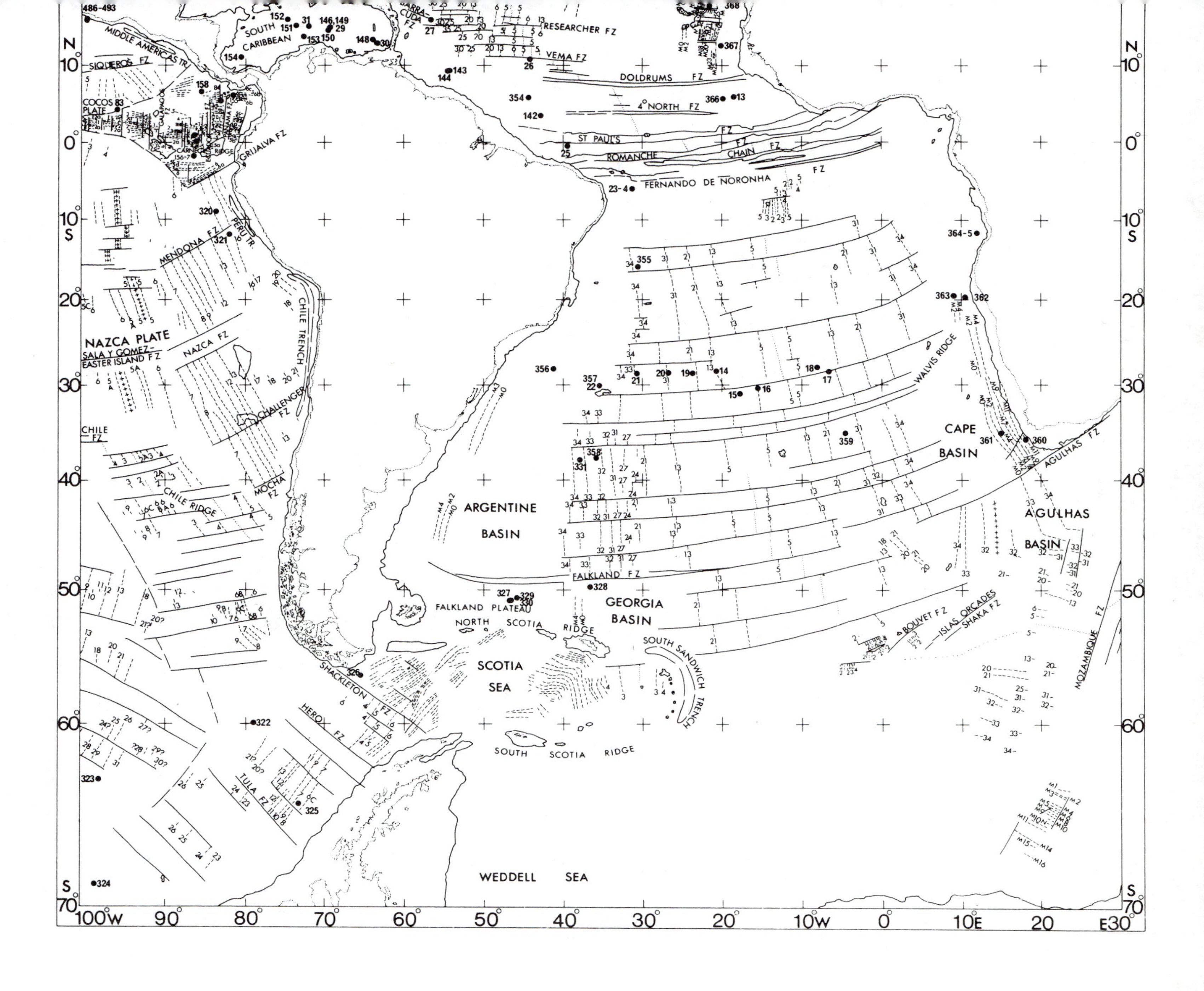

MIDDLE AMERICAS TR.
SIQUEROS FZ
COCOS PLATE
SOUTH CARIBBEAN
GRIJALVA FZ
PERU TR.
MENDONA FZ
NAZCA PLATE
SALA Y GOMEZ-EASTER ISLAND FZ
NAZCA FZ
CHILE TRENCH
CHALLENGER FZ
CHILE FZ
MOCHA FZ
CHILE RIDGE
SHACKLETON FZ
HERO FZ
TULA FZ
RESEARCHER FZ
VEMA FZ
DOLDRUMS FZ
4° NORTH FZ
ST PAUL'S FZ
ROMANCHE FZ
CHAIN FZ
FERNANDO DE NORONHA FZ
WALVIS RIDGE
CAPE BASIN
AGULHAS FZ
AGULHAS BASIN
ARGENTINE BASIN
FALKLAND FZ
GEORGIA BASIN
FALKLAND PLATEAU
NORTH SCOTIA RIDGE
SCOTIA SEA
SOUTH SANDWICH TRENCH
SOUTH SCOTIA RIDGE
BOUVET FZ
ISLAS ORCADES
SHAKA FZ
MOZAMBIQUE FZ
WEDDELL SEA
100W
90°
80°
70°
60°
50°
40°
30°
20°
10W
0°
10E
20
E30°
N 10°
0°
10° S
20°
30°
40°
50°
60°
S 70°

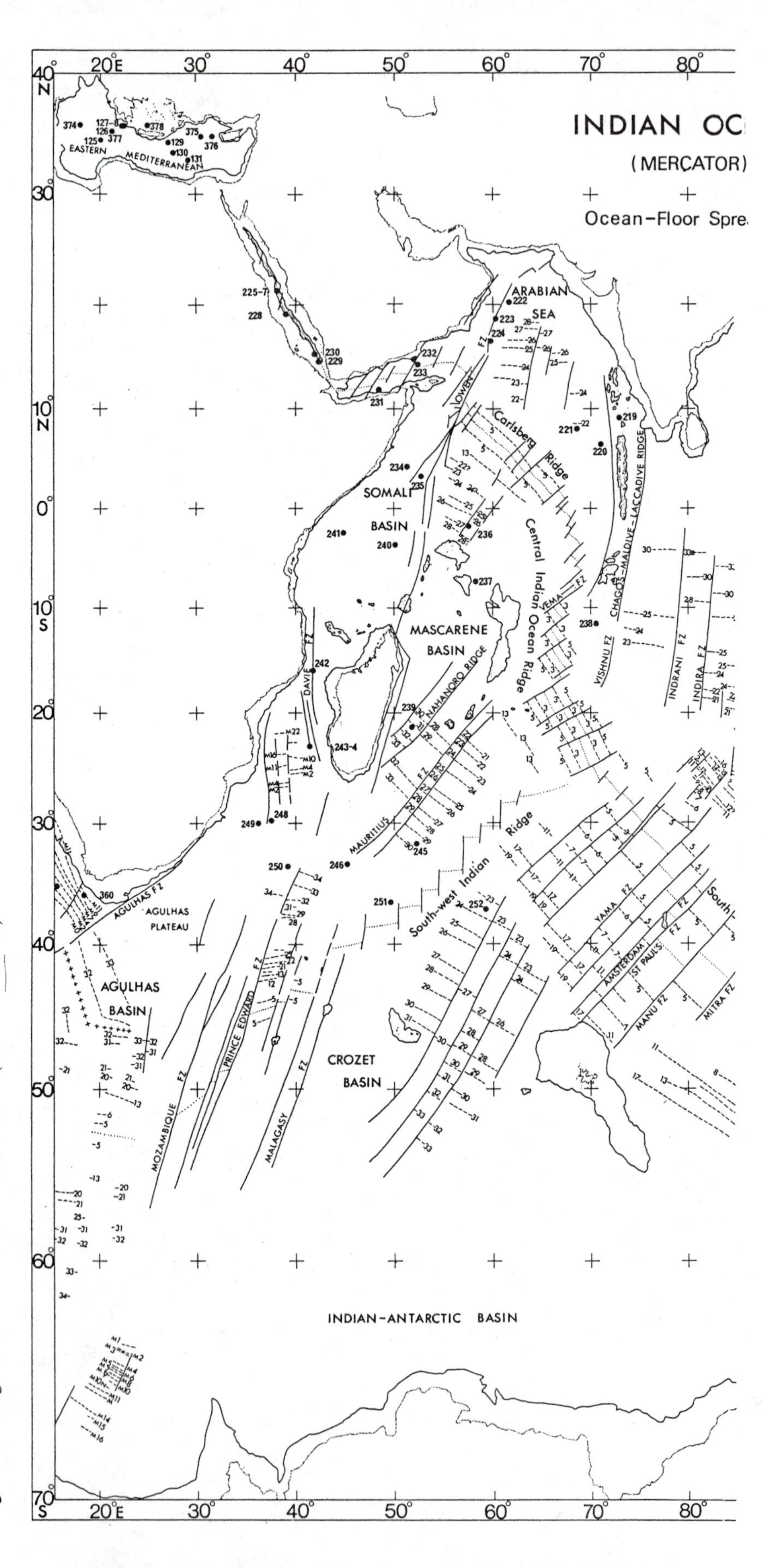

Figure 3.20. Magnetic anomalies over the Indian Ocean. (From Owen 1983.)

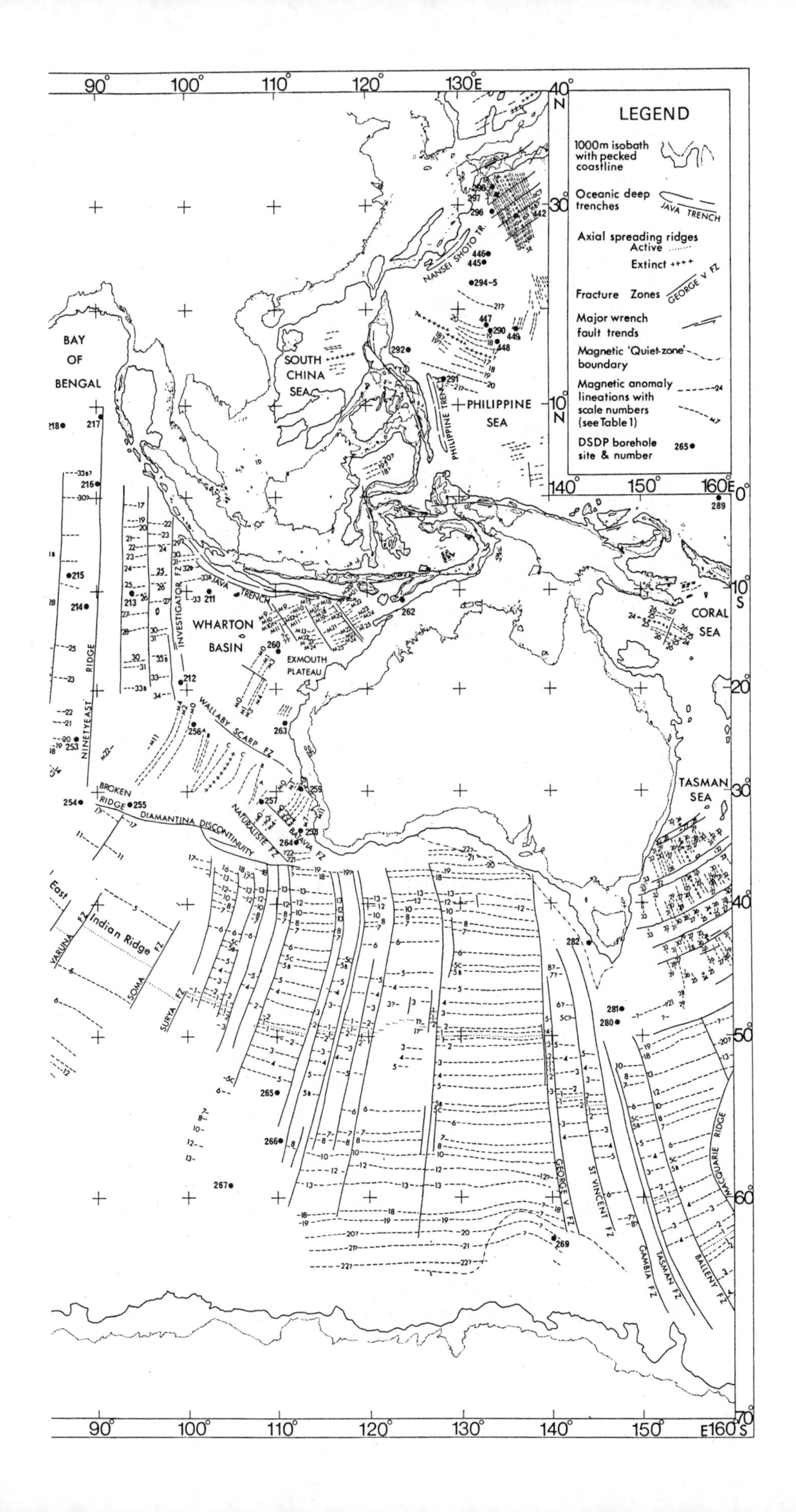
LEGEND
1000m isobath with pecked coastline
Oceanic deep trenches
JAVA TRENCH
Axial spreading ridges
Active
Extinct
Fracture Zones
GEORGE V FZ
Major wrench fault trends
Magnetic 'Quiet-zone' boundary
Magnetic anomaly lineations with scale numbers (see Table 1)
DSDP borehole site & number
BAY OF BENGAL
SOUTH CHINA SEA
PHILIPPINE SEA
NANSEI SHOTO TR.
PHILIPPINE TRENCH
WHARTON BASIN
EXMOUTH PLATEAU
JAVA TRENCH
INVESTIGATOR FZ
NINETYEAST RIDGE
WALLABY SCARP FZ
BROKEN RIDGE
DIAMANTINA DISCONTINUITY
NATURALISTE FZ
CORAL SEA
TASMAN SEA
East Indian Ridge
VARUNA FZ
SOMA FZ
SURYA FZ
GEORGE V FZ
ST VINCENT FZ
GAMBIA FZ
TASMAN FZ
BALLENY FZ
MACQUARIE RIDGE
90°
100°
110°
120°
130°E
140°
150°
160°E
40° N
30°
10° N
0°
10° S
20°
30°
40°
50°
60°
70° S
E160°S

ranean region because Eurasia and North America did not start to separate until about 90 Ma; anomaly 34 is identified on both sides of the North Atlantic. The NW–SE anomalies in the Labrador Sea between Canada and Greenland (anomalies 24–19) and extending northwards into Baffin Bay (anomalies 24–13) indicate that there was an active ridge there from about 55 Ma until 43 or 35 Ma. Since this time, Greenland has not moved independently and has been part of the North American Plate; the spreading which started with anomaly 24 has continued only along the Reykjanes Ridge.

3.3.3 The Indian Ocean

The magnetic anomaly map of the Indian Ocean (Fig. 3.20) is considerably more complex than that of the Atlantic Ocean. The three present-day mid-ocean ridges – Central Indian Ridge, Southwest Indian Ridge and the Southeast Indian Ridge – intersect at the Indian Ocean (or Rodriguez) Triple Junction (Fig. 2.2), a ridge–ridge–ridge triple junction. The Southeast Indian Ridge is spreading fairly fast (3 cm yr^{-1} half-rate) and has smooth topography, whereas the Southwest Indian Ridge is spreading very slowly (less than 1 cm yr^{-1} half-rate) and has rough topography and many long fracture zones. The Carlsberg Ridge starts in the Gulf of Aden and trends southeast. At the equator it is intersected and offset by many transform faults, so the net strike of the plate boundary between Africa and India becomes almost north–south. This part of the plate boundary is called the Central Indian Ridge. The Southwest Indian Ridge, which extends from the Bouvet Triple Junction in the South Atlantic to the Indian Ocean Triple Junction, is the boundary between the African and Antarctic plates. The position of the present-day rotation pole for Africa relative to Antarctica at about 6°N, 39°W (Table 2.1) means that this ridge is offset by a series of very long transform faults. One of the main faults, the Prince Edward Fracture Zone, offsets the ridge axis by over 1000 km. The half-spreading rate in this region is about 0.6 cm yr^{-1}. The other major bathymetric features of the Indian Ocean are the Ninety-East Ridge and the Chagos–Maldive–Laccadive Ridge system, both of which are linear north–south submarine mountain chains.

The oldest magnetic anomalies in the Indian Ocean occur in the Wharton Basin (M9–M25), which lies between Australia and the Java Trench, west of Western Australia (M0–M22), between Madagascar and east Africa (M2–M22), in the Somali Basin (M13–M21), and north of Antarctica at 20°E (M1–M16). These anomalies indicate that the separation between Africa and Antarctica began at about the time of anomaly M21 (150 Ma). At this time, east Antarctica and Madagascar (then joined) moved southwards away from Africa (Fig. 3.21a). In so doing, they generated the symmetric anomalies in the Somali Basin, as well as those off the east Antarctic coast and west of Madagascar, which appear to be the northern and southern halves of a symmetric pattern. At about the time of anomaly M0, sea-floor spreading in the Somali Basin north of Madagascar stopped and a new ridge was initiated between Madagascar and Antarctica: The proto Southwest Indian Ridge formed, separating Antarctica from Africa, Madagascar and India. From the time of anomaly 34 (~ 90 Ma) until

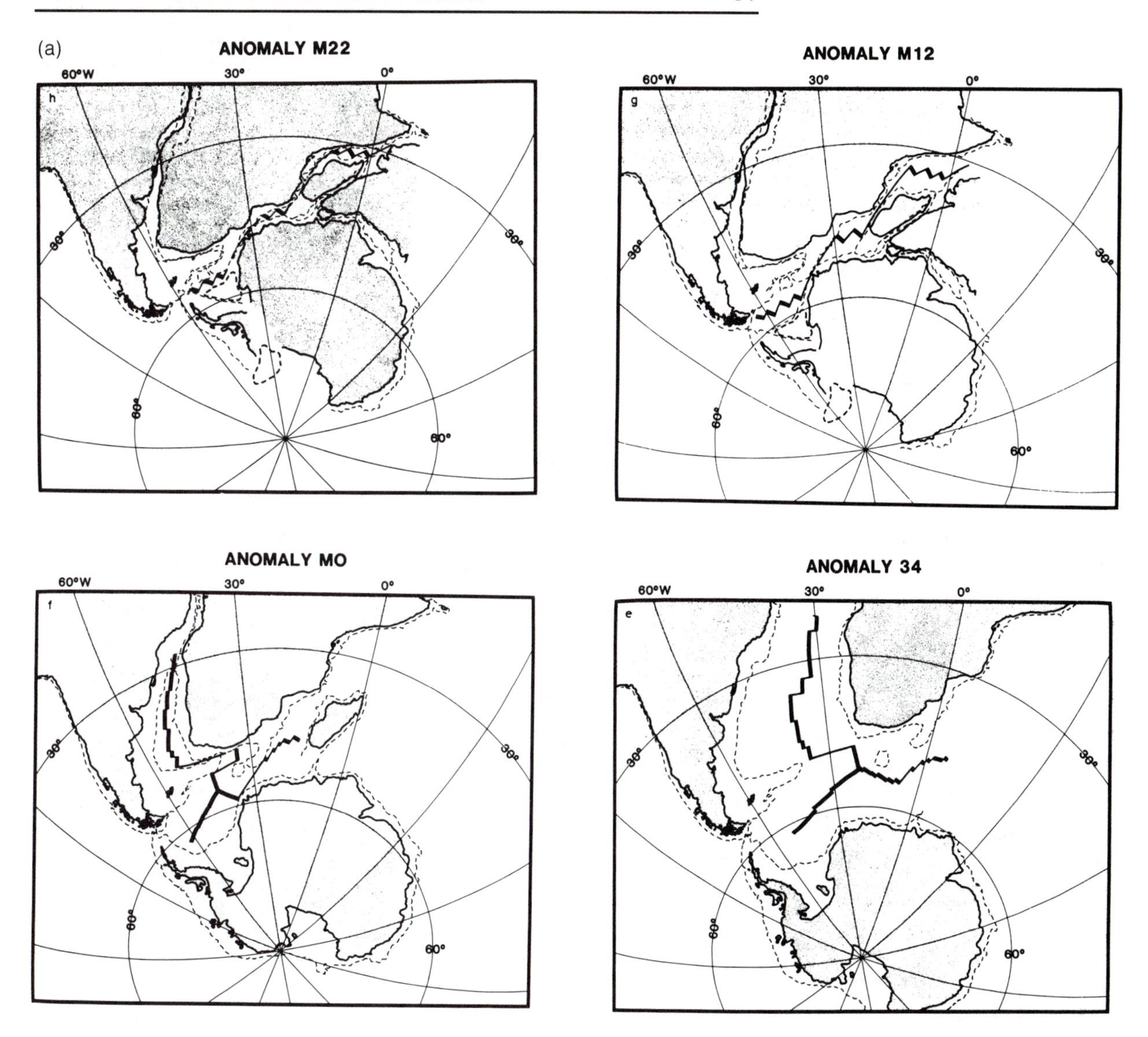

Figure 3.21. (a) Reconstructions of the positions of South America, Africa and Antarctica at the time of anomalies M22, M12, M0 and 34. (From Lawver et al. 1985.)

the present, spreading along this ridge appears to have been fairly constant, but magnetic anomaly data from this ridge are very scarce.

The anomalies north of Australia, to the north and east of the Exmouth Plateau, M9–M25, are only the southern half of a symmetric pattern; the northern part has been subducted by the Java Trench which is currently active there. The oldest anomaly close to the continental margin north of Australia, M25, gives an estimate of the date of opening of this ocean as 155 Ma. West of Western Australia a symmetric pattern of anomalies M0–M10 is present, indicating that a ridge was active here by M10 time (134 Ma) until about 96 Ma. This ridge was the plate boundary between the Indian Plate and the Australia–Antarctic Plate. Australia and Antarctica were still joined at this stage, though continental extension had been taking place since the separation of India (Fig. 3.21b).

At about 96 Ma a major change occurred in the plate motions; and the major part of the Indian Ocean has been created since then. At this time (or

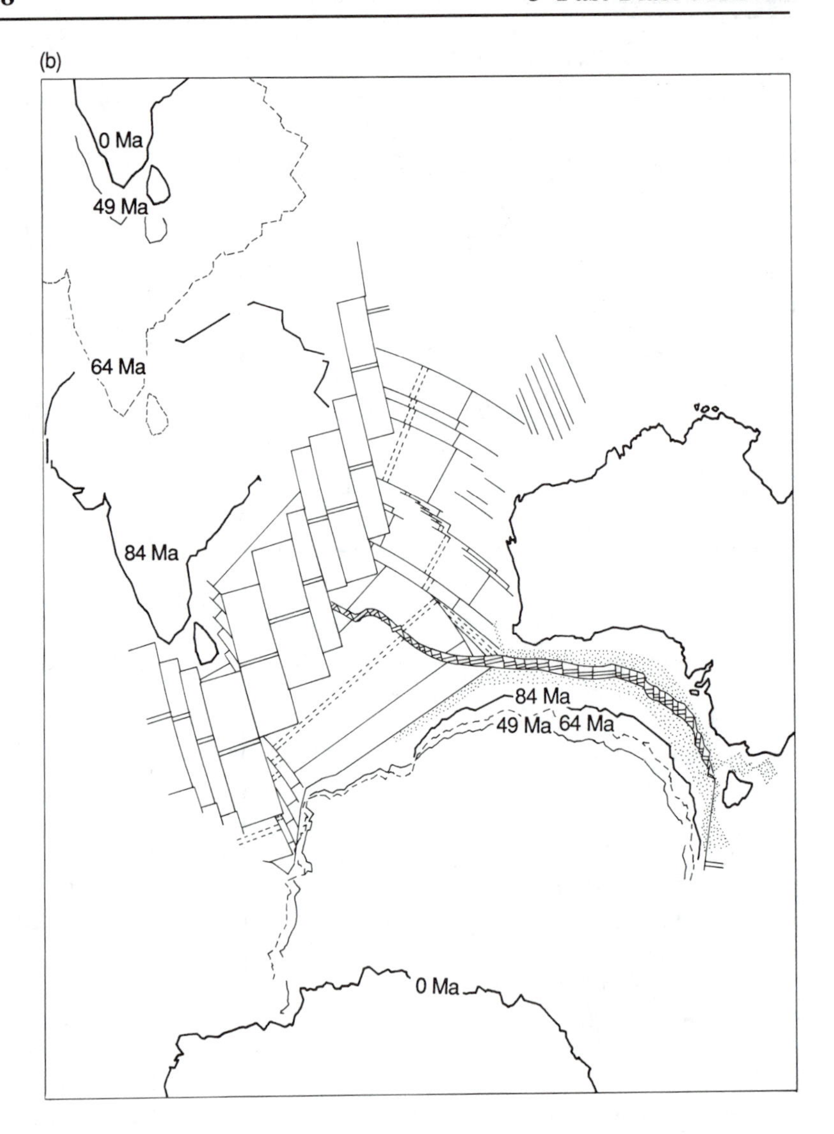

Figure 3.21 (*cont.*). (b) Reconstructions of the positions of Indian and Antarctica relative to Australia 84, 64 and 49 Ma ago and at the present. The midocean ridge system 84 Ma ago is shown with solid lines. The abandoned 96Ma midocean ridge system is shown with dashed lines. The continental extension between Australia and Antarctica is stippled. (After Powell, Roots and Veevers 1988.)

shortly before, as anomaly information is not available for the Magnetic Quiet Zone), the Southeast Indian Ridge started spreading. This began the splitting of Australia from Antarctica. At the same time, India began to move rapidly northwards from Antarctica. The Carlsberg–Central Indian Ridge also started spreading, separating Africa from India. The half-spreading rate of the portion of the Southeast Indian Ridge between Australia and Antarctica was slow (perhaps only 0.5 cm yr^{-1}), whereas the spacing of the same anomalies to the south and east of India indicate half-spreading rates of 10 cm yr^{-1} or more for the ridge between India and Antarctica.

It is apparent in Figure 3.20 that a second major change of spreading direction must have occurred on the Southeast Indian Ridge at about the time of anomaly 19 (43 Ma). Older anomalies strike approximately east–west, whereas the younger anomalies strike northeast–southwest. Therefore, prior to the time of anomaly 19, the Southeast Indian Ridge was striking east–west. Thus, the east–west magnetic anomalies in the Arabian

Sea and south of India and the Bay of Bengal were all formed by the same ridge. All the lineations formed on the north side of the section of the ridge east of the Ninety East Ridge (which is believed to be a hot spot trace) and the extinct ridge itself have been subducted by the Java Trench, the only subduction zone in the Indian Ocean. The ridge between Australia and Antarctica continued to spread very slowly.

At about the time of anomaly 19 (43 Ma), there was a major reorganization in the plate motions. Spreading on the Carlsberg–Central Indian and Southeast Indian ridges changed from approximately north–south to the present northeast–southwest. This was the time when the Indian and Asian continents collided (refer to Sect. 9.2.3 for details of continent–continent collisions). The fate of the ridge between Australia and India is not clear. It is possible that India and Australia then lay on the same plate. There has been slight motion between India and Australia since about the time of anomaly 13.

The fine details of the motions of the plates in this region and the exact positioning of the continents as they were prior to anomaly M25 when they formed the megacontinent Gondwanaland remain to be firmly established, but on a broad scale the motions are fairly well understood.

3.3.4 The Pacific Ocean

The magnetic anomaly pattern off the west coast of North America was the first magnetic anomaly pattern to be studied in detail (Fig. 3.7). An

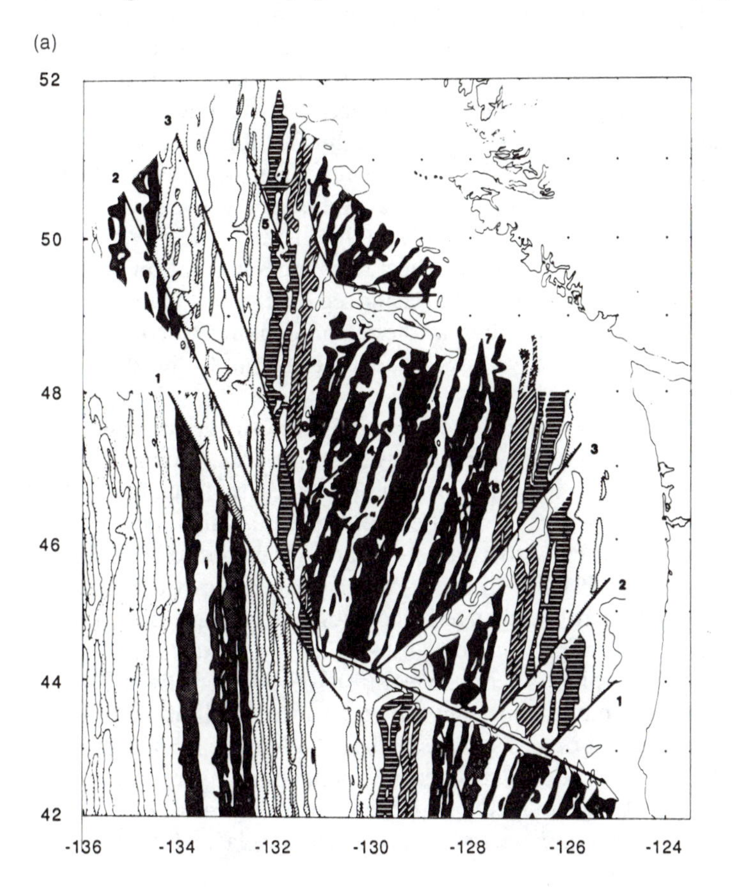

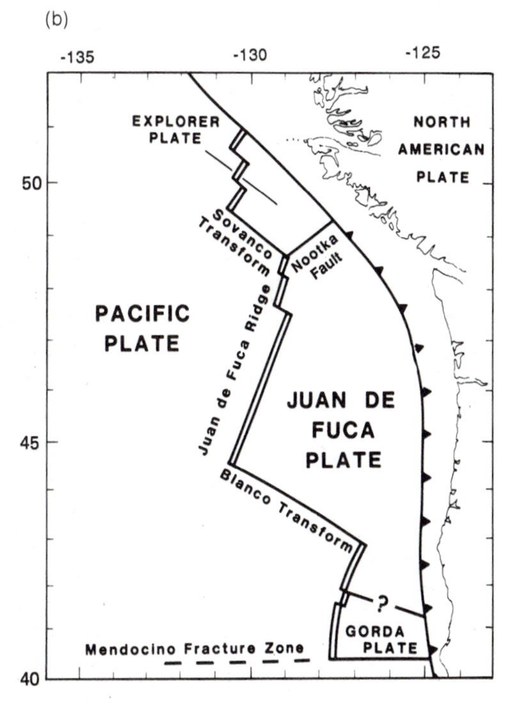

Figure 3.22. (a) Magnetic anomaly data for the northwestern continental margin of the North American Plate. Solid black lines are major offsets in anomaly pattern (termed *pseudo faults*). This figure includes the data shown in Figure 3.7 and has been shaded so that the ages of the various anomalies stand out. (b) Location map for (a) showing the plates and plate boundaries. Juan de Fuca and Explorer plates are undergoing oblique subduction beneath the North American Plate. There is left lateral strike-slip motion along the Nootka Fault and right lateral strike-slip motion on the Queen Charlotte Fault (Fig. 3.27). (After Wilson et al. 1984.)

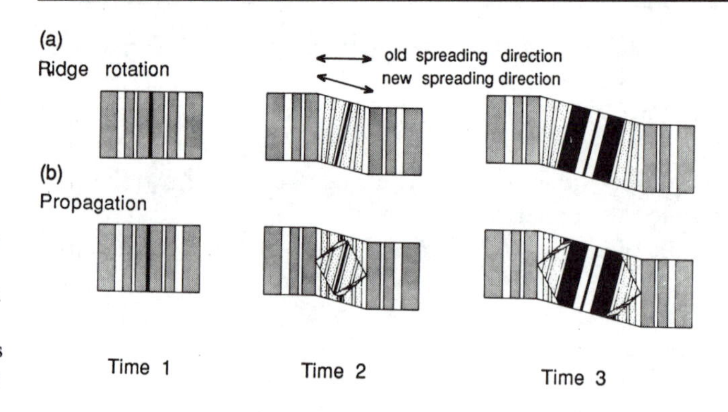

Figure 3.23. Magnetic anomaly patterns for two possible models by which a midocean ridge can adjust to changes in its spreading direction caused by a change in the rotation pole. (a) Ridge rotation. In this model, during the adjustment time, the spreading at the ridge is asymmetric. Half-spreading rates are not equal on both sides of the ridge and change along the length of the ridge. (b) Propagating rifting. In this model, a section of ridge is assumed to jump to its new orientation. This 'propagating' ridge segment then lengthens pseudofaults at oblique angles to the ridge, as observed in Figure 3.22a. (From Wilson et al. 1984.)

interpretation of these anomalies, when combined with other available data on plate motions and earthquakes in the area, is shown in Figure 3.22. The details of the plate boundaries and their relative motions and ages are much clearer. Although the ridges in this region are spreading relatively slowly, the plates are small, so the lithosphere is only approximately 10 Ma old when it is subducted beneath the North American Plate. Another feature of the magnetic anomalies in this region is the difference of some 20° between the present-day trend of the Juan de Fuca Ridge and the strike of the older anomalies on the Pacific Plate (southwest part of Fig. 3.22a). This difference is explained by changes in the rotation pole and subsequent reorientation of the ridge. The diagonal *pseudofaults* which offset the magnetic anomalies in this region are also thought to be due to the adjustment of the ridge system to changes in the rotation poles (Fig. 3.23). Thus, when an area is studied in detail, the original questions may well be answered and theories validated, but usually new questions are also raised (in this instance, the new problem is the exact method by which ridges adjust to changes in rotation poles, or vice versa). In some other parts of the Pacific region, the past plate motions are more difficult to interpret than in the region flanking North America where the presence of an active ridge system means that both sides of the anomaly pattern are preserved.

Further to the south and west on the Pacific Plate, the oceanic anomaly pattern is, on a broad scale, fairly simple (Fig. 3.24 and 3.28). Anomalies strike almost north–south and are offset by fracture zones (see Sect. 8.5.1). The central part of the ocean was formed during a period which included the Magnetic Quiet Zone. Thus, there are not many anomalies to be observed over this part of the Pacific. However, much farther north towards the Aleutian islands the pattern changes. The anomalies change direction so that they are striking approximately east–west. This feature is called the Great Magnetic Bight. The other main feature of the northern Pacific is that the north–south anomalies represent only the western half of the pattern, and except for the short ridge segments such as the Juan de Fuca Ridge, the midocean ridge that created the oceanic plate no longer exists. This vanished ridge has been subducted under the North American Plate along with much of the Farallon Plate, the name given to the plate which once was

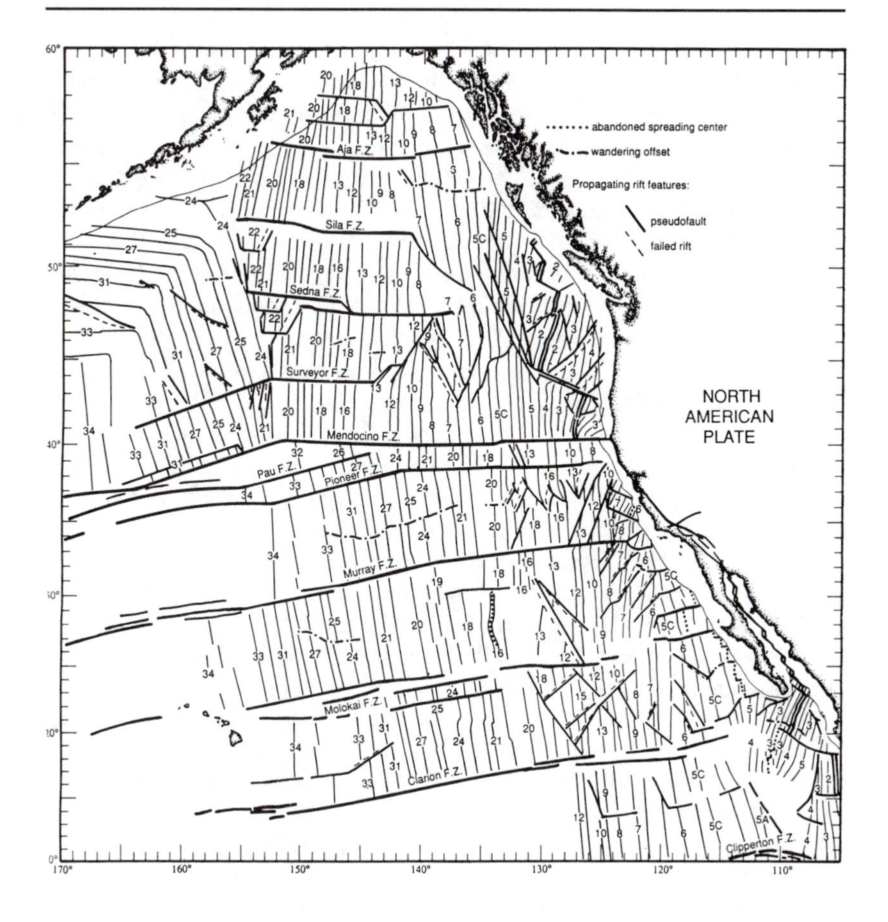

Figure 3.24. Magnetic anomalies in the northeastern Pacific. Numbers are anomaly numbers (Fig. 3.14) not ages. (From Atwater 1989.)

to the east of the ridge and had the matching half of the symmetrical anomaly pattern (Fig. 3.24). For the Farallon Plate and the ridge to have been subducted in this manner, the rate of subduction must have been greater than the rate at which the ridge was spreading. All that remains are short segments of the Pacific–Farallon Ridge and fragments of the Farallon Plate, now the Juan de Fuca, Cocos and Nazca plates.

The change in direction of the magnetic anomalies in the Great Magnetic Bight region indicates that a third plate was involved. This plate, which no longer exists as such, though a very small piece may be trapped on the Pacific Plate in the western Aleutian arc, has been named the Kula Plate. Examination of Figure 2.16 is a reminder that three ridges meeting at a triple junction produce a magnetic anomaly pattern as observed here. Therefore, the Pacific, Farallon and Kula plates are assumed to have met at an RRR triple junction (the Kula Triple Junction). The spreading rates and directions of these ridges have to be determined from the anomalies and fracture zones in the vicinity of the Great Magnetic Bight. The Kula Plate has been subducted beneath the North American Plate, as have both the other ridges.

Putting all of this information together to determine the motions of the plates in the northern Pacific region for the last 80 Ma involves much spherical geometry and computing. An idealized flat–plate model of this region was shown in Chapter 2, Problem 2. This illustrates the main features of the evolution of the northeastern Pacific. Figure 3.25 shows a

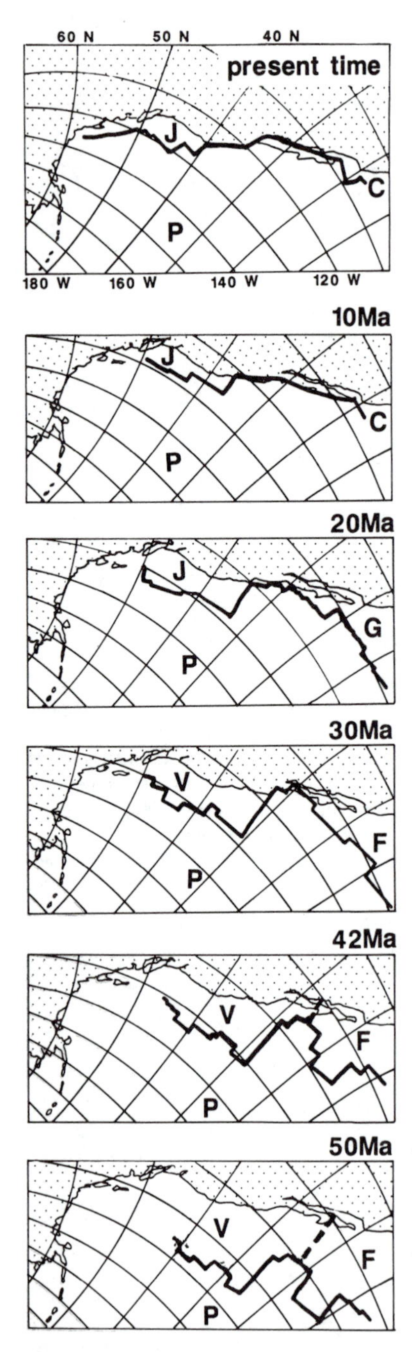

Figure 3.25. Reconstructions of the positions of the Pacific Plate and the Farallon Plate with respect to the North American Plate (shaded). P, Pacific Plate; F, Farallon Plate; V, Vancouver Plate; J, Juan de Fuca Plate; G, Guadalupe Plate; C, Cocos Plate. (From Atwater 1989 after Stock and Molnar 1988.)

reconstruction of the evolution of this region from 50 Ma until the present. From 80 Ma until about 55 Ma there were four plates in this northern Pacific region (North America, Kula, Farallon and Pacific), and two of them were being subducted beneath the North American Plate: the Kula Plate in the north and the Farallon Plate farther south. About 55 Ma ago the northern part of the Farallon Plate broke off to form the Vancouver Plate. The present Juan de Fuca Plate is the remnant of this Farallon Plate. The location of the boundary between the Vancouver and Farallon Plates was rather variable but the relative motion was slow oblique compression. Subduction of the Vancouver Plate beneath North America continued to take place. At about 30 Ma, the situation changed when the ridge between the Farallon and Pacific plates first reached the subduction zone. At this time, subduction ceased on that part of the North American Plate boundary because there the North American Plate was adjacent to the Pacific Plate, and the relative motion between the Pacific and North American plates was parallel to the plate boundary. Thus, the San Andreas Fault and the Mendocino Triple Junction were born. By 10 Ma, the Farallon/Guadalupe/Cocos Plate was very small, and the San Andreas Fault system had lengthened. At about 9 Ma and 5 Ma, as discussed earlier the strike of the Juan de Fuca Ridge changed by some 20° in total, resulting in the present configuration of the plates. The geological evolution of western North America was controlled by the motions of these oceanic plates. If the relative motion between the Pacific and North American plates is assumed to have been parallel to the subduction zone between the Farallon and North American plates (the present-day Cascade Subduction Zone), the Mendocino Triple Junction, where the Farallon, North American and Pacific plates meet, must have been stable (Figs. 2.16 and 3.26). However, since the Cascade Subduction Zone is not, at present, co-linear with the San Andreas Fault, the triple junction is unstable and may well have always been unstable (Fig. 3.27). The evolution of this unstable triple junction is shown in Figure 3.26e. If the three plates are assumed to be rigid, a hole must develop. Such a hole would presumably fill with rising mantle material from below and sediments from above. It would become in effect a microplate. Alternatively, if the continental North American Plate is allowed to deform, then the triple junction evolves as shown in Figure 3.26f, with internal deformation involving both extension and rotation over a wide zone to the east of both the subduction zone and fault. The deformations which would be produced by this process are identical to those indicated by continental palaeomagnetic data and would account for the regional extension in the western United States (which began about 30 Ma ago when the Mendocino Triple Junction was formed), as well as for the eastward stepping of the San Andreas Fault with time, which has transferred parts of coastal California from the North American Plate to the Pacific Plate. Details of the past plate motions and their effect on the interior of the North American Plate will no doubt continue to be revised for many years, but the main features of the model as presented here are probably nearly correct.

Farther south, to the west of the Middle Americas Trench, magnetic lineations striking east–west are sandwiched between north–south anomalies (Fig. 3.28). This region has undergone a series of plate reorganizations since 30 Ma when the Pacific–Farallon Ridge first intersected the North

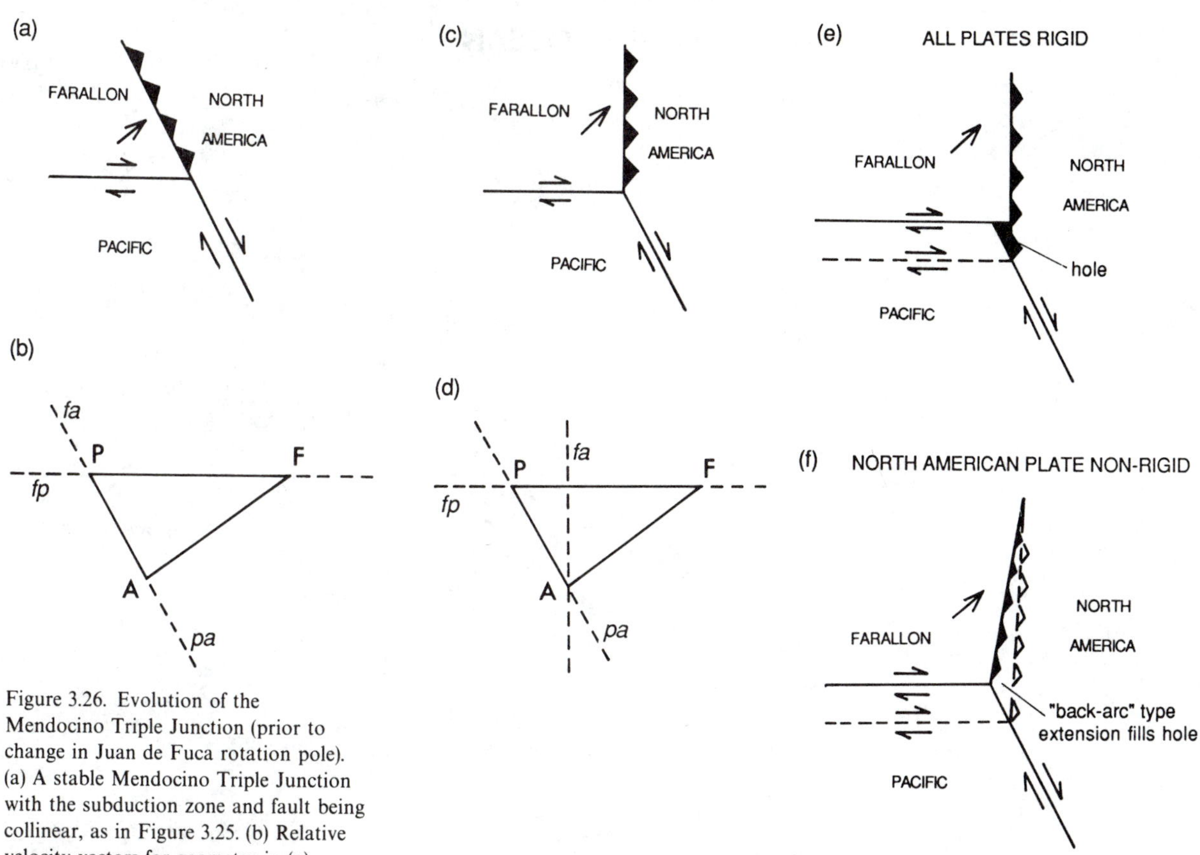

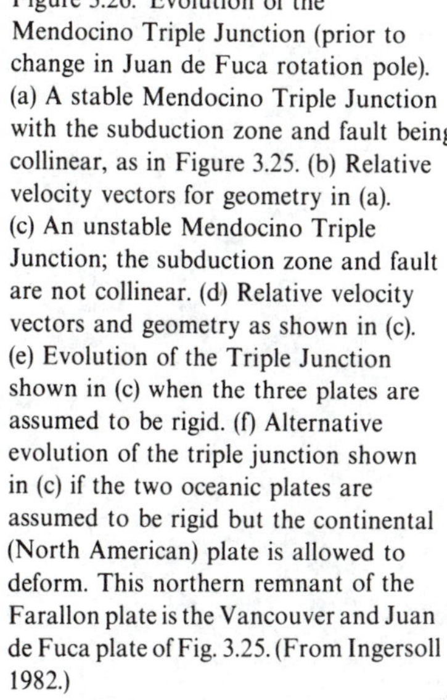

Figure 3.26. Evolution of the Mendocino Triple Junction (prior to change in Juan de Fuca rotation pole). (a) A stable Mendocino Triple Junction with the subduction zone and fault being collinear, as in Figure 3.25. (b) Relative velocity vectors for geometry in (a). (c) An unstable Mendocino Triple Junction; the subduction zone and fault are not collinear. (d) Relative velocity vectors and geometry as shown in (c). (e) Evolution of the Triple Junction shown in (c) when the three plates are assumed to be rigid. (f) Alternative evolution of the triple junction shown in (c) if the two oceanic plates are assumed to be rigid but the continental (North American) plate is allowed to deform. This northern remnant of the Farallon plate is the Vancouver and Juan de Fuca plate of Fig. 3.25. (From Ingersoll 1982.)

American Subduction Zone. In general terms these reorganizations can best be described as the breaking up of large plates into smaller plates as the ridge was progressively subducted. At 30 Ma one plate (Farallon) lay to the east of the ridge. At 25 Ma the east–west anomalies indicate that the Cocos–Nazca Ridge (otherwise known as the Galapagos spreading centre) started spreading; so two plates, the Nazca Plate and the Cocos or Guadalupe Plate, lay to the east of the ridge. At about 12 Ma, the Cocos Plate subdivided, spawning the Rivera Plate; and now three plates lie to the east of the ridge, now called the East Pacific Rise (Fig. 2.2). A schematic flat model of the plates in this region was presented in Chapter 2, problem 10.

The magnetic lineations in the northwestern Pacific region are much

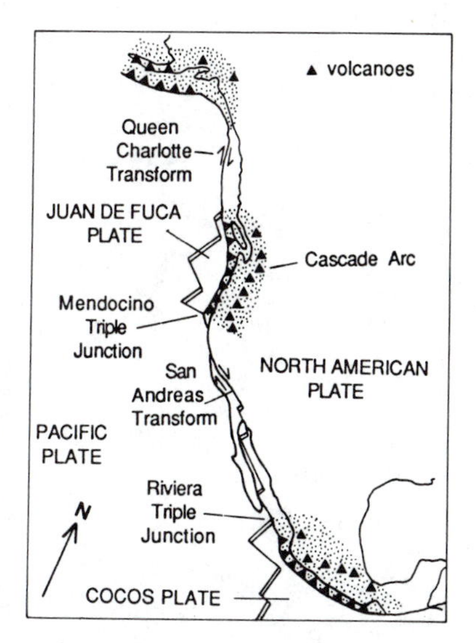

Figure 3.27. Present-day tectonic map of the western part of the North American Plate. Stippled regions mark extent of expected arc magmatism. (After Dickinson 1979.)

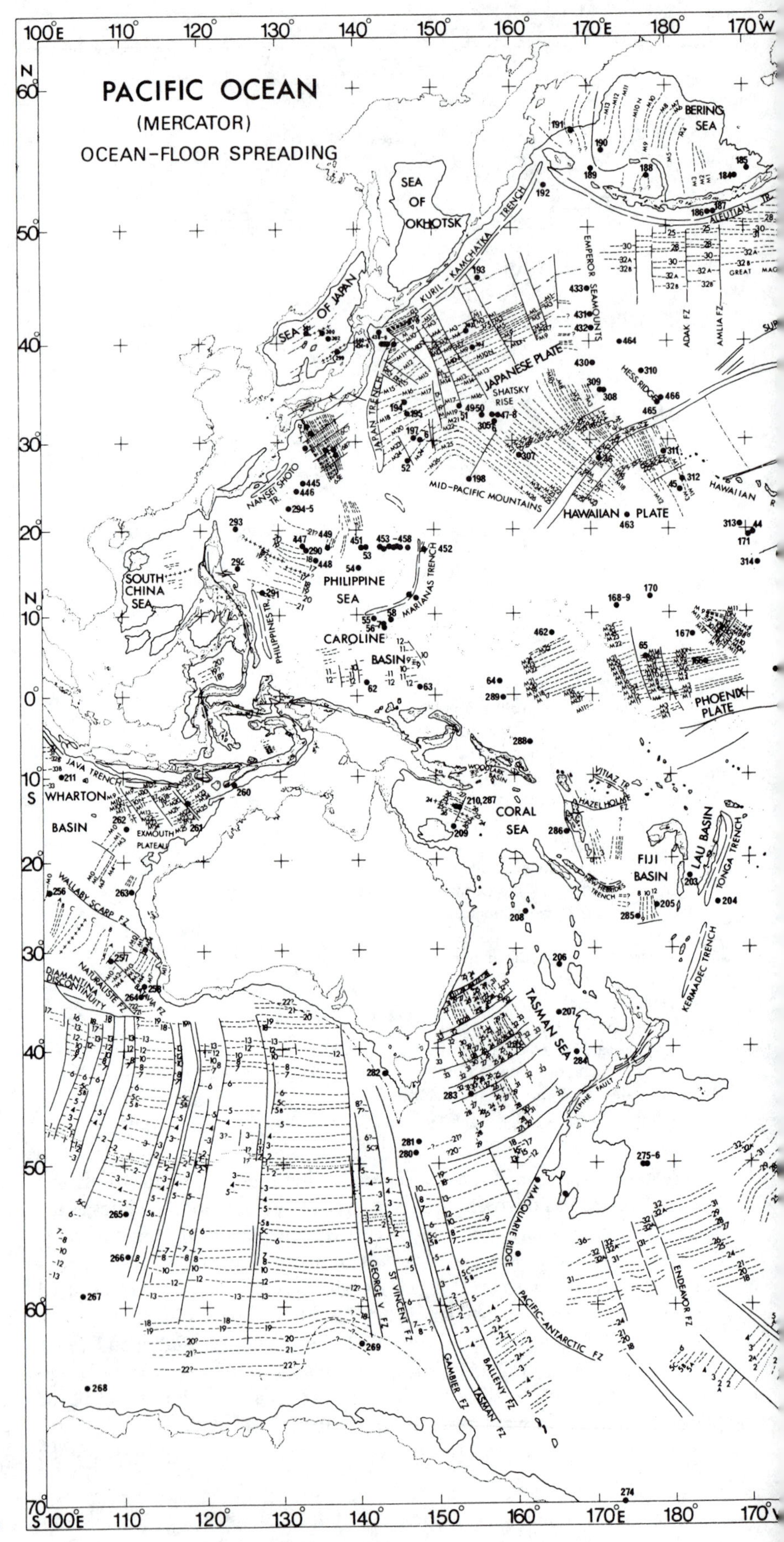

Figure 3.28. Magnetic anomalies over the Pacific Ocean. (From Owen 1983.)

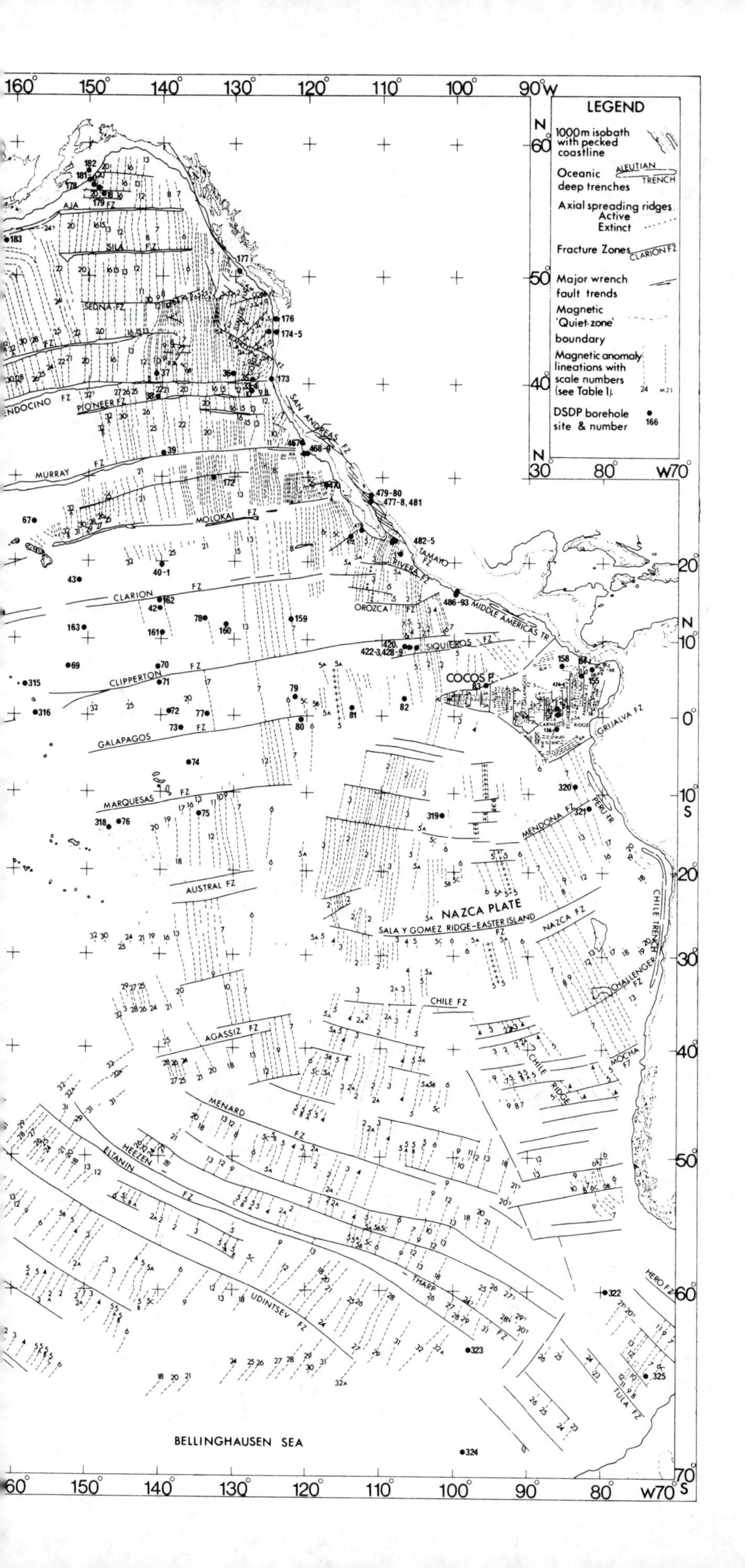
LEGEND
1000m isobath with pecked coastline
Oceanic deep trenches
ALEUTIAN TRENCH
Axial spreading ridges
Active
Extinct
Fracture Zones
CLARION FZ
Major wrench fault trends
Magnetic 'Quiet-zone' boundary
Magnetic anomaly lineations with scale numbers (see Table 1).
DSDP borehole site & number
166
160° 150° 140° 130° 120° 110° 100° 90°W
80° W70°
60° 50° 40° 30° 20° 10° 0° 10° 20° 30° 40° 50° 60° 70°
N
S
AJA
SILA FZ
SEDNA FZ
MENDOCINO FZ
PIONEER FZ
SAN ANDREAS FZ
MURRAY FZ
MOLOKAI FZ
TAMAYO
RIVERA FZ
CLARION FZ
OROZCA FZ
MIDDLE AMERICAS TR
SIQUEIROS FZ
CLIPPERTON FZ
COCOS P.
GALAPAGOS FZ
GRIJALVA FZ
MARQUESAS FZ
MENDONA FZ
PERU TR
AUSTRAL FZ
NAZCA PLATE
SALA Y GOMEZ RIDGE-EASTER ISLAND FZ
NAZCA FZ
CHILE TRENCH
CHALLENGER FZ
CHILE FZ
AGASSIZ FZ
CHILE RIDGE
MOCHA FZ
MENARD FZ
HEEZEN
ELTANIN FZ
THARP FZ
UDINTSEV FZ
HERO FZ
TULA FZ
BELLINGHAUSEN SEA

older and more complex than those of the northeastern Pacific (Fig. 3.28). As plates get older, tectonic reconstructions generally become increasingly difficult and subject to error and, frequently, to speculation. The southwest–northeast lineations extending from the Japan trench towards the Aleutian trench are called the Japanese lineations. The northwest–southeast lineations to the south and west of the Emperor Seamounts and the Hawaiian Ridge are called the Hawaiian lineations. Both the Japanese and Hawaiian lineations are identified as M1 to M29 inclusive. The east–west lineations that straddle the equator and extend from 160°E to 170°W are the Phoenix lineations (so named because of their proximity to the Phoenix Islands).

The Japanese and Hawaiian lineations form a well-defined magnetic bight, older but otherwise very similar to the Great Magnetic Bight in the northeastern Pacific. This older magnetic bight is thought to have been formed by an RRR triple junction, where the Farallon and Pacific Plates met a third plate, the Izanagi Plate (Fig. 3.29).

The Phoenix lineations are from anomalies M1 to M25 inclusive and were produced, between about 127 and 155 Ma, by an east–west striking Phoenix–Pacific Ridge possibly some 40° south of their present latitude. It has been suggested that this ridge extended far to the west and joined the ridge system of the Indian Ocean that gave rise to the magnetic anomalies north of the Exmouth Plateau (Sect. 3.3.3). Any younger lineations and the symmetric southern half of these Phoenix lineations are no longer present here.

There are two magnetic bights in the southern Pacific, both less well defined than their northern counterparts. The first is at the eastern end of the Phoenix lineations where the (M12–M4, 137–131 Ma) lineations bend from approximately east–west to northwest–southeast. These northwest–southeast anomalies were probably produced by the Pacific–Farallon Ridge. The second magnetic bight is at about 40°S, 145°W, where anomalies 32–20 (45–70 Ma) bend from approximately southwest–northeast to northwest–southeast. Unfortunately, much of the evidence which would enable the details of these western Pacific plates and their relative motions to be better determined has been swallowed by the hungry western Pacific subduction zones.

Figure 3.29 shows a series of reconstructions of the plates in the Pacific from 110 Ma to the present. The Magnetic Quiet Zone lasted from 124 Ma to 84 Ma, so not much can be deduced about the motions of the plates during that period. The Kula Plate came into existence then. It is not possible for the Kula Plate to be the older Izanagi Plate; rather it is believed to have been a piece which broke off either the Farallon Plate or the Pacific Plate. The Farallon Plate was very large indeed: Between about 85 and 55 Ma ago the Pacific–Farallon Ridge extended for some 10,000 km.

The evolution of the western and southern Pacific was clearly very complex with a number of changes in spreading centres. The process probably included the creation and subduction of whole plates. No doubt further detailed mapping of the lineations and palaeomagnetic measurements and dating of drilled and dredged basement samples will refine and improve the picture of the history of the Pacific back to Jurassic time.

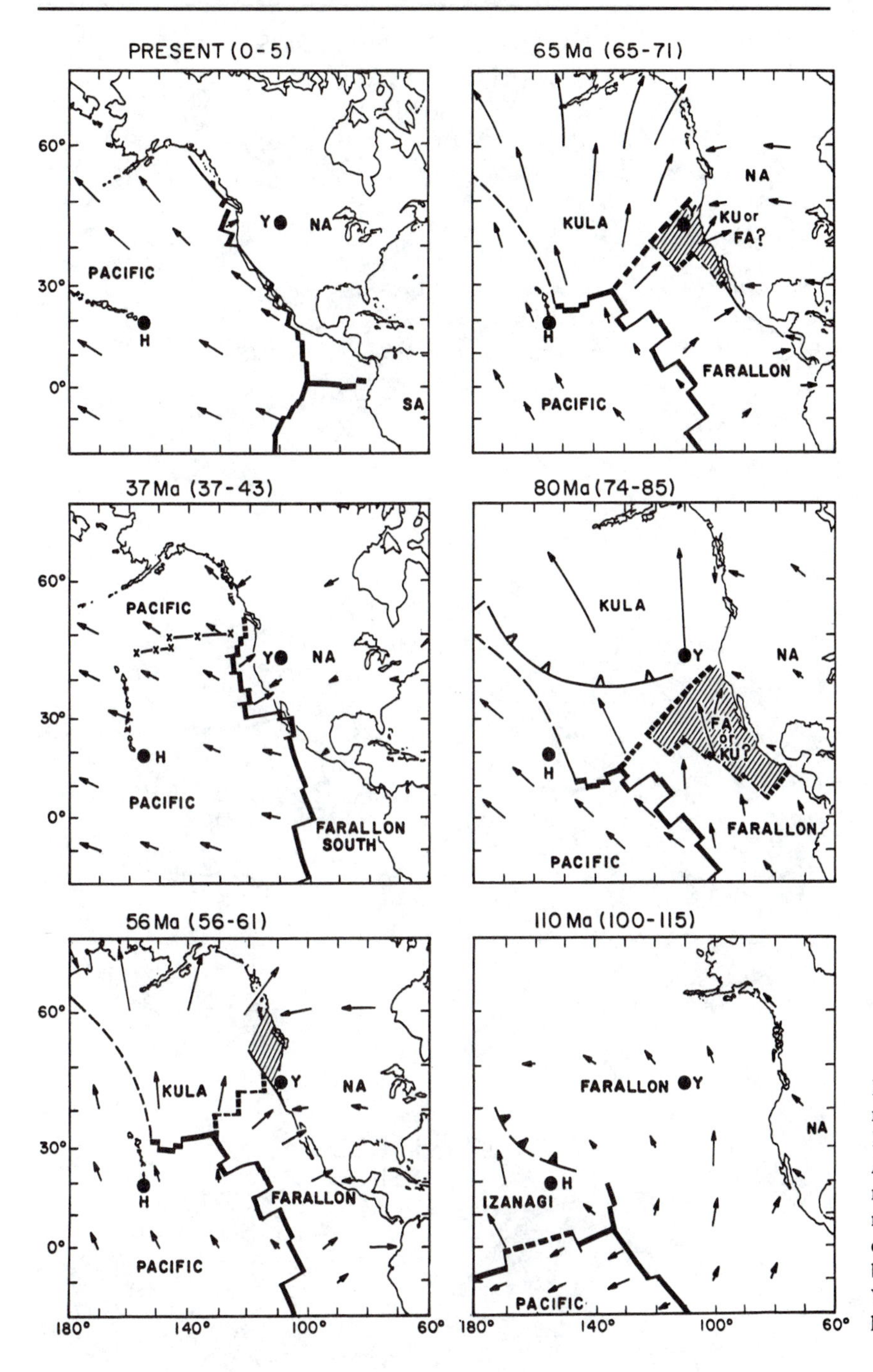

Figure 3.29. Reconstruction of the motions of the Pacific, Izanagi, Farallon, Kula and North American (NA) Plates. Arrows indicate motion of the plates with respect to the hotspots. The hatched region at 56–80 Ma indicates the range of possible locations for the ridge between the Kula and Farallon Plates. Y, Yellowstone hotspot; H, Hawaüan hotspot. (From Atwater 1989.)

3.3.5 The Continents

Figure 3.30, a series of snapshots of the continents, shows how they have moved relative to each other during the last 280 million years. By the late Carboniferous to earliest Permian, the continents were all joined together and formed one supercontinent which we call *Pangaea* (Greek, 'all lands').

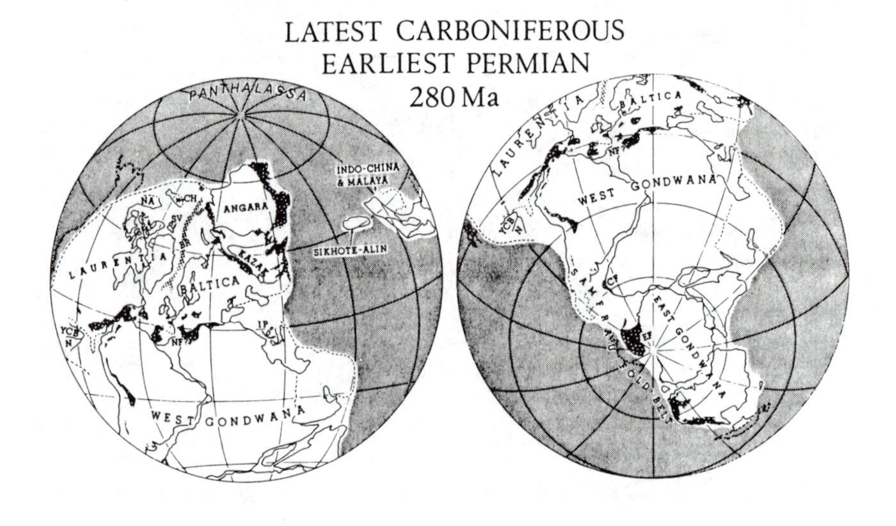

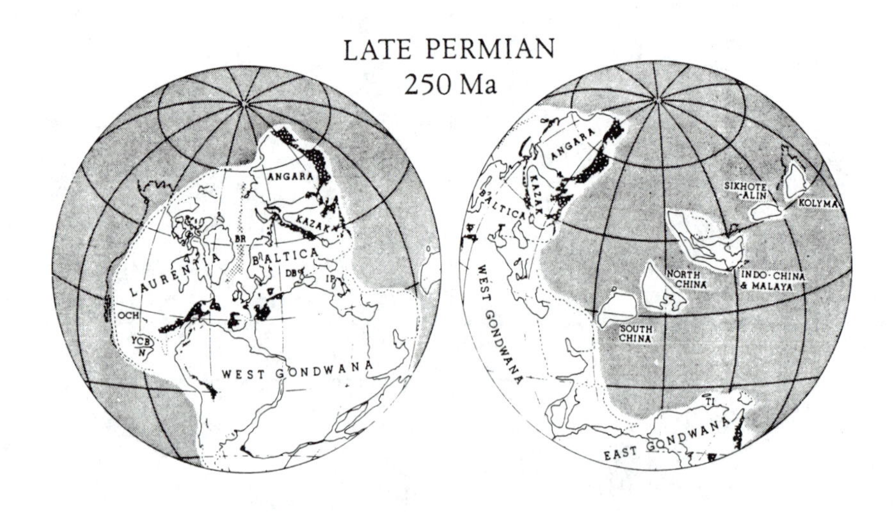

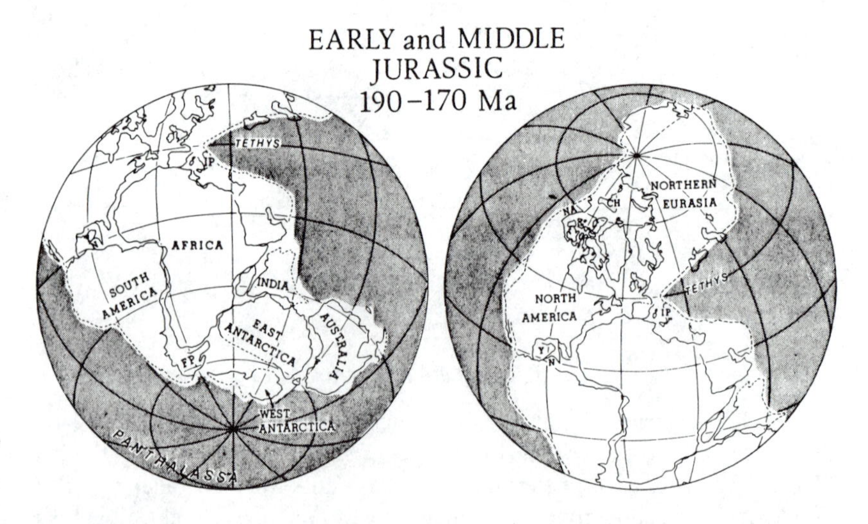

Figure 3.30. Palaeocontinental maps showing the present-day continents in their previous positions. BB, Baffin Bay; BR, Boreal rift; CF, Cape foldbelt; CH, Chukotsk peninsula; DB, Donbass; EF, Ellesworth foldbelt; FP, Falkland Plateau; GM, Gulf of Mexico; IM, Indochina–Malaya block; IP, Italian promontory; KAZAK, Kazakhstan; KY, Kolyma block; LS, Labrador Sea; N, Nicaragua; NA, North Alaska; NF, NW African foldbelt; OCH, Ouchita foldbelt; ST, Stikinia; SV, Svalbad; TI, Timor; TT, Tibet; Y, Yucatan; YCB, Yucatan and Campeche Bank. (From Irving 1983.)

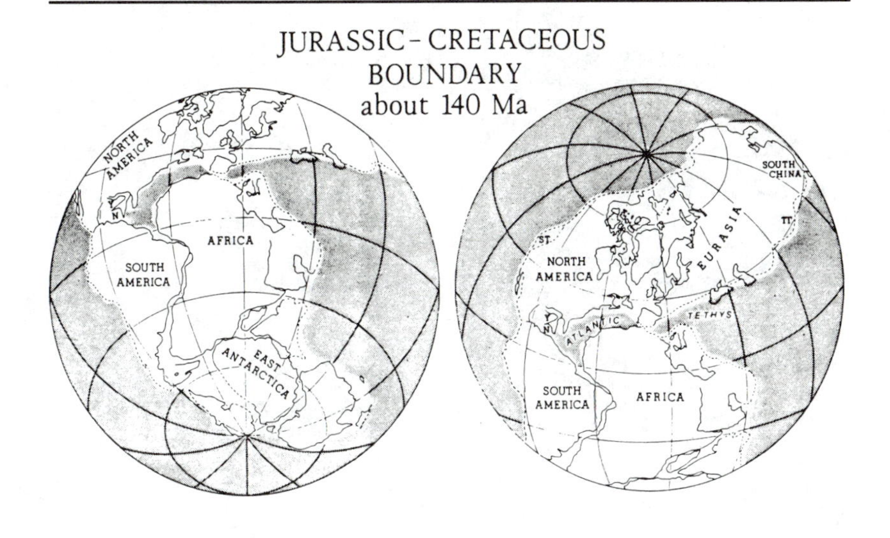
JURASSIC-CRETACEOUS
BOUNDARY
about 140 Ma
NORTH AMERICA
AFRICA
SOUTH AMERICA
EAST ANTARCTICA
SOUTH CHINA
NORTH AMERICA
EURASIA
ATLANTIC
TETHYS
SOUTH AMERICA
AFRICA

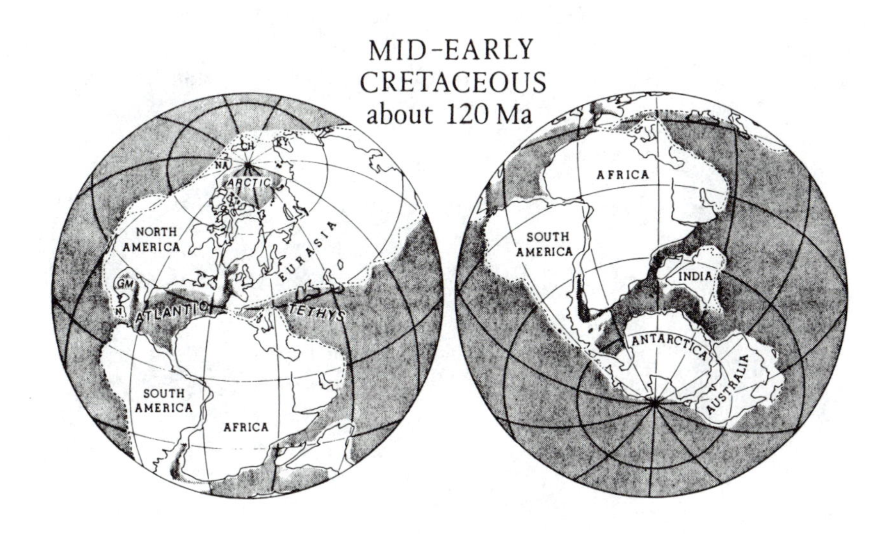
MID-EARLY
CRETACEOUS
about 120 Ma
ARCTIC
NORTH AMERICA
EURASIA
ATLANTIC
TETHYS
SOUTH AMERICA
AFRICA
AFRICA
SOUTH AMERICA
INDIA
ANTARCTICA
AUSTRALIA

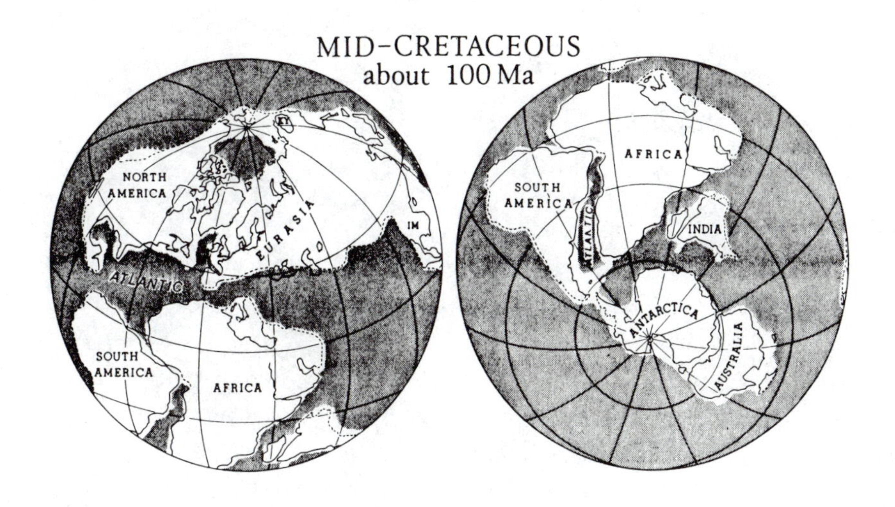
MID-CRETACEOUS
about 100 Ma
NORTH AMERICA
EURASIA
ATLANTIC
SOUTH AMERICA
AFRICA
AFRICA
SOUTH AMERICA
ATLANTIC
INDIA
ANTARCTICA
AUSTRALIA

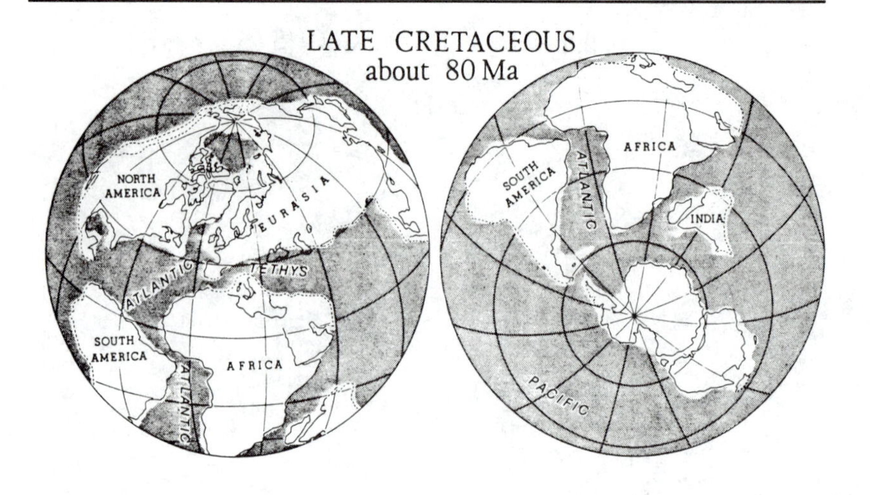

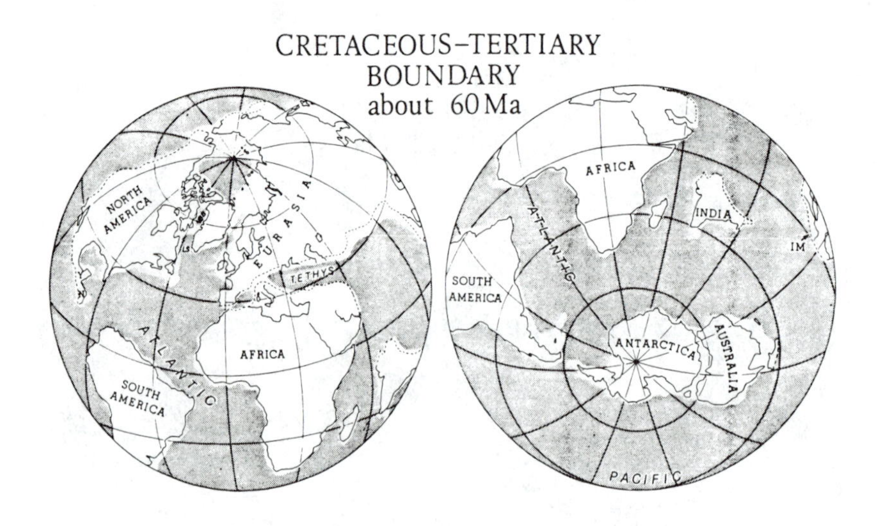

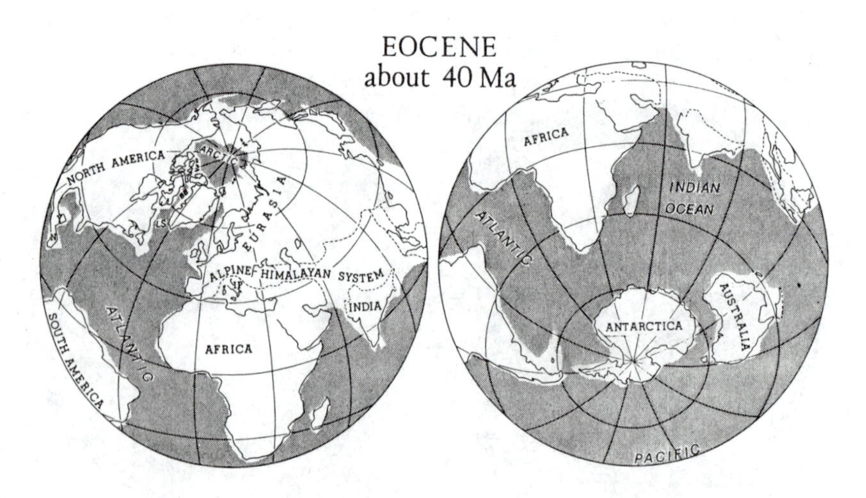

Figure 3.30 (*cont.*)

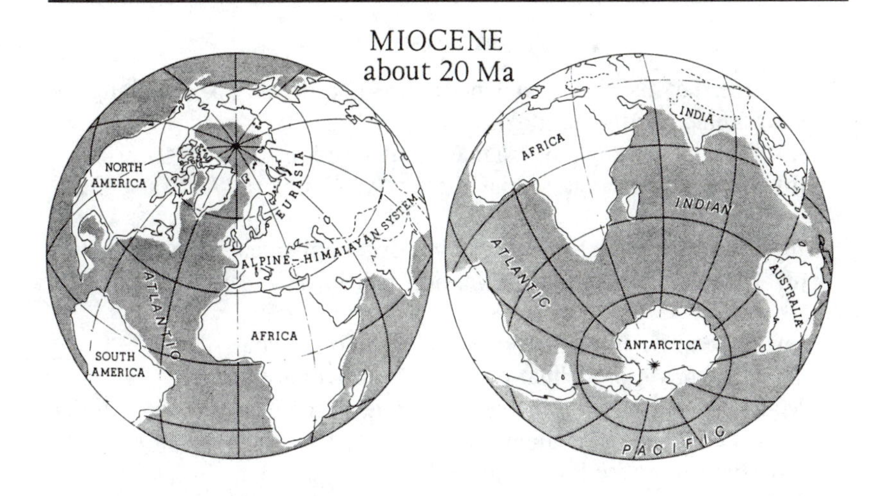

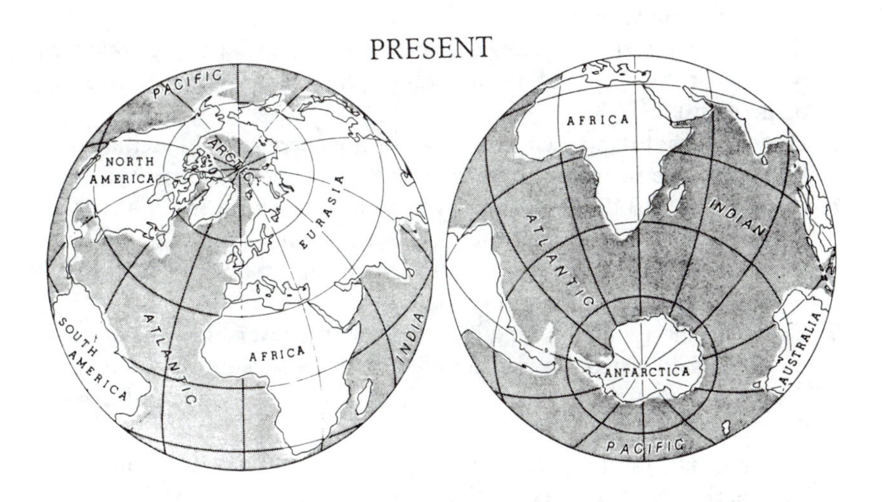

The northern part of Pangaea, comprised of today's North America and Eurasia, has been named *Laurasia*, and the southern part of the continent, comprised of South America, Africa, India, Antarctica and Australia, is called *Gondwanaland*. Laurasia and Gondwanaland became distinct during the Jurassic when Pangaea rifted in two. The wedge-shaped ocean between Laurasia and Gondwanaland is the *Tethys* Sea. (Tethys was the wife of Oceanus in Greek mythology.) It is this sea which has presumably been subducted beneath Laurasia as India and Africa have moved northwards. The Mediterranean, Caspian and Black seas are the last vestiges of this ancient ocean, the completion of whose subduction resulted in the building of the Alpine, Carpathian and Himalayan mountain chains. All these regions contain scattered outcrops of ophiolites (a suite of rocks which have chemical and lithological similarities to the oceanic crust and which may be examples of crust from ancient back-arc basins).

PROBLEMS

1. Assume that the earth's magnetic field is a dipole aligned along the geographic north–south axis.
   (a) What is the angle of inclination at London, U.K. (51°N, 0°E)?
   (b) What is the angle of inclination at Canberra, Australia (35°S, 149°E)?
   (c) What is the angle of inclination on Spitzbergen (78°N, 16°E)?
   (d) What is the angle of inclination at Rio de Janeiro, Brazil (23°S, 43°W)?
   (e) If the angle of inclination is 76°, where are you?
   (f) If the angle of inclination is − 36°, where are you?
2. Magnetic measurements have been made on some lavas found at 60°N, 90°W. The angle of inclination is measured at 37°. At what magnetic latitude were these lavas erupted?
3. If the direction of magnetization of the lavas of problem 2 is due west, calculate the position of the pole of the earth's magnetic field at the time the lavas were erupted. What does this indicate about the continent on which these lavas occur?
4. If the measurement of the angle of inclination of the lavas of problem 2 is in error by 5°, what is the subsequent error in the calculated palaeolatitude?
5. Down to what depth can (a) oceanic lithosphere and (b) continental lithosphere be permanently magnetized? (*Hint*: see Chapter 7.)
6. (a) Use Figure 3.12 to calculate a spreading rate for the South Atlantic. Is this a half-spreading rate or a plate separation rate?
   (b) Using Figure 3.12, estimate minimum and maximum spreading rates for the Pacific during the last 80 Ma.
7. Calculate the minimum length of time that a reversal of the earth's magnetic field lasted if it was detected from sea-surface magnetic data collected in (a) the Atlantic Ocean and (b) the equatorial Pacific Ocean.
8. Where is the oldest ocean floor, and what age is it? Discuss what may have happened to the rest of the ocean floor of this age and why.
9. At any given latitude, what orientation of ridge gives rise to the largest amplitude magnetic anomalies?
10. At what latitude do magnetic anomalies due to magnetized oceanic crust have minimum amplitude?

BIBLIOGRAPHY

Atwater, T. 1970. Implications of plate tectonics for the Cenozoic tectonic evolution of western North America. *Bull. Geol. Soc. Am.*, *81*, 3513–36.

Plate tectonic history of the northeast Pacific and western North America. In E. L. Winterer, D. M. Hussong and R. W. Decker, eds., *The geology of North America, Vol. N, The eastern Pacific Ocean and Hawaii*, Geol. Soc. Am., 21–7.

Atwater, T., and Sverınghaus, J. 1989. Tectonic maps of the northeast Pacific. In E. L. Winterer, D. M. Hussong, and R. W. Decker, eds., *The geology of North America, Vol. N, The eastern Pacific Ocean and Hawaii*, Geol. Soc. Am., 15–20.

Bambach, R. K., Scotese, C. R., and Ziegler, A. M. 1980. Before Pangea: The geographies of the Paleozoic world. *Am. Sci.*, *68*, 26–38.

Briden, J. C., Hurley, A. M., and Smith, A. G. 1981. Paleomagnetism and Mesozoic–Cenozoic paleocontinental world maps. *J. Geophys. Res.*, *86*, 11631–56.

Brunhes, B. 1906. Récherches sur la direction d'aimentation des roches volcaniques (1). *J. Physique, 4e Ser.*, *5*, 705–24.

Bullard, E. C., and Mason, R. G. 1963. The magnetic field over the oceans. *In* M. N. Hill, ed., *The sea, Vol. 3*, Wiley, New York, 175–217.

Cande, S. C., and Mutter, J. C. 1982. A revised identification of the oldest sea-floor spreading anomalies between Australia and Antarctica. *Earth Planet. Sci. Lett.*, *58*, 151–60.

Dickinson, W. R. 1979. Cenozoic plate tectonic setting of the Cordilleran region in the United States. *In* J. M. Armentrout, M. R. Cole and H. TerBest, Jr., eds., *Cenozoic Paleography of the western United States*, Society of Economic Palaeontologists and Mineralogists, Pacific Section, Pacific Coast Paleogeography Symposium 3, 1–13.

1979. Plate tectonic evolution of North Pacific Rim. *In* S. Uyeda et al., eds., *Geodynamics of western Pacific, Advances in Earth and Planetary Sciences, 6*, Centre for Academic Publications Japan, 1–19.

Ellis, R. M., Spence, G. D., Clowes, R. M., Waldron, D. A., Jones, I. F., Green, A. G., Forsyth, D. A., Mair, J. A., Berry, M. J., Mereu, R. F., Kanasewich, E. R., Cumming, G. L., Hajnal, Z., Hyndman, R. D., McMechan, G. A., and Loncarevic, B. D. 1983. The Vancouver Island Seismic Project: A COCRUST onshore–offshore study of a convergent margin. *Can. J. Earth Sci.*, *20*, 719–41.

Engebretson, D. C., Cox, A., and Gordon, R. G. 1985. Relative Motion between Oceanic and Continental Plates in the Pacific Basin. Geol. Soc. Am. Special Paper 206.

Falvey, D. A., and Mutter, J. C. 1981. Regional plate tectonics and evolution of Australia's passive continental margins. *Bur. Miner. Resour. J. Aust. Geol. Geophys.*, *6*, 1–29.

Farrar, E., and Dixon, J. M. 1981. Early Tertiary rupture of the Pacific plate: 1700 km of dextral offset along the Emperor Trough–Line Islands lineament. *Earth Planet. Sci. Lett.*, *53*, 307–22.

Fisher, R. L., and Sclater, J. G. 1983. Tectonic evolution of the southwest Indian Ocean ridge system since the mid-Cretaceous: Plate mobility and stability of the pole of Antarctica/Africa for at least 80 My. *Geophys. J. R. Astr. Soc.*, *73*, 553–76.

Garland, D. G. 1979. *Introduction to geophysics: Mantle, core and crust*, 2nd ed. Saunders, Philadelphia.

Grantz, A., Johnson, L. and Sweeney, J.F. eds. 1990, *The Arctic Ocean Region*, The Geology of North America vol. L, *Geol. Soc. Am.*

Handschumacher, D. W. 1976. Post-Eocene plate tectonics of the Eastern Pacific. *In* G. H. Sutton et al., eds., *The geophysics of the Pacific Ocean Basin and its margin.* Vol. 19 of *Geophys. Monogr. Am. Geophys. Union*, 177–202.

Harland, W. B., Cox A. V., Llewellyn, P. G., Pickton, C. A. G., Smith, A. G., and Walters, R. 1990. *A geologic time scale 1989*. Cambridge Univ. Press.

Harrison, C. G. A. 1987. Marine magnetic anomalies – the origin of the stripes. *Ann. Rev. Earth Planet. Sci.*, *15*, 505–43.

Heirtzler, J. R., Dickson, G. O., Herron, E. M., Pitman, W. C., III, and Le Pichon X. 1968. Marine magnetic anomalies, geomagnetic field reversals and motions of the ocean floor and continents. *J. Geophys. Res.*, *73*, 2119–36.

Hilde, T. W. C., Isezaki, N., and Wageman, J. M. 1976. Mesozoic sea-floor spreading in the North Pacific. *In* G. H. Sutton et al., eds., *The geophysics of the Pacific Ocean Basin and its margin.* Vol. 19 of Geophys. Monogr Am. Geophys. Union, 205–26.

Hutton, V. R. S. 1976. The electrical conductivity of the earth and planets. *Rep. Prog. Phys., 39*, 487–572.

Ingersoll, R. V. 1982. Triple-junction instability as cause for late Cenozoic extension and fragmentation of the western United States. *Geology, 10*, 621–4.

Irving, E. 1983. Fragmentation and assembly of the continents, mid-Carboniferous to present. *Geophysical Surveys, 5*, 299–333.

1988. The paleomagnetic confirmation of continental drift. *EOS Trans. Am. Geophys. Un., 69*, 44, 994–1014.

Irving, E., and Sweeney, J. F. 1982. Origin of the Arctic Basin. *Trans. Roy. Soc. Canada, Ser. IV, XX*, 409–16.

Klitgord, K. D., and Mammerickx, J. 1982. Northern East Pacific Rise: Magnetic anomaly and bathymetric framework. *J. Geophys. Res., 87*, 6725–50.

Larson, R. L., and Chase, C. G. 1972. Late Mesozoic evolution of the western Pacific Ocean. *Bull. Geol. Soc. Am., 83*, 3627–44.

Lawver, L. A., Sclater, J. G., and Meinke, L. 1985. Mesozoic and Cenozoic reconstructions of the South Atlantic. *Tectonophys., 114*, 233–54.

Malin, S. R. C., and Bullard, E. 1981. The direction of the earth's magnetic field at London, 1570–1975. *Phil. Trans. R. Soc., 299A*, 357–423.

Mammerickx, J., and Klitgord, K. D. 1982. Northern East Pacific Rise: Evolution from 25 mybp to the present. *J. Geophys. Res., 87*, 6751–9.

Matuyama, Motonori. 1929. On the direction of magnetisation of basalt in Japan, Tyôsen and Manchuria. *Japan Academy Proceedings, 5*, 203–5.

Maxwell, A. E., Von Herzen, R. P., Hsü, K. J., Andrews, J. E., Saito, T., Percival, S. F., Jr., Milow, E. D., and Boyce, R. E. 1970. Deep sea drilling in the South Atlantic. *Science, 168*, 1047–59.

McElhinny, M. W. 1973. *Palaeomagnetism and plate tectonics.* Cambridge Univ. Press, Cambridge.

McKenzie, D. P., and Sclater, J. G. 1971. The evolution of the Indian Ocean since the late Cretaceous. *Geophys. J. R. Astr. Soc., 25*, 437–528.

1973. The evolution of the Indian Ocean. *Sci. Am., 228*, 5, 62–72.

Molnar, P., Pardo-Casas, F., and Stock, J. 1988. The Cenozoic and late Cretaceous evolution of the Indian Ocean Basin: Uncertainties in the reconstructed positions of the Indian, African and Antarctic plates. *Basin Research, 1*, 23–40.

Morel, P., and Irving, E. 1981. Paleomagnetism and the evolution of Pangaea. *J. Geophys. Res., 86*, 1858–72.

Mutter, J. C., Hegarty, K. A., Cande, S. C., and Weissel, J. K. 1985. Breakup between Australia and Antarctica: A brief review in the light of new data. *Tectonophys., 114*, 255–79.

Ness, G., Levi, S., and Couch, R. 1980. Marine magnetic anomaly timescales for the Cenozoic and late Cretaceous: A précis, critique and synthesis, *Rev. Geophys. Space Phys., 18*, 753–70.

Nevanlinna, H., Pesonen, L. J., and Blomster, K. 1983. *Earth's magnetic field charts (IGRF 1980).* Geological Survey of Finland, Report Q19/22.0/ World/1983/1.

Norton, I. O., and Sclater, J. G. 1979. A model for the evolution of the Indian Ocean and the breakup of Gondwanaland. *J. Geophys. Res., 84*, 6803–30.

Owen, H. G. 1983. *Atlas of continental displacement: 200 million years to the present.* Cambridge Univ. Press.

Parkinson, W. D. 1983. *Introdution to geomagnetism.* Scottish Academic Press, Edinburgh.

Peddie, N. W. 1982. IGRF 1980: A report by IAGA Division Working Group 1. *Geophys. J. R. Astr. Soc., 68*, 265–8.

Pitman, W. C., III, and Heirtzler, J. R. 1966. Magnetic anomalies over the Pacific-Antarctic ridge. *Science, 154*, 1164–71.

Powell, C. McA., Roots, S. R., and Veevers, J. J. 1988. Pre-breakup continental extension in East Gondwanaland and the early opening of the eastern Indian Ocean. *Tectonophys., 155*, 261–83.

Raff, A. D., and Mason, R. G. 1961. Magnetic survey off the west coast of North America, 40°N latitude to 50°N latitude. *Geol. Soc. Am. Bull., 72*, 1267–70.

Schouten, H., and Denham, C. R. 1979. Modelling the oceanic source layer. *In* M. Talwani et al., eds., *Implications of deep drilling results in the Atlantic Ocean: Ocean crust*, Vol. 2 of Maurice Ewing Series, Am. Geophys. Un., 151–9.

Schouten, H., Denham, C., and Smith, W. 1982. On the quality of marine magnetic anomaly sources and sea-floor topography. *Geophys. J. R. Astr. Soc., 70*, 245–59.

Schouten, J. A. 1971. A fundamental analysis of magnetic anomalies over oceanic ridges. *Mar. Geophys. Res., 1*, 111–44.

Sclater, J. G., Fischer, R. L., Patriat, P., Tapscott, C., and Parsons, B. 1981. Eocene to recent development of the southwest Indian ridge, a consequence of the evolution of the Indian Ocean triple junction. *Geophys. J. R. Astr. Soc., 64*, 587–604.

Sclater, J. G., Jaupart, C., and Galson, D. 1980. The heat flow through oceanic and continental crust and the heat loss of the earth. *Rev. Geophys. Space Phys., 18*, 269–311.

Sclater, J. G., Parsons, B., and Jaupart, C. 1981. Oceans and continents: Similarities and differences in the mechanisms of heat loss. *J. Geophys. Res., 86*, 11535–52.

Scotese, C. R. 1984. An introduction to this volume: Paleozoic paleomagnetism and the assembly of Pangea. *In* R. Van der Voo, C. R. Scotese, and N. Bonhommet, eds. *Plate reconstruction from Paleozoic paleomagnetism*, Geodynamics Series, Vol. 12, Am. Geophys. Un. 1–10.

Scotese, C. R., Bambach, R. K., Barton, C., Van der Voo, R., and Ziegler, A. M. 1979. Paleozoic Base Maps. *J. Geology, 87*, 217–77.

Scotese, C. R., Gahagan, L. M., and Larson, R. L. 1988. Plate tectonic reconstructions of the Cretaceous and Cenozoic ocean basins. *Tectonophys., 155*, 27–48.

Scotese, C. R., and Sager, W. W., eds. 1988. Mesozoic and Cenozoic plate reconstructions. *Tectonophys., 155*.

Smith, A. G., Hurley, A. M., and Briden, J. C. 1981. *Phanerozoic Paleocontinental World Maps*. Cambridge Univ. Press.

Stock, J., and Molnar, P. 1988. Uncertainties and implications of the Late Cretaceous and Tertiary position of the North American relative to Farallon, Kula and Pacific plates. *Tectonics, 8*, 1359–70.

Van der Voo, R., Peinado, J., and Scotese, C. R. 1984. A paleomagnetic re-evaluation of Pangea reconstructions. *In* R. Van der Voo, C. R. Scotese, and N. Bonhommet, eds., *Plate reconstruction from Paleozoic paleomagnetism*, Geodynamics Series, Vol. 12, Am. Geophys. Un., 11–26.

Veevers, J. J. 1984. *Phanerozoic earth history of Australia*. Clarendon, Oxford.

Veevers, J. J. 1986. Breakup of Australia and Antarctica estimated as mid-Cretaceous (95 ± 5 Ma) from magnetic and seismic data at the continental margin. *Earth Planet. Sci. Lett., 77*, 91–9.

Vine, F. J. 1966. Spreading of the ocean floor: New evidence. *Science, 154*, 1405–15.

Vine, F. J., and Matthews, D. H. 1963. Magnetic anomalies over oceanic ridges. *Nature, 199*, 947–9.

Vine, F. J., and Wilson, J. T. 1965. Magnetic anomalies over a young oceanic ridge off Vancouver Island. *Science, 150*, 485–9.

Wilson, D. S., Hey, R. N., and Nishimura, C. 1984. Propagation as a mechanism of reorientation of the Juan de Fuca Ridge. *J. Geophys. Res., 89*, 9215–25.

4

Seismology
Measuring the Interior

4.1 Waves through the Earth

4.1.1 Introduction

Seismology is the study of the passage of elastic waves through the earth. It is arguably the most powerful method available for studying the structure of the interior of the earth, especially the crust and mantle. There are a variety of other geophysical techniques, including the study of gravity, magnetism and the electrical properties of the earth, which can be applied on scales ranging from the planet as a whole to large regions or small areas or even individual rock samples (Telford et al. 1990; Dobrin and Savit 1988); but seismology is probably the most widely used and the most informative. This chapter discusses the methods by which we obtain information about the interior of the planet from the study of elastic waves passing through the earth.

Earthquake seismology is perhaps the best tool for investigating the earth's interior. The study of earthquakes was of major significance in giving us our understanding of plate tectonics: earthquake foci have very accurately delineated the boundaries of the tectonic plates. It has also helped us to map the internal structure of our planet. The distribution of earthquakes shows us where the earth is active (mostly near the surface), and the passage of seismic waves through the earth allows us, as it were, to CAT-scan its interior.

When an earthquake or an explosion occurs within the earth, part of the energy released takes the form of elastic waves which are transmitted through the earth. These waves can be detected by instruments called *seismometers*, which measure, amplify and record (on paper, magnetic tape or disc) the motion of the ground on which they are positioned. The speed with which these elastic waves travel depends on the density and elastic moduli of the rocks through which the waves pass. There are two types of elastic waves: *body waves* and *surface waves.*

4.1.2 Body Waves

Body waves are seismic waves which travel through the body of the earth. The propagation of body waves is similar to that of light: Body waves are reflected and transmitted at interfaces where the seismic velocity and/or density change, and they obey Snell's law (see Sect. 4.4.2). There are two types of body waves:

1. P-Waves (P stands for primary or pressure or push-pull). These waves involve compression and rarefaction of the material as the wave passes through it but not rotation. P-waves are most correctly called dilatational or irrotational waves. They are the analogue in a solid of sound waves in air.
2. S-Waves (S stands for secondary or shear or shake). These waves involve shearing and rotation of the material as the wave passes through it but not volume change. S-waves are most correctly called rotational or equivoluminal waves.

Figure 4.1 shows the deformation undergone by a block of material when when P- and S-waves pass through it. The P-wave *particle motion* is longitudinal, meaning that the particles making up the medium through which the P-wave is passing vibrate about an equilibrium position, in the same direction as the direction in which the P-wave is travelling. In contrast, the particle motion of S-waves is transverse, that is, perpendicular to the direction of motion of the S-wave. The S-wave motion can be split into a horizontally polarized motion termed SH and a vertically polarized motion termed SV.

The *wave equation* is derived and discussed in Appendix 2. The three-dimensional *compressional wave equation* (Eq. A2.48) is

$$\frac{\partial^2 \phi}{\partial t^2} = \alpha^2 \nabla^2 \phi \tag{4.1}$$

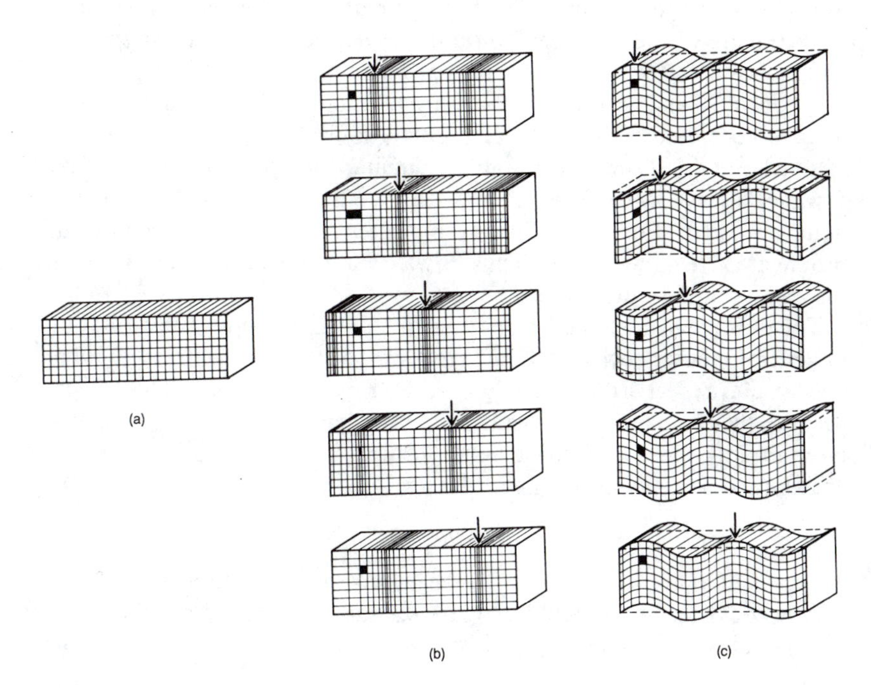

Figure 4.1. Successive stages in the deformation of (a) a block of material by P-waves and SV-waves. The sequences progress in time from top to bottom and the seismic wave is travelling through the block from left to right. Arrow marks the crest of the wave at each stage. (b) For P-waves, both the volume and the shape of the marked region change as the wave passes. (c) For S-waves, the volume remains unchanged and the region undergoes rotation only. The particle motion and deformation for SH-waves (horizontally polarized S-waves) is the same as for SV-waves (vertically polarized S-waves) but is entirely in the horizontal plane instead of the vertical plane; imagine (c) rotated through 90°. (From Phillips 1968.)

where ϕ is the *scalar displacement potential* and α the speed at which dilatational waves travel. The three-dimensional *rotational wave equation* (Eq. A2.49) is

$$\frac{\partial^2 \boldsymbol{\psi}}{\partial t^2} = \beta^2 \nabla^2 \boldsymbol{\psi} \tag{4.2}$$

where $\boldsymbol{\psi}$ is the *vector displacement potential* and β the speed at which rotational waves travel. The *displacement* $\mathbf{u}$ of the medium by any wave is given by Eq. A2.45:

$$\mathbf{u} = \nabla\phi + \nabla \wedge \boldsymbol{\psi} \tag{4.3}$$

The speeds at which dilatational and rotational waves travel, α and β, are termed the *P-wave* and *S-wave seismic velocities.* Often the symbols v_P and v_S are used instead of α and β.

The P-wave and S-wave velocities depend on the physical properties of the material through which the wave travels (Eqs. A2.31, A2.39 and A2.44):

$$\alpha = \sqrt{\frac{K + \frac{4}{3}\mu}{\rho}} \tag{4.4}$$

$$\beta = \sqrt{\frac{\mu}{\rho}} \tag{4.5}$$

where K is the bulk modulus or incompressibility, μ the shear modulus or rigidity and ρ the density.

The bulk modulus K, which is defined as the ratio of the increase in pressure to the resulting fractional change in volume, is a measure of the force per unit area required to compress material. The shear modulus μ is a measure of the force per unit area needed to change the shape of a material. Since P-waves involve change of both volume and shape, α is a function of K and μ, whereas β is only a function of μ because S-waves involve no volume change. Since the bulk modulus K must be positive, Eqs. 4.4 and 4.5 show that α is always greater than β, or, in other words, P-waves always travel faster than S-waves. The rigidity modulus μ for a liquid is zero; Eq. 4.5 therefore indicates that S-waves cannot be propagated through liquids. Thus, S-waves are not transmitted through the earth's liquid outer core.

The dependence of α and β on density is not immediately obvious, but, in general, denser rocks have higher seismic velocities, contrary to what one would expect from a first glance at Eqs. 4.4 and 4.5. This occurs because the elastic moduli K and μ are also dependent on ρ and increase more rapidly than ρ^1. A linear relationship, termed *Birch's law*, between density and seismic velocity is of the form

$$v = a\rho + b \tag{4.6}$$

where a and b are constants; it fits measurements from many crustal and mantle rocks fairly reasonably. Figure 4.2 shows examples of such linear relationships. The Nafe–Drake curve in Figure 4.2d is another example

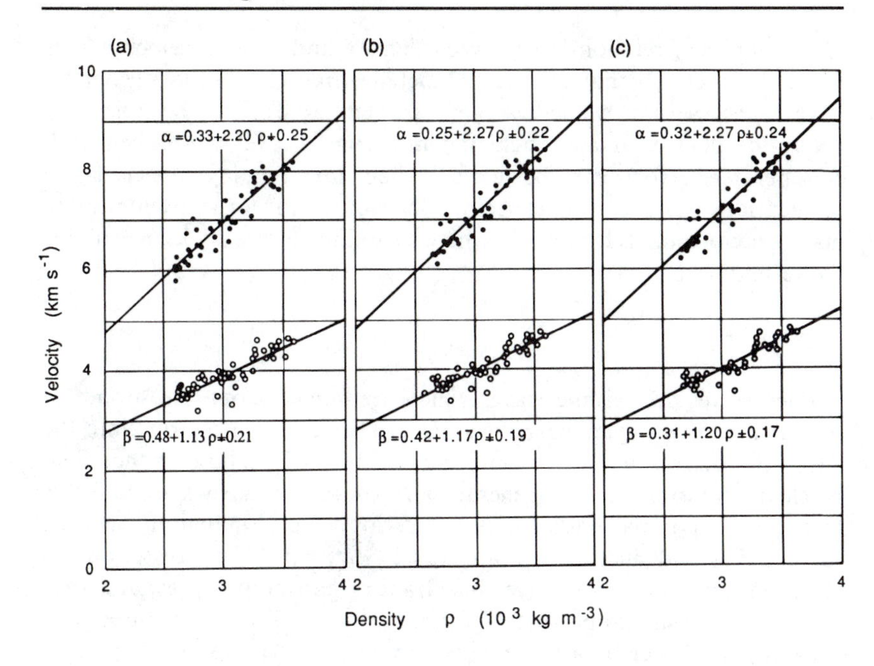

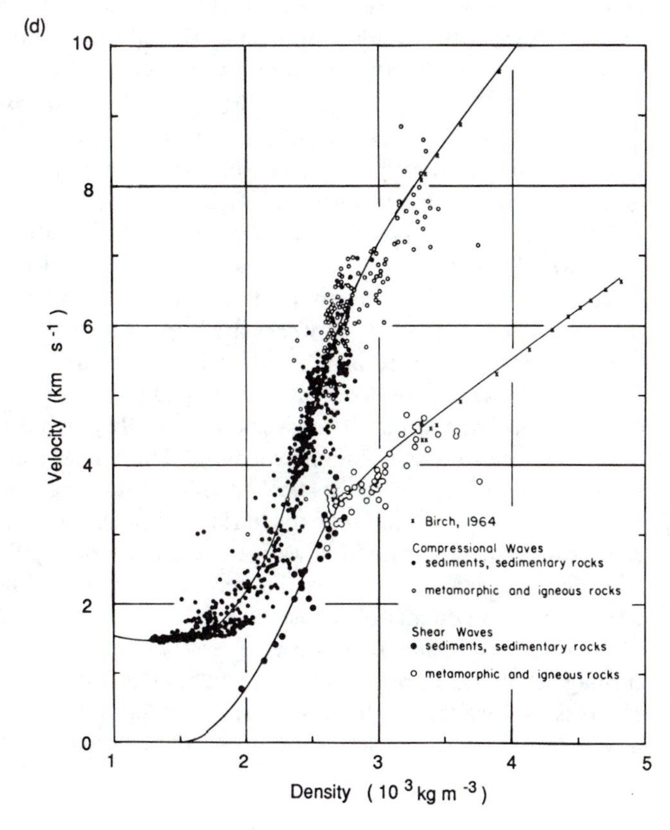

Figure 4.2. (a)–(c) Examples of the essentially linear relationship between seismic velocities and density known as Birch's law. Laboratory measurements made on crustal rocks at confining pressures of (a) 0.2 GPa, (b) 0.6 GPa and (c) 1.0 GPa, which correspond to approximate depths of 6, 18 and 30 km. (1 GPa is 10^9 pascal. The pascal, a unit of pressure, is equal to 1 newton per square metre: $1\ \text{Pa} = 1\ \text{N m}^{-2}$. A unit of pressure still often used today is the bar: $1\ \text{bar} = 10^6\ \text{dyne cm}^{-2}$. Thus, $1\ \text{GPa} = 10^4\ \text{bar} = 10\ \text{kbar}$.) (After Gebrande 1982.) (d) Relationship between seismic velocities and density referred to as the Nafe–Drake curve after its originators. Open circles, values for igneous and metamorphic rocks; solid circles, values for sedimentary rocks; crosses show a linear model from Birch 1964. (After Ludwig et al. 1970.)

of an empirical relationship between density and seismic velocity. This figure shows clearly that igneous and metamorphic crustal rocks generally have higher seismic velocities than sedimentary rocks. The ability to determine density at a particular depth from the velocity (albeit with considerable error) can be useful when investigating the isostatic implications of structures determined seismologically, or in attempting to make gravity models for regions where something is already known about the seismic velocity structure.

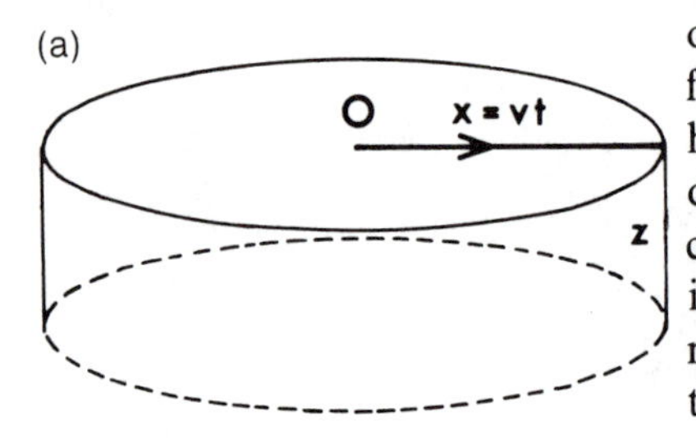

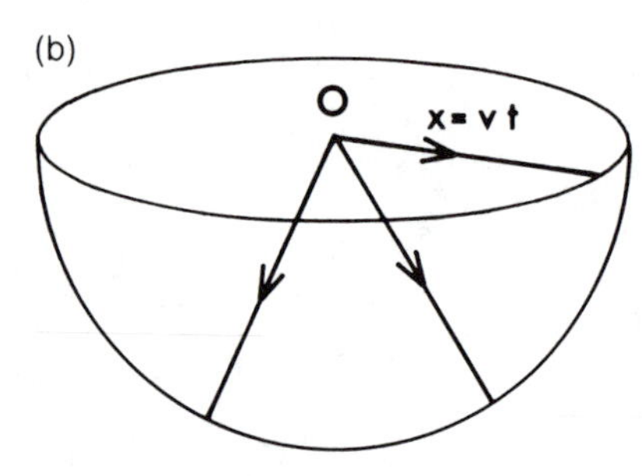

Figure 4.3. Amplitudes of surface waves and body waves. (a) A surface wave is travelling from a source O at speed v. After a time t the area of the cylindrical wavefront is $2\pi xz = 2\pi vtz$. Energy is proportional to the square of amplitude (Eq. 4.26). By conservation of energy, the amplitude at distance x is proportional to $x^{-1/2}$. (b) A body wave is travelling from a source O at speed v. After a time t the area of the spherical wavefront is $4\pi x^2 = 4\pi v^2 t^2$. By conservation of energy, the amplitude at distance x is proportional to x^{-1}.

4.1.3 Surface Waves

Surface waves are seismic waves which are guided along the surface of the earth and the layers near the surface. They do not penetrate into the deep interior. Surface waves are generated best by shallow earthquakes. Nuclear explosions do not generate comparable surface waves, and this fortunate fact is the basis for one criterion of discrimination between earthquakes and nuclear explosions (see Sect. 4.2.3). Surface waves are larger in amplitude and longer in duration than body waves, and they arrive at a seismograph after the main P- and S-waves because their velocities are lower than those of body waves. The reason for the larger amplitude of surface waves is easy to understand (Fig. 4.3). The area of the cylindrical wavefront is proportional to x, the distance from its source, which means that the amplitude of the surface wave is inversely proportional to $\sqrt{x}$. In contrast, the wavefront of P- and S-waves at any time is spherical; therefore, the area of the wavefront is proportional to x^2, the square of the distance from the source. The amplitude of the body wave is therefore inversely proportional to x.

There are two types of surface waves,* both named after famous physicists: *Rayleigh waves*, sometimes descriptively called 'ground roll' in exploration seismology, are named after Lord Rayleigh, who predicted their existence in 1887; *Love waves* are named after A. E. H. Love, who predicted their existence in 1911. Rayleigh waves are denoted by LR or R, and Love waves are denoted by LQ or Q (L for long; R for Rayleigh; Q for *Querwellen*, German, 'transverse waves'). Rayleigh waves occur close to the surface of a semi-infinite medium. The particle motion for these waves is confined to a vertical plane containing the direction of propagation (Fig. 4.4a). Near the surface of a uniform half-space it is a retrograde (anticlockwise for a wave travelling to the right) vertical ellipse. Rayleigh waves can therefore be recorded by both the vertical and horizontal components of a seismometer. In contrast, Love waves occur when there is a general increase of S-wave velocity with depth. They propagate by multiple internal reflections of horizontally polarized S-waves (SH-waves) in this near-surface medium and thus propagate in a waveguide. The particle motion of these waves is transverse and

* A third type of surface wave, the Stonely wave, propagates along an interface between two media and is more correctly an interface wave. Stonely waves are not dispersive; thus they decrease in amplitude with distance from the interface and have a velocity between the lesser S-wave velocity and the greater Rayleigh-wave velocity.

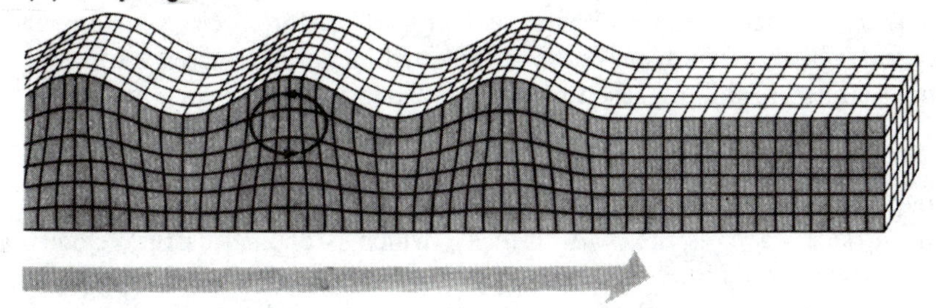

Figure 4.4. The particle motion for surface waves. (a) Rayleigh waves and (b) Love waves. (After Bolt 1976.)

horizontal, so they can only be recorded by horizontal seismometers (Fig. 4.4b).

The amplitude of Rayleigh waves decays exponentially with depth beneath the surface. The amplitude of any particular frequency component is dependent upon the ratio:

$$\frac{\text{depth beneath the surface}}{\text{wavelength}}$$

Thus, for each frequency, the amplitude decreases by the same factor when the depth increases by a wavelength. This means that the longer-wavelength (longer-period, lower-frequency) surface waves contain more information about the deep velocity structure, and the shorter-wavelength (shorter-period, higher-frequency) surface waves yield information about the shallow structure. A good rule of thumb is that surface waves sample to a depth of their wavelength divided by three. Surface waves with periods less than about 50 s are used to determine gross crustal shear-wave velocity structures. Because their wavelength (velocity/frequency) is large, even the highest-frequency surface waves cannot resolve the fine structure of the crust. Surface waves with longer periods are used to determine the mantle shear-wave velocity structure.

Both Rayleigh waves and Love waves are, in practice, *dispersive*, which means that their velocities depend on frequency (different frequencies travel at different velocities). Dispersion means that a wave train changes shape as it travels. For details of the mathematics of the equations of motion for surface waves the reader is referred to Officer 1974 (chapter 7). The first surface-wave energy to arrive at any seismometer is of those frequencies that have the greatest velocities. The other frequencies will arrive later according to their velocities. Therefore, seismometers at increasingly greater distances from an earthquake (measured along the great circle arc linking seismometer and earthquake) record wave trains which are increasingly spread out.

The velocity with which surface wave energy associated with a particular frequency travels is called the *group velocity*. The other velocity used in the surface waves is the *phase velocity*, which is defined as the velocity with which any particular phase (i.e., peak or trough) travels. Both the group velocity and the phase velocity are a function of frequency. Consider Figure 4.5, which illustrates a record section of surface waves. Notice that on each record the lower-frequency waves arrive before the higher-frequency waves. The first phase to arrive on each record is peak A. Notice also that the frequency of the A peak is not a constant from one record to the next: The frequency decreases (period increases) as the distance increases. The dashed curve linking peak A on the records defines the phase velocity for peak A. The phase velocity for the frequency of peak A at any particular distance is the slope of this dashed curve at that distance. The dashed lines linking the subsequent peaks B, C and D also determine the phase velocity as a function of frequency. The slopes of all these dashed lines indicate that, in this example, the phase velocity decreases as the frequency of these surface waves increases (the phase velocity increases with the period of the surface waves). The group velocity of surface waves is a constant for a given frequency. Therefore, the group velocity for a particular frequency is a straight line passing through the origin and through the signal of that particular frequency on each successive record. Such lines for the three frequencies f_1, f_2 and f_3, where $f_1 < f_2 < f_3$, are shown on the record section. The corresponding group velocities are U_1, U_2 and U_3, where $U_1 > U_2 > U_3$. In this example, the phase velocity also decreases as the frequency of these surface waves

Figure 4.5. (a) Seismic records of dispersive surface waves obtained at increasing distances from the source (at origin). The phases A, B, C and D are each associated with a different frequency from one record to the next – frequency decreases with increasing distance in this example. The inverse slope of the dashed line linking the phases is the phase velocity, which is a function of frequency. The straight solid lines have inverse slope U_1, U_2 and U_3, the group velocities for frequency f_1, f_2 and f_3. The group velocity is a constant for each frequency and, in this example, decreases with increasing frequency. (From Officer 1974.) (b) Group and phase velocity dispersion curves for the example shown in (a).

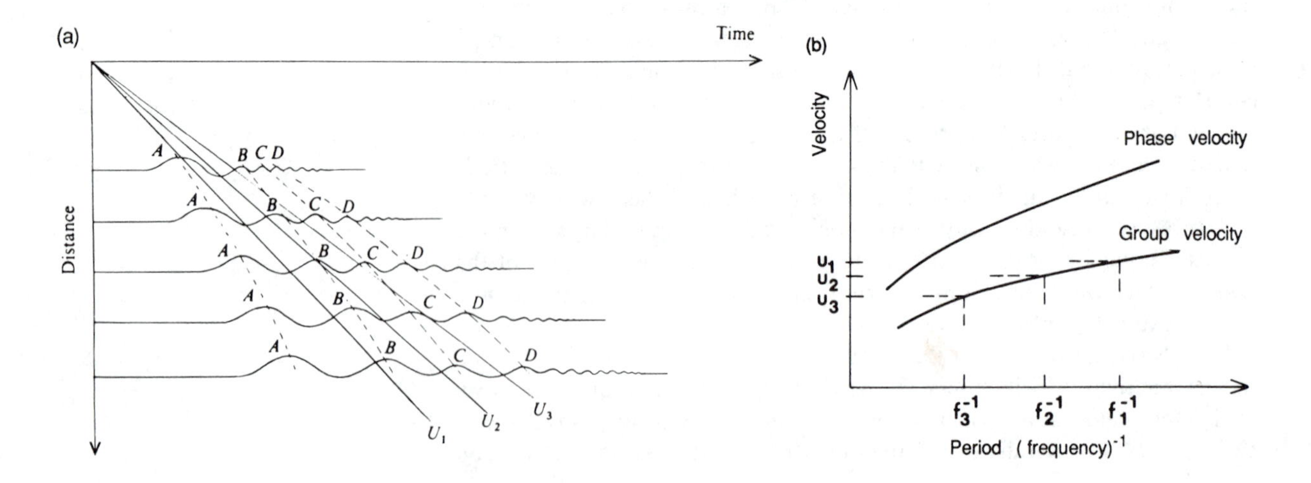

increases (and increases with their period). A plot of velocity against period, called a *dispersion curve*, is the usual way of presenting this velocity–frequency information. Figure 4.5b illustrates dispersion curves appropriate for the records of Figure 4.5a. Notice that in this example the group velocity is less than the phase velocity. To repeat, energy travels with the group velocity not the phase velocity.

Theoretically, group velocity $U(f)$ and phase velocity $V(f)$ are linked by

$$U = V + f\frac{dV}{df} \tag{4.7}$$

where f is frequency. In our example, dV/df is negative (phase velocity decreases as frequency increases), so $U < V$.

First, consider Rayleigh waves. In the ideal theoretical situation, in which the elastic properties of the earth are constant with depth, Rayleigh waves are not dispersive and travel at a velocity of approximately 0.92β (i.e., slower than S-waves). However, the real earth is layered, and when the equations of motion for Rayleigh waves in a layered earth are solved (far beyond the scope of this text), then Rayleigh waves are indeed found to be dispersive.

Next, consider Love waves. Love waves can exist only when the shear-wave velocity increases with depth or, in the layered case, when the shear-wave velocity of the overlying layer β_1 is less than that of the substratum β_2 (i.e., $\beta_1 < \beta_2$). Love waves are always dispersive. The phase velocity is always between β_1 and β_2 in magnitude. The low-frequency (long-period) limit of the phase velocity is β_2, and its high-frequency (short-period) limit is β_1. Generally, Love wave group velocities are greater than Rayleigh wave group velocities, which means that on seismograms Love waves usually arrive before Rayleigh waves.

Dispersion curves contain much information about the velocity structure of the crust and upper mantle, but it is no simple matter to extract it. A *linearized inversion technique* (an iterative scheme which obtains a best fit between actual dispersion curves and model curves which would be generated by particular velocity–depth structures), calculated by computer, is often used to obtain a velocity–depth structure appropriate for particular dispersion curves. Figure 4.6a shows standard continental and oceanic dispersion curves. Notice that over much of the frequency range, oceanic paths are faster than continental paths. In Figure 4.6b notice the low S-wave velocities starting at approximately 50–100 km depth. This low-velocity region is generally identified with the asthenosphere, the zone beneath the lithospheric plates, where the temperatures approach the solidus of mantle material and thus where the material ceases to behave rigidly, becoming more ductile and able to creep.

4.1.4 *Free Oscillations of the Earth*

Any mechanical system has a natural oscillation which can be excited, and the earth is no exception. The earth can oscillate in an indefinite number of *normal modes* of oscillation, rather like a giant bell. Although such oscillations had been predicted theoretically at the beginning of this century, the first definite measurement did not take place until May 22,

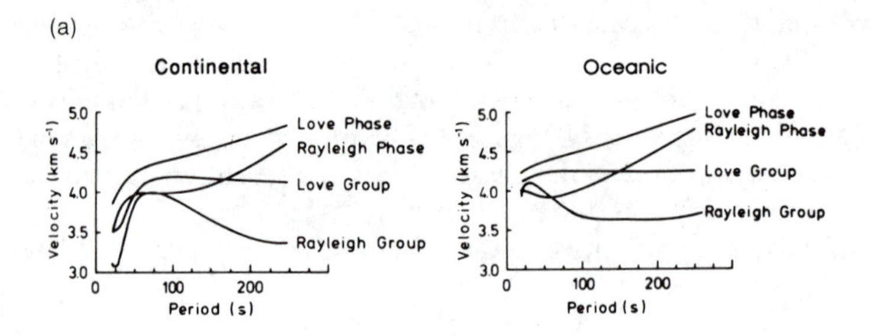

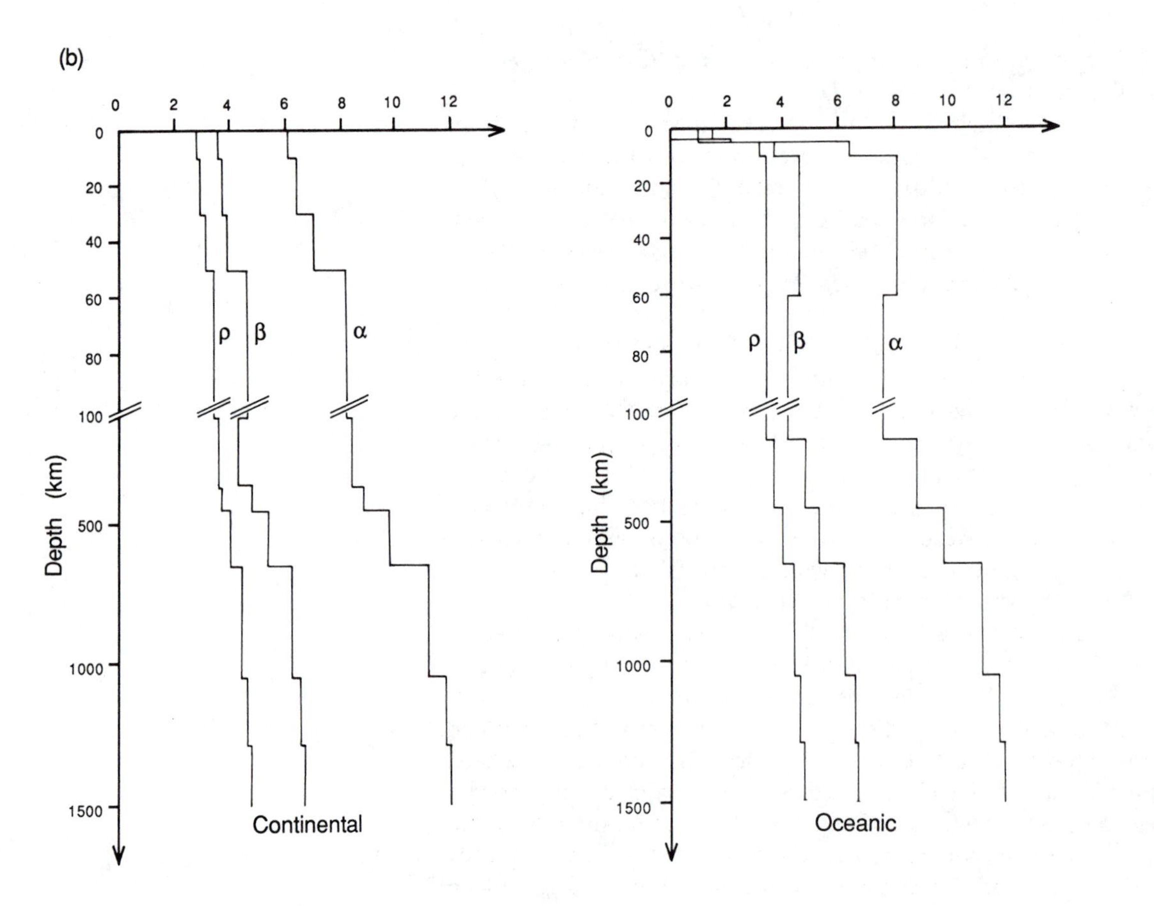

Figure 4.6. (a) Standard surface wave dispersion curves for continental and oceanic paths. (b) Seismic velocity and density models for the dispersion curves in (a). Note change in depth scale at 100 km. (After Knopoff and Chang 1977.)

1960, when a large earthquake in Chile excited the oscillations sufficiently for them to be detected.

There are two independent types of free oscillations: *toroidal*, or *torsional*, oscillations (T) and *spheroidal* oscillations (S). The displacement for toroidal oscillations is always perpendicular to the radius vector and so is confined to the surfaces of concentric spheres within the earth. Such oscillations involve only the crust and mantle. The general displacement for spheroidal oscillations has both radial and tangential components. The simplest spheroidal oscillation is a purely radial motion. Both types of oscillation have an infinite number of *modes* (or, as in music, overtones). Figure 4.7 shows the modes ${}_0T_2$, ${}_1T_2$, ${}_0S_2$ and ${}_0S_3$. Free oscillations can be detected by strain meters and gravimeters used for measuring earth tides. Seismometers are not suitable for such very long-period signals.

The toroidal oscillations are equivalent to all the Love waves, whereas the spheroidal oscillations are equivalent to all the Rayleigh waves. This means that measurements of the periods of free oscillations can be used to extend the dispersion curves for surface waves out to the maximum periods of ~2500 s (T mode) and ~3200 s (S mode).

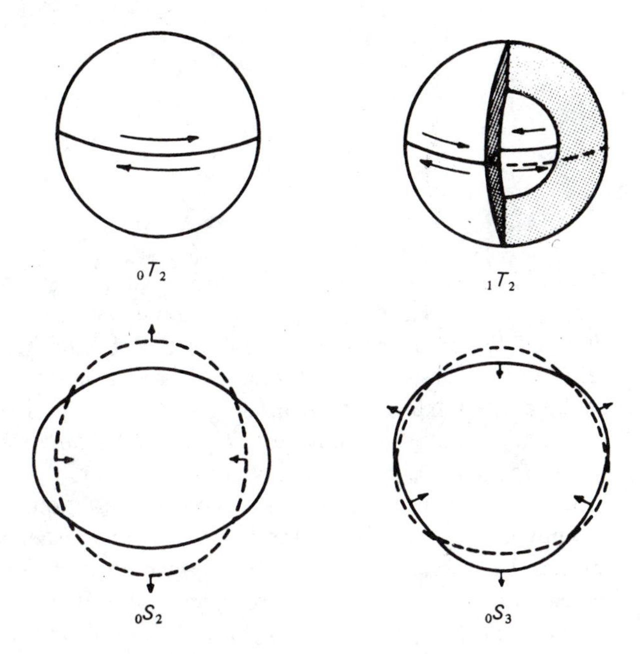

Figure 4.7. Normal modes $_0T_2$, $_1T_2$, $_0S_2$ and $_0S_3$. In mode $_0T_2$, the opposite hemispheres twist in opposite directions. In mode $_1T_2$ the outside part of the sphere twists as in mode $_0T_2$, but the inner sphere twists in the opposite direction. (From Bullen and Bolt 1985.)

4.2 Earthquake Seismology

4.2.1 Location of Earthquakes

The earthquake focus or hypocentre is the point in the earth where the earthquake nucleated. It is specified by latitude, longitude and depth beneath the surface. The *earthquake epicentre* is the point on the earth's surface vertically above the focus. For an earthquake beneath California, the focus might be at 37°N, 122°W and 10 km depth. The epicentre for this earthquake would be 37°N, 122°W. In fact, earthquakes do not occur exactly at points; rather the stress releases occur within small volumes or along fault planes.

Let us say that an earthquake occurs at the surface at time t_0. There are three unknowns: time of origin, latitude and longitude. To determine these three unknowns, we need to know the arrival times of seismic waves at three nearby seismometers. If the shallow P-wave velocity is α and the shallow S-wave velocity is β, then the time for P-waves to travel from the focus to seismometer 1 at distance r_1 is r_1/α. Similarly, the time for the S-waves is r_1/β. The arrival time of P-waves at seismometer 1 is then $t_0 + r_1/\alpha$, and the arrival time of S-waves at seismometer 1 is $t_0 + r_1/\beta$. The difference between the P- and S-wave arrival times at seismometer 1,

$t_{1,S-P}$, is given by

$$t_{1,S-P} = \frac{r_1}{\beta} - \frac{r_1}{\alpha} \tag{4.8}$$

and similarly for seismometers 2 and 3. If we assume values for α and β, we now have three linear equations with three unknowns, r_1, r_2 and r_3, which can easily be solved:

$$\begin{aligned} t_{1,S-P} &= \frac{r_1}{\beta} - \frac{r_1}{\alpha} \\ t_{2,S-P} &= \frac{r_2}{\beta} - \frac{r_2}{\alpha} \\ t_{3,S-P} &= \frac{r_3}{\beta} - \frac{r_3}{\alpha} \end{aligned} \tag{4.9}$$

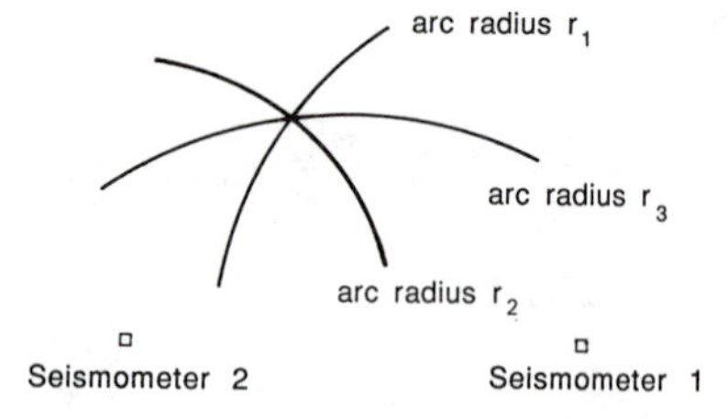

Seismometer 3

Figure 4.8. Map showing the location of seismometers 1, 2 and 3. The earthquake focus is distance r_1 from seismometer 1, r_2 from seismometer 2 and r_3 from seismometer 3 (r_1, r_2 and r_3 are given by Eqs. 4.9). The focus can be located by drawing an arc of a circle of radius r_1 about seismometer 1 and then repeating this for seismometers 2 and 3 (radius r_2 and r_3, respectively). The point at which the three arcs intersect is the focus.

It would be easiest to solve this simple example graphically by drawing a map of the area, marking the seismometers and then drawing an arc of a circle radius r_1 about seismometer 1, another radius r_2 about seismometer 2 and another radius r_3 about seismometer 3. The focus of the earthquake is then at the intersection of the three arcs (Fig. 4.8).

In reality, things are not quite so simple because the earth is not flat, nor homogeneous with globally constant P- and S-wave velocities, and errors in arrival times easily occur. In addition, there is a fourth unknown, focal depth. Earthquakes do not conveniently occur at the surface. In practice, therefore, several seismometers are required to locate an earthquake. Locations are routinely calculated using as many travel times from as many stations as possible. The distance between the epicentre and the seismometer is called the *epicentral distance*. The shortest distance on the surface of a sphere between any two points on the sphere is along the great circle which intersects the two points. The epicentral distance is therefore the length of this great circle arc and is usually measured in degrees (except for local earthquake studies, in which case it is measured in kilometers).

Focal depths can also be determined from measurement of the difference in travel time between the P phase and the pP phase (pP is a P-wave reflected at the earth's surface in the vicinity of the earthquake, as in Figure 4.9). The P phase travels along path FS and has arrival time t_{pP}, whereas the pP phase travels along path FRS and arrives at time t_{pP}. At teleseismic distances FR is small compared with FS; and

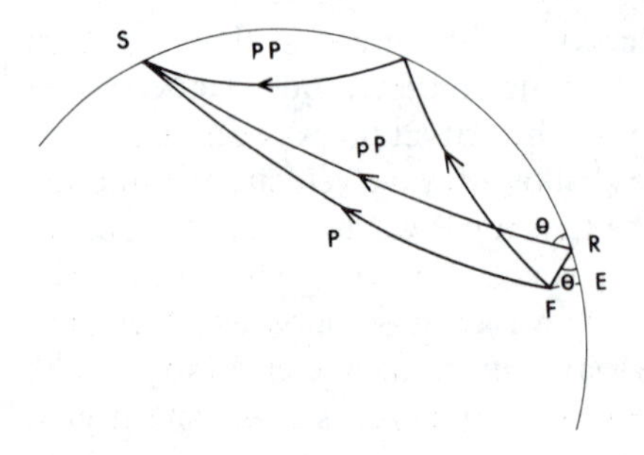

Figure 4.9. Two points on the earth's surface satisfy the condition for a reflection of the P-wave (or S-wave) travelling through the mantle. (At a reflection point the angle of incidence equals the angle of reflection.) The first reflection point is close to the focus, and the second is midway between the focus and the seismometer. The first reflected ray is coded pP (or sS), and the second is coded PP (or SS). The difference in time between the pP arrival and P (the direct P-wave without reflection) or between the sS arrival and S can be used to determine the focal depth (EF) of the earthquake. E is the epicentre, F the focus, S the seismometer and R the reflection point close to the focus. The diagram is purely schematic and not drawn to scale.

Table 4.1. *Variation of earthquake frequency with focal depth for earthquakes recorded during the period 1918–1946*

	Depth (km)	Number of earthquakes
Shallow	< 100	800
Intermediate	75–125	139
	125–175	56
	175–225	38
	225–275	15
	275–325	8
Deep	325–375	11
	375–425	12
	425–475	4
	475–525	7
	525–575	8
	575–625	12
	625–675	7
	675–725	3

Source: After Gutenberg and Richter (1954).

so, to a first approximation, the length FR is given by

$$FR = \frac{\alpha}{2}(t_{\mathrm{pP}} - t_{\mathrm{P}}) \tag{4.10}$$

For shear waves (using phases sS and S), the corresponding distance would be

$$FR = \frac{\beta}{2}(t_{\mathrm{sS}} - t_{\mathrm{S}}) \tag{4.11}$$

The focal depth h is then given by

$$h = EF = FR \sin\theta \tag{4.12}$$

In practice, travel times of all the earthquake wave phases are tabulated: The Jeffreys–Bullen (1988) tables (often called J–B tables) are probably the most complete (they are a revision of the first tabulation of travel times to be published). Even with a record from only one station, if the phases P, S and pP have been identified and the S – P time (Eq. 4.9) and the pP – P time have been calculated, it is possible to use these tables to estimate both the epicentral distance and the focal depth.

Table 4.1 shows how earthquake occurrence vary with focal depth. Most earthquakes are shallow, and none have been detected from depths greater than about 700 km. This fact is used in discussing the assimilation of subducted plates into the surrounding mantle at subduction zones.

4.2.2 *Aftershocks, Foreshocks and Swarms*

An earthquake occurs as the result of a slow buildup c" strain (deformation) in rock, usually caused by the relative motion of adjacent plates. When a fault or volume of rock can no longer resist movement, the stored strain energy is suddenly released, causing an earthquake.

Frequently, a strong earthquake is followed by a sequence of *aftershocks*, which can continue for months. The aftershocks occur during a period of readjustment following the main shock, in which small localized strains on the fault are released. Deep focus earthquakes usually do not have aftershocks. The aftershocks are usually much smaller than the main earthquake and gradually decrease in magnitude. The largest aftershock, which often occurs within hours of the main shock, usually reaches a magnitude about 1.2 units lower than the main shock (Båth 1979). (Earthquake magnitude is discussed in Sect. 4.2.3.)

Sometimes the main earthquake is preceded by a small *foreshock*. Unfortunately, it is not usually possible to identify it as a foreshock until after the main event has occurred.

Earthquake *swarms* are yet another sequence of activity. In a swarm there is no main earthquake but instead a large number of small shocks, often with many occurring every day. The numbers of shocks build up slowly to a maximum and then die down again. Most earthquake swarms occur in volcanic areas, especially along the midocean ridges, and these events usually have very shallow foci.

4.2.3 Earthquake Magnitude

The concept of earthquake magnitude was introduced by C. F. Richter in 1935. He was studying local earthquakes in southern California and proposed a particular logarithmic magnitude scale for them. Since then, measurements of magnitude have been extended to earthquakes at all depths and distances and to both body and surface waves. Richter's name is known around the world because press releases on earthquakes invariably quote magnitude as being 'on the Richter scale'.

A number of other logarithmic *magnitude scales* for earthquakes exist, all of which are based on measurements of the amplitude of the seismic waves. In addition there are *intensity scales* such as the Mercalli scale, which are subjective and are based on the shaking of buildings, breaking of glass, ground cracking, people running outside, and so on.

All the magnitude scales are of the form

$$M = \log_{10}\left(\frac{A}{T}\right) + q(\Delta, h) + a \tag{4.13}$$

where M is the magnitude, A the maximum amplitude of the wave (in 10^{-6} metres), T the period of the wave (in seconds), q a function correcting for the decrease of amplitude of the wave with distance from the epicentre and focal depth, Δ the angular distance from seismometer to epicentre, h the focal depth of the earthquake, and a an empirical constant.

It is important in all magnitude work to use the period ranges for which the particular equation has been determined; waves of different frequencies have different amplitude behavior. Many determinations of the constant a and function q of Eq. 4.13 have been made, for various frequency bands, focal depths, geographic locations and so on. Those that are given and discussed here are the equations generally accepted and used today. Båth

(1981) provides a detailed review of the present state of research on earthquake magnitudes.

For shallow focus (<50 km) teleseismic earthquakes ($20° < \Delta < 160°$) a *surface-wave magnitude* M_S can be defined as

$$M_S = \log_{10}\left(\frac{A}{T}\right)_{\max} + 1.66\log_{10}\Delta + 3.3 \quad (4.14)$$

The amplitude A used is the maximum of the horizontal component of the Rayleigh wave in the period range 17–23 s. (If the vertical component is used instead, the empirical constant in the equation depends on the seismic layering beneath the recording station; and if there is no layering, then a constant of 3.1 is appropriate.) Because earthquakes occurring at greater depths are not so efficient at generating surface waves, Eq. 4.14 gives too small a value for their magnitude. To allow for this, a correction must be applied to Eq. 4.14 for earthquakes with a focal depth greater than 50 km:

$$(M_S)_{\text{corrected}} = M_S + \delta M_S(h) \quad (4.15)$$

where M_S is calculated from Eq. 4.14 and $\delta M_S(h)$ is the correction for focal depth h. The maximum value of $\delta M_S(h)$, 0.4, is used for any earthquake with a focal depth greater than 90 km. At ranges shorter than 20°, a correction, $\delta M_S(\Delta)$, must also be made to Eq. 4.14 to account for differences in absorption, scattering, geometrical spreading and dispersion:

$$(M_S)_{\text{corrected}} = M_S + \delta M_S(\Delta) \quad (4.16)$$

Estimates of this correction vary between 0.6 and 0.1. These corrections illustrate the fact that magnitude, although easy to measure, is not an exact description of an earthquake. Many variables are associated with the recording site and passage of the seismic signal through the earth, as well as with the earthquake itself, all of which can have a large effect on magnitude. Differences of 0.2–0.3 units in calculated magnitudes are not uncommon even under the most favourable conditions.

Because deep-focus earthquakes are not effective at generating surface waves, a better magnitude scale for them is based on body waves (P- and S-waves). A first suggestion by Gutenberg (1945) for *body-wave magnitude* m_b of shallow-focus earthquakes for P, PP and S waves of 12-s period was

$$m_b = \log_{10}\left(\frac{A}{12}\right) + 0.01\Delta + 5.9 \quad (4.17)$$

Many more recent determinations of the calibration function $q(\Delta, h)$ in Eq. 4.13 have been made since then, but the one generally used today is still the Gutenberg–Richter (1956) calibration function:

$$m_b = \log_{10}\left(\frac{A}{T}\right)_{\max} + q(\Delta, h) \quad (4.18)$$

In this tabulation, $q(\Delta, h)$ for P-waves is not strongly dependent on focal depth and increases from ~6.0 at $\Delta = 10°$ to ~6.5 at $\Delta = 80°$ to ~8.0 at $\Delta = 110°$.

When m_b and M_s are calculated for an earthquake, they do not usually

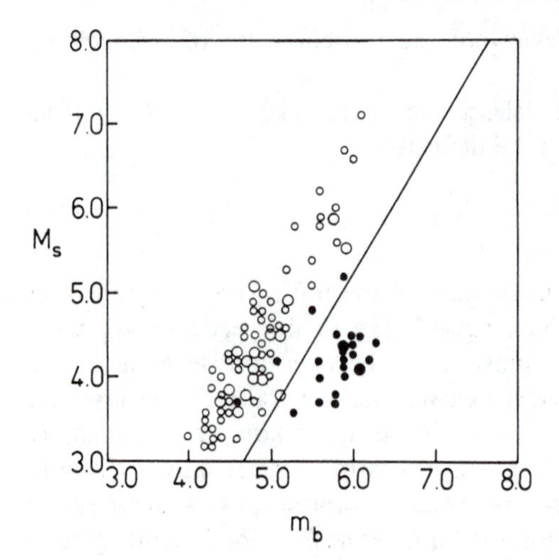

Figure 4.10. A plot of surface wave magnitude M_s against body wave magnitude m_b for 26 nuclear explosions (solid circles) and 99 earthquakes (open circles), all in Eurasia. Solid line is the discriminant line $m_b = 2.87 + 0.60\, M_s$. Nuclear explosions are less efficient at generating surface waves than earthquakes with the same body-wave magnitude. (After Nowroozi 1986.)

have the same value. In other words, the two magnitude scales do not give the same value for the magnitude of an earthquake. However, the two scales are related, as can be seen from Figure 4.10. The data are scattered, but m_b and M_s appear to be linearly related. The m_b:M_s data plots vary appreciably from one earthquake region to the next, but a worldwide average of m_b:M_s relations is

$$m_b = 2.94 + 0.55 M_s \tag{4.19}$$

The two magnitude scales coincide at magnitude 6.5. For smaller magnitudes, m_b is greater than M_s, and for large magnitudes, m_b is less than M_s.

As mentioned earlier, nuclear explosions are not such good generators of surface waves as earthquakes are. In general, nuclear explosions have m_b values 1.0 to 1.5 magnitude units greater than earthquakes with the same M_s values. The straight line in Figure 4.10,

$$M_s = 1.68 m_b - 4.82 \tag{4.20}$$

or

$$m_b = 2.87 + 0.60 M_s \tag{4.21}$$

separates most of the explosions and earthquakes. To be certain that an event is classified correctly, it is necessary to include other parameters such as location, depth and shape of the waveforms. For example, an event with a focal depth of 200 km is exceedingly unlikely to be a nuclear explosion, even assuming the maximum errors in depth calculation, whereas an event with a very shallow focal depth of 5 km would be a candidate for a nuclear explosion.

4.2.4 Seismic Moment

The *seismic moment* M_0 of an earthquake is defined by

$$M_0 = \mu A u \tag{4.22}$$

where μ is the shear modulus, A the area of the fault and u the average displacement on the fault. The seismic moment can be determined by observations and estimates of the fault-plane area and displacement. The seismic moment can also be expressed in terms of the low-frequency amplitude spectra of surface waves. It does not yield an accurate estimate of the drop in stress, $\Delta\sigma$, for the earthquake. Plots of $\log M_0$ against M_s and of M_0 against $\log A$ are shown in Figure 4.11.

Both the body wave and the surface wave magnitude scales saturate (do not give large enough values) for very large earthquakes and are strongly dependent on the frequency of the seismic wave. Therefore, to estimate the size of big earthquakes, long-period (low-frequency) waves are used. A more suitable magnitude scale called the *moment magnitude* M_W is obtained from the seismic moment. The moment magnitude has been defined as

$$M_W = \tfrac{2}{3}\log_{10} M_0 - 6.0 \tag{4.23}$$

and has the important advantage over the other magnitude scales that it does not saturate towards the top end of the scale.

A few examples of the magnitude of some famous earthquakes are the 1906 San Francisco, California, earthquake, $M_s = 8.3$ and $M_W = 7.9$; the 22 May 1960 Chile earthquake, $M_s = 8.3$ and $M_W = 9.5$; the 1963 earthquake in Skopje, Yugoslavia, $M_s = 6.0$; the 1964 Alaska earthquake, $M_s = 8.4$ and $M_W = 9.2$; the 1971 San Fernando, California, earthquake, $M_s = 6.6$; the 1976 Tangshan, China, earthquake, which killed 650,000

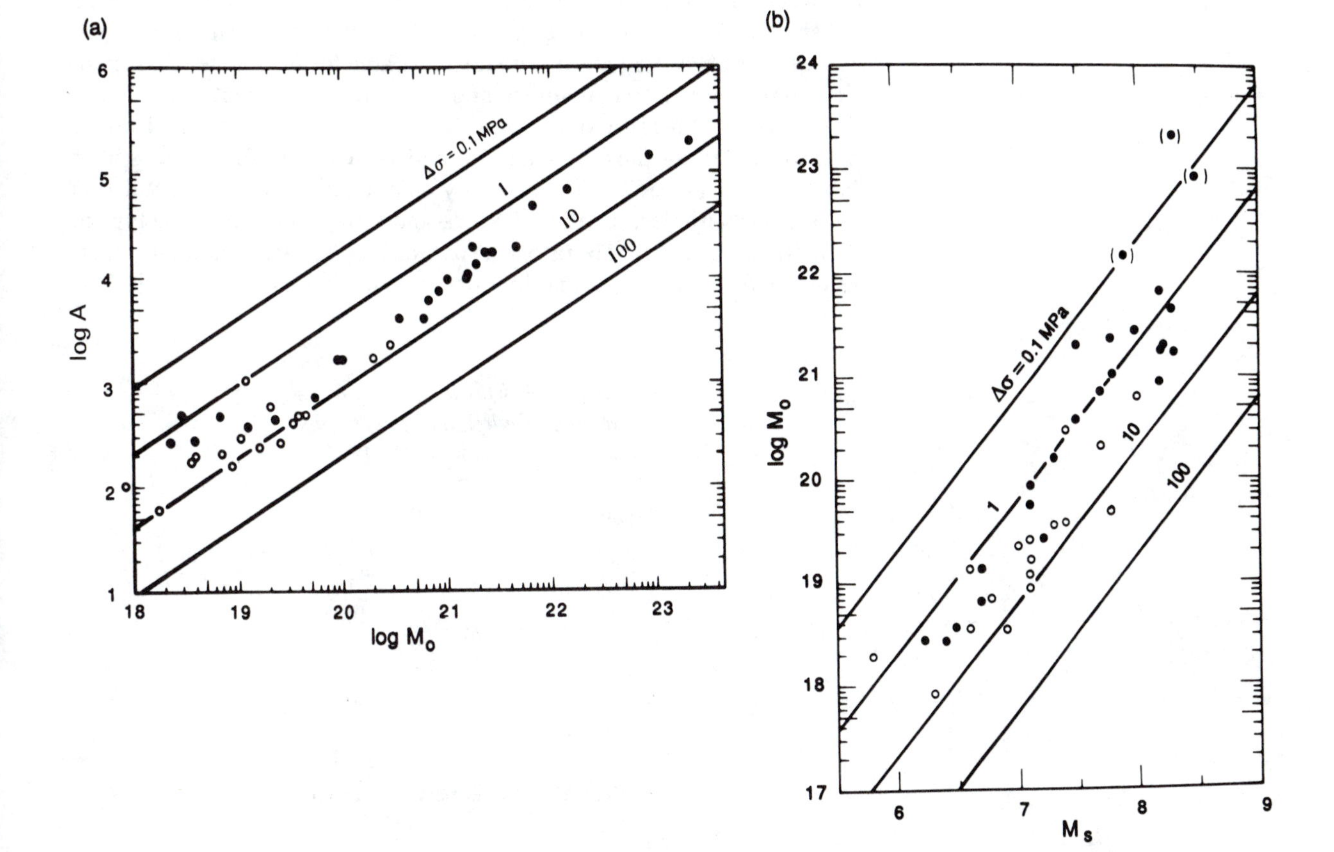

Figure 4.11. (a) Relationship between A (surface area of the fault in km^2) and M_0 (seismic moment in N m). (b) Relationship between M_0 and M_s (surface wave magnitude). Straight lines are for circular cracks with a constant stress drop $\Delta\sigma$ (in MPa). Both (a) and (b) show some difference in stress drop for interplate (●) and intraplate (○) earthquakes. (After Kanamori and Anderson 1975.)

people, $M_s = 7.6$; the 1985 Mexico earthquake had a magnitude of 7.8; and the estimates of the magnitude of the 1988 Armenian earthquake range between 6.7 and 7.0. The 1989 Loma Prieta, California, earthquake had a surface wave magnitude of 7.1. The largest earthquake ever recorded in Japan occurred in 1933 and had $M_s = 8.9$. The amount of damage done by an earthquake depends on factors such as population density, soil conditions and local building standards in the epicentral region, in addition to the focal depth and magnitude of the earthquake. In general, a shallow earthquake with an M_s of 6 is very destructive, one with an M_s of 5 produces moderate damage, and earthquakes with magnitudes less than about 3 are usually not be felt by many people, though they will, of course, be detected by local seismometers.

4.2.5 Magnitude–Frequency Relation

The mean annual frequency of earthquakes is tabulated in Table 4.2. As might be expected, small earthquakes are much more abundant than large ones. It is typically found that there is a straight-line relationship between the surface wave magnitude M_s and the logarithm of N, where N is the number of earthquakes with magnitude $M_s \pm \delta M_s/2$ occurring in an area per unit of time:

$$\log N = \text{constant} - bM_s \tag{4.24}$$

and where the value of b is a measure of the relative abundance of large and small earthquakes. A large value for b in an area indicates that small earthquakes occur frequently, and a small value for b indicates that small earthquakes are not so frequent and that large earthquakes are more likely to occur. An approximate worldwide average for b is 1; small values for b are generally those less than 1, and large values for b are greater than 1. Small values for b are usually characteristic of continental rifts and regions with deep earthquake foci, and large values for b are typical of midocean ridges. Laboratory experiments indicate that high stress results in low values for b, and low stress results in high values for b. This

Table 4.2. *Worldwide mean annual frequency of earthquakes recorded during the period 1918–1945*

Magnitude	Number per year
>8.0	1
7–7.9	18
6–6.9	108
5–5.9	800
4–4.9	6,200
3–3.9	49,000
2–2.9	300,000

Source: After Gutenberg and Richter (1954).

observation is consistent with most of the regional determinations that have been made, but it would be incorrect and misleading to say that b is understood. Much more research is needed.

4.2.6 Energy Released by Earthquakes

Few determinations of the total energy released as seismic waves by earthquakes have been made, probably because the energy must be computed by an integration over both time and space and must include all frequencies. To do this, the signal must be recorded with broad-band instruments* (instruments which record both the high and low frequencies).

Consider a simple plane wave, a simple harmonic with displacement u:

$$u = A\cos(\kappa x - \omega t) \tag{4.25}$$

where ω is the angular frequency, κ the wave number (ω/wave-velocity), A the maximum amplitude and t time. The kinetic energy per unit volume associated with this wave is

$$\begin{aligned} \text{K.E.} &= \tfrac{1}{2}\rho\left(\frac{\partial u}{\partial t}\right)^2 \\ &= \tfrac{1}{2}\rho A^2\omega^2 \sin^2(\kappa x - \omega t) \end{aligned} \tag{4.26}$$

where ρ is the density of the material. The kinetic energy varies from a maximum value of $\rho A^2\omega^2/2$ to a minimum value of zero during each cycle. For a perfectly elastic medium, energy is conserved (the sum of potential energy and kinetic energy is constant), so the total energy per unit volume is equal to the maximum of the kinetic energy per unit volume $\rho A^2\omega^2/2$. To calculate the total seismic energy released by an earthquake it is necessary to integrate or sum an expression such as this over time, frequency and position around the epicentre. Amplitude and magnitude are related logarithmically in Eq. 4.13, so the expression of energy in terms of magnitude is also logarithmic (constants are determined from integrations as well as assumptions about density, etc.). The relation between energy and magnitude generally used by seismologists (Båth 1966) is

$$\log_{10} E = 5.24 + 1.44 M_s \tag{4.27}$$

where E (measured in joules J) is the total energy of the seismic waves. This relationship was determined for, and is therefore reliable for, earthquakes with $M_s > 5$ only. An earthquake with $M_s = 7.3$ would release 5.6×10^{15} J as seismic energy.

In comparison, the Hiroshima atomic bomb was approximately equivalent in terms of energy to an earthquake of magnitude 5.3. A one

* Ideally, for local earthquakes frequencies 0.1–500 Hz should be recorded, whereas for teleseismic recordings of surface waves 0.01–20 Hz would be appropriate. In contrast, for vibroseis (Sect. 4.5.1) deep reflection data 5–200 Hz should be recorded and for explosion seismology (refraction lines, *not* nuclear explosion detection) 2–200 Hz. These frequency ranges are those suggested by the National Research Council (U.S.) 1983.

Table 4.3 *Number of large earthquakes occurring in various magnitude and depth ranges, 1918–1955*

Magnitude range	Shallow focus (0–70 km)	Intermediate focus (70–300 km)	Deep focus (> 300 km)
⩾8.6	9	1	0
7.9–8.5	66	8	4
7.0–7.8	570	214	66

Source: Richter (1958).

megaton nuclear explosion would release about the same amount of energy as an earthquake of magnitude 6.5.

By applying Eq. 4.27 to Table 4.2, we can calculate the total energy released each year by earthquakes of various magnitudes, and by applying it to Table 4.3, we can calculate the energy released in the various depth ranges. About three-quarters of the energy is released by shallow focus events.

4.2.7 The Observation of Earthquakes: Seismic Phases

The various seismic ray paths within the earth are coded as follows:*

P	a P-wave in the mantle
S	an S-wave in the mantle
K	a P-wave through the outer core
I	a P-wave through the inner core
J	an S-wave through the inner core
c	a reflection from the mantle–outer core boundary
i	a reflection from the outer core–inner core boundary
p	a P-wave reflected from the surface of the earth close to the earthquake focus
s	an S-wave reflected from the surface of the earth close to the earthquake focus
LR	a Rayleigh wave
LQ	a Love wave

The first arrival at a seismometer is always P, closely followed by pP, the P-wave reflected once near the focus (Fig. 4.9). The S-waves arrive next and finally the surface waves.

When a body wave is incident on a discontinuity in the Earth, some of the energy is reflected as a P-wave, some is reflected as an S-wave, some is transmitted as a P-wave and some is transmitted as an S-wave. This is termed *mode conversion.* The amplitudes of the various waves are discussed in Section 4.4.5. Some examples of the coding of seismic

* The terminology P (primary), S (secondary) and L was introduced by van der Borne (1904).

waves follow:

PKP	a P-wave which passed down through the mantle and the outer core and then up through the mantle
PKIKP	a P-wave which passed down through the mantle and outer core and then through the inner core and up through the outer core and mantle
PKJKP	a wave that may exist, but which has not been unequivocably observed
PKiKP	a P-wave which travelled down through the mantle and outer core, was reflected at the outer core–inner core boundary and travelled back up through the outer core and mantle
sSP	a wave which travelled from the focus as an S-wave, was reflected at the earth's surface close to the focus, then travelled through the mantle as an S-wave, was reflected for a second time at the earth's surface, converted to a P-wave and travelled as a P-wave through the mantle

Figure 4.12 shows some ray paths. Figure 4.13 is a travel-time plot for all the main earthquake phases. Identifying and recognizing the phases on an earthquake record is a skilled task (Fig. 4.14). Plotting earthquake records side by side, as a record section (Fig. 4.15) greatly facilitates identification of phases. This record section should be compared with the travel-time plot of Figure 4.13, the main mantle phases and also some of the core phases are immediately apparent even to the untrained eye. Few surface waves, which have greater amplitude and arrive later than the main body wave phases, are plotted (see top left-hand corner of the section for the start of surface waves out to 80°).

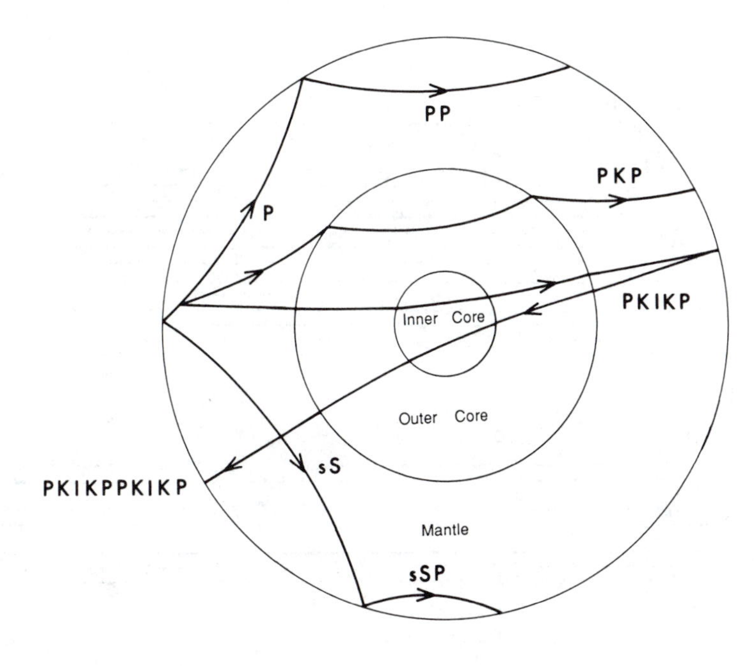

Figure 4.12. Some of the possible ray paths for seismic waves penetrating the earth. In the mantle and core, the velocities increase with depth, so the ray paths bend away from the normal. The decrease in velocity at the mantle–core boundary causes those rays refracted into the core to bend towards the normal.

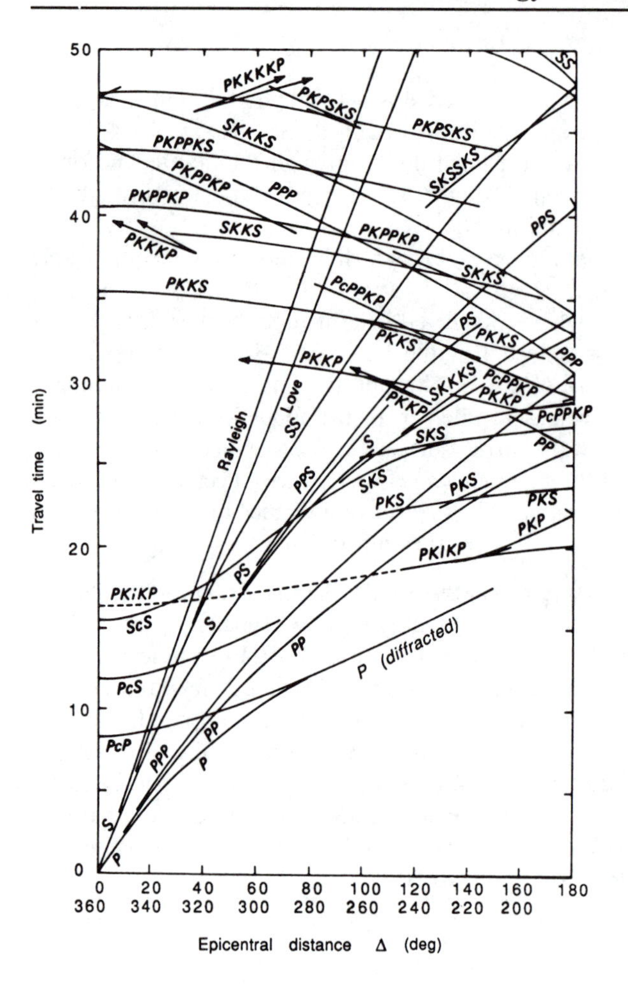

Figure 4.13. Jeffreys–Bullen (J–B) travel-time curves for earthquake focus at the surface, with some modifications. Such curves are usually accurate to a few seconds. Epicentral distance is the angle subtended by the earthquake epicentre and seismometer at the centre of the earth. (After Bullen and Bolt 1985.)

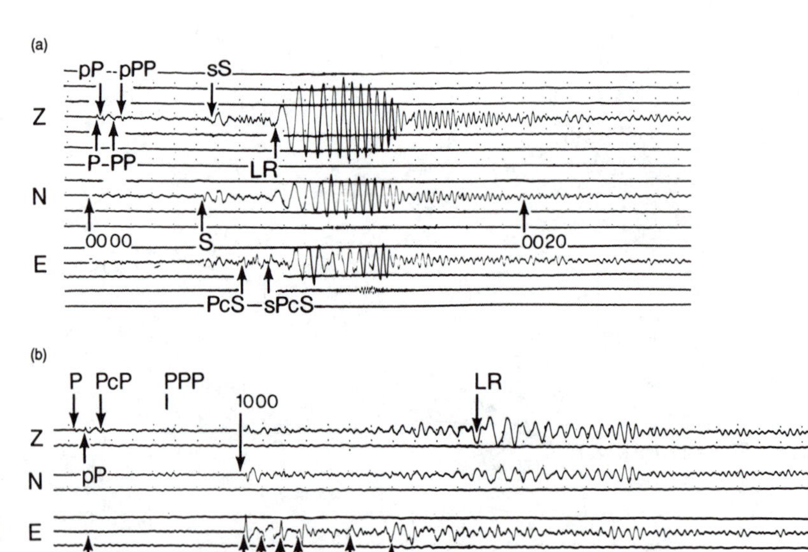

Figure 4.14. Examples of long-period earthquake records. Z is the vertical component and N (north) and E (east) the two horizontal components. Time marks (dots) are every minute. The recordings are made in ink on a rotating drum – hence the other traces above and below the earthquake. (a) Offshore Nicaragua, focal depth 70 km, surface wave magnitude 5.7, recorded in New York. Note the clear dispersion of the Rayleigh waves. (b) Northern Chile, focal depth 129 km, surface wave magnitude 6.5, recorded in New York. (From Simon 1981.)

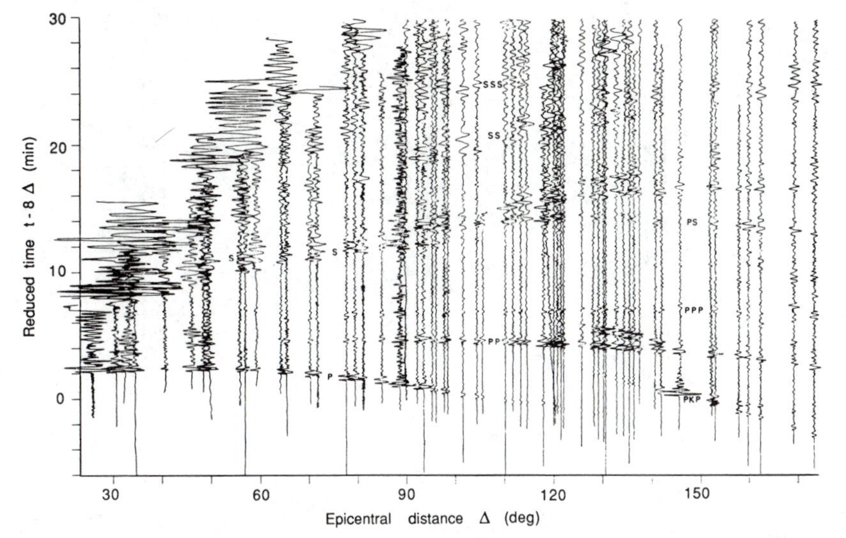

Figure 4.15. Vertical component seismogram section of an earthquake near Sumatra, recorded by long-period stations. All traces have the same amplitude scale. The reduction velocity for this plot is 13.9 km s^{-1}. This means that phases with velocities of 13.9 km s^{-1} appear horizontal, lower phase velocities have a positive slope and higher phase velocities have negative slope. (From Müller and Kind 1976.)

4.2.8 Earthquake Fault-Plane Solutions

The relative motions between plates on the earth's surface, although regular over geological time, are not continuous over a daily or yearly period. Stress builds up along a fault, or in a region, over a period of years before it reaches some critical level; then an earthquake and, perhaps, aftershocks relieve it (Fig. 4.16). The length of the *fault plane*, along which the rocks are displaced, varies from metres for a very small earthquake, to about 1000 km for a very large earthquake. The 1960 Chile earthquake had a fault plane 1000 km in length, and the aftershock zone of the 1957 Aleutian earthquake ($M_w = 9.1$) was some 1200 km long, the longest aftershock zone known. Figure 4.17 illustrates the fundamental types of faults and the various names by which they are known. Most earthquakes occur along plate boundaries as a direct result of plate motions; these are *interplate* (between plate) earthquakes. *Intraplate* (within plate) earthquakes are only a small proportion of the total number occurring, but they can be large and can produce considerable damage. Examples of these include the great 1811 and 1812 earthquakes in New Madrid, Missouri, U.S.A. which must have exceeded magnitude 8.

To determine the location of any particular earthquake, the seismic waves it produces must be observed at a number of recording stations around the world. The arrival time of the waves at the stations can then be used to locate the source (Sect. 4.2.1). In 1961, the United States Coast and Geodetic Survey established the World-Wide Standardized Seismograph Network (WWSSN). As a result, earthquakes occurring since 1961 have improved epicentre locations. Figure 2.1 shows the 1961–1967 distribution of epicentres. It is apparent that these seismic belts accurately delineate the plate boundaries and that there are very few intraplate events.

An earthquake occurring along the San Andreas Fault in California, U.S.A., is likely to have a different mechanism from one occurring on a subduction zone beneath Japan. By studying direction of movement, or

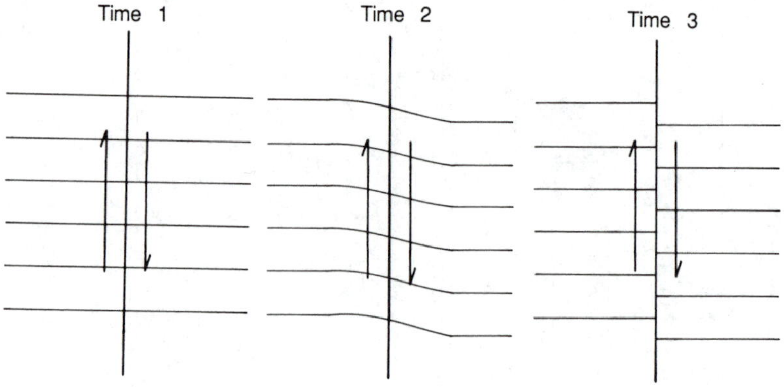

Figure 4.16. A simple model of the *elastic rebound theory* of earthquake source mechanisms. Over time, one side of the fault is displaced relative to the other side (1, 2). The deformation continues until the stresses on the fault are large enough to overcome the friction between the two blocks of material; then an earthquake (sudden displacement or rupture) occurs, and the strain is released (3). Thus, the size of the earthquake is directly related to the friction in the fault.

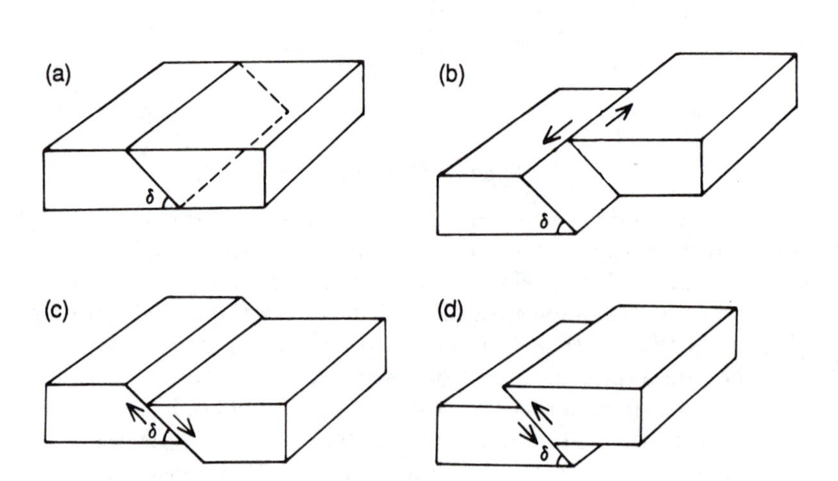

Figure 4.17. Fundamental types of faults and their various names. (a) Block model of a fault before any motion has taken place; δ is the angle of dip. (b) Strike-slip fault, transcurrent fault, lateral fault, tear fault, wrench fault; δ is often near 90°. Fault shown is left-handed or sinistral; a strike-slip fault with the opposite offset would be called right-handed or dextral. (c) Normal fault, dip-slip fault, normal-slip fault, tensional fault, gravity fault; generally, $45° < \delta < 90°$. (d) Thrust fault, dip-slip fault, reverse-slip fault, reverse fault, compressional fault; often, $0° < \delta < 45°$. Arrows show the slip vector, the relative motion between the two sides of the fault. The slip vector always lies in the fault plane. (From Båth 1979.)

polarity, of the first seismic waves from an earthquake arriving at a variety of seismograph stations distributed over the earth's surface, one usually can determine both the type of earthquake and the geometry of the fault plane. To understand this, consider a simple strike-slip fault as shown in Figure 4.18. Imagine that the world is flat and that seismograph stations A, B, C, D, E and F are located some distance from the fault where the earthquake occurred.

The first P-wave motion to arrive at any particular station from an earthquake on this fault will either be compressional (first motion is a push) or dilatational (first motion is a pull). At each of these five seismograph stations the polarity of the first-arriving energy is noted. Thus, the polarities at stations A, C and E are positive (compressional) and at stations B and F negative (dilatational), whereas station D receives no P-wave energy. If the distribution of compressional and dilatational first motions for many seismograph stations around the earthquake are plotted, it is seen that they fall into four quadrants (Fig. 4.18b) alternately positive and negative. The length of the arrows in each lobe represents the relative magnitude of the first P-wave at any location.

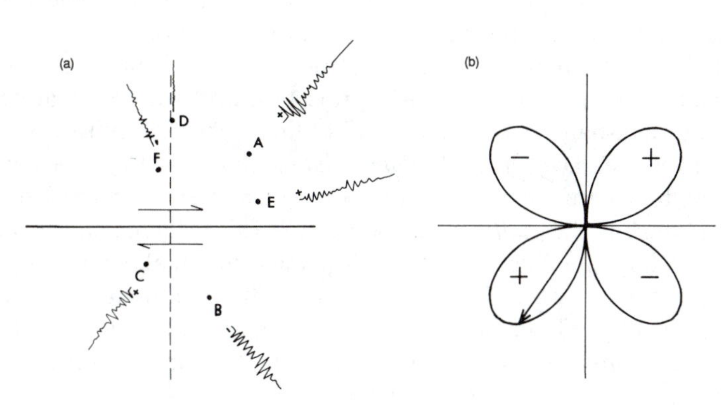

Figure 4.18. (a) Plane view of an earthquake strike-slip fault. A, B, C, D, E and F are six seismograph stations which recorded the earthquake. The first P-wave recorded at stations A, C and E would be compressional (positive, up); the first P-wave at stations B and F would be dilatational (negative, down); station D would record no first P-wave arrival. (b) The distribution of polarity of the first P-wave motion falls into four quadrants. The lobes indicate the relative magnitude of the first motion at any location. Arrow shows the magnitude at location C.

Of course, the earth is spherical, and so in reality we must work in spherical coordinates, which complicates the geometry slightly but not the results. Imagine a small sphere centred on and surrounding the focus of an earthquake. This sphere is known as the *focal sphere*. The ray that first arrives at each receiver intersects the *lower focal hemisphere* at an appropriate angle from the vertical, *i*, and azimuth A (Fig. 4.19). To calculate the angle from the vertical at which the rays intersect the lower focal hemisphere, knowledge of the gross P-wave structure of the earth is required, and a ray is traced back from receiver to source (see Sect. 4.3.1 and Appendix 3). This angle of departure can be obtained from standard seismological tables. This lower focal hemisphere is then projected onto a horizontal plane using an equal area or a stereographic projection. It is obvious from Figure 4.19 that the first arriving P-wave for a seismograph close to the earthquake focus must travel almost horizontally. Thus, this ray intersects the lower focal hemisphere almost at its equator $i = 90°$. In contrast, the direct P-wave to a station on the opposite side of the earth to the earthquake travels almost vertically down from the focus and intersects the lower focal hemisphere near its centre (pole) $i = 0°$. This means that nearby seismograph stations plot around the edge of the projection, and distant stations plot towards the centre. The azimuth of each seismograph station is easily measured geographically. The polarity

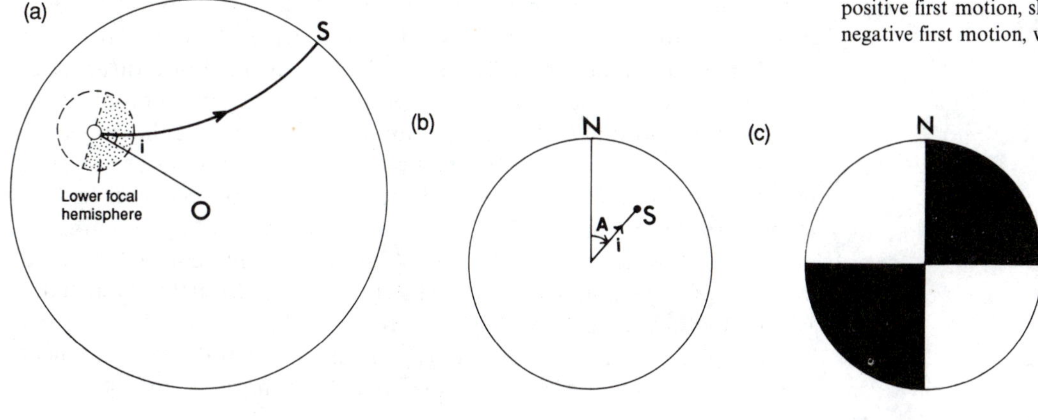

Figure 4.19. (a) Schematic cross section through the earth, centre 0. The focal sphere is an imaginary small sphere centred on the earthquake focus (○). The lower focal hemisphere is stippled. The ray path for the first arrival at seismograph S intersects the lower focal hemisphere at an angle *i* from the vertical. (b) Projection of the lower focal hemisphere onto a horizontal plane. N is north. Seismograph S plots at (*i*, *A*), where *i* is the angle of incidence shown in (a), and *A* is the geographic azimuth of S from the earthquake focus. The polarity of the first motion recorded at S is then plotted at (*i*, *A*). (c) Fault-plane solution for the strike-slip earthquake shown in Figure 4.18a. Compression, positive first motion, shaded; dilatation, negative first motion, white.

(positive = compressional, or negative = dilatation) of the first motion recorded by each seismograph station is then plotted on this projection. In this way, data from seismograph stations around the world can be plotted on a graph which is a projection of the lower focal hemisphere. The right-handed or dextral strike-slip earthquake illustrated in Figure 4.18 would have the *fault-plane solution* shown in Figure 4.19c. The four quadrants are separated by two orthogonal planes, or nodal surfaces, one of which is the *fault plane*. The other is called the *auxiliary plane*. There is no way of telling from the fault-plane solution alone which plane is the fault plane and which is the auxiliary plane. The radiation pattern shown in Figure 4.19c could have been generated either by dextral strike-slip motion on a vertical fault plane striking 090° or by a sinistral strike-slip motion on a vertical plane striking 000°.

If the fault plane is not vertical, the fault-plane solution for strike-slip motion still has four quadrants, but the fault plane and auxiliary plane do not plot as orthogonal straight lines passing through the origin. Instead, they plot as orthogonal great circles offset from the origin by $90° - \delta$ where δ is their dip. An example is shown in Figure 4.20. Again the

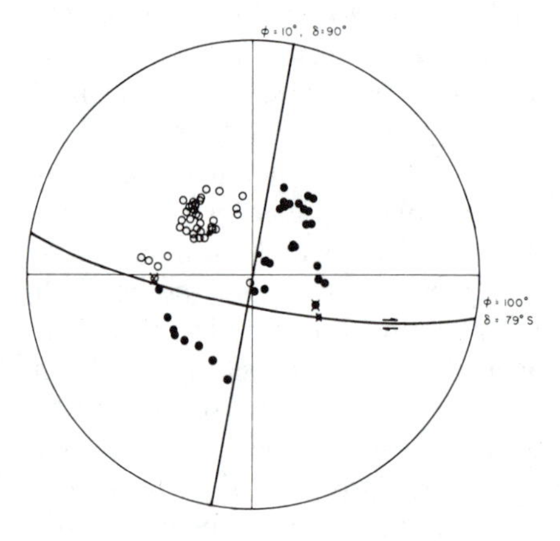

Figure 4.20. A fault-plane solution for a strike-slip earthquake. The strike and dip of the nodal planes are given by ϕ and δ. The east-striking nodal plane was chosen as the fault plane because the earthquake epicentre lies on an east-striking transform fault. The fault plane is not vertical in this instance; it is dipping at 79°. (After Sykes 1967.)

fault-plane solution does not distinguish between the fault plane and the auxiliary plane. A normal faulting earthquake (Fig. 4.21a) would have the fault-plane solution shown in Figure 4.21b. In contrast, a thrust faulting earthquake and its fault-plane solution are shown in Figures 4.21c,d. In these cases, there is no ambiguity about the strike of the nodal planes, but there is ambiguity about the dip of the fault plane. For a thrust fault, the first motion at the centre of the projection is always compressional, whereas for a normal fault the first motion at the centre of the projection is always dilatational. When the fault plane is vertical, it plots as a straight line passing through the origin, and the auxiliary plane plots around the circumference. Figure 4.22a shows an example of a predominantly normal-faulting event from the East African Rift. Figure 4.22b shows an example of a predominantly thrust-faulting event from the Macquarie Ridge (southwest of New Zealand). Both of these events had a small strike-slip component in addition to the main normal or thrust component. Note that motion on a fault can have both normal and strike-slip

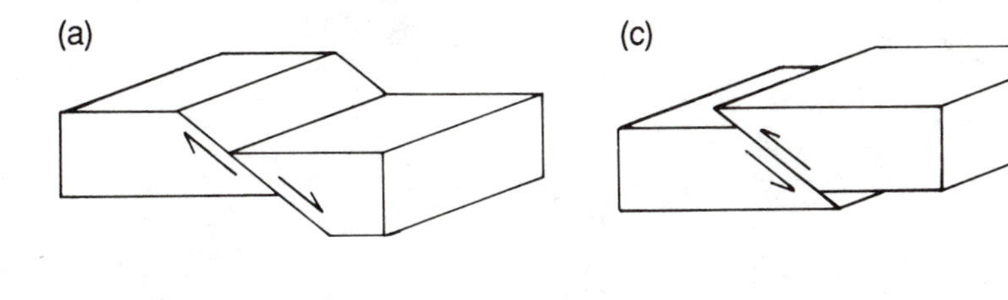

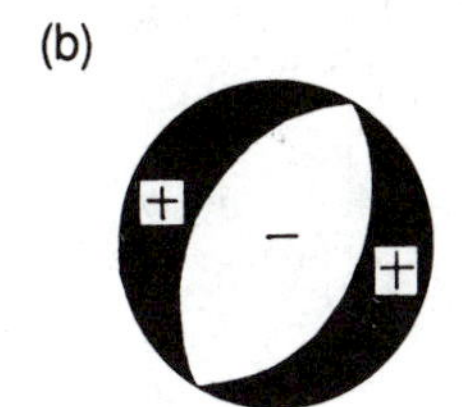

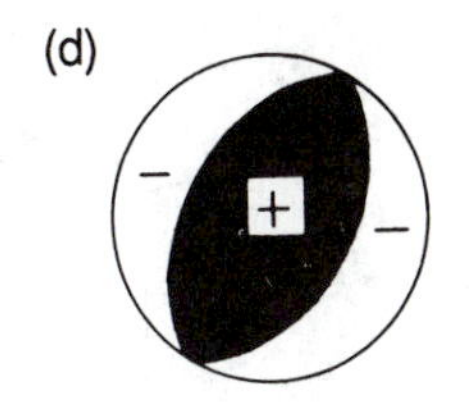

Figure 4.21. (a) Normal faulting earthquake. (b) Fault-plane solution for the normal faulting earthquake in (a). White region represents locations at which the first motion is dilatational (negative); black region represents locations at which the first motion is compressional (positive). (c) Thrust faulting earthquake. (d) Fault-plane solution for the thrust faulting earthquake in (c).

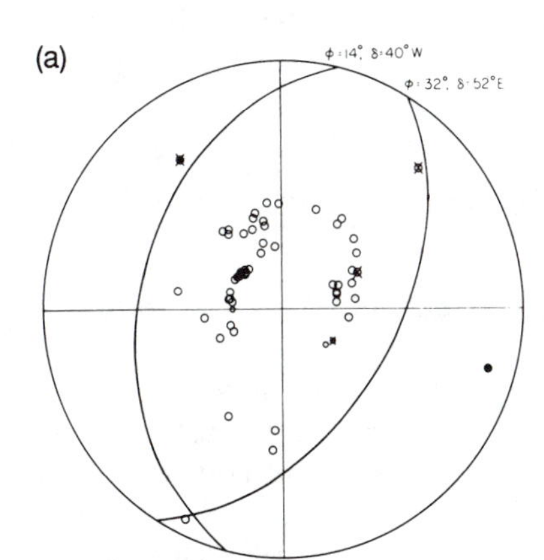

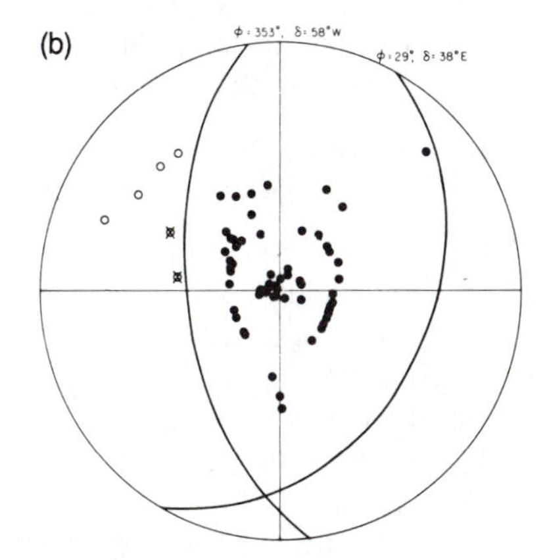

Figure 4.22. (a) A fault-plane solution for a predominantly normal faulting earthquake which occurred on the East African Rift. (b) A fault-plane solution for a predominantly thrust faulting earthquake which occurred on the Macquarie Ridge. Closed circles, first motion is compressional (positive); open circles, first motion is dilatational (negative); crosses, no clear first motion. (After Sykes 1967.)

components or thrust and strike-slip components but never normal and thrust components.

The *slip vector* **u** of the earthquake is the relative motion which occurred between the two sides of the fault plane. It always lies in the fault plane (Fig. 4.17). The horizontal component of the slip vector u_h gives the azimuth of the relative horizontal motion occurring at the epicentre (Fig. 4.23). Although, for a pure strike-slip event such as that illustrated in Figure 4.17b and 4.20, u_h is parallel to the fault plane, this is not the case in general. Consider the thrust-faulting earthquake shown in Figure 4.21a; in this case, u_h is perpendicular to the strike of the fault plane. For any fault-plane solution, the strike of u_h is normal to the strike of the auxiliary plane. Thus, the strike of u_h can be found by adding 90° to the strike of the auxiliary plane. The fault-plane solution cannot tell us anything about the magnitude of u_h, simply its direction.

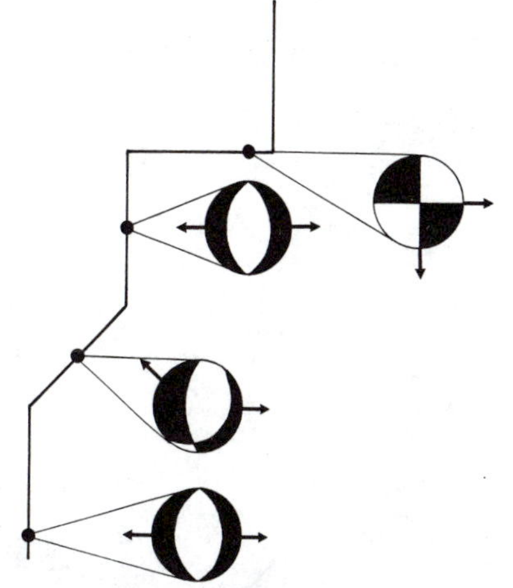

Figure 4.23. Fault-plane solutions for earthquakes (●) occurring along a plate boundary (solid line). Arrows indicate the azimuth of the possible horizontal components of the slip vector. The consistent arrows are the east–west set. The region is in extension, and the area is characterized by a mixture of strike-slip and normal faulting.

How to use earthquake first-motion directions to determine a fault-plane solution

To use earthquake first-motion directions you need to have tracing paper, pencil, eraser, pin and an equal-area projection net (also known as

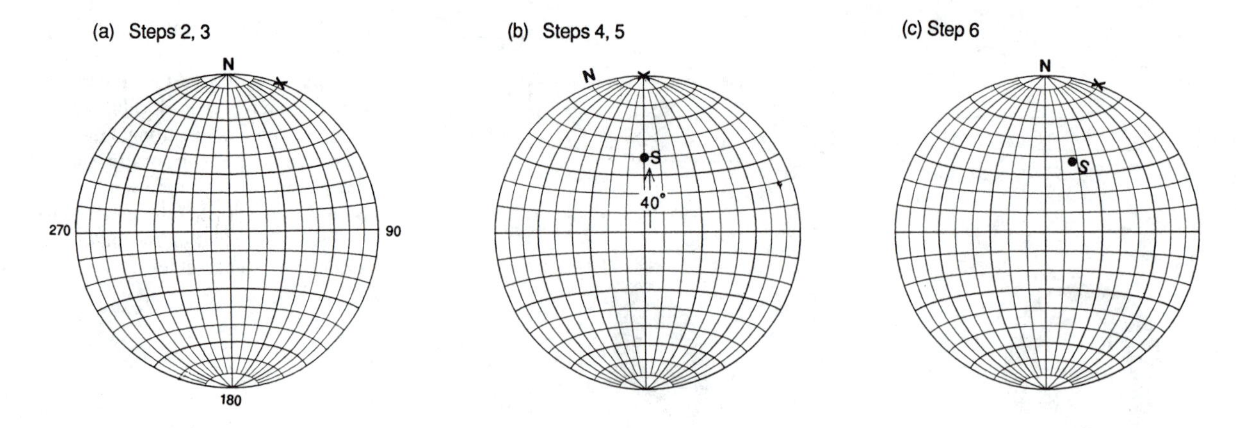

Figure 4.24. How to plot a seismograph station on an equal-area projection of the lower focal hemisphere.

Lambert or Schmidt equal-area projection net) (Figs. 4.24a and 4.59c). On that base projection, great circles are the north–south lines of longitude; azimuth, or strike, is measured by counting lines of latitude clockwise around the edge of the projection.

Example 1 Plot seismograph station S (azimuth N20°E from the earthquake; angle of first-motion ray 40° from the vertical) on an equal-area projection.

1. Pin the tracing paper onto the base projection through its centre.
2. Mark N, the north axis.
3. Mark the azimuth N20°E as *x* at the edge of the projection (Fig. 4.24a.)
4. Rotate the tracing paper so that *x* is at the top of the projection.
5. Mark the point S 40° from the centre of the projection towards *x* (Fig. 4.24b.)
6. Rotate the tracing paper so that N is again at the top of the projection (Fig. 4.24c.)

S is now correctly plotted on the projection of the lower focal hemisphere.

Example 2 Find the great circle which joins the two points R and S.

1. Plot the points R and S as described in Example 1 (Fig. 4.25a.)
2. Rotate the tracing paper until R and S both lie on the same great circle (line of longitude.)
3. Trace that great circle (Fig. 4.25b.)
4. Rotate the tracing paper so that N is again at the top of the projection (Fig. 4.25c.)

The great circle which joins points R and S is now drawn.

Example 3 Determine the fault-plane, auxiliary plane and slip vector for an earthquake, given the direction of first motion at several seismograph stations.

The fault plane and the auxiliary plane plot on the projection of the lower focal hemisphere as great circles. The fault plane and the auxiliary plane are orthogonal. This means that we need to find two orthogonal great circles which separate the positive first motions from the negative first motions.

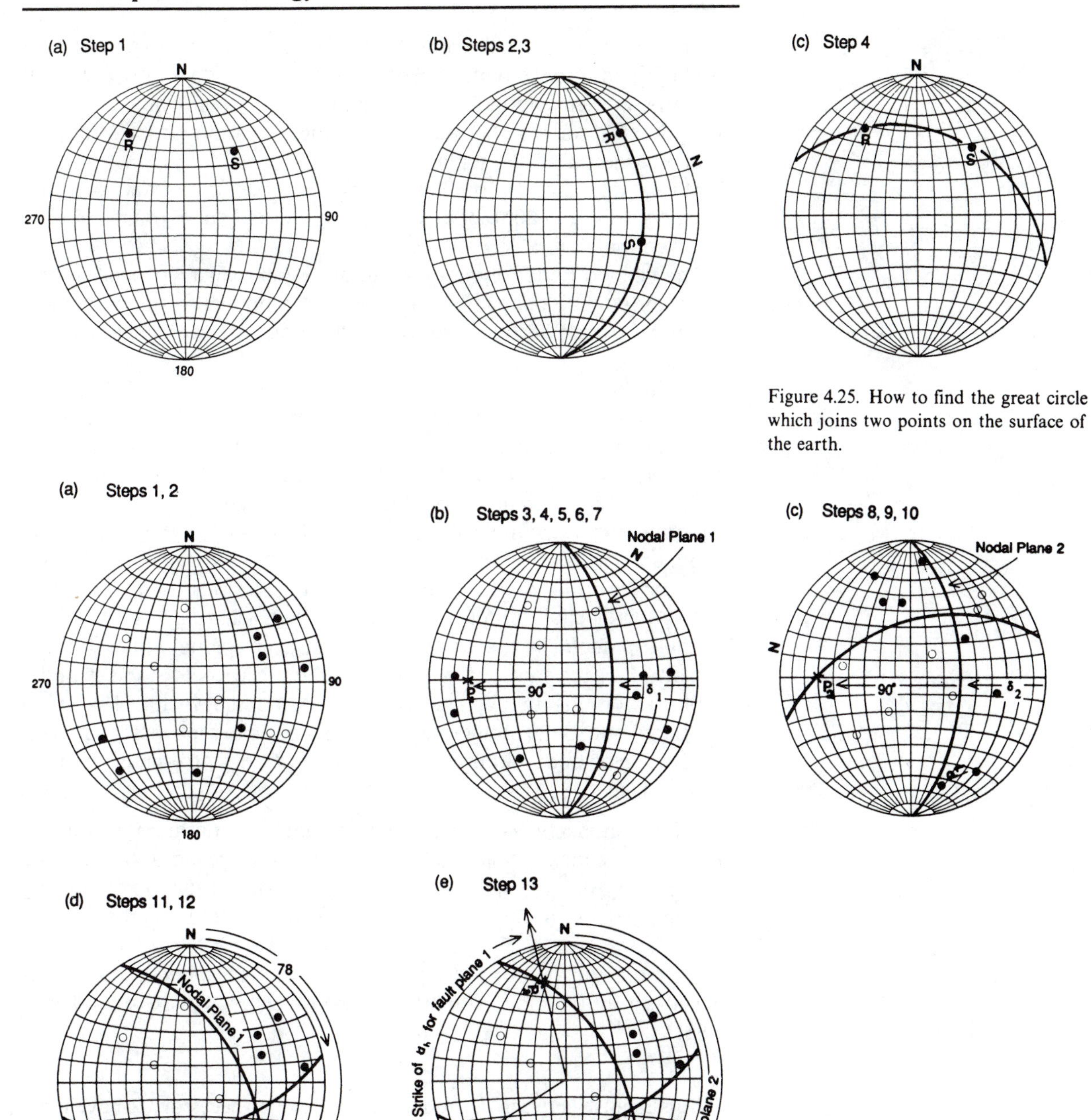

Figure 4.25. How to find the great circle which joins two points on the surface of the earth.

Figure 4.26. How to determine the fault plane, auxiliary plane and azimuth of the slip vector for an earthquake.

1. Pin the tracing paper onto the base projection through its centre.
2. Put the first motions onto the tracing paper (Fig. 4.26a).
3. Rotate the tracing paper until you find a great circle which separates positive and negative first motions.
4. Trace that great circle; it is nodal plane 1.
5. The dip of nodal plane 1, δ_1, is measured along the equator of the

base projection from the outside edge to the great circle. (A horizontal plane with zero dip plots around the edge of the projection. A vertical plane, with 90° dip plots as a straight north—south line.)

6. Count 90° along the equator of the base projection from its intersection with nodal plane 1.
7. Mark that point P_1 on the tracing paper; it is normal to nodal plane 1 (Fig. 4.26b). (Sometimes the normal to a great circle is called the pole.)
8. The second nodal plane must separate the remaining positive and negative first motions, and since it is also normal to nodal plane 1, point P_1 must lie on it. So, rotate the tracing paper until you find such a great circle.
9. Trace that great circle; it is nodal plane 2 (Fig. 4.26c). If you cannot find nodal plane 2, go back to step 3 and check that nodal plane 1 was correct.
10. Repeat steps 5, 6 and 7 to find the dip of nodal plane 2, δ_2, and point P_2, the normal to nodal plane 2 (Fig. 4.26c).
11. Rotate the tracing paper so that N is again at the top of the projection.
12. The strike of the nodal planes is measured clockwise around the outside of the projection from N, 78° and 147° (Fig. 4.26d).
13. The slip vector is the normal to the auxiliary plane. Thus, if nodal plane 2 is the fault plane, point P_1 is the slip vector; and if nodal plane 1 is the fault plane, point P_2 is the slip vector. The strike of the horizontal component of the possible slip vector is measured clockwise around the outside of the projection from N (Fig. 4.26e).

This earthquake was therefore a combination of normal faulting and strike-slip on either a fault plane striking 78° and dipping at 60° with the strike of horizontal component of slip 238°, or a fault plane striking 147° and dipping at 60° with the strike of the horizontal component of slip 348°.

For more details on using lower focal hemisphere projections, consult an introductory structural geology textbook or Cox and Hart (1986).

To distinguish between the nodal planes, it is often necessary to have some additional information. Sometimes local geology can be used to decide which nodal plane is most likely to have been the fault plane. On other occasions, the earthquake may show a surface break or fault, in which case the strike of that fault and the offset along it should agree with one of the nodal planes. Often (and of course for all oceanic earthquakes), this information is unavailable. McKenzie and Parker (1967) faced this problem in determining earthquakes in the Pacific. They plotted both of the possible horizontal components of the slip vector for a set of earthquakes which occurred along the northern boundary of the Pacific Plate. They found that one set changed slowly and systematically in direction and the others showed no consistency. The correct horizontal components were the consistent set, and so the fault and auxiliary planes were determined. An example of this method is shown in Figure 4.23.

4.3 The Interior of the Earth

4.3.1 Seismic Velocities for the Whole Earth

After the travel times of seismic waves have been measured and travel-time curves (Fig. 4.13) plotted, it is possible to use the results to determine the variation of seismic velocities in the earth. There are two approaches to this determination: One is an inverse method and the other is forward. In the inverse problem used in the early determinations, such as that of Jeffreys (see Appendix 3 for details), the velocities are obtained directly from the travel times. In the forward method, a velocity–depth model is assumed and travel times calculated and compared with observations. The model is then adjusted until agreement at the desired level is attained. A recent velocity model is shown in Figure 4.27. The two zones in which refinement of the structure is still taking place are the low-velocity zone in the upper mantle and the transition zone between the inner core and the outer core.

Crust The seismic structure of the continental crust is variable, but it has an average thickness of 35 km and an average P-wave velocity of about 6.5 km s^{-1} (see Sect. 9.1.2). The oceanic crust is thinner, 7–8 km thick, with an average P-wave velocity of more than 6 km s^{-1} (see Sect. 8.2.1).

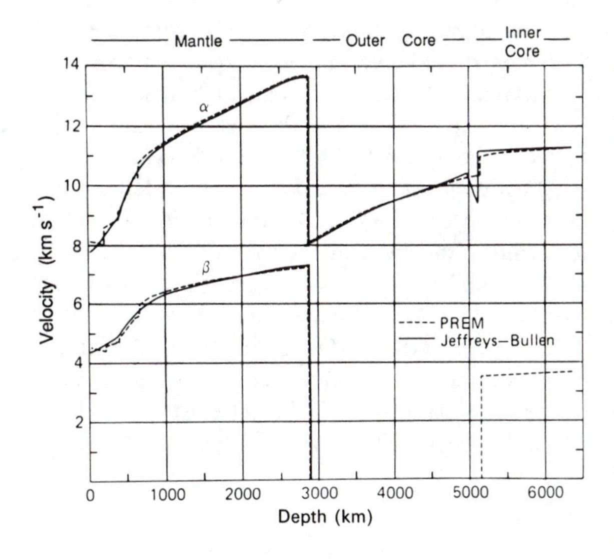

Figure 4.27. Seismic velocity–depth models for the whole earth. Since the early determination by Jeffreys (1939), which was based on the Herglotz–Wiechert inversion of the Jeffreys–Bullen (J–B) compilation of travel-time and angular distance data, there have been many revisions, but the agreement among them is good. The two regions where the models have been most revised and refined are the low-velocity zone in the upper mantle (asthenosphere) and the inner core–outer core transition zone. The preliminary reference earth model (PREM) of Dziewonski and Anderson (1981) was determined by a joint inversion of the free oscillation periods of the earth, its mass and moment of inertia as well as the travel-time–distance data. (After Bullen and Bolt 1985.)

Mantle The discontinuity between the crust and mantle, which is compositional (Sect. 4.3.5), is called the Mohorovičić discontinuity. The normal P-wave velocity at the top of the mantle is 8.1 km s^{-1}.

A low-velocity zone for S-waves in the upper mantle is well established by the surface-wave dispersion data. In contrast, the P-wave low-velocity zone in the upper mantle is less well established. The inference that there is a low-velocity zone for P-waves is based on a shadow zone effect for P-waves out to about 15° (Fig. 4.28) and on a matching of waveforms of

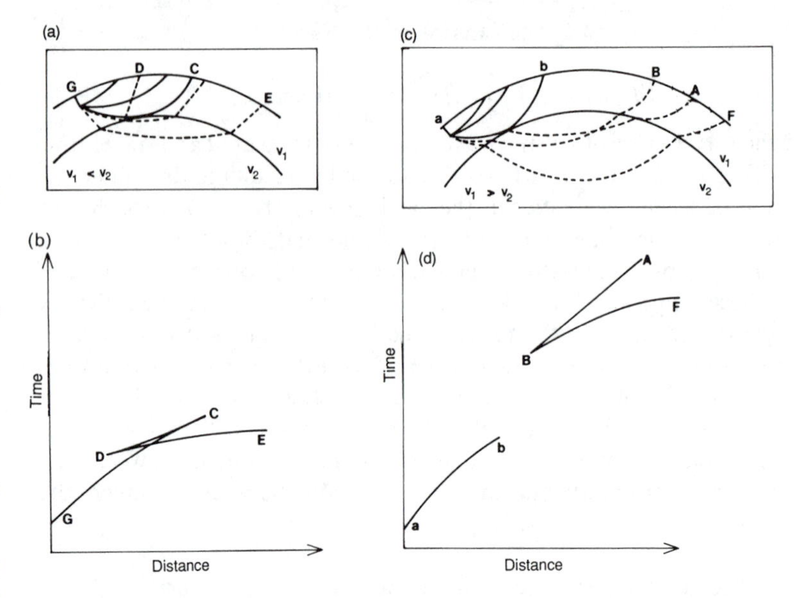

Figure 4.28. The 'shadow zone' resulting from a low-velocity zone. As an example, consider a two-layered sphere for which the seismic velocity increases gradually with depth in each layer. The seismic velocity immediately above the discontinuity in the upper layer is V_1 and that immediately below the discontinuity is V_2. The ray paths for the case $V_2 > V_1$ (the velocity increases at the discontinuity) are shown in (a). If $V_2 < V_1$ (the velocity decreases at the discontinuity resulting in a low-velocity zone at the top of the second layer), then the rays refracted into the inner layer bend towards the normal (Snell's law), yielding the ray paths shown in (c). The travel-time–distance curves for (a) and (c) are shown in (b) and (d). When $V_2 > V_1$, arrivals are recorded at all distances, but when $V_2 < V_1$, there is a distance interval over which no arrivals are recorded. This is the shadow zone. The angular extent of the shadow zone (b to B) and the corresponding delay in travel time (b to B) are dependent on the depth, extent and reduction of velocity in the low-velocity zone. (After Gutenberg 1959.)

P-wave arrivals with synthetic seismograms computed for possible velocity structures. Beneath this low-velocity zone, P- and S-wave velocities increase markedly to about 400 km. At 400 km and 670 km, there are sharp changes in velocity; both P- and S-wave velocities increase. These increases have been independently verified by computing synthetic seismograms to match earthquake and nuclear explosion amplitudes and waveforms. The approximate depth at which earthquake activity in subduction zones ceases is also 670 km, and this depth is commonly taken as the boundary between the upper and lower mantle. The entire region between 400 and 1000 km depth is often called the mantle transition zone. The mantle at depths between 1000 and 2700 km is referred to as the D′ shell. The lowermost 200 km of the mantle (depth ~ 2700–2900 km) is referred to as the D″ shell. Velocities increase slowly with depth through the lower mantle. The direct P-wave through the mantle can be observed out to 103°. At epicentral distances between 103° and 120°, a weak P-wave is diffracted (Sect. 4.5.4) at the core–mantle boundary in what is called the shadow zone (Fig. 4.29). Evidence exists that velocity gradients are reduced in the D″ shell. This could be due to chemical interaction between the

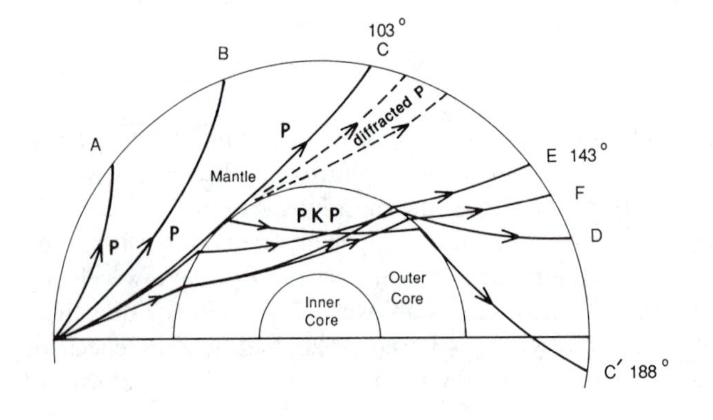

Figure 4.29. Ray paths for PKP, the direct P-wave passing through the mantle and outer core. The mantle P-wave (C) which has grazing incidence on the core has an epicentral distance of 103°. Beyond this distance, there can be no direct P-waves although PP and PPP can be recorded at greater distances, and a weak diffracted P can be recorded out to about 120°. Because there is a sharp velocity decrease for rays refracted into the core, all these rays bend towards the normal at the mantle–core boundary, causing a shadow zone for P-waves. The PKP ray with the shallowest angle of incidence on the outer core (C′) is refracted and finally emerges at an epicentral distance of 188°. With increasing angle of incidence (C′, D, E, F), PKP rays emerge at epicentral distances decreasing to 143° and then increasing to about 155°. Each ray penetrates deeper into the outer core than its predecessor. Thus, no direct P-waves are recorded at epicentral distances between 103° and 143°; this is the shadow zone. At 143° there is a *caustic*; the amplitude of PKP is large (this shows clearly in Fig. 4.15). (Based on Gutenberg 1959.)

core and mantle or to a thermal boundary layer which would conduct, and not convect, heat (refer to Sects. 4.3.5 and 7.9).

Core At the mantle–core boundary (also known as the Gutenberg discontinuity after its discoverer, or the CMB) the P-wave velocity drops sharply from about 13.6 to about 8.0 $\mathrm{km\,s^{-1}}$, and the S-wave velocity drops from about 7.3 $\mathrm{km\,s^{-1}}$ to zero. This structure is determined by the strong reflections PcP, ScS and so on. The P-wave velocity increases slowly in the outer core until the inner core boundary. This is determined mainly by the rays PKP and SKS (Fig. 4.29). However, since PKP does not sample the outermost core, the velocities there are based on SKS. The inner core was discovered in 1936 by Inge Lehmann, a Danish seismologist (who celebrated her hundredth birthday in 1988), using seismograms from an earthquake near Buller on the Southern Alpine Fault in New Zealand. She realized that to explain particular phases (observed at epicentral distances greater than about 120° with travel times of 18–20 min), the core must contain a distinct inner region. The phases she identified are then explained as being refractions through the higher-velocity inner core (PKIKP in today's notation) which then arrive earlier than the PKP phase. There is a transition zone at the outer core–inner core boundary: The P-wave velocity increases from about 10.2 to about 11.2 $\mathrm{km\,s^{-1}}$. The P-wave and S-wave velocities (3.5 $\mathrm{km\,s^{-1}}$) are both constant through the inner core. There is still some uncertainty about the exact velocity structure of the outer core–inner core transition, but a velocity gradient as shown in Figure 4.27 is the most favoured current model.

A zero S-wave velocity for the outer core, consistent with its being liquid, is in agreement with studies on the tides, which require a liquid core. This conclusion is supported by all other seismological evidence and, indeed, is essential if the earth's magnetic field (and its secular variation) is to be accounted for by convection currents in the outer core (Sect. 7.9.2). The structure of the inner core is mainly determined by using ray path PKIKP (Fig. 4.30). The depth of the inner core–outer core transition can be determined from the travel times of PKiKP (the reflection from this transition), and the velocity increase/velocity gradient occurring there follows from the amplitude of this reflected arrival. It has been suggested that the boundary between the inner and outer core should be termed the Lehmann discontinuity. The name has been used for a discontinuity at ~ 220 km depth beneath North America, but it seems most appropriate in the core.

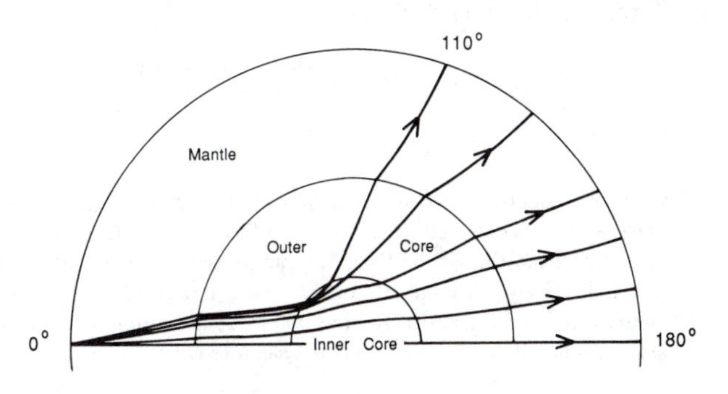

Figure 4.30. Ray paths for PKIKP, the direct P-wave passing through the mantle, outer core and inner core. (Based on Gutenberg 1959.)

4.3.2 *Density and Elastic Moduli for the Whole Earth*

The variation of seismic velocity with depth has been discussed in the previous section. To further understand the internal structure of the earth and its composition, it is also necessary to know how density and the elastic moduli vary with depth.

We have already seen (Eqs. 4.4 and 4.5) that the bulk or compressibility modulus K, shear modulus μ and density ρ are related to the P-wave and S-wave velocities by

$$\alpha = \sqrt{\frac{K + \frac{4}{3}\mu}{\rho}} \tag{4.28}$$

$$\beta = \sqrt{\frac{\mu}{\rho}} \tag{4.29}$$

Even if we know that α and β vary with depth in the earth, these two equations alone cannot tell how K, μ and ρ vary with depth because they contain three unknowns. A third equation which assists us to determine these three unknowns is the *Adams–Williamson equation.*

Let us assume that the earth is made up of a series of infinitesimally thin, spherical shells, each with uniform physical properties. The increase in pressure δP which results during the descent from radius $r + \delta r$ to radius r is due only to the weight of the shell thickness δr:

$$\delta P = -g(r)\rho(r)\,\delta r \tag{4.30}$$

where $\rho(r)$ is the density of that shell and $g(r)$ the acceleration due to gravity* at radius r.

Writing Eq. 4.30 in the form of a differential equation, we have

$$\frac{dP}{dr} = -g(r)\rho(r) \tag{4.31}$$

where dP/dr is simply the gradient of the hydrostatic pressure. There is a minus sign in Eqs. 4.30 and 4.31 because the pressure P decreases as the radius r increases. The gravitational acceleration at radius r can be written in terms of the gravitational constant G and M_r, the mass of the earth within radius r:

$$g(r) = \frac{GM_r}{r^2} \tag{4.32}$$

Therefore, Eq. 4.31 becomes

$$\frac{dP}{dr} = -\frac{GM_r\rho(r)}{r^2} \tag{4.33}$$

* Outside a spherical shell the gravitational attraction of that shell is the same as if all its mass were concentrated at its centre. Within a spherical shell there is no gravitational attraction from that shell. Together, the preceding statements mean that at a radius r in the earth, the gravitational attraction is the same as if all the mass inside r were concentrated at the centre of the earth. All the mass outside radius r makes no contribution to the gravitational attraction and so can be ignored. This is proved in Section 5.2.

To determine the variation of density with radius it is necessary to determine dP/dr.

From Eq. 4.33, we can write

$$\frac{d\rho}{dr} = \frac{dP}{dr}\frac{d\rho}{dP} \tag{4.34}$$

$$= -\frac{GM_r\rho(r)}{r^2}\frac{d\rho}{dP} \tag{4.35}$$

The compressibility or bulk modulus for adiabatic compression K is used to obtain $d\rho/dP$, the variation of density with pressure:

$$K = \frac{\text{increase in pressure}}{\text{fractional change in volume}} = -\frac{dP}{dV/V} \tag{4.36}$$

There is a minus sign in Eq. 4.36 because volume decreases as pressure increases. Since density ρ is the ratio of mass to volume,

$$\rho = \frac{m}{V} \tag{4.37}$$

we can write

$$\frac{d\rho}{dV} = -\frac{m}{V^2} = -\frac{\rho}{V} \tag{4.38}$$

Substituting this into Eq. 4.36 gives

$$K = \rho\frac{dP}{d\rho} \tag{4.39}$$

Equation 4.35 can therefore be written

$$\frac{d\rho}{dr} = -\frac{GM_r\rho(r)}{r^2}\frac{\rho(r)}{K} \tag{4.40}$$

Combining Eqs. 4.28 and 4.29 gives

$$\frac{K}{\rho} = \alpha^2 - \tfrac{4}{3}\beta^2 \tag{4.41}$$

so that

$$\frac{d\rho}{dr} = -\frac{GM_r\rho(r)}{r^2(\alpha^2 - \frac{4}{3}\beta^2)} = -\frac{GM_r\rho(r)}{r^2\Phi} \tag{4.42}$$

where Φ is equal to $\alpha^2 - \frac{4}{3}\beta^2$.

Equation 4.42 is the *Adams–Williamson* equation and is used to determine density as a function of radius. To use the equation, it is necessary to start at the earth's surface and to work inwards, applying the equation successively to shells of uniform composition. It is important to remember that M_r is the mass within radius r:

$$M_r = 4\pi\int_{a=0}^{a=r}\rho(a)a^2\,da \tag{4.43}$$

$$M_r = M_E - 4\pi \int_{a=r}^{a=R} \rho(a) a^2 \, da \tag{4.44}$$

where M_E is the mass of the earth (known from study of periods of rotation for satellites and from direct measurements of gravity, see Chapter 5), and R is the radius of the earth. Thus, at each stage of the calculation, working from the earth's surface inwards, all the terms on the right-hand side of equation refer to material outside or at the radius r and so have already been determined.

A density structure for the whole earth obtained by using the Adams–Williamson equation is called a *self-compression model* because the density at each point is assumed to be due to compression by the material above it. In practice, it is pointless to assume that the whole earth has uniform composition since we know this to be untrue, so the density determination is usually begun at the top of the mantle, with a chosen density (the crustal thickness and density vary widely). Then densities can be calculated all the way to the base of the mantle using the Adams–Williamson equation. The core clearly has a different composition from the mantle since the dramatic seismic velocity changes occurring at the core–mantle boundary could hardly be due to pressure alone. A new starting density is therefore chosen for the top of the core, and the densities down through the core can be calculated using the Adams–Williamson equation. Because the total mass in the model must equal the mass of the earth, successive guesses at the density at the top of the core must be made until this constraint is satisfied.

Although such a self-compression density model for the earth satisfies the seismic velocity data from which it was derived, it does not satisfy data on the rotation of the earth. In particular, the earth's moment of inertia, which is sensitive to the distribution of mass in the earth (test the difference between opening the refrigerator door when all the heavy items in the door compartments are next to the hinge and when they are all next to the handle) is significantly greater than the moment of inertia for the self-compression model. There must be more mass in the mantle than the self-compression model allows. To determine the reasons for this discrepancy, it is necessary to reexamine the assumptions made in determining the Adams–Williamson equation. First, it was assumed that the temperature gradient in the earth is adiabatic (Sect. 7.7). However, since we know that convection is occurring in both the mantle (Sect. 7.8) and liquid outer core (Sect. 7.9), the temperature gradients there must be superadiabatic. Equation 4.42 can be modified to include a nonadiabatic temperature gradient:

$$\frac{d\rho}{dr} = -\frac{GM_r\rho(r)}{r^2\Phi} + \alpha\rho(r)\tau \tag{4.45}$$

where α is the coefficient of thermal expansion and τ the difference between the actual temperature gradient and the adiabatic temperature gradient. This modification means that in the case of a superadiabatic gradient ($\tau > 0$), the density increases more slowly with depth. Conversely, in the case of a subadiabatic temperature gradient, the density increases more rapidly with depth. This means that corrections for the temperature gradient, which, in practice, are found to be fairly small, act in the opposite direction to that required to explain the missing mantle mass. Another explanation must be found.

The second assumption made in deriving the Adams–Williamson equation was that there were no chemical or phase changes in the earth (other than differences between crust, mantle and core, which have already been included in the model). This assumption provides the answer to the problem of the missing mantle mass. In the mantle transition zone (400–1000 km), a number of jumps in seismic velocity occur which seem to be due to changes to state (phase changes). An example of a change of state is the change from liquid to solid such as occurs when water freezes. This is not the only type of change of state possible; there are also solid–solid phase changes in which the atoms in a solid rearrange and change the crystal structure. Examples of this are the change of carbon from graphite to diamond under increasing pressure and the changes which take place in the transformation from basalt to greenschist to amphibolite to pyroxene granulite (at high temperatures) or to blueschist (at low temperatures), and finally to eclogite under increasing pressure and temperature. The phase changes thought to occur in the transition zone are olivine to spinel and pyroxene to garnet at about 400 km, and spinel to postspinel forms at about 700 km. Density increases of about 10% are associated with these phase changes and are sufficient to account for the self-compression model's low estimate of mantle mass.

Figure 4.31 shows a density model for the earth which is based on the Adams–Williamson equation and the additional constraints provided by free oscillations, moment of inertia and total mass. These models are continually being updated and modified, but the densities are unlikely to change substantially from those shown here, though the details of the model, particularly in the transition zone and inner core, may alter. After a density model has been determined, it is straightforward to work backwards using Eqs. 4.28 and 4.29 to determine the elastic moduli (Fig. 4.31).

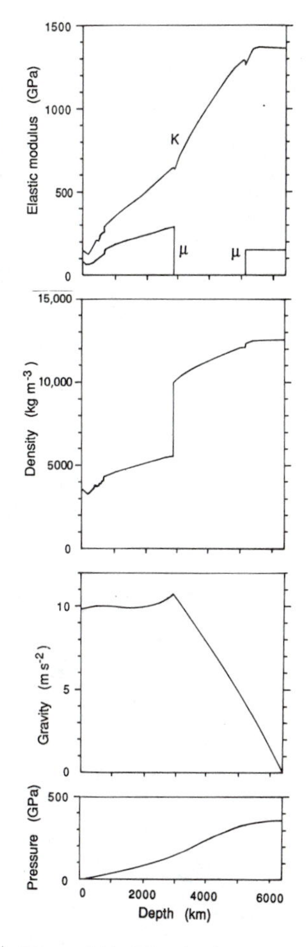

Figure 4.31. The elastic moduli K (bulk), μ (rigidity), density ρ, gravity g and pressure P in the interior of the earth. The elastic moduli and pressure are given in GPa. To convert GPa to kilobars (kb) multiply by 10. Thus, pressure at the centre of the earth is 361.7 GPa or 3617 kb. (Model of Hart et al. 1977 and Anderson and Hart 1976.)

$$\mu = \rho\beta^2 \tag{4.46}$$

$$K = \rho\alpha^2 - \tfrac{4}{3}\rho\beta^2 \tag{4.47}$$

Table 4.4 provides an interesting comparison of the volume, mass and density of the earth taken region by region. We know most about the structure of the crust, yet it represents only 0.5% of the total by volume and 0.3% by mass. Uncertainty increases with depth and mass.

4.3.3 Attenuation of Seismic Waves

In a perfectly elastic medium no elastic energy would be lost during the passage of a seismic wave. However, in practice the earth is not perfectly elastic, and some energy is dissipated (i.e., turned into heat) as a seismic wave passes. The amount of energy lost as a seismic wave passes through any medium is used to define a parameter Q for that medium. The *quality factor* Q is defined as

$$Q = \frac{2\pi\,(\text{elastic energy stored in the wave})}{\text{energy lost in one cycle or wavelength}} \tag{4.48}$$

Table 4.4. *Volume, mass and density of the earth*

	Depth (km)	Volume (10^{18} m^3)	%	Mass (10^{21} kg)	%	Density range[a] (10^3 kg m^{-3})
Crust	0–Moho	10	0.9	28	0.5	2.60–2.90
Upper mantle	Moho–670	297	27.4	1064	17.8	3.38–3.99
Lower mantle	670–2891	600	55.4	2940	49.2	4.38–5.56
Outer core	2891–5150	169	15.6	1841	30.8	9.90–12.16
Inner core	5150–6371	8	0.7	102	1.7	12.76–13.08
Whole earth		1083	100	5975	100	

[a] After Dziewonski and Anderson (1981).

Thus, for a perfectly elastic material Q is infinite, whereas for a totally dissipative medium Q is zero. A highly attenuative region in the earth is often referred to as a low-Q region.

Equation 4.48 can be written in differential form as

$$Q = -\frac{2\pi E}{T\dfrac{dE}{dt}}$$

$$\frac{dE}{dt} = -\frac{2\pi E}{QT} \tag{4.49}$$

where E is energy, t time and T the period of the seismic wave. Integrating Eq. 4.49 gives

$$E = E_0 e^{-2\pi t/QT} \tag{4.50}$$

where E_0 was the energy of the wave at time t ago. Alternatively, since the amplitude of the wave A is proportional to the square root of its energy E (Sect. 4.2.6), Eq. 4.50 can be written

$$A = A_0 e^{-\pi t/QT}$$

or

$$A = A_0 e^{-\omega t/2Q} \tag{4.51}$$

where ω is the angular frequency and A_0 the amplitude of the wave at time t ago. Performing similar calculations on the spatial form of Eq. 4.48, one obtains

$$A = A_0 e^{-\pi x/Q\lambda} \tag{4.52}$$

where x is the distance travelled by the wave from the point at which it had amplitude A_0, and λ is the wavelength. Thus, after one allows for geometrical spreading, Q can be estimated by taking the ratio of the amplitude of a body wave of a particular frequency at various distances or times. The quality factor determined by using Eqs. 4.51 and 4.52 is for one particular wave type (P or S) only. Q for P-waves, Q_p, is higher than Q for S-waves, Q_s; in general Q_p is approximately twice Q_s.

Estimates of Q in the earth are 10^2 for the inner core, 10^4 for the outer core, 10^3 for the lower mantle, 10^2 for the upper mantle and 200 for the lithosphere.

4.3.4 Seismic Tomography

Much more detailed velocity models of the mantle can be obtained by using tomography, a technique which is similar in method to the whole-body scanning method that medical physicists use. First, the phase and/or group velocities and/or waveforms of surface waves are measured for hundreds of earthquakes and recording stations. To do this, a network of digital long-period seismic stations is required. The scarcity of such stations (compared with the WWSSN) means that, at present, the method is not capable of resolving structures on a horizontal scale of less than about 2000 km in the earth. A best-fitting three-dimensional model of the shear-wave velocity structure of the upper mantle is then constructed; the methods are very complex and are described in Dziewonski and Woodhouse (1987), Woodhouse and Dziewonski (1984) and Nataf et al. (1984). Figure 4.32 shows a shear-wave velocity perturbation model, plotted at increasing depths in the upper mantle. The standard velocity model used here was the PREM from Figure 4.27. It is clear that at shallow depths the continental shields are defined as regions of high velocity, whereas the midocean ridge systems are associated with low velocities. By 350 km depth, the midocean ridge system is (except for the South Pacific region) no longer well delineated, much being underlain by

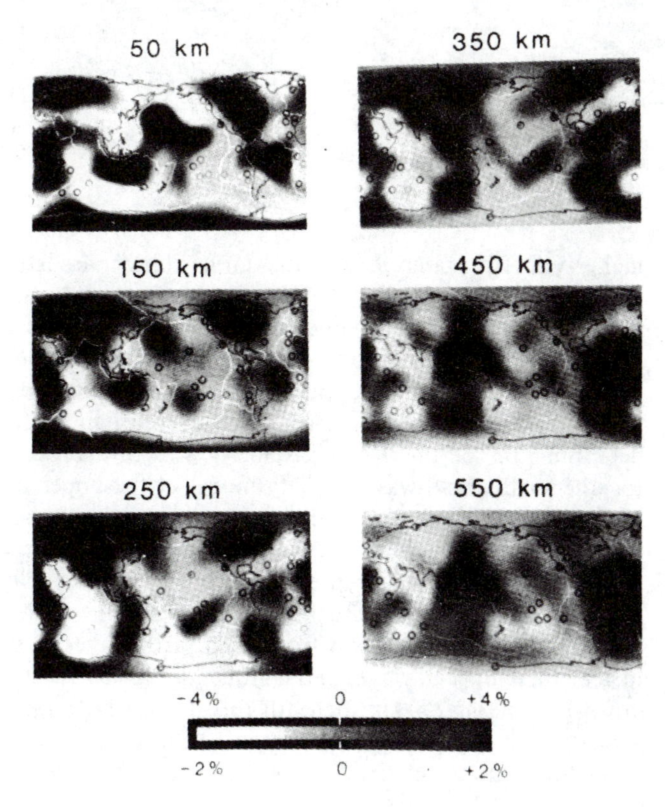

Figure 4.32. Plots of perturbation in shear-wave velocity from a standard whole-earth model (PREM), shown at depths of 50, 150, 250, 350, 450 and 550 km. The graduation in shading indicates increasing perturbation from the standard model. For depths of 50 and 150 km, the scale range is $\pm 4\%$. For all greater depths, the scale range is $\pm 2\%$. (From Woodhouse and Dziewonski 1984.)

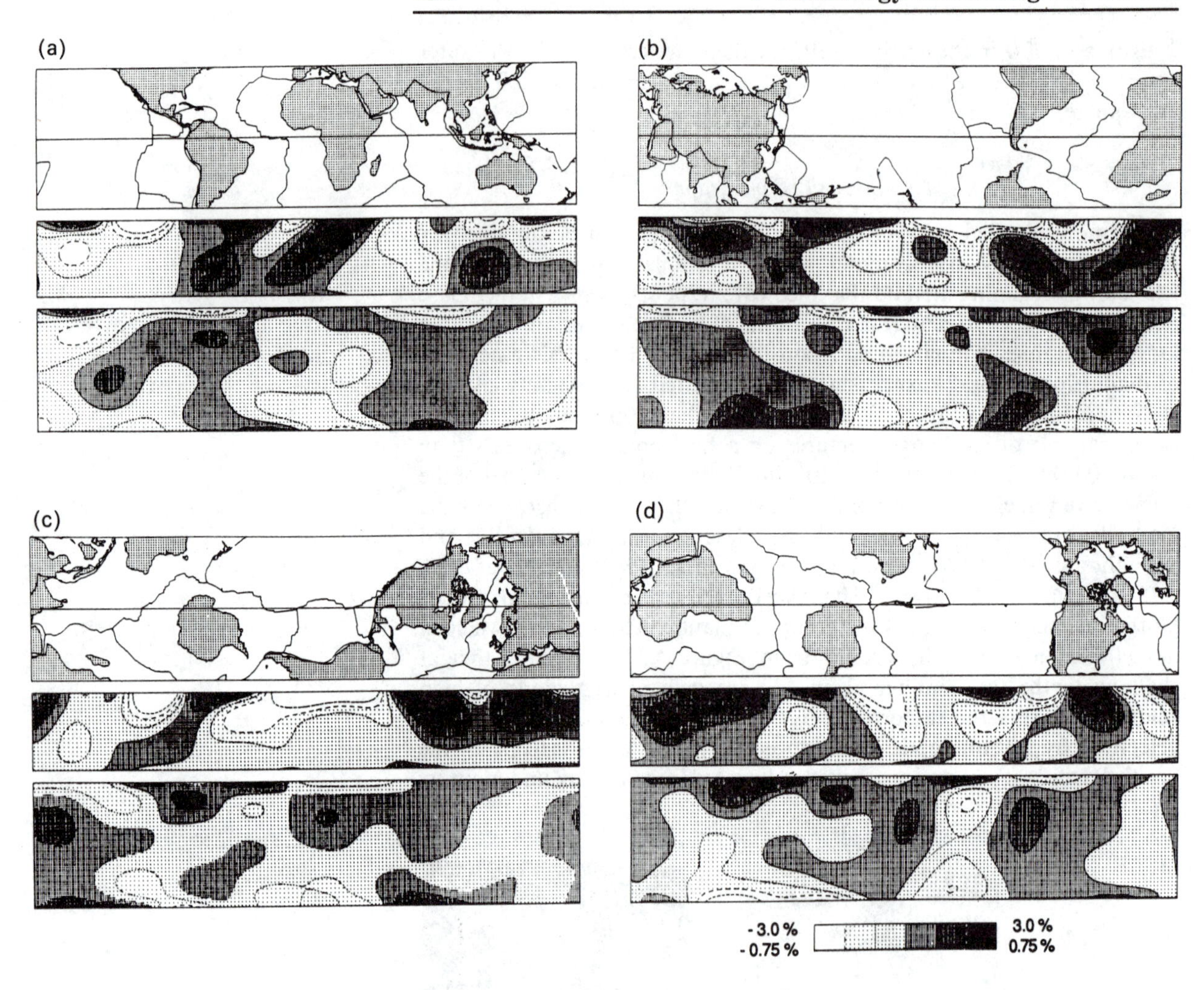

Figure 4.33. Four vertical sections through the mantle showing perturbations from the PREM shear-wave velocity model of the upper mantle above perturbations from the PREM P-wave model of the lower mantle (Dziewonski 1984). Above each pair of sections the map shows the great circle along which the sections are taken. Unfortunately, the velocity resolution for upper and lower mantle is poorest near 670 km depth, which means that the details of the models at this boundary should be treated with some caution. Upper mantle: depth 25–670 km, vertical exaggeration 8:1, scale of perturbation ±3%. Lower mantle: Depth 670–2891 km, vertical exaggeration 4:1, scale of perturbation ±0.75%. (From Woodhouse and Dziewonski 1984.)

higher-velocity material, but the large shields are still underlain by high-velocity material. By 450 km depth, the mantle beneath both the midocean ridges and the subduction zones has, for the most part, high velocities. The perturbations are of the same magnitude as the velocity jumps in the upper-mantle transition zone (220, 400 and 670 km).

Figure 4.33 shows four cross sections through a three-dimensional determination of the P-wave velocity structure of the lower mantle, together with the S-wave velocity model of the upper mantle shown in Figure 4.32. In the lower-mantle determination, travel-time residuals for teleseismic P-waves were used, and the velocity model, expanded in spherical harmonics, is a representation of P-wave velocity perturbations from a standard earth model. This lower-mantle model was based on 500,000 travel-time residuals from 5000 earthquakes. We see again that, in general, midocean ridges and continental shields are the major features down to at least 250 km. Beneath this depth, the correlation of mantle velocity with the surface plate tectonic features decreases. There is some indication that anomalies or roots can cross the boundary between the upper and lower mantle. Figure 4.33 shows how the low-velocity anomaly associated with the Mid-Atlantic Ridge separates the high-velocity roots of the South American and African shields, which link at depth and then appear to continue into the lower mantle.

The velocity of seismic waves through olivine (which is a major constituent of the mantle) is greater when parallel to the *a* axis of the olivine crystal than when perpendicular to the *a* axis. Such dependence of seismic velocity on direction is called *velocity anisotropy* (i.e., the material is not perfectly isotropic). Anisotropy is not the same as *inhomogeneity*, which refers to a localized change in physical parameters within a larger medium. Any flow in the mantle tends to align the olivine crystals with their *a* axes parallel to the direction of flow. For this reason, measurement of anisotropy in the mantle can indicate whether any flow is vertical or horizontal. Plots of the difference between vertically and horizontally polarized shear-wave velocities, as determined by tomography, do indeed suggest that in the upper mantle there is horizontal flow beneath the shields and vertical flow beneath the midocean ridges and subduction zones.

4.3.5 Composition of the Earth

Our knowledge of the composition of the deep interior of the earth is largely constrained by the seismic and density models. Chemistry, of course, is also important, but unfortunately we do not have in situ measurements. Although this is a chapter on seismology, it is instructive and useful at this point to pause and consider the gross composition of the crust, mantle and core.

The continental crust varies greatly in the variety of its igneous, sedimentary and metamorphic rocks. However, on average, the continental crust is silica-rich and, very loosely speaking, of granitoid composition (see Sect. 9.1.3). In contrast, the oceanic crust is basaltic and richer in mafic (Mg, Fe-rich) minerals (see Sect. 8.1 and 9.1.3).

The compositions of the mantle and core are much more difficult to determine because no one has yet drilled that deeply. Our chemical knowledge of the deep earth has to be inferred from the chemistry of the volcanic and intrusive rocks derived from liquids which originated in the mantle, from structurally emplaced fragments of mantle, from nodules brought up during volcanism and from geochemical models of the various seismic and density models discussed in the previous sections.

This lack of direct evidence might suggest that the mantle composition is pretty much unknown, but geochemistry is a sophisticated and powerful branch of earth science and has developed many techniques which use the compositions of rocks available to us on the surface to model the composition at depth. In addition, experiments at high temperatures and pressures have enabled the behaviour of minerals thought to exist in the deep mantle to be studied in some detail. There are many compositional models of the upper mantle, some of which are more popular than others. These models include imaginary rocks such as pyrolite, which is a mixture of basalt and residual mantle material. The main constituent of the mantle is believed to be magnesian silicate, mostly in the form of olivine. The core is believed to be iron-rich, and the core-mantle boundary is thought to be a very major compositional boundary.

Olivine in the mantle lies between two end members: forsterite (Fo), which is Mg_2SiO_4, and fayalite (Fa), which is Fe_2SiO_4. Normal mantle olivine is very forsteritic, probably in the range Fo_{91}–Fo_{94}, where 91 and 94 represent percentages of Mg in $(Mg, Fe)_2SiO_4$. Other trace components of mantle olivine include nickel (Ni) and chromium (Cr). The other major mantle minerals include orthopyroxene and clinopyroxene. Orthopyroxene (Opx) varies between the end members enstatite ($MgSiO_3$) and ferrosilite ($FeSiO_3$), with about the same Mg to Fe ratio (94–91) in the mantle as olivine. Clinopyroxene (Cpx) also contains calcium, as $Ca(Mg, Fe)Si_2O_6$. Figure 4.34 illustrates the relationships among the common ultramafic rocks.

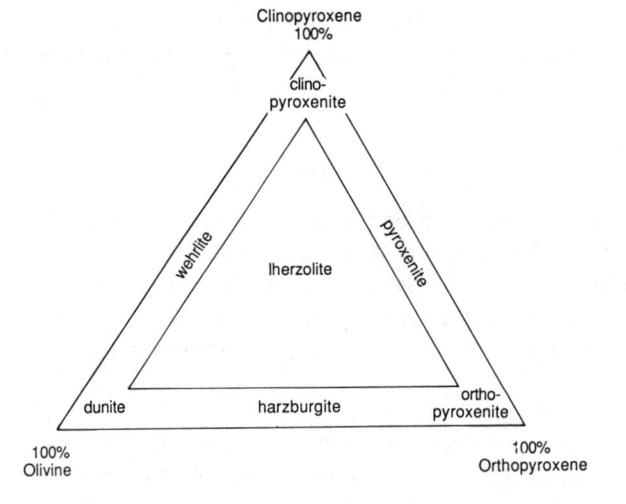

Figure 4.34. Common ultramafic rocks. The three outer corners of the triangle represent 100% compositions: Dunite is an ultramafic rock with close to 100% olivine and almost no clinopyroxene or orthopyroxene; wehrlite is made up of approximately equal parts of olivine and clinopyroxene but no orthopyroxene. The interior triangle is the 90% contour. The classification of igneous rocks is discussed in Section 8.1.

Experimental work on olivine has shown that it undergoes phase changes to denser structures at pressures equivalent to depths of 390–450 km and about 700 km. A phase change does not involve a change in chemical composition but a reorganization of the atoms into a different crystalline structure. With increasing pressure, these phase changes involve closer packing of the atoms into denser structures. At 390–450 km, olivine changes to a spinel structure with a resultant 10% density increase (pyroxene also changes to a garnet structure at this depth). This change of olivine to spinel structure is accompanied by a release of heat (i.e., the reaction is *exothermic*). At about 700 km, the spinel structure undergoes another change to minerals with a postspinel structure, such as perovskite and magnesium oxide. These reactions, which also involve density increases of about 10%, are thought to be *endothermic* (heat is absorbed during the reaction). The silicon atoms in olivine and the spinel structure are both surrounded by four oxygen atoms, but in the perovskite structure the silicon is surrounded by six oxygen atoms. Figure 4.35 shows these phase changes alongside the S-wave velocity profile of the upper mantle. The depths at which the phase changes take place coincide with those at which the seismic velocity increases more rapidly.

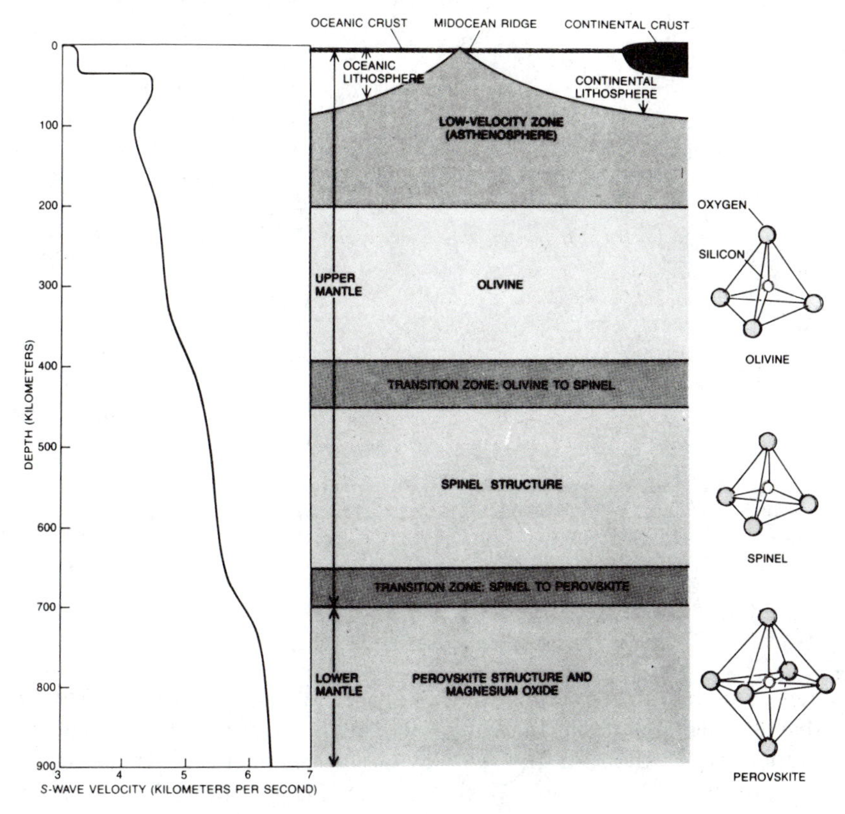

Figure 4.35. The S-wave velocity profile of the upper mantle compared with the phases and transition zones for olivine in the upper mantle. (From The Earth's Mantle, McKenzie. Copyright (©) 1983 by Scientific American, Inc. All rights reserved.)

The situation for the core is far more difficult: No one has ever had a sample of the core to analyse. The closest we can get to sampling the core is to consider the abundance of elements in the sun and in meteorites. The earth is believed to have formed from an accretion of meteoritic material. Meteorites are classified into two types: *stony* and *iron* (discussed in Sects. 6.8 and 6.10). The stony meteorites are similar to the mantle in composition, and perhaps the iron meteorites are similar to the core. If so, the core should be rich in iron with a small proportion of nickel. The major problem with theories of core composition is that they depend on theories of the origin of the earth and its chemical and thermal evolution, which are also poorly understood. Solar abundances of iron are slightly higher than that in stony meteorites. If the solar model is taken, the lower mantle may have as much as 15% FeO. The core is very iron rich and may in bulk be roughly Fe_2O in composition.

More direct evidence for the composition of the core comes from its seismic velocity and density structure. Pressures appropriate for the core can be momentarily attained in experiments using shock waves and diamond anvils. Thus, laboratory measurements can be made on test samples at core pressures. When corrections for temperature are made, such laboratory velocity measurements can be compared with seismic models. Figure 4.36 shows a plot for a number of metals of $\sqrt{\Phi} = \sqrt{K/\rho} = \sqrt{\alpha^2 - \frac{4}{3}\beta^2}$ (Eq. 4.47) against density as obtained from shock-wave experiments. The range of values appropriate for the mantle

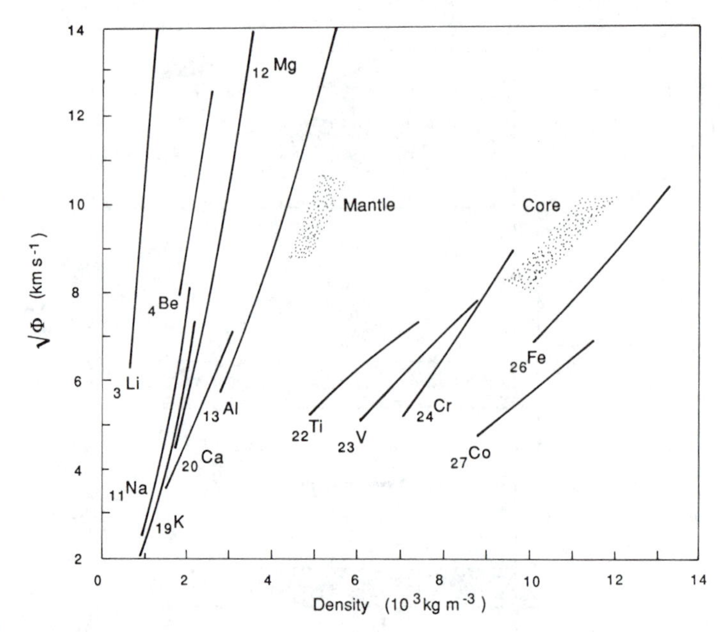

Figure 4.36. Seismic parameter ϕ, $\sqrt{\phi} = \sqrt{K/\rho} = \sqrt{\alpha^2 - 4/3\beta^2}$ plotted against density for metals. These values were obtained from shock-wave experiments. The shaded regions show the range of values for the mantle and core as given by the seismic models. (After Birch 1968.)

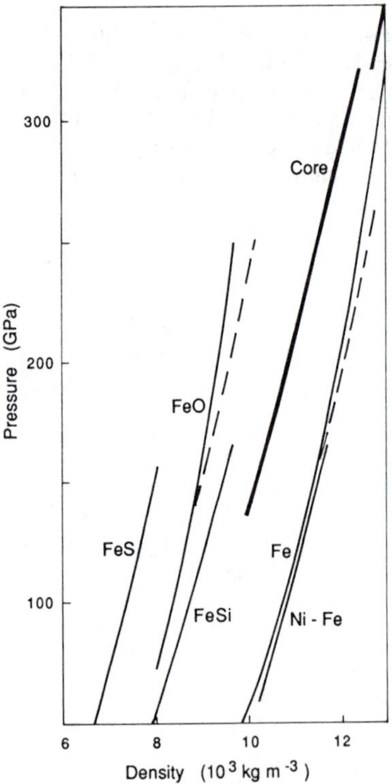

Figure 4.37. Pressure and density as measured in shock-wave experiments for iron and the iron compounds which may be present in the core. The heavy line shows values for the core calculated from a seismic velocity model. (Based on Jeanloz and Ahrens 1980 and Jeanloz 1983.)

and core, indicated by the seismic velocity and density models, indicate that although magnesium and aluminum are likely candidates for a major proportion of the mantle, such low-atomic-number metals are quite inappropriate for the core. All the evidence on the properties of iron at high temperatures and pressures points unequivocably to a core which is predominantly composed of iron. The data for the inner core indicate that it may well be virtually pure iron. The outer core is probably an iron alloy: iron with a small percentage, 10% by weight, of lighter elements. Amongst the favoured candidates for the minor alloying element(s) are oxygen, sulphur, silicon, and nickel. Figure 4.37 is a plot of density against pressure (obtained from shock-wave experiments) for pure iron (molten and solid) and possible iron compounds compared with the in situ seismic values for the core. This plot shows that the outer core cannot be composed of either pure iron or the nickel–iron compound found in meteorites: Both of these materials are too dense. However, either are possible candidates for the inner core. Each alloying element has its advantages and disadvantages, with oxygen and sulphur the strongest candidates.

At low temperatures, FeO is nonmetallic and forms an immiscible liquid with Fe. In the past this led to doubts about the presence of oxygen in the core. High-pressure and temperature experiments on iron oxide, however, have shown that it becomes metallic at pressures greater than 70 GPa and temperatures greater than 1000 K. This means that, in the core, oxygen can alloy with iron. Further experiments at temperatures and pressures appropriate for the core–mantle boundary have shown that liquid iron and iron alloys react vigorously with solid oxides and solid silicates. These experiments suggest that oxygen is very probably a constituent of the core. They also suggest that an iron-rich core would react chemically with the silicate mantle and that oxygen in the core may be derived from the mantle. This may well be the explanation for the seismic complexity of the core–mantle boundary (Sect. 4.3.1): The mantle and core are not in chemical equilibrium. Most probably the core contains oxygen and has the gross overall composition of Fe_2O. However, the seismic and density models are not yet accurate enough to be certain.

4.4 Refraction Seismology

4.4.1 Refraction Experiments

Earthquake seismology is able to reveal the broad details of the velocity and density structure of the earth, but to look in greater detail at the structure of the crust and uppermost mantle, especially in those regions well away from active seismic zones, it becomes impracticable to use earthquakes as the energy source. In refraction seismology, portable seismometers are deployed in the regions to be studied, and explosives are used as the energy source. Figure 4.38 shows some examples of the positioning of sources and receivers for such refraction experiments.

Refraction seismology on land is expensive and messy because tons of explosive are needed to record energy at distances of 300 km or so, the distance necessary to determine a continental crust and uppermost mantle structure. It is necessary to have a seismometer at least every 5 km along each line, and preferably much closer together. Because the instruments are also expensive, this can entail detonating one shot at the shotpoint, moving the seismometers along the line to the next positions and then detonating another shot at the shotpoint, and so on. Thus, large refraction experiments on land are usually carried out cooperatively by several universities or institutions so that enough people and recording instruments are available.

The situation at sea is different. In marine work, the source and receiver locations (Fig. 4.38) are usually exchanged so that a small number of seismometers or hydrophones (pressure recorders) are deployed and then a large number of shots are fired at ever increasing distances. Seismometers for use at sea are very expensive because they are necessarily enclosed in waterproof pressure vessels, have a homing detection device to ensure their subsequent recovery and, if laid on the seabed, must be equipped with a release device. Many fewer people are needed for an experiment, however, because the research ship can steam along, firing charges as it goes (no drilling needed here). At shorter ranges, an air gun (an underwater source which discharges a volume of very high pressure air) can be used as an energy source. Because the oceanic crust is much thinner than the continental crust, marine refraction lines need to be only 50 km or so in length to determine the crustal structure and upper mantle velocity.

Various textbooks on exploration geophysics (e.g., Telford et al. 1990 and Dobrin and Savit 1988) give the details of the field procedures and the corrections to be applied to seismic refraction data. These will not be discussed further here.

The coding used for crustal and uppermost mantle phases is

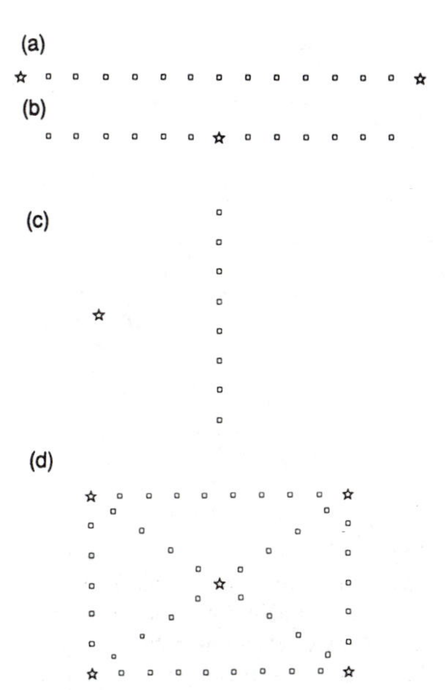

Figure 4.38. Examples of some layouts of energy sources (usually explosives) and seismometers used in refraction experiments: (a) Reversed refraction line. (b) Split profile. (c) Fan shooting. (d) Refraction experiment designed to determine the three-dimensional structure, including velocity anisotropy. For refraction experiments on land: ☆ is the explosive source and □ the seismometer locations. For refraction experiments at sea, the source and receiver locations are exchanged: ☆ is the seismometer/hydrophone location and □ the explosive sources.

P_g	P-wave through the upper continental crust*
S_g	S-wave through the upper continental crust
P_mP	P-wave reflection from the Moho
S_mS	S-wave reflection from the Moho
P_n	upper mantle P-head wave
S_n	upper mantle S-head wave

4.4.2 A Two-Layered Model

Let us assume that the crust beneath a refraction line consists of two horizontal layers, with distinct and constant P-wave velocities α_1 and α_2 such that $\alpha_2 > \alpha_1$ (Fig. 4.39). Energy from the source can then reach the seismometer by a variety of paths: directly through the top layer, by reflection from the interface between the two layers, by multiple reflections within the top layer or by travelling along the interface as a critically refracted wave or 'head' wave. The head wave, which is often called a refraction or refracted wave, has a travel time corresponding to a ray which travelled down to the interface at the *critical angle* i_c,† then along the interface with the velocity of the lower layer and then back up to the seismometer, again at the critical angle. (Head waves are second-order waves not predicted by geometrical ray theory.)

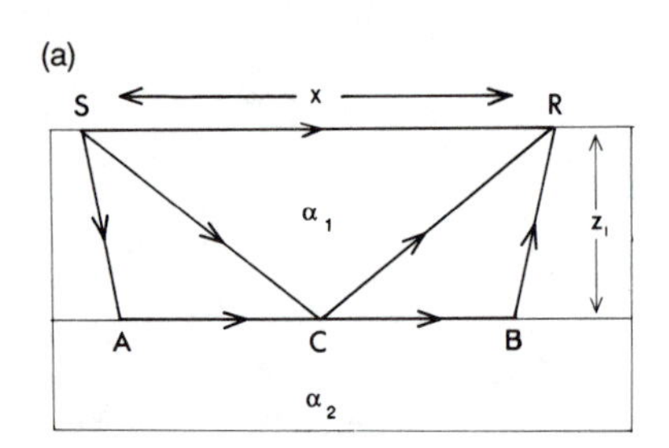

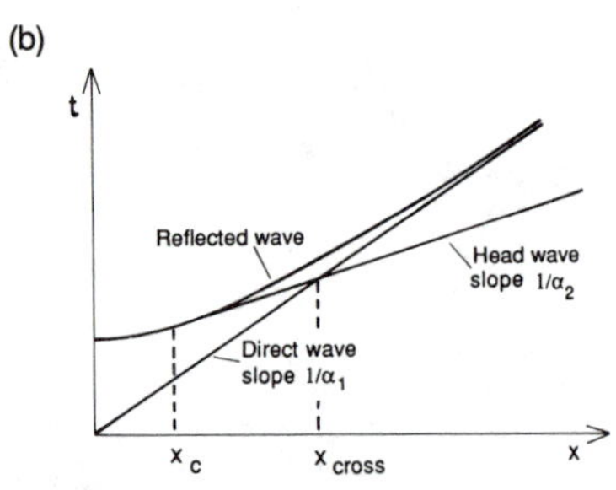

Figure 4.39. (a) Ray paths for seismic energy travelling from source S to receiver R in a simple two-layer model. The P-wave velocity is α_1 for the upper layer and α_2 for the lower layer, where $\alpha_2 > \alpha_1$. The direct wave takes ray path *SR* in the upper layer. The reflected wave takes ray path *SCR* in the upper layer. The head wave takes ray path *SABR*. (b) Travel-time–distance plots for the model in (a).

Direct Wave The time taken for energy to reach the receiver directly through the top layer is simply

$$t = \frac{x}{\alpha_1} \tag{4.53}$$

This is the equation of a straight line when time is plotted against distance.

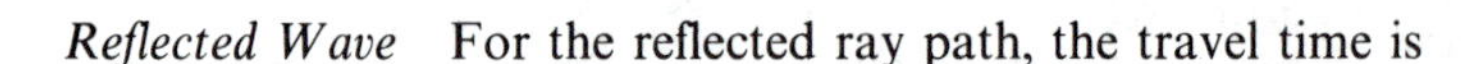

Reflected Wave For the reflected ray path, the travel time is

$$t = \frac{SC}{\alpha_1} + \frac{CR}{\alpha_1} \tag{4.54}$$

where *SC* and *CR* are the lengths of the ray paths shown in Figure 4.39. Since the top layer is uniform, the reflection point *C* is midway between source *S* and seismometer *R*. Using Pythagoras' theorem, we can write

$$SC = CR = \sqrt{z_1^2 + \frac{x^2}{4}} \tag{4.55}$$

where z_1 is the thickness of the top layer.

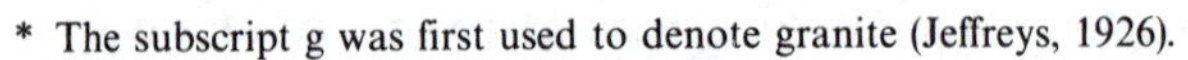

* The subscript g was first used to denote granite (Jeffreys, 1926).

† Snell's law for the interface between two media where *i* is the angle of incidence and *r* the angle of refraction is

$$\frac{\sin i}{\sin r} = \frac{\alpha_1}{\alpha_2}$$

When *r* is 90°, the angle of incidence $i = i_c = \sin^{-1}(\alpha_1/\alpha_2)$ is called the *critical angle*. For angles of incidence greater than the critical angle, no energy is refracted into the second layer.

The travel time t can then be expressed in terms of distance by substituting Eq. 4.55 into Eq. 4.54:

$$t = \frac{2}{\alpha_1}\sqrt{z_1^2 + \frac{x^2}{4}} \tag{4.56}$$

or

$$\alpha_1^2 t^2 = 4z_1^2 + x^2$$

which is the equation of a hyperbola.

Head Wave or Refracted Wave The travel time for the head wave is

$$t = \frac{SA}{\alpha_1} + \frac{AB}{\alpha_2} + \frac{BR}{\alpha_1} \tag{4.57}$$

By symmetry and using Snell's law ($\sin i_c = \alpha_1/\alpha_2$),

$$\begin{aligned} SA &= BR \\ &= \frac{z_1}{\cos i_c} \end{aligned} \tag{4.58}$$

and

$$AB = x - 2z_1 \tan i_c \tag{4.59}$$

Substituting these expressions into Eq. 4.57 gives

$$\begin{aligned} t &= \frac{2z_1}{\alpha_1 \cos i_c} + \frac{x}{\alpha_2} - \frac{2z_1}{\alpha_2}\tan i_c \\ &= \frac{2z_1}{\alpha_1 \cos i_c}\left[1 - \frac{\alpha_1}{\alpha_2}\sin i_c\right] + \frac{x}{\alpha_2} \\ &= \frac{2z_1}{\alpha_1}\cos i_c + \frac{x}{\alpha_2} \\ &= \frac{2z_1}{\alpha_1}\sqrt{1 - \frac{\alpha_1^2}{\alpha_2^2}} + \frac{x}{\alpha_2} \end{aligned} \tag{4.60}$$

On a time–distance (t:x) graph this is the equation of a straight line with slope $1/\alpha_2$ and intercept on the time axis $(2z_1/\alpha_1)\sqrt{1 - \alpha_1^2/\alpha_2^2}$.

Critical Distance Notice that the shortest range at which a head wave can be recorded is x_c, the critical distance, where

$$\begin{aligned} x_c &= 2z_1 \tan i_c \\ &= \frac{2z_1\alpha_1}{\sqrt{\alpha_2^2 - \alpha_1^2}} \end{aligned} \tag{4.61}$$

At ranges shorter than this, there is no head wave, only the precritical reflection from the interface. At the critical distance x_c, the travel time for the head wave is the same as the travel time for the reflected wave since at this range the distance AB (Fig. 4.39) is zero, and these two ray paths are identical. The slope of the reflection hyperbola (Eq. 4.56) is

$$\frac{dt}{dx} = \frac{d}{dx}\left(\frac{1}{\alpha_1}\sqrt{4z_1^2 + x^2}\right)$$

$$= \frac{x}{\alpha_1}\frac{1}{\sqrt{4z_1^2 + x^2}} \tag{4.62}$$

At the critical distance ($x_c = 2z_1 \tan i_c$), this slope is

$$\left.\frac{dt}{dx}\right|_{x=x_c} = \frac{2z_1 \tan i_c}{\alpha_1}\frac{1}{\sqrt{4z_1^2 + 4z_1^2 \tan^2 i_c}}$$

$$= \frac{\sin i_c}{\alpha_1}$$

$$= \frac{1}{\alpha_2} \tag{4.63}$$

So, at the critical distance, the head wave is the tangent to the reflection hyperbola.

Crossover Distance The crossover distance x_{cross} is the range at which the direct wave and the head wave have the same travel time. It can be obtained from Eqs. 4.53 and 4.60:

$$\frac{x_{\text{cross}}}{\alpha_1} = \frac{x_{\text{cross}}}{\alpha_2} + \frac{2z_1}{\alpha_1}\sqrt{1 - \frac{\alpha_1^2}{\alpha_2^2}} \tag{4.64}$$

Rearranging, we obtain

$$x_{\text{cross}} = 2z_1\sqrt{\frac{\alpha_2 + \alpha_1}{\alpha_2 - \alpha_1}} \tag{4.65}$$

Time–Distance Graph Figure 4.39 shows the travel-time versus distance graphs for this simple two-layer model. At short ranges, the first arrival is the direct wave, followed by the reflected wave; and at long ranges, the first arrival is the head wave, followed by the direct wave, followed by the reflected wave.

To determine an initial velocity–depth structure from a refraction experiment, it is necessary to display the data on a time–distance graph. If we had a record section or first-arrival travel times from a seismic experiment shot over this model, we would determine α_1, α_2 and z_1, in that order, as follows:

1. α_1 is determined as the inverse of the slope of the direct wave time–distance plot for distances less than x_{cross}.
2. α_2 is determined as the inverse of the slope of the head wave time–distance plot for distances greater than x_{cross}.
3. z_1 is determined from the intercept of the head wave time–distance line on the t axis; z_1 could also be calculated from the crossover distance (Eq. 4.65), but this distance is not usually well enough defined to give an accurate value for z_1.

Note that although all of the equations have used P-wave velocities α_1

and α_2, they all hold for shear waves as well (velocities would then be β_1 and β_2). Shear waves are not usually studied in detail in crustal seismology, partly because of the extra expense of recording horizontal components as well as the vertical component but mainly because they are second arrivals; it is often very difficult to pick travel times accurately enough to make a good shear wave velocity–depth model.

4.4.3 A Multilayered Model

The travel times for a model consisting of *n* uniform horizontal layers of thickness z_j and P-wave velocity α_j are determined in exactly the same way as for the two-layer model. The only extra matter to be remembered is that the rays bend according to Snell's law as they cross each interface [$\sin(i)/\alpha$ is a constant along each ray]. The travel time for a wave refracted along the top of the *m*th layer is

$$t = \sum_{j=1}^{m-1} \left(\frac{2z_j}{\alpha_j} \sqrt{1 - \frac{\alpha_j^2}{\alpha_m^2}} \right) + \frac{x}{\alpha_m} \tag{4.66}$$

provided $\alpha_j < \alpha_m$.

It is possible to determine correctly the velocities and thickness of all layers in such an *n*-layer structure only by using the first-arrival travel times *if* refracted arrivals from each interface are first arrivals over some distance range. If this is not the case and the refractions from a layer are always second arrivals, this layer is a *hidden layer*, and first arrivals alone do not give the correct structure (Fig. 4.40a, b). Another structure which cannot be determined uniquely by using first arrivals alone is illustrated in Figure 4.40c, d. A low-velocity layer cannot give rise to any head wave at its upper surface because the refracted ray bends towards the normal ($\alpha_{j-1} > \alpha_j$). The only indication that a low-velocity layer is present comes

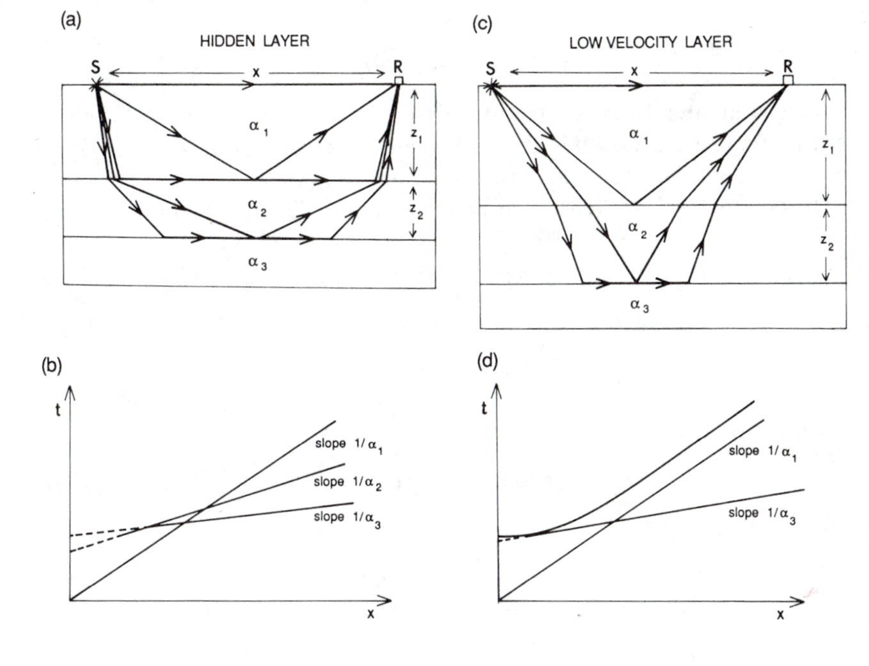

Figure 4.40. (a) Three-layer model in which $\alpha_3 > \alpha_2 > \alpha_1$. (b) Travel-time–distance plot for the model in (a). Only the direct wave (slope $1/\alpha_1$) and the two head waves (slopes $1/\alpha_2$ and $1/\alpha_3$) are shown. For distances less than the critical distance, the head wave arrivals are extended back to the time axis by dashed lines. For this particular structure, the head wave from the second layer is always a second arrival; so this layer cannot be detected using first-arrival times alone. It is a hidden layer. (c) Three-layer model in which $\alpha_3 > \alpha_1 > \alpha_2$. (d) Travel-time–distance plot for the model in (c). Notice that the reflection from the interface at the base of the low-velocity layer is delayed with respect to the direct wave. When there is no low-velocity zone, this reflection and the direct wave are asymptotic (compare with Fig. 4.39b).

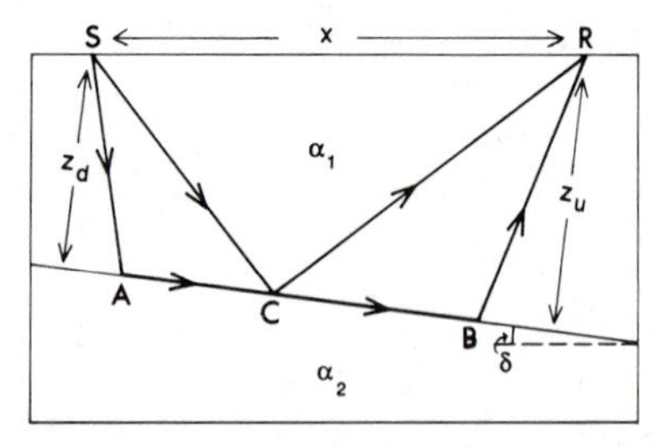

Figure 4.41. Ray paths for seismic energy travelling from source S to receiver R in a two-layer model in which the interface between the two layers dips at an angle δ. P-Wave velocity for the upper layer is α_1, and for the lower layer is α_2, where $\alpha_2 > \alpha_1$.

from the reflections from its upper surface, and the reflections and head wave from its lower surface. The critical distance for this lower interface is less than anticipated, and arrivals are greater in amplitude due to the large velocity contrast occurring there. This is exactly the same as the shadow-zone effect illustrated for the spherical earth in Figure 4.28.

Dipping Layers The real earth is far more complex than Figure 4.39a. If we allow the first interface to dip at an angle δ (Fig. 4.41) instead of being horizontal ($\delta = 0$) as in Figure 4.39, then the travel time for the head wave is

$$t_d = \frac{2z_d}{\alpha_1}\sqrt{1 - \frac{\alpha_1^2}{\alpha_2^2}} + \frac{x}{\alpha_1}\sin(i_c + \delta) \tag{4.67}$$

where z_d is the perpendicular distance from the shotpoint S to the interface. This is called shooting *down-dip*. Equation 4.67 is the equation of a straight line, but in this case the apparent head wave velocity is α_d:

$$\alpha_d = \frac{\alpha_1}{\sin(i_c + \delta)} \tag{4.68}$$

α_d is less than $\alpha_2 (= \alpha_1/\sin i_c)$. The fact that the interface is dipping cannot be determined from this time–distance graph alone. However, if the refraction line is 'reversed', that is, the shotpoint is placed at R and the receiver positions from R towards S, the travel times are

$$t_u = \frac{2z_u}{\alpha_1}\sqrt{1 - \frac{\alpha_1^2}{\alpha_2^2}} + \frac{x}{\alpha_1}\sin(i_c - \delta) \tag{4.69}$$

where z_u is the perpendicular distance from the new shotpoint to the interface. This is called shooting *up-dip*. In this case, the apparent velocity is α_u:

$$\alpha_u = \frac{\alpha_1}{\sin(i_c - \delta)} \tag{4.70}$$

which is greater than the true velocity of the lower layer α_2. When a refraction line is reversed in this way, the true velocity of the dipping layer α_2 can be determined from Eqs. 4.68 and 4.70.

In one method, we rearrange Eqs. 4.68 and 4.70 and obtain

$$\sin(i_c + \delta) = \frac{\alpha_1}{\alpha_d}$$

and

$$\sin(i_c - \delta) = \frac{\alpha_1}{\alpha_u}$$

Hence

$$i_c + \delta = \sin^{-1}\left(\frac{\alpha_1}{\alpha_d}\right) \tag{4.71}$$

and

$$i_c - \delta = \sin^{-1}\left(\frac{\alpha_1}{\alpha_u}\right) \tag{4.72}$$

Therefore, i_c and δ are easily obtained by adding and subtracting Eqs. 4.71 and 4.72:

$$i_c = \frac{1}{2}\left[\sin^{-1}\left(\frac{\alpha_1}{\alpha_d}\right) + \sin^{-1}\left(\frac{\alpha_1}{\alpha_u}\right)\right] \tag{4.73}$$

$$\delta = \frac{1}{2}\left[\sin^{-1}\left(\frac{\alpha_1}{\alpha_d}\right) - \sin^{-1}\left(\frac{\alpha_1}{\alpha_u}\right)\right] \tag{4.74}$$

The velocity α_2 is known when i_c is determined since i_c has been defined as the angle for which $\sin i_c = \alpha_1/\alpha_2$.

An alternative method which does not involve the use of inverse sines can be used in situations in which the dip angle δ is small enough for the approximation $\sin\delta = \delta$ and $\cos\delta = 1$ to be made (δ measured in radians). In this case, expanding $\sin(i_c + \delta)$ and $\sin(i_c - \delta)$ in Eqs. 4.68 and 4.70, we obtain

$$\begin{aligned} \frac{\alpha_1}{\alpha_d} &= \sin(i_c + \delta) \\ &= \sin i_c \cos\delta + \cos i_c \sin\delta \\ &= \sin i_c + \delta\cos i_c \end{aligned} \tag{4.75}$$

and

$$\begin{aligned} \frac{\alpha_1}{\alpha_u} &= \sin(i_c - \delta) \\ &= \sin i_c \cos\delta - \cos i_c \sin\delta \\ &= \sin i_c - \delta\cos i_c \end{aligned} \tag{4.76}$$

Now, i_c and δ can be obtained by adding and subtracting Eqs. 4.75 and 4.76:

$$\begin{aligned} \sin i_c &= \frac{1}{2}\left[\frac{\alpha_1}{\alpha_d} + \frac{\alpha_1}{\alpha_u}\right] \\ &= \frac{\alpha_1}{2}\left[\frac{1}{\alpha_d} + \frac{1}{\alpha_u}\right] \end{aligned} \tag{4.77}$$

$$\begin{aligned} \delta &= \frac{1}{2\cos i_c}\left[\frac{\alpha_1}{\alpha_d} - \frac{\alpha_1}{\alpha_u}\right] \\ &= \frac{\alpha_1}{2\cos i_c}\left[\frac{1}{\alpha_d} - \frac{1}{\alpha_u}\right] \end{aligned} \tag{4.78}$$

for small δ, α_2 is therefore given by

$$\begin{aligned} \frac{1}{\alpha_2} &= \frac{\sin i_c}{\alpha_1} \\ &= \frac{1}{2}\left[\frac{1}{\alpha_d} + \frac{1}{\alpha_u}\right] \end{aligned} \tag{4.79}$$

If the dipping interface is not the first but a deeper interface, the velocities and thicknesses are determined exactly as above except that contributions from the overlying layers must be included in Eqs. 4.67 and 4.69. (The algebra can get somewhat tedious; see Slotnick 1959 for detailed derivations.)

4.4.4 Seismic Record Sections

When seismic refraction travel-time data are plotted on a time and distance graph, straight line segments can be fitted to the points, and then velocities and layer thickness can be calculated. However, as has been mentioned, many structures are most unlikely to be detected by such a simple travel-time interpretation alone.

A good way to display all the data available is to plot a *record section* such as those shown in Figures 4.15, 4.46, 8.6, 8.7, 8.22, 9.16 and 9.20. The recording (*record*) from each seismometer is plotted at its appropriate range on a time–distance plot. The great advantage of plotting the data on a record section is that all of the amplitude and travel-time information is displayed together. Phases can be easily correlated from trace to trace, which means that second arrivals and reflections, which often do not have a clear starting time and so would not necessarily be identified on a single record, stand out clearly. Hidden layers and low-velocity zones, which cannot be detected from the first-arrival travel times (Fig. 4.40), should be resolved when a record section is used in the interpretation of a refraction line.

To avoid having to plot record sections on exceedingly large pieces of paper or with a very small time scale, *reduced record sections* are used. To do this, the time (t) axis of the plot is replaced by a reduced time ($t - x/v$) axis, where x is the offset distance (horizontal axis) and v is the *reduction velocity*. This means that phases arriving with a velocity v line up horizontally on the reduced time–distance plot, and that phases with velocity less than v have a positive slope and phases with velocity greater than v a negative slope. Reduced record sections greatly facilitate correlation of phases. Refraction data are generally plotted with a reduction velocity appropriate for the first-arrival velocities: Long-range crustal/upper mantle refraction lines would use a reduction velocity of perhaps 6.6 km s^{-1} or 8 km s^{-1}, whereas the example of Figure 4.15, for which the first arrivals come from the deep mantle, is plotted with a reduction velocity of 13.9 km s^{-1}.

4.4.5 Amplitudes

Reflection and Refraction Coefficients Although we can determine a velocity–depth structure by using only the first-arrival travel times and the offset distances, much more information clearly is available in seismograms. Using amplitude information and the travel times of secondary phases can help immensely to minimize errors and to reduce the number of possible structures that could fit the first-arrival time–distance data. The examples of Figure 4.40 show common situations for which the first-arrival times do not tell the whole story.

To be able to use the information in the amplitudes and waveforms of seismic waves, it is necessary as a start to determine the amplitudes of reflected and refracted waves at interfaces. Figure 4.42 shows the rays reflected and transmitted at an interface between two media. Consider first the case of an incident P-wave, which gives rise to reflected and transmitted P-waves but can also generate S-waves when it strikes the

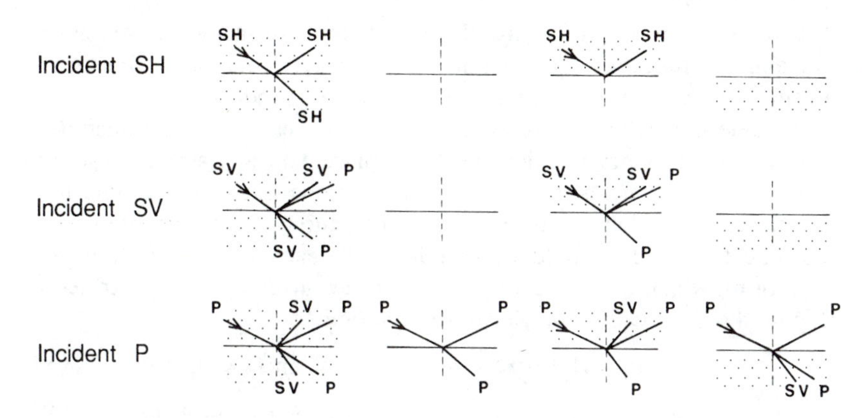

Figure 4.42. Reflection and transmission (refraction) of P-waves, vertically polarized shear waves (SV) and horizontally polarized shear waves (SH) incident from above on the boundary between two media (solid, shaded; liquid, white). Remember that S-waves cannot propagate in a liquid. Notice that incident P-waves can only give rise to reflected and transmitted P- and SV-waves and that incident SH-waves can only give rise to reflected and transmitted SH-waves. (After Båth 1979.)

interface. Since the particle motion for P-waves is longitudinal, there is no motion in the direction perpendicular to the plane of Figure 4.43. This means that the S-waves generated by the incident P-wave cannot be SH-waves and must be SV-waves. Similarly, an incident SV-wave can generate reflected and refracted SV- and P-waves, but in the case of an incident SH-wave, only SH-waves can be transmitted and reflected.

The angles made by the various rays with the normal to the interface are defined by Snell's law. The constant along each ray path sin i/velocity is often called p, the *ray parameter*. For the case of the incident P-wave (Fig. 4.43), the angles for the reflected and transmitted P- and SV-waves are therefore determined from

$$\frac{\sin i}{\alpha_1} = \frac{\sin e_1}{\alpha_1} = \frac{\sin f_1}{\beta_1} = \frac{\sin e_2}{\alpha_2} = \frac{\sin f_2}{\beta_2} = p \tag{4.80}$$

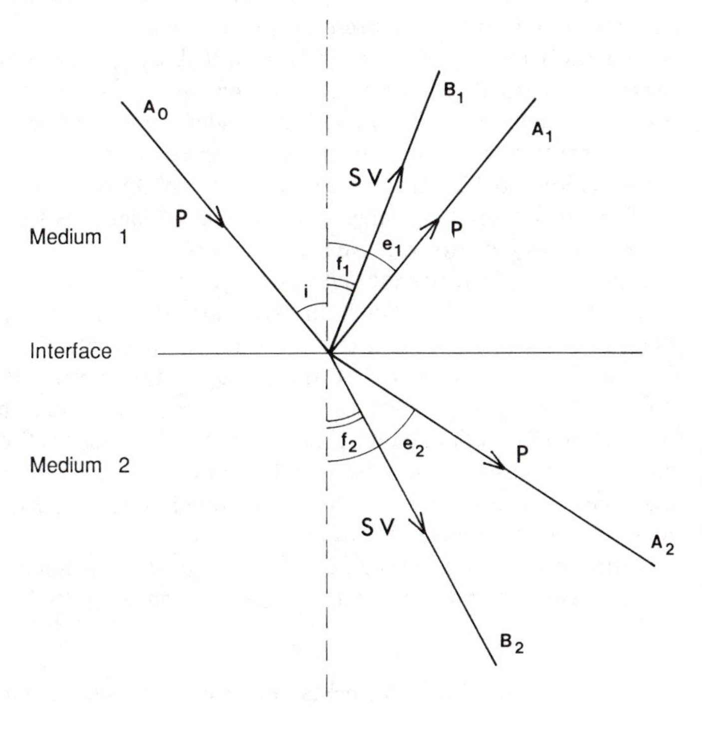

Figure 4.43. Waves generated at an interface between two elastic media by an incident P-wave. The incident P-wave has amplitude A_0 and angle of incidence i. The reflected P- and SV-waves have angles of reflection e_1 and f_1 and amplitude A_1 and B_1, respectively. The transmitted P- and SV-waves have angles of refraction e_2 and f_2 and amplitude A_2 and B_2, respectively.

Clearly, for P-waves in the first layer, the angle of incidence is equal to the angle of reflection ($i = e_1$), and the other rays bend according to the seismic velocities of the media and the angle of incidence.

To determine the relative amplitudes of the reflected and transmitted waves, it is necessary to calculate the displacement and stresses resulting from the wave fields and to equate these values at the interface. Displacement and stress must both be continuous across the interface or else the two media would move relative to each other. The following equations, which relate the amplitudes of the various waves illustrated in Figure 4.43, are called the *Zoeppritz equations*:

$$A_1 \cos e_1 - B_1 \sin f_1 + A_2 \cos e_2 - B_2 \sin f_2 = A_0 \cos i \tag{4.81}$$

$$A_1 \sin e_1 + B_1 \cos f_1 - A_2 \sin e_2 - B_2 \cos f_2 = -A_0 \sin i \tag{4.82}$$

$$A_1 Z_1 \cos 2f_1 - B_1 W_1 \sin 2f_1 - A_2 Z_2 \cos 2f_2 + B_2 W_2 \sin 2f_2 = -A_0 Z_1 \cos 2f_1 \tag{4.83}$$

$$A_1 \gamma_1 W_1 \sin 2e_1 + B_1 W_1 \cos 2f_1 + A_2 \gamma_2 W_2 \sin 2e_2 + B_2 W_2 \cos 2f_2 = A_0 \gamma_1 W_1 \sin 2i \tag{4.84}$$

where

$$Z_1 = \rho_1 \alpha_1, \qquad Z_2 = \rho_2 \alpha_2$$

$$W_1 = \rho_1 \beta_1, \qquad W_2 = \rho_2 \beta_2$$

$$\gamma_1 = \frac{\beta_1}{\alpha_1}, \qquad \gamma_2 = \frac{\beta_2}{\alpha_2}$$

Equations 4.81 and 4.82 come from requiring continuity of vertical and horizontal displacement, and Eqs. 4.83 and 4.84 are required by continuity of vertical and horizontal stress on the boundary.

The ratio of the reflected or transmitted amplitude to the incident amplitude is called the *reflection* or *transmission coefficient*. (Unfortunately, and confusingly, the fractions of the incident energy which are reflected or transmitted are also sometimes referred to as the reflection and transmission coefficients.) Calculation of reflection and transmission coefficients for anything other than normal incidence is lengthy because the equations have to be solved numerically.

Figure 4.44 illustrates the reflection and transmission of energy for an incident P-wave on a solid–solid boundary. For this example there are two critical angles, 30° and 60°. Beyond the first critical angle, that for P-waves ($\sin 30° = \alpha_1/\alpha_2 = 0.5$), there are no transmitted P-waves. The reflected P-wave energy increases greatly as the angle of incidence increases towards the first critical angle; these are the wide-angle reflections which are used extensively in seismic refraction work to determine critical distances. Similarly, beyond the critical angle ($\sin^{-1} \alpha_1/\beta_2$) for S-waves there are no transmitted S-waves.

Reflection and transmission coefficients take a very much simpler form for the case of normal incidence on the boundary ($i = 0$). In this case,

$$B_1 = B_2 = 0$$

since the tangential displacements and stresses on the boundary are zero

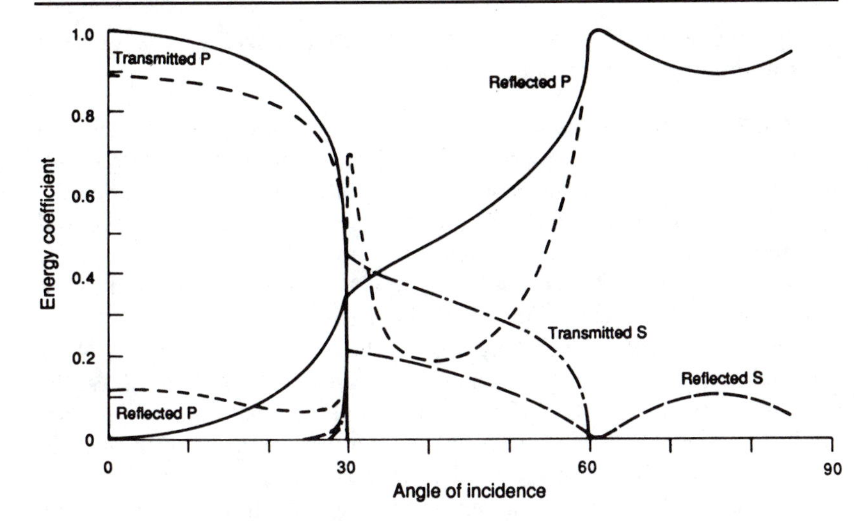

Figure 4.44. Reflected and transmitted energy (energy/incident energy) for the situation illustrated in Figure 4.43: a P-wave incident on an interface between two solid media. For this example, $\alpha_1/\alpha_2 = 0.5$, $\rho_1/\rho_2 = 2.0$, $\alpha_1/\beta_1 = 1.87$ and $\alpha_2/\beta_2 = 1.73$. Solid lines, reflected and transmitted P-wave; long dashed line, reflected S-wave; the dashed and dotted line, transmitted S-wave; the short dashed lines, reflected and transmitted P-wave when $\rho_1/\rho_2 = 1.0$. (After Tooley et al. 1965.)

for a vertically travelling P-wave. Equations 4.81–4.84 reduce to

$$A_1 + A_2 = A_0 \tag{4.85}$$

$$Z_1 A_1 - Z_2 A_2 = -Z_1 A_0 \tag{4.86}$$

The amplitude ratios A_1/A_0 and A_2/A_0 are the reflection and transmission coefficients. Equations 4.85 and 4.86 are solved for the amplitude ratios:

$$\frac{A_1}{A_0} = \frac{Z_2 - Z_1}{Z_2 + Z_1} = \frac{\rho_2\alpha_2 - \rho_1\alpha_1}{\rho_2\alpha_2 + \rho_1\alpha_1} \tag{4.87}$$

$$\frac{A_2}{A_0} = \frac{2Z_1}{Z_2 + Z_1} = \frac{2\rho_1\alpha_1}{\rho_2\alpha_2 + \rho_1\alpha_1} \tag{4.88}$$

If Figure 4.40 is studied again in the light of the foregoing discussion of reflection and transmission coefficients, it should be clear that although the first arrivals should be the easiest phases to pick accurately, the largest amplitude events are the reflection from any interface from the critical distance onwards. Thus, in attempts to determine a crustal structure, the reflections from major discontinuities in the crust are very important. Normally, the main reflection seen on refraction profiles is P_mP, the P-wave reflection from the Moho (the crust–mantle boundary). At the critical distance for P_n (the mantle head wave), the amplitude of P_mP is large, which often helps to constrain the mantle velocity since P_n does not usually become a first arrival on continental lines until perhaps 200 km, at which distance from the source accurate picking of arrival times may be difficult due to low amplitudes relative to background noise. At long range, the reflection P_mP is asymptotic to the head wave from the layer above the mantle, and so the large amplitude P_mP at long ranges is very useful in determining the lower crustal velocity.

Amplitude–Distance Relations In addition to reflection and transmission coefficients it is useful to understand how the amplitude of waves decreases with increasing distance from the source. We have already seen in Section 4.1.3 that the amplitude of body waves varies as x^{-1} with distance and that the amplitude of surface waves varies only as $x^{-1/2}$.

For an interface between two uniform elastic media, the amplitude of a head wave decreases rapidly with increasing distance as $L^{-3/2}x^{-1/2}$ where $L = x - x_c$, x is the distance and x_c the critical distance (Fig. 4.45a). This relationship holds for values of L greater than 5–6 times the predominant wavelength of the seismic signal (i.e., for $8\,\mathrm{km\,s^{-1}}$ material and a 5 Hz signal, $L > 10$ km). At distances much greater than the critical distance, the amplitude of the head wave varies as x^{-2}.

In the case of continuous refraction in a velocity gradient (Fig. 4.45b), the refracted wave is a body wave, and so its amplitude varies as x^{-1}. This type of arrival therefore dominates head waves at long distances.

However, if the velocity beneath the interface $\alpha(z)$ continues to increase with depth z as

$$\alpha(z) = \alpha_0(1 + bz) \tag{4.89}$$

the resultant head wave has an interference character (Fig. 4.45c). The first arrival is no longer the head wave but a wave continuously refracted below the interface. This is closely followed by a succession of continuously refracted waves whch have been reflected from the undersurface of the interface. Those that arrive within ΔT of each other, where ΔT is the duration of the incident signal, interfere and are termed the *interference head wave*. For $Lb^{2/3}\lambda^{-1/3} > 1$, where λ is the wavelength just below the interface, the amplitude of this interference head wave increases with range as $L^{3/2}x^{-1/2}$. At long ranges, when the first arrival arrives more than ΔT before the interference packet, the amplitude of the first arrival varies as x^{-1}. It should be clear, therefore, that introduction of velocity gradients can have a considerable effect on amplitude–distance behaviour.

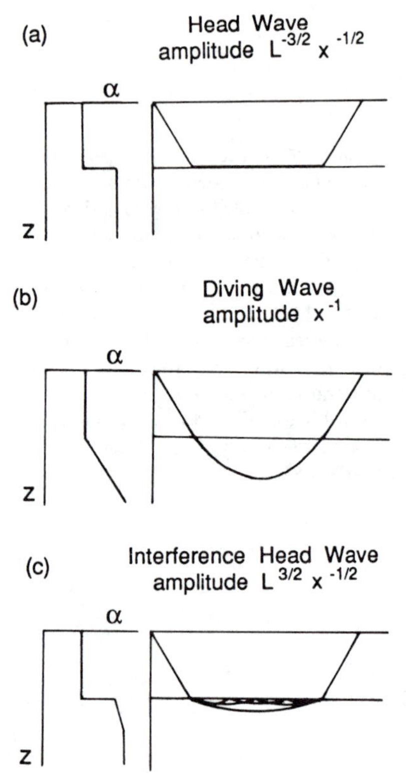

Figure 4.45. Amplitude–distance behaviour for head waves, diving waves and interference head waves. (From Kennett 1977.)

Synthetic refraction seismograms

The best way to determine a velocity structure from seismic refraction data is to compute synthetic (theoretical) seismograms and to compare these with those recorded in the refraction experiment. The velocity model can then be adjusted, velocities and thickness changed, or layers inserted or removed, and the computations repeated until the synthetic seismograms match the recorded seismograms to the desired degree. It is not usually possible to match every detail of a seismogram, but it is possible to match the arrival times and relative amplitudes of the main phases in a refraction record section and probably to match a few of the waveforms well. It should be stressed that this is still not an easy, quick or cheap process, even with increasing computer power.

Figure 4.46 shows examples of velocity models which agree with the first-arrival travel times but for which the synthetic seismograms do not match the recorded seismograms. For the final velocity model, both travel

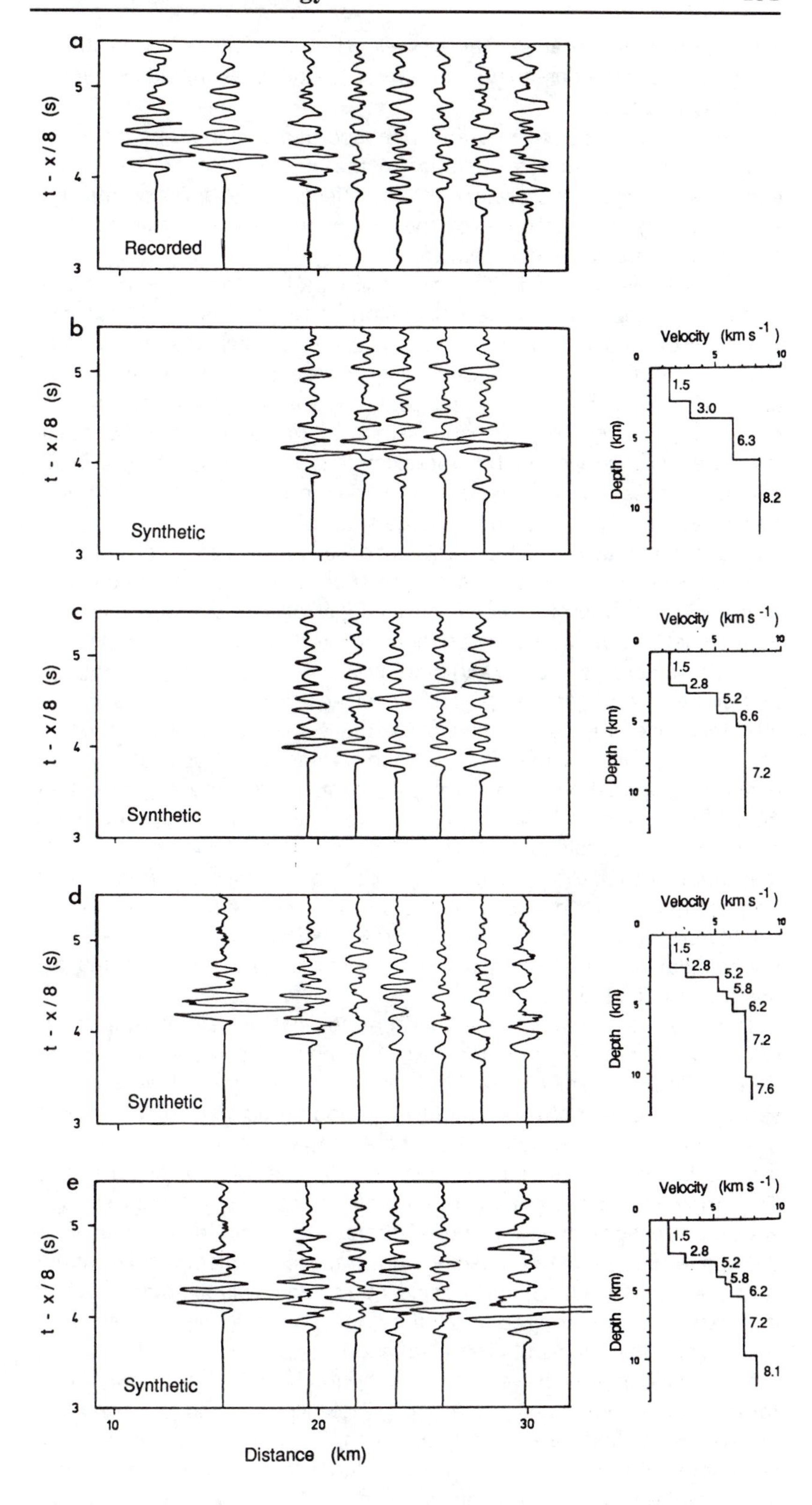

Figure 4.46. Use of synthetic seismograms in the determination of oceanic crustal structure. Synthetic seismograms were calculated using the 'reflectivity' method for four oceanic crustal models, all of which have first-arrival travel times which fit the observed times. Each seismogram is calculated for the charge size used in the original experiment – hence the variable amplitude–distance behaviour. All record sections are reduced to $6\,\mathrm{km\,s^{-1}}$. All models include 2.6 km of sea water ($\alpha = 1.5\,\mathrm{km\,s^{-1}}$) and a thin seabed layer of fractured material ($\alpha = 2.8$–$3.0\,\mathrm{km\,s^{-1}}$). (a) Recorded record section is part of a refraction line shot along the median valley of the Mid-Atlantic Ridge at 37°N; these were some of the data used for the model in Figure 8.17. (b) Model with a layer 3 and normal upper mantle. The large amplitude arrivals at 4 s are the reflections from the $6.3/8.2\,\mathrm{km\,s^{-1}}$ interface. (c) The amplitude and complexity of the first arrivals have been increased. There are no large amplitude reflections because the velocity contrast at the interfaces is much less than in (b). (d) These seismograms match the recorded record section (a) best; model was obtained from model (c) by making slight changes in layer thicknesses and velocities and by including $7.6\,\mathrm{km\,s^{-1}}$ material at depth. (e) Replacing the $7.6\,\mathrm{km\,s^{-1}}$ material of model (d) with material with a normal upper mantle velocity of $8.1\,\mathrm{km\,s^{-1}}$ dramatically changes the seismograms. The large-amplitude arrival is the reflection from the $7.2/8.1\,\mathrm{km\,s^{-1}}$ interface.

times and seismograms match well. The importance of synthetic seismograms in the determination of velocity structures should be evident from this example.

To match synthetic seismograms with recorded seismograms, one must compute the synthetics to include (a) the source function (explosion, earthquake, airgun or vibrator); (b) the effects of transmission through the earth; (c) the response of the detection systems (seismometers, amplifiers, filters, tape recorders etc.).

The most difficult and time-consuming part of the computation of synthetic seismograms is the calculation of the effects on the seismic waveform caused by its passage through the earth. There are undoubtedly lateral variations in the crust and upper mantle, but provided that these are gradual and small compared with the wavelength of the incident seismic signal and the horizontal scale of the proposed or possible structure, they can often be approximated by a model in which the structure varies only with depth. This variation of α, β and ρ with depth can always be approximated by a large stack of uniform horizontal layers. Because computation of the total response, which means including all interconversions and reverberations of such a stack of layers, is costly, a number of approximations have been developed. Some of the approximations allow lateral variations to be included in the model.

In the *generalized* (or exact) and *WKBJ* (Wenzel, Kramers, Brillouin and Jeffreys) *ray methods* (e.g., Helmberger 1968; Chapman 1978), the total response of a stack of layers is expressed as an infinite sum of the impulse responses of partial rays. Each partial ray is associated with a different propagation path through the structure, and together they include all interconversions and reverberations. The final synthetic seismogram is obtained by convolving the impulse response sum with the source and receiver functions. The problem with these methods comes from the termination of this infinite sum of impulse responses: Rays which would make a significant contribution may inadvertently be excluded from the calculation. Usually only primary reflections are considered, making this a cheap method.

The *classical ray method* allows the inclusion of lateral variation in velocity and of curved interfaces (e.g., Červený 1979). In this method, the particle displacement is expressed as an infinite sum, known as a ray series. For seismic body waves (reflected and refracted P- and S-waves), it is often sufficient to consider only the first term in the ray series: This is the solution obtained according to the principles of geometrical optics. This approximation is not valid near critical points, caustics or shadow zones; corrections must be applied there. Ray paths to be included in the computation must be specified (in practice, frequently only primary rays are considered, multiples being neglected). This method is not as accurate as the generalized ray method but has the significant advantage of including the effects of lateral variation.

The most comprehensive method, although also the most costly, is the *reflectivity method* (Fuchs and Müller 1971; Müller 1985). In this technique, all the calculations are performed in the frequency domain, and all reflections and interconversions are included by using a matrix formulation for the reflection and transmission coefficients. Attenuation Q can also be specified for each layer (Kennett 1975) as well as α, β and ρ. The computation is reduced by calculating only arrivals within a specified phase–velocity window, and by including only the multiples and interconversions for the 'reflection zone', the lower part of the model. The layers above the reflection zone, which need not be the same beneath source and receiver, are assumed to introduce only transmission losses; no reflections are calculated for these upper layers. The source and receiver responses are included in the frequency domain by multiplication, and the synthetic seismogram is then obtained as a time series by Fourier transform. The disadvantage of this method is the larger amount of computing time required and the fact that lateral variations cannot be included in the reflection zone.

4.5 Reflection Seismology

4.5.1 The Reflection Method

Although earthquake seismology and refraction seismology enable scientists to determine gross earth structures and crustal and upper mantle structures, reflection seismology is the method used to determine fine details of the shallow structures, usually over small areas. The resolution obtainable with reflection seismology makes it the main method used by oil exploration companies to map subsurface sedimentary structures. The method has also been increasingly used to obtain new information on the fine structures within the crust and at the crust–mantle boundary.

For land profiles, explosives can be used as a source. Other sources include the *gas exploder*, in which a gas mixture is exploded in a chamber which has a moveable bottom plate resting on the ground, and the *vibrator*, in which a steel plate pressed against the ground is vibrated at increasing frequency (in the range 5–60 Hz) for several seconds (up to 30 s for deep crustal reflection profiling). Vibrators require an additional step in the data processing to extract the reflections from the recordings: The cross-correlation of the recordings with the source signal.

Of the many marine sources, the two most frequently used for deep reflection profiling are the *air gun*, in which a bubble of very high pressure air is released into the water, and the *explosive cord*. Several air guns are usually used in an array which is towed behind the shooting ship.

Deconvolution is the process which removes the effects of the source and receiver from the recorded seismograms and allows direct comparison of data recorded with different sources and/or receivers. For the details of the methods of obtaining and correcting seismic reflection profiles, the reader is again referred to the textbooks on exploration geophysics (e.g., Telford et al. 1990; Dobrin and Savit, 1988; Yilmaz 1987; Claerbout 1985).

The basic assumption of seismic reflection is that there is a stack of horizontal layers in the crust and mantle, each with a distinct seismic P-wave velocity. Dipping layers, faults and so forth can be included in the method (see Sect. 4.5.4). P-waves from a surface energy source, which are almost normally incident on the interfaces between these layers, are reflected and can be recorded by *geophones* (vertical component seismometers) close to the source. Because the rays are close to normal incidence, effectively no S-waves are generated (Fig. 4.44). The P-waves reflected at close to normal incidence are very much smaller in amplitude than the wide-angle reflections near to and beyond the critical distance. This fact means that normal incidence reflections are less easy to recognize than wide-angle reflections and more likely to be obscured by background noise, and that sophisticated averaging and enhancement techniques must be used to detect reflecting horizons.

4.5.2 *A Two-Layered Model*

Consider the two-layer model in Figure 4.39. By application of Pythagoras' theorem, the travel time t for the reflection path SCR is given by Eqs. 4.54 and 4.55 as

$$t = \frac{SC}{\alpha_1} + \frac{CR}{\alpha_1}$$

$$= \frac{2}{\alpha_1}\sqrt{z_1^2 + \frac{x^2}{4}}$$

or

$$t^2 = \frac{4z_1^2}{\alpha_1^2} + \frac{x^2}{\alpha_1^2} \tag{4.90}$$

which is the equation of a hyperbola. At normal incidence ($x = 0$), the travel time is $t = t_0$, where

$$t_0 = \frac{2z_1}{\alpha_1} \tag{4.91}$$

This is the two-way normal incidence time. At large distances ($x \gg z_1$) the travel time (Eq. 4.90) can be approximated by

$$t \approx \frac{x}{\alpha_1} \tag{4.92}$$

This means that at large distances the travel-time curve is asymptotic to the travel time for the direct wave, as illustrated in Figure 4.39b. In reflection profiling, since we are dealing with distances much shorter than the critical distance, the travel-time distance plot is still curved. Notice that with increasing values of α_1 the hyperbola (Eq. 4.90) becomes flatter. If travel-time distance data were obtained from a reflection profile shot over such a model, one way to determine α_1 and z_1 would be to plot, not t against x but t^2 against x^2. Equation 4.90 is then the equation of a straight line with slope $1/\alpha_1^2$ and an intercept on the t^2 axis of $t_0^2 = 4z_1^2/\alpha_1^2$.

The normal incidence reflection coefficient for P-waves is given by Eq. 4.87. Since normal incidence reflections have small amplitudes, it is advantageous to average the signals from nearby receivers to enhance the reflections and reduce the background noise. This averaging process is called *stacking*. *Common Depth Point* (CDP) *stacking*, which combines all the recordings of reflections from every subsurface point, is the method

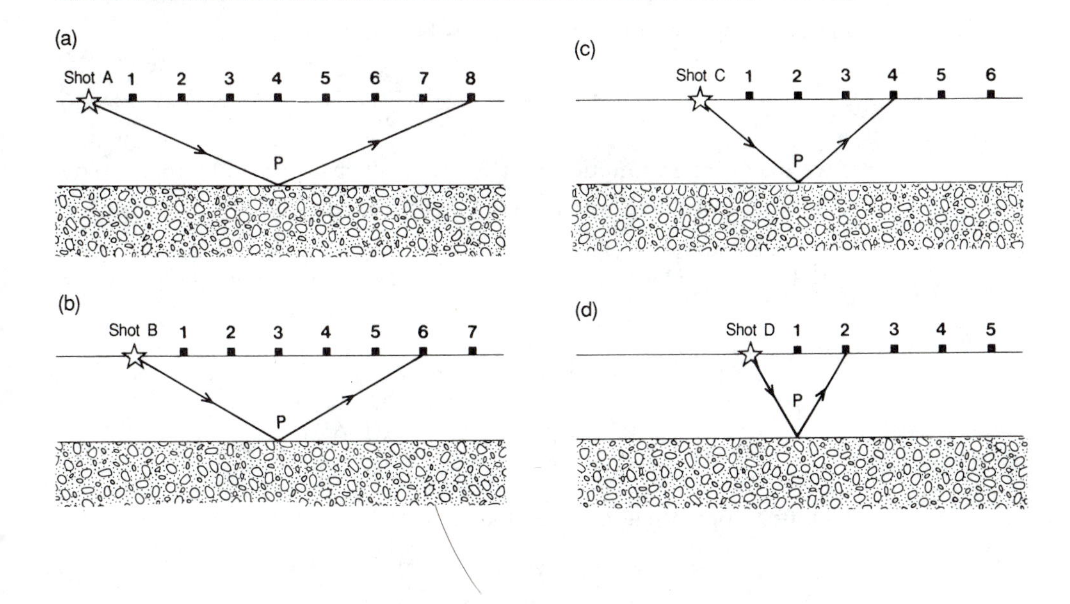

Figure 4.47. Common depth point (CDP) reflection profiling. In this example, eight geophones (■) record each shot (*). In (a), shot A is fired and a reflection from a particular point *P* on the reflector (interface between the two layers) is recorded by geophone 8. In (b), all the geophones and the shotpoint have been moved one step to the right, and shot B is fired; the reflection from point *P* is recorded by geophone 6. Similarly, a reflection from *P* is recorded (c) by geophone 4 when shot C is fired and (d) by geophone 2 when shot D is fired. The four reflections from point *P* can be stacked (added together after time corrections are made). In this example, because there are four reflections from each reflecting point on the interface (fewer at the two ends of the profile), there is said to be four-fold coverage. Alternatively, the reflection profile can be described as a four-fold CDP profile.

usually used. Common-offset stacking, which combines all the recordings with a common offset distance, is less popular. Figure 4.47 shows the layout of shots (or vibrators) and receivers used in CDP reflection profiling. The coverage obtained by any profile is

$$\text{coverage} = \frac{\text{number of receivers}}{\text{twice the shot spacing}} \tag{4.93}$$

where the shot spacing is in units of receiver spacing. In the example of Figure 4.47, the number of receivers is eight and the shot spacing is one which results in the four-fold coverage. Reflection profiling systems usually have 48 or 96 recording channels (and hence receivers), which means that 24-, 48-, or 96-fold coverage is possible. The greater the multiplicity of coverage, the better the system is for imaging weak and deep reflectors and the better the final quality of the record section. In practice, receiver spacing of a few tens to hundreds of metres is used, in contrast to the several-kilometre spacing of refraction surveys.

To be able to add all these recordings to produce a signal reflected from the common depth point, one must first correct them for their different travel times which are due to their different offset distances. This correction to the travel times is called the *normal moveout* (NMO) *correction.*

The travel time for the reflected ray in the simple two-layered model of Figure 4.39 is given by Eq. 4.90. The difference in the travel time t at two distances is called the *moveout* Δt. The moveout can be written

$$\Delta t = \frac{2}{\alpha_1}\sqrt{z_1^2 + \frac{x_a^2}{4}} - \frac{2}{\alpha_1}\sqrt{z_1^2 + \frac{x_b^2}{4}} \tag{4.94}$$

where x_a and x_b, ($x_a > x_b$), are the distances of the two geophones a and b from the shotpoint. The normal moveout Δt_{NMO} is the moveout for the special case when geophone b is at the shotpoint (i.e., $x_b = 0$). In this case, and dropping the subscript a, Eq. 4.94 becomes

$$\Delta t_{\mathrm{NMO}} = \frac{2}{\alpha_1}\sqrt{z_1^2 + \frac{x^2}{4}} - \frac{2z_1}{\alpha_1} \tag{4.95}$$

If we make the assumption that $2z_1 \gg x$, which is generally appropriate for reflection profiling, we can use a binomial expansion for $\sqrt{z_1^2 + x^2/4}$:

$$\begin{aligned}\sqrt{z_1^2 + \frac{x^2}{4}} &= z_1\sqrt{1 + \frac{x^2}{4z_1^2}} \\ &= z_1\left[1 + \left(\frac{x}{2z_1}\right)^2\right]^{1/2} \\ &= z_1\left[1 + \frac{1}{2}\left(\frac{x}{2z_1}\right)^2 - \frac{1}{8}\left(\frac{x}{2z_1}\right)^4 + \frac{1}{16}\left(\frac{x}{2z_1}\right)^6 + \cdots\right]\end{aligned} \tag{4.96}$$

To a first approximation, therefore,

$$\sqrt{z_1^2 + \frac{x^2}{4}} = z_1\left[1 + \frac{1}{2}\left(\frac{x}{2z_1}\right)^2\right] \tag{4.97}$$

Substituting this value into Eq. 4.95 gives a first approximation for the normal moveout Δt_{NMO}.

$$\begin{aligned}\Delta t_{\mathrm{NMO}} &= \frac{2z_1}{\alpha_1}\left[1 + \frac{1}{2}\left(\frac{x}{2z_1}\right)^2\right] - \frac{2z_1}{\alpha_1} \\ &= \frac{x^2}{4\alpha_1 z_1}\end{aligned} \tag{4.98}$$

Using Eq. 4.91 for t_0, the two-way normal incidence time, we obtain an alternative expression for Δt_{NMO}:

$$\Delta t_{\mathrm{NMO}} = \frac{x^2}{2\alpha_1^2 t_0} \tag{4.99}$$

This illustrates again the fact that the reflection time–distance curve is flatter (Δt_{NMO} smaller) for large velocities and large normal incidence times. This NMO correction must be subtracted from the travel times for the common-depth point recordings. The effect of this correction is to line up all the reflections from each point P with the same arrival time t_0 so that they can be stacked (added together) to produce one trace. This procedure works well when we are using a model for which α_1 and z_1 are known, but in practice we do not know them: They are precisely the unknowns which we would like to determine from the reflections! This difficulty is overcome by the bootstrap technique illustrated in Figure 4.48. A set of arrivals is identified as reflections from point P if their travel times fall on a hyperbola. Successive values of α_1 and t_0 are tried until a combination defining a hyperbola which gives a good fit to the travel times is found. These values α_1 and t_0 then define the model. In the example of Figure 4.48, it is obvious that curve 2 is correct, but with real data it is not sufficient to rely on the eye alone to determine velocities – a numerical criterion must be used. In practice, the power (or some similar

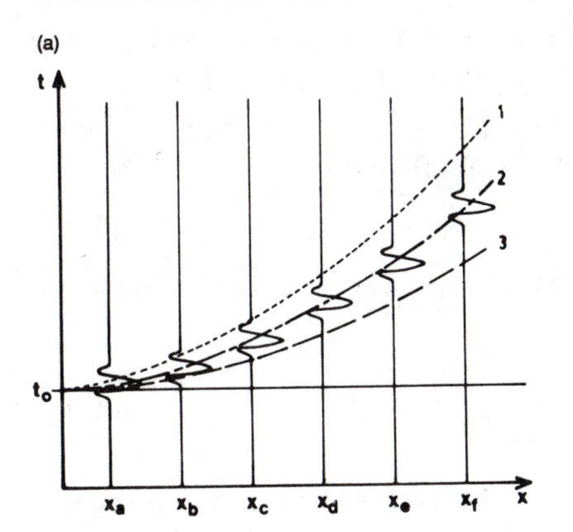

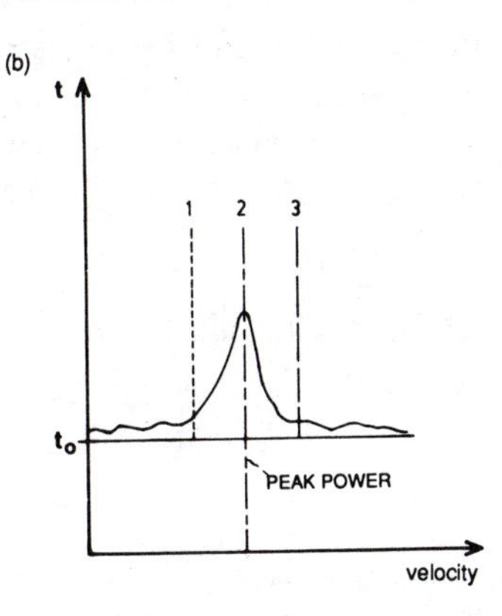

Figure 4.48. (a) Reflections from an interface, recorded at distances x_a, x_b, x_c, x_d, x_e and x_f. Three travel-time curves (1, 2, 3) are shown for two-way normal incidence time t_0 and increasing values of velocity. Clearly, curve 2 is the best fit to the reflections. To stack these traces, the NMO correction (Eq. 4.99) for curve 2 is subtracted from each trace so that the reflections line up with a constant arrival time of t_0. Then the traces can be added to yield a final trace with increased signal-to-noise ratio. (b) The power in the stacked signal is calculated for each value of the stacking velocity and displayed on a time–velocity plot. The velocity which gives the peak value for the power in the stacked signal is then the best stacking velocity for that particular value of t_0. For (a), velocity 2 is best; velocity 1 is too low and velocity 3 too high. (After Taner and Koehler 1969.)

entity) in the stacked signal is calculated for each value of α_1 and t_0. The maximum value of the power is then used to determine the best velocity for any value of t_0. A plot of power against both velocity and time, as in Figure 4.48b, is usually called a *velocity spectrum* display.

A common-depth point record section shows the travel times as if shots and receivers were coincident. This is achieved by the stacking process, which necessarily involves some averaging over fairly short horizontal distances, but which has the considerable advantage that the signal-to-noise ratio of the stacked traces is increased by a factor of $\sqrt{n}$ over the signal-to-noise ratio of the n individual traces.

4.5.3 A Multilayered Model

A two-layer model is obviously not a realistic approximation to a pile of sediments or to the earth's crust, but it serves to illustrate the principle of the reflection method. A multiple stack of layers is a much better model than a single layer. Travel times through a stack of multiple layers are calculated in the same way as for two layers, with the additional constraint that Snell's law ($\sin i/\alpha = p$, a constant for each ray) must be applied at each interface (Eq. 4.80). The travel times and distances for a model with n layers, each with thickness z_j and velocity α_j are best expressed in the parametric form:

$$x = 2\sum_{j=1}^{n} \frac{z_j p \alpha_j}{\sqrt{1 - p^2\alpha_j^2}}$$

$$t = 2\sum_{j=1}^{n} \frac{z_j}{\alpha_j\sqrt{1 - p^2\alpha_j^2}} \tag{4.100}$$

where $p = \sin i_j/\alpha_j$. It is unfortunately not usually possible to eliminate p from these equations in order to express the distance curve as one equation. In this multilayered case, the time–distance curve is *not* a hyperbola as it is when $n = 1$. However, it can be shown that the square of the travel time, t^2, can be expressed as an infinite series in x^2:

$$t^2 = c_0 + c_1x^2 + c_2x^4 + c_3x^6 + \cdots \tag{4.101}$$

where the coefficients c_0, c_1, $c_2, \ldots$ are constants dependent on the layer thicknesses z_j and velocities α_j. In practice, it has been shown that use of just the first two terms of Eq. 4.101 (c_0 and c_1) gives travel times to an accuracy of about 2%, which is good enough for most seismic reflection work. This means that Eq. 4.101 can be simplified to

$$t^2 = c_0 + c_1 x^2 \tag{4.102}$$

which is, after all, like Eq. 4.90, a hyperbola. The value of the constant c_0 is given by

$$c_0 = \left\{ \sum_{m=1}^{n} t_m \right\}^2 \tag{4.103}$$

where $t_m = 2z_m/\alpha_m$ is the two-way vertical travel time for a ray in the mth layer. The two-way normal incidence travel time from the nth interface $t_{0,n}$ is the sum of all the t_m.

$$t_{0,n} = \sum_{m=1}^{n} t_m = \sum_{m=1}^{n} \frac{2z_m}{\alpha_m} \tag{4.104}$$

Equation 4.103 can therefore be more simply written as

$$c_0 = (t_{0,n})^2 \tag{4.105}$$

The second constant of Eq. 4.102, c_1, is given by

$$c_1 = \frac{\sum_{m=1}^{n} t_m}{\sum_{m=1}^{n} t_m \alpha_m^2} \tag{4.106}$$

We can define a time-weighted root mean square (rms) velocity $\bar{\alpha}_n$ as

$$\bar{\alpha}_n^2 = \frac{\sum_{m=1}^{n} t_m \alpha_m^2}{\sum_{m=1}^{n} t_m} \tag{4.107}$$

With these expressions for c_0 and c_1, Eq. 4.102 becomes

$$t^2 = (t_{0,n})^2 + \frac{x^2}{\bar{\alpha}_n^2} \tag{4.108}$$

This now has exactly the same form as the equation for the two-layer case (Eq. 4.90), but instead of t_0 we have $t_{0,n}$, and instead of the constant velocity above the reflecting interface α_1, we now have $\bar{\alpha}_n$, the time-weighted rms velocity above the nth interface. This means that NMO corrections can be calculated and traces stacked as described in Section 4.5.2. The only difference is that in this multilayered case the velocities determined are not the velocity above the interface but the rms velocity above the interface.

Figure 4.49 shows a typical velocity spectrum display for real data. By stacking reflection records and using such a velocity spectrum display, one can estimate both $t_{0,n}$ and $\bar{\alpha}_n$ for each reflector. However, determining $t_{0,n}$ and $\bar{\alpha}_n$ is not enough; to relate these values to the rock structure over which the reflection line was shot, we have to be able to calculate the layer thicknesses and seismic velocities and obtain an estimate of the accuracy of such values.

Let us suppose that the rms velocity and normal incidence times have been determined for each of two successive parallel interfaces (i.e., $t_{0,n-1}$,

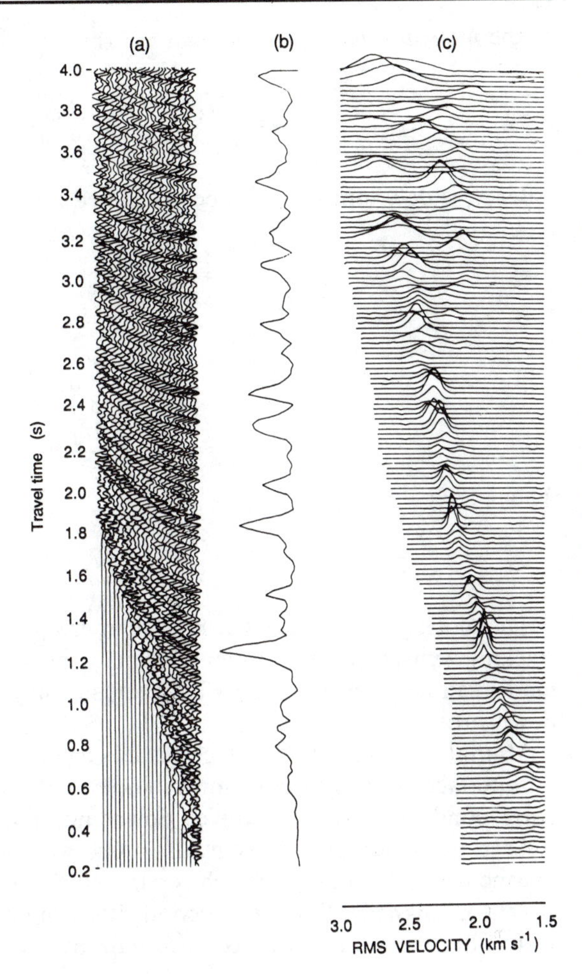

Figure 4.49. Steps in the computation of a velocity analysis. (a) The twenty-four individual reflection records used in the velocity analysis. (b) Maximum amplitude of the stacked trace shown as a function of t_0, two-way time along the trace. Notice that the reflections are enhanced compared with the original traces. Although the main reflections at $t_0 = 1.2$ and 1.8 s do stand out on the original traces, these intercept times are clearly defined by the stacked trace, and subsequent deeper reflections which were not clear on the original traces can now be identified with some confidence on the stacked trace. (c) The velocity spectrum for the traces in (a). The peak power at each time served to identify the velocity which would best stack the data. The stacking velocity clearly increases steadily with depth down to about 3 s. After this, some strong multiples (rays which have bounced twice or more in the upper layers and therefore need a smaller stacking velocity) confuse the velocity display. (From Taner and Koehler 1969.)

$t_{0,n}$, $\bar{\alpha}_{n-1}$ and $\bar{\alpha}_n$ are known). Then by using Eqs. 4.104 and 4.107, we can determine the velocity of the nth layer α_n:

$$\bar{\alpha}_n^2 \sum_{m=1}^{n} t_m = \sum_{m=1}^{n} \alpha_m^2 t_m$$

$$= \sum_{m=1}^{n-1} \alpha_m^2 t_m + \alpha_n^2 t_n \tag{4.109}$$

and

$$\bar{\alpha}_{n-1}^2 \sum_{m=1}^{n-1} t_m = \sum_{m=1}^{n-1} \alpha_m^2 t_m \tag{4.110}$$

Subtracting Eq. 4.110 from Eq. 4.109 gives

$$\bar{\alpha}_n^2 \sum_{m=1}^{n} t_m - \bar{\alpha}_{n-1}^2 \sum_{m=1}^{n-1} t_m = \alpha_n^2 t_n \tag{4.111}$$

or

$$\bar{\alpha}_n^2 t_{0,n} - \bar{\alpha}_{n-1}^2 t_{0,n-1} = \alpha_n^2 (t_{0,n} - t_{0,n-1}) \tag{4.112}$$

Rearranging this equation gives α_n, the velocity of the nth layer (also known as the *interval velocity*), in terms of the rms velocities:

$$\alpha_n = \sqrt{\frac{\bar{\alpha}_n^2 t_{0,n} - \bar{\alpha}_{n-1}^2 t_{0,n-1}}{t_{0,n} - t_{0,n-1}}} \tag{4.113}$$

After α_n is determined, z_n can be calculated from Eq. 4.104.

$$\begin{aligned} t_{0,n} &= \sum_{m=1}^{n} \frac{2z_m}{\alpha_m} \\ &= \sum_{m=1}^{n-1} \frac{2z_m}{\alpha_m} + \frac{2z_n}{\alpha_n} \\ &= t_{0,n-1} + \frac{2z_n}{\alpha_n} \end{aligned} \tag{4.114}$$

Thus,

$$z_n = \frac{\alpha_n}{2}(t_{0,n} - t_{0,n-1}) \tag{4.115}$$

Therefore, given the two-way normal intercept times and corresponding stacking (rms) velocities from a velocity analysis, the velocity–depth model can be determined layer by layer, starting at the top and working downwards.

Multiples are reflections which have been reflected more than once at an interface. The most common multiple is the surface multiple, which corresponds to a ray which travels down and up through the layers twice. Reflections with multiple ray paths in one or more layers also occur. In marine work, the multiple which is reflected at the sea surface and seabed is very strong (Fig. 8.18). The periodicity of multiple reflections enables us to filter them out of the recorded data by deconvolution.

Example: calculation of layer thickness and seismic velocity from a normal incidence reflection line

A velocity analysis of reflection data has

$$t_{0,1} = 1.0\,\text{s}, \qquad \bar{\alpha}_1 = 3.6\,\text{km s}^{-1}$$

$$t_{0,2} = 1.5\,\text{s}, \qquad \bar{\alpha}_2 = 4.0\,\text{km s}^{-1}$$

Calculate a velocity–depth model from these values, assuming constant velocity in each layer. Since $\bar{\alpha}_1$ must be equal to α_1, the velocity of the top layer z_1 can be calculated from Eq. 4.115:

$$z_1 = \frac{3.6 \times 1.0}{2.0} = 1.8\,\text{km}$$

Now α_2 can be calculated from Eq. 4.113:

$$\begin{aligned} \alpha_2 &= \sqrt{\frac{4.0^2 \times 1.5 - 3.6^2 \times 1.0}{1.5 - 1.0}} \\ &= \sqrt{22.08} = 4.7\,\text{km s}^{-1} \end{aligned}$$

Finally, z_2 is then calculated from Eq. 4.115:

$$z_2 = \frac{4.7 \times 0.5}{2.0} = 1.175 \text{ km}$$

Notice that the velocity α_2 is larger than the rms velocity $\bar{\alpha}_2$ in this example.

Unfortunately, the interval velocities and depths frequently cannot be determined accurately by these methods (see Problem 21). In exploration work the inaccuracy of velocity–depth information can usually be made up for by detailed measurements from drill holes. However, when reflection profiling is used to investigate structures deep in the crust and upper mantle, such direct velocity information is not available. There is an added problem in this situation: When the depths of horizons are much larger than the maximum offset, the NMO correction is insensitive to velocity. This means that any velocity which is approximately correct stacks the reflections adequately, and so the interval velocities calculated from the stacking velocities will not be very accurate. As an example, consider a two-layered model:

$$\alpha_1 = 6 \text{ km s}^{-1}, \qquad z_1 = 20 \text{ km}, \qquad \alpha_2 = 8 \text{ km s}^{-1}$$

The reflection hyperbola (Eq. 4.90) for this interface is

$$t = \frac{1}{3}\sqrt{400 + \frac{x^2}{4}}$$

and the two-way normal incidence time t_0 is 6.667 s. The NMO correction is (Eq. 4.98)

$$\Delta t_{\text{NMO}} = \frac{x^2}{480}$$

The maximum offset used in deep reflection profiling is rarely greater than about 5 km and often less. At this offset, the correct NMO correction for the reflection is 0.052s. But this correction has to be estimated from the data. Using the correct value for t_0, we obtain $\Delta t_{\text{NMO}} = 0.062$ s when $\alpha_1 = 5.5 \text{ km s}^{-1}$ and $\Delta t_{\text{NMO}} = 0.044$ s when $\alpha_1 = 6.5 \text{ km s}^{-1}$. Clearly, a signal with predominant frequency of 20 Hz or less could not give an accurate value for α_1. Higher-frequency signals give more accurate velocity analyses, as discussed in Section 4.5.4. The best way to obtain reliable interval velocity measurements in such cases is to supplement the reflection profiles with wide-angle reflection profiles. Such profiles, perhaps 50–80 km in length, enable deep wide-angle reflections to be recorded. Wide-angle reflections have larger amplitudes than normal incidence reflections (Fig. 4.44), and their travel times(Eqs. 4.90 and 4.92) give more reliable values for the interval velocities .

4.5.4 Diffraction and Migration

All the preceding sections have dealt with a horizontally layered model. Although some structures can be approximated by horizontal layers, structures such as faults, dykes and intrusions cannot. It is also necessary

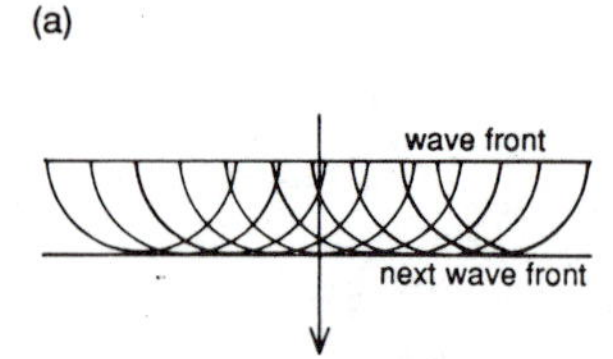

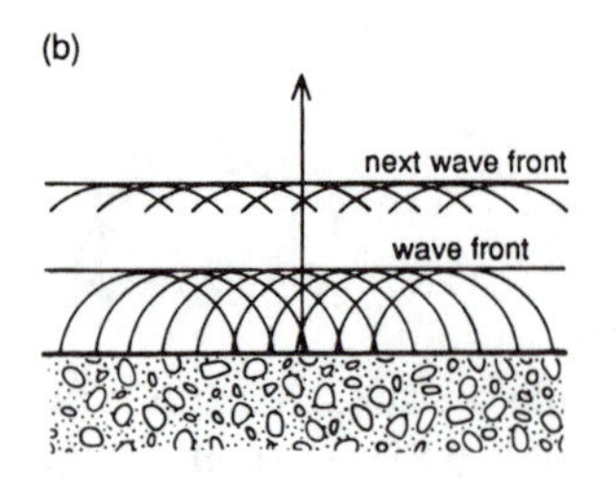

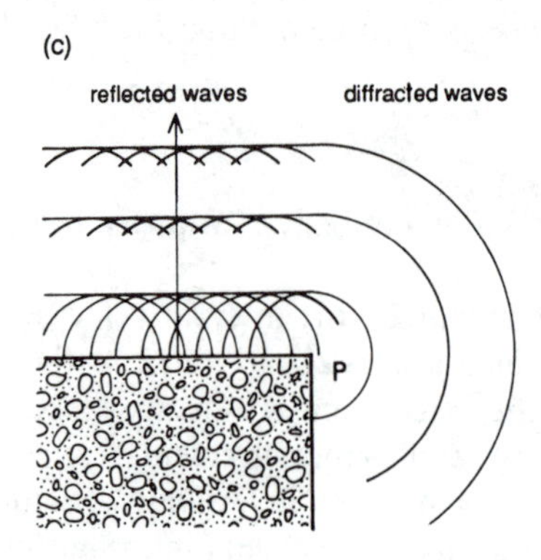

Figure 4.50. (a) A plane wave propagating vertically downwards. Each point on the wavefront acts as a point source and generates spherical waves. The next wavefront, which is the sum of all these spherical waves, is also a plane wave. All the lateral disturbances cancel each other out. (b) A plane wave travelling vertically downwards is reflected at an interface (lower material is shaded). Each point on the interface acts as a point source and generates spherical waves. Thus, the reflected wave is a plane wave travelling upwards (all lateral disturbances again cancel out). (c) A plane wave reflected from an interface which terminates at a corner P. The reflected wave well to the left of P is unaffected. Since there are no point sources to the right of P, the reflected wave close to P is reduced in amplitude. In addition, the spherical wave generated by P is not cancelled out. It is called the diffracted wave.

to allow for the effects of dipping layers. These complications bring into question the minimum size or contrast in physical properties which can be resolved by using the reflection profiling method.

Diffraction Huygens' principle states that every point on a wavefront acts as a point source generating spherical waves. The wavefront at a later time is then the sum of all the waves from these point sources. The case of a plane wave is illustrated in Figure 4.50a: The new wavefront is also planar because the lateral disturbances from the point sources cancel each other out. When a plane wave is reflected from an interface, the same principle applies: Each point on the interface acts as a point source and generates spherical waves. When summed together these spherical waves give a reflected plane wave (Fig. 4.50b). When an interface is not infinitely long but terminates at a point P (Fig. 4.50c), the reflected wave far to the left of P is not affected. However, since there are no more point sources to the right of point P to ensure cancellation of all but the plane wave, the wave field in the region around point P is affected. The spherical waves generated by point P which are not cancelled out combine to form the *diffracted* wave. This wave is detected to both the right and the left of P. Huygens' principle also applies to spherical waves: A spherical wave incident on the interface which terminates at P also gives rise to a diffraction.

Diffractions are not restricted to the ends of reflectors; a diffracted wave is produced by any irregularity in a structure that is comparable in scale to the wavelength of the signal. [Recall the diffraction of light through a hole, the diffraction of water waves around the end of a breakwater and diffraction at the core–mantle boundary (Sect. 4.3.1).]

Consider the horizontal reflector which terminates abruptly at point P (Fig. 4.51a). Normal-incidence reflections are recorded from this reflector only at loctions betwen A and B. However, the corner P acts as a point source when a seismic wave hits it and so radiates seismic waves in all directions in accordance with Huygens' principle. At B this diffracted wave arrives at the same time as the reflection, but with increasing distance from B (left or right) it arrives progressively later.

The travel time t from point C to the diffraction point P and back to point C, where C is a horizontal distance x from P, is given by

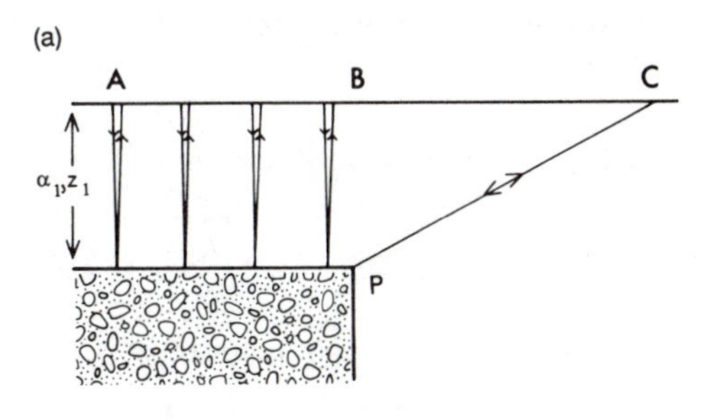

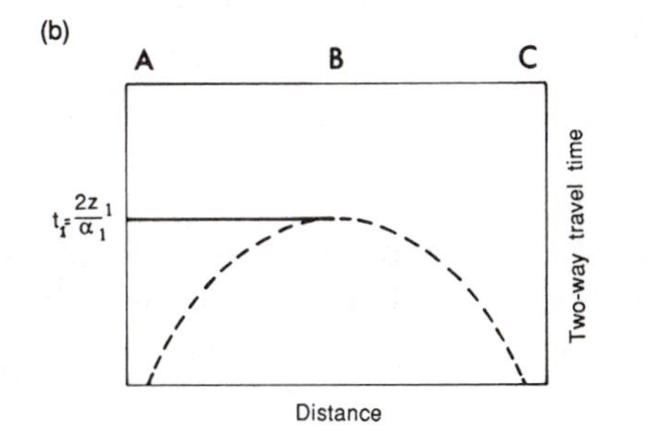

Figure 4.51. (a) A reflection profile over a horizontal reflector which terminates at point P. The overlying material has a constant P-wave velocity of α_1, and the reflector is at a depth z_1. Normal incidence reflections are recorded from A to B, but no reflections can be recorded to the right of location B. A diffracted wave produced by the point P is recorded at all locations. The ray path for the diffracted wave arriving at location C is the straight line CPC. The amplitude of diffracted waves is smaller than that of the reflections (amplitude being distance dependent). (b) Reflection record section for the model in (a). The reflection is a solid line and the diffraction a dashed line.

$$t = \frac{2}{\alpha_1}\sqrt{x^2 + z_1^2}$$

or

$$\alpha_1^2 t^2 = 4(x^2 + z_1^2) \tag{4.116}$$

This is the equation of a hyperbola (with a greater curvature than the reflection hyperbola of Eq. 4.90). The reflection and diffraction travel times are plotted on the record section in Fig. 4.51b. The curvature of this diffracted arrival on the record section decreases with the depth z_1 of the diffracting point P; diffractions from a deep fault are flatter than those from a shallow fault.

In the immediate vicinity of B, the amplitude of the reflection decreases as point B is approached until at B the amplitude of the reflection is only half its value well away from the reflector edge. (See discussion of Fresnel zone later in this section.) The diffraction amplitude decreases smoothly with increasing distance from the reflector edge. Since the reflection is tangent to the diffraction hyperbola at distance B, there is no discontinuity in either travel time or amplitude to mark the edge of the reflector. However, the diffraction branch recorded to the left of B has opposite polarity to the branch recorded to the right of B (they are 180° out of phase). Detection of both diffraction branches can assist in the location of the edge of the reflector.

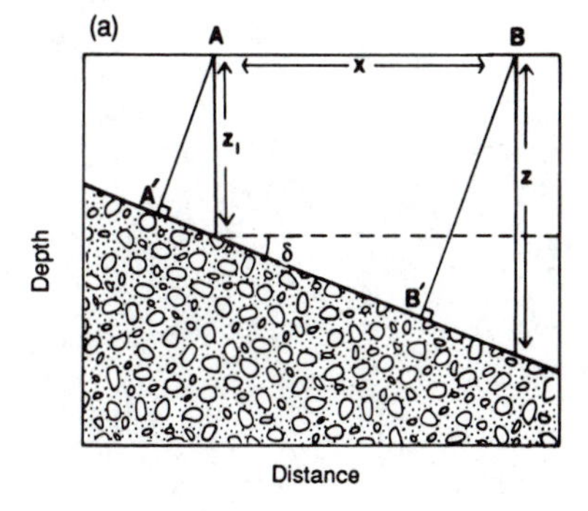

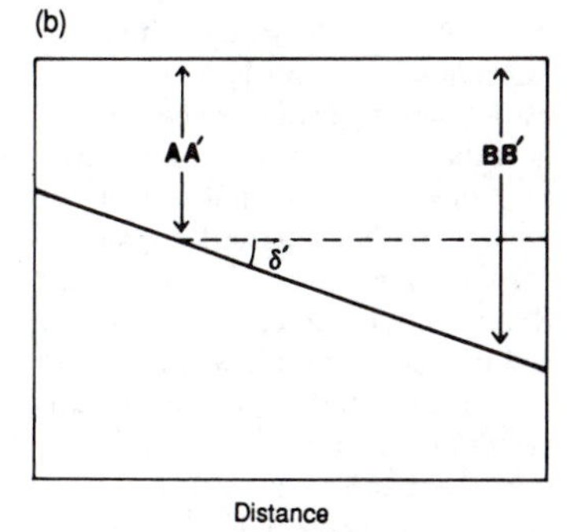

Figure 4.52. (a) Normal incidence reflections from an interface with dip δ. With the source and receiver at A, the reflection point is A' (likewise B and B'). The upper layer has a P-wave velocity of α_1. (b) Record section for the structure in (a). On a section with equal horizontal and vertical distance scales, the apparent dip δ' is less than the real dip δ (Eq. 4.124).

Dipping Layers On a CDP record section, the reflection point is shown as being vertically below the shot/receiver location. Dipping interfaces are therefore distorted on CDP record sections because the reflection points are not vertically below the shot/receiver location. The situation is illustrated in Figure 4.52: The normal-incidence reflection recorded at location A is reflected from point A' and that recorded at location B is reflected from point B'. If z_1 is the vertical depth of the reflector beneath

A and δ is the dip of the reflector, then the length AA' is given by

$$AA' = z_1 \cos \delta \tag{4.117}$$

and the two-way travel time for this reflection is

$$t = \frac{2z_1}{\alpha_1} \cos \delta \tag{4.118}$$

Similarly, the length BB' is given by

$$BB' = z \cos \delta \tag{4.119}$$

and the two-way travel for this reflection is

$$t = \frac{2z}{\alpha_1} \cos \delta \tag{4.120}$$

where z is the vertical depth of the reflector below B. The depths z_1 and z, dip angle δ and horizontal distance x between A and B, are related by

$$\tan \delta = \frac{z - z_1}{x} \tag{4.121}$$

Since the reflections from A' and B' plot on the record section as though they were vertically beneath A and B, respectively, the reflector appears to have a dip of δ' where

$$\tan \delta' = \frac{BB' - AA'}{x} \tag{4.122}$$

Substituting from Eqs. 4.117 and 4.119, this becomes

$$\tan \delta' = \frac{z \cos \delta - z_1 \cos \delta}{x} \tag{4.123}$$

Further substitution from Eq. 4.121 yields

$$\tan \delta' = \sin \delta \tag{4.124}$$

Thus, the apparent dip δ' is less than the actual dip δ, except for very small angles for which the approximation $\sin \delta \approx \tan \delta$ is valid. Reflection record sections are rarely plotted at true scale: The distortion is due to differing horizontal and vertical scales as well as to the vertical axis frequently being two-way travel time.

Migration A dipping reflector is not the only structure which is incorrectly located on a CDP record section. A reflection from any variation in structure in the third dimension, conveniently assumed to be nonexistent here, could also appear on the section. Diffractions can be mistaken for reflections. Curved interfaces can also produce some rather strange reflections (Fig. 4.53).

Migration is the name given to the process which attempts to deal with all of these problems and to move the reflectors into their correct spatial location. Thus, diffractions are removed. They are recognized as being diffractions by a method similar to the method of estimating the best stacking velocity; that is, if they lie on a hyperbola, they are a diffraction

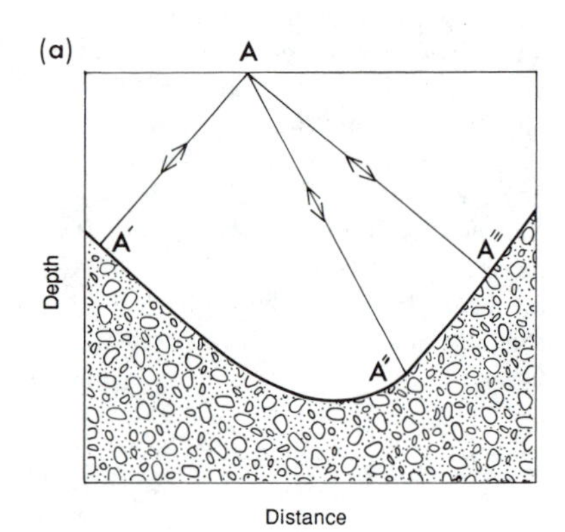

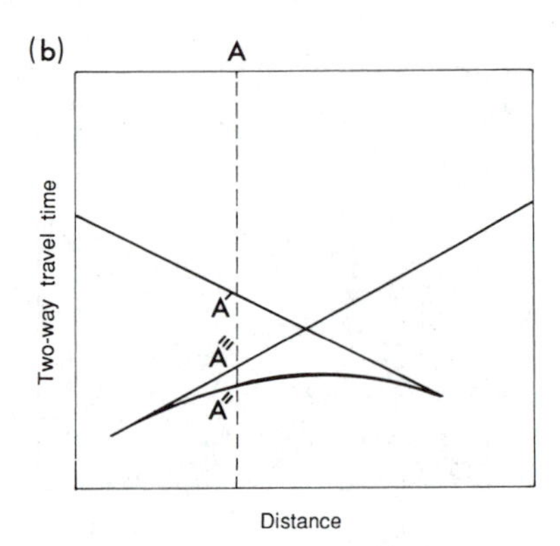

Figure 4.53. (a) A syncline with a centre of curvature (focus) which is below the earth's surface gives three normal incidence reflections for shot/receiver locations over the central part of the syncline. A shot/receiver location at A has reflections from points A', A'' and A''', one from each flank of the syncline and the third from the centre. The extreme example of a syncline is two planes intersecting – there is diffraction from the intersection. (b) Travel times for the syncline shown in (a). The deeper the centre of curvature (the tighter the syncline), the broader the convex reflection from the central portion of the syncline. Thus, on a record section, a syncline can look much like an anticline. The process of migration uses these three reflection branches to determine the shape of the reflecting surface, the syncline.

and are coalesced into a point at the apex. Dipping layers are plotted correctly, multiples (reflections bouncing twice or more in a layer) are removed and the reflections from curved interfaces are inverted. There are many methods by which migration is performed, all of them complex; the interested reader is referred to a reflection processing text such as Robinson and Treitel 1980, Hatton, Worthington and Makin 1985 or Yilmaz 1987 for more details.

Resolution of Structure The resolution of structure obtainable with reflection profiling is controlled by the wavelength of the source signal. A wave of frequency 20 Hz travelling through material with a P-wave velocity of 4.8 km s^{-1} has a wavelength of 240 m. When the same wave is travelling through material with a P-wave velocity of 7.2 km s^{-1}, the wavelength is increased to 360 m.

Consider a fault with a vertical throw of h (Fig. 4.54). For the fault to be detected, it must be greater than about one-eighth to one-quarter of the wavelength of the seismic wave. This means that the reflected wave from the down-faulted side is delayed one-quarter to one-half a wavelength. A smaller time delay than this is unlikely to be resolved as being due to a fault since it would appear as a slight time error. Thus, a 20 Hz signal could resolve faults with throws in excess of about 30 m in 4.8 km s^{-1} material. However, in 6.4 km s^{-1} material, the same 20 Hz signal could only resolve faults with throws of more than 40 m; in 7.2 km s^{-1} material, the minimum resolvable throw would be 45 m. Clearly, shorter wavelengths (higher frequencies) give better resolution of structures and greater detail than are obtainable from long wavelengths. Unfortunately, high-frequency vibrations do not travel as well through rock as low-frequency vibrations do (Sect. 4.3.3), and so high frequencies are usually absent in reflections from deep structures.

The lateral extent that a reflector must have to be detected is also controlled by wavelength. To understand the lateral resolution obtainable with a given source signal, we need to introduce the concept of the *Fresnel zone.* The first Fresnel zone is that part of a reflecting interface which returns energy back to the receiver within half a cycle of the first reflection (Fig. 4.55). In plan view, the first Fresnel zone is circular; the second Fresnel zone is a ring (concentric with the first Fresnel zone) from which the reflected energy is delayed by one-half to one cycle; and so on for the third, fourth and nth Fresnel zones. For an interface to be seen as such, it must be at least as wide as the first Fresnel zone (usually referred to as *the* Fresnel zone). If a reflecting zone is narrower than the first Fresnel zone it effectively appears to be a diffractor.

The width of the Fresnel zone w can be calculated by applying Pythagoras' theorem to triangle SOQ of Figure 4.55b.

$$\left(d + \frac{\lambda}{4}\right)^2 = d^2 + \left(\frac{w}{2}\right)^2 \tag{4.125}$$

Rearranging this, we obtain

$$w^2 = 2d\lambda + \frac{\lambda^2}{4} \tag{4.126}$$

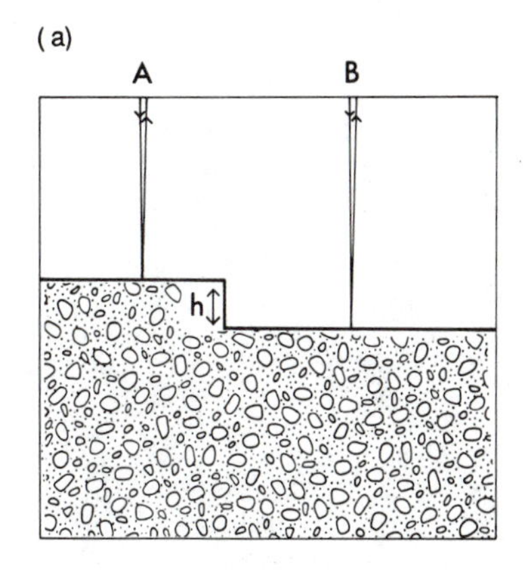

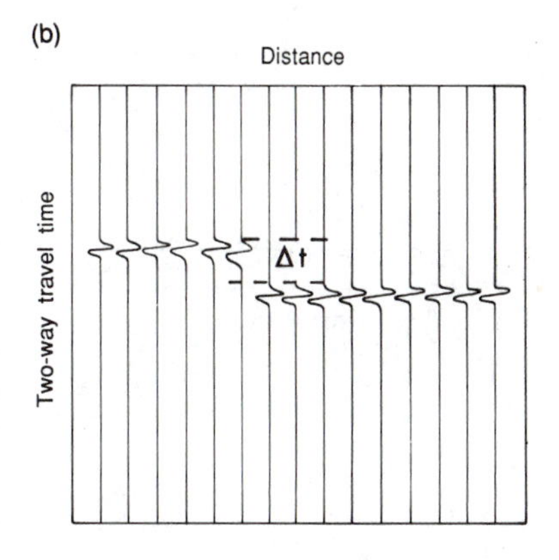

Figure 4.54. (a) Normal incidence reflection profile over an interface which has a fault with a vertical throw of h. (b) Record section for (a). The time difference between the two-way travel times on either side of the fault is Δt. For the fault to be resolved, its throw h must be greater than about $\lambda/8$ to $\lambda/4$, where λ is the wavelength of the signal. [In terms of Δt, Δt must be greater than $1/(4 \times \text{frequency}) - 1/(2 \times \text{frequency})$.] The corners at the top and bottom of the fault would both give rise to diffractions.

Table 4.5 *Widths of the first Fresnel zone (in km) for various wavelengths and depths*

Depth (km)	Wavelength (km) 0.10	0.20	0.30	0.40	0.50
2	0.63	0.90	1.11	1.28	1.44
5	1.00	1.42	1.74	2.01	2.25
10	1.42	2.00	2.45	2.84	3.17
30	2.45	3.46	4.24	4.90	5.48
50	3.16	4.47	5.48	6.33	7.08

Note: Calculated using Eq. 4.126.

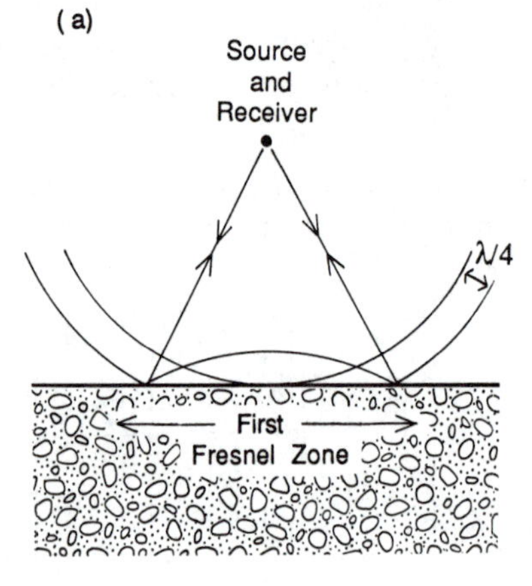

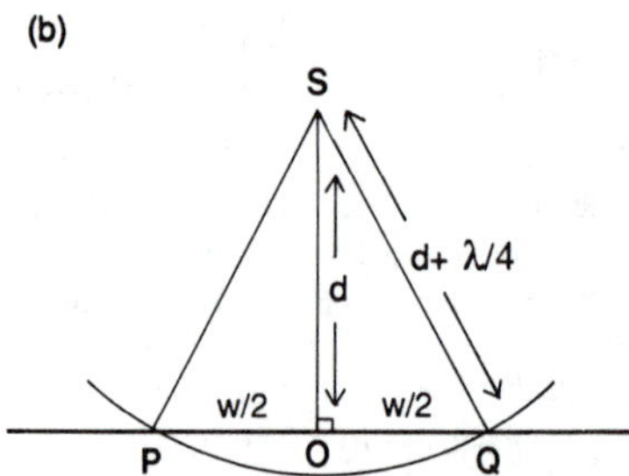

Figure 4.55. (a) Spherical waves from a point source reflected from a plane interface. The first Fresnel zone is that part of the interface which reflects energy back to the receiver within one-half a cycle of the first (normal-incidence) reflection. Because the wave must travel from source to interface and back to the receiver, energy from the wavefront one-quarter of a wavelength behind the first wavefront when reflected back to the receiver is delayed by half a cycle. (b) Geometry of the first Fresnel zone. The width w of this zone can be calculated by applying Pythagoras' theorem to the right triangle SOQ; d is the depth of the interface beneath the source and λ the wavelength of the signal. The upper material is assumed to have constant P-wave velocity α_1.

Table 4.5 gives values of w for various interface depths and wavelengths. Even at 2 km depth and with a wavelength of 0.10 km (which could correspond to $\alpha_1 = 3\,\text{km s}^{-1}$ and frequency = 30 Hz, etc.) the width of the Freznel zone is 0.63 km. If depths appropriate for the middle or deep crust the considered, the Fresnel zone is several kilometres in width, which is large on a geological scale. Deep-reflection profiling is therefore not able to give as clear a picture of the crust as one might at first hope.

4.5.5 Deep Seismic Reflection Profiling

The reflection data shown in Fig. 4.49a have two-way travel times of up to 4 s. Such recording times are used primarily when attempting to determine shallow sedimentary structures in oil exploration (seismic reflection profiling is the main prospecting tool of the petroleum industry). However, it is also possible to use reflection profiles to probe the deeper structures in the earth's crust. To do this, recording times of 15–30 s or more are necessary. In addition, large energy sources are required for deep crustal or uppermost mantle reflections to be detected. As mentioned in Section 4.5.3, the best method of obtaining detailed velocity information on the lower crust and uppermost mantle is not normal incidence reflection profiling but wide-angle reflection profiles. Recall from Eq. 4.90 that as the curvature of the reflection hyperbola decreases with increasing velocity (and thus increasing depth), deep interval velocities can only be rather poorly determined. However, normal incidence reflection and wide-angle reflection methods used together enable velocities and depths to be much better determined.

Deep seismic reflection profiling in the United States started with the COCORP (Consortium for Continental Reflection Profiling) profile shot in Hardman County, Texas, U.S.A., in 1975. Since then, many profiles have been shot by many organizations around the world. The method is used increasingly to relate surface and shallow geology to structures at depth, to trace thrusts and to map plutons. Deep seismic reflection profiling has dramatically increased our knowledge of the

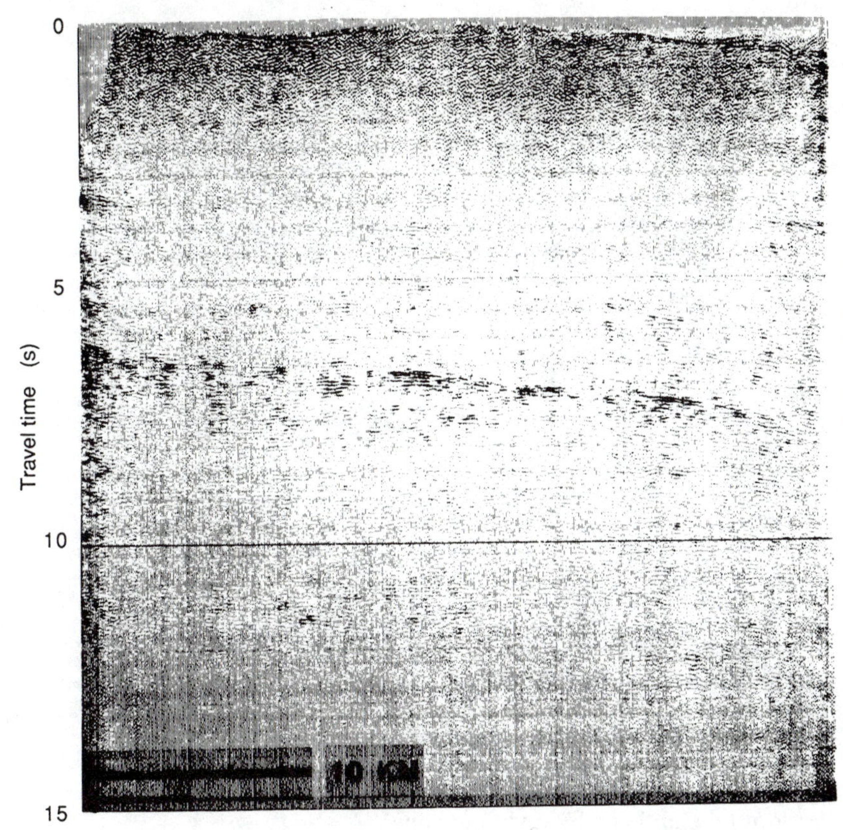

Figure 4.56. The Consortium for Continental Reflection Profiling (COCORP) has shot reflection profiles over many types of continental structures in the United States: accreted terranes, extensional rifts, thrust zones and so on. This record section is part of a series of profiles shot across the Rio Grande Rift, New Mexico, an active extensional rift (discussed in Sect. 9.4.3). The high-amplitude reflections at about 7 s two-way time in the midcrust are inferred to come from the lid of a magma chamber. The solid–liquid interface gives a large reflection coefficient. (From Brown 1986, reprinted by permission.)

structure of the continental crust. The Moho is frequently a reflector, although not unequivocably observed on every reflection line. The resolution obtainable with deep reflection profiling far exceeds that with seismic refraction or wide angle reflection, or magnetic, electrical or gravity methods. However, as with every method, the ultimate questions which remain unanswered concern the interpretation of the physical parameters (in this case, reflections and velocities) in terms of geology. For example, some interpretations ascribe lower crustal reflections to lithology (rock type) and changes in metamorphic facies, whereas others ascribe reflections to physical factors, perhaps the presence of fluids trapped in the lower crust, or to sheared zones.

As an example, a reflection coefficient of about 0.10 is necessary to explain the strong lower crustal and upper mantle reflections seen on some deep seismic reflection profiles. Such values are not easily achieved by juxtaposition of proposed materials or measured velocities. A P-wave velocity contrast of 8.1–8.3 $\mathrm{km\,s^{-1}}$ in the mantle gives a reflection coefficient (Eq. 4.87) of 0.012, assuming no density change. Even if somewhat larger velocity contrasts are considered and a density change is included, the reflection coefficient could hardly increase by an order of magnitude to 0.1. Mantle reflections of the correct magnitude could, however, be produced by regions of partially hydrated peridotite (i.e., water acting on mantle peridotite to produce serpentine, which has a very

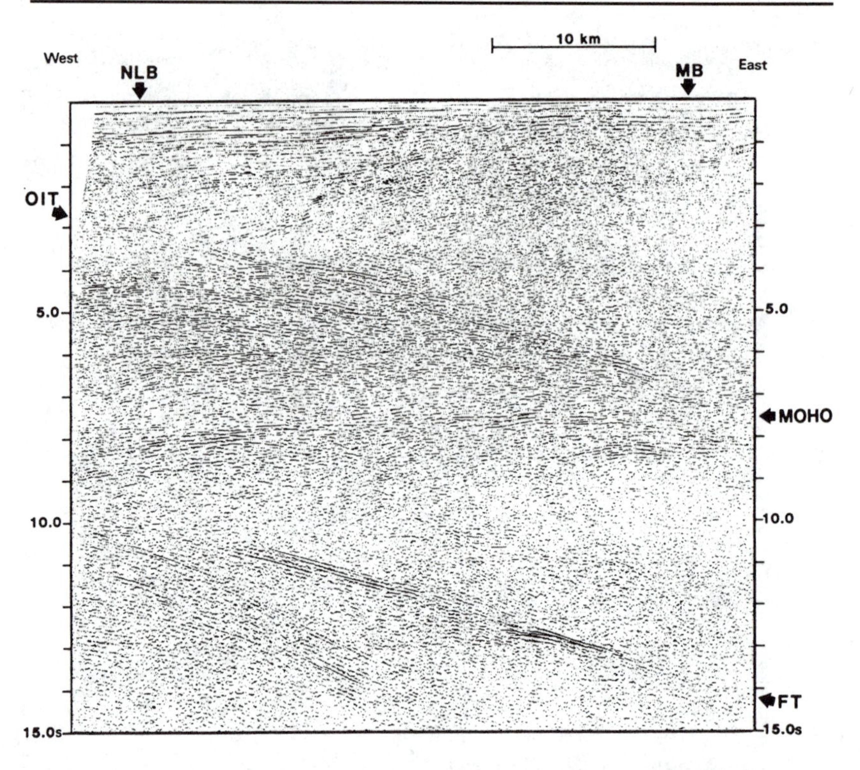

Figure 4.57. The British Institutes Reflection Profiling Syndicate (BIRPS), which started data acquisition in 1981, has collected a large amount of data on the continental shelf around the British Isles. The short segment of unmigrated 30-fold data shown here (taken from the WINCH profile) is from an east–west line off the north coast of Scotland. NLB, North Lewis Basin; MB, Minch Basin; OIT, Outer Isles Thrust; FT, Flannan Thrust. The Outer Isles Thrust can be traced from the surface down into the lower crust with a dip of approximately 25°. The basins are Mesozoic half-grabens. The North Lewis Basin and other basins apparently formed as a result of extension associated with the opening of the North Atlantic. The structure of the basins was controlled by the fabric of the existing Caledonian basement, in this instance the Outer Isles Thrust. The Moho on this section is a very clear reflector at 8–9 s two-way time (depth about 27 km). The Flannan Thrust cuts the Moho and extends into the upper mantle as a very clear reflector. Further data from this area are shown in Figures 9.23 and 9.24. (From Brewer et al. 1983.)

much lower P-wave velocity), by bodies of gabbro or eclogite in a peridotitic mantle or by mylonite zones.

Details of the structures as revealed by reflection profiling are discussed in Chapter 9; here we point out two illustrations of the type of data which can be obtained: Figure 4.56, which is a record section obtained by COCORP over the Rio Grande Rift in New Mexico, U.S.A. and Figure 4.57, which is part of a record section obtained by BIRPS (British Institutions Reflection Profiling Syndicate) off the north coast of Scotland.

PROBLEMS

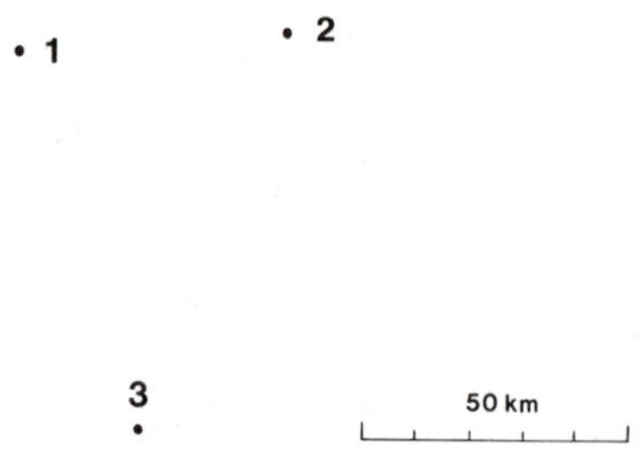

1. How would you distinguish between Love waves and Rayleigh waves if you were given an earthquake record from a WWSSN station?
2. What is the wavelength of a surface wave with a period of (a) 10 s, (b) 100s and (c) 200s? Comment on the use of surface waves in resolving small-scale lateral inhomogeneities in the crust and mantle.
3. How much greater is a nuclear explosion body-wave amplitude likely to be than an earthquake body-wave amplitude if both have the same M_s value and are recorded at the same distance?
4. (a) Which types of seismic wave can propagate in an unbounded solid medium?
   (b) Which types of seismic waves can be detected by a vertical component seismometer?
5. During a microearthquake survey in central Turkey, an earthquake was recorded by three seismometers (Fig. 4.58). Calculate the origin

• 1
• 2
3 •
50 km

Figure 4.58. Map showing location of seismometers 1, 2 and 3.

time for this earthquake and locate the epicentre relative to the seismometers by using information that follows:

	Hours	Minutes	Seconds
Seismometer 1			
P-wave arrival time	13	19	58.9
S-wave arrival time	13	20	04.7
Seismometer 2			
P-wave arrival time	13	20	02.6
S-wave arrival time	13	20	10.8
Seismometer 3			
P-wave arrival time	13	19	54.5
S-wave arrival time	13	19	57.4

Assume that the focus was the surface and use P- and S-wave velocities of 5.6 and 3.4 km s^{-1}, respectively.

6. Calculate the amount by which the seismic energy released by an earthquake increases when the surface-wave magnitude increases by one unit. Repeat the calculation for an increase of one unit in the body-wave magnitude.
7. The daily electrical consumption of the United States in 1985 was about 7×10^9 kilowatt hours. If this energy were released by an earthquake what would its magnitude be?
8. It is sometimes suggested that small earthquakes can act as safety valves by releasing energy in small amounts and so averting a damaging large earthquake. Assuming that an area (perhaps California) can statistically expect a large earthquake ($M_s > 8$) once every 50–100 years, calculate how many smaller earthquakes with (a) $M_s = 6$, (b) $M_s = 5$ or (c) $M_s = 4$ would be required during these years to release the same amount of energy. If it were possible to trigger these smaller earthquakes, would it help?
9. (a) Using Table 4.3, calculate the total amount of energy released each year by earthquakes. Suggest an estimate of the amount by which this figure may be in error.
   (b) Compare the total amount of energy released each year by earthquakes with the energy lost each year by heat flow (Chapter 7).
10. Using the earthquake travel-time curves in Figure 4.13 and the two earthquake records in Figure 4.14, determine for each earthquake (a) the epicentral distance (in degrees) and (b) the origin time.
11. Explain how the epicentre, focal mechanism and slip vector of an earthquake are determined. (Cambridge University Natural Sciences Tripos IB, 1974.)
12. Figures 4.59a, b give the directions of first motion for two earthquakes, plotted on equal-area projections of the lower focal hemisphere. For each earthquake, do the following:
    (a) Draw the two perpendicular planes that divide the focal sphere into positive and negative regions.

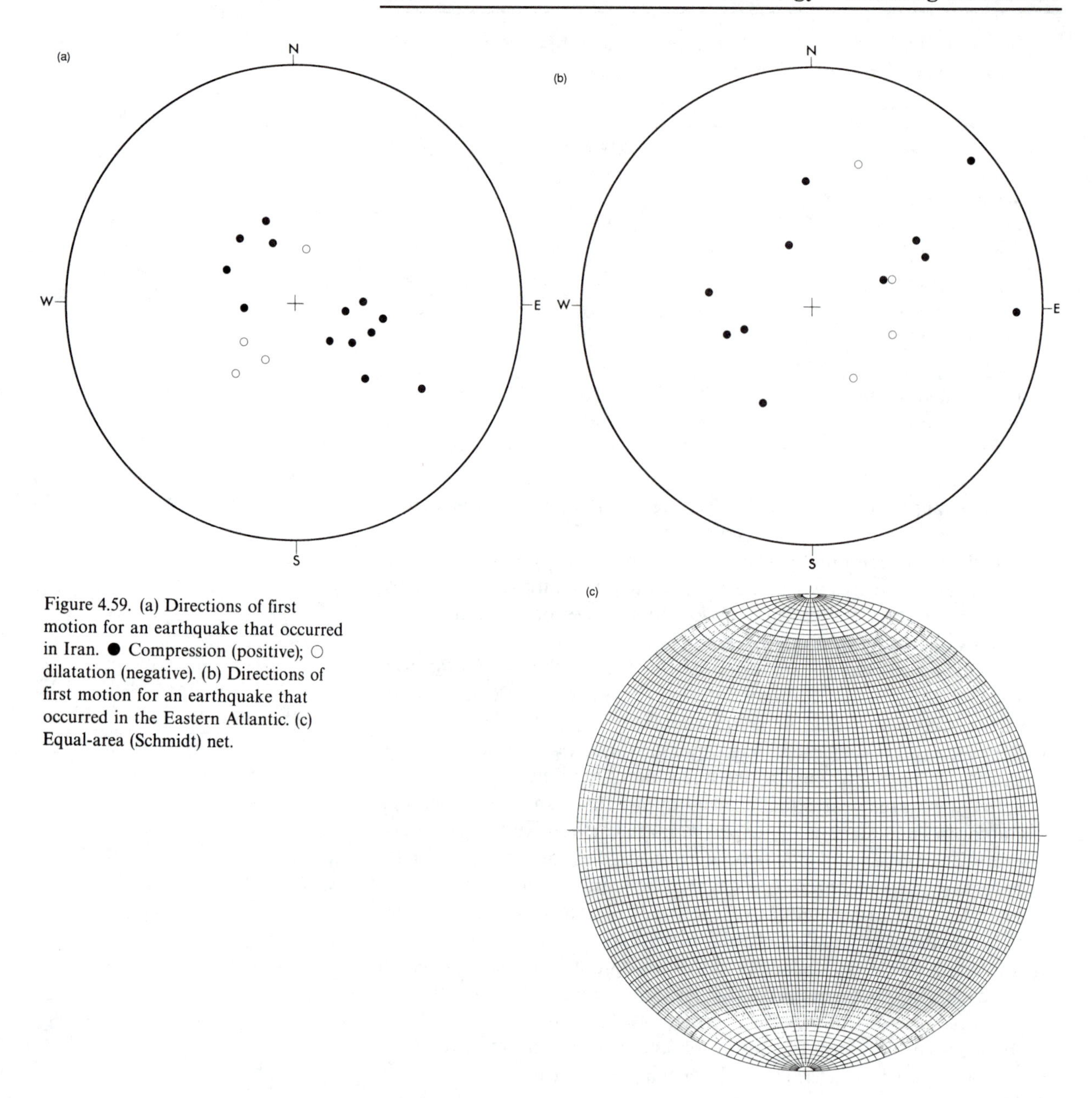

Figure 4.59. (a) Directions of first motion for an earthquake that occurred in Iran. ● Compression (positive); ○ dilatation (negative). (b) Directions of first motion for an earthquake that occurred in the Eastern Atlantic. (c) Equal-area (Schmidt) net.

(b) Find the strike and dip of these planes.
(c) Plot the horizontal component of strike.
(d) Discuss the mechanism of the earthquake.

13. Contrast the distribution of earthquake foci in and around the Atlantic and Pacific oceans and describe the mechanism of earthquakes on the midocean ridges. What light do these facts throw on the plate motions? (Cambridge University Natural Sciences Tripos IB, 1976.)
14. The outline map in Figure 4.60a shows the positions of two earthquakes recorded on the Mid-Atlantic ridge. Focal spheres of earthquakes 1 and 2 are also shown (Fig. 4.60b, c).

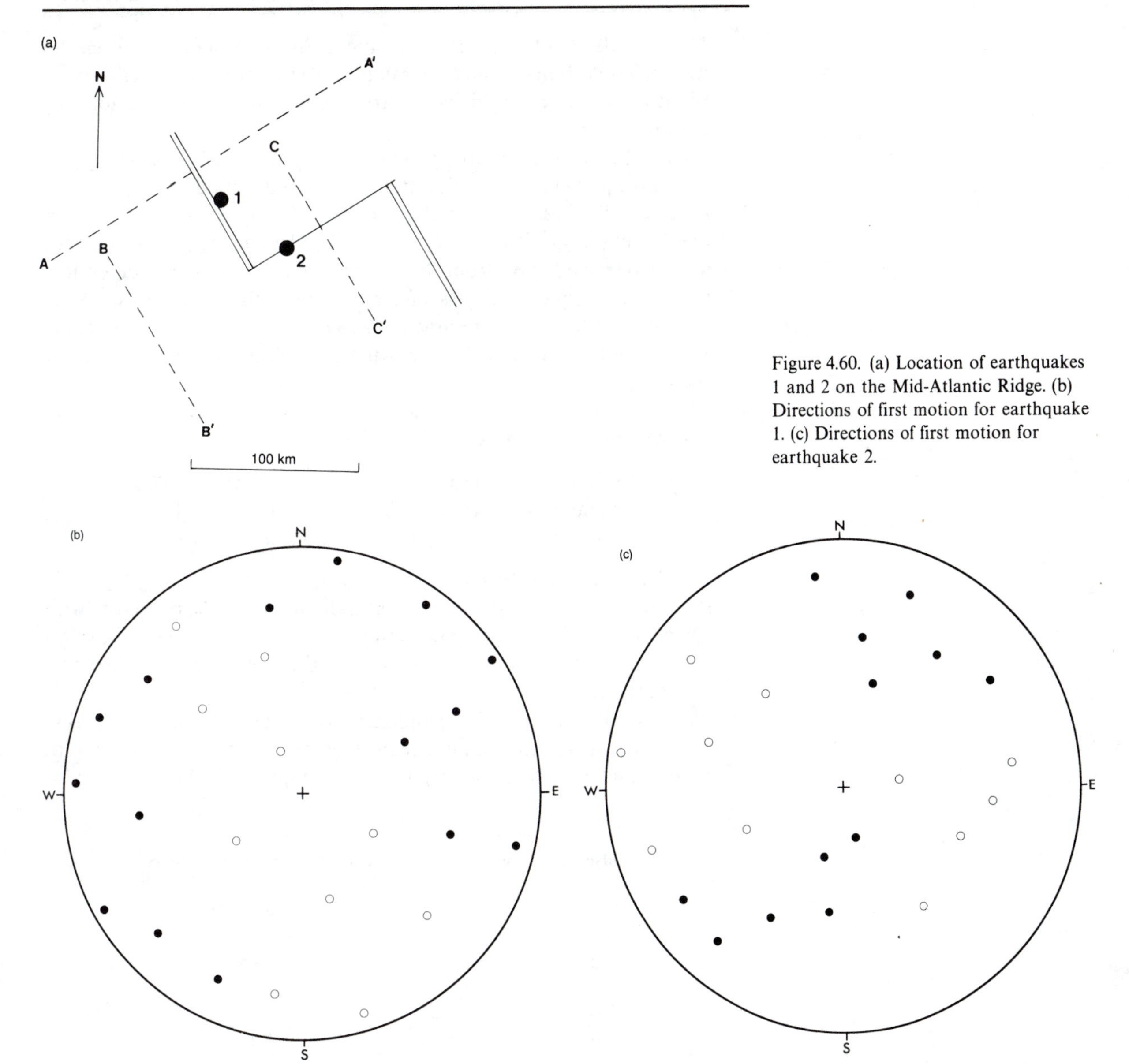

Figure 4.60. (a) Location of earthquakes 1 and 2 on the Mid-Atlantic Ridge. (b) Directions of first motion for earthquake 1. (c) Directions of first motion for earthquake 2.

(a) Draw the fault-plane solutions on the focal spheres for earthquakes 1 and 2.
(b) Indicate the fault types which generated earthquakes 1 and 2. From the tectonic setting of each earthquake, explain how it is possible to infer which plane is likely to be the fault plane. List the strikes and dips of the probable fault planes and the probable slip vectors.
(c) Sketch the bathymetric profiles along sections A–A', B–B' and C–C'.
(d) In this tectonic setting, over what depth range would you expect the earthquakes to occur?

(Cambridge University Natural Sciences Tripos IB, 1983.)

15. The boundary between the African and Eurasian plates, between the Azores Triple Junction and Gibraltar, strikes approximately east–west. What is the nature of this boundary? Sketch fault-plane solutions for the following:
    (a) Earthquakes along this plate boundary west of Gibraltar
    (b) Earthquakes east of Gibraltar in the Mediterranean
16. A 200 m thick layer of sediment with P-wave velocity $2\,\mathrm{km\,s^{-1}}$ overlies material with velocity $3\,\mathrm{km\,s^{-1}}$. Calculate the critical distance for the refracted wave from the second layer. If you had to determine the velocities and thickness of a structure such as this by shooting a seismic refraction experiment, indicate how many seismometers you would require and over what distances you would deploy them.
17. The first-arrival time–distance data from a seismic refraction experiment are as follows:

    refraction velocity $6.08\,\mathrm{km\,s^{-1}}$, intercept 1.00 s
    refraction velocity $6.50\,\mathrm{km\,s^{-1}}$, intercept 1.90 s
    refraction velocity $8.35\,\mathrm{km\,s^{-1}}$, intercept 8.70 s

    (a) Calculate the velocity–depth structure.
    (b) Now assume that there is a hidden layer: A refractor with velocity $7.36\,\mathrm{km\,s^{-1}}$ and intercept 5.50 s is a second-arrival at all ranges. By how much is the total crustal thickness increased?
18. The first-arrival travel times and distances obtained from a refraction experiment shot along a midocean ridge axis are tabulated here. Plot them on a time–distance graph and determine a crustal structure for

Offset distance (km)	First-arrival travel time (s)
2.0	4.10
3.0	4.45
4.0	4.85
5.0	5.20
6.0	5.40
7.0	5.60
8.0	5.80
9.0	5.95
10.0	6.15
12.0	6.50
14.0	6.75
16.0	7.00
18.0	7.35
20.0	7.50
22.0	7.90
24.0	8.10
26.0	8.40
28.0	8.70
30.0	8.90

the line, assuming that the structure can be approximated by uniform horizontal layers. The water is 3 km deep and has a P-wave velocity of 1.5 km s^{-1}.
If the upper mantle has a normal P-wave velocity of 8.1 km s^{-1} in this region, what is its minimum depth? Repeat the calculation assuming that the upper mantle has a velocity of only 7.6 km s^{-1}. Ideally, how long should the refraction line have been to make a measurement of the upper mantle velocity and depth? Explain how a record section would assist in determining the velocity structure.

19. Using the velocities for Love waves along continental ray paths (Fig. 4.6a), make a rough estimate of the minimum time over which Love waves arrive at a seismometer at an epicentral distance of (a) 20° and (b) 100°. Assume that the seismometer records waves with periods 20–250 s.
20. Calculate a velocity–depth model for the data shown in Fig. 4.49. Assume that the structure is horizontally layered and that all layers have constant velocity. Use only the reflections at 1.2, 1.5, 1.8, 2.0, 2.3 and 2.45 s.
21. For the simple example used in the text,

$$t_{0,1} = 1.0\,\text{s}, \qquad \bar{\alpha}_1 = 3.6\,\text{km s}^{-1}$$

$$t_{0,2} = 1.5\,\text{s}, \qquad \bar{\alpha}_2 = 4.0\,\text{km s}^{-1}$$

calculate the error in velocity and thickness which results in the following cases:
(a) The travel times have an error of ± 0.05 s.
(b) The travel times have an error of ± 0.01 s.
(c) The rms velocities have an error of ± 0.1 km s^{-1}.
Comment on the accuracy of interval velocity–depth models obtained from rms velocities and two-way travel times.

22. Calculate the normal incidence P-wave reflection coefficients for the following interfaces:
(a) Sandstone ($\alpha = 3.0$ km s^{-1}, $\rho = 2.2 \times 10^3$ kg m^{-3}) above a limestone ($\alpha = 4.1$ km s^{-1}, $\rho = 2.2 \times 10^3$ kg m^{-3})
(b) A possible lower crustal interface of basalt ($\alpha = 6.8$ km s^{-1}, $\rho = 2.8 \times 10^3$ kg m^{-3}) over granulite ($\alpha = 7.3$ km s^{-1}, $\rho = 3.2 \times 10^3$ kg m^{-3})
(c) A possible crust–mantle interface of granulite ($\alpha = 7.3$ km s^{-1}, $\rho = 3.2 \times 10^3$ kg m^{-3}) over peridotite ($\alpha = 8.1$ km s^{-1}, $\rho = 3.3 \times 10^3$ kg m^{-3})
23. Calculate the wavelength of a 30 Hz seismic P-wave in these rocks:
(a) Sandstone ($\alpha = 3.0$ km s^{-1})
(b) Gabbro ($\alpha = 7.0$ km s^{-1})
What do these calculations indicate about the resolution of shallow and deep crustal structures by seismic reflection profiling.
24. Given the following time–distance data from a seismic reflection survey, calculate the depth of the reflector and a velocity for the material above it. What type is this velocity?

Distance (km)	Time (s)
0.5	1.01
1.0	1.03
1.5	1.07
2.0	1.11
2.5	1.17
3.0	1.25
3.5	1.33

25. If normal incidence reflection lines were shot over the structure shown in Fig. 4.61, mark on record sections the arrival times of the phases that would be recorded. What would be the relative amplitudes of the phases and by what physical factors are they controlled?

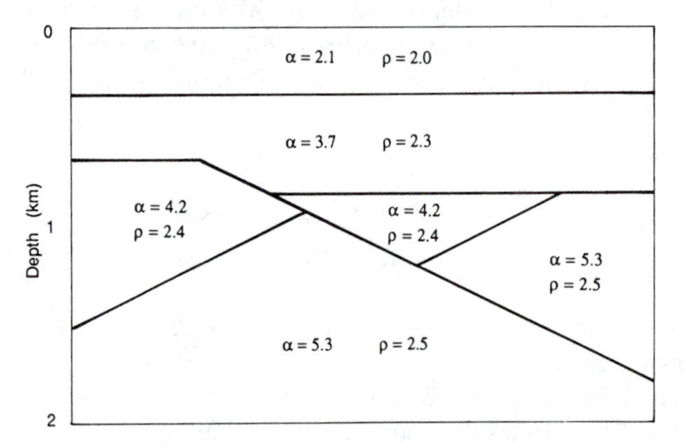

Figure 4.61. Vertical cross section. P-Wave velocities in km s^{-1}; densities in 10^3 kg m^{-3}.

26. Using the parametric form of the travel-time and epicentral distance equations for earthquakes (see Appendix 3), obtain the following relation:

$$t = p\Delta + 2\int_{r_{\min}}^{r_{\max}} \frac{(\eta^2 - p^2)^{1/2}}{r}\,dr$$

27. A region is believed to have the following horizontally layered structure:

Depth (km)	P-Wave velocity (km s^{-1})
0.0–0.5	1.0
0.5–2.0	3.0
2.0–5.0	5.5
5.0–∞	7.0

(a) Plot the time–distance curve for the structure.
(b) How would you position shots and receivers to test whether the assumed structure is correct? Illustrate with simple layout diagrams and outline your reasoning.

(c) How can you exclude the possibility that the structure has dipping interfaces?
(d) Give two methods by which a velocity inversion in the structure could be detected.
(Cambridge University Natural Science Tripos II, 1986)

28. Ewing Oil decides to shoot two marine seismic surveys. For the first survey, water guns are fired every 17 m, the hydrophone streamer is 2.45 km long and consists of 72 separate hydrophone sections and both the source and streamers are towed at 5 m depth. In the second survey, air guns are fired every 50 m, the streamer is 5 km long and consists of 100 sections and both the source and receivers are towed at 15 m depth.
(a) If the velocity of sea water is 1.5 km s^{-1}, what is the peak frequency for each survey?
(b) What is the 'fold' of each survey (i.e., the number of separate traces stacked together in the final section)?
(c) One survey was intended to be a high-resolution survey of a shallow sedimentary section, and the other was a deep survey to image the whole crust. Which survey is which? Give your reasoning.

29. The shallow survey of problem 28 revealed a simple sedimentary sequence, shown in the seismic section of Fig. 4.62. The normal moveouts observed for each reflector when the source-to-receiver offset was 1000 m are listed in the table.

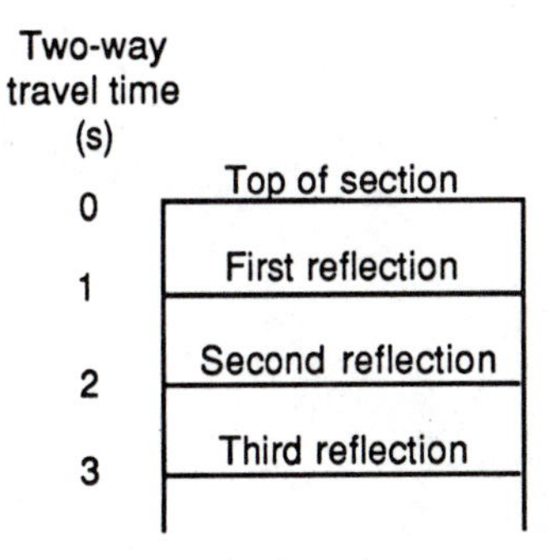

Figure 4.62. Seismic section.

	Two-way time (s)	Δt_{NMO} (ms)
First reflection	1	222
Second reflection	2	77
Third reflection	3	27

(a) What is the appropriate stacking velocity for each reflector?
(b) Calculate the corresponding interval velocities, and from these determine the depth to the deepest reflector.
(c) Suggest a plausible stratigraphy for this area.

30. The deep survey of problem 28 was run over the cross section shown in Fig. 4.63. Velocities and densities are given in the table.

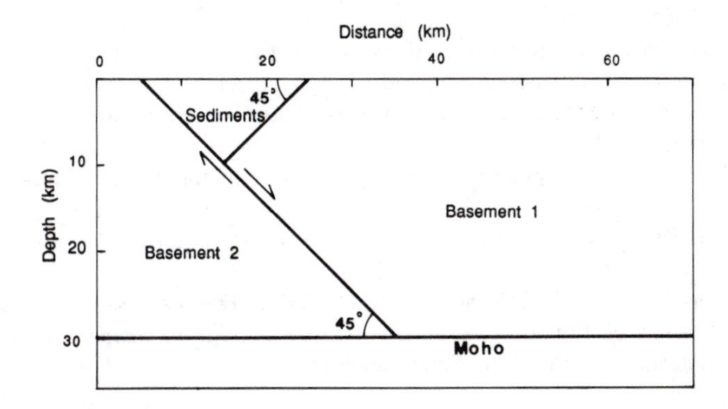

Figure 4.63. Vertical cross section.

	Sedimentary rocks	Basement 1	Basement 2
Velocity ($km\,s^{-1}$)	3.0	6	6
Density ($\times 10^3\,kg\,m^{-3}$)	2.2	2.6	2.7

(a) By considering rays perpendicular to the reflectors, draw the corresponding time section. Use a scale of $2\,cm \equiv 5\,km$ and $1\,cm \equiv 1\,s$. Ignore diffractions.

(b) Draw a rough sketch of the shape of the Moho after migration if this section had been mistakenly migrated at a constant velocity of $6\,km\,s^{-1}$. Ignore diffractions.

31. What is the reflection coefficient of the fault, shown in Figure 4.63, (a) at the sediment–basement contact and (b) within the basement?
32. Desperate to find something of interest in the data of problems 29 and 30, J. R. Ewing authorized heat-flow measurements over the sedimentary basin of Figure 4.63. If the surface heat flow is $0.75\,mW\,m^{-2}$ and the thermal conductivity is $2.5\,W\,m^{-1}\,°C^{-1}$, what is the temperature (a) at the base of the sediments and (b) at the Moho? State your assumptions and comment on your answer (see Chapter 7). (Problems 28–32, from Cambridge Natural Science Tripos II, 1986.)
33. Your foot is trapped under a railway sleeper. You feel the rail vibrate and then 20 s later hear a faint whistle. Estimate how much time you have to free yourself. State your assumptions.

BIBLIOGRAPHY

Ahrens, T. J. 1982. Constraints on core composition from shock-wave data. *Phil. Trans. R. Soc. Lond. A, 306*, 37–47.

Aki, K., and Richards, P. G. 1980. *Quantitative seismology – Theory and methods.* Freeman, San Francisco.

Anderson, D. L. 1986. Properties of iron at the earth's core conditions. *Geophys. J. R. Astr. Soc., 84*, 561–80.

Anderson, D. L., and Hart, R. S. 1976. An earth model based on free oscillations and body waves. *J. Geophys. Res., 81*, 1461–75.

1978. Q of the Earth. *J. Geophys. Res., 83*, 5869–82.

Anderson, D. L., and Dziewonski, A. M. 1984. Seismic tomography. *Sci. Am., 251*, 4, 60–8.

Båth, M. 1966. Earthquake energy and magnitude. *Phys. Chem. Earth, Pergamon*, 7, 115–65.

1977. Teleseismic magnitude relations. *Ann. Geofis., 30*, 299–327.

1979. *Introduction to seismology*, 2nd ed. Birkhauser Verlag, Basel.

1981. Earthquake magnitude – recent research and current trends. *Earth Science Reviews, 17*, 315–98.

Birch, F. 1968. On the possibility of large changes in the earth's volume. *Phys. Earth Planet. Int., 1*, 141–7.

Bolt, B. A. 1976. *Nuclear explosions and earthquakes: The parted veil.* Freeman, San Francisco.

1982. *Inside the earth: Evidence from earthquakes.* Freeman, San Francisco.

1987. 50 years of studies on the inner core. *EOS Trans. Am. Geophys. Un.*, **68**, 73.

1988. *Earthquakes.* Freeman, San Francisco.

Brewer, J. A. Matthews, D. H., Warner, M. R., Hall, J., Smythe, D. K., and Whittington, R. J. 1983. BIRPS deep seismic reflection studies of the British Caledonides. *Nature, 305*, 206–10.

Brown, L. D. 1986. Aspects of COCORP deep seismic profiling. *In* M. Barazangi and L. D. Brown, eds., *Reflection seismology: A global perspective*, Vol. 13 of Geodynamics Series, Am. Geophys. Un., 209–22.

Brown, L. D., Kaufman, S., and Oliver, J. E. 1983. COCORP seismic traverse across the Rio Grande Rift. In A. W. Bally, ed., *Seismic expression of structural styles*, Amer. Assoc. Petrol. Geol. Studies in Geol. No. 15, 2.2.1–1–6.

Bullen, K. E., and Bolt, B. A. 1985. *An introduction to the theory of seismology*, 4th ed., Cambridge Univ. Press.

Cerv̌eny, V. 1979. Ray theoretical seismograms for laterally inhomogeneous structures. *J. Geophys. Res., 46*, 335–42.

Chapman, C. H. 1978. A new method for computing synthetic seismograms. *Geophys. J. R. Astr. Soc., 54*, 481–518.

Claerbout, J. F. 1985. *Imaging the earth's interior*. Blackwell Scientific, Oxford.

Clark, S. P. 1966. *Handbook of physical constants*, Vol. 97 of Mem. Geol. Soc. Am.

Cox, A., and Hart, R. B. 1986. *Plate tectonics: How it works*. Blackwell Scientific, Oxford.

Davies, D., and McKenzie, D. P. 1969. Seismic travel-time residuals and plates. *Geophys. J. R. Astr. Soc., 18*, 51–63.

Davies, J., Sykes, L., House, L., and Jacob, K. 1981. Shumagin seismic gap, Alaska peninsula: History of great earthquakes, tectonic setting and evidence for high seismic potential. *J. Geophys. Res., 86*, 3821–55.

Dey-Sarkar, S. K., and Chapman, C. H. 1978. A simple method for the computation of body wave seismograms. *Bull. Seism. Soc. Am., 68*, 1577–93.

Dobrin, M. B., and Savit, C. H. 1988. *Introduction to geophysical prospecting*, 4th ed. McGraw-Hill, New York.

Dziewonski, A. M. 1984. Mapping the lower mantle: Determination of lateral heterogeneity in P velocity up to degree and order 6. *J. Geophys. Res., 89*, 5929–52.

Dziewonski, A. M., and Anderson, D. L. 1981. Preliminary reference earth model. *Phys. Earth Planet. Inter., 25*, 297–356.

Dziewonski, A. M., and Woodhouse, J. H. 1987. Global images of the earth's interior. *Science, 236*, 37–48.

Frolich, C. 1989. Deep earthquakes. *Sci. Am., 259*, 1, 48–55.

Fuchs, K., and Müller, G. 1971. Computation of synthetic seismograms with the reflectivity method and comparison with observations. *Geophys. J. R. Astr. Soc., 23*, 417–33.

Garland, G. D. 1979. *Introduction to geophysics – Mantle, core and crust*, 2nd ed. Saunders, Philadelphia.

Gebrande, H. 1982. *In* "Landolt-Bornstein" ed. K. H. Hellwege Group 5, vol. 16, "Physical Properties of Rocks" ed. G. Angenheister. Springer-Verlag, Berlin 1–96.

Gubbins, D. 1989. *Seismology and plate tectonics*. Cambridge Univ. Press, Cambridge.

Gutenberg, B. 1945. Amplitudes of P, PP and S and magnitude of shallow earthquakes. *Bull. Seism. Soc. Am.*, **35**, 57–69.

1945. Magnitude determination for deep focus earthquakes. *Bull. Seism. Soc. Am., 35*, 117–30.

1959. *Physics of the earth's interior*. Academic, New York.

Gutenberg, B., and Richter, C. F. 1954. *Seismicity of the earth and associated phenomena*. Princeton Univ. Press, Princeton.

1956. Magnitude and energy of earthquakes. *Ann. Geoȷis., 9*, 1–15.

Hart, R. S., Anderson, D. L., and Kanamori, H. 1977. The effect of attenuation on gross earth models. *J. Geophys. Res.*, *82*, 1647–54.

Hatton, L., Worthington, M. H., and Makin, J. 1985. *Seismic data processing: Theory and practice.* Blackwell Scientific, Oxford.

Helmberger, D. W. 1968. The crust–mantle transition in the Bering Sea. *Bull. Seism. Soc. Am.*, *58*, 179–214.

Herrin, E. 1968. Introduction to 1968 seismological tables for P phases. *Bull. Seism. Soc. Am.*, *58*, 1193–241.

Jacobs, J. A. 1987. *The earth's core*, 2nd ed. Academic, London.

Jeanloz, R. 1983. The earth's core. *Sci. Am.*, *249*, 56–65.

1988. High-pressure experiments and the earth's deep interior. *Physics Today* (Jan.), S44–5.

Jeanloz, R., and Ahrens, T. J. 1980. Equations of state of FeO and CaO. *Geophys. J. R. Astr. Soc.*, *62*, 505.

Jeffreys, H. 1926. On near earthquakes. *Mon. Not. Roy. Astr. Soc. Geophys. Suppl.*, *1*, 385.

1939. The times of P, S and SKS and the velocities of P and S. *Mon. Not. R. Astr. Soc. Geophys. Suppl.*, *4*, 498–533.

1939. The times of PcP and ScS. *Mon. Not. R. Astr. Soc. Geophys. Suppl.*, *4*, 537–47.

1976. *The earth*, 6th ed. Cambridge Univ. Press.

Jeffreys, H., and Bullen, K. E. 1988. *Seismological tables.* British Association Seismological Investigations Committee, Black Bear Press, Cambridge.

Kanamori, H., and Anderson, D. L. 1975. Theoretical basis of some empirical relations in seismology. *Bull. Seism. Soc. Am.*, *65*, 1073–95.

Kennett, B. L. N. 1975. The effects of attenuation on seismograms. *Bull. Seism. Soc. Am.*, *65*, 1643–1651.

1977. Towards a more detailed seismic picture of the oceanic crust and mantle. *Marine Geophys. Res.*, *3*, 7–42.

Kerr, R. A. 1979. Earthquake – Mexican earthquake shows the way to look for the big ones. *Science*, *203*, 860–2.

Knopoff, L., and Chang, F.-S. 1977. The inversion of surface wave dispersion data with random errors. *J. Geophys.*, *43*, 299–309.

Liu, L.-G. 1974. Birch's diagram: Some new observations. *Phys. Earth Planet. Int.*, *8*, 56–62.

Love, A. E. H. 1911. *Some problems of geodynamics.* Cambridge Univ. Press.

Ludwig, W. J., Nafe, J. E., and Drake, C. L. 1970. Seismic refraction. *In* A. E. Maxwell, ed., *The sea*, Vol. 4, Part 1. Wiley-Interscience, New York, 53–84.

McGeary, S., and Warner, M. R. 1985. DRUM: Seismic profiling of the continental lithosphere. *Nature*, *317*, 795–7.

McKenzie, D. P. 1983. The earth's mantle. Sci. Am., *249*, 3, 66–113.

McKenzie, D. P., and Parker, R. L. 1967. The North Pacific: An example of tectonics on a sphere. *Nature*, *216*, 1276–80.

Meissner, R. 1986. *The continental crust, a geophysical approach.* Vol. 34 of International Geophysics Series, W. L. Donn, ed., Academic, Orlando, Florida.

Morelli, A., and Dziewonski, A. M. 1987. Topography of the core–mantle boundary and lateral homogeneity of the outer core. *Nature*, *325*, 678.

Müller, G. 1970. Exact ray theory and its applications to the reflection of elastic waves from vertically homogeneous media. *Geophys. J. R. Astr. Soc.*, *21*, 261–84.

1985. The reflectivity method: A tutorial. *J. Geophys.*, *58*, 153–74.

Müller, G., and Kind, R. 1976. Observed and computed seismogram sections for the whole earth. *Geophys. J. R. Astr. Soc.*, *44*, 699–716.

Nataf, H.-C., Nakanishi, I., and Anderson, D. L. 1984. Anisotropy and shear wave velocity heterogeneities in the upper mantle. *Geophys. Res. Lett.*, *11*, 109.

National Research Council (U.S.). 1983. Workshop on guidelines for instrumentation design in support of a proposed lithospheric seismology program.

Nowroozi, A. A. 1986. On the linear relation between m_b and M_s for discrimination between explosions and earthquakes. *Geophys. J. R. Astr. Soc.*, *86*, 687–99.

Officer, C. B. 1974. *Introduction to theoretical geophysics.* Springer-Verlag, Berlin and New York.

Ohtani, E., and Ringwood, A. E. 1984. Composition of the Core, I. Solubility of oxygen in molten iron at high temperatures. *Earth Planet. Sci. Lett.*, *71*, 85–93.

Ohtani, E., Ringwood, A. E., and Hibberson, W. 1984. Composition of the core, II. Effect of high pressure on solubility of FeO in molten iron. *Earth Planet. Sci. Lett.*, *71*, 94–103.

Panza, G. F. 1981. *In* R. Cassinus, ed., *The solution of the inverse problem in geophysical interpretation*, Plenum, New York, 39–78.

Phillips, O. M. 1968. *The heart of the earth.* Freeman, Cooper & Co., San Francisco.

Press, F. 1966. Seismic velocities. *In* S. P. Clark, ed., *Handbook of physical constants*, Vol. 97 of Geol. Soc. Am. Mem., 195–218.

1975. Earthquake prediction. *Sci. Am.*, *232*, May, 14–23.

Rayleigh (J. W. Strutt). 1887. On waves propagated along the plane surface of an elastic solid. *London Math. Soc. Proc.*, *17*, 4–11.

Richter, C. F. 1958. *Elementary seismology.* Freeman, New York.

Robinson, E. A., and Treitel, S. 1980. *Geophysical signal analysis.* Prentice-Hall, Englewood Cliffs, N.J.

Silver, P. G., Carlson, R. W., and Olson, P. 1988. Deep slabs, geochemical heterogeneity and the large-scale structure of mantle convection: Investigation of an enduring paradox. *Ann. Rev. Earth Planet. Sci.*, *16*, 477–541.

Simon, R. B. 1981. *Earthquake interpretations: a manual for reading seismograms.* Kaufmann, Los Altos, California.

Slotnick, M. M. 1959. Lessons in seismic computing. R. A. Geyer, ed., Society Exploration Geophysicists.

Sykes, L. R. 1967. Mechanism of earthquakes and faulting on the mid-ocean ridges. *J. Geophys. Res.*, *72*, 2131–53.

Taner, T. G., and Koehler, F. 1969. Velocity-spectra-digital computer derivation and applications of velocity functions. *Geophysics*, *34*, 859–81.

Telford, W. M., Geldart, L. P., and Sheriff, R. E. 1990. *Applied geophysics*, 2nd ed. Cambridge Univ. Press, New York.

Tooley, R. D., Spencer, T. W., and Sagoci, H. F. 1965. Reflection and transmission of plane compressional waves. *Geophysics*, *30*, 552–70.

van der Borne, G. 1904. *Nachr. Ges. Wiss. Göttingen*, 1–25.

Woodhouse, J. H., and Dziewonski, A. M. 1984. Mapping the upper mantle: Three dimensional modelling of earth structure by inversion of seismic waveforms. *J. Geophys. Res.*, *89*, 5953–86.

Yilmaz, O. 1987. *Seismic data processing.* Society of Exploration Geophysicists.

Zoback, M. D. 1980. Recurrent intraplate tectonism in the New Madrid seismic zone. *Science*, *209*, 971–6.

5

Gravity

5.1 Introduction

Seismic methods for determining the internal structure of the earth require the recording and analysis of energy which has passed through the earth, energy which was produced either by earthquakes or artificially by exploration teams. Gravimetric and magnetic methods are different in that they utilize measurements of existing potential fields, which are physical properties of the earth itself.

5.2 Gravitational Potential and Acceleration

Two point masses m_1 and m_2 at distance r apart attract each other with a force F,

$$F = \frac{Gm_1m_2}{r^2} \tag{5.1}$$

where G is the gravitational or Newtonian constant (Fig. 5.1). In the SI system of units, G has a value of $6.67 \times 10^{-11}\,\mathrm{m^3\,kg^{-1}\,s^{-2}}$. This inverse square law of gravitational attraction was deduced by Sir Isaac Newton in 1666: The legend is that an apple falling from a tree gave him the revolutionary idea that the same force that attracted the apple downwards could also account for the moon's orbit of the earth.

The acceleration of the mass m_1 due to the presence of the mass m_2 is Gm_2/r^2 towards m_2 (since force equals mass multiplied by acceleration), and the acceleration of the mass m_2 is Gm_1/r^2 towards m_1.

The *gravitational potential* V due to mass m_1 can be defined* as

$$V = -\frac{Gm_1}{r} \tag{5.2}$$

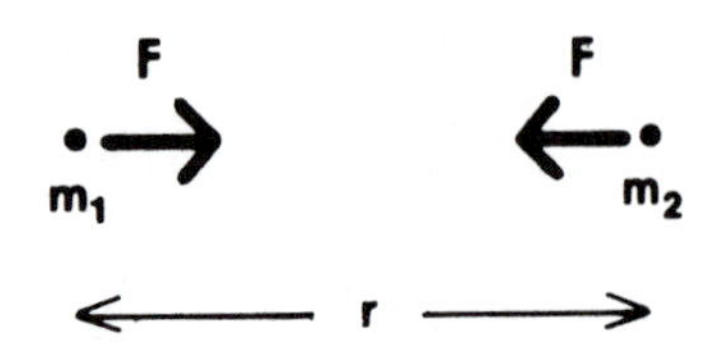

Figure 5.1. Two masses m_1 and m_2, at distance r apart, attract each other with force $F = Gm_1m_2/r^2$.

The *gravitational potential energy* of a mass m_2 at distance r from mass m_1 is $-Gm_1m_2/r$. If mass m_2 moves to a new position, r' from mass m_1, the gravitational energy that is released is $Gm_1m_2/r' - Gm_1m_2/r$. The *gravitational acceleration* a of mass m_2 towards m_1 (taken to be positive in the

* The standard convention is to define $V = Gm_1/r$.

outward radial direction) can be written in terms of the potential V by using Eq. 5.2:

$$a = -\frac{Gm_1}{r^2}$$
$$= -\frac{\partial}{\partial r}\left(-\frac{Gm_1}{r}\right)$$
$$= -\frac{\partial V}{\partial r} \qquad (5.3)$$

If we generalize to three dimensions, Eq. 5.3 is written as

$$a = -\operatorname{grad} V \qquad (5.4)$$

or

$$a = -\nabla V$$

(See Appendix 1 for discussion of *grad* or ∇.)

If instead of just one mass m_1, we imagine a distribution of masses, we can then define a potential V as

$$V = -G\sum_i \frac{m_i}{r_i} \qquad (5.5)$$

or

$$V = -G\int_m \frac{dm}{r} \qquad (5.6)$$

where (in Eq. 5.5) each mass m_i is at position $\mathbf{r}_i$ and (in Eq. 5.6) the integral over m is summing all the infinitesimal masses dm, each at its position $\mathbf{r}$. The gravitational acceleration due to either of these distributions is then again given by Eq. 5.4.

We can now use Eq. 5.6 to calculate the potential of a spherical shell (Fig. 5.2). Let us calculate the potential at a point P at distance r from the centre O of the shell. Consider the thin strip of shell, half of which is shown (stippled) in the figure. This circular strip has an area of

$$(2\pi b \sin\theta)(b\, d\theta)$$

If we assume the shell to be t in thickness (and very thin) and to be of uniform density ρ, the total mass of the strip is

$$\rho t 2\pi b^2 \sin\theta\, d\theta$$

Because every point on the strip is the same distance D from point P, Eq. 5.2 gives the potential at P due to the strip as

$$-\frac{G\rho t 2\pi b^2 \sin\theta\, d\theta}{D}$$

Applying the cosine formula to triangle OQP gives D in terms of r, b and θ as

$$D^2 = r^2 + b^2 - 2br\cos\theta \qquad (5.7)$$

The potential of the entire spherical shell can now be evaluated from Eq. 5.6

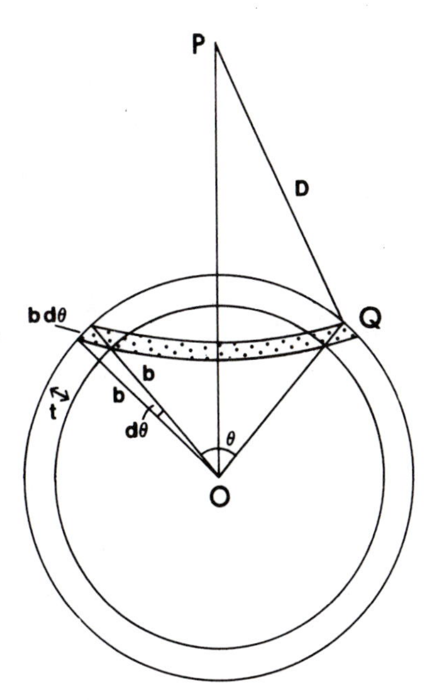

Figure 5.2. A strip of a thin, spherical shell with centre O. P is an arbitrary point at distance r from O.

by integrating the potential of the strip

$$V = -G\rho t 2\pi b^2 \int \frac{\sin\theta \, d\theta}{(r^2 + b^2 - 2br\cos\theta)^{1/2}} \tag{5.8}$$

To evaluate this integral, we need to change it from an integral over θ to an integral over D. Differentiation of Eq. 5.7 gives

$$D\,dD = br\sin\theta\,d\theta \tag{5.9}$$

Equation 5.8 is then written

$$V = -G\rho t 2\pi b^2 \int_{D_{\min}}^{D_{\max}} \frac{dD}{br} \tag{5.10}$$

When the point P is *external* to the spherical shell, as in Figure 5.2, the limits $D_{\min}$ and $D_{\max}$ are $r-b$ and $r+b$, respectively, and the potential at point P is

$$\begin{aligned} V &= -G\rho t 2\pi b^2 \left[\frac{D}{br}\right]_{r-b}^{r+b} \\ &= -\frac{G\rho t 4\pi b^2}{r} \end{aligned} \tag{5.11}$$

The total mass of the spherical shell is $4\mu b^2 \rho t$, and so, at the point P the potential of the shell is the same as that due to an equal mass placed at O, the centre of the shell. The gravitational acceleration a due to the spherical shell can be calculated from Eq. 5.4 (in spherical polar coordinates) as

$$a = -\frac{\partial V}{\partial r} = -\frac{G\rho t 4\pi b^2}{r^2} \tag{5.12}$$

This acceleration is the same as the acceleration for the situation when the entire mass of the shell $4\pi b^2 \rho t$ is concentrated at O, the centre of the shell.

When the point P is *inside* the spherical shell, the limits $D_{\min}$ and $D_{\max}$ in Eq. 5.10 are $b-r$ and $b+r$, respectively. In this case, the potential at P is

$$\begin{aligned} V &= -G\rho t 2\pi b^2 \left[\frac{D}{br}\right]_{b-r}^{b+r} \\ &= -G\rho t\, 4\pi b \end{aligned} \tag{5.13}$$

This potential is a constant, independent of the position of point P inside the shell. The gravitational acceleration, being the negative gradient of the potential, is therefore zero inside the shell.

Using these results for the gravitational potential and acceleration of a spherical shell, we can immediately see that at any external point the gravitational potential and the acceleration due to a sphere are the same as the values due to an equal mass placed at the centre of the sphere. In addition, at any point within a sphere the gravitational acceleration is that due to all the material closer to the centre than the point itself. The contribution from all the material outside the point is zero. Imagine the sphere being made of a series of thin uniform spherical shells; then from Eq. 5.13 we can see that none of the shells surrounding the point make any contribution to the acceleration. The gravitational acceleration a at a distance r from a sphere of radius b ($b < r$) and density ρ is, therefore,

$$a = -\frac{G\rho\frac{4}{3}\pi b^3}{r^2} = -\frac{GM}{r^2} \tag{5.14}$$

where M is the mass of the sphere. The minus sign in Eqs. 5.12 and 5.14 arises because gravitational acceleration is positive inwards, whereas r is positive outwards. A radial variation of density within the sphere does not affect these results, but any lateral variations within each spherical shell do render them invalid.

5.3 Gravity of the Earth

5.3.1 Reference Gravity Formula

We can apply Eq. 5.14 to the earth if we assume it to be perfectly spherical. The gravitational acceleration towards the earth is then given by

$$-a = \frac{GM_E}{r^2} \tag{5.15}$$

where M_E is the mass of the earth. The value of the gravitational acceleration at the *surface* (denoted by g which is taken to be positive inwards) of a spherical earth is, therefore, GM_E/R^2 where R is the radius of the earth. At the earth's surface, gravity has a value of about 9.81 m s^{-2}.

The first person to measure the earth's gravity was Galileo. Another celebrated legend is that he conducted his experiments by dropping objects from the top of the leaning tower in Pisa and timing their fall to the ground. (In fact, he slid objects down inclined planes, which reduced their acceleration g to $g \sin\delta$, where δ is the dip angle of the plane; thus, they moved more slowly and he could time them more accurately.) In his honour a gravitational unit, the *gal*, was named: 1 gal $= 10^{-2}$ m s^{-2}. The gravitational acceleration at the earth's surface is therefore about 981 gal.

If the earth were perfectly spherical and not rotating, the gravitational acceleration would have the same value at every point on its surface. However, the earth is not a perfect sphere (it bulges at the equator and is flattened at the poles, and it is rotating. The earth's shape can be approximated by an *oblate spheroid*, the surface that is generated by revolving an ellipse about its minor (shorter) axis. The *ellipticity*, or *polar flattening*, f of an ellipse is defined as

$$f = \frac{R_e - R_p}{R_e} \tag{5.16}$$

where R_e and R_p are the equatorial (longer) and polar (shorter) radii, respectively. Ellipticity should not be confused with eccentricity, which is defined as $\sqrt{R_e^2 - R_p^2}/R_e$. The oblate spheroid that best approximates the earth's shape has an ellipticity of 1/298·247. The 'radius' of an oblate spheroid is given, to first order in f, by

$$r = R_e(1 - f\sin^2\lambda) \tag{5.17}$$

where f is the ellipticity, λ the latitude and R_e the equatorial radius.

Centrifugal acceleration means that the gravitational acceleration on a sphere rotating with angular frequency ω, g_{rot}, is *less* than that on a nonrotating sphere, g, and is dependent on latitude, λ:

$$g_{\mathrm{rot}} = g - \omega^2 R_{\mathrm{e}} \cos^2 \lambda \tag{5.18}$$

The gravitational acceleration on a rotating oblate spheroid can also be calculated mathematically. The *reference gravity formula* adopted by the International Association of Geodesy in 1967 is

$$g(\lambda) = g_{\mathrm{e}}(1 + \alpha \sin^2 \lambda + \beta \sin^4 \lambda) \tag{5.19}$$

where the gravitational acceleration at the equator g_{e} is $9.7803185\,\mathrm{m\,s^{-2}}$ and the constants are $\alpha = 5.278895 \times 10^{-3}$ and $\beta = 2.3462 \times 10^{-5}$. (There are relations between α and β and f and ω.) About 40% of this variation of gravity with latitude λ is a result of the difference in shape between the spheroid with the best-fitting ellipticity and a perfect sphere; the remaining 60% of the variation is due to the earth's rotation. Gravity observations are expressed as deviations from Eq. 5.19.

5.3.2 *Orbits of Satellites*

The orbit of a small object about a point mass is an ellipse. Consider a satellite orbiting the earth, and for simplicity consider a circular orbit with radius r (many satellite orbits are almost circular; a circle is an ellipse with an ellipticity of zero). The gravitational force of the earth acting on the satellite is $GM_{\mathrm{E}}m/r^2$ (from Eq. 5.15), and this is balanced by the outward centrifugal force $m\omega^2 r$ (where ω is the angular velocity and m the mass of the satellite). Thus,

$$\frac{GM_{\mathrm{E}}m}{r^2} = m\omega^2 r \tag{5.20}$$

Rearranging this equation gives the angular velocity ω:

$$\omega = \left(\frac{GM_{\mathrm{E}}}{r^3}\right)^{1/2} \tag{5.21}$$

Alternatively, the period of the satellite orbit $T = 2\pi/\omega$ is given by

$$T = \left(\frac{4\pi^2 r^3}{GM_{\mathrm{E}}}\right)^{1/2} \tag{5.22}$$

Equations 5.21 and 5.22 are Kepler's third law. Johann Kepler (1571–1630) spent years studying the motions of the planets and discovered that the square of the period of the planet's orbit was proportional to the cube of the orbit's radius. This relationship was explained later when Isaac Newton developed his laws of motion and gravitation, which led to the exact relation derived and given in Eq. 5.22.

5.4 The Shape of the Earth

The earth is neither a perfect sphere nor a perfect oblate spheroid. Clearly, mountains and deep oceanic trenches are deviations of several kilometres.

Geodesists use the surface of the oceans as the reference surface, which is sensible since a liquid surface is necessarily an equipotential;* if it were not, the liquid would adjust until the surface *was* an equipotential. The earth's reference surface is called the *geoid.* Over the oceans the geoid is the mean sea level, and over the continents it can be visualized as the level at which water would lie if imaginary canals were cut through the continents. All navigation and all surveying are referenced to the geoid. The surveyor's plumb bob, for example, does not point 'down', it points perpendicular to the local equipotential surface which, if not too far above sea level, means perpendicular to the geoid.

The oblate spheroid (discussed in the previous section) approximates the geoid. This reference oblate spheroid, known as the *reference spheroid* or *reference ellipsoid,* is a mathematical figure whose surface is an equipotential of the theoretical gravity field of a symmetric spheroidal earth model with realistic radial variations in density, plus the centrifugal potential. The international gravity formula (Eq. 5.19) gives the value of g on this spheroid. Figure 5.3a shows the deviation of the geoid from the reference spheroid. The largest feature is the 'hole' south of India. A ship sailing across that hole would drop by almost 100 m and then rise again by the same amount, all without doing work against gravity! Figure 5.3b shows the 'average' shape of the earth (roughly pear-shaped) compared with the reference spheroid.

The fine details of the shape of the earth and its gravity field have been determined from artificial satellites (Fig. 5.10). The results have come from studies of small changes in their orbital parameters and from direct radar altimetry measurements, in which a radar pulse from the satellite is reflected at the ocean surface and its time of arrival back at the satellite measured. Two of the satellites which have used radar altimetry to measure the geoid are GEOS3 and SEASAT, launched in 1975 and 1978, respectively. These were so successful that another dedicated altimetric satellite GEOSAT was launched, and future missions are planned. This radar altimetry has defined the marine geoid to perhaps 10 cm. The gravimetrically defined geoid and the geoid defined by radar altimetry agree to better than ± 50 cm, indicating that the theories and assumptions are correct.

The shape of the geoid can be determined from gravity observations, but to determine the geoid at a point we effectively need worldwide gravity observations. Since it is difficult to measure gravity at sea due to accelerations of the ship, prior to the statellite era we had a relatively poor knowledge of the geoid, particularly over the oceans. Within weeks of the launch of the first earth-orbiting satellite by the Soviet Union in 1957, geophysicists had begun to reevaluate the geoid.

Altimetry satellites, of course, only measure the geoid over the oceans; thus, within months of the launch we had much more detailed information on the geoid over oceans than over land, a reversal of the previous situation. Satellites now measure the geoid so accurately that we can detect the slow rebound of the earth, to a more spherical shape following the removal of the

* An equipotential is a surface over which the potential has a constant value. The gradient of the potential is therefore perpendicular to this surface (see Appendix 1, *Gradient*). Thus, gravity ($-$grad V) is *always normal* to the mean sea-level surface.

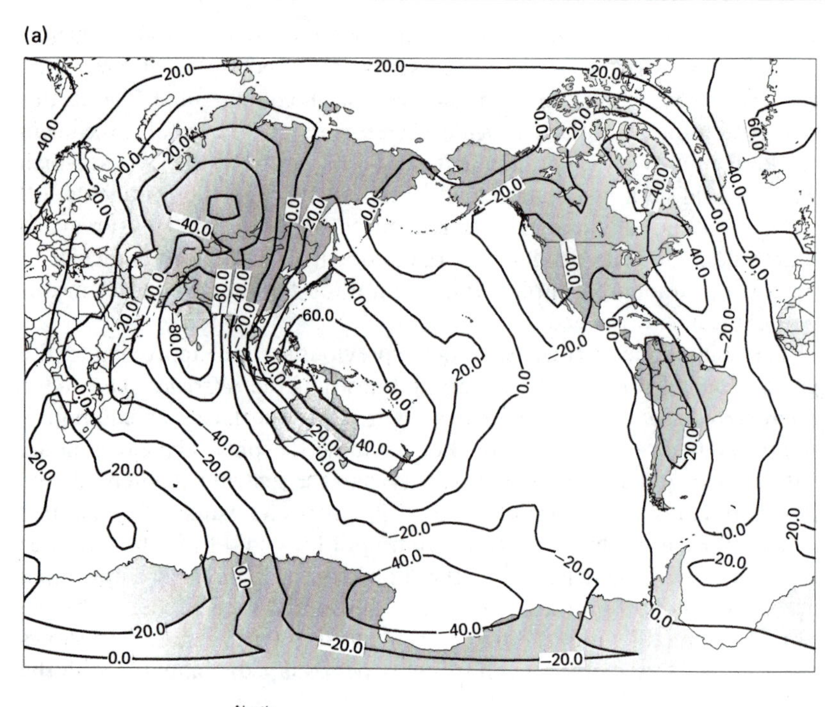

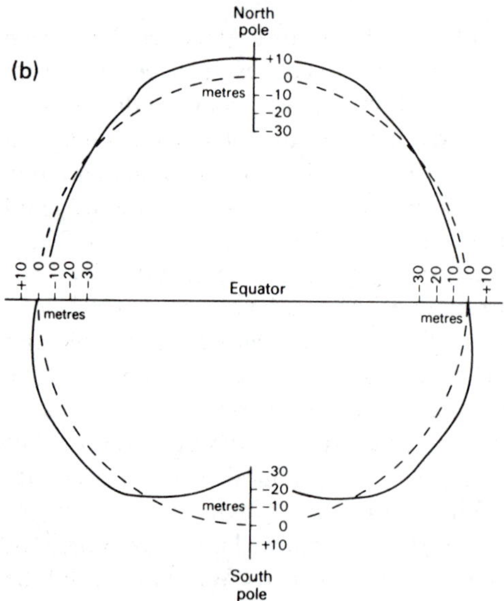

Figure 5.3. (a) Geoid height anomalies: the height of the geoid above (positive) or below (negative) the spheroid in metres. (b) The averaged shape of the earth, calculated by assuming that the earth is symmetric about its rotation axis (solid line), compared with a spheroid of flattening 1/298.25 (dashed line). (From King-Hele 1969.)

Pleistocene glaciers. This represents a relative change in moment of inertia of about $10^{-8}\,yr^{-1}$.

5.5 Gravity Anomalies

5.5.1 Introduction

Measurements of the gravitational attraction of the earth are not only useful in discovering the exact shape of the earth and its rotational properties; they also provide information about the structure of the lithosphere and mantle. Gravity anomalies are very small compared with the mean surface gravity value of 9.81 m s^{-2} and are therefore often quoted in a more convenient unit, the *milligal*, which is 10^{-5} m s^{-2} (10^{-3} gal). Another unit which is sometimes used is the *gravity unit*, gu, which is 10^{-6} m s^{-2} (1 gu = 10^{-1} mgal). Gravimeters have a sensitivity of about 10^{-5} gal (10^{-2} mgal), or about 10^{-8} of the surface gravitational acceleration. John Milne (1906) gave an interesting example of an anomaly:

> When a squad of 76 men marched to within 16 or 20 feet of the Oxford University Observatory it was found that a horizontal pendulum inside the building measured a deflection in the direction of the advancing load....

5.5.2 *Isostasy*

Between 1735 and 1745 a French expedition under the leadership of Pierre Bouguer made measurements of the length of a degree of latitude in Peru and near Paris in order to determine the shape of the earth. They realized that the mass of the Andes mountains would attract their plumb line but were surprised that this deflection was much less than they had estimated. In the nineteenth century, the Indian survey under Sir George Everest (for whom the mountain is named) found the same reduced deflection near the Himalayas. This is now known to be a fairly common attribute of surface features. In 1855, Archdeacon J. H. Pratt and Sir George Airy proposed two separate hypotheses to explain these observations; and in 1889, the term *isostasy* was first used to describe them. For both the Himalayas and the Andes, the lack of mass or mass deficiency beneath the mountain chains, which is required to account for the reduced deflection of the plumb lines, was found to be approximately equal to the mass of the mountains themselves. This is an alternative statement of Archimedes' principle of hydrostatic equilibrium: A floating body displaces its own weight of water. A mountain chain can therefore be compared with an iceberg or cork floating in water. Thus, isostasy requires the surface layers of the earth to be rigid and to 'float' on, or in, a denser substratum. The rigid surface layer is termed the *lithosphere* and the region beneath, the *asthenosphere* (Sect. 2.1); recall that these layers are distinct from the compositional layers, crust and mantle, which have been discussed previously.

The depth below which all pressures are hydrostatic is termed the *compensation depth*. At or below the compensation depth, the weight of (imaginary) vertical columns with the same cross-sectional area must be the same. A mountain in *isostatic equilibrium* is therefore *compensated* by a mass deficiency beneath it but above the compensation depth. In contrast, an ocean basin in isostatic equilibrium is compensated by extra mass at depth but above the compensation depth.

Airy's Hypothesis In this hypothesis the rigid upper layer and the substratum are assumed to have constant density, ρ_u and ρ_s, respectively. Isostatic compensation is achieved by mountains having deep roots (exactly like an iceberg). Figure 5.4 illustrates this hypothesis. Taking an arbitrary compensation depth that is deeper than the deepest mountain

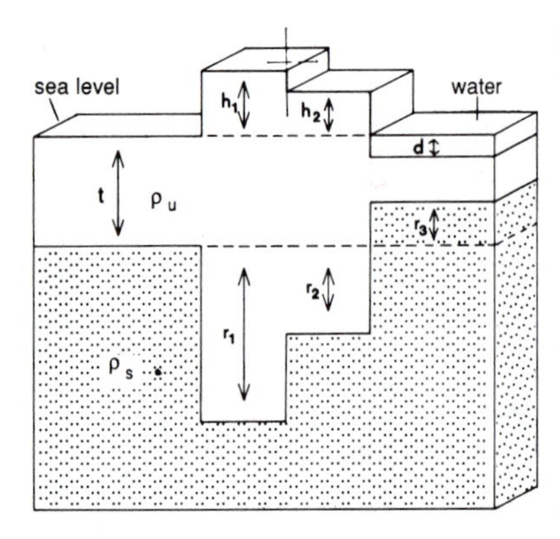

Figure 5.4. Illustration of Airy's hypothesis of isostasy. The upper layer has density ρ_u, and the substratum has density ρ_s. Isostatic compensation is achieved by variation in the thickness of the upper layer: Mountains have deep roots and ocean basins have antiroots.

root in the substratum and equating the masses above that depth in each vertical column of unit cross-sectional area, one obtains

$$\begin{aligned} t\rho_u + r_1\rho_s &= (h_1 + t + r_1)\rho_u \\ &= (h_2 + t + r_2)\rho_u + (r_1 - r_2)\rho_s \\ &= d\rho_w + (t - d - r_3)\rho_u + (r_1 + r_3)\rho_s \end{aligned} \tag{5.23}$$

A mountain of height h_1 would therefore have a root r_1 given by

$$r_1 = \frac{h_1\rho_u}{\rho_s - \rho_u} \tag{5.24}$$

Similarly, a feature at a depth d beneath sea level would have an anti root r_3 given by

$$r_3 = \frac{d(\rho_u - \rho_w)}{\rho_s - \rho_u} \tag{5.25}$$

The rigid upper layer (lithosphere) has density ρ_u, but Eqs. 5.23–5.25 apply equally well when ρ_u is replaced by ρ_c (the density of the crust) and ρ_s is replaced by ρ_m (the density of the mantle). This is because the crust–mantle boundary is embedded in and is part of the lithosphere, and so loading at the surface and subsequent deflection of the base of the lithosphere deflects the crust–mantle boundary. Furthermore, the difference between the density of the mantle at the crust–mantle boundary and the density of the mantle at the lithosphere–asthenosphere boundary may be very small (see Sect. 4.3.1). Therefore, Eqs. 5.24 and 5.25 are often applied in the following forms:

$$r_1 = \frac{h_1\rho_c}{\rho_m - \rho_c} \tag{5.26}$$

and

$$r_3 = \frac{d(\rho_c - \rho_w)}{\rho_m - \rho_c} \tag{5.27}$$

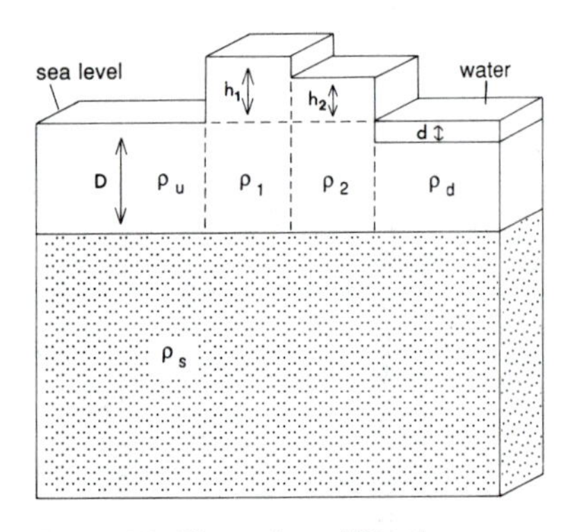

Figure 5.5. Illustration of Pratt's hypothesis of isostasy. Density of sea water, ρ_w; density of substratum (hatched), ρ_s; densities of upper layer beneath mountains of heights h_2 and h_1, ρ_2 and ρ_1; density of upper layer beneath ocean of depth d, ρ_d; density of upper layer beneath land at sea level, ρ_u; compensation depth, D. Isostatic compensation is achieved by a lateral variation of density beneath surface features: Material beneath mountains is less dense; material beneath ocean basins is more dense.

Pratt's Hypothesis Pratt assumed that the depth of the base of the upper layer is a constant and that isostatic equilibrium is achieved by allowing this upper layer to be composed of columns of constant density (Fig. 5.5). Taking the base of the upper layer as the compensation depth and equating the masses above this level in each column of unit cross-sectional area gives

$$\begin{aligned} \rho_u D &= (h_1 + D)\rho_1 \\ &= (h_2 + D)\rho_2 \\ &= \rho_w d + \rho_d(D - d) \end{aligned} \tag{5.28}$$

Thus, in this model, compensation is achieved by mountains consisting of and being underlain by material of low density,

$$\rho_1 = \rho_u\left(\frac{D}{h_1 + D}\right) \tag{5.29}$$

and the oceans being underlain by material of higher density,

$$\rho_d = \frac{\rho_u D - \rho_w d}{D - d} \tag{5.30}$$

These two hypotheses of Pratt and Airy are very different; but determining whether one, the other or a combination of both actually operates in a particular part of the earth is not a simple matter. Determining whether or not a surface feature is in isostatic equilibrium (compensated) is easier.

Example: isostasy and seismic reflection

The many deep continental reflection lines from around the British Isles produced a result that was initially very surprising. The results of the lines showed that although the structure of the upper crust varies from sedimentary basins to outcropping crystalline rocks, nevertheless, the clear reflection identified as the Moho is generally horizontal and arrives at about 10 s two-way time on unmigrated sections. The reflection typically shows neither any depression (pull down) beneath basins nor elevation (pull up) beneath the crystalline rocks. It is possible that this feature may be just a desire of the eye of the seismic interpreter to find a suitable candidate for the Moho reflection at about this depth, but it is more likely that the feature is real and a consequence of isostasy.

Let us modify the Airy and Pratt isostatic compensation models to include a depth-dependent density $\rho(z)$ (Warner 1987). For the British reflection lines, which were shot in relatively shallow water, Airy-type isostatic compensation (Eq. 5.23) requires

$$\int_0^{t_1} \rho_1(z)\, dz = \int_0^{t_2} \rho_2(z)\, dz + (t_1 - t_2)\rho_m \tag{5.31}$$

where $\rho_1(z)$ and $\rho_2(z)$ are the density functions for crust with total thickness t_1 and t_2, respectively, and ρ_m is the density of the mantle. When density and seismic velocity, v_1, v_2, are related by

$$\rho_1(z) = \rho_m - \frac{k}{v_1(z)}$$

$$\rho_2(z) = \rho_m - \frac{k}{v_2(z)} \tag{5.32}$$

where k is a constant, the two-way normal incidence travel time T_1 and T_2 for the two structures, given by

$$T_1 = \int_0^{t_1} \frac{2}{v_1(z)}\, dz$$

$$T_2 = \int_0^{t_2} \frac{2}{v_2(z)}\, dz \tag{5.33}$$

are equal. Although the density–velocity relationship does not exactly fit Eq. 5.32, values of k between 3×10^6 and 4×10^6 kg m^{-2} s^{-1} approximate the Nafe–Drake density–velocity curve (Fig. 4.2d).

Pratt-type isostatic compensation (Eq. 5.28) with a depth-dependent density requires that

$$\int_0^{t_1} \rho_1(z)\,dz = \int_0^{t_2} \rho_2(z)\,dz \tag{5.34}$$

In this case, the two two-way travel times (Eqs. 5.33) are equal if the density and seismic velocity are related by

$$\begin{aligned} \rho_1(z) &= \rho_k - \frac{k}{v_1(z)} \\ \rho_2(z) &= \rho_k - \frac{k}{v_2(z)} \end{aligned} \tag{5.35}$$

where k and ρ_k are arbitrary constants. Thus, the Airy restriction on the density–velocity relationship (Eq. 5.32) is just a special case of Eq. 5.35.

It is therefore possible that observation of a nearly horizontal Moho on time sections may just be an indication that the observed structures are isostatically compensated. In an isostatically compensated region, if the density–velocity relationship approximates Eq. 5.32 or 5.35, a structure on the Moho would not be seen on an unmigrated seismic section. Structures on the Moho would, however, be seen after corrections for velocity were made.

5.5.3 Calculation of Gravity Anomalies

Before any gravity measurements can be used, a number of corrections have to be made. First, allowance must be made for the fact that the earth is not a perfect sphere but is flattened at the poles and is rotating. The reference gravity formula of 1967 (Eq. 5.19) includes these effects and expresses gravity g as a function of latitude λ. This enables a correction for the latitude of the measurement point to be made by subtracting the reference value (Eq. 5.19) from the actual gravity measurement.

The second correction which must be made to any gravity measurement allows for the fact that the point at which the measurement was made was at an elevation h and not at sea level on the spheroid. This correction, known as the *free-air correction*, makes no allowance for any material between the measurement point and sea level: It is assumed to be air. Therefore, using the inverse-square law and assuming that the earth is a perfect sphere, we find that gravity at elevation h is

$$g(h) = g_0\left(\frac{R}{R+h}\right)^2 \tag{5.36}$$

where R is the radius of the earth and g_0 is gravity at sea level. Since $h \ll R$, Eq. 5.36 can be written as

$$g(h) \simeq g_0\left(1 - \frac{2h}{R}\right) \tag{5.37}$$

The free-air correction δg_F, which is to be added to the measured value to correct it to a sea-level value, is then

$$\delta g_{\mathrm{F}} = g_0 - g(h) = \frac{2h}{R} g_0 \tag{5.38}$$

As gravity decreases with height above the surface, points above sea level are corrected to sea level by adding $2hg_0/R$. This correction amounts to 3.1 $\times 10^{-6}\,\mathrm{m\,s^{-2}}$ per metre of elevation. A more accurate value of this correction can be made by using McCullagh's formula for the gravitational attraction of a rotating spheroid (e.g., Cook 1973, pp. 280–282.) The *free-air anomaly* g_{F} is then the measured gravity value g_{obs} with these two corrections applied:

$$\begin{aligned} g_{\mathrm{F}} &= g_{\mathrm{obs}} - g(\lambda) + \delta g_{\mathrm{F}} \\ &= g_{\mathrm{obs}} - g(\lambda)\left(1 - \frac{2h}{R}\right) \end{aligned} \tag{5.39}$$

Two other corrections are frequently applied to gravity measurements. The first is the *Bouguer correction*, which allows for the gravitational attraction of the rocks between the measurement point and sea level, assuming that these rocks are of infinite horizontal extent. The Bouguer correction is given by

$$\delta g_{\mathrm{B}} = 2\pi G \rho h \tag{5.40}$$

where G is the constant of gravitation, ρ the density of the material between the measurement point and sea level, and h the height of the measurement point above sea level. Taking G to be $6.67 \times 10^{-11}\,\mathrm{m^3\,kg^{-1}\,s^{-2}}$ and assuming an average crustal density of $2.7 \times 10^3\,\mathrm{kg\,m^{-3}}$, we obtain a Bouguer correction of $1.1 \times 10^{-6}\,\mathrm{m\,s^{-2}}$ per metre of elevation. Alternatively, the Bouguer correction for a 1 km thick layer with density $10^3\,\mathrm{kg\,m^{-3}}$ is 42 mgal. The second correction is a *terrain correction*, δg_{T}, which allows for deviations of the surface from an infinite horizontal plane. This correction can be calculated graphically by using a set of templates and a topographic map. The terrain correction is small and, except for areas of mountainous terrain, can often be ignored in crustal studies.

The *Bouguer anomaly* g_{B} is the free-air anomaly with these two extra corrections applied:

$$\begin{aligned} g_{\mathrm{B}} &= g_{\mathrm{F}} - \delta g_{\mathrm{B}} + \delta g_{\mathrm{T}} \\ &= g_{\mathrm{obs}} - g(\lambda) + \delta g_{\mathrm{F}} - \delta g_{\mathrm{B}} + \delta g_{\mathrm{T}} \end{aligned} \tag{5.41}$$

This Bouguer anomaly is the observed value of gravity minus the theoretical value at the latitude and elevation of the observation point. Since we have allowed for the attraction of all the rock above sea level, the Bouguer anomaly represents the gravitational attraction of the material below sea level.

The free-air anomaly is usually used for gravity measurements at sea. It is comparable to the Bouguer anomaly over continents since the measurements are then all corrected to the sea level datum. If a Bouguer anomaly is required for oceanic gravity measurements, it must be calculated by 'replacing' the sea water with rocks of average crustal density. A terrain correction must then also be applied to account for the seabed topography.

We can use gravity measurements to determine whether an area is in isostatic equilibrium. If a region is in isostatic equilibrium, there should be no gravity anomaly and hence no excess or lack of mass above the compensation depth. However, in practice this becomes a rather convoluted problem. Take, for example, the mountains illustrated in Figure 5.4 and assume them to be in isostatic equilibrium with the left-hand column of crust (t in thickness). The Bouguer anomaly across these mountains is negative since below sea level there is a mass deficiency under the mountains. The fact that this mass deficiency balanced the excess mass of the mountains themselves has been removed from this anomaly. In contrast, the free-air anomaly over the mountains is positive and much smaller in magnitude. The free-air anomaly is positive over the mountains because the mountains are closer to the measurement point than the deep compensating structure. Even though the compensating structure has the same mass as the mountains, it gives a smaller negative contribution to the gravity at the measurement point because it is further away from the measurement point.

The simplest way to determine whether a large-scale structure such as a mountain chain or large sedimentary basin is in isostatic equilibrium is to use the free-air anomaly. If a structure or region is totally compensated, the free-air anomaly is very small away from the edges of the structure, provided the structure is about ten times or more wider than the compensation depth. If the structure is only partially compensated, or not compensated at all, then the free-air anomaly is positive, perhaps up to several hundred milligals in magnitude depending on the structure and degree of compensation. For a totally or partially compensated structure the Bouguer anomaly is negative, whereas for a noncompensated structure the Bouguer anomaly is zero. Free-air anomalies are almost isostatic anomalies. They do not assume any specific mechanism for compensation but are small if compensation is complete.

Another way to determine whether a structure or region is in isostatic equilibrium is to propose a series of density models and then calculate the Bouguer anomaly that each would give. The *isostatic anomaly* for the region is then the actual Bouguer anomaly minus the computed Bouguer anomaly for the proposed density model. Thus, each density model for the region has a different isostatic anomaly.

The effect that isostatic compensation has on the gravity anomalies is illustrated in Figure 5.6. Figure 5.6a shows a schematic wide mountain range which is totally compensated. The Bouguer anomaly across this model is therefore very large and negative, whereas the free-air anomaly is small and positive in the centre of the model and large and positive at the edge of the mountains. Also shown in Figure 5.6a are isostatic anomalies for three models made to test whether the structure is in isostatic equilibrium. All three isostatic anomalies are very close to zero and the anomaly calculated for Airy-type compensation with $D = 30$ km is exactly zero. The fact that the other two anomalies are almost zero indicates that the structure is in isostatic equilibrium.

Figure 5.6b shows the same mountain range, but this time it is only 75% compensated. Now the free-air anomaly is large and positive, whereas the Bouguer anomaly is large and negative; all the indications are that the

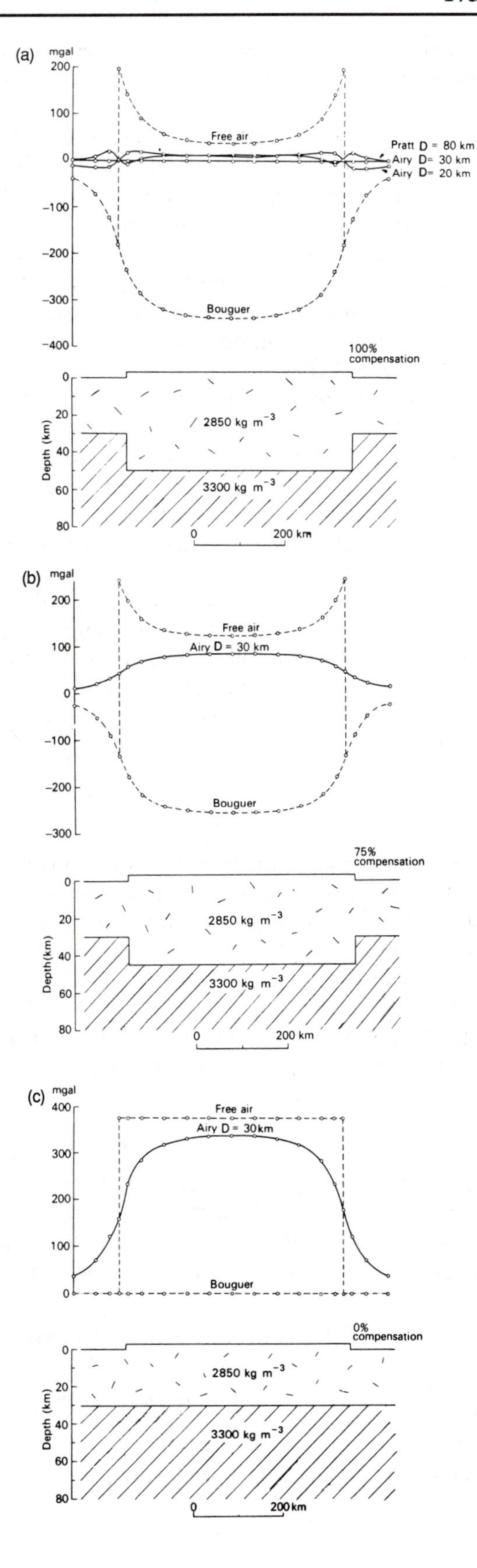

Figure 5.6. Gravity anomalies over a schematic mountain range. (a) Mountain range is 100% compensated (Airy-type). (b) Mountain range is 75% compensated (Airy-type). (c) Mountain range is uncompensated. Dashed lines, the free-air and Bouguer anomalies that would be measured over the mountain. Solid lines, isostatic anomalies calculated for the density models (Pratt compensation depth $D = 80$ km; Airy $D = 20$ km and $D = 30$ km). Densities used to calculate the isostatic anomalies are (fortuitously) those shown in the model. (From Bott 1982.)

mountains are not in isostatic equilibrium. The isostatic anomaly calculated for Airy-type compensation for $D = 30$ km confirms this.

Figure 5.6c shows the case in which the mountains are totally uncompensated. Then the Bouguer anomaly is exactly zero since all the excess gravitational attraction is provided by the material above sea level, whereas the free-air anomaly is very large and positive. The isostatic anomaly is, in this case, also very large and positive.

To determine what form the compensation takes, the gravity anomaly must be calculated for a number of possible subsurface density structures and various compensation depths. A zero isostatic anomaly indicates that a correct density distribution and compensation depth have been determined (as in Fig. 5.6). Unfortunately, it is often not possible to distinguish unequivocably between the various hypotheses or compensation depths because gravity is insensitive to minor changes in the deep density structure. In addition, small shallow structures can easily hide the effects of the deeper structure. To determine the extent and shape that any compensating mass takes, it is helpful to have additional information on the structure of the crust such as that given by seismic refraction and reflection experiments.

The continents and oceans are in broad isostatic equilibrium. This is mainly achieved by variations in crustal thickness (Airy's hypothesis) although, for example, the midocean ridges are partially compensated by lower-density rocks in the mantle (Pratt's hypothesis). Gabbro, a typical oceanic rock, is denser than granite, a typical continental rock, so Pratt's hypothesis also plays a role in this isostatic balancing of oceanic and continental crust.

5.5.4 Gravity Anomalies Due to Some Buried Bodies

The gravitational attraction of some simple shapes can be calculated analytically by using Eqs. 5.4 and 5.6. The attraction due to more complex shapes, however, must be calculated numerically by computer. To illustrate the magnitude and type of gravitational anomaly caused by subsurface bodies, we consider a few simple examples.

Figure 5.7 shows one problem. A sphere of density ρ_1 and radius b is buried with its centre at depth h in a medium with density ρ. The density contrast of the sphere with respect to the surrounding medium, $\Delta\rho$, is given by

$$\Delta\rho = \rho_1 - \rho \tag{5.42}$$

From the calculations in Section 5.2 we know that the gravitational acceleration g due to a sphere of mass m is Gm/r^2. However, that is the acceleration at point P in the radial direction r, and in this particular case we need to determine the vertical component of gravity, g_z:

$$g_z = g\cos\theta = \frac{Gm}{r^2}\cos\theta$$

$$= \frac{Gm}{r^2}\frac{h}{r} = \frac{Gmh}{(x^2+h^2)^{3/2}} \tag{5.43}$$

The gravity anomaly δg_z is therefore given by

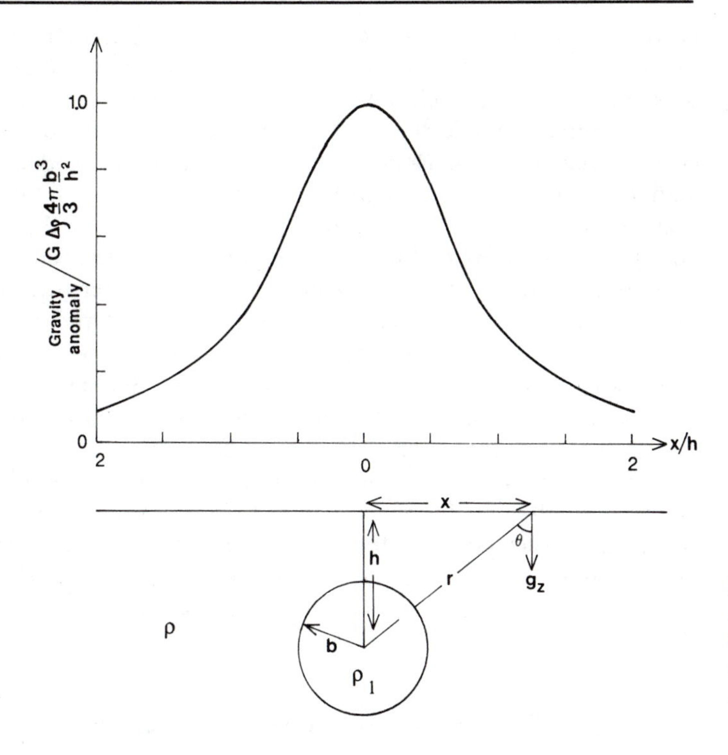

Figure 5.7. Gravity anomaly due to a sphere of radius b buried at a depth h. Density of surrounding medium ρ, density of sphere ρ_1. Density contrast $\Delta\rho = \rho_1 - \rho$.

$$\delta g_z = \frac{4G\Delta\rho\pi b^3 h}{3(x^2 + h^2)^{3/2}} \tag{5.44}$$

In SI units, Eq. 5.44 is

$$\delta g_z = 2.79 \times 10^{-10} \frac{\Delta\rho\, b^3 h}{(x^2 + h^2)^{3/2}} \tag{5.45}$$

The anomaly due to this buried sphere is therefore symmetrical about the centre of the sphere and essentially confined to a width of about two to three times the depth of the sphere (Fig. 5.7).

This is the simplest gravity anomaly to calculate. All the others involve more tedious algebra (for many detailed examples the reader is referred to Telford et al. 1990). Here we merely note that the gravity anomaly for an infinitely long cylinder with anomalous density $\Delta\rho$ and radius b, buried at a depth d beneath the surface, is

$$\delta g_z = \frac{G\,\Delta\rho\, 2\pi b^2 d}{(x^2 + d^2)} \tag{5.46}$$

and the gravity anomaly for a semi-infinite (extending to infinity in the positive x direction) horizontal sheet with anomalous density $\Delta\rho$, t in thickness and buried at depth d beneath the surface, is

$$\delta g_z = 2G\,\Delta\rho\, t\left[\frac{\pi}{2} + \tan^{-1}\left(\frac{x}{d}\right)\right] \tag{5.47}$$

Each particular buried body gives rise to its own anomaly. In many cases, the shape of the anomalous body can be determined from the shape of the gravity anomaly (e.g., the gravity anomaly due to a sphere is narrower than that due to an infinite horizontal cylinder and not stretched out in the y-direction). Gravity models are, however, unfortunately *not* unique.

These three examples illustrate the fact that to determine a shallow density structure, one must make a detailed local survey. Anomalies due to bodies buried at depth d can only be detected at distances out to perhaps $2d$ from the body. To resolve details of the density structure of the lower crust (at say 20–40 km), gravity measurements must be made over an extensive area, as the width of these anomalies are much greater (say, five to ten times) than the anomalies due to the shallow crustal structure. Likewise, gravity anomalies due to anomalous bodies in the mantle are of a much longer wavelength (hundreds of kilometers). Thus, although at first sight it might seem impossible to extract mantle density information from surface gravity measurements (which cannot fail to be affected by near-surface density anomalies), application of a wavelength filter or smoothing of the gravity data allow such deep structures to be studied. The opposite is, of course, also true: To study the shallow structure, one must remove the regional long-wavelength anomaly which is of deeper origin. However, no amount of processing can ensure such a separation.

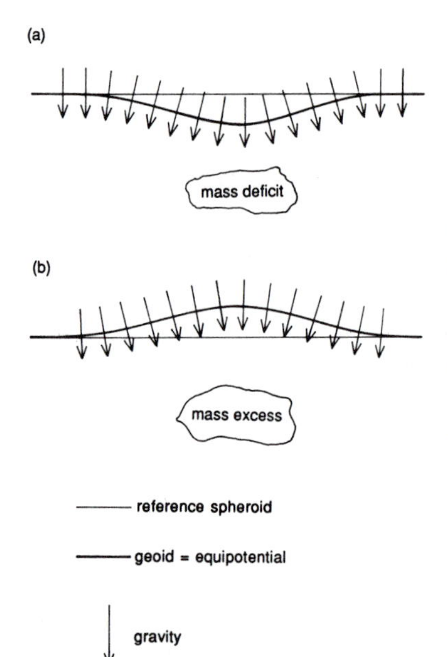

Figure 5.8. (a) A trough in the geoid, or negative geoid height anomaly, occurs over a region of mass deficit (such as a depression in the seabed). A negative free-air gravity anomaly also occurs over such a mass deficit. (b) A bulge in the geoid, or positive geoid height anomaly, occurs over regions of excess mass (such an elevated region of the seabed). A positive free-air gravity anomaly also occurs over such a mass excess.

5.6 Observed Gravity and Geoid Anomalies

5.6.1 Gravity Anomalies

A gravity profile across the Mid-Atlantic Ridge is shown in Figure 8.11. The maximum free-air anomaly on this profile is about 100 mgal. The four density models in the figure yield anomalies that adequately match the measurements, despite being very different. Note that, as is to be expected, model (c), the deepest model, has only a very small $40\,\mathrm{kg\,m^{-3}}$ density anomaly, whereas the shallower models (e), (f) and (g) have density anomalies of $300\,\mathrm{kg\,m^{-3}}$ and more. Gravity profiles across oceanic trenches (Figs. 5.10, 8.45, 8.46 and 8.47) show very large magnitude anomalies. Gravity measurements over the continents and their implications are discussed in Chapter 9.

5.6.2 Geoid Height Anomalies

The lateral variations in density distribution within the earth, although resulting in gravity anomalies, also result in deviations of the geoid from the theoretical spheroid. The *geoid height anomaly* Δh (geoid radius minus spheroid radius) is directly related to the anomaly in the gravitational potential ΔV as measured on the spheroid:

$$g\,\Delta h = -\Delta V \tag{5.48}$$

where g is the gravitational acceleration due to the spheroid (Eq. 5.19). Since the gravitational acceleration is normal to the equipotential surface (which is by definition the geoid), a trough in the geoid is present wherever there is a negative gravity anomaly (mass deficit), and likewise there is a bulge in the geoid wherever there is a positive gravity anomaly (mass excess)

(Fig. 5.8). Thus, a trough in the geoid, or negative geoid height anomaly, is the result of a negative gravity anomaly or a positive potential anomaly; and a bulge in the geoid, or positive geoid height anomaly, is the result of a positive gravity anomaly or a negative potential anomaly.

Overall, the deviations from the geoid are on a small scale, indicating that isostatic equilibrium is a normal attribute of large-scale surface features. In turn, this indicates that the mantle is, on a long time scale, not particularly strong; it must have a finite viscosity to be able to flow and adjust to changing surface loads. In contrast, we find that small-scale surface topography may not be compensated. This observation is explained by the notion of the strong lithosphere, which is able to support small mass anomalies but which bends or flexes under very large-scale mass anomalies, causing flow and readjustment in the weaker underlying asthenosphere. Erosion also occurs. Isostatic balance is thus achieved.

The geoid height anomaly resulting from an isostatic density distribution is not zero and can be calculated. It can be shown (see Turcotte and Schubert, 1982, p. 223) that the geoid height anomaly at any point P is given by

$$\Delta h = -\frac{2\pi G}{g}\int_0^D \Delta\rho(z) z \, dz \tag{5.49}$$

where g is the reference gravity value, $\Delta\rho(z)$ the anomalous density at depth z beneath point P, and D the compensation depth. Depth z is measured positively downwards with $z = 0$ corresponding to the spheroid. Equation 5.49 therefore gives the geoid height anomaly due to long-wavelength isostatic density anomalies. Geoid anomalies can be used to estimate the variation of density with depth. In practice, we need to work in reverse, by first calculating the geoid height anomaly for an isostatic density model and then comparing the calculations with the observed measurements.

As an example, consider the Airy compensation model illustrated in Figure 5.4. The reference structure is an upper layer of density ρ_u and thickness t and a substratum of density ρ_s. All density anomalies are *with respect to this reference structure.* The geoid height anomaly over a mountain range height h_1, calculated by using Eq. 5.49, is

$$\begin{aligned}\Delta h &= -\frac{2\pi G}{g}\left[\int_{-h_1}^{0} \rho_u z \, dz + \int_t^{t+r_1} (\rho_u - \rho_s) z \, dz\right] \\ &= -\frac{\pi G}{g}\left[-h_1^2\rho_u + (\rho_u - \rho_s)(2tr_1 + r_1^2)\right]\end{aligned} \tag{5.50}$$

After substituting for r_1 from Eq. 5.24 and rearranging terms, we finally obtain

$$\Delta h = \frac{\pi G}{g}\rho_u h_1\left[2t + \frac{\rho_s h_1}{\rho_s - \rho_u}\right] \tag{5.51}$$

Therefore, for crustal and mantle densities of 2.8 and 3.3×10^3 kg m^{-3}, respectively, and a reference crust 35 km thick, the geoid height anomaly is

$$\Delta h \simeq 6h_1(0.7 + 0.066h_1) \quad \text{m} \tag{5.52}$$

where h_1 is in kilometers. Thus, a compensated mountain range 3 km high would result in a positive geoid height anomaly of about 16 m.

Likewise, the geoid height anomaly for an Airy-compensated ocean basin of depth d (Fig. 5.4) is

$$\begin{aligned}\Delta h &= -\frac{2\pi G}{g}\left[\int_0^d (\rho_w - \rho_u)z\,dz + \int_{t-r_3}^{t} (\rho_s - \rho_u)z\,dz\right] \\ &= -\frac{\pi G}{g}\left[(\rho_w - \rho_u)d^2 + (\rho_s - \rho_u)(2tr_3 - r_3^2)\right] \end{aligned} \tag{5.53}$$

Substituting for r_3 from Eq. 5.25 and rearranging terms, we obtain

$$\Delta h = -\frac{\pi G}{g}(\rho_u - \rho_w)d\left[2t - d\left(\frac{\rho_s - \rho_w}{\rho_s - \rho_u}\right)\right] \tag{5.54}$$

Thus, with the numerical values given for Eq. 5.52, the geoid height anomaly is

$$\Delta h \simeq -3.85d(0.7 - 0.046d) \quad \text{m} \tag{5.55}$$

where d is the ocean depth in kilometers. A compensated ocean basin 5 km deep would result in a negative geoid height anomaly of about 9 m.

Geoid height anomalies calculated for the Pratt compensation model are larger than those for the Airy model. The difference is particularly significant for seabed topography: The Pratt model gives a geoid height anomaly of about twice that of the Airy model. This difference can be used to estimate the type and depth of compensation for major features.

In Chapter 7, the topic of convection in the mantle is discussed in some detail. Here, we assume that this process occurs and see what can be discovered about it by using measurements of the geoid and the earth's gravity field.

The complex flow of material within the mantle gives rise to gravity and geoid height anomalies just as mountains and ocean basins do. Figure 5.9 shows a numerical simulation of convection in the mantle (refer to Sect. 7.8 for details of these models). The columns of rising material are hotter and therefore less dense than the columns of sinking mantle material. Calculations indicate that there are small, positive gravity seabed and geoid height anomalies above the rising hot regions. These anomalies occur because the deflection of the surface by the rising current produces a larger anomaly than the negative anomaly which results from the density deficit. Similarly, there are negative anomalies above the sinking columns. Detailed studies of long-wavelength gravity, geoid and bathymetric anomalies may therefore be able to give a powerful insight into the planform of convection in the mantle.

Figure 5.11 shows the bathymetry and the free-air gravity anomaly north and south of the Hawaiian Island of Oahu. The central 200 km wide Hawaiian ridge 'mountain' whose pinnacle is Oahu acts as a load on the Pacific Plate and bends it, resulting in the depression on either side of the mountain. The bending of the plate and the associated gravity anomaly are confined to distances within about 200 km of the load. The plates are too thin and too flexible for local loads to result in any long-wavelength

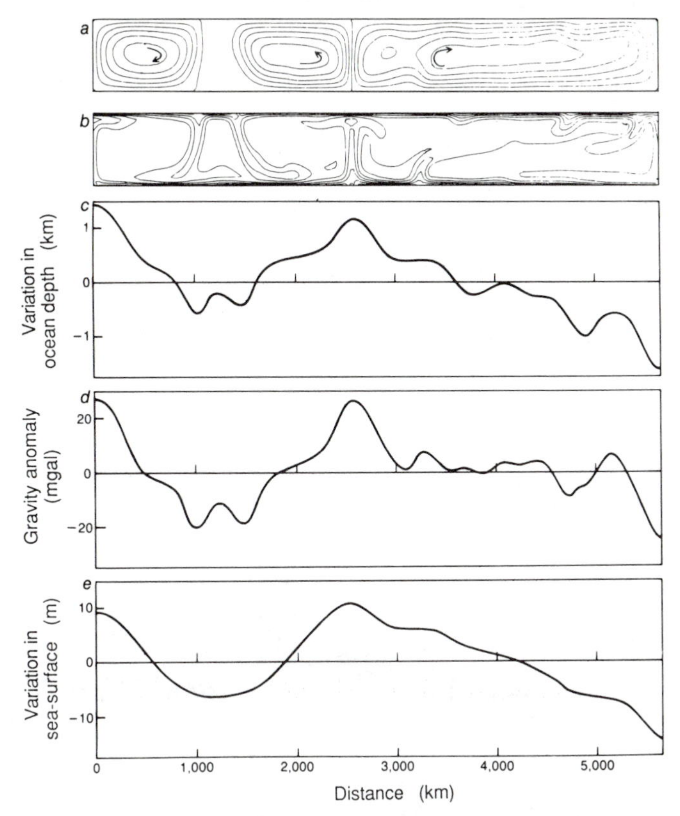

Figure 5.9. Computer modelling of convection in the upper mantle. This model assumes that the mantle has constant viscosity and is heated from below and cooled from above. (a) Circulation of the mantle material in convection cells. (b) Temperature in the model mantle contoured at 100°C intervals. There are two rising limbs at the left-hand edge and the centre and two regions of sinking material at centre left and right. (c) Variation in the depth of the ocean caused by the convection. (d) Variation in the gravitational acceleration (gravity anomaly) caused by the convection. (e) Variation in the height of the sea surface (geoid height anomaly) caused by the convection. (From McKenzie et al. 1980.)

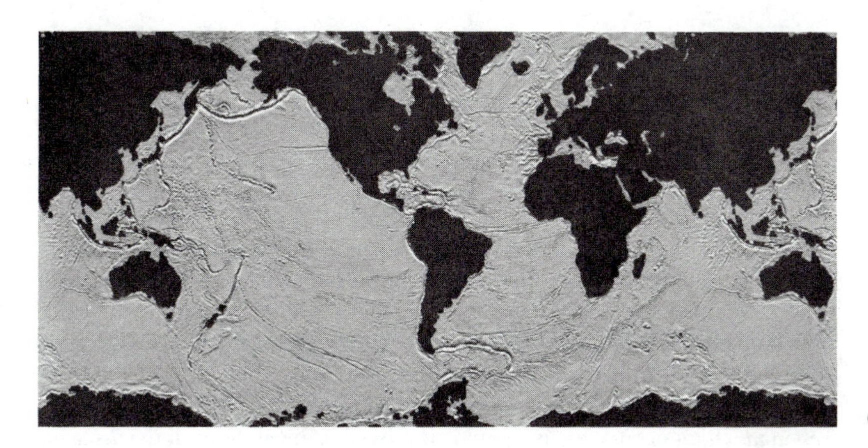

Figure 5.10. SEASAT gravity map. (Courtesy of W. Haxby)

bathymetric or gravity anomalies. The long-wavelength bulge evident in both the bathymetry and the gravity is, however, thought to be the surface expression of a hot upwelling region in the mantle as discussed previously. Again, the fortunate difference in wavelength between the elastic deformation of the plate to a surface load and the apparent wavelength of mantle convection allows these two deviations to be distinguished.

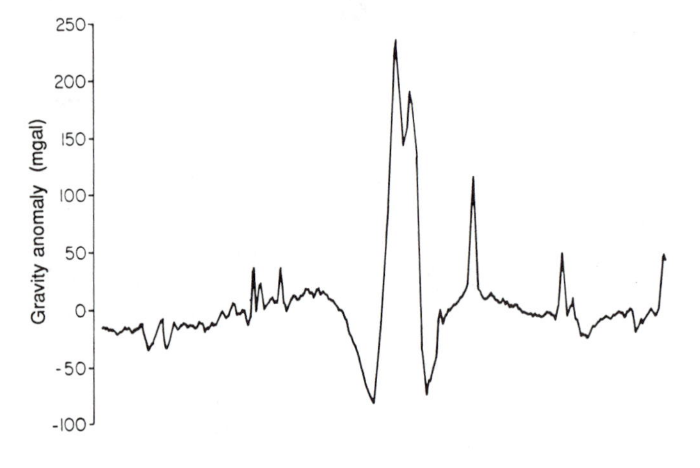

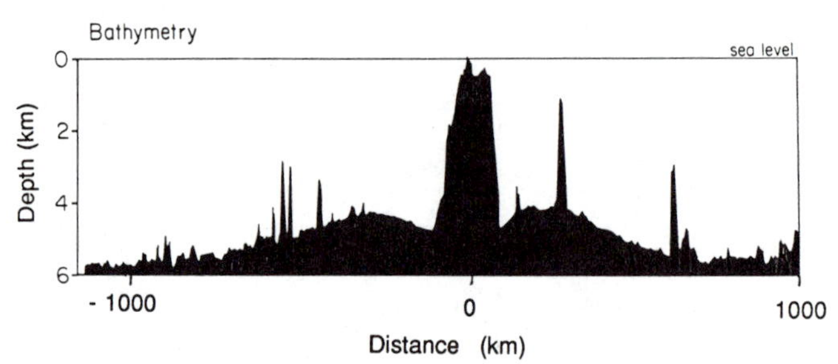

Figure 5.11. Free-air gravity and bathymetry anomalies along a north–south line centred on the Hawaiian island of Oahu. The volcanic islands act as a load on the Pacific Plate bending it downwards and resulting in the symmetric trough in both the bathymetry and gravity anomaly. (After Watts and Daly 1981. Reproduced with permission from the *Annual Review of Earth and Planetary Sciences*, vol. 9, © 1981 by Annual Reviews Inc.)

5.7 Flexure of the Lithosphere and the Viscosity of the Mantle

5.7.1 The Lithosphere as an Elastic Plate

In the theory of plate tectonics the thin lithospheric plates are assumed to be rigid and to float on the underlying mantle. On a geological time scale the lithosphere behaves elastically and the mantle behaves as a viscous fluid, whereas on the very short seismic time scale both behave as elastic solids. The study of the bending or *flexure* of the lithosphere which results from its loading by mountain chains, volcanoes and so on enables us to estimate the elastic properties of the lithosphere. Additionally, the rate of recovery or *rebound* which occurs when a load is removed is dependent on the viscosity of the underlying mantle as well as the elastic properties of the lithosphere. Thus, given suitable loads, we can make estimates of mantle viscosity.

The general fourth-order differential equation governing the equilibrium deflection of an elastic plate as a function of horizontal distance x is well known in engineering:

$$D\frac{d^4w}{dx^4} = V(x) - H\frac{d^2w}{dx^2} \tag{5.56}$$

where $w(x)$ is the deflection of the plate, $V(x)$ a vertical force per unit length applied to the plate, H a constant horizontal force per unit length applied to the plate and D the *flexural rigidity* of the plate (Fig. 5.12). The flexural rigidity is defined by

$$D = \frac{Eh^3}{12(1-\sigma^2)} \tag{5.57}$$

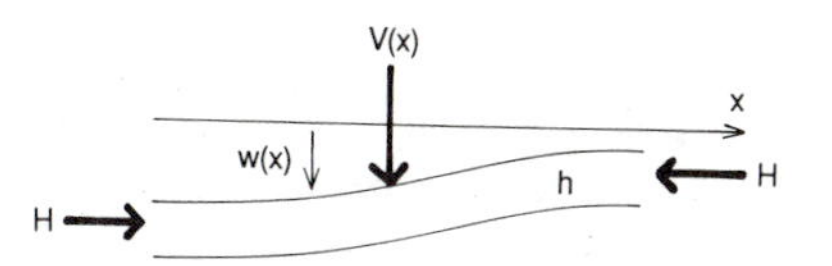

Figure 5.12. A thin plate of thickness h is deflected by $w(x)$ as a result of an imposed variable vertical force per unit area $V(x)$ and a constant horizontal force H per unit length.

where E is Young's modulus (see Appendix 2), h the thickness of the plate and σ Poisson's ratio (see Appendix 2).

Consider the oceanic lithosphere deforming under an applied vertical load $V(x)$ and no horizontal force. Water fills the resulting depression in the seabed. However, there is a net hydrostatic restoring force of $(\rho_m - \rho_w)gw$ per unit area. This restoring force acts because the deformed lithosphere is not in isostatic equilibrium: A thickness w of mantle with density ρ_m has been replaced by water with density ρ_w. Thus, for the oceanic lithosphere, Eq. 5.56 is

$$D\frac{d^4w}{dx^4} = V(x) - (\rho_m - \rho_w)gw \tag{5.58}$$

In the case of the deformation of continental lithosphere, when the depression is filled with sediment, the hydrostatic restoring force is $(\rho_m - \rho_c)gw$, since mantle with density ρ_m has been replaced by crust with density ρ_c. For continental lithosphere Eq. 5.56 is

$$D\frac{d^4w}{dx^4} = V(x) - (\rho_m - \rho_c)gw \tag{5.59}$$

These differential equations must be solved for given loads and boundary conditions to give the deflection of the plate as a function of horizontal distance. In the particular case in which the load is an island chain (assumed to be at $x = 0$), Eq. 5.58 is

$$D\frac{d^4w}{dx^4} + (\rho_m - \rho_w)gw = 0 \tag{5.60}$$

The solution to this equation for a line load V at $x = 0$ is

$$w(x) = w_0 e^{(-x/\alpha)}[\cos(x/\alpha) + \sin(x/\alpha)], \qquad x \geqslant 0 \tag{5.61}$$

where

$$w_0 = \frac{V\alpha^3}{8D} \tag{5.62}$$

and

$$\alpha = \left[\frac{4D}{(\rho_m - \rho_w)g}\right]^{1/4} \tag{5.63}$$

The parameter α is often known as the *flexural parameter*. Figure 5.13 shows the deflection given by Eq. 5.61 as a function of x. Notice the clear arch, or forebulge, on either side of the central depression. Estimating V can be difficult but is fortunately not necessary because the width of the depression defines α. Therefore, Eq. 5.63 provides a better method of determining the elastic thickness of the lithosphere h. Consider, for example, a depression with a half-width of 150 km. Using Figure 5.13, we

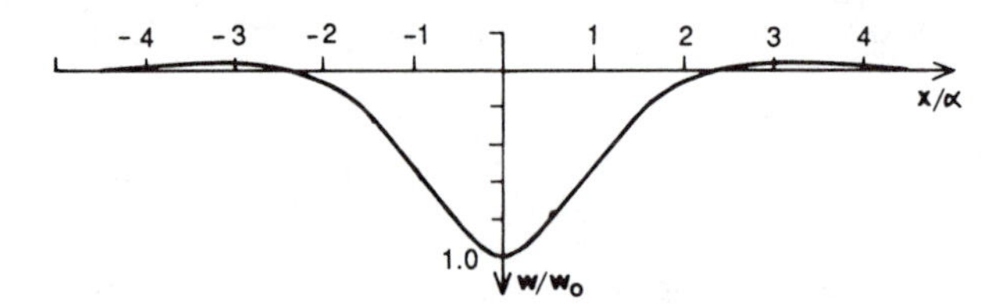

Figure 5.13. Deflection of an elastic plate by a line load at $x=0$. The deflection is normalized to w_0, the deflection at $x=0$, which is determined by the load and the physical properties of the plate (Eq. 5.62).

can estimate the flexural parameter α as 64 km. Then, with Eqs. 5.63 and 5.57, the elastic thickness of the lithosphere is calculated as 25 km when $\rho_m = 3.3 \times 10^3\,\mathrm{kg\,m^{-3}}$, $\rho_w = 1.0 \times 10^3\,\mathrm{kg\,m^{-3}}$, $g = 10\,\mathrm{m\,s^{-2}}$, $D = 9.6 \times 10^{22}\,\mathrm{N\,m}$, $E = 70\,\mathrm{GPa}$ and $\sigma = 0.25$.

Similar but more complex analyses have been used to estimate the elastic thickness of the Pacific Plate under the Hawaiian–Emperor island chain (Fig. 5.11). (The problem is more complex than the simple solution given here because the island chain has a finite width and so cannot be treated as a line force acting at $x=0$. Also, the age of the Pacific Plate and therefore its thickness change along the length of the island chain.)

The bending of the oceanic lithosphere at a subduction zone can also be modelled by Eq. 5.56. In this case, it is necessary to include a load at one end ($x=0$) of the plate V and a horizontal bending moment M per unit length. The deflection of plate is then given by

$$w(x) = \frac{\alpha^2}{2D}\exp\left(-\frac{x}{\alpha}\right)\left[-M\sin\left(\frac{x}{\alpha}\right) + (V\alpha + M)\cos\left(\frac{x}{\alpha}\right)\right], \qquad x \geqslant 0 \tag{5.64}$$

This topographic profile has, like the island chain profile, a pronounced forebulge, as is observed for subduction zones. The parameters M and V cannot be reliably estimated in practice, but fortunately the width and height of the forebulge can be, and these can be used to calculate α and hence the elastic thickness of the lithosphere. Figure 5.14a shows a topographic profile across the Mariana Trench and a theoretical profile for $\alpha = 70$ km. The two profiles are clearly in good agreement. From Eqs. 5.63 and 5.57 and the same values as in the previous example, the elastic thickness of the subducting oceanic plate is then 28 km.

Although the topography of many trenches can be explained simply as the bending of an elastic plate, there are exceptions. The Tonga Trench, for

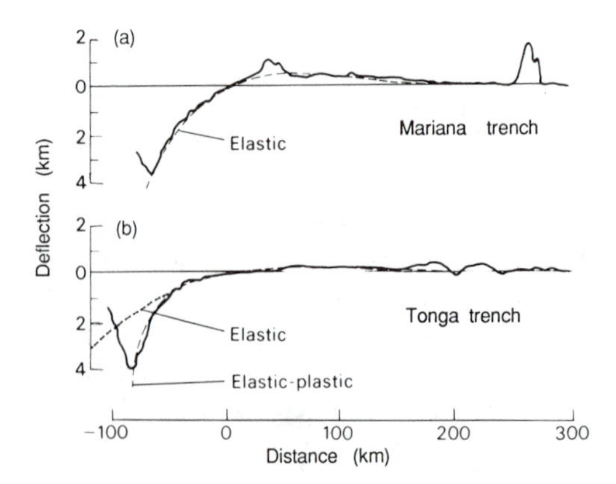

Figure 5.14. Observed (solid line) and theoretical (dashed line) bathymetric profiles across two subduction zones. (a) The bathymetry of the Mariana Trench is fitted by the flexure of a 28 km-thick elastic plate. (b) the bathymetry of the Tonga Trench is not fitted well by flexure of an elastic plate but is fitted by a flexure of an elastic and then perfectly plastic plate 32 km thick with a yield stress of 1000 MPa. (After Turcotte et al. 1978.)

example, bends more steeply than can be explained by an elastic model of the plate. However, if the plate is assumed to behave elastically up to some critical yield stress (e.g., 1000 MPa) and then to behave perfectly plastically* above that stress, the observed deformation of the Tonga Trench can be explained. Figure 5.14b shows observed and theoretical topographic profiles across the Tonga Trench, which indicate that an elastic and perfectly plastic lithosphere is the more appropriate model there.

The values for the elastic thickness of the oceanic plates determined in this section are considerably less than the values determined from seismic and thermal data (Sects. 4.1.3 and 7.5). A 100 km thick elastic plate could not deform as the oceanic lithosphere is observed to bend, bending would be much more gradual. These apparent contradictions are a consequence of the thermal structure of the lithosphere (discussed in Sect. 7.5). Figure 5.15 shows that the long-term elastic thickness of the Pacific Plate apparently increases with age and approximately corresponds to the 450°C isotherm. The elastic thickness of the continental lithosphere is considerably thicker than that of the oceanic lithosphere: 80–100 km is typical.

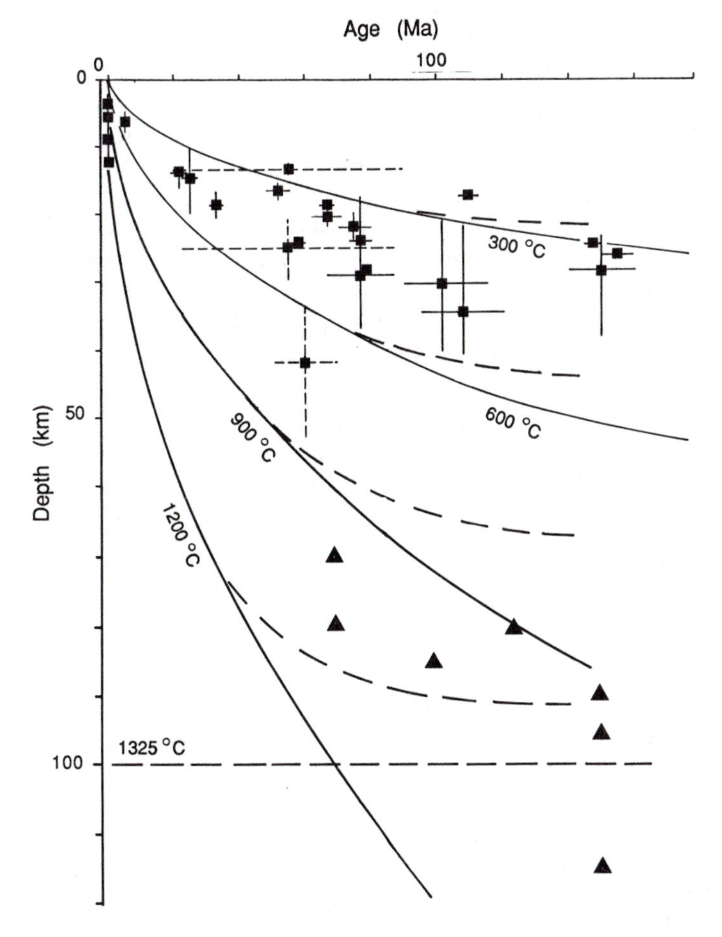

Figure 5.15. The long-term elastic thickness of the lithosphere estimated from studies of flexure (squares). The short-term elastic thickness estimated from surface waves (triangles). Also shown are isotherms for the plate model (dashed curves) and a cooling half-space model (solid curves) (Sect. 7.5) with a mantle temperature of 1325°C, specific heat 1.25×10^3 J kg^{-1} °C^{-1} and thermal conductivity 3 W m^{-1} °C^{-1}. (After Watts and Daly 1981. Reproduced with permission from the *Annual Review of Earth and Planetary Sciences*, vol. 9, © 1981 by Annual Reviews Inc.)

* If a material deforms plastically (or anelastically), the deformation is irreversible, unlike elastic deformation, which is reversible.

5.7.2 Isostatic Rebound

The examples just discussed assume an equilibrium situation in which the load has been in place for a long time and deformation has occurred. A study of the rate of deformation after the application or removal of a load, however, shows that the rate of deformation is dependent on both the flexural rigidity of the lithosphere and the viscosity* of the mantle. Mountain building and subsequent erosion can be so slow that the viscosity of the underlying mantle is not important; the mantle can be assumed to be in equilibrium at all times. However, the ice caps – which during the late Pleistocene covered much of Greenland, northern North America and Scandinavia – provide loads of both the right magnitude and age to enable the viscosity of the mantle to be estimated. Figure 5.16 illustrates the deformation and rebound which occurs as the lithosphere is first loaded and then unloaded.

To determine the viscosity of the uppermost mantle, one must find a narrow load. An example of such a load was the water of Lake Bonneville in Utah, U.S.A., the ancestor of the present Great Salt Lake. The old lake, which existed during the Pleistocene and had a radius of about 100 km and a central depth of about 300 m, dried up about 10,000 years ago. As a result of the drying of this lake, the ground is now domed: The centre of the old lake has risen 65 m relative to the margins. This doming is a result of the isostatic adjustment which took place after the water load was removed. Estimates of the viscosity of the asthenosphere which would permit this amount of uplift to occur in the time available range from 10^{20} Pa s for a 250 km thick asthenosphere to 4×10^{19} Pa s for a 75 km thick asthenosphere. Postglacial uplift of a small region of northeastern Greenland and of some of the Arctic Islands indicate that the lower value is a better estimate. This low-viscosity channel appears to correspond to the low-velocity zone in the upper mantle (Fig. 4.27), but since the geographical distribution of the viscosity determination is very limited, the correlation is speculative.

Wider and much more extensive loads must be used to determine the viscosity of the upper and lower mantle. The last Fennoscandian (Finland plus Scandinavia) ice sheet, which melted some 10,000 years ago, was centred on the Gulf of Bothnia. It covered an area of approximately 4×10^6 km^2 with a maximum average thickness of ~2.5 km. The maximum present-day rate of uplift, which is more than 0.9 cm yr^{-1}, is occurring in the Gulf of Bothnia close to the centre of the ancient ice sheet (Fig. 5.17).

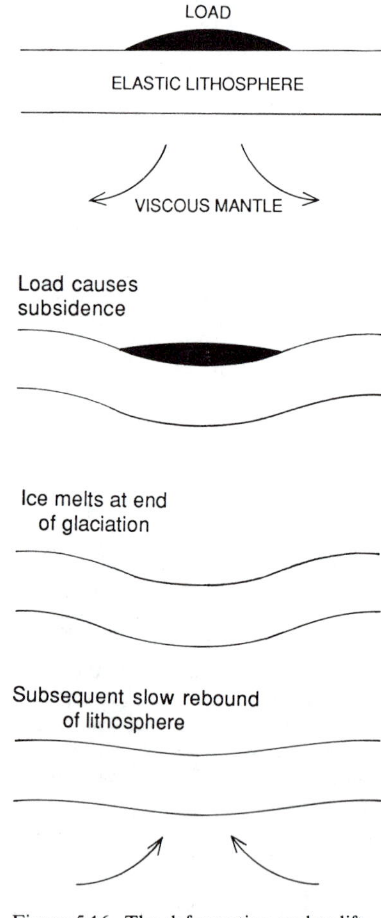

Figure 5.16. The deformation and uplift which occur as a result of loading and unloading of an elastic lithospheric plate overlying a viscous mantle.

* *Newtonian viscosity* is defined as the ratio of shear stress to strain rate and is therefore essentially a measure of the internal friction of a fluid. Although many fluids have Newtonian viscosity, some, such as most paints and egg white, do not. For these non-Newtonian fluids, the strain rate increases much more rapidly with increasing shear stress. There is debate about the exact properties of the mantle. However, here we simply assume it to be a Newtonian fluid. The unit of viscosity is the Pascal second (Pa s), in which 1 Pa s $\equiv$ 1 N m^{-2} s. The viscosity of water at 20°C is 10^{-3} Pa s; at 100°C, 0.3×10^{-3} Pa s. Castor oil at 0°C has a viscosity of 5.3 Pa s; at 20°C, 1 Pa s; and at 100°C, 2×10^{-2} Pa s. The viscosity of most fluids and of rock decreases rapidly with increasing temperature. However, within the earth, pressure tends to counteract the effects of temperature.

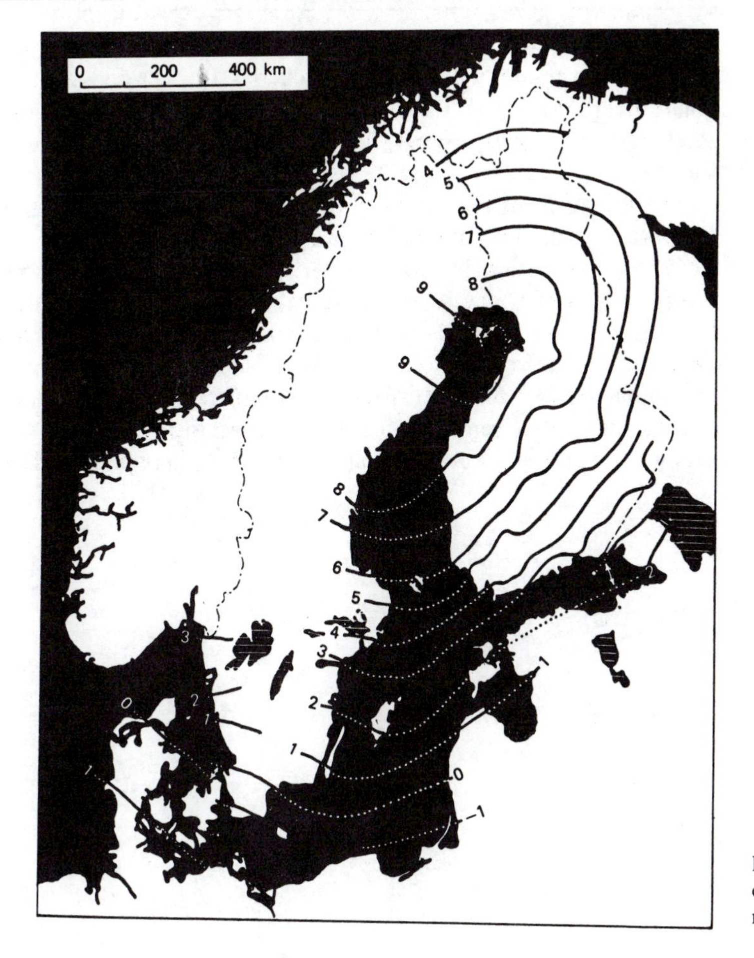

Figure 5.17. Present-day rate of uplift occurring in Fennoscandia. Contours in millimetres per year. (After Flint 1971.)

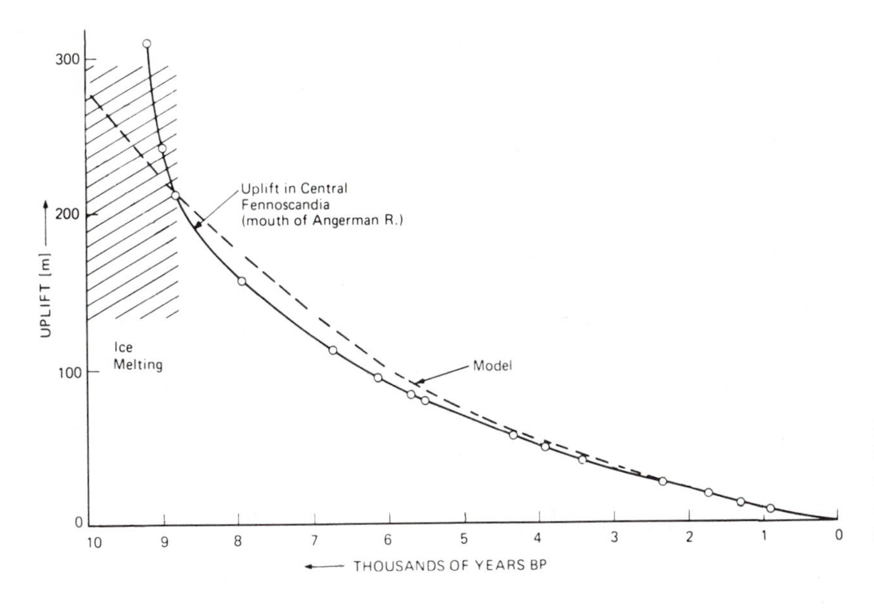

Figure 5.18. Uplift in central Fennoscandia calculated for a constant viscosity (10^{21} Pa s) mantle (dashed line) and geological observations (○) from the nothern Gulf of Bothnia. (After Cathles 1975.)

Figure 5.18 shows the uplift calculated to occur in the northern Gulf of Bothnia for a model with a 75 km thick, 4×10^{19} Pa s viscosity asthenosphere overlying a 10^{21} Pa s viscosity mantle. This predicted uplift is in good agreement with observations. It is predicted that 30 m of uplift remains.

Determination of the viscosity of the lower mantle requires loads of very large extent. The Wisconsin ice sheet, formed during the most recent Pleistocene glaciation in North America, was perhaps 3.5 km thick and covered much of Canada as well as part of the northern United States. Melting of the ice resulted in rebound of the continent but also loading of the oceans. Thus, melting of an extensive ice sheet provides both loading and unloading data for study. Figure 5.19 illustrates the calculated uplift at various times after removal of a model Wisconsin ice sheet which had attained isostatic equilibrium on an elastic lithosphere plate underlain by a

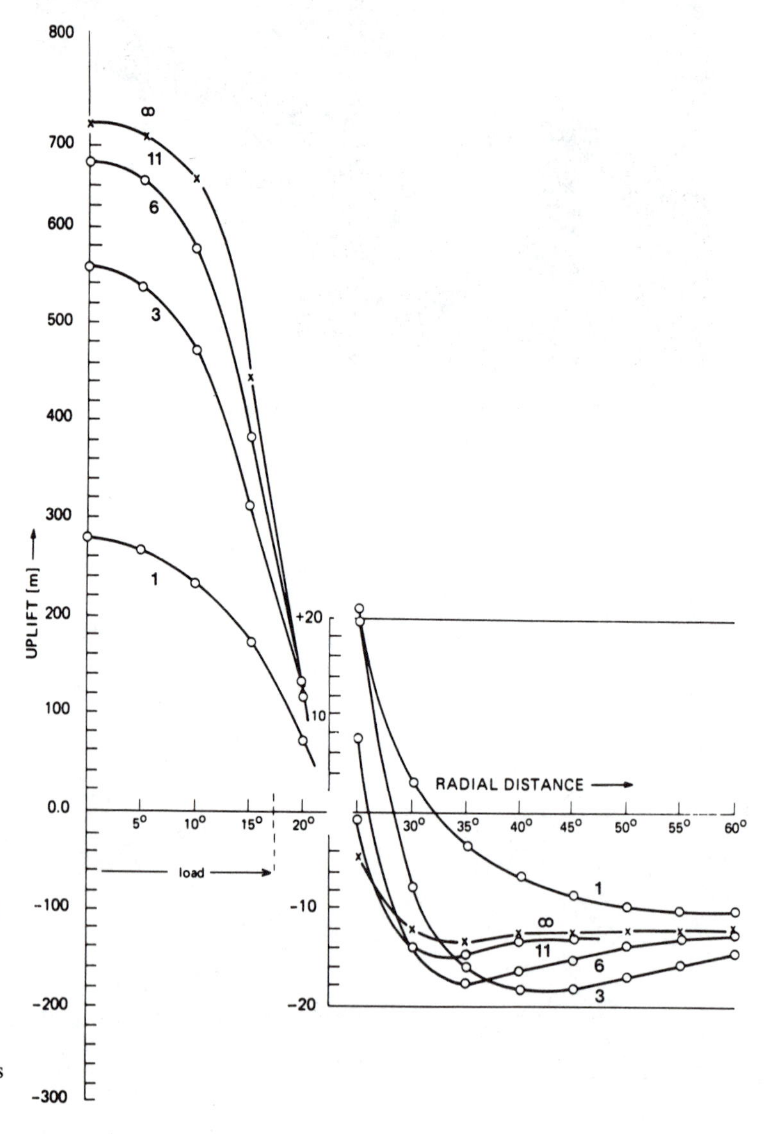

Figure 5.19. The uplift at 1, 3, 6, 11 (circles) and ∞ (crosses) thousand years after the removal of a 2.5 km thick model Wisconsin ice sheet from an elastic lithospheric plate overlying a uniform 10^{21} Pa s viscosity mantle. Note that for clarity the vertical scale at distances greater than 25° (1° = 110 km) is magnified by a factor of 10. (From Cathles 1975.)

Table 5.1. *Estimates of the viscosity of the mantle as determined from studies of postglacial rebound*

	Depth (km)	Viscosity (Pa s)
Lithosphere	0–100	Elastic (rigidity = 5×10^{24} N m)
Asthenosphere	100–175	4×10^{19}
Mantle	175–2885	10^{21}

Source: From Cathles (1975).

constant viscosity mantle. Uplift is greatest at the centre of the load, and uplift rate decreases with time as expected. Regions peripheral to the load also undergo initial uplift before undergoing subsidence. Such uplift, followed by subsidence, is documented in ancient sea-level changes along the east coast of the United States. (The melting of the ice sheet immediately causes sea level to rise, but the oceanic plate also deforms in response to the increased water load, thus resulting in a further change in sea level.) The uplift documented in Canada (present-day rates around Hudson Bay are about 1 cm yr^{-1}) and the sea-level changes along the east coast of North America are all in reasonable agreement with a mantle of reasonably constant viscosity: The models imply an upper mantle with viscosity 10^{21} Pa s and a lower mantle with viscosity in the range 10^{21}–10^{23} Pa s.

Table 5.1 gives a viscosity model for the mantle as determined by these postglacial rebound studies.

PROBLEMS

1. Calculate the mass of the earth.
2. Calculate the maximum gravitational attraction between Newton and the apple, assuming (a) it passed 30 cm from his centre of mass and (b) it hit him.
3. Calculate the weight of a 60 kg astronaut on the moon. How much less is this than the astronaut's weight on the earth? (Moon mass = 7.4×10^{22} kg and radius = 1738 km.)
4. (a) Calculate the radius of orbit of a geostationary satellite (a satellite whose orbit is such that it remains above the same point on the earth's surface).
   (b) Calculate the period of a satellite orbiting 200 km above the earth.
   (c) Calculate the period of a satellite orbiting 200 km above the moon.
5. What is the radius of the moon's orbit about the earth? Use the mass of the earth as calculated in Problem 1 and assume that the period of rotation is 28 days. Explain why it is not necessary to know the mass of the moon to perform the calculation.
6. Calculate the mass of the sun, assuming the earth's orbit to be circular with a radius of 1.5×10^{8} km.
7. A small moon of mass m and radius a orbits a planet of mass M, keeping the same face towards the planet. Show that if the moon approaches the planet closer than $r_c = a(3M/m)^{1/3}$, loose rocks lying on the surface of the moon will be lifted off. (From Thompson 1987.)

8. What is the difference in length, in kilometres, between a degree of latitude at the equator and at the pole?
9. Explain (qualitatively and quantitatively) how the source of the Mississippi River can be about 5 km closer to the centre of the earth than its mouth. (From Thompson 1987.)
10. It is observed that the acceleration g due to gravity is greater down a mine than at the earth's surface. Show that this result can be explained if the earth's density increases sufficiently rapidly with depth. (From Thompson 1987.)
11. What is the difference between the value of the earth's gravitational acceleration at the equator and the poles? In what units would this difference usually be quoted?
12. (a) Verify the isostatic equilibrium calculation for continents and ocean basins, using densities of sea water, crust and mantle of 1.03, 2.9 and $3.3 \times 10^3\,kg\,m^{-3}$, respectively, an ocean-basin depth of 5 km and an oceanic crustal thickness of 6.6 km.
    (b) What would happen if the ocean-basin depth were to change to (i) 1 km, (ii) 3 km or (iii) 8 km?
13. Calculate the depths and densities beneath a 5 km high mountain chain in isostatic equilibrium with a 35 km thick, $2.8 \times 10^3\,kg\,m^{-3}$ density continental crust and a $3.3 \times 10^3\,kg\,m^{-3}$ mantle by using the hypothesis of (a) Pratt and (b) Airy.
14. A mountain range 4 km high is in isostatic equilibrium.
    (a) During a period of erosion, a 2 km thickness of material is removed from the mountains. When the new isostatic equilibrium is achieved, how high are the mountains?
    (b) How high would they be if 10 km of material were eroded away?
    (c) How much material must be eroded to bring the mountains down to sea level? (Use crustal and mantle densities of $2.8 \times 10^3\,kg\,m^{-3}$ and $3.3 \times 10^3\,kg\,m^{-3}$.)
15. A 500 m deep depression on the earth's surface fills with (a) sandstone of density $2.2 \times 10^3\,kg\,m^{-3}$ and (b) ironstone of density $3.4 \times 10^3\,kg\,m^{-3}$. Assuming that isostatic equilibrium is attained, calculate the thickness of sediment that will be deposited. (Use crustal and mantle densities of 2.8 and $3.3 \times 10^3\,kg\,m^{-3}$, respectively.)
16. If subduction doubled the thickness of the continental crust, calculate the elevation of the resulting plateau. Assuming that all plateaux are eroded down to sea level, calculate the total thickness of material that would be eroded.
17. Assume that the oceanic regions are in Airy-type isostatic equilibrium. If the lithosphere has a uniform density, show that the depth of the seabed is given by

$$d + (L(t) - L(0))\left(\frac{\rho_1 - \rho_a}{\rho_a - \rho_w}\right)$$

    where $L(t)$ is the thickness of the lithosphere age t, d is the depth of water at the ridge axis and ρ_w, ρ_l and ρ_a are the densities of water, lithosphere and asthenosphere, respectively.
18. (a) Calculate the maximum gravity anomaly due to a sphere of radius 1 km and a density contrast $300\,kg\,m^{-3}$ buried at a depth of (i) 1 km, (ii) 2 km and (iii) 15 km.

(b) Calculate the gravity anomalies at distances of 1 km, 5 km and 10 km from these spheres.

19. (a) What is the maximum gravity anomaly due to a cylinder of radius 1 km with density contrast 200 kg m^{-3} which is buried at a depth of 1 km.
    (b) A cylinder, radius 50 km and buried at a depth of 100 km, yields the same maximum gravity anomaly as that in (a). Calculate the density contrast of this deep cylinder.
    (c) Can gravity measurements resolve deep anomalous mantle density structures?
20. (a) Calculate the geoid height anomaly due to a mountain height h for Pratt-type compensation. What is this anomaly for $h = 2$ km?
    (b) Calculate the same anomaly for an ocean depth d. What is this anomaly for $d = 5$ km?
21. What is the anomaly in the gravitational potential as measured on the spheroid if the geoid height anomaly is (a) 3 m, (b) – 5 m and (c) 8 m?
22. Calculate the depression of the land surfaces beneath (a) the Wisconsin ice sheet and (b) the Fennoscandia ice sheet, assuming that prior to and after the emplacement of the ice sheets, they were in isostatic equilibrium.
23. What gravitational arguments can you put forward to counter the proposal that the oceans formed when huge continental areas sunk beneath the sea?
24. Estimate the gravitational effect of the 76 men mentioned in Section 5.2.1.

BIBLIOGRAPHY

Airy, G. B. 1855. On the computations of the effect of the attraction of the mountain masses as disturbing the apparent astronomical latitude of stations in geodetic surveys. *Phil. Trans. R. Soc. Lond.*, *145*, 101–4.

Beaumont, C. 1978. The evolution of sedimentary basins on a viscoelastic lithosphere: theory and examples. *Geophys. J. R. Astr. Soc.*, *55*, 471–98.

Bott, M. H. P. 1982. *The interior of the earth*, 2nd ed. Elsevier, Amsterdam.

Caldwell, J. E., Haxby, W. F., Kang, D. E., and Turcotte, D. L. 1976. On the applicability of a universal elastic trench profile. *Earth Planet. Sci. Lett.*, *31*, 239–46.

Cathles, L. M., III. 1975. *The viscosity of the earth's mantle.* Princeton Univ. Press, Princeton, N.J.

Chapman, M. E., and Talwani, M. 1979. Comparison of gravimetric geoids with Geos 3 altimetric geoid. *J. Geophys. Res.*, *84*, 3803–16.

Cook, A. H. 1973. *Physics of the earth and planets.* Wiley, New York.

De Bremaecker, J.-C. 1985. *Geophysics: The earth's interior.* Wiley, New York.

Dobrin, M. B., and Savit, C. H. 1988. *Introduction to geophysical prospecting*, 4th ed. McGraw-Hill, New York.

Flint, R. F. 1971. *Glacial and quarternary geology.* Wiley, New York.

Garland, G. D. 1979. *Introduction to geophysics: Mantle, core and crust*, 2nd ed. Saunders, Philadelphia.

Hoffman, N. R. A., and McKenzie, D. P. 1985. The destruction of geochemical heterogeneities by differential fluid motions during mantle convection. *Geophys. J. R. Astr. Soc.*, *82*, 163–206.

King-Hele, D. G. 1969. Royal Aircraft Establishment Technical Memorandum Space 130, Farnborough, U.K.

King-Hele, D. G., Brookes, C. J., and Cook, G. E. 1981. Odd zonal harmonics in the geopotential from analysis of 28 satellite orbits. *Geophys. J. R. Astr. Soc., 64*, 3–30.

Lerch, F. J., Klosko, S. M., Laubscher, R. E., and Wagner, C. A. 1979. Gravity model improvement using GEOS 3 (GEM 9 and 10). *J. Geophys. Res., 84*, 3897–916.

McKenzie, D. P. 1977. Surface deformation, gravity anomalies and convection. *Geophys. J. R. Astr. Soc., 48*, 211–38.

1983. The earth's mantle. *Sci. Am.*, 249, 3, 66–113.

McKenzie, D.P., Roberts, J. M., and Weiss, N. O. 1974. Convection in the earth's mantle: Towards a numerical simulation. *J. Fluid Mech., 62*, 465–538.

McKenzie, D. P., Watts, A. B., Parsons, B., and Roufosse, M. 1980. Planform of mantle convection beneath the Pacific Ocean. *Nature, 288*, 442–6.

Milne, J. 1906. Bakerian Lecture: Recent advances in seismology. *Proc. Roy. Soc. A, 77*, 365–76.

Morner, N. A., ed. 1980. *Earth rheology, isostacy and eustacy.* Wiley, New York.

Nakada, M., and Lambeck, K. 1987. Glacial rebound and relative sea-level variations: A new appraisal. *Geophys. J. R. Astr. Soc., 90*, 171–224.

Peltier, W. R. 1983. Constraint on deep mantle viscosity from Lageos acceleration data. *Nature, 304*, 434–6.

1984. The thickness of the continental lithosphere. *J. Geophys. Res., 89*, 11303–16.

1985. New constraint on transient lower mantle rheology and internal mantle buoyancy from glacial rebound data. *Nature, 318*, 614–17.

Pratt, J. H. 1855. On the attraction of the Himalaya Mountains, and of the elevated regions beyond them, upon the plumb line in India. *Phil. Trans. R. Soc. Lond., 145*, 53–100.

Sandwell, D. T., and Renkin, M. L. 1988. Compensation of swells and plateaus in the North Pacific: No direct evidence for mantle convection. *J. Geophys. Res., 93*, 2775–83.

Telford, W. M., Geldart, L. P., and Sheriff, R. E. 1990. *Applied geophysics*, 2nd ed. Cambridge Univ. Press, New York.

Thompson, N., ed. 1987. *Thinking like a physicist.* Adam Hilger, Bristol.

Turcotte, D. L. 1979. Flexure. *Adv. Geophys., 21*, 51–86.

Turcotte, D. L., McAdoo, D. C., and Caldwell, J. G. 1978. An elastic–perfectly plastic analysis of the bending of the lithosphere at a trench. *Tectonophysics, 47*, 193–205.

Turcotte, D. L., and Schubert, G. 1982. *Geodynamics: Applications of continuum physics to geological problems.* Wiley, New York.

Warner, M. R. 1987. Seismic reflections from the Moho: The effect of isostasy. *Geophys. J. R. Astr. Soc., 88*, 425–35.

Watts, A. B. 1978. An analysis of isostasy in the world's oceans 1. Hawaiian–Emperor seamount chain. *J. Geophys. Res., 83*, 5989–6004.

1979. On geoid heights derived from Geos 3 altimeter data along the Hawaiian–Emperor seamount chain. *J. Geophys. Res., 84*, 3817–26.

Watts, A. B., Bodine, J. H., Ribe, N. M. 1980. Observations of flexure and the geological evolution of the Pacific Ocean. *Nature, 283*, 532–7.

Watts, A. B., and Daly, S. F. 1981. Long wavelength gravity and topography anomalies. *Ann. Rev. Earth Planet. Sci., 9*, 415–48.

Watts, A. B., McKenzie, D. P., Parsons, B. E., and Roufosse, M. 1985. The relationship between gravity and bathymetry in the Pacific Ocean. *Geophys. J. R. Astr. Soc., 83*, 263–98.

6

Geochronology

6.1 Introduction

Radioactivity was discovered in 1896 by Henri Becquerel. Over the following decades the new understanding of the atom that came from the work of Rutherford, Soddy, Boltwood and others had a major impact on geology. Before this work, the age of the earth was unknown. In the last century, Lord Kelvin (William Thompson) attempted to calculate the age of the earth by assuming that the planet was a hot body cooling by conduction. He obtained a young age which conflicted with the observations of geologists, who had concluded that the earth must be at least several hundred million years old. The geological reasoning was based on rather qualitative evidence such as the observation of sedimentary deposition rates, calculations about the amount of salt in the sea and guesses of evolutionary rates. Not unnaturally, Kelvin's apparently more rigorous and quantitative physical calculation was regarded as much sounder by most scientists. In 1904, while at McGill University in Montreal, Ernest Rutherford realized that radioactive heat could account for some of the apparent discrepancy. Kelvin was sceptical to the extent that he bet Rayleigh (Hon R. J. Strutt) five shillings on the matter, but later he paid up. Rutherford gave a lecture at the Royal Institution in London in 1904 about which he wrote:

> To my relief Kelvin fell fast asleep but as I came to the important point, I saw the old bird sit up, open an eye and cock a baleful glance at me. Then a sudden inspiration came and I said Lord Kelvin had limited the age of the earth, provided no new source (of heat) was discovered.... Behold! the old boy beamed upon me.

In fact, Kelvin's calculation is now known to be too simple; not only does radioactive decay provide an extra source of heat, but more importantly, heat is transferred within the mantle by convection and not conduction (see Sect. 7.4). Had Kelvin carried out a convection calculation, as he did for the sun, he would have obtained a more reasonable answer. Soon after this lecture, Rutherford and his colleague Boltwood developed radioactive dating, which was first applied to show that even the geological estimates of the age of the earth were too modest. Rutherford's first dating method was to measure the accumulation of helium (α particles) from the radioactive decay of uranium but he later suggested that measuring the accumulation of lead would be a better method. In 1907, Boltwood published the first uranium–lead dates.

Precision describes the reproducibility with which measurements are made. *Accuracy* describes the truth of these measurements. The geological tools of stratigraphy and palaeontology provide a very precise method of measuring *relative ages*. In good cases, a palaeontological resolution of better than 0.25 Ma can be attained. These tools also enable us to discover the order in which rocks were laid down and the order in which tectonic events and sea transgressions occurred. The geologic time scale is made up of stratigraphic divisions based on observed rock sequences. However, palaeontology is unable to give any accurate estimate of the *absolute ages* of geological events. The use of radioactive isotopes to date rocks, being a measurement of physical properties, is quite separate from any intuitive method and thus can provide an independent, fairly accurate and sometimes very precise date. There are, of course, many assumptions and inherent problems in these dating methods, as in any other, of which anyone using such dates should be aware. The principal radioactive dating methods are discussed in the next section.

A standard geologic time scale, based on stratigraphy, palaeontology, geochronology and geomagnetic polarity reversals is shown in Table 6.1.

6.2 General Theory

The disintegration of any radioactive atom is an entirely random event, independent of neighbouring atoms, physical conditions and the chemical state of the atom. Disintegration depends only on the structure of the nucleus. This means that every atom of a given type has the same probability of disintegrating in unit time. This probability is called the *decay constant*, λ, a different constant for each isotope. Suppose that at time t there are N atoms and that at time $t + \delta t$, δN of these have disintegrated, then δN can be expressed as

$$\delta N = -\lambda N \, \delta t \tag{6.1}$$

In the limit as δN and $\delta t \to 0$, Eq. 6.1 becomes

$$\frac{dN}{dt} = -\lambda N \tag{6.2}$$

Thus, the rate of disintegration dN/dt is proportional to the number of atoms present. Integration of Eq. 6.2 gives

$$\log_e N = -\lambda t + c \tag{6.3}$$

where c is a constant. If at time $t = 0$ there are N_0 atoms present, then

$$c = \log_e N_0 \tag{6.4}$$

Thus, Eq. 6.3 can be written

$$N = N_0 e^{-\lambda t} \tag{6.5}$$

The rate of disintegration dN/dt is sometimes called the activity A.

Table 6.1. *A geological time scale*

Aeon	Era	Subera, period	Subperiod, epoch	Age (Ma)
Phanerozoic Ph	Cenozoic Cz	Quaternary Q	Pleistocene	1.64
		Tertiary TT	Pliocene	5.2
			Miocene	23.3
			Oligocene	35.4
			Eocene	56.5
			Palaeocene	65.0
	Mesozoic Mz	Cretaceous K	Late	97.0
			Early	146
		Jurassic J	Malm	157
			Dogger	178
			Lias	208
		Triassic Tr	Late	235
			Middle	241
			Early	245
	Palaeozoic Pz	Permian P	Late	256
			Early	290
		Carboniferous C	Pennsylvanian	323
			Mississppian	363
		Devonian D	Late	377
			Middle	386
			Early	408
		Silurian S	Pridoli	411
			Ludlow	424
			Wenlock	430
			Llandovery	439
		Ordovician O	Ashgill	443
			Caradoc	464
			Llandeilo	469
			Llanvirn	476
			Arenig	493
			Tremadoc	510
		Cambrian Ꞓ	Merioneth	517
			St. David's	536
			Caerfai	570
PRECAMBRIAN: Proterozoic Pt		Hadrynian		
		Helikian		1000
		Aphebian (Canada)		1800
Archaean Ar		Kenoran Witwatersrand	Shamvaian	2500
			Bulawayan	
		Pongola	Belingwean	
	Isua (Greenland)	(Canada) Barberton (S. Africa)	Sebakwian (Zimbabwe) Pilbara (Australia)	4000
	Hadean			4600

Note: There is continuing debate about the ages assigned to various boundaries. Precambrian names have no international status and are illustrative examples only.
Source: Modified after Harland et al. (1990).

Equation 6.2 can be rewritten in terms of A instead of N, giving

$$A = A_0 e^{-\lambda t} \tag{6.6}$$

which is then an alternative expression for Eq. 6.5. Suppose that at some starting time N_0 radioactive atoms are present. The *half-life*, $T_{1/2}$, is then the length of time required for one-half of those original atoms to undergo disintegration. Therefore, putting $N = N_0/2$ into Eq. 6.5 gives

$$\frac{N_0}{2} = N_0 e^{-\lambda T_{1/2}} \tag{6.7}$$

or

$$T_{1/2} = \frac{1}{\lambda} \log_e 2 \simeq \frac{0.693}{\lambda} \tag{6.8}$$

Consider the case of a radioactive *parent* atom disintegrating to a stable atom called the *daughter*. After time t, $N = N_0 - D$ parent atoms remain, and Eq. 6.5 can be written

$$N_0 - D = N_0 e^{-\lambda t} \tag{6.9}$$

where D is the number of daughter atoms (all of which have come from disintegration of the parent) present at time t. Thus,

$$D = N_0(1 - e^{-\lambda t}) \tag{6.10}$$

However, since it is not possible to measure N_0, but only N, Eq. 6.5 must be used to rewrite Eq. 6.10:

$$D = N(e^{\lambda t} - 1) \tag{6.11}$$

This equation expresses the number of daughter atoms D in terms of the number of parent atoms N, both measured at time t, and it means that t can be calculated by taking the natural logarithm:

$$t = \frac{1}{\lambda} \log_e \left(1 + \frac{D}{N} \right) \tag{6.12}$$

These equations give the probability that any atom will survive a time t. The actual proportion that survives in any particular case is subject to statistical fluctuations, which are usually very small because N_0 and N are very large numbers. In practice, measurements of the ratio D/N are made by using a mass spectrometer.

Age determination is not always as simple as applying Eq. 6.12. In most cases there may have been an initial concentration of the daughter in the sample, and thus the assumption that all the measured daughter is a product of the parent is not necessarily valid. Also, all systems are not *closed*, as has been assumed here; in other words, over time there may have been some exchange of the parent and/or daughter atoms with surrounding material. If more than one age method is used to estimate t and the resulting ages are within analytical error, then these problems can be neglected; in this case we say that the various dates are *concordant*. In the case of *discordant* ages, the possibility of a nonclosed, open system or the initial presence of the daughter must be considered and, when possible, appropriate corrections made. The date t that is estimated by these methods is not necessarily the date of formation of a rock; it may be the date the rock crystallized or the date of some metamorphic event which heated the rock to such a degree that chemical changes took place (see Sect. 7.11.5 for discussion). The half-life of a radioactive isotope to be used to estimate the age of a rock must be comparable to the age of the sample because if they are greatly different, the measurement of D/N in Eq. 6.12 becomes impracticable. The fitting of straight lines to data points and the calculation of errors are described in Appendix 4.

Table 6.2 shows the decay products, half-lives and decay constants for those radioactive isotopes most often used in dating geological samples. Also given in the table are the *heat generation rates* for the four isotopes that are responsible for effectively all the radioactive heating in the earth (see Chapter 7). All the radiations produced, except the neutrinos that accompany every decay, contribute heat as they are absorbed within about 30 cm of their origin. The neutrinos interact so little with matter that they penetrate the whole earth and escape. (A remarkable illustration of the neutrino's penetrating ability is that the burst of neutrinos detected on February 23, 1987, which came from the supernova explosion in the Greater Magellanic Cloud high in the southern sky, was detected in the northern hemisphere, in Japan and the United States, after they had passed through the earth.)

Radioactive decay schemes are not simple. Generally, a series of disintegrations takes place until a stable daughter nuclide is formed. Of those isotopes that are commonly used for geological dating, ^{87}Rb and ^{147}Sm undergo a simple decay to a stable daughter (β decay and α decay, respectively); many of the others undergo a sequence of decays. Thus, Eq. 6.12 can be applied directly only to the decays of ^{87}Rb to ^{87}Sr and of ^{147}Sm to ^{143}Nd.

The long series of decays undergone by the isotopes of uranium and thorium can be described by a series of equations like Eq. 6.2. Suppose that the radioactive isotope decays in series to $X_1, X_2, \ldots, X_n$ and the isotope X_n is the stable daughter. Let the number of atoms of each isotope present at time t be $N, N_1, N_2, \ldots, N_{n-1}, D$, and let the decay constants be $\lambda, \lambda_1, \lambda_2, \ldots, \lambda_{n-1}$.

Then Eq. 6.2 can be applied to each disintegration in the series:

$$
\begin{aligned}
\frac{dN}{dt} &= -\lambda N \\
\frac{dN_1}{dt} &= -\lambda_1 N_1 + \lambda N \\
\frac{dN_2}{dt} &= -\lambda_2 N_2 + \lambda_1 N_1 \\
&\vdots \\
\frac{dD}{dt} &= \lambda_{n-1} N_{n-1}
\end{aligned}
\tag{6.13}
$$

The general solution to these equations is more complicated than Eq. 6.5. The solution for, say, the ith isotope in the chain involves exponentials for

Table 6.2. *Decay schemes for radioactive isotopes primarily used in geochronology*

Parent isotope	Daughter isotope	Decay products	Decay constant (a^{-1})	Half-life (a)	Present rate of heat generation ($W\,kg^{-1}$)
^{238}U	^{206}Pb	$8\alpha + 6\beta$	1.55×10^{-10}	4468×10^6	9.4×10^{-5}
^{235}U	^{207}Pb	$7\alpha + 4\beta$	9.85×10^{-10}	704×10^6	5.7×10^{-4}
^{232}Th	^{208}Pb	$6\alpha + 4\beta$	4.95×10^{-11}	14010×10^6	2.7×10^{-5}
^{87}Rb	^{87}Sr	β	1.42×10^{-11}	48800×10^6	
^{147}Sm	^{143}Nd	α	6.54×10^{-12}	106000×10^6	
^{40}K	^{40}Ca	β	4.96×10^{-10}	1400×10^6 } 1250×10^6	2.8×10^{-5}
	^{40}Ar	Electron capture	5.81×10^{-11}	11900×10^6 }	
^{39}Ar	^{39}K	β	2.57×10^{-3}	269	
^{176}Lu	^{176}Hf	β	1.94×10^{-11}	35000×10^6	
^{187}Re	^{187}Os	β	1.52×10^{-11}	45600×10^6	
^{14}C	^{14}N	β	1.21×10^{-4}	5730	

Note: Annum (a) is the SI unit for year: $a^{-1} = yr^{-1}$.
Decay constants and half-lives based on Steiger and Jaeger (1977).

all the decays from the parent to isotope i inclusive. (In physics, these equations are known as the Bateman relations.) Fortunately, for the decay schemes of uranium and thorium, the half-life of the first decay is many orders of magnitude greater than those of the subsequent disintegrations ($\lambda \ll \lambda_1, \ldots, \lambda_{n-1}$). In this case, the solution to Eqs. 6.13 is much simplified:

$$N = N_0 e^{-\lambda t}$$

$$N_i = \frac{\lambda}{\lambda_i} N \qquad (i = 1, \ldots, n-1) \tag{6.14}$$

$$D = N_0(1 - e^{-\lambda t})$$

Thus, just as for the simple decay scheme, the time t is given by Eq. 6.12.

The decay of ^{40}K is also complex. It can decay either by β decay to ^{40}Ca or by electron capture to ^{40}Ar. At time t, let N be the number of ^{40}K atoms and D_C and D_A the number of ^{40}Ca and ^{40}Ar atoms, respectively. Then Eq. 6.2 can be written as

$$\frac{dN}{dt} = -(\lambda_A + \lambda_C)N \tag{6.15}$$

where λ_A and λ_C are the decay rates to ^{40}Ar and ^{40}Ca, respectively. Equation 6.5 then becomes

$$N = N_0 e^{-(\lambda_A + \lambda_C)t} \tag{6.16}$$

The rate of increase of ^{40}Ar atoms dD_A/dt is given by

$$\frac{dD_A}{dt} = \lambda_A N \tag{6.17}$$

Similarly, the rate of increase of ^{40}Ca atoms is

$$\frac{dD_C}{dt} = \lambda_C N \tag{6.18}$$

Substituting the expression for N from Eq. 6.16 into Eq. 6.17 gives

$$\frac{dD_A}{dt} = \lambda_A N_0 e^{-(\lambda_A + \lambda_C)t} \tag{6.19}$$

and so

$$D_A = -\frac{\lambda_A N_0}{(\lambda_A + \lambda_C)} e^{-(\lambda_A + \lambda_C)t} + c \tag{6.20}$$

which c is a constant of integration.

If we assume that $D_A = 0$ at $t = 0$, $c = \lambda_A N_0/(\lambda_A + \lambda_C)$, and so

$$D_A = \frac{\lambda_A N_0}{(\lambda_A + \lambda_C)} [1 - e^{-(\lambda_A + \lambda_C)t}] \tag{6.21}$$

Using Eq. 6.16 to eliminate N_0, we can finally express the ratio of the number of atoms of ^{40}Ar to ^{40}K at time t in terms of the decay rates:

$$\frac{D_A}{N} = \frac{\lambda_A}{(\lambda_A + \lambda_C)} [e^{(\lambda_A + \lambda_C)t} - 1] \tag{6.22}$$

The age of the sample t is then given by

$$t = \frac{1}{(\lambda_A + \lambda_C)} \log_e \left[1 + \left(\frac{\lambda_A + \lambda_C}{\lambda_A} \right) \frac{D_A}{N} \right] \tag{6.23}$$

If the decay to ^{40}Ca were used, Eq. 6.19–6.23 would be exactly the same but with D_C and the decay rate to calcium λ_C, instead of D_A and the decay rate to argon λ_A. The decay of ^{40}K to ^{40}Ca is not used in geochronology because calcium is so common that the ^{40}Ca resulting from ^{40}K decays cannot be accurately determined.

The choice and application of these radioactive dating methods to actual samples require knowledge of a number of factors concerning both the methods and the samples.

In the first instance, as has already been mentioned, the choice of a particular dating method depends on the probable age of the sample. Ideally, the decay scheme used should have a half-life of about the same order of magnitude as the age of the sample. This ensures that the number of daughter atoms is of the same magnitude as the number of parent atoms (see Eq. 6.11). If the half-life is very much greater or smaller than the age of the sample, the ratio of daughter to parent atoms is either very small or very large, which, in either case, is difficult to measure accurately. Hence, in this respect, the uranium–lead method appears to be the best radioactive dating method for younger rocks. The half-lives of the other decays are not ideal, but they are used for lack of better methods. Very small quantities of the daughter atoms have to be measured, and careful analysis is required in all isotope work. For recent samples, particularly archaeological samples, short half-life methods such as carbon-14 are used.

Two other important factors affect the choice of dating method: the amounts of parent and daughter elements present in the rock, and whether or not the sample was a closed system. Clearly a sample which never contained any rubidium could not be a candidate for rubidium–strontium dating (rubidium is a trace element). The advantage of potassium–argon dating is that it can be used on most rocks because potassium is a widespread element. Table 6.3 gives rough estimates of the concentrations of radioactive elements in average rocks.

In the formulation of Eqs. 6.1–6.23, we have assumed that there has been no loss or gain of parent or daughter atoms except by radioactive decay. In

Table 6.3. *Broad estimates of concentrations of radioactive and common daughter elements in rocks*

	U (ppm)	Th (ppm)	Pb (ppm)	K %	Rb (ppm)	Sr (ppm)	Sm (ppm)	Nd (ppm)
Granitoid	4	15	20	3.5	200	300	8	44
Basalt	0.5	1	4	0.8	30	470	10	40
Ultramafic	0.02	0.08	0.1	0.01	0.5	50	0.5	2
Shale	4	12	20	2.7	140	300	10	50

Source: After York and Farquhar (1972) and Faure (1986).

other words, we have assumed that our sample was a closed system. For most rocks this is clearly a false assumption. For example, radioactively derived lead (termed *radiogenic* lead) is sometimes deposited along with uranium in cooling rocks (both are transported in circulating hydrothermal water). The extra amount of daughter lead present in the rocks would result in an overestimation of their age. As another example consider argon, a gas, which is the daughter product of ^{40}K decay. Although argon is not likely to have been present in the rock initially, it cannot be retained in minerals until they have cooled to below their *closure temperatures* (about 280 ± 40°C for biotite and 530 ± 40°C for hornblende). These are informal estimates, actual temperatures depend on local conditions. Thus, a potassium–argon date of, say, a granite is not the data of its intrusion but the time at which the minerals in the granite cooled below their particular closure temperatures. If, after cooling, the granite were reheated to temperatures above the closure temperatures, then argon would be lost. A potassium–argon date, therefore, dates the last time that the sample cooled below the closure temperature. Table 6.4 is an informal compilation of closure temperatures for various dating methods.

The relationship between the closure temperature of a mineral and its cooling history can be put onto a more rigorous footing by the use of some thermodynamics. Since the diffusion of any species in a solid mineral is

Table 6.4 *An informal compilation of closure temperatures*

Mineral	Closure temperature (°C)
Potassium–argon	
Hornblende	530 ± 40
Biotite	280 ± 40
Muscovite	~ 350
Microcline	130 ± 15 (plateau segment)
	110 (0% release intercept)
Uranium–lead	
Zircon	> 750
Monazite	> 650
Sphene	> 600
Allanite	> 600
Apatite	~ 350
Rubidium–strontium	
Biotite	320
Muscovite	500 or more
Apatite, feldspar	~ 350
Fission tracks	
Zircon	175–225
Sphene	290 ± 40
Apatite	105 ± 10

Source: Ghent et al. (1988).

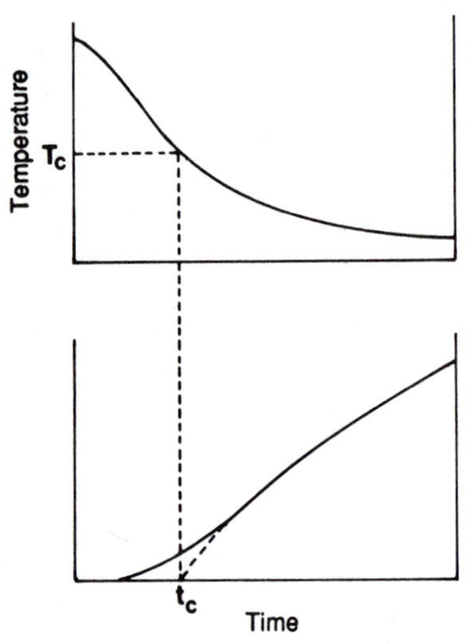

Figure 6.1. Relationship between geochronological closure of a mineral and its cooling history. Upper graph shows the cooling history of a mineral. The vertical axis in the lower graph is related to the diffusion coefficient D and approximates the rate of accumulation of the radiogenic daughter product. T_c is the closure temperature, which is attained at time t_c. Each mineral in a rock has a different set of diffusion coefficients resulting in a variety of closure temperatures and times, as discussed in the text. (From Dodson 1973.)

controlled by temperature, the *diffusion coefficient* D can be defined as

$$D = D_0 e^{(-E/RT)} \tag{6.24}$$

where D_0 is the diffusion coefficient of the particular species and mineral involved at infinitely high temperature, E the activation energy of the diffusion process, R the gas constant and T the temperature. This is the *Arrhenius equation.* Clearly, D is very dependent on temperature; a small change in T can produce an order of magnitude change in D (Fig. 6.1). An expression for the closure temperature T_c (sometimes referred to as the blocking temperature) is

$$\frac{E}{RT_c} = \log_e \left[-\frac{AD_0RT_c^2}{a^2E \left.\dfrac{\partial T}{\partial t}\right|_{T=T_c}} \right] \tag{6.25}$$

where T_c is the closure temperature, a the characteristic dimension of the crystal, A a geometrical factor that depends on the geometry of the system and $\partial T/\partial t|_{T=T_c}$ is the rate of change of temperature at the closure temperature (the slope of the upper graph at T_c). The solution of this equation can only be found by substitution; it cannot be inverted to give an explicit relation between T_c and the other variables.

To interpret the isotopic systematics of a mineral, all of these factors have to be taken into account, which is not always an easy matter. In the following sections, we discuss ways to get around some of the problems, and even to turn some problems around to obtain information on the chemical and thermal history of the crust and mantle. Rubidium–strontium dating is discussed first because it is the most important and commonly used method, as well as one of the simplest.

6.3 Rubidium–Strontium

For the decay of ^{87}Rb to ^{87}Sr, Eq. 6.11 is

$$[^{87}\text{Sr}]_{\text{now}} = [^{87}\text{Rb}]_{\text{now}}(e^{\lambda t} - 1) \tag{6.26}$$

where $[^{87}Sr]_{now}$ is the number of ^{87}Sr atoms and $[^{87}Rb]_{now}$ is the number ^{87}Rb atoms measured now. Since strontium occurs naturally in rocks independently of rubidium, it is not reasonable to assume that all of $[^{87}Sr]_{now}$ is a result of the decay of ^{87}Rb. Equation 6.26 must therefore be modified:

$$[^{87}\text{Sr}]_{\text{now}} = [^{87}\text{Sr}]_0 + [^{87}\text{Rb}]_{\text{now}}\ (e^{\lambda t} - 1) \tag{6.27}$$

where $[^{87}Sr]_0$ is the amount of originally occurring ^{87}Sr. Natural strontium possesses four isotopes with masses 84, 86, 87 and 88, which have fractional abundances of 0.6%, 10%, 7% and 83%, respectively. Rubidium itself has just two isotopes, ^{85}Rb and ^{87}Rb; ^{87}Rb has on average a fractional abundance of ~ 28%.

Since strontium-86 is not a product of any radioactive decay, the amount of strontium-86 present now should be the amount that was originally present:

$$[^{86}\text{Sr}]_{\text{now}} = [^{86}\text{Sr}]_0 \tag{6.28}$$

Normalizing Eq. 6.27 by $[^{86}Sr]_{now}$ gives

$$\frac{[^{87}Sr]_{now}}{[^{86}Sr]_{now}} = \frac{[^{87}Sr]_0}{[^{86}Sr]_{now}} + \frac{[^{87}Rb]_{now}}{[^{86}Sr]_{now}}(e^{\lambda t} - 1)$$

$$= \frac{[^{87}Sr]_0}{[^{86}Sr]_0} + \frac{[^{87}Rb]_{now}}{[^{86}Sr]_{now}}(e^{\lambda t} - 1) \qquad (6.29)$$

This is the equation of a straight line with intercept $[^{87}Sr]_0/[^{86}Sr]_0$ and slope $(e^{\lambda t} - 1)$. If the ratios $[^{87}Sr]_{now}/[^{86}Sr]_{now}$ and $[^{87}Rb]_{now}/[^{86}Sr]_{now}$ are measured for various minerals in a rock and plotted against each other, then the slope of the resulting line is $e^{\lambda t} - 1$, from which t can be determined. Alternatively, from any locality, a set of rock samples having varying amounts of rubidium and strontium can be used. The slope and intercept and their errors can be obtained by least-squares fitting, as discussed in Appendix 4. The intercept $[^{87}Sr]_0/[^{86}Sr]_0$ is called the *initial ratio.* The straight line is called a *whole-rock isochron.* An example of such an isochron is shown in Figure 6.2a. This method presumes that the initial strontium isotope-ratio was the same for all the minerals in the rock and means that the age t can be determined without having to make any assumptions about $[^{87}Sr]_0$, the original amount of strontium-87.

The rubidium–strontium method became popular in the 1960's because the measurements were accessible to early mass spectrometer techniques. The main disadvantages are that rubidium and strontium are often mobile – that is, they can be transported in or out of the rock by geochemical processes – and that rubidium does not occur in abundance (in particular, not in limestones or ultramafic rocks); thus, not all rocks are suitable for dating by this method. In addition, because the half-life of ^{87}Rb is very long, very young rocks cannot be accurately dated.

The initial ratio $[^{87}Sr]_0/[^{86}Sr]_0$ is an indicator of whether the samples have been derived from the remelting of crustal rocks, or otherwise reset, or are of deeper origin. Since λ is small $(1.42 \times 10^{-11} yr^{-1})$, λt is very much less than unity for all geological ages. Therefore, a good approximation to Eq. 6.29 is

$$\frac{[^{87}Sr]_{now}}{[^{86}Sr]_{now}} = \frac{[^{87}Sr]_0}{[^{86}Sr]_0} + \frac{[^{87}Rb]_{now}}{[^{86}Sr]_{now}}\lambda t \qquad (6.30)$$

When $[^{87}Sr]_{now}/[^{86}Sr]_{now}$ is plotted against t (Fig. 6.2b), Eq. 6.30 is the equation of the straight line with slope $\lambda[^{87}Rb]_{now}/[^{86}Sr]_{now}$ and intercept $[^{87}Sr]_0/[^{86}Sr]_0$, describing the increase in the strontium ratio with time. Rocks derived from the Archaean mantle have an initial ratio of roughly 0.700, and it is believed that when the earth formed, the ratio was 0.699 (the value for some meteorites). The ratio for modern oceanic basalts (which are mantle-derived) is 0.704, and this increase of the initial ratio for strontium is attributed to the accumulation of ^{87}Sr from the decay of ^{87}Rb. These ratios are consistent with a gross rubidium–strontium ratio for the mantle of about 0.03. The line OA on Figure 6.2b is known as the *mantle growth curve.* Melts, now forming crustal rocks, which were extracted from mantle can be assumed to have the same initial strontium isotope ratio as the mantle at the time of their extraction. Because the gross rubidium–strontium ratio for crustal rocks is greater than for the mantle (generally between 0.05 and 1.0, compared with 0.03), the strontium isotope ratio for

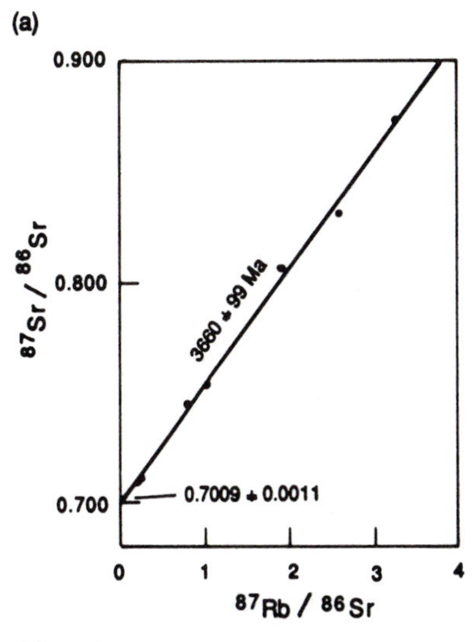

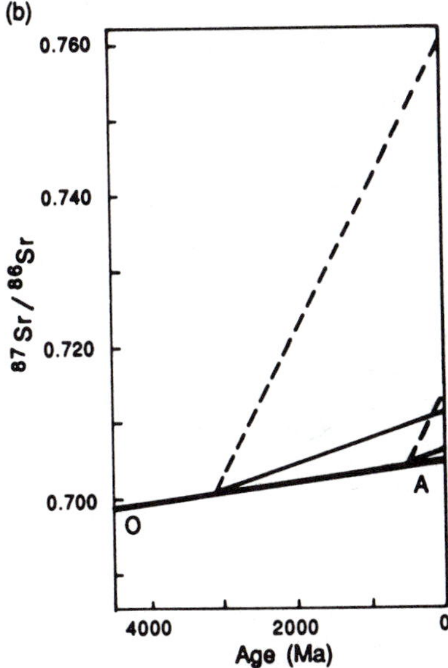

Figure 6.2. (a) Whole-rock rubidium–strontium isochron for the Amitsôq gneiss of the Godthaab district of western Greenland. These rocks are some of the oldest found on the earth. They were originally granite before undergoing metamorphism to gneiss. The 3660 Ma age determined by this method probably represents the time of metamorphism. The original granite was therefore older. (After Moorbath et al. 1972.) (b) Strontium isotope growth curves for the mantle and crust. Heavy line OA is the mantle growth curve. Crustal growth curves are illustrated for rocks extracted from the mantle 3000 and 500 Ma ago. Gross rubidium–strontium ratios for crustal rocks are 0.10 (solid lines) and 0.50 (dashed lines).

crustal rocks, the *crustal growth curve*, is much steeper than the mantle growth curve. Equation 6.30 enables us to calculate crustal growth curves, given the gross rubidium–strontium ratio and assuming approximate isotopic proportions. Figure 6.2b shows crustal growth curves for rocks derived from the mantle at 3000 Ma and 500 Ma with rubidium–strontium ratios of 0.10 and 0.5. The strontium isotope ratio measured today for these rocks would therefore be considerably more than 0.704.

If a crustal rock is remelted, then the subsequent rock has an initial $^{87}Sr/^{86}Sr$ ratio corresponding to the $^{87}Sr/^{86}Sr$ ratio of the source rock at the time of melting. Clearly this is *always* considerably *greater* than 0.704, often greater than 0.710. However, rocks derived directly from the mantle have an initial ratio of *less* than 0.704. Thus, the initial ratio provides an excellent method of determining the origin of some rocks.

6.4 Uranium–Lead

For the decay of ^{238}U to ^{206}Pb, Eq. 6.11 is

$$[^{206}\mathrm{Pb}]_{\mathrm{now}} = [^{238}\mathrm{U}]_{\mathrm{now}}(e^{\lambda_{238}t} - 1) \tag{6.31}$$

and from Eq. 6.12 the age t is given by

$$t = \frac{1}{\lambda_{238}} \log_e \left[1 + \frac{[^{206}\mathrm{Pb}]_{\mathrm{now}}}{[^{238}\mathrm{U}]_{\mathrm{now}}} \right] \tag{6.32}$$

For the decay of ^{235}U to ^{207}Pb, these equations are

$$[^{207}\mathrm{Pb}]_{\mathrm{now}} = [^{235}\mathrm{U}]_{\mathrm{now}}(e^{\lambda_{235}t} - 1) \tag{6.33}$$

and

$$t = \frac{1}{\lambda_{235}} \log_e \left[1 + \frac{[^{207}\mathrm{Pb}]_{\mathrm{now}}}{[^{235}\mathrm{U}]_{\mathrm{now}}} \right] \tag{6.34}$$

Therefore, provided the samples come from closed systems which had no initial lead, measurement of the lead–uranium ratios will yield an age. An alternative method of calculating the age is to take the ratio of Eqs. 6.33 and 6.31,

$$\frac{[^{207}\mathrm{Pb}]_{\mathrm{now}}}{[^{206}\mathrm{Pb}]_{\mathrm{now}}} = \frac{[^{235}\mathrm{U}]_{\mathrm{now}}}{[^{238}\mathrm{U}]_{\mathrm{now}}} \frac{(e^{\lambda_{235}t} - 1)}{(e^{\lambda_{238}t} - 1)} \tag{6.35}$$

where $[^{207}\mathrm{Pb}]_{\mathrm{now}}/[^{206}\mathrm{Pb}]_{\mathrm{now}}$ is the present-day ratio of radiogenic ^{207}Pb to radiogenic ^{206}Pb. The present-day ratio $[^{235}\mathrm{U}]_{\mathrm{now}}/[^{238}\mathrm{U}]_{\mathrm{now}}$ is 1/137.88 and is assumed to be independent of the age and history of the sample. Equation 6.35 can therefore be used to estimate t from just one sample if the lead ratio is measured. The mineral zircon is the best candidate for dating by this method because it is retentive to uranium and its daughter products, crystallizes with almost no initial lead and is widely distributed.

6.4.1 Concordia Diagram

Many rocks do not come from ideal locations with no initial lead and no loss or gain of lead or uranium and so require alternative dating methods.

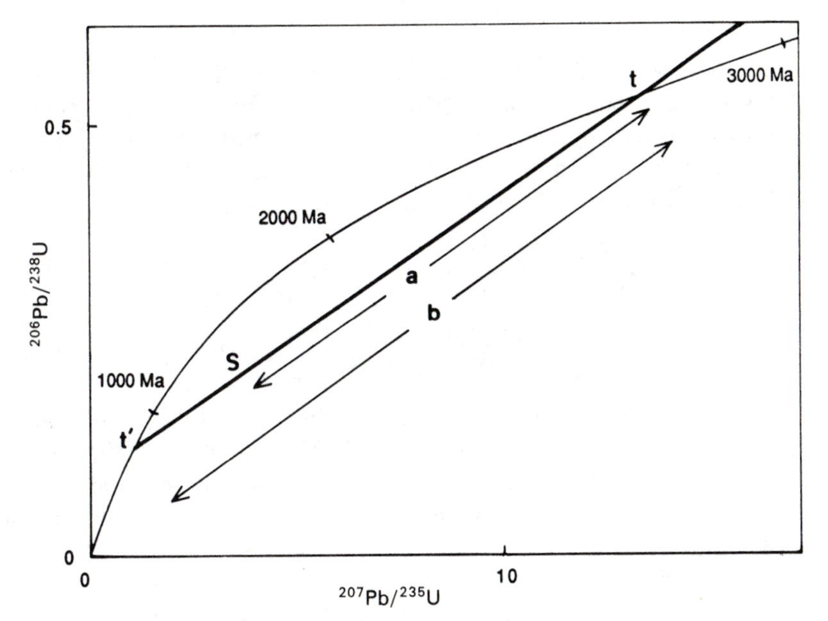

Figure 6.3. The concordia curve as defined by Eq. 6.35. The solid straight line joining points t and t' is the discordia. A sample which crystallized at time t ago and then lost lead and closed at time t' ago will have lead–uranium ratios which plot on this line. A sample which plots at point S must have lost a fraction a/b of its lead in the disturbing event.

An ideal plot of $[^{206}Pb]_{now}/[^{238}U]_{now}$ against $[^{207}Pb]_{now}/[^{235}U]_{now}$ from Eq. 6.35 is known as the *concordia* curve (Fig. 6.3) because all the points along the curve have concordant U–Pb ages (i.e. the ^{238}U–^{206}Pb age is the same as the ^{235}U–^{207}Pb age). If a system has been closed to uranium and lead and the other elements in the decay scheme since its formation, and if correction is made for initial lead, then the ^{238}U–^{206}Pb and ^{235}U–^{207}Pb methods generally give concordant ages and the isotope ratios plot on the concordia. In the event of discordant ages, a concordia plot is often able to yield good estimates of both the ages of the samples and the time at which the disturbance occurred.

Suppose crystallization of the samples occurred at time t ago and that at time t' ago a disturbing event caused the system to lose lead. Since time t', the system has been closed. Then, if this disturbing event removed ^{207}Pb and ^{206}Pb in the same proportion in which they are present in the rock, the $[^{206}Pb]_{now}/[^{238}U]_{now}$ and $[^{207}Pb]_{now}/[^{235}U]_{now}$ ratios will lie on a straight line called *discordia*, joining points t' and t on the theoretical concordia curve. The position of any particular sample on this straight line depends on the amount of lead the sample lost. Any loss or gain of uranium by the sample during the disturbing event changes the position of the sample on the line (uranium loss plots on the extension of the line above the concordia curve and uranium gain below) but *does not* alter the intersection of the line with the concordia curve and therefore does not alter the values t' and t. If lead is lost continuously (e.g., by diffusion) as opposed to loss in a single disturbing event, the isotope ratios still plot on a straight line. However, in this cast t still represents the time of crystallization but t' is fictitious and has no meaning. Thus, the stratigraphic and erosional/depositional setting of samples must be known fully to understand their history. This concordia/discordia method provides a powerful way of dating altered rocks.

6.4.2 Isochrons

Lead isotopes 204, 206, 207 and 208 all occur naturally, but only isotope-204 is nonradiogenic. Equations 6.31–6.34 assume that all the lead present is a decay product of uranium. However, we can normalize the equations and correct for an initial, unknown amount of lead, in the same manner as for initial strontium in the rubidium–strontium method:

$$\frac{[^{206}\mathrm{Pb}]_{\mathrm{now}}}{[^{204}\mathrm{Pb}]_{\mathrm{now}}} = \frac{[^{206}\mathrm{Pb}]_0}{[^{204}\mathrm{Pb}]_0} + \frac{[^{238}\mathrm{U}]_{\mathrm{now}}}{[^{204}\mathrm{Pb}]_{\mathrm{now}}}(e^{\lambda_{238}t} - 1)$$

$$\frac{[^{207}\mathrm{Pb}]_{\mathrm{now}}}{[^{204}\mathrm{Pb}]_{\mathrm{now}}} = \frac{[^{207}\mathrm{Pb}]_0}{[^{204}\mathrm{Pb}]_0} + \frac{[^{235}\mathrm{U}]_{\mathrm{now}}}{[^{204}\mathrm{Pb}]_{\mathrm{now}}}(e^{\lambda_{235}t} - 1) \tag{6.36}$$

Thus, *uranium–lead isochrons* can be plotted in exactly the same manner as for rubidium–strontium. The problem with this method is that the isotope ratios are difficult to measure accurately. In addition, this method is often unsuccessful because extensive losses of uranium are not uncommon (Fig. 6.4b).

Taking the ratio of Eqs. 6.36 gives

$$\frac{[^{207}\mathrm{Pb}]_{\mathrm{now}}/[^{204}\mathrm{Pb}]_{\mathrm{now}} - [^{207}\mathrm{Pb}]_0/[^{204}\mathrm{Pb}]_0}{[^{206}\mathrm{Pb}]_{\mathrm{now}}/[^{204}\mathrm{Pb}]_{\mathrm{now}} - [^{206}\mathrm{Pb}]_0/[^{204}\mathrm{Pb}]_0} = \frac{[^{235}\mathrm{U}]_{\mathrm{now}}}{[^{238}\mathrm{U}]_{\mathrm{now}}}\frac{(e^{\lambda_{235}t} - 1)}{(e^{\lambda_{238}t} - 1)}$$

$$= \frac{1}{137.88}\frac{(e^{\lambda_{235}t} - 1)}{(e^{\lambda_{238}t} - 1)} \tag{6.37}$$

This is the equation of a straight line which passes through the point

$$\frac{[^{206}\mathrm{Pb}]_0}{[^{204}\mathrm{Pb}]_0} \quad \text{and} \quad \frac{[^{207}\mathrm{Pb}]_0}{[^{204}\mathrm{Pb}]_0}$$

and has a slope of

$$\frac{1}{137.88}\frac{(e^{\lambda_{235}t} - 1)}{(e^{\lambda_{238}t} - 1)}$$

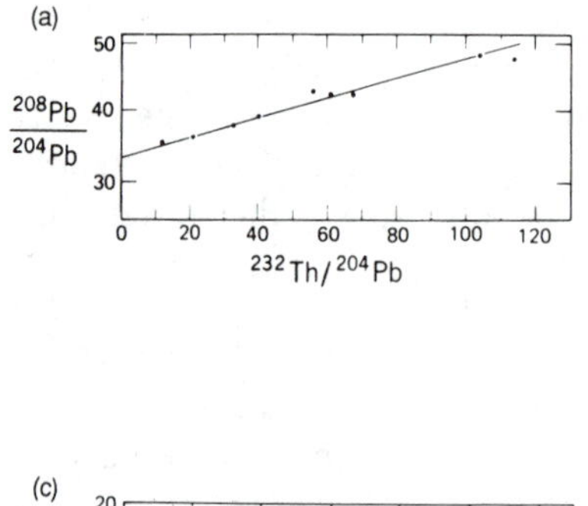

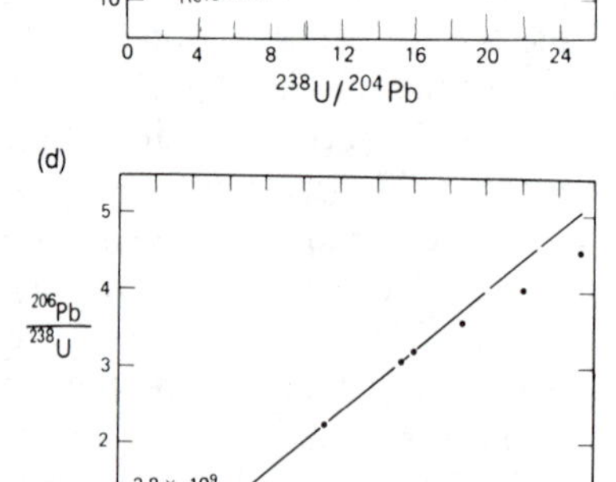

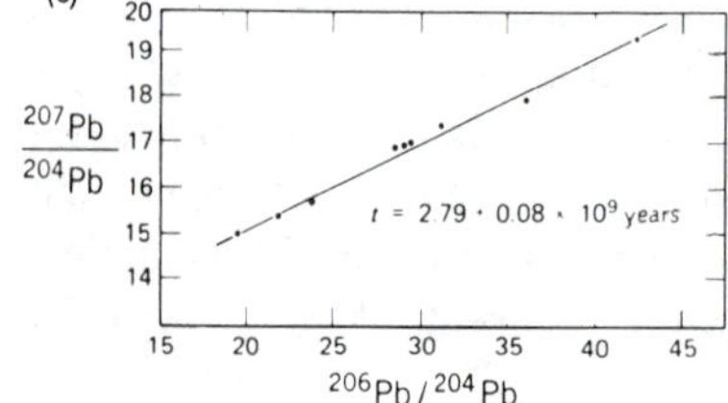
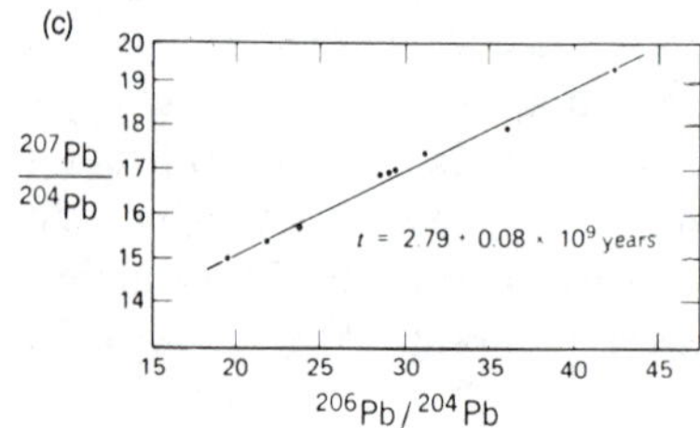

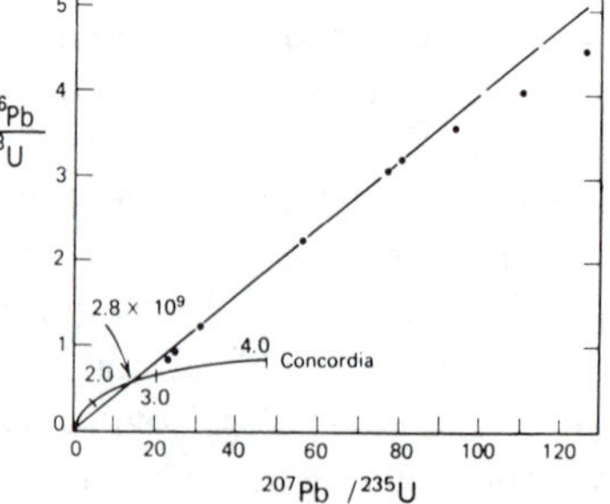

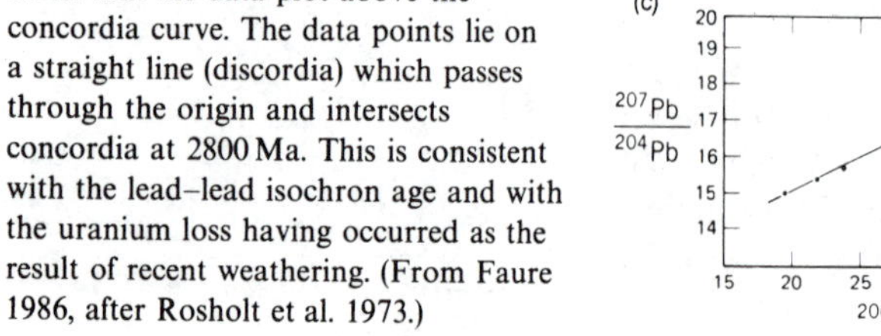

Figure 6.4. Lead–lead, thorium–lead and uranium–lead whole-rock isochrons and the concordia plot for a granite. (c) Lead–lead isochron. The points fall on a straight line with slope 0.1911, which corresponds to an age of 2790 Ma. (a) Thorium–lead isochron. The slight scatter in the data indicates that the rocks may have lost or gained variable amounts of thorium. However, since the points are not very scattered, the actual quantity of thorium lost or gained was small. (b) Uranium–lead isochron. These isotope ratios are very scattered due to loss of uranium. Had no uranium been lost, the points would fall on the 2790 Ma reference isochron. However, loss of uranium causes the rocks data to plot to the left of their position on the reference isochron, as indicated by the arrow. The amount of uranium lost by each rock can be estimated by assuming that the $^{206}\mathrm{Pb}/^{204}\mathrm{Pb}$ ratio and the reference isochron are correct. The reference $^{238}\mathrm{U}/^{204}\mathrm{Pb}$ ratio can then be compared with the actual value: This gives the fraction of uranium retained by each rock. Plots a, b, c show some of the problems which are encountered when using the uranium and thorium decay schemes and some of the ways in which information can be obtained despite the mobility of uranium and thorium. (d) Concordia plot. The losses of uranium mean that the data plot above the concordia curve. The data points lie on a straight line (discordia) which passes through the origin and intersects concordia at 2800 Ma. This is consistent with the lead–lead isochron age and with the uranium loss having occurred as the result of recent weathering. (From Faure 1986, after Rosholt et al. 1973.)

Thus, plotting $[^{207}\text{Pb}]_{\text{now}}/[^{204}\text{Pb}]_{\text{now}}$ against $[^{206}\text{Pb}]_{\text{now}}/[^{204}\text{Pb}]_{\text{now}}$ gives a straight line, a *lead–lead isochron* (Fig. 6.4c). A lead–lead isochron yields a reliable value for t provided all the samples have the same initial isotope ratios and were closed to uranium and lead at least until recent time.

6.5 Thorium–Lead

Applying Eq. 6.11 to the thorium–lead decay (^{232}Th–^{208}Pb) gives

$$[^{208}\text{Pb}]_{\text{now}} = [^{232}\text{Th}]_{\text{now}}(e^{\lambda t} - 1) \tag{6.38}$$

Using ^{204}Pb, the nonradiogenic isotope of lead, in the same way as for the uranium–lead method and including an initial amount of ^{208}Pb, we can write Eq. 6.38 as

$$\frac{[^{208}\text{Pb}]_{\text{now}}}{[^{204}\text{Pb}]_{\text{now}}} = \frac{[^{208}\text{Pb}]_0}{[^{204}\text{Pb}]_0} + \frac{[^{232}\text{Th}]_{\text{now}}}{[^{204}\text{Pb}]_{\text{now}}}(e^{\lambda t} - 1) \tag{6.39}$$

The construction of an isochron therefore allows t and $[^{208}\text{Pb}]_0/[^{204}\text{Pb}]_0$ to be calculated (Fig. 6.4a). Alternatively, with an estimate for the initial lead-isotope ratio, Eq. 6.39 gives a value for t directly with one sample. Thorium–lead isochrons can be more successful than uranium–lead isochrons because thorium and lead tend to be less easily lost than uranium. However, because thorium has a much longer half-life than uranium, this is not such a good method for dating younger rocks.

6.6 Potassium–Argon

Potassium has three naturally occurring isotopes, ^{39}K, ^{40}K and ^{41}K. Their relative abundances are 93%, 0.012% and 6.7%. Equation 6.22 gives the amount of ^{40}Ar produced by ^{40}K decay:

$$[^{40}\text{Ar}]_{\text{now}} = [^{40}\text{K}]_{\text{now}}\left(\frac{\lambda_{\text{A}}}{\lambda_{\text{A}} + \lambda_{\text{C}}}\right)[e^{(\lambda_{\text{A}} + \lambda_{\text{C}})t} - 1] \tag{6.40}$$

Including an initial amount of ^{40}Ar means that the amount of ^{40}Ar measured at time t is

$$[^{40}\text{Ar}]_{\text{now}} = [^{40}\text{Ar}]_0 + [^{40}\text{K}]_{\text{now}}\left(\frac{\lambda_{\text{A}}}{\lambda_{\text{A}} + \lambda_{\text{C}}}\right)[e^{(\lambda_{\text{A}} + \lambda_{\text{C}})t} - 1] \tag{6.41}$$

The nonradiogenic isotope ^{36}Ar is used to normalize this equation:

$$\frac{[^{40}\text{Ar}]_{\text{now}}}{[^{36}\text{Ar}]_{\text{now}}} = \frac{[^{40}\text{Ar}]_0}{[^{36}\text{Ar}]_0} + \frac{[^{40}\text{K}]_{\text{now}}}{[^{36}\text{Ar}]_{\text{now}}}\left(\frac{\lambda_{\text{A}}}{\lambda_{\text{A}} + \lambda_{\text{C}}}\right)[e^{(\lambda_{\text{A}} + \lambda_{\text{C}})t} - 1] \tag{6.42}$$

An isochron can be constructed using this equation, which allows t and the initial argon ratio to be estimated (Fig. 6.5). It is often possible to assume that the initial argon ratio $[^{40}\text{Ar}]_0/[^{36}\text{Ar}]_0$ was 295.5, its present-day value in the atmosphere, and thus to obtain an age from a single, whole rock or mineral, which makes this a very attractive dating method. In addition, the shorter half-life of ^{40}K and its relative abundance compared with the

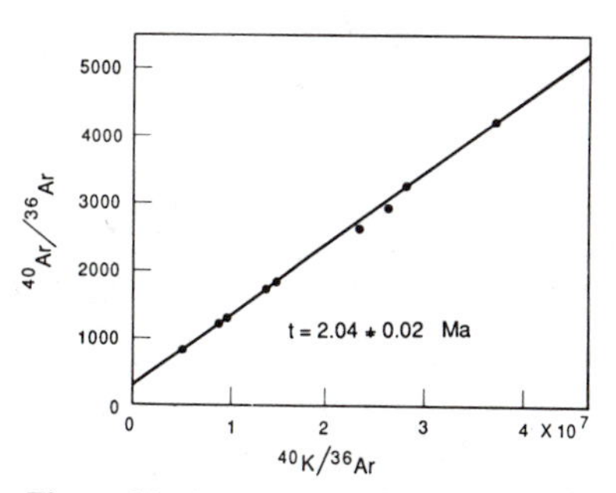

Figure 6.5. A potassium–argon whole-rock isochron for tuffs (volcanic ash) from the Olduvai Gorge in Tanzania. The age of these rocks is very important because it is used to date both recent geomagnetic reversals (as discussed in Sect. 3.2.1) and the early hominoid remains which have been discovered in the gorge by Leakey and co-workers. (After Fitch et al. 1976.)

previously discussed elements means that this method is good for dating young rocks. However, problems arise because argon is a gas and thus is easily lost from the system. Closure temperatures (see Sects. 6.2 and 7.11.5) for argon depend on the mineral involved, ranging between approximately 100 and 600°C; thus, each mineral gives a different age (see Fig. 7.37 for an example). However, as discussed in Section 7.11.5, this leads to a potentially powerful method of unravelling the thermal history of the rocks.

6.7 Argon–Argon

The three naturally occurring isotopes of argon are ^{36}Ar, ^{38}Ar and ^{40}Ar, which are present in the atmosphere in proportions of 0.34%, 0.06% and 99.6%, respectively. The argon–argon dating method depends on bombarding ^{39}K with fast neutrons in a nuclear reactor to produce ^{39}Ar. The reaction is

$$^{39}\mathrm{K} + \mathrm{n} \rightarrow {}^{39}\mathrm{Ar} + \mathrm{p} \tag{6.43}$$

where n is a neutron and p a proton. To date a rock sample it is first irradiated with neutrons and then heated step by step in a vacuum. At each temperature step, the $^{40}Ar/^{39}Ar$ ratio of the gas given off is measured in a mass spectrometer. Argon-39 is an unstable isotope and decays to potassium-39, but its half-life is long enough (compared with the time taken to make the analysis) for ^{39}Ar to be treated as though it were stable.

The amount of ^{39}Ar present in the sample after it has been irradiated is proportional to the amount of ^{39}K present before it was irradiated in the nuclear reactor:

$$[^{39}\mathrm{Ar}]_{\mathrm{now}} = c[^{39}\mathrm{K}]_{\mathrm{now}} \tag{6.44}$$

where c, the constant of proportionality, is a function of the length of time for which the sample was irradiated, the neutron energies, and the capture cross section for the reaction.

The number of radiogenic ^{40}Ar atoms present in the sample is given by Eq. 6.40. Therefore, by combining Eqs. 6.44 and 6.40, we can write the $^{40}Ar/^{39}Ar$ ratio as

$$\frac{[^{40}\mathrm{Ar}]_{\mathrm{now}}}{[^{39}\mathrm{Ar}]_{\mathrm{now}}} = \frac{1}{c}\frac{[^{40}\mathrm{K}]_{\mathrm{now}}}{[^{39}\mathrm{K}]_{\mathrm{now}}}\left(\frac{\lambda_{\mathrm{A}}}{\lambda_{\mathrm{A}} + \lambda_{\mathrm{C}}}\right)[e^{(\lambda_{\mathrm{A}} + \lambda_{\mathrm{C}})t} - 1]$$

$$= \frac{(e^{(\lambda_{\mathrm{A}} + \lambda_{\mathrm{C}})t} - 1)}{D} \tag{6.45}$$

where

$$D^{-1} = \frac{1}{c}\frac{[^{40}\mathrm{K}]_{\mathrm{now}}}{[^{39}\mathrm{K}]_{\mathrm{now}}}\left(\frac{\lambda_{\mathrm{A}}}{\lambda_{\mathrm{A}} + \lambda_{\mathrm{C}}}\right)$$

If a standard sample of known age t_s is irradiated at the same time as the sample to be studied, this term D can be determined. Measurement of the $^{40}Ar/^{39}Ar$ ratio for this standard sample gives

$$\frac{[^{40}\mathrm{Ar}]^{\mathrm{s}}_{\mathrm{now}}}{[^{39}\mathrm{Ar}]^{\mathrm{s}}_{\mathrm{now}}} = \frac{(e^{(\lambda_{\mathrm{A}} + \lambda_{\mathrm{C}})t_{\mathrm{s}}} - 1)}{D} \tag{6.46}$$

where the superscript s refers to measurements on the standard. Rearranging Eq. 6.46 gives

$$D = (e^{(\lambda_A + \lambda_C)t_s} - 1)\frac{[^{39}\mathrm{Ar}]^s_{now}}{[^{40}\mathrm{Ar}]^s_{now}} \tag{6.47}$$

Substituting this value for D back into Eq. 6.45, we obtain

$$\frac{[^{40}\mathrm{Ar}]_{now}}{[^{39}\mathrm{Ar}]_{now}} = \frac{(e^{(\lambda_A + \lambda_C)t} - 1)\,[^{40}\mathrm{Ar}]^s_{now}}{(e^{(\lambda_A + \lambda_C)t_s} - 1)\,[^{39}\mathrm{Ar}]^s_{now}} \tag{6.48}$$

Corrections for the presence of atmospheric ^{40}Ar can be made to Eq. 6.48. An age t can now be calculated from this equation.

As already mentioned, the isotope ratios are measured at each temperature step, and so an age can be calculated for every temperature. If the sample had been closed to both argon and potassium since its formation, then the ages determined at every temperature should be the same. However, loss of argon from some minerals and not others results in a spectrum of ages, which can then be interpreted to give an idea of the thermal history of the sample.

Problems with this dating method can arise if excess ^{40}Ar is present, in which case anomalously old dates may be obtained. Another problem with very fine-grained samples is that ^{39}Ar can be lost during the irradiation because it recoils out of the sample during the reaction.

6.8 Samarium–Neodymium

Samarium and neodymium are rare earth elements which occur in a wide range of rocks. Although the decay constant for ^{147}Sm–^{143}Nd is very small (see Table 6.2), which makes the decay most useful for dating very old rocks, the system can also be used for younger samples. Applying Eq. 6.11 to this decay and including an initial concentration of ^{143}Nd gives

$$[^{143}\mathrm{Nd}]_{now} = [^{143}\mathrm{Nd}]_0 + [^{147}\mathrm{Sm}]_{now}(e^{\lambda t} - 1) \tag{6.49}$$

One of the seven nonradiogenic isotopes of neodymium, ^{144}Nd, is used as a standard:

$$\frac{[^{143}\mathrm{Nd}]_{now}}{[^{144}\mathrm{Nd}]_{now}} = \frac{[^{143}\mathrm{Nd}]_0}{[^{144}\mathrm{Nd}]_0} + \frac{[^{147}\mathrm{S_M}]_{now}}{[^{144}\mathrm{Nd}]_{now}}(e^{\lambda t} - 1) \tag{6.50}$$

An isochron can therefore give t and the initial Nd ratio. Instrumentation of high precision is required since neodymium and samarium are frequently present in abundances of less than 10 ppm. Values of $[^{147}\mathrm{Sm}]_{now}/[^{144}\mathrm{Nd}]_{now}$ are typically 0.1–0.2. In the same manner as was discussed for the rubidium–strontium method, the $[^{143}\mathrm{Nd}]_0/[^{144}\mathrm{Nd}]_0$ initial ratio provides an indication of the origin of the samples. The advantage of this method is that samarium and neodymium are similar in their chemical characteristics and little affected by processes such as weathering or metamorphism; so even when a system is not closed to rubidium and strontium, the Sm–Nd method is ofter successful. However, Sm–Nd results for some volcanic rocks can reflect contamination of the igneous melt as it ascended through country rock; such results can be very difficult to interpret.

The beta decays of rhenium to osmium (^{187}Re to ^{187}Os) and lutetium to hafnium (^{176}Lu to ^{176}Hf) mean that these isotopes can be used as geochronometers in the same way that the decays ^{87}Rb to ^{87}Sr and ^{147}Sm to ^{143}Nd are used. Isochrons and isotopic evolution diagrams are constructed in an analogous manner to the rubidium–strontium and samarium–neodymium methods. These isotopes, rarely used as geochronometers, are useful in the study of mantle evolution and the origin of magmas.

The abundance of terrestrial ^{143}Nd has increased with time since the earth's formation because of the decay of ^{147}Sm to ^{143}Nd. The ^{143}Nd/^{144}Nd ratio has therefore also increased with time. We can model the time dependence of the ^{143}Nd/^{144}Nd ratio by using Eq. 6.50 and by assuming that the ^{147}Sr/^{144}Nd ratio of the earth is the same as that of the chondritic meteorites (Sect. 6.10). For chondritic meteorites, the average present-day ^{147}Sm/^{144}Nd ratio is 0.1967, and the average present-day ^{143}Nd/^{144}Nd ratio is 0.512638.* Therefore, we can rewrite Eq. 6.50 for such chondrites as

$$\frac{[^{143}\mathrm{Nd}]_0}{[^{144}\mathrm{Nd}]_0} = \frac{[^{143}\mathrm{Nd}]_{\mathrm{now}}}{[^{144}\mathrm{Nd}]_{\mathrm{now}}} - \frac{[^{147}\mathrm{Sm}]_{\mathrm{now}}}{[^{144}\mathrm{Nd}]_{\mathrm{now}}}(e^{\lambda t} - 1)$$
$$= 0.512638 - 0.1967(e^{\lambda t} - 1) \qquad (6.51)$$

where the subscript *now* represents measurements made now and the subscript *o* refers to values time t ago.

The model of the primitive mantle called the *chondritic uniform reservoir* (CHUR) is used as a standard for mantle isotopic geochemistry. CHUR is described by Eq. 6.51, and since the half-life of ^{147}Sm is so long, a plot of the ^{143}Nd/^{144}Nd ratio against time is a straight line (Fig. 6.6a). The initial ^{143}Nd/^{144}Nd ratios for samarium–neodymium isochrons are often quoted as deviations from the CHUR model, where the parameter $\varepsilon_{\mathrm{Nd}}$ (epsilon Nd) is the deviation in parts per ten thousand:

$$\varepsilon_{\mathrm{Nd}} = \left[\frac{[^{143}\mathrm{Nd}]_0/[^{144}\mathrm{Nd}]_0}{[^{143}\mathrm{Nd}]_0^{\mathrm{CHUR}}/[^{144}\mathrm{Nd}]_0^{\mathrm{CHUR}}} - 1\right] \times 10^4 \qquad (6.52)$$

where $[^{143}\mathrm{Nd}]_0/[^{144}\mathrm{Nd}]_0$ is the initial ratio of the whole-rock samarium–neodymium isochron and $[^{143}\mathrm{Nd}]_0^{\mathrm{CHUR}}/[^{144}\mathrm{Nd}]_0^{\mathrm{CHUR}}$ is the initial neodymium ratio of the CHUR model (calculated from Eq. 6.51) at the age t given by the whole-rock isochron.

As shown in Figure 6.6, a rock which was derived directly from CHUR would have an initial ratio equal to the CHUR ratio for that age and hence a zero epsilon value ($\varepsilon_{\mathrm{Nd}} = 0$). However, partial melts of CHUR have negative epsilon values, and so a rock derived by the melting of older crustal rocks has a negative epsilon value ($\varepsilon_{\mathrm{Nd}} < 0$). Likewise, rocks derived from the residue left after CHUR had been partially molten and the melt

* This isotope ratio is normalized to the reference isotope ratio ^{146}Nd/^{144}Nd = 0.7219. Various laboratories use slightly different values due to varying methods and calibrations of their chemical methods; this is a problem for isotope geologosts using the Sm–Nd method.

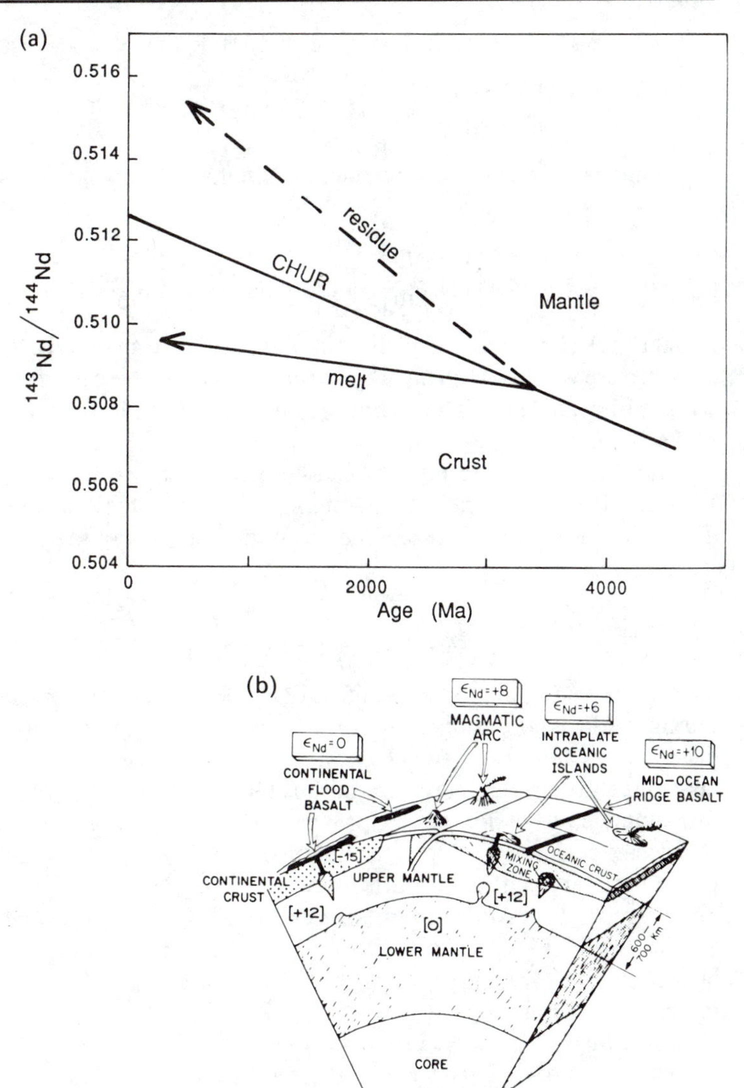

Figure 6.6. (a) The isotopic evolution of neodymium for the CHUR model for the primitive mantle. The straight-line CHUR is defined by Eq. 6.51. If CHUR undergoes partial melting at, say, 3400 Ma the melt fraction (solid line) has a lower Sm/Nd ratio (and hence a lower $^{143}Nd/^{144}Nd$ ratio) than CHUR, whereas the solid residue (dashed line) has a higher Sm/Nd ratio (and hence a higher $^{143}Nd/^{144}Nd$ ratio) than CHUR. Thus, the residual mantle plots above CHUR (positive epsilon) as it evolves, and the crust plots below (negative epsiton). (b) A model of the earth based on neodymium isotopic data. The present ε_{Nd} value of the continental crust is about -15, and the value for the upper mantle is $+12$. The lower mantle is chondritic (i.e, $\varepsilon_{Nd} = 0$ there). (From DePaolo 1981.)

withdrawn will have positive epsilon values ($\varepsilon_{Nd} > 0$). Such a source is generally referred to as *depleted mantle* because it is depleted in the large ion lithophile (LIL) elements such as rubidium. Equations 6.50 and 6.51 can be used together to calculate a model age T, which represents the time that has elapsed since the neodymium in the rock sample had the same isotopic ratio as the CHUR model (i.e., since the rock separated from a CHUR mantle). Such an age only has a meaning provided the samarium–neodymium ratio of the rock has not been altered since the rock separated from the CHUR mantle. However, the samarium–neodymium ratio is relatively insensitive to disturbing events such as metamorphism, and so the model age T can be used, with caution, as a time estimate. Equation 6.50 gives the initial neodymium ratio for a rock as

$$\frac{[^{143}\mathrm{Nd}]_0}{[^{144}\mathrm{Nd}]_0} = \frac{[^{143}\mathrm{Nd}]_{\mathrm{now}}}{[^{144}\mathrm{Nd}]_{\mathrm{now}}} - \frac{[^{147}\mathrm{Sm}]_{\mathrm{now}}}{[^{144}\mathrm{Nd}]_{\mathrm{now}}}(e^{\lambda T} - 1) \tag{6.53}$$

From Eq. 6.51, the neodymium ratio in CHUR time T ago was

$$\frac{[^{143}\mathrm{Nd}]_0^{\mathrm{CHUR}}}{[^{144}\mathrm{Nd}]_0^{\mathrm{CHUR}}} = 0.512638 - 0.1967(e^{\lambda T} - 1) \tag{6.54}$$

Equating these two expressions and rearranging terms to obtain a value for T gives

$$T = \frac{1}{\lambda}\log_e\left\{\frac{[^{143}\mathrm{Nd}]_{\mathrm{now}}/[^{144}\mathrm{Nd}]_{\mathrm{now}} - 0.512638}{[^{147}\mathrm{Sm}]_{\mathrm{now}}/[^{144}\mathrm{Nd}]_{\mathrm{now}} \quad - 0.1967} + 1\right\} \tag{6.55}$$

If the rock came from a depleted mantle source instead of CHUR, such an age estimate would be too low. Age estimates such as these need to be made with caution and with as much knowledge of the origin and chemistry of the rocks as possible.

A notation similar to Eq. 6.52 can be used to express deviations of the lutetium–hafnium or rhenium–osmium systems from their bulk reservoirs. These deviations can then be used in an analogous way to $\varepsilon_{\mathrm{Nd}}$.

6.9 Fission Track Dating

As well as undergoing a series of radioactive decays to stable lead-206, uranium-238 is also subject to *spontaneous fission*. That is, its nucleus disintegrates into two large but unequal parts, releasing two or three neutrons and considerable energy (about 150 MeV). The decay constant for this spontaneous fission of ^{238}U is $8.46 \times 10^{-17}\ a^{-1}$, very much less than the decay constant for the α emission of ^{238}U, which eventually leads to ^{206}Pb after a chain of fourteen decays. Thus, fission of ^{238}U occurs only rarely; the ratio of spontaneous fission to α-particle emission is $8.46 \times 10^{-17}/1.55 \times 10^{-10} = 5 \times 10^{-7}$ only. The decay products from the fission of ^{238}U are of such energy that they are able to travel through minerals for about 10 μm (1 $\mu\mathrm{m} = 10^{-6}$ m).

The passage of a charged particle through a solid results in a damaged zone along its path. This is one of the ways by which *cosmic rays** can be studied. If a singly charged particle passes through a photographic emulsion which is subsequently developed, the track of the particle can be seen under the microscope as a trail of grains of silver, but in the case of, say, a nucleus of iron (charge 26), one sees a hairy sausagelike cylinder penetrating the emulsion.

Another way of detecting particles uses solid-state track detectors. Tracks can be registered in many important mineral crystals and in a number of commercially available plastic sheets. A highly charged particle

* Cosmic rays are very high-velocity nuclei which constantly bombard the earth. Their flux is about 1 particle $\mathrm{cm}^{-2}\,\mathrm{s}^{-1}$, about the same energy flux as starlight. Most cosmic rays are protons, and 10% are ^{3}He and ^{4}He. The remainder are heavier elements (with B, C, O, Mg, Si and Fe being prominent); in fact, iron has an abundance of $\sim 3 \times 10^{-4}$ of the protons by number or $\sim$1.7% by mass. The abundance of the heaviest nuclei, with charge greater than 70, is $\sim 3 \times 10^{-9}$ of the protons. The boron and similar nuclei are fragments of the original carbon and oxygen that underwent collision with interstellar hydrogen in their 10^7 year journey to earth at $\sim$90% of the speed of light.

which passes through the plastic sheet produces sufficient damage for later etching (usually in NaOH) to reveal the damaged zone, which is dissolved more rapidly than the undamaged material. Under the microscope, one sees two cones, one on each side of the plastic, marking the entry and exit points of the particle. An important feature of all solid-state track detectors is that they have a threshold damage level below which no track is produced; this enables one to detect a minute number of, say, fission particles among a very large number of particles which leave no tracks. This is relevant in detection of spontaneous fission in uranium-containing mineral crystals. One of the most noteworthy results of this technique was the discovery of tracks of spontaneous fission of plutonium-244 in crystals in certain meteorites. Plutonium was probably initially present on earth with an abundance $\sim 10\%$ of uranium. It decays principally by α-particle emission, but in $1:10^4$ cases it decays by spontaneous fission. The activity on earth of plutonium-244 is now essentially extinct since it has a half-life of only 8×10^7 years; thus, the proportion remaining is $\sim 2^{-60} = 10^{-20}$.

When a surface of a rock or mineral is cut and polished and then etched in a suitable solvent, tracks of these fission products of ^{238}U, or *fission tracks*, are visible under a microscope because the very numerous α particles do not register. Thus, it is possible to use fission tracks to date geological samples. For dating meteorite samples, one has to be concerned about the now-extinct ^{244}Pu; in principle, a correction could be needed for Archaean terrestrial samples also.

Consider a small polished sample of a mineral and assume that at present it has $[^{238}\text{U}]_{\text{now}}$ atoms of ^{238}U distributed evenly throughout its volume. The number of radioactive decays of uranium 238, D_R, during time t is given by Eq. 6.11 as

$$D_R = [^{238}\text{U}]_{\text{now}}(e^{\lambda t} - 1) \tag{6.56}$$

where λ is the decay constant for ^{238}U decay.

The number of decays of uranium 238 by spontaneous fission, D_S, which occur in time t is then

$$D_S = \frac{\lambda_S}{\lambda}[^{238}\text{U}]_{\text{now}}(e^{\lambda t} - 1) \tag{6.57}$$

where λ_s is the decay constant for the spontaneous fission of ^{238}U. To use Eq. 6.57 to calculate a date t, we must count the visible fission tracks. In addition, we must estimate what proportion of fission tracks produced in the sample have crossed the polished surface and will be visible and therefore countable after the sample is etched. We must also measure $[^{238}\text{U}]_{\text{now}}$.

Fortunately, it is not necessary to carry out the analysis in an absolute manner because another isotope of uranium, ^{235}U, can be made to undergo fission by the absorption of slow neutrons. (Such an *induced fission* is the heart of the generation of power in nuclear reactors and atomic bombs.) The induced fission of ^{235}U is achieved by putting the sample in a reactor and bombarding it with slow neutrons for a specified time (hours). This provides us with a standard against which to calibrate the number of tracks per unit area (track density).

The analysis that follows assumes that neither of the other two isotopes of uranium that occur naturally (^{234}U and ^{235}U) contributes significantly to the spontaneous fissions; we can make this assumption because their fission branching ratios and the isotope abundances are very low relative to those of ^{238}U. The analysis also assumes that there has been no previous interaction with neutrons, which would have contributed neutron-induced fission tracks from ^{235}U. This is almost always a safe assumption unless the uranium has been associated with any of the natural thermal nuclear reactors that occurred in uranium mineral deposits in the early Precambrian (an example is the set of natural nuclear reactors at Oklo in Gabon). At that time, the ^{235}U was relatively more abundant than now since it has a shorter half-life than ^{238}U.

The number of induced fissions of ^{235}U, D_I, is defined as

$$D_I = [^{235}\text{U}]_{\text{now}}\,\sigma n \tag{6.58}$$

where σ is the known neutron-capture cross section (the probability that capture of a neutron by ^{235}U will occur) and n is the neutron dose in the reactor (number of neutrons crossing a square centimetre). Since the fission products of ^{235}U have almost exactly the same average kinetic energy as the fission products of ^{238}U, we can assume that, if uranium-235 is distributed throughout the sample in the same even way as uranium-238, the same proportion of both fission products will cross the sample's polished surface and be counted. This being the case, we can combine Eqs. 6.57 and 6.58 to obtain

$$\frac{\lambda_s[^{238}\text{U}]_{\text{now}}}{\lambda[^{235}\text{U}]_{\text{now}}}\frac{(e^{\lambda t}-1)}{\sigma n} = \frac{D_s}{D_I} = \frac{N_s}{N_I} \tag{6.59}$$

where N_s and N_I are the numbers of spontaneous and induced fission tracks counted in a given area. The time t can then be determined by rearranging Eq. 6.59 and using the uranium isotopic ratio $[^{238}\text{U}]_{\text{now}}/[^{235}\text{U}]_{\text{now}} = 137.88$:

$$t = \frac{1}{\lambda}\log_e\left[1 + \frac{N_s}{N_I}\frac{\lambda}{\lambda_s}\frac{\sigma n}{137.88}\right] \tag{6.60}$$

In practice, after the number of spontaneous fission tracks N_s has been counted, the sample is placed in the reactor for a specified time and then etched again (which enlarges the original spontaneous fission tracks as well as etching the newly induced fission tracks) so that the number of induced ^{235}U fission tracks N_I can be counted.

One major advantage of using these ancient fission tracks as a geological dating method is that their stability is temperature dependent (Eq. 6.25). At high temperatures over geological periods of time, the damaged zones in the crystals along the particle track 'heal', and the rate of healing differs for every mineral and is also temperature dependent. At room temperatures the tracks are stable. Thus, two minerals of the same age which have been at the same high temperature for the same length of time can yield two different fission track ages. For example, after 1 Ma at 50°C a small number of fission tracks in apatite will have heated, but to heal all fission tracks within 1 Ma the apatite must be at 175°C. If the heating time is only 10,000 years, the healing temperatures will increase to 75 and 190°C, respectively. Tracks in the mineral *sphene* can withstand much higher temperatures: Healing will start if the mineral has been at 250°C for 1 Ma, but not all tracks will completely heal unless the mineral has been at 420°C for 1 Ma. The corresponding temperatures for 10,000 years are 295 and 450°C. This means that although fission track dates can be completely reset by heating, the temperature history of a particular sample or set of samples can be determined by measuring dates in various minerals.

For the two minerals discussed above, ages determined from fission tracks in sphene are always greater than ages determined from fission tracks in apatite. The ages are interpreted as representing the last time the mineral cooled below its closure temperature (which depends on the cooling rate!) The differences among the closure temperatures of minerals do not, however, depend on the cooling rate. Thus, the difference between the sphene and apatite ages indicates the length of time taken for the sample to cool between the two closure temperatures, and so the cooling rate can be determined. Closure temperatures for fission tracks in a wide range of minerals which cool at various rates are shown in Figure 6.7.

The temperature dependence of fission tracks provides an excellent method of determining the details of the cooling history of rock samples. This method has been used in the analysis of the erosional history of sedimentary basins.

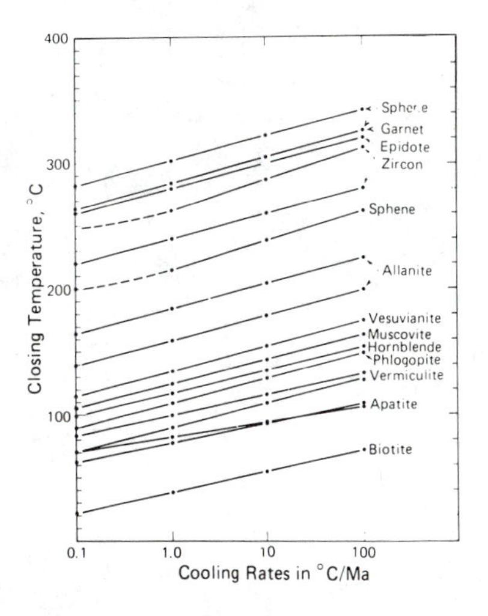

Figure 6.7. Closure temperatures for the retention of fission tracks as a function of cooling rate, for a variety of minerals. (From Faure 1986.)

6.10 The Age of the Earth

Some of the early estimates of the age of the earth were discussed in Section 6.1. Radioactivity provided the tool with which accurate estimates of the earth's age could be made as well as providing an 'unknown' source of heat which helped to make sense of the early thermal models (see Sect. 7.4).

The first radioactive dating method used to limit the age of the earth was the accumulation of α particles (helium nuclei) in minerals as the result of the decay of uranium. In 1905, Rutherford obtained ages of around 500 Ma for the uranium mineral he tested. Also in 1905, Boltwood, as a result of an idea of Rutherford, used the relative proportion of lead and uranium in a rock sample to obtain a date. Measurements on a variety of samples gave dates of between 92 and 570 Ma with the radioactive production rates then available. (Isotopes were not yet understood in 1905 and neither was the thorium–lead decay.)

The oldest rocks on the surface of the earth are to be found in the ancient cratons which form the hearts of the continents. Amongst these oldest rocks are the deformed and metamorphosed Isua supracrustal rocks in Greenland, for which the igneous activity is dated at 3770 ± 42 Ma by the samarium–neodymium method and 3769^{+11}_{-8} Ma by the uranium–lead method on zircons. Even older rocks occur in the Yellowknife region of Canada. Felsic volcanic rocks from the Duffer Formation of the Pilbara supergroup in Western Australia are dated at 3452 ± 16 Ma by uranium–lead work on zircons; and most interesting of all, dates of 4100–4300 Ma have been obtained for zircons from the Jack Hills region of the Yilgarn block in the south of Western Australia. The conclusion to be drawn from these ancient rocks is that although they are rare, they do indicate that perhaps by 4200 Ma and without doubt by 3500 Ma, continents were in existence. The earth itself is certainly older than these oldest rocks.

Our present knowledge of the age of the earth comes from a study of the isotopes of lead and from meteorites. First, consider a general model of the lead evolution of the earth, usually known as the *Holmes–Houtermans model*, after its two independent creators, who built on earlier work by Holmes and by Rutherford. They assumed that when the earth was formed it was homogeneous with a uniform internal distribution of U, Pb and Th. Very soon afterwards, the earth separated (*differentiated*) into a number of subsystems (e.g., mantle and core), each of which had its own characteristic U/Pb ratio. After this differentiation, the U/Pb ratio in each subsystem changed only as a result of the radioactive decay of uranium and thorium to lead (i.e., each subsystem was closed). Finally, when any lead mineral formed (a common one is galena), its lead separated from all uranium and thorium; and so its lead isotopic ratios now are the same as they were at its formation. Applying Eq. 6.37 to this model gives

$$\frac{[^{207}\mathrm{Pb}]_{\mathrm{now}}/[^{204}\mathrm{Pb}]_{\mathrm{now}} - [^{207}\mathrm{Pb}]_0/[^{204}\mathrm{Pb}]_0}{[^{206}\mathrm{Pb}]_{\mathrm{now}}/[^{204}\mathrm{Pb}]_{\mathrm{now}} - [^{206}\mathrm{Pb}]_0/[^{204}\mathrm{Pb}]_0} = \frac{1}{137.88}\frac{(e^{\lambda_{235}T} - e^{\lambda_{235}t})}{(e^{\lambda_{238}T} - e^{238t})} \tag{6.61}$$

where T is the age of the earth, t the time since the formation of the lead mineral, the subscript now refers to the isotope ratio of the lead mineral measured now, and the subscript 0 refers to the primordial isotope ratio of the earth time T ago. This is the *Holmes–Houtermans equation.* There are three unknowns in the equation: T, $[^{207}\mathrm{Pb}]_0/[^{204}\mathrm{Pb}]_0$ and $[^{206}\mathrm{Pb}]_0/[^{204}\mathrm{Pb}]_0$.

Thus, having at least three lead minerals of known age and lead-isotope ratios from different subsystems should enable us to determine the age of the earth and the primordial isotope ratios from Eq. 6.61. Unfortunately, the complex history of the crust and the fact that rocks are frequently not closed to uranium means that in practice T cannot be determined satisfactorily using terrestrial samples. However, meteorites satisfy the criteria of the Holmes–Houtermans model. They are thought to have had a common origin with the planets and asteroids and to have remained a separate subsystem since their separation at the time of formation of the earth.

Meteorites are fragments of comets and asteroids which hit the earth. They vary widely in size from dust upwards and can be classified into three main types: *chondrites, achondrites* and *iron.* The most common type is the chondrites, comprising about 90% of those meteorites observed to fall on earth. Chondrites are characterized by chondrules (small glassy spheres of silicate) which indicate that the material was heated then rapidly cooled and later coalesced into larger bodies. Achondrites are crystalline silicates containing no chondrules and almost no metal phases. Chondrites and achondrites together are termed *stony meteorites.* Some of the chondritic meteorites, termed *carbonaceous chondrites* are the least metamorphosed of the meteorites and still retain significant amounts of water and other volatiles. Their composition is believed to be close to the original composition of the solar nebula from which the solar system formed, and thus they provide an initial composition to use for chemical models, such as

CHUR, of the earth. (see Sect. 6.8). An asteroid which had undergone partial melting and chemical differentiation into crust, silicate mantle and iron core would fragment into stony and iron meteorites. The stony meteorites, composed primarily of the silicate minerals olivine and pyroxene, are thus similar to the earth's crust and mantle, whereas the iron meteorites are made up of alloys of iron and nickel, which have been postulated to be present in the core (see Sect. 4.3.5). A particular iron sulphide (FeS) phase known as troilite is present in meteorites. Because troilite contains lead but almost no uranium or thorium, its present lead isotopic composition must be close to its primordial composition. Thus, lead-isotope ratios of meteorites can be used as in Eq. 6.37 (Eq. 6.61 with $t = 0$ because meteorites are still closed systems) to construct a lead–lead isochron (as in Fig. 6.4c):

$$\frac{[^{207}\mathrm{Pb}]_{\mathrm{now}}/[^{204}\mathrm{Pb}]_{\mathrm{now}} - [^{207}\mathrm{Pb}]_0/[^{204}\mathrm{Pb}]_0}{[^{206}\mathrm{Pb}]_{\mathrm{now}}/[^{204}\mathrm{Pb}]_{\mathrm{now}} - [^{206}\mathrm{Pb}]_0/[^{204}\mathrm{Pb}]_0} = \frac{1}{137.88}\frac{(e^{\lambda_{235}T} - 1)}{(e^{\lambda_{238}T} - 1)} \tag{6.62}$$

This is the equation of a straight line passing through the point $[^{206}\mathrm{Pb}]_0/[^{204}\mathrm{Pb}]_0$, $[^{207}\mathrm{Pb}]_0/[^{204}\mathrm{Pb}]_0$, with a slope of

$$\frac{1}{137.88}\frac{(e^{\lambda_{235}T} - 1)}{(e^{\lambda_{238}T} - 1)}$$

Thus, plotting the lead-isotope ratios of meteorites $[^{207}\mathrm{Pb}]_{\mathrm{now}}/[^{204}\mathrm{Pb}]_{\mathrm{now}}$ against $[^{206}\mathrm{Pb}]_{\mathrm{now}}/[^{204}\mathrm{Pb}]_{\mathrm{now}}$ enables the time T to be determined from the slope of the best-fitting straight lines. The first determination (made by Patterson in 1956) using three stony meteorites and two iron meteorites yielded a value for T of 4540 Ma (Fig. 6.8). Many subsequent measurements have been made, all giving an age for the meteorites, and by inference for the earth, of between 4550 and 4570 Ma.

Using a value for the primordial lead-isotope ratios obtained from meteorites (the lead in the iron meteorite from Canyon Diablo is frequently used because it is the least radiogenic of all the meteorite lead) means that Eq. 6.61 can be applied to terrestrial samples. If a sample's age t is known from other dating methods, T can then be obtained from the present lead-isotope ratios. Such estimates of T are in agreement with meteorite ages.

If the meteorites and the earth are of the same age and initially contained lead of the same isotopic composition, the isotope ratios of average terrestrial lead should lie on the meteorite lead isochron. Average terrestrial lead is not a particularly easy sample to obtain, but recent oceanic sediments, originating as they do from such varied sources, provide a reasonable estimate. That the isotope ratios of these oceanic sediments do lie on the meteorite lead isochron (Fig. 6.8) confirms that the age of the earth is indeed the same as the age of the meteorites and that they have the same primordial lead-isotope ratios. Dating stony meteorites by the rubidium–strontium method gives ages of about 4550 Ma (their initial ratio is 0.699). Dates from iron meteorites are similar.

This concept that the earth and meteorites are of the same age indicates that by 'age of the earth' we do not mean the age of the solid earth but the time when the parts of the solar system became separate bodies.

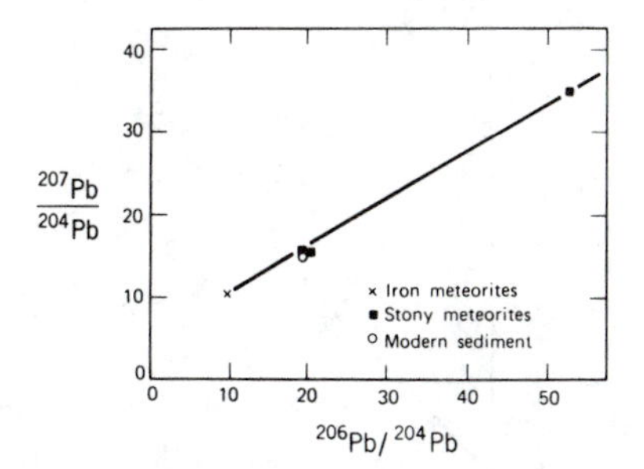

Figure 6.8. Lead–Lead isochron for meteorites and recent oceanic sediments. The slope of the straight line gives an age T of 4540 ± 70 Ma for the meteorites. That lead isotopic ratios of recent oceanic sediments also fall on this line indicates that meteorites and the earth are of the same age and initially contained lead of the same isotopic composition. (From Faure 1986, after Patterson 1956.)

PROBLEMS

1. Six samples of granodiorite from a pluton in British Columbia, Canada, have strontium and rubidium isotopic compositions as follows:

$^{87}Sr/^{86}Sr$	$^{87}Rb/^{86}Sr$
0.7117	3.65
0.7095	1.80
0.7092	1.48
0.7083	0.82
0.7083	0.66
0.7082	0.74

(a) Find the age of the intrusion.
(b) Find the initial $^{87}Sr/^{86}Sr$ ratio of the magma at the time of the intrusion.
(c) Assuming an $^{87}Sr/^{86}Sr$ ratio of 0.699 and an $^{87}Rb/^{86}Sr$ ratio of 0.1 for the undifferentiated earth 4550 Ma ago, comment on the possibility that this batholith originated in the mantle.

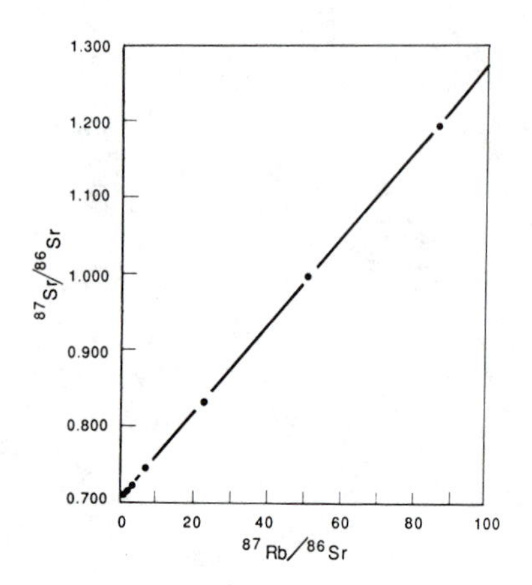

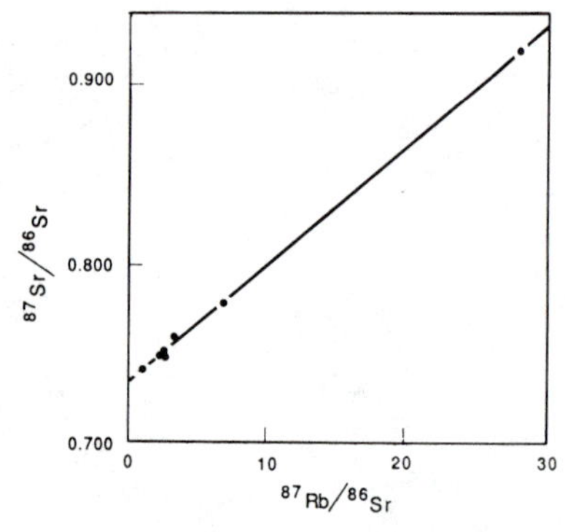

Figure 6.9. Rubidium–strontium whole-rock isochrons for two plutons. (From Fullagar et al. 1971 and Gunner 1974.)

2. Whole-rock rubidium–strontium isochrons for two plutons are shown in Figure 6.9. Calculate the age of each pluton and comment on the source of the magma.
3. Figure 6.10 shows a rubidium–strontium isochron for some unmetamorphosed sediments and clay minerals, illite and chlorite found in the sediment. Calculate an age from this isochron and comment on its meaning.
4. (a) Given that $N = N_0 e^{-\lambda t}$, where N is the number of surviving radioactive atoms at time t, N_0 the initial number and λ the decay constant, and given that today $^{235}U/^{238}U = 0.007257$ in the earth, moon and meteorites, estimate the date of sudden nucleosynthesis for the two estimated production ratios of $^{235}U/^{238}U = 1.5$ and 2.0. Show all steps in your work clearly. *Note:* $\lambda(^{235}U) = 9.8485 \times 10^{-10}\,yr^{-1}$. $\lambda(^{238}U) = 1.55125 \times 10^{-10}\,yr^{-1}$.
   (b) Given that the solar system had an $^{87}Sr/^{86}Sr$ ratio of 0.699 4.6 Ga ago, what should be the initial $^{87}Sr/^{86}Sr$ ratio of the earliest continental crust that formed at 3.8 Ga, given that the mantle had an $^{87}Rb/^{86}Sr$ ratio of 0.09 at that time and that $\lambda(^{87}Rb) = 1.42 \times 10^{-11}\,yr^{-1}$? Show all steps in your work clearly.
   (Cambridge University Natural Science Tripos IB 1980.)

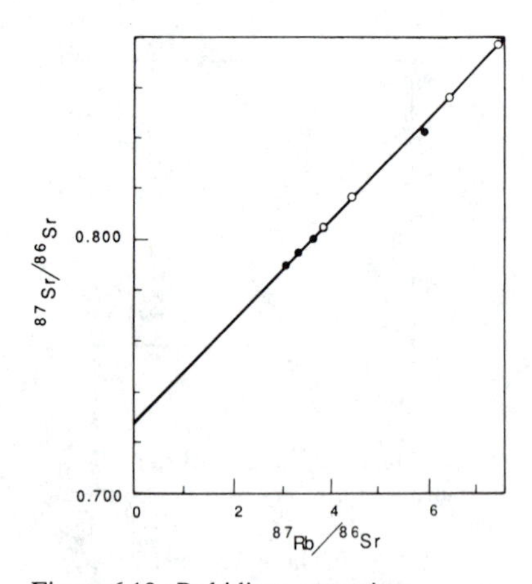

Figure 6.10. Rubidium–strontium isochron for an unmetamorphosed sediment. Open circles, whole rock; solid circles, illite and chlorite. (Data from Gorokhov et al. 1981.)

5. A granite was extracted from the mantle 2500 Ma ago. Later, at 500 Ma, this granite was remelted. What is the present-day strontium isotope ratio ($^{87}Sr/^{86}Sr$) in the young pluton? Assume that the granites have gross rubidium–strontium ratios of 0.5 and 0.7, respectively.
6. A sample is being dated by the $^{40}Ar/^{39}Ar$ ratio method. Would a delay of one month between the irradiation of the sample and the spectrographic measurements adversely affect the results?
7. Discuss which radioactive dating methods are most appropriate for dating basalt, granite, shale and ultramafic samples (use Table 6.3).
8. Four samples from a meteorite have neodymium and strontium

isotope ratios as follows:

$^{143}Nd/^{144}Nd$	$^{147}Sm/^{144}Nd$
0.5105	0.12
0.5122	0.18
0.5141	0.24
0.5153	0.28

(a) Find the age of the meteorite.
(b) Find the initial $^{143}Nd/^{144}Nd$ ratio for this meteorite.
(c) Discuss the relevance of the initial $^{143}Nd/^{144}Nd$ ratio of meteorites.

9. Fission track dating was performed on two minerals in a sample. The track date from garnet was 700 Ma and the date from muscovite was 540 Ma.
(a) Determine the cooling rate for these minerals in $°C\ Ma^{-1}$.
(b) Determine the closure temperatures for these minerals.
(c) Assuming that cooling continued at this rate, calculate the track date that would be given by apatite and its closure temperature.
(d) Would a biotite fission track date be useful?
10. Fission track dating on four minerals from one rock sequence yielded the following dates: zircon, 653 Ma; sphene, 646 Ma; hornblende, 309 Ma; apatite, 299 Ma. Comment on the cooling history of this rock. If you could have one radiometric date determined for these rocks, which method would you choose and why?

BIBLIOGRAPHY

Badash, L. 1968. Rutherford, Boltwood and the age of the earth: The origin of radioactive dating techniques. *Proc. Am. Phil. Soc., 112*, 157–169.

1969. *Rutherford and Boltwood: Letters on radioactivity.* Yale Univ. Press, New Haven.

Curtis, G. H., and Hay, R. L. 1972. Further geological studies and potassium–argon dating at Olduvai Gorge and Ngorongoro Crater. In W. W. Bishop and J. A. Miller, eds., *Calibration of hominoid evolution*, Scottish Academic Press, Edinburgh, 289–301.

DePaolo, D. J. 1981. Nd Isotopic studies: Some new perspectives on earth structure and evolution, *EOS, 62*, 137–40.

Dodson, M. A. 1973. Closure temperature in cooling geochronological and petrological systems. *Contrib. Mineral. Petrol., 40*, 259–74.

Eicher, D. L. 1976. *Geologic time*, 2nd ed. Prentice-Hall, Englewood Cliffs, N.J.

Eve, A. S. 1939. *Rutherford, being the life and letters of the Rt. Hon. Lord Rutherford, O.M.* Cambridge Univ. Press.

Faul, H. 1966. *Ages of rocks, planets and stars.* McGraw-Hill, New York.

Faure, G. 1986. *Principles of isotope geology*, 2nd ed. Wiley, New York.

Fitch, F. J., Miller, J. A., and Hooker, P. J. 1976. Single whole rock K-Ar isochrons. *Geol. Mag., 113*, 1–10.

Fleischer, R. L., Price, B., and Walker, R. M. 1975. *Nuclear tracks in solids.* Univ. California Press, Berkeley.

Fullagar, P. D., Lemmon, R. E., and Ragland, P. C. 1971. Petrochemical and geochronological studies of plutonic rocks in the southern Appalachians: Part I. The Salisbury pluton. *Geol. Soc. Am. Bull., 82*, 409–16.

Ghent, E. D., Stout, M. Z., and Parrish, R. R. 1988. Determination of metamorphic pressure–temperature–time (*P–T–t*) paths. *In* E. G. Nisbet and C. M. R. Fowler, eds., *Heat, metamorphism and tectonics*, Min. Assoc. Can. Short Course, 14, 155–88.

Gorokhov, I. M., Clauer, N., Varshavskaya, S., Kutyavin, E. P., and Drannik, A. S. 1981. Rb–Sr Ages of Precambrian sediments from the Ovruch Mountain Range, northwestern Ukraine (USSR). *Precambrian Research, 16*, 55–65.

Gunner, J. D. 1974. Investigations of lower Paleozoic granites in the Beardmore Glacier region. *Ant. J.U.S., 9*, 76–81.

Harland, W. B., Armstrong, R. L., Cox, A. V., Craig, L. E., Smith, A. G., and Smith, D. G. 1990. *A geologic time scale 1989*. Cambridge Univ. Press, Cambridge.

Harrison, T. M. 1987. Comment on 'Kelvin and the age of the Earth'. *J. Geology, 94*, 725–7.

Jeffreys, H. 1976. *The earth*, 6th ed. Cambridge Univ. Press.

McNutt, R. H., Crocket, J. H., Clark, A. H., Caelles, J. C., Farrar, E., Haynes, S. J., and Zentilli, M. 1975. Initial $^{87}Sr/^{86}Sr$ ratios of plutonic and volcanic rocks of the central Andes between latitudes 26° and 29° South. *Earth Planet. Sci. Lett., 27*, 305–13.

Moorbath, S., O'Nions, R. K., Pankhurst, R. J., Gale, N. H., and McGregor, V. R. 1972. Further rubidium–strontium age determinations on the very early Precambrian rocks of the Godthaab district, West Greenland. *Nature, Phys. Sci., 240*, 78–82.

Patchett, P. J., White, W. M., Feldmann, H., Kielinczuk, S., and Hofmann, A. W. 1984. Hafnium/rare earth element fractionation in the sedimentary system and crustal recycling into the earth's mantle. *Earth Planet. Sci. Lett., 69*, 365–75.

Patterson, C. C. 1956. Age of meteorites and the earth. *Geochem. Cosmochim. Acta, 10*, 230–7.

Richter, F. M. 1986. Kelvin and the age of the earth. *J. Geology, 94*, 395–401.

Rosholt, J. N., Zartman, R. E., and Nkomo, I. T. 1973. Lead isotope systematics and uranium depletion in the Granite Mountains, Wyoming. *Geol. Soc. Am. Bull., 84*, 989–1002.

Rutherford, E. 1907. Some cosmical aspects of radioactivity. *J. R. Astr. Soc. Canada*, May–June, 145–65.

Steiger, R. H., and Jaeger, E. 1977. Subcommission on geochemistry: Convention on the use of decay constants in geo- and cosmochronology. *Earth Planet. Sci. Lett., 36*, 359–62.

Strutt, Hon. R. J. 1906. On the distribution of radium in the earth's crust and on the earth's internal heat. *Proc. Roy. Soc. A, 77*, 472–85.

Wasserburg, G. J., and DePaolo, D. J. 1979. Models of earth structure inferred from neodymium and strontium isotopic abundances. *Proc. Natl. Acad. Sci., U.S.A., 76*, 3594–8.

7

Heat

7.1 Introduction

Volcanoes, intrusions, earthquakes, mountain building and metamorphism are all controlled by the transfer and generation of heat. The earth's thermal budget controls the activity of lithosphere and asthenosphere as well as the development of the innermost structure of the earth.

Heat arrives at the earth's surface from its interior and from the sun. Virtually all the heat comes from the sun, as any sunbather knows, but is all eventually radiated back into space. The rate at which heat is received by the earth, and reradiated, is about 2×10^{17} W or, averaged over the surface, about $4 \times 10^{2}\ \mathrm{W\,m^{-2}}$. Compare this value with the mean rate of loss of internal heat from the earth, 4×10^{13} W (or $8 \times 10^{-2}\ \mathrm{W\,m^{-2}}$); the approximate rate at which energy is released by earthquakes, 10^{11} W; the rate at which heat is lost by a clothed human body on a very cold (−30°C), windy ($10\ \mathrm{m\,s^{-1}}$) Canadian winter day, $2 \times 10^{3}\ \mathrm{W\,m^{-2}}$. From a geological perspective, the sun's heat is important because it drives the surface water cycle, the rainfall and, hence, erosion. However, the heat source for igneous intrusion, metamorphism and tectonics is in the earth, and it is this internal source which accounts for most geological phenomena. The sun and the biosphere have kept the surface temperature within the range of the stability of liquid water, probably 15–25°C averaged over geological time. Given that constraint, the movement of heat derived from the interior has governed the geological evolution of the earth, controlling plate tectonics, igneous activity, metamorphism, the evolution of the core and hence the earth's magnetic field.

Heat moves by conduction, convection, radiation and advection. *Conduction* is the transfer of heat through a material by atomic or molecular interaction within the material. In *convection*, heat transfer occurs because the molecules themselves are able to move from one location to another within the material; it is important in liquids and gases. In a room with a hot fire, air currents are set up which move the light, hot air upwards and away from the fire while dense cold air moves in. Convection is a much faster way of transferring heat than conduction. As an example, when we boil a pan of water on the stove, the heat is transferred through the metal saucepan by conduction but through the water primarily by convection. *Radiation* involves direct transfer of heat by electromagnetic radiation (e.g., from the sun or an electric bar heater). Within the earth, heat moves predominantly by conduction through the lithosphere (both oceanic and continental) and the solid inner core. Although convection cannot take place in rigid solids, over geological times the earth's mantle appears to behave as a very high-viscosity liquid, which means that slow convection is possible in the mantle (see Sects. 6.1, 7.4 and 7.8); in fact, heat is generally thought to be transferred by convection through most of the mantle as well as through the liquid outer core. Although hot lava radiates heat, as do crystals at deep, hot levels in the mantle, radiation is a minor factor in the transfer of heat within the earth. *Advection* is a special form of convection. If a hot region is uplifted by tectonic events or by erosion and isostatic rebound, heat (called advected heat) is physically lifted up with the rocks.

It is not possible to directly measure temperatures deep in the earth. Temperatures and temperature gradients can only be measured close to the earth's surface, usually in boreholes or mines or in oceanic sediments. The deeper thermal structure must be deduced by extrapolation, by inference from seismic observations, from knowledge of the behaviour of materials at high temperatures and pressures, from metamorphic rocks and from models of the distribution of heat production and of the earth's thermal evolution.

7.2 Conductive Heat Flow

7.2.1 The Heat Conduction Equation

Heat, as everyone knows, flows from a hot body to a cold body, and not the other way around. The *rate* at which heat is conducted through a solid is proportional to the temperature gradient (the difference in temperature per unit length). Heat is conducted faster when there is a large temperature gradient than when there is a small temperature gradient (all other things remaining constant). Imagine an infinitely long and wide solid plate, d in thickness, with its lower surface kept at temperature T_1 and its upper surface at temperature T_2 ($T_2 > T_1$). The rate of flow of heat per unit area *down* through the plate is proportional to

$$\frac{T_2 - T_1}{d} \tag{7.1}$$

The rate of flow of heat per unit area *up* through the plate, Q, is therefore

$$Q = -k\left(\frac{T_2 - T_1}{d}\right) \tag{7.2}$$

where k, the constant of proportionality, is called the *thermal conductivity*. The thermal conductivity is a physical property of the material of which the plate is made and is a measure of its physical ability to conduct heat. The rate of flow of heat per unit area Q is measured in units of watts per square metre ($\mathrm{W\,m^{-2}}$), and thermal conductivity k is measured in watts per metre per degree centigrade ($W\,m^{-1}\,{}^{\circ}C^{-1}$).* Values of the thermal conductivity of solids vary widely: 418 $W\,m^{-1}\,{}^{\circ}C^{-1}$ for silver; 159 $W\,m^{-1}\,{}^{\circ}C^{-1}$ for magnesium; 1.2 $W\,m^{-1}\,{}^{\circ}C^{-1}$ for glass; 1.7–3.3 $W\,m^{-1}\,{}^{\circ}C^{-1}$ for rock; 0.1 $W\,m^{-1}\,{}^{\circ}C^{-1}$ for wood.

To express Eq. 7.2 as a differential equation, let us assume that the temperature of the lower surface (at z) is T and that the temperature of the upper surface (at $z + \delta z$) is $T + \delta T$ (Fig. 7.1). Substituting these values into Eq. 7.2 then gives

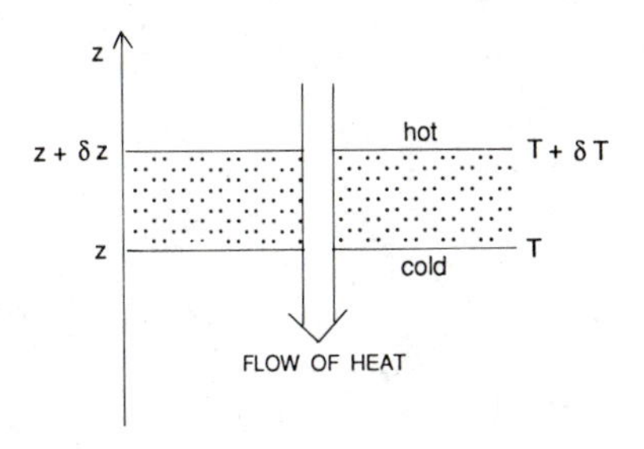

Figure 7.1. Conductive transfer of heat through an infinitely wide and long plate δz in thickness. Heat flows from the hot side of the slab to the cold side (in the negative z direction).

$$Q(z) = -k\left(\frac{T + \delta T - T}{\delta z}\right)$$

$$= -k\frac{\delta T}{\delta z} \tag{7.3}$$

In the limit as $\delta z \to 0$, Eq. 7.3 is written

$$Q(z) = -k\frac{\partial T}{\partial z} \tag{7.4}$$

The minus sign in Eq. 7.4 arises because the temperature is increasing in the positive z direction; since heat flows from a hot region to a cold region, it flows in the negative z direction.

If we consider Eq. 7.4 in the context of the earth, z denotes depth beneath the surface. As z increases downwards, a positive temperature gradient (temperature increases with depth) means that there is a net flow of heat upwards out of the earth. Measurement of temperature gradients and conductivities in near-surface boreholes and mines can provide estimates of the rate of loss of heat from the earth.

Consider a small volume element of height δz and cross-sectional area a (Fig. 7.2). Any change in temperature δT of this small volume element in time δt depends on

1. flow of heat across the element's surface (net flow is in or out),
2. heat generated in the element and
3. thermal capacity (specific heat) of the material.

The heat per unit of time entering the element across its face at z is $aQ(z)$, whereas the heat per unit time leaving the element across its face at $z + \delta z$ is $aQ(z + \delta z)$. Expanding $Q(z + \delta z)$ in a Taylor series gives

$$Q(z + \delta z) = Q(z) + \delta z\frac{\partial Q}{\partial z} + \frac{(\delta z)^2}{2}\frac{\partial^2 Q}{\partial z^2} + \cdots \tag{7.5}$$

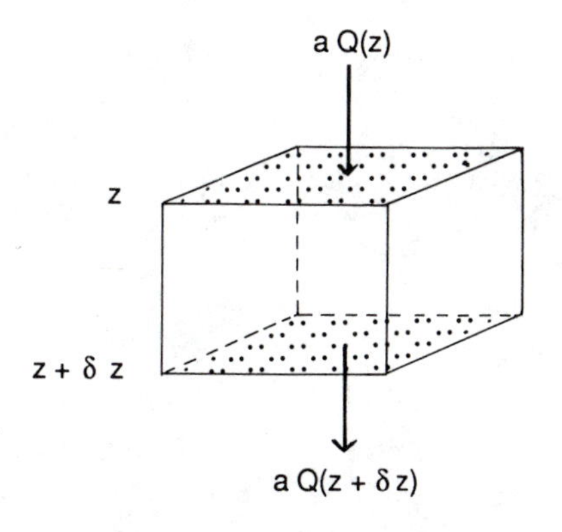

Figure 7.2. Volume element, height δz, cross-sectional area a. Heat is conducted into and out of the element across the shaded faces only. We assume there is no heat transfer across the other four faces.

In the Taylor series, the $(\delta z)^2$ term and those of higher order are very small and can be ignored. From Eq. 7.5 the net gain of heat per unit time is

* Until recently, the c.g.s. system was used in heat flow work. In that system, 1 hgu (heat generation unit) = 10^{-13} cal $cm^{-3}\,s^{-1} = 4.2 \times 10^{-7}\,W\,m^{-3}$; 1 hfu (heat flow unit) = 10^{-6} cal $cm^{-2}\,s^{-1} = 4.2 \times 10^{-2}\,W\,m^{-2}$; and conductivity, 0.006 cal $cm^{-1}\,s^{-1}\,{}^{\circ}C^{-1}$ = 2.52 $W\,m^{-1}\,{}^{\circ}C^{-1}$.

$$\text{heat entering across } z - \text{heat leaving across } z + \delta z$$

$$= aQ(z) - aQ(z + \delta z)$$

$$= -a\,\delta z \frac{\partial Q}{\partial z} \qquad (7.6)$$

Suppose heat is generated in this volume element at a rate A per unit volume per unit time. The total amount of heat generated per unit time is then

$$Aa\,\delta z \qquad (7.7)$$

Radioactive heat is the main internal heat source for the earth as a whole; however, local heat sources and sinks include latent heat, shear heating and endothermic/exothermic chemical reactions. Radioactive heat generation is discussed in Section 7.2.2. Combining expressions 7.6 and 7.7 gives the total gain in heat per unit time to first order in δz as

$$Aa\,\delta z - a\,\delta z \frac{\partial Q}{\partial z} \qquad (7.8)$$

The specific heat c_P of the material of which the element is made determines the temperature rise due to this gain in heat since *specific heat is defined as the amount of heat necessary to raise 1 kg of the material by 1°C*. Specific heat is measured in units of $W\,kg^{-1}\,°C^{-1}$.

If the material has density ρ and specific heat c_p, and undergoes a temperature increase δT in time δt, the rate at which heat is gained is

$$c_p a \delta z \rho \frac{\delta T}{\delta t} \qquad (7.9)$$

Thus, equating the expressions 7.8 and 7.9 for the rate at which heat is gained by the volume element gives

$$c_p a \delta z \rho \frac{\delta T}{\delta t} = Aa\delta z - a\delta z \frac{\partial Q}{\partial z}$$

$$c_p \rho \frac{\delta T}{\delta t} = A - \frac{\partial Q}{\partial z} \qquad (7.10)$$

In the limiting case when $\delta z, \delta t \to 0$, Eq. 7.10 becomes

$$c_p \rho \frac{\partial T}{\partial t} = A - \frac{\partial Q}{\partial z} \qquad (7.11)$$

Using Eq. 7.4 for Q (heat flow per unit area), we can write

$$c_p \rho \frac{\partial T}{\partial t} = A + k \frac{\partial^2 T}{\partial z^2} \qquad (7.12)$$

or

$$\frac{\partial T}{\partial t} = \frac{k}{\rho c_p} \frac{\partial^2 T}{\partial z^2} + \frac{A}{\rho c_p} \qquad (7.13)$$

This is the one-dimensional heat conduction equation.

In the derivation of this equation, temperature was assumed to be a function only of time t and depth z. It was assumed not to vary in the x and y

directions. If temperature were assumed to be a function of x, y, z and t, a three-dimensional heat conduction equation could be derived in the same way as this one-dimensional equation. It is not necessary to go through the algebra again: We can generalize Eq. 7.13 to a three-dimensional Cartesian coordinate system as

$$\frac{\partial T}{\partial t}=\frac{k}{\rho c_{\mathrm{p}}}\left[\frac{\partial^2 T}{\partial x^2}+\frac{\partial^2 T}{\partial y^2}+\frac{\partial^2 T}{\partial z^2}\right]+\frac{A}{\rho c_{\mathrm{p}}} \tag{7.14}$$

Using differential operator notation (see Appendix 1), we write Eq. 7.14 as

$$\frac{\partial T}{\partial t}=\frac{k}{\rho c_{\mathrm{p}}}\nabla^2 T+\frac{A}{\rho c_{\mathrm{p}}} \tag{7.15}$$

Equations 7.14 and 7.15 are both known as the *heat conduction equation.* The term $k/\rho c_{\mathrm{p}}$ is known as the *thermal diffusivity* κ. Thermal diffusivity expresses the ability of a material to lose heat by conduction. Although we have derived this equation for a Cartesian coordinate system, we can use it in any other coordinate system (e.g., cylindrical or spherical) provided we remember to use the definition of the Laplacian operator, ∇^2 (Appendix 1), which is appropriate for the desired coordinate system.

For a steady-state situation when there is no change in temperature with time, Eq. 7.15 becomes

$$\nabla^2 T=-\frac{A}{k} \tag{7.16}$$

In the absence of any heat generation, Eq. 7.15 becomes

$$\frac{\partial T}{\partial t}=\frac{k}{\rho c_{\mathrm{p}}}\nabla^2 T \tag{7.17}$$

This is the *diffusion equation.*

So far we have assumed that there is no relative motion between the small volume of material and its immediate surroundings. Now consider how the temperature of the small volume changes with time if it is in relative motion through a region where the temperature varies with depth. This is an effect not previously considered, and so Eq. 7.13 and its three-dimensional analogue, Eq. 7.15, must be modified. Assume that the volume element is moving with velocity u_z in the z direction. It is now no longer fixed at depth z; instead, at any time t, its depth is $z+u_z t$. The $\partial T/\partial t$ in Eq. 7.13 must therefore be replaced by $\partial T/\partial t+(dz/dt)\cdot(\partial T/\partial z)$. The first term is the variation of temperature with time at a fixed depth z in the region. The second term $(dz/dt)\cdot(\partial T/\partial z)$ is equal to $u_z(\partial T/\partial z)$ and accounts for the effect of the motion of the small volume of material through the region where the temperature varies with depth. Equations 7.13 and 7.15 become, respectively,

$$\frac{\partial T}{\partial t}=\frac{k}{\rho c_{\mathrm{p}}}\frac{\partial^2 T}{\partial z^2}+\frac{A}{\rho c_{\mathrm{p}}}-u_z\frac{\partial T}{\partial z} \tag{7.18}$$

and

$$\frac{\partial T}{\partial t}=\frac{k}{\rho c_{\mathrm{p}}}\nabla^2 T+\frac{A}{\rho c_{\mathrm{p}}}-\mathbf{u}\cdot\nabla T \tag{7.19}$$

In Eq. 7.19, **u** is the three-dimensional velocity of the material. The term $\mathbf{u}\cdot\nabla T$ is the *advective transfer* term.

Relative motion between the small volume and its surroundings can occur for various reasons. The difficulty involved in solving Eqs. 7.18 and 7.19 depends on the cause of this relative motion. If material is being eroded from above the small volume or deposited on top of it, then the volume is getting nearer to or farther from the cool surface of the earth. In these cases, u_z is the rate at which erosion or deposition is taking place. This is the process of advection referred to earlier. On the other hand, the volume element may form part of a thermal convection cell driven by temperature-induced density differences. In this latter case, the value of u_z depends on the temperature field itself rather than on an external factor such as erosion rates. The fact that, for convection, u_z is a function of temperature makes Eqs. 7.18 and 7.19 nonlinear and significantly more difficult to solve.

7.2.2 *Radioactive Heat Generation*

Heat is produced by the decay of radioactive isotopes (Table 6.2). Those radioactive elements which contribute most to the internal heat generation of the earth are uranium, thorium and potassium. These elements are present in the crust in very small quantities, parts per million for uranium and thorium and of the order of a percent for potassium; in the mantle they are some two orders of magnitude less abundant. Nevertheless, these radioactive elements are important in determining the temperature and tectonic history of the earth. Other radioactive isotopes, such as aluminium 26 and plutonium 244, may have been important in the earliest history of the planet.

The radioactive isotopes producing most of the heat generation in the crust are ^{238}U, ^{235}U, ^{232}Th and ^{40}K. The uranium in the crust can be considered to be ^{238}U and ^{235}U, with present-day relative abundances of 99.28% and 0.72%, respectively; but ^{40}K is only present at a level $1{:}10^4$ of total potassium (Chapter 6). Table 7.1 gives the radioactive heat generation of some average rock types. It is clear from this table that, on average, the uranium and thorium contributions to heat production are larger than the potassium contribution. On average, granite has a greater internal heat generation than mafic igneous rocks, and the heat generation of undepleted mantle is very low.

The heat generated by these radioactive isotopes as measured today can be used to calculate the heat generated at earlier times. At time t ago, a radioactive isotope with a decay constant λ would have been a factor $e^{\lambda t}$ more abundant than today (Eq. 6.5). Table 7.2 shows the changes in abundance of isotopes and consequent higher heat generation in the past relative to the present.

Although the heat generation of the crust is some two orders of magnitude greater than that of the mantle, the rate at which the earth as a whole produces heat is influenced by the mantle because the volume of the mantle is so much greater than the total crustal volume. About one-fifth of radioactive heat is generated in the crust. The mean abundances of potassium, thorium and uranium, for the crust and mantle taken together,

Table 7.1. *Typical concentrations of radioactive elements and heat production of some rock types*

	Granite	Tholeiitic basalt	Alkali basalt	Peridotite	Average continental upper crust	Average oceanic crust	Undepleted mantle
Concentration by weight							
U (ppm)	4	0.1	0.8	0.006	1.6	0.9	0.02
Th (ppm)	15	0.4	2.5	0.04	5.8	2.7	0.10
K (%)	3.5	0.2	1.2	0.01	2.0	0.4	0.02
Heat generation (10^{-10} *W kg*$^{-1}$)							
U	3.9	0.1	0.8	0.006	1.6	0.9	0.02
Th	4.1	0.1	0.7	0.010	1.6	0.7	0.03
K	1.3	0.1	0.4	0.004	0.7	0.1	0.007
Total	9.3	0.3	1.9	0.020	3.9	1.7	0.057
Density (10^3 kg m^{-3})	2.7	2.8	2.7	3.2	2.7	2.9	3.2
Heat generation (μ*Wm*$^{-3}$)	2.5	0.08	0.5	0.006	1.0	0.5	0.02

Table 7.2. *Relative abundance of isotopes and crustal heat generation in the past relative to the present*

Age (Ma)	Relative abundance					Heat generation	
	^{238}U	^{235}U	U[a]	Th	K	Model A[b]	Model B[c]
Present	1.00	1.00	1.00	1.00	1.00	1.00	1.00
500	1.08	1.62	1.10	1.03	1.31	1.13	1.17
1000	1.17	2.64	1.23	1.05	1.70	1.28	1.37
1500	1.26	4.30	1.39	1.08	2.34	1.48	1.64
2000	1.36	6.99	1.59	1.10	2.91	1.74	1.98
2500	1.47	11.4	1.88	1.13	3.79	2.08	2.43
3000	1.59	18.5	2.29	1.16	4.90	2.52	3.01
3500	1.71	29.9	2.88	1.19	6.42	3.13	3.81

[a]This assumes a present isotopic composition of 99.2886% ^{238}U and 0.7114% ^{235}U.
[b]Model A based on Th/U = 4, K/U = 20,000.
[c]Model B based on Th/U = 4, K/U = 40,000.
Source: Jessop and Lewis (1978).

are in the ranges 150–260 ppm, 80–100 ppb and 15–25 ppb, respectively. These abundances result in a total radioactive heat production for the crust and mantle of $1.4–2.7 \times 10^{13}$ W, with a best guess value of 2.1×10^{13} W.

7.3 Calculation of Simple Geotherms

7.3.1 *Equilibrium Geotherms*

As can be seen from Eq. 7.18, the temperature in a column of rock is controlled by several parameters, some internal and some external to the rock column. Internal parameters are the conductivity, specific heat, density and radioactive heat generation. External factors include heat flow into the column, the surface temperature and the rate at which material is removed from or added to the top of the column (erosion or deposition). Temperature–depth profiles within the earth are called *geotherms.* If we consider a one-dimensional column with no erosion or deposition and a constant heat flux, the column may eventually reach a state of thermal equilibrium in which the temperature at any point is steady. In that case, the temperature–depth profile is called an *equilibrium geotherm.* In this equilibrium situation, $\partial T/\partial t = 0$, and Eq. 7.16 applies:

$$\frac{\partial^2 T}{\partial z^2} = -\frac{A}{k} \tag{7.20}$$

Since this is a second-order differential equation, it can be solved given two boundary conditions. Assume that the surface is at $z = 0$ and that z increases downwards. Let us consider two pairs of boundary conditions. One possible pair is

(i) $T=0$ on $z=0$ and
(ii) a surface heat flow $Q=-k\,\partial T/\partial z=-Q_0$ on $z=0$.

The surface heat flow $Q=-Q_0$ is negative because heat is assumed to be flowing upwards out of the medium and z is positive downwards. Integrating Eq. 7.20 once gives

$$\frac{\partial T}{\partial z}=-\frac{Az}{k}+c_1 \tag{7.21}$$

where c_1 is the constant of integration. Because $\partial T/\partial z=Q_0/k$ on $z=0$ is boundary condition (ii), the constant c_1 is given by

$$c_1=\frac{Q_0}{k} \tag{7.22}$$

Substituting Eq. 7.22 into Eq. 7.21 and then integrating the second time gives

$$T=-\frac{A}{2k}z^2+\frac{Q_0}{k}z+c_2 \tag{7.23}$$

where c_2 is the constant of integration. However, since $T=0$ on $z=0$ was specified as boundary condition (i), c_2 must equal zero. The temperature within the column is therefore given by

$$T=-\frac{A}{2k}z^2+\frac{Q_0}{k}z \tag{7.24}$$

An alternative pair of boundary conditions could be

(i) $T=0$ on $z=0$ and
(ii) $Q=-Q_{\mathrm{d}}$ on $z=d$.

This could, for example, be used to estimate equilibrium crustal geotherms if d was the depth of the crust/mantle boundary and Q_{d} was the mantle heat flow into the base of the crust. For these boundary conditions, integrating Eq. 7.20 gives, as before,

$$\frac{\partial T}{\partial z}=-\frac{A}{k}z+c_1 \tag{7.25}$$

where c_1 is the constant of integration. Because $\partial T/\partial z=Q_{\mathrm{d}}/k$ on $z=d$ is boundary condition (ii), c_1 is given by

$$c_1=\frac{Q_{\mathrm{d}}}{k}+\frac{Ad}{k} \tag{7.26}$$

Substituting Eq. 7.26 into Eq. 7.25 and then integrating again gives

$$T=-\frac{A}{2k}z^2+\left(\frac{Q_{\mathrm{d}}+Ad}{k}\right)z+c_2 \tag{7.27}$$

where c_2 is the constant of integration. Because $T=0$ on $z=0$ was boundary condition (i), c_2 must equal zero. The temperature in the column

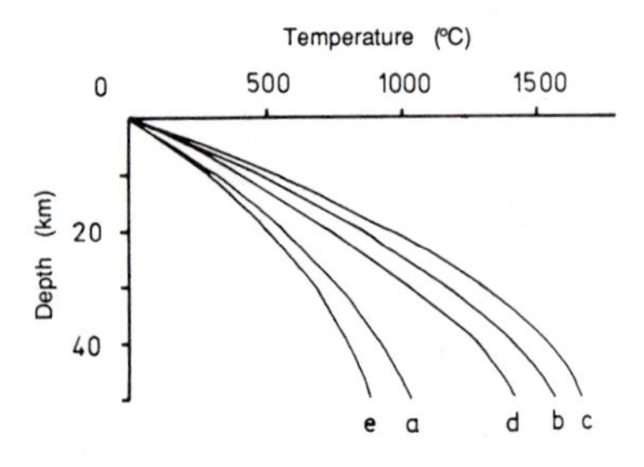

Figure 7.3. Equilibrium geotherms calculated from Eq. 7.28 for a 50 km thick column of rock. Curve a: *standard model* with conductivity 2.5 $W m^{-1} °C^{-1}$, radioactive heat generation 1.25 $\mu W m^{-3}$ and basal heat flow $21 \times 10^{-3} W m^{-2}$. Curve b: standard model with conductivity reduced to 1.7 $W m^{-1} °C^{-1}$. Curve c: standard model with radioactive heat generation increased to 2.5 $\mu W m^{-3}$. Curve d: standard model with basal heat flow increased to $42 \times 10^{-3} W m^{-2}$. Curve e: standard model with basal heat flow reduced to $10.5 \times 10^{-3} W m^{-2}$. (From Nisbet and Fowler 1982.)

$0 \leqslant z \leqslant d$ is therefore given by

$$T = -\frac{A}{2k}z^2 + \left(\frac{Q_d + Ad}{k}\right)z. \tag{7.28}$$

Comparison of the second term in Eqs. 7.24 and 7.28 shows that a column of material of thickness d and radioactive heat generation A makes a contribution to the surface heat flow of Ad. Similarly, the mantle heat flow Q_d contributes $Q_d z/k$ to the temperature at depth z.

7.3.2 One-Layer Models

Figure 7.3 illustrates how the equilibrium geotherm for a model rock column changes when the conductivity, radioactive heat generation and basal heat flow are varied. This model column is 50 km thick, has conductivity 2.5 $W m^{-1} °C^{-1}$, radioactive heat generation 1.25 $\mu W m^{-3}$ and a heat flow into the base of the column of $21 \times 10^{-3} W m^{-2}$. The equilibrium geotherm for this model column is given by Eq. 7.28 and is shown in Figure 7.3a; at shallow levels the gradient is approximately 30 $°C km^{-1}$, whereas at deep levels the gradient is 15 $°C km^{-1}$ or less.

Conductivity Reducing the conductivity of the whole column to 1.7 $W m^{-1} °C^{-1}$ has the effect of increasing the shallow-level gradient to about 45 $°C km^{-1}$ (see Fig. 7.3b). Increasing the conductivity to 3.4 $W m^{-1} °C^{-1}$ would have the opposite effect of reducing the gradient to about 23 $°C km^{-1}$ at shallow levels.

Heat Generation Increasing the heat generation from 1.25 $\mu W m^{-3}$ to 2.5 $\mu W m^{-3}$ raises the shallow-level gradient to over 50 $°C km^{-1}$ (Fig. 7.3c); in contrast, reducing the heat generation to 0.4 $\mu W m^{-3}$ reduces this shallow-level gradient to about 16 $°C km^{-1}$.

Basal Heat Flow If the basal heat flow is doubled from 21 to $42 \times 10^{-3} W m^{-2}$ the gradient at shallow level is increased to about 40 $°C km^{-1}$ (Fig. 7.3d). If the basal heat flow is halved to $10.5 \times 10^{-3} W m^{-2}$, the shallow-level gradient is reduced to about 27 $°C km^{-1}$ (Fig. 7.3e).

7.3.3 Two-Layer Models

The models described so far have been very simple with a 50 km thick surface layer of uniform composition. This is not appropriate for the real earth but is a mathematically simple illustration. More realistic models have a layered crust with the heat generation concentrated towards the top (see, e.g., Sect. 7.6.1). The equilibrium geotherm for such models is calculated exactly as described in Eqs. 7.20–7.28 except that each layer must be considered separately and temperature and temperature gradients must be matched across the boundaries.

Consider a two-layer model:

$$A = A_1 \quad \text{for} \quad 0 \leqslant z < z_1$$

$$A = A_2 \quad \text{for} \quad z_1 \leqslant z < z_2,$$

$$T = 0 \quad \text{on} \quad z = 0$$

with a basal heat flow $Q = -Q_2$ on $z = z_2$. In the first layer, $0 \leqslant z < z_1$, the equilibrium heat conduction equation is

$$\frac{\partial^2 T}{\partial z^2} = -\frac{A_1}{k} \tag{7.29}$$

In the second layer, $z_1 \leqslant z < z_2$, the equilibrium heat conduction equation is

$$\frac{\partial^2 T}{\partial z^2} = -\frac{A_2}{k} \tag{7.30}$$

The solution to these two differential equations, subject to the boundary conditions and matching both temperature T and temperature gradient $\partial T/\partial z$ on the boundary $z = z_1$, is

$$T = -\frac{A_1}{2k}z^2 + \left[\frac{Q_2}{k} + \frac{A_2}{k}(z_2 - z_1) + \frac{A_1 z_1}{k}\right]z \quad \text{for} \quad 0 \leqslant z < z_1 \tag{7.31}$$

$$T = -\frac{A_2}{2k}z^2 + \left[\frac{Q_2}{k} + \frac{A_2 z_2}{k}\right]z + \left[\frac{A_1 - A_2}{2k}\right]z_1^2 \quad \text{for} \quad z_1 \leqslant z < z_2 \tag{7.32}$$

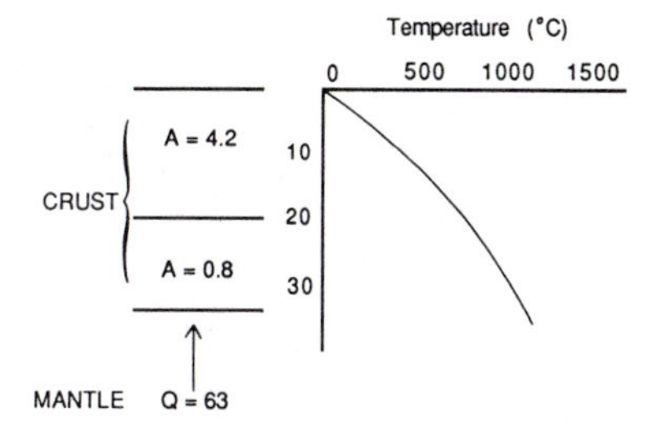

Figure 7.4. Two-layer model Archaen crust and equilibrium geotherm. Heat generation A in $\mu\mathrm{W\,m^{-3}}$; heat flow from mantle Q in $10^{-3}\,\mathrm{W\,m^{-2}}$. (After Nisbet and Fowler 1982.)

Figure 7.4 shows an equilibrium geotherm calculated for a model Archaean crust. The implication is that Archaean crustal temperatures may have been relatively high (compare with Fig. 7.3.).

7.3.4 The Time-Scale of Conductive Heat Flow

Geological structures such as young mountain belts are not usually in thermal equilibrium because the thermal conductivity of rock is so low that equilibrium takes many millions of years to attain. For example, consider the model rock column with the geotherm shown in Figure 7.3a. If the basal heat flow were suddenly increased from 21 to $42 \times 10^{-3}\,\mathrm{W\,m^{-2}}$, the temperature of the column would increase until the new equilibrium temperatures (Fig. 7.3d) were attained. That this process is very slow can be illustrated by considering a rock at depth 20 km. The initial temperature at 20 km would be 567°C, and 20 Ma after the basal heat flow increased, conduction would have raised the temperature at 20 km to about 580°C. Only after 100 Ma would the temperature at 20 km be over 700°C and close to the new equilibrium value of 734°C. This can be estimated quantitatively from Eq. 7.17:

$$\frac{\partial T}{\partial t} = \kappa \frac{\partial^2 T}{\partial z^2}$$

The *characteristic time* $\tau = l^2/\kappa$ gives an indication of the amount of time necessary for a temperature change to propagate a distance of l in a medium having thermal diffusivity κ. Likewise, the *characteristic thermal diffusion distance*, $l=\sqrt{\kappa\tau}$, gives an indication of the distance that temperature changes propagate in a time τ. To give a geological example, it would take many tens of millions of years for thermal transfer from a subduction zone at 100 km depth to have a significant effect on the temperatures at shallow depth if all heat transfer were by conduction alone. Hence, melting and intrusion are important mechanisms for heat transfer above subduction zones. As a second example, a metamorphic belt caused by a deep-seated heat source is characterized by abundant intrusions, often of mantle-derived material; this is the dominant factor in heat transfer to the surface. Magmatism occurs because large increases in the deep heat flow cause large-scale melting at depth long before the heat can penetrate very far towards the surface by conduction.

When a rock column is assembled by some process such as sedimentation, overthrusting or intrusion, the initial temperature gradient is likely to be very different from the equilibrium gradient.

Example: periodic variation of surface temperature

Because the earth's surface temperature is not constant but varies periodically (daily, annually, ice ages) it is necessary to ensure that temperature measurements are made deep enough so that distortion due to these surface periodicities is minimal.

We can model this periodic contribution to the surface temperature as $T_0e^{i\omega t}$, where ω is 2π multiplied by the frequency of the temperature variation, i is the square root of -1 and T_0 is maximum variation of the mean surface temperature. The temperature $T(z,t)$ is then given by Eq. 7.13 (with $A=0$) subject to the two boundary conditions:

(i) $T(0,t)=T_0\,e^{i\omega t}$ and
(ii) $T(z,t)\to 0$ as $z\to\infty$.

We can use the separation of variables technique to solve this problem. Let us assume that the variables z and t can be separated and that the temperature can be written as

$$T(z,t)=V(z)W(t) \tag{7.33}$$

This supposes that the periodic nature of the temperature variation is the same at all depths as it is at the surface, but it allows the magnitude and phase of the variation to be depth dependent, which seems reasonable. Substitution into Eq. 7.13 (with $A=0$) then yields

$$V\frac{dW}{dt}=\frac{k}{\rho c_P}W\frac{d^2V}{dz^2} \tag{7.34}$$

which, upon rearranging, becomes

$$\frac{1}{W}\frac{dW}{dt}=\frac{k}{\rho c_P}\frac{1}{V}\frac{d^2V}{dz^2} \tag{7.35}$$

Because the left-hand side of this equation is a function of z alone and the right-hand side is a function of t alone, it follows that each must equal a constant, say, c_1. However, substitution of Eq. 7.33 into the boundary condition (i) and (ii) yields, respectively,

$$W(t) = e^{i\omega t} \tag{7.36}$$

and

$$V(z) \to 0 \quad \text{as} \quad z \to \infty \tag{7.37}$$

Boundary condition (i) therefore means that the constant c_1 must be equal to $i\omega$ (differentiate Eq. 7.36 to check this). Substituting Eq. 7.36 into Eq. 7.35 gives the equation to be solved for $V(z)$:

$$\frac{d^2 V}{dz^2} = \frac{i\omega\rho c_p V}{k} \tag{7.38}$$

This has two solutions:

$$V(z) = c_2 e^{-qz} + c_3 e^{qz} \tag{7.39}$$

where $q = (1 + i)\sqrt{\omega\rho c_p/2k}$ [remember that $\sqrt{i} = (1 + i)/\sqrt{2}$] and c_2 and c_3 are constants. Equation 7.37, boundary condition (ii), indicates that the positive exponential solution is not allowed; the constant c_3 must be zero. Boundary condition (i) indicates that the constant c_2 is T_0; so finally, $T(z, t)$ is given by

$$\begin{aligned} T(z, t) &= T_0 \exp(i\omega t) \exp\left[-(1 + i)\sqrt{\frac{\omega\rho c_p}{2k}}\, z \right] \\ &= T_0 \exp\left(-\sqrt{\frac{\omega\rho c_p}{2k}}\, z \right) \exp\left[i\left(\omega t - \sqrt{\frac{\omega\rho c_p}{2k}}\, z \right) \right] \end{aligned} \tag{7.40}$$

For large z this periodic variation dies out. Thus, temperatures at great depth are unaffected by the variations in surface temperatures, as required by boundary condition (ii).

At a depth of

$$L = \sqrt{\frac{2k}{\omega\rho c_p}} \tag{7.41}$$

the periodic distrubance has an amplitude $1/e$ of the amplitude at the surface. This depth L is called the *skin depth*. Taking $k = 2.5\,\mathrm{W\,m^{-1}\,{}^\circ C^{-1}}$, $c_p = 10^3\,\mathrm{J\,kg^{-1}\,{}^\circ C^{-1}}$ and $\rho = 2.3 \times 10^3\,\mathrm{kg\,m^{-3}}$, which are reasonable values for a sandstone, then for the daily variation ($\omega = 7.27 \times 10^{-5}\,\mathrm{s^{-1}}$), L is approximately 17 cm; for the annual variation ($\omega = 2 \times 10^{-7}\,\mathrm{s^{-1}}$), L is 3.3 m; and for an ice age (with period of the order of 100,000 years), L is greater than 1 km. Therefore, provided temperature measurements are made at depths greater than 10–20 m, the effects of the daily and annual surface temperature variation are negligible. The effects of ice ages cannot be so easily ignored and must be considered when borehole measurements are made. Measurement of temperatures in ocean sediments are not usually subject to these constraints, the ocean bottom temperature being comparatively constant.

Equation 7.40 shows that there is a phase difference ϕ between the surface temperature variation and that at depth z,

$$\phi = \sqrt{\frac{\omega \rho c_{\rm p}}{2k}}\, z \tag{7.42}$$

At the skin depth, this phase difference is one radian. When the phase difference is π, the temperature at depth z is exactly half a cycle out of phase with the surface temperature.

7.3.5 *Instantaneous Cooling or Heating*

Assume that there is a semi-infinite solid with an upper surface at $z = 0$, no heat generation ($A = 0$) and an initial temperature throughout the solid of $T = T_0$. For $t > 0$, let the surface be kept at temperature $T = 0$. We want to determine how the interior of the solid cools with time.

The differential equation to be solved is Eq. 7.13 with $A = 0$, the *diffusion equation*:

$$\frac{\partial T}{\partial t} = \kappa \frac{\partial^2 T}{\partial z^2} \tag{7.43}$$

where $\kappa = k/\rho c_{\rm p}$ is the thermal diffusivity.

Derivation of the solution to this problem is beyond the scope of this book, and the interested reader is referred to Carslaw and Jaeger 1959, chapter 2, or Turcotte and Schubert 1982, chapter 4. Here we merely state that the solution of this equation which satisfies the boundary conditions is given by an *error function* (Fig. 7.5 and Appendix 5):

$$T = T_0 \operatorname{erf}\left(\frac{z}{2\sqrt{\kappa t}}\right) \tag{7.44}$$

The error function is defined by

$$\operatorname{erf}(x) = \frac{2}{\sqrt{\pi}} \int_0^x e^{-y^2}\, dy \tag{7.45}$$

You can check that Eq. 7.44 is a solution to Eq. 7.43 by differentiating with respect to t and z. Equation 7.44 shows that the time taken to reach a given temperature is proportional to z^2 and inversely proportional to κ.

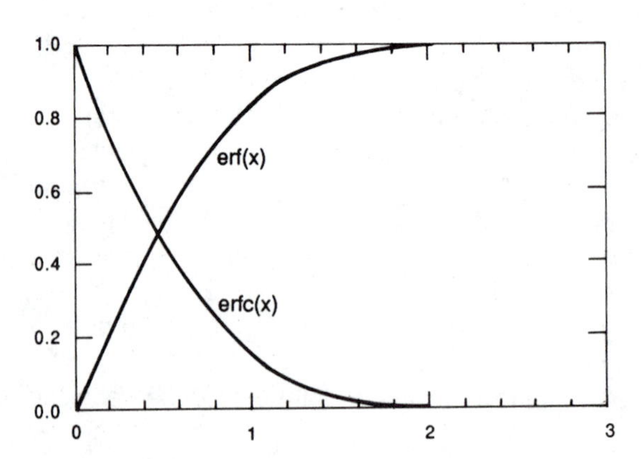

Figure 7.5. Error function erf(x) and complementary error function erfc(x).

The temperature gradient is given by differentiating Eq. 7.44 with respect to z:

$$\begin{aligned}\frac{\partial T}{\partial z} &= \frac{\partial}{\partial z}\left(T_0 \operatorname{erf}\left(\frac{z}{2\sqrt{\kappa t}}\right)\right) \\ &= T_0 \frac{2}{\sqrt{\pi}} \frac{1}{2\sqrt{\kappa t}} e^{-z^2/4\kappa t} \\ &= \frac{T_0}{\sqrt{\pi \kappa t}} e^{-z^2/4\kappa t} \end{aligned} \tag{7.46}$$

This error function solution to the heat conduction equation can be applied to many geological situations. For solutions to these problems, and numerous others, the reader is again referred to Carslaw and Jaeger 1959.

For example, imagine a dyke of width $2w$ and of infinite extent in the y and z directions. If we assume that there is no heat generation and that the dyke has an initial temperature of T_0, and if we ignore latent heat of solidification, then the differential equation to be solved is

$$\frac{\partial T}{\partial t} = \kappa \frac{\partial^2 T}{\partial z^2}$$

with initial conditions

(i) $T = T_0$ at $t = 0$ for $-w \leqslant x \leqslant w$ and
(ii) $T = 0$ at $t = 0$ for $|x| > w$.

The solution of this equation which satisfies the initial conditions is

$$T(x,t) = \frac{T_0}{2}\left[\operatorname{erf}\left(\frac{w-x}{2\sqrt{\kappa t}}\right) + \operatorname{erf}\left(\frac{w+x}{2\sqrt{\kappa t}}\right)\right] \tag{7.47}$$

If the dyke were 2 m in width ($w = 1$ m) and intruded at a temperature of 1000°C and if κ were $10^{-6}\ \mathrm{m^2\,s^{-1}}$, then the temperature at the centre of the dyke would be about 640°C after one week, 340°C after one month and only about 100°C after one year! Clearly, a small dyke cools very rapidly.

For the general case, the temperature in the dyke is about $T_0/2$ when $t = w^2/\kappa$ and about $T_0/4$ when $t = 5w^2/\kappa$. High temperatures outside the dyke are confined to a narrow contact zone: At a distance w away from the edge of the dyke the highest temperature reached is only about $T_0/4$. Temperatures close to $T_0/2$ are reached only within about $w/4$ of the edge of the dyke.

7.4 Worldwide Heat Flow: Total Heat Loss from the Earth

The total present-day worldwide rate of heat loss by the earth is estimated at 4.2×10^{13} W. Table 7.3 shows how this heat loss is distributed by area: 73% of this heat loss occurs through the oceans (which cover 60% of the earth's surface). Thus, most of the heat loss results from the creation and

Table 7.3. *Heat loss and heat flow from the earth*

	Area (10^6 km^2)	Mean heat flow (10^{-3} W m^{-2})	Heat loss (10^{12} W)
Continents (including volcanoes)	149		8.8
Continental shelves	52		2.8
Total continents and continental shelves	201	57	11.6
Deep oceans	282		27.4
Marginal basins	27		3.0
Conductive contribution		66	20.3
Hydrothermal contribution		33	10.1
Total oceans and basins	309	99	30.4
Worldwide total	510	82	42.0

Note: Estimate of convective heat transport by plates is ~65% of total heat loss; this includes lithospheric creation on oceans and magmatic activity on continents. Estimate of heat loss as a result of radioactive decay in the crust is ~17% of total heat loss. Estimate of the heat loss of the core 10^{12}–10^{13} W; this is a major heat source for the mantle.
Source: Sclater et al. (1980, 1981).

cooling of oceanic lithosphere as it moves away from the midocean ridges. Plate tectonics is a primary result of a cooling earth. Conversely, it seems clear that the mean rate of plate generation is determined by some balance between the total rate at which heat is generated within the earth and the rate of heat loss at the surface. Some models of the thermal behaviour of the earth in Archaean time (before 2500 Ma) suggest that the plates were moving around the surface of the earth an order of magnitude faster than today. Other models suggest less marked differences from the present. The heat generated within the Archaean earth by long-lived radioactive isotopes was probably three to four times greater than that generated now (see Table 7.2). A large amount of heat also has been left over from the gravitational energy that was dissipated during accretion of the earth (see Problem 23) and from short-lived but energetic isotopes such as ^{26}Al which decayed during the first few million years of the earth's history. Evidence from Archaean lavas that were derived from the mantle suggests that the earth has probably cooled several hundred degrees since the Archean as the original inventory of heat has dissipated. The earth is gradually cooling, and the plates and rates of plate generation may be slowing to match. Presumably, after many billion years all plate motion will cease.

Measured values of heat flow depend on the age of the underlying crust, be it oceanic or continental (Figs. 7.6 and 7.11). Over the oceanic crust the heat flow generally decreases with age: The highest and very variable measurements occur over the midocean ridges and young crust, and the lowest values are over the deep ocean basins. In continental regions the highest heat flows are measured in regions which are subject to the most recent tectonic activity, and the lowest values occur over the oldest, most stable regions of the Precambrian Shield. These continental and oceanic heat flow observations and their implications are discussed in Sects. 7.5 and 7.6.

To apply the heat conduction equation 7.15 to the earth as a whole, we need to use spherical polar coordinates (r, θ, ϕ) (refer to Appendix 1). If temperature is not a function of θ or ϕ but only of radius r, Eq. 7.15 is

$$\frac{\partial T}{dt} = \frac{k}{\rho c_p} \frac{1}{r^2} \frac{\partial}{\partial r}\left(r^2 \frac{\partial T}{dr}\right) + \frac{A}{\rho c_p} \tag{7.48}$$

First let us assume that there is no internal heat generation. The equilibrium temperature is then the solution to

$$\frac{1}{r^2} \frac{\partial}{\partial r}\left(r^2 \frac{\partial T}{\partial r}\right) = 0 \tag{7.49}$$

Integrating once we obtain

$$r^2 \frac{\partial T}{\partial r} = c_1 \tag{7.50}$$

and integrating the second time gives

$$T = -\frac{c_1}{r} + c_2 \tag{7.51}$$

where c_1 and c_2 are the constants of integration.

Now impose the boundary conditions for a *hollow sphere* $b < r < a$:

(i) zero temperature $T = 0$ at the surface $r = a$ and
(ii) constant heat flow $Q = -k(\partial T/\partial r) = Q_b$ at $r = b$.

The temperature in the spherical shell $b < r < a$ is then given by

$$T = -\frac{Q_b b^2}{k}\left(\frac{1}{a} - \frac{1}{r}\right) \tag{7.52}$$

An expression such as this could be used to estimate a steady temperature for the lithosphere. However, since the thickness of the lithosphere is very small compared with the radius of the earth, $(a - b)/a \ll 1$, this solution is the same as the solution to the one-dimensional equation 7.28 with $A = 0$.

There is no nonzero solution to Eq. 7.49 for the whole sphere, which has a finite temperature at the origin $(r = 0)$. However, there is a steady-state solution to Eq. 7.48 with constant internal heat generation A within the sphere:

$$0 = \frac{k}{r^2} \frac{\partial}{\partial r}\left(r^2 \frac{\partial T}{\partial r}\right) + A$$

$$\frac{\partial}{\partial r}\left(r^2 \frac{\partial T}{\partial r}\right) = -\frac{Ar^2}{k} \tag{7.53}$$

Integrating twice, the temperature is given by

$$T = -\frac{Ar^2}{6k} - \frac{c_1}{r} + c_2 \tag{7.54}$$

where c_1 and c_2 are the constants of integration.

Let us impose the two boundary conditions:

(i) T finite at $r = 0$ and
(ii) $T = 0$ at $r = a$.

Then Eq. 7.54 becomes

$$T = \frac{A}{6k}(a^2 - r^2) \tag{7.55}$$

and the heat flux is given by

$$-k\frac{dT}{dr} = \frac{Ar}{3} \tag{7.56}$$

The surface heat flow (at $r = a$) is therefore equal to $Aa/3$.

If we model the earth as a solid sphere with constant thermal properties and uniform heat generation, Eqs. 7.55 and 7.56 yield the temperature at the centre of this model earth, given a value for the surface heat flow. Assuming values for surface heat flow $= 80 \times 10^{-3}\,\mathrm{W\,m^{-2}}$, $a = 6370\,\mathrm{km}$ and $k = 4\,\mathrm{W\,m^{-1}\,°C^{-1}}$, we obtain a temperature at the centre of this model solid earth of

$$T = \frac{80 \times 10^{-3} \times 6370 \times 10^3}{2 \times 4}$$

$$= 63{,}700°\mathrm{C}$$

This temperature is clearly too high for the real earth because the temperature at the surface of the sun is only about 5700°C. The model is unrealistic since, in fact, heat is not conducted but convected through the mantle, and the heat-generating elements are concentrated in the upper crust and are not uniformly distributed throughout the earth. These facts mean that the actual temperature at the centre of the earth is much lower than this estimate. Convection is important because it allows the surface heat flow to exploit the entire internal heat of the earth, instead of just the surface portions of a conductive earth.

Another fact that we have neglected to consider is the decrease of the radioactive heat generation with time. Equation 7.48 can be solved for an exponential time decay and a nonuniform distribution of the internal heat generation; the temperature solutions are rather complicated (Carslaw and Jaeger 1959, sect. 9.8, give some examples) and still are not applicable to the earth because heat is convected through mantle and outer core and not conducted.

It is thought that the actual temperature at the centre of the earth is about 7000°C, based on available evidence: thermal and seismic data, laboratory behaviour of solids at high temperatures and pressures and laboratory melting of iron-rich systems at high pressures.

7.5 Oceanic Heat Flow

7.5.1 Heat Flow and the Depth of the Oceans

Figure 7.6 shows the mean heat flow as a function of age for the five major oceans and the ocean basins. The average heat flow is higher over young oceanic crust but shows much greater standard deviation than that over the older ocean basins. This decrease of heat flow with increasing age is to be expected if we consider hot volcanic material rising along the axes of the midocean ridges and plates cooling as they move away from the spreading centres. The very scattered heat-flow values measured over young oceanic crust are a consequence of the hydrothermal circulation of sea water through the crust (also discussed in Sect. 8.4.4). Heat flow is locally high in the vicinity of hot water vents and low where cold sea water enters the crust. Water temperatures approaching 400°C have been measured at the axes of spreading centres by submersibles, and the presence of hot springs on Iceland (which is located on the Reykjanes Ridge) and other islands or regions proximal to spreading centres is well known. As will be discussed in Chapter 8, the oceanic crust is formed by the intrusion of basaltic magma from below. Contact with sea water causes rapid cooling of the new crust, and many cracks form in the lava flows and dykes. Convection of sea water through the cracked crust occurs, and it is probable that this circulation penetrates through most of the crust, providing an efficient cooling mechanism (unless you drive a Volkswagen, your car's engine is cooled in the same manner). As the newly formed plate moves away from the ridge, sedimentation occurs. Deep sea sediments have a low permeability* and, in sufficient thickness, are impermeable to sea water. In well-sedimented and therefore generally older crust, measurements of conductive heat flow yield reliable estimates of the actual heat flow. In addition, in older crust the pores and cracks may tend to become plugged with mineral deposits. As a result, hydrothermal circulation may cease. Loss of heat due to hydrothermal circulation is difficult to measure, and so heat flow estimates for young crust are generally very scattered and also significantly lower than the theoretical estimates of heat loss. When we use only reliable heat flow measurements from well-sedimented areas in the Pacific and Atlantic oceans, the scatter in the plot of heat flow versus age is much reduced, and measurements agree well with theoretical estimates (Fig. 7.7 and Sect. 7.5.2).

As an oceanic plate moves away from the ridge axis and cools, it contracts and thus increases in density. If we assume the oceanic regions to be compensated (see Sect. 5.5.2), the depth of the oceans should increase with increasing age (and thus plate density). For any model of the cooling lithosphere, the expected ocean depth can be simply calculated (see Sect. 7.5.2).

* Permeability and porosity are not the same. Sediments have a higher porosity than crustal rocks but lack the connected pore spaces needed for high permeability.

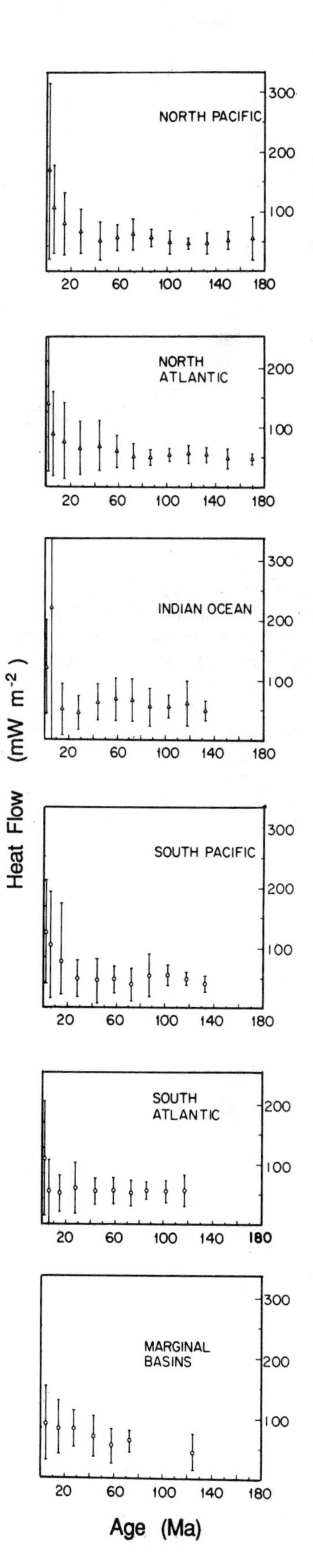

Figure 7.6. Mean heat flow and standard deviation plotted against age for the five major oceans and the marginal basins. (After Sclater et al. 1980.)

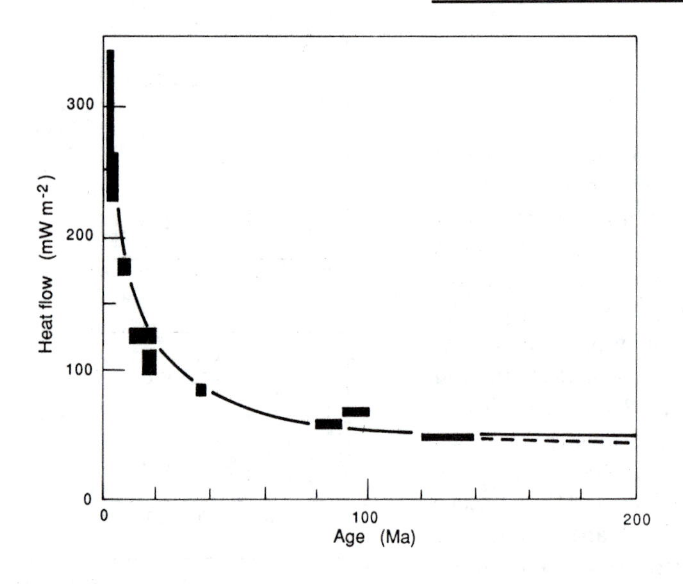

Figure 7.7. Mean heat flow for well-sedimented areas in the North Pacific and North Atlantic plotted against age. Solid curve is the heat flow predicted by the plate model; dashed curve, the heat flow predicted by the boundary layer model. (After Sclater et al. 1980.)

Figure 7.8. Mean depth for the North Atlantic (squares) and North Pacific (circles) plotted against age. Shaded region represents the scatter in the data. Solid curve is the depth predicted by the plate model; dashed line, the depth predicted by the boundary layer model. (After Sclater et al. 1980.)

Figure 7.8 shows the depths of the North Atlantic and North Pacific sea beds plotted against age. For ages less than 70 Ma a simple empirical relation is observed between bathymetric depth d (km) and age t (Ma):

$$d = 2.5 + 0.35t^{1/2} \tag{7.57a}$$

Depth increases linearly with the square root of age. For ages greater than 70 Ma this simple relation does not hold; depth increases much more slowly with increasing age and approximates a negative exponential:

$$d = 6.4 - 3.2e^{-t/62.8} \tag{7.57b}$$

A similar empirical relation is observed between heat flow $Q(10^{-3}\,\mathrm{W\,m^{-2}})$ and age t (Ma) for crust younger than 120 Ma:

$$Q = 473t^{-1/2} \tag{7.58a}$$

Heat flow decreases linearly with the inverse square root of age. For ages greater than 120 Ma this simple relation does not hold; heat flow decreases more slowly with increasing age and follows a negative exponential:

$$Q = 33.5 + 67e^{-t/62.8} \tag{7.58b}$$

7.5.2 Models of Plate Formation and Cooling

The creation of a lithospheric plate at the axis of a midocean ridge and the subsequent cooling of the plate as it moves away from the ridge axis give rise to the type of problem that can be solved by using the two-dimensional version of the heat conduction equation in a moving medium (Eq. 7.19). The boundary conditions can be specified in a number of ways: These necessarily lead to different solutions and thus to different estimates of heat flow and bathymetric depth. The bathymetric depth is calculated from the temperature by assuming that the plate is in isostatic equilibrium, and the heat flow is calculated from the temperature gradient at the surface of the lithosphere. In this way, an understanding of the thermal structure and formation of the plates has been built up; as in all scientific work, the best model is the one which best fits the observations, in this case variations of bathymetric depth and heat flow with age.

A Simple Model The simplest thermal model of the lithosphere is to assume that the lithosphere is cooled asthenospheric material which at the ridge axis had a constant temperature T_a and no heat generation. If we assume the ridge to be infinite in the y direction and the temperature field to be in equilibrium, then the differential equation to be solved is

$$\frac{k}{\rho c_p}\left(\frac{\partial^2 T}{\partial x^2}+\frac{\partial^2 T}{\partial z^2}\right)=u\frac{\partial T}{\partial x} \tag{7.59}$$

where u is the horizontal velocity of the plate and the term on the right-hand side of the equation is due to advection of heat with the moving plate. A further simplification can be introduced by the assumption that horizontal conduction of heat is insignificant in comparison with horizontal advection and vertical conduction of heat. In this case, we can disregard the $\partial^2 T/\partial x^2$ term, leaving the equation to be solved as

$$\frac{k}{\rho c_p}\frac{\partial^2 T}{\partial z^2}=u\frac{\partial T}{\partial x} \tag{7.60}$$

This equation, however, is identical to Eq. 7.43 if we write $t = x/u$, which means that we reintroduce time through the spreading of the ridge. Approximate initial and boundary conditions are $T = T_a$ at $x = 0$ and $T = 0$ at $z = 0$. According to Eq. 7.44, the solution to Eq. 7.60 is

$$T(z,t) = T_a \operatorname{erf}\left(\frac{z}{2\sqrt{\kappa t}}\right) \tag{7.61}$$

The surface heat flow at any distance (age) from the ridge axis is then obtained by differentiating Eq. 7.61:

$$\begin{aligned} Q(t) &= -k\left.\frac{\partial T}{\partial z}\right|_{z=0} \\ &= -\frac{kT_a}{\sqrt{\pi\kappa t}} \end{aligned} \tag{7.62}$$

The observed $t^{-1/2}$ relationship between heat flow and age is thus a feature of this model called a half-space cooling model.

We can estimate the lithospheric thickness L from Eq. 7.61 by specifying a temperature for the base of the lithosphere. For example, if we assume the temperature of the asthenosphere at the ridge axis to be 1300°C and the temperature at the base of the lithosphere to be 1100°C, then we need to find the combination of L and t such that

$$1100 = 1300 \operatorname{erf}\left(\frac{L}{2\sqrt{\kappa t}}\right) \tag{7.63}$$

In other words, we need the inverse error function of 0.846. Using

Figure 7.5 (or Appendix 5), we can write

$$1.008 = \frac{L}{2\sqrt{\kappa t}} \tag{7.64}$$

Thus, for $\kappa = 10^{-6}\,\mathrm{m^2\,s^{-1}}$,

$$L = 2.016 \times 10^{-3}\sqrt{t} \tag{7.65a}$$

when L is in metres and t in seconds, or

$$L = 11\sqrt{t} \tag{7.65b}$$

when L is in kilometres (km) and t in millions of years (Ma). At 10 Ma this lithosphere would be 35 km thick, and at 80 Ma it would be 98 km thick.

The depth of the seabed at any given age can be calculated by using the principle of isostasy (see Chapter 5) and the gradual increase in density of the lithosphere as it cools. If we take the compensation depth D to be in the mantle beneath the base of the lithosphere, the total mass in a vertical column extending down to D is

$$\int_0^D \rho(z)\,dz$$

Isostatic compensation requires that this mass be constant for all columns whatever their age. At the ridge axis the lithosphere has zero thickness, and so, taking $z = 0$ to be at sea level, the mass of the column is

$$\int_0^{d_r} \rho_w\,dz + \int_{d_r}^D \rho_a\,dz$$

where ρ_w is the density of sea water, ρ_a the density of the asthenosphere (at temperature T_a) and d_r the depth of the water over the ridge axis. The mass of a column aged t is then

$$\int_0^d \rho_w\,dz + \int_d^{d+L} \rho(z)\,dz + \int_{d+L}^D \rho_a\,dz$$

where d is the water depth and L the thickness of the lithosphere. Because the mass in this column must be the same as the mass in the column at the ridge axis, we obtain the equation

$$\int_0^{d_r} \rho_w\,dz + \int_{d_r}^D \rho_a\,dz = \int_0^d \rho_w\,dz + \int_d^{d+L} \rho(z)\,dz + \int_{d+L}^D \rho_a\,dz \tag{7.66}$$

Rearranging yields

$$(d - d_r)(\rho_a - \rho_w) = \int_0^L (\rho(z) - \rho_a)\,dz \tag{7.67}$$

To determine $\rho(z)$ we must use the expression for density as a function of temperature and α the coefficient of thermal expansion,

$$\rho(T) = \rho_a[1 - \alpha(T - T_a)] \tag{7.68}$$

and Eq. 7.61 for the temperature structure of the lithosphere. Substituting these two equations into Eq. 7.67 gives

$$(d-d_r)(\rho_a-\rho_w)=\rho_a\alpha T_a\int_0^L \operatorname{erfc}\left(\frac{z}{2\sqrt{\kappa t}}\right)dz \tag{7.69}$$

where erfc is the complementary error function (see Appendix 5). The integral on the right-hand side of this equation can be simply calculated or looked up in a set of mathematical tables. However, for our purposes it is sufficient to change the upper limit of integration from L to ∞ (the error introduced by this approximation is about 5%). This integral of erfc(x) between $x=0$ and infinity is $1/\sqrt{\pi}$ (Appendix 5). When this approximation is made, Eq. 7.69 becomes

$$(d-d_r)(\rho_a-\rho_w)=2\rho_a\alpha T_a\sqrt{\frac{\kappa t}{\pi}} \tag{7.70}$$

Rearranging Eq. 7.70 gives

$$d=d_r+\frac{2\rho_a\alpha T_a}{(\rho_a-\rho_w)}\sqrt{\frac{\kappa t}{\pi}} \tag{7.71}$$

If we assume values for ρ_a and ρ_w of 3.3×10^3 and $1.0\times10^3\,\mathrm{kg\,m^{-3}}$; α, $3\times10^{-5}\,^\circ\mathrm{C}^{-1}$; κ, $10^{-6}\,\mathrm{m^2\,s^{-1}}$; T_a, 1200°C; d_r, 2.5 km; t in millions of years and d in kilometres, then Eq. 7.71 is

$$d=2.5+0.33\sqrt{t} \tag{7.72}$$

When T_a is taken to be 1300°C, Eq. 7.71 is

$$d=2.5+0.36\sqrt{t} \tag{7.73}$$

Such dependence of ocean depth on age is in agreement with the depths observed for oceanic plates less than 70 Ma old (Eq. 7.57a). Thus, as the lithosphere moves away from the ridge axis, it cools, contracts and so subsides.

The Boundary Layer Model The boundary layer model is a modification of the simple half-space model: The rigid lithosphere is assumed to be cooled asthenosphere, and the base of the lithosphere is defined by an isotherm (Fig. 7.9). The boundary condition at the base of the lithosphere is that the heat flux from the mantle is specified. The bathymetric depth predicted by this model increases as $t^{1/2}$ for all t. It is apparent from Figure 7.8 that although this model fits the observed depths out to 70 Ma, for greater ages the predicted depth is too great. The heat flow values predicted by this model decrease as $t^{-1/2}$ for all t. These heat flow predictions are hardly different from the observed values. For ages greater than a few million years, the temperatures and lithospheric thicknesses predicted by the simple model are very close to the values for the boundary layer model.

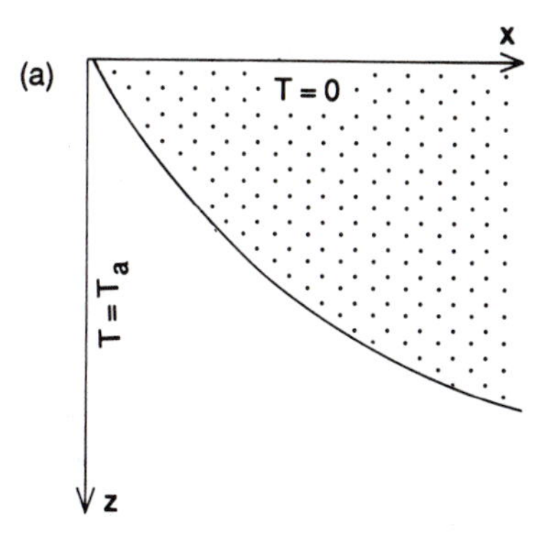

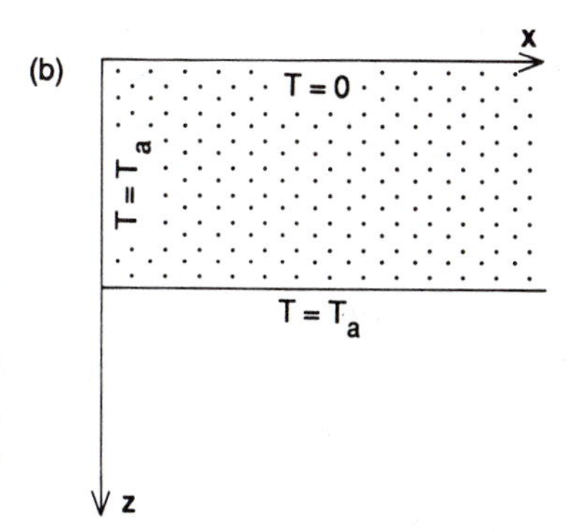

Figure 7.9. Schematic of (a) the half-space cooling and boundary layer models and (b) the plate model of the oceanic lithosphere. Lithosphere is stippled.

The Plate Model In the plate model, the oceanic lithosphere is taken to be of constant thickness L, the base of the lithosphere is at the same constant temperature T_a as the vertical ridge axis and the top surface of the lithosphere and the seabed is another isotherm, usually put at 0°C

(Fig. 7.9). The solution to Eq. 7.59 with these boundary conditions, $T = T_a$ on $z = L$, $T = 0$ on $z = 0$ and $T = T_a$ on $t = 0$, is

$$T = T_a\left(\frac{z}{L} + \sum_{n=1}^{\infty} \frac{2}{n\pi} \sin\left(\frac{n\pi z}{L}\right) \exp\left[\left(\frac{uL}{2\kappa} - \sqrt{\frac{u^2L^2}{4\kappa^2} + n^2\pi^2}\right)\frac{ut}{L}\right]\right) \tag{7.74}$$

When the horizontal conduction of heat is ignored, this simplifies to

$$T = T_a\left(\frac{z}{L} + \sum_{n=1}^{\infty} \frac{2}{n\pi} \sin\left(\frac{n\pi z}{L}\right) \exp\left(-\frac{n^2\pi^2\kappa t}{L^2}\right)\right)$$

A thermal time constant can be defined as $t_0 = L^2/\pi^2\kappa$.

Isotherms for a 100-km-thick plate model with a temperature of 1325°C at the base of the lithosphere and at the ridge axis were shown in Figure 5.15. As the lithosphere ages and moves away from the ridge axis, the isotherms descend until, far from the ridge, they reach essential equilibrium. Figures 7.7 and 7.8 show that the heat flow and bathymetric depths predicted by this model are in good agreement with the observations. The differences between ocean depths predicted by the plate model and by the boundary layer model become apparent beyond about 70 Ma. Since there is no limit to how cool the upper regions of the boundary layer model can get, there is no limit to its predicted ocean depths. The plate model has a uniformly thick lithosphere, and so temperatures in the lithosphere, as well as ocean depths, predicted by that model approach equilibrium as age increases. For the same reason, differences in the surface heat predicted by the two types of model become apparent for ages greater than 140 Ma or so – the boundary layer model keeps on cooling and the plate model approaches equilibrium.

Thermal Structure of the Oceanic Lithosphere Both the plate and boundary layer models of the lithosphere provide heat-flow values that are in reasonable agreement with the measured values, but the ocean depths predicted by the plate model are in much better agreement with observed ocean depths than those predicted by the boundary layer model. However, there is other geophysical evidence on the structure of the oceanic lithosphere which can be used to constrain these thermal models. Surface wave dispersion studies show that the lithosphere thickens away from the ridge axis in approximately the same way as predicted by the boundary layer model. This is a strong constraint on the thermal models.

The observations and the boundary layer model could be reconciled if some mechanism were found to slow the cooling of the boundary layer model for ages greater than 70 Ma so that it would resemble the plate model. Two mechanisms of maintaining the heat flux at the base of the lithosphere have been proposed: shear stress heating caused by a differential motion between lithosphere and asthenosphere or increased heat production rates in the upper mantle. These mechanisms are both somewhat unlikely; perhaps a better proposal is that small-scale convection occurs in the asthenosphere at the base of the older lithosphere. This would increase the heat flux into the base of the rigid lithosphere and maintain a more constant lithospheric thickness.

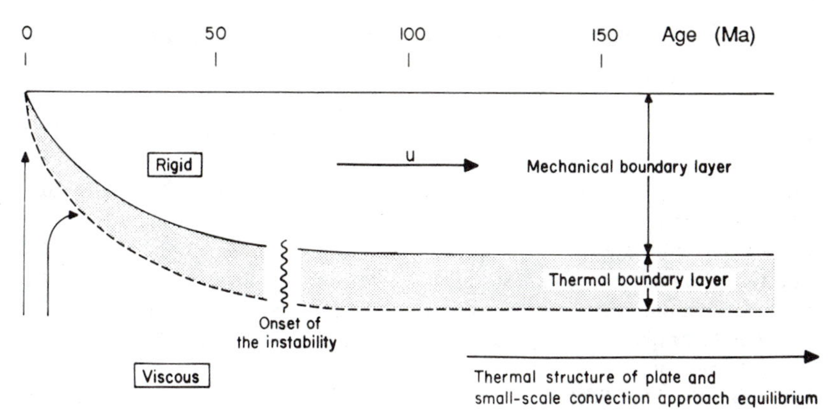

Figure 7.10. Schematic of the oceanic lithosphere, showing the proposed division of the lithospheric plate. The base of the mechanical boundary layer is the isotherm chosen to represent the transition between rigid and viscous behaviour. The base of the thermal boundary layer is another isotherm, chosen to represent correctly the temperature gradient immediately beneath the base of the rigid plate. In the upper mantle beneath these boundary layers, the temperature gradient is approximately adiabatic. At about 60–70 Ma the thermal boundary layer becomes unstable, and small-scale convection starts to occur. With a mantle heat flow of about 38×10^{-3} W m^{-2} the equilibrium thickness of the mechanical boundary layer is approximately 90 km. (From Parsons and McKenzie 1978.)

The lithospheric plate is thought to consist of two parts: an upper rigid layer and a lower viscous thermal boundary layer (Fig. 7.10). At about 60 Ma this thermal boundary layer becomes unstable, and small-scale convection develops within it (see Sect. 7.8), resulting in increased heat flow to the base of the rigid layer and a thermal structure similar to that predicted by the plate model. Very detailed, accurate measurements of heat flow, bathymetry and the geoid on old oceanic crust and across fracture zones may improve our knowledge of the thermal structure of the lithosphere.

7.6 Continental Heat Flow

7.6.1 Mantle Contribution to Continental Heat Flow

Continental heat flow is harder to understand than oceanic heat flow and harder to fit into general theory of thermal evolution of the continents or of the earth. Continental heat-flow values are affected by many factors including erosion, deposition, glaciation, the length of time since any tectonic events, local concentrations of heat-generating elements in the crust, presence or absence of aquifers and the drilling of the hole in which the measurements were made. Nevertheless, it is clear that the measured heat-flow values decrease with increasing age (Fig. 7.11). This is an indication that the continental lithosphere is cooling and slowly thickening with time like the oceanic lithosphere. Alternatively, it could indicate that radioactive elements are lost by erosion.

That all erosional, depositional, tectonic and magmatic processes occurring in the continental crust affect the measured surface heat-flow values is shown in the examples of Sections 7.3 and 7.11. The particularly scattered heat-flow values measured at ages less than about 800 Ma are evidence of strong influence by these transient processes and are therefore very difficult to interpret in terms of the deeper thermal structure of the continents.

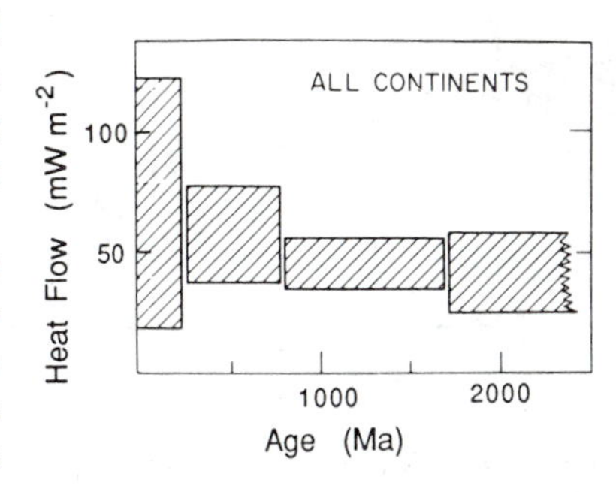

Figure 7.11. Heat flow versus age for the continents. The height of the boxes represents the scatter in heat flow, and the width represents the age range. (From Sclater et al. 1981.)

In some specific areas known as *heat-flow provinces*, a linear relationship exists between surface heat flow and surface radioactive heat generation (Fig. 7.12). Using this relationship, one can make an approximate estimate

Table 7.4. *Some continental heat flow provinces*

Province	Mean Q_0 ($10^{-3}\,W\,m^{-2}$)	Q_r ($10^{-3}\,W\,m^{-2}$)	D (km)
Basin and Range (U.S.)	92	59	9.4
Sierra Nevada (U.S.)	39	17	10.1
Eastern United States	57	33	7.5
Superior Province (Canadian Shield)	39	21	14.4
United Kingdom	59	24	16.0
Western Australia	39	26	4.5
Central Australia	83	27	11.1
Ukrainian Shield (USSR)	37	25	7.1

Source: Sclater et al. (1980).

of the contribution of the heat-generating elements in the continental crust to the surface heat flow. In these heat-flow provinces, some of which are listed in Table 7.4, the surface heat flow Q_0 can be expressed in terms of the measured surface radioactive heat generation A_0 as

$$Q_0 = Q_r + A_0 D \tag{7.75}$$

where Q_r and D are constants for each heat-flow province.

We consider here two extreme models of the distribution of the radioactive heat generation in the crust which both yield a surface heat flow in agreement with this observed linear observation.

1. *Heat generation is uniformly concentrated within a slab with thickness D.* In this case, using Eq. 7.16, we obtain

$$\frac{\partial^2 T}{\partial z^2} = -\frac{A_0}{k} \quad \text{for} \quad 0 \leqslant z \leqslant D$$

Integrating once gives

$$\frac{\partial T}{\partial z} = -\frac{A_0}{k} z + c \tag{7.76}$$

where c is the constant of integration. At the surface, $z = 0$, the upward heat flow $Q(0)$ is

$$Q(0) = Q_0 = k \left.\frac{\partial T}{\partial z}\right|_{z=0}$$

$$= kc \tag{7.77}$$

Therefore, the constant c is given by

$$c = \frac{Q_0}{k}$$

At depth D, the upward heat flow is

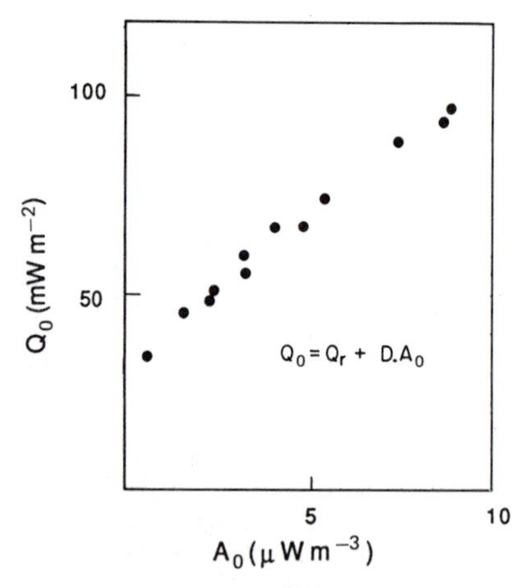

Figure 7.12. Measured heat flow Q_0 versus internal heat generation A_0 for the Eastern United States heat–flow province. The straight line $Q_0 = Q_r + DA_0$ fitted to these measurements has $Q_r = 33 \times 10^{-3}\,W\,m^{-2}$ and $D = 7.5$ km. (After Roy et al. 1968.)

$$\begin{aligned} Q(D) &= k\left(\frac{-A_0 D}{k} + \frac{Q_0}{k}\right) \\ &= -A_0 D + Q_0 \\ &= Q_r \end{aligned} \tag{7.78}$$

Thus, in this case, the heat flow $Q(D)$ into the base of the uniform slab (and the base of the crust, since all the heat generation is assumed to be concentrated in the slab) is the Q_r of Eq. 7.75.

2. *Heat generation is an exponentially decreasing function of depth within a slab with thickness* z^*. Equation 7.16 then becomes

$$\frac{\partial^2 T}{\partial z^2} = \frac{-A(z)}{k} \tag{7.79}$$

where

$$A(z) = A_0 e^{-z/D} \qquad \text{for} \qquad 0 \leqslant z \leqslant z^*$$

Integrating Eq. 7.79 once gives

$$\frac{\partial T}{\partial z} = \frac{A_0}{k} D e^{-z/D} + c \tag{7.80}$$

where c is the constant of integration. At the surface, $z = 0$, the heat flow is $Q(0)$:

$$\begin{aligned} Q(0) = Q_0 &= k\left(\frac{A_0 D}{k} + c\right) \\ &= A_0 D + kc \end{aligned} \tag{7.81}$$

The constant c is given by

$$c = \frac{Q_0 - A_0 D}{k} \tag{7.82}$$

At depth z^* (which need not be uniform throughout the heat-flow province), the heat flow is

$$\begin{aligned} Q(z^*) &= k\left(\frac{A_0 D}{k} e^{-z^*/D} + \frac{Q_0 - A_0 D}{k}\right) \\ &= A_0 D e^{-z^*/D} + Q_0 - A_0 D \end{aligned} \tag{7.83}$$

Thus, by rearranging we obtain

$$Q_0 = Q(z^*) + A_0 D - A_0 D e^{-z^*/D} \tag{7.84}$$

Equation 7.84 is the same as Eq. 7.75 if we write

$$\begin{aligned} Q_r &= Q(z^*) - A_0 D e^{-z^*/D} \\ &= Q(z^*) - A(z^*) D \end{aligned} \tag{7.85}$$

Thus, the linear relation is valid for this model if the heat generation $A(z^*)$ at depth z^* is constant throughout the heat-flow province. Unless $A(z^*)D$ is small, the observed value of Q_r may be very different from the actual heat flow $Q(z^*)$ into the base of the z^* thick layer. However, it can be shown (for details see Lachenbruch 1970) that for some heat-flow

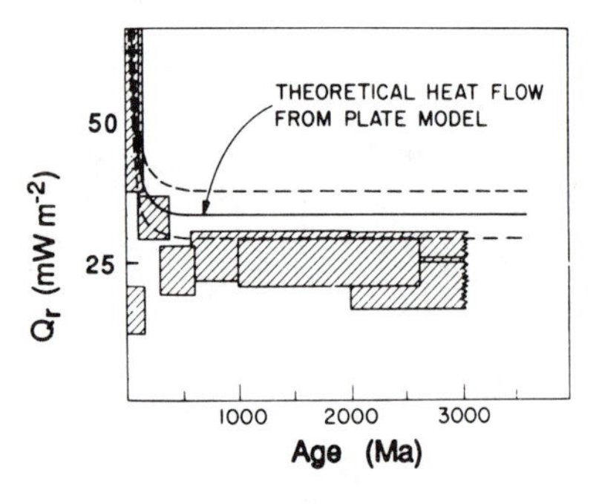

Figure 7.13. Reduced heat flow Q_r versus age for the continental heat flow provinces. The size of the shaded boxes represents an estimate of the scatter in the data. The solid line is the reduced heat flow predicted by the plate model. The dashed lines show the $\pm 4 \times 10^{-3}$ W m^{-2} error about this reduced heat flow. (From Sclater et al. 1981.)

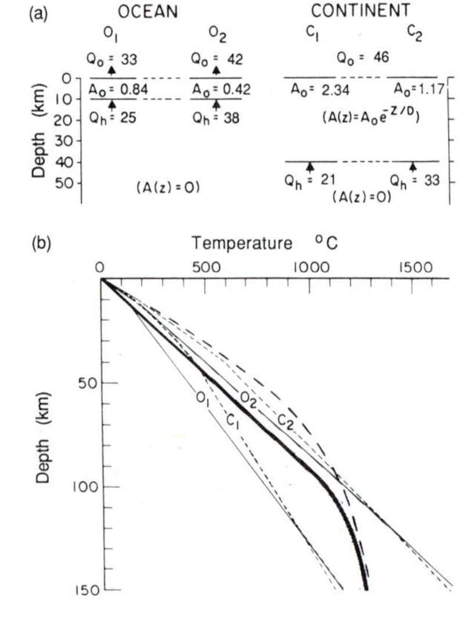

Figure 7.14. (a) Extreme thermal models used to calculate geotherms beneath an equilibrium ocean, O_1 and O_2, and beneath an old stable continent, C_1 and C_2. Heat flows Q_o and Q_h are in 10^{-3} W m^{-2}; heat generation A_o, in μW m^{-3}. (b) Predicted geotherms for these models. The heavy solid line is the equilibrium geotherm for the plate model, taking into account the small-scale convection occurring in the thermal boundary layer (see Fig. 7.10). The heavy dashed line is an error function for the age 70 Ma geotherm (see Sect. 7.5.2). Mantle temperature T_a is taken as 1300°C. (After Parsons and McKenzie 1978 and Sclater et al. 1981.)

provinces, $A(z^*)D$ is small, and thus Q_r is a reasonable estimate of $Q(z^*)$. This removes the constraint that $A(z^*)$ must be the same throughout the heat-flow province. Additionally, for those provinces in which $A(z^*)D$ is small, it can be shown that z^* must be substantially greater than D. Thus, the exponential distribution of heat production satisfies the observed linear relationship between surface heat flow and heat generation and does so even in cases of differential erosion. In this model, D is a measure of the upwards migration of the heat-producing radioactive isotopes (which can be justified geochemically), and Q_r is approximately the heat flow into the base of the crust (because z^* is probably approximately the thickness of the crust).

Neither of these models of the distribution of heat generation within the crust allows for different vertical distributions among the various radioactive isotopes. There is some evidence for such variation. Nevertheless, it is clear that much of the variation in measured surface heat flow is caused by the radioactive heat generation in the crust and that the reduced heat flow Q_r is a reasonable estimate of the heat flow into the base of the crust. Figure 7.13 shows this reduced heat flow plotted against age. After about 300 Ma, the reduced heat flow shows no variation and attains a value of $25 \pm 8 \times 10^{-3}$ W m^{-2}. This is within experimental error of the value predicted by the plate model of the oceanic lithosphere and suggests that there should be no significant difference between models of the thermal structure of the oceanic and continental lithospheres. The present thermal differences are a consequence of the age disparity between oceanic and continental lithospheres.

7.6.2 Temperature Structure of the Continental Lithosphere

Figure 7.14 shows two extreme temperature models of the equilibrium oceanic lithosphere, O_1 and O_2, and two extreme models of the old stable continental lithosphere, C_1 and C_2. These have been calculated by using the one-dimensional heat conduction equation. The extensive region of overlap between these four geotherms indicates that, on the basis of surface measurements, for depths greater than about 80 km there need be little difference in equilibrium temperature structure beneath oceans and continents. The solid line is the geotherm for the oceanic plate model in Figure 7.10. The heavy dashed line is the geotherm for the simple error function model of Section 7.5.2. Figure 7.15 shows thermal models of oceanic and old continental lithospheres.

7.7 The Adiabat and Melting in the Mantle

The previous sections have dealt in some detail with the temperatures in the continental and oceanic lithosphere and with attempts to estimate the temperatures in the mantle and core, assuming that heat is transferred by conduction. For the mantle and outer core, however, where conduction is not the primary method of heat transfer, the methods and estimates of the previous sections are not appropriate. In the mantle and the outer core, where convection is believed to be occurring, heat is transported as

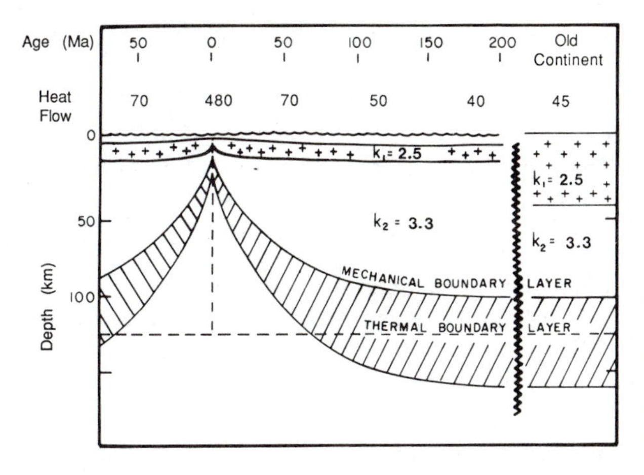

Figure 7.15. Thermal models of the lithospheric plates beneath oceans and continents. The dashed line is the plate thickness predicted by the plate model; k is the conductivity in $\mathrm{W\,m^{-1}\,^{\circ}C^{-1}}$; heat flow in $10^{-3}\,\mathrm{W\,m^{-2}}$. (After Sclater et al. 1981.)

hot material moves; thus, the rate of heat transfer is much greater than by conduction alone, and as a result the temperature gradient and temperatures are much lower. In the interior of a vigorously convecting fluid, the mean temperature gradient is approximately adiabatic. Hence, the temperature gradient in the mantle is approximately adiabatic.

To estimate the adiabatic temperature gradient in the mantle, we need to use some thermodynamics. Consider *adiabatic expansion*, an expansion in which entropy is constant for the system (the system can be imagined to be in a sealed and perfectly insulating rubber bag). Imagine that a rock unit which is initially at depth z and at temperature T is suddenly raised up to depth z'. Assume that the rock unit is a closed system, and let us consider the change in temperature that the unit undergoes. When it reaches its new position z', it is hotter than the surrounding rocks; but because it was previously at a higher pressure, it expands and, in so doing, cools. If the temperature to which it cools as a result of this expansion is the temperature of the surrounding rocks, then the temperature gradient in the rock pile is *adiabatic*. Thus, an adiabatic gradient is essentially the temperature analogue of the self-compression density model discussed in Section 4.3.1. Temperature gradients in a convecting system are close to adiabatic.

To determine the adiabatic gradient, we need to determine the rate of change of temperature T (in K not °C) with pressure P at constant entropy S. Using the reciprocal theorem (a mathematical trick), we can write $(\partial T/\partial P)_S$ as

$$\left(\frac{\partial T}{\partial P}\right)_S = -\left(\frac{\partial T}{\partial S}\right)_P\left(\frac{\partial S}{\partial P}\right)_T \tag{7.86}$$

However, we know from Maxwell's thermodynamic relations that

$$\left(\frac{\partial S}{\partial P}\right)_T = -\left(\frac{\partial V}{\partial T}\right)_P \tag{7.87}$$

where V is volume. Thus, Eq. 7.86 becomes

$$\left(\frac{\partial T}{\partial P}\right)_S = \left(\frac{\partial T}{\partial S}\right)_P\left(\frac{\partial V}{\partial T}\right)_P \tag{7.88}$$

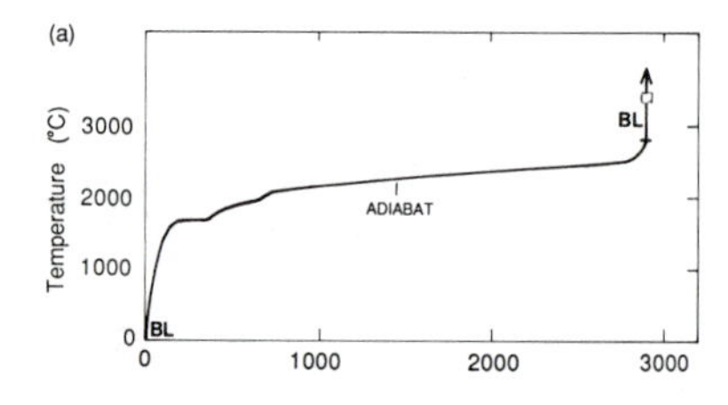

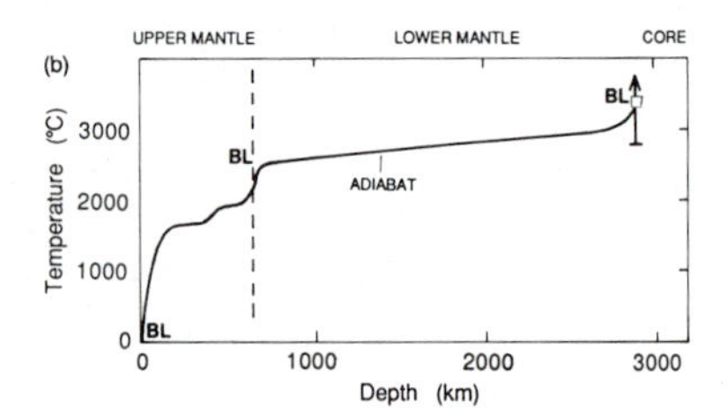

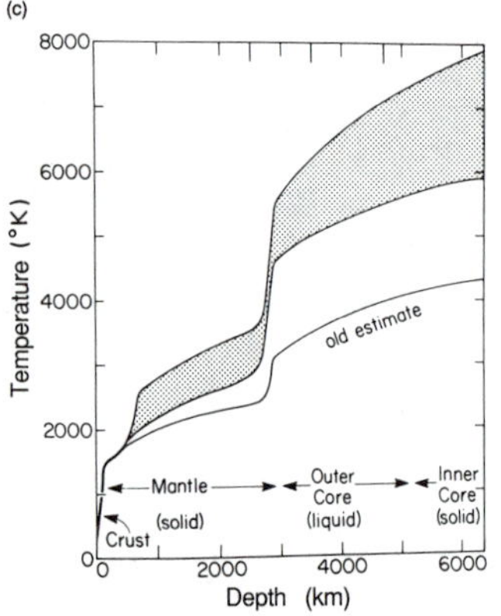

Figure 7.16. Models of temperature profiles in the earth. (a) Mantle adiabat with a thermal boundary layer (BL) at the surface and at the core–mantle boundary. (b) Mantle adiabat with a thermal boundary layer (BL) at both the top and the bottom of the lower mantle. The dashed line indicates a chemical and dynamic boundary between the upper and lower mantle, which are assumed to be separate systems. For (a) and (b) the temperature at the core–mantle boundary is assumed to be in the range 2900–3200°C. (c) An alternative estimate of the temperature in the earth based on high-pressure (over 1000 GPa) and high-temperature (700–6700°C) experiments on iron. The shaded region reflects the uncertainty in the temperature of the core (6600 ± 1000°C at the centre of the earth). The old estimate is similar to profiles shown in (a) and (b) and is typical of temperature profiles proposed prior to 1987. (From Jeanloz and Richter 1979 and Jeanloz 1988.)

The definition of α, the coefficient of thermal expansion, is

$$\alpha = \frac{1}{V}\left(\frac{\partial V}{\partial T}\right)_P \tag{7.89}$$

The definition of specific heat at constant pressure c_p is

$$mc_p = T\left(\frac{\partial S}{\partial T}\right)_P \tag{7.90}$$

where m is the mass of the material.

Using Eqs. 7.89 and 7.90, we can finally write Eq. 7.88 as

$$\left(\frac{\partial T}{\partial P}\right)_S = \frac{T\alpha V}{mc_p} \tag{7.91}$$

Since $m/V = \rho$, density, Eq. 7.91 further simplifies to

$$\left(\frac{\partial T}{\partial P}\right)_S = \frac{T\alpha}{\rho c_p} \tag{7.92}$$

In the earth, we can write

$$\frac{dP}{dr} = -g\rho \tag{7.93}$$

where g is the acceleration due to gravity. The change in temperature with radius r is therefore given by

$$\begin{aligned}\left(\frac{\partial T}{\partial r}\right)_S &= \left(\frac{\partial T}{\partial P}\right)_S \frac{dP}{dr}\\ &= -\frac{T\alpha}{\rho c_p} g\rho\\ &= -\frac{T\alpha g}{c_p}\end{aligned} \tag{7.94}$$

For the uppermost mantle, the adiabatic temperature gradient given by Eq. 7.94 is about 5×10^{-4} °C m^{-1} (0.5°C km^{-1}) assuming the following values: T, 1573 K (1300°C); α, 3×10^{-5} °C^{-1}; g, 9.8 m s^{-2}; c_p, 10^3 J kg °C^{-1}. At greater depths in the mantle where the coefficient of thermal expansion is somewhat less, the adiabatic gradient is reduced to about 3×10^{-4} °C m^{-1} (0.3°C km^{-1}). Figure 7.16 illustrates two possible models of the temperature through the mantle.

The adiabatic gradient can also tell us much about melting in the mantle. For most rocks, the melting curve is very different from the adiabatic gradient (Fig. 7.17), and the two curves intersect at some depth. Imagine a mantle rock rising along an adiabat. At the depth where the two curves intersect, the rock begins to melt and then rises along the melting curve. At some point, the melted material segregates from the solid residue and, being less dense, rises to the surface. Since melt is liquid, it has a coefficient of thermal expansion greater than that of the solid rock. The adiabat along which the melt rises is therefore considerably different from the mantle adiabat (perhaps 1°C km^{-1} instead of 0.5°C km^{-1}). One way of comparing the thermal states of rising melts is to define a *potential temperature* which is the temperature an adiabatically rising melt would have at the earth's surface. Integrating Eq. 7.94 gives the potential temperature T_p of a melt at depth z and temperature T:

$$T_p = Te^{-(\alpha g z / c_p)} \quad (7.94a)$$

T_P is unaffected by adiabatic upwelling.

Equation 7.94 can also be used to estimate temperature gradients in the outer core (see also Sect. 7.9). Using the following values – T, perhaps 5773 K (5500°C) estimated from high-pressure melting experiments on iron compounds (see Fig. 7.16c); α, perhaps $10^{-5}\,°C^{-1}$; g, $5\,m\,s^{-2}$; c_p $5 \times 10^2\,J\,kg^{-1}\,°C^{-1}$ – gives an adiabatic gradient of $6 \times 10^{-4}\,°C\,m^{-1}$ ($0.6°C\,km^{-1}$). However, as estimates of the ratio α/c_p in the core decrease with depth, the adiabatic gradient in the outer core may decrease with depth from perhaps $0.8°C\,km^{-1}$ to $0.2°C\,km^{-1}$. These estimates are just that, however, and are reliable perhaps to $\pm 0.3°C\,km^{-1}$; such is the uncertainty in physical properties of the outer core.

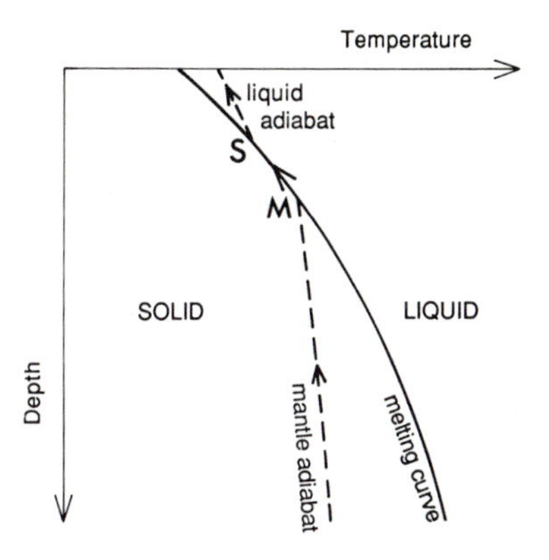

Figure 7.17. Temperatures in the mantle are approximately adiabatic. Rising mantle material is initially too cool to melt. At point M, where the mantle adiabat and the melting curve intersect, melting starts. The melting rock then follows the melting curve until the melt separates from the residue S; it then rises to the surface as a liquid along a liquid adiabat.

7.8 Convection in the Mantle

7.8.1 Rayleigh–Bénard Convection

The horizontal movements of the lithospheric plates require a driving mechanism. The most generally favoured mechanism is thermal convection within the mantle. Convection in liquids occurs when the density distribution deviates from equilibrium. When this occurs, buoyancy forces cause the liquid to flow until it returns to equilibrium. Density disturbances in the earth could be brought about by chemical stratification, such as Fe-Ni in the core; but in the mantle, such disturbances are most likely caused by temperature differences. The simplest illustration of thermal convection is probably a saucepan of water, or soup heating on the stove.

For a *Newtonian viscous fluid*, stress is proportional to strain rate, with the constant of proportionality being the dynamic viscosity of the fluid. Rayleigh–Bénard convection occurs when a tank of Newtonian viscous fluid is uniformly heated from below and cooled from above. Initially, heat is transported by conduction, and there is no lateral variation. As heat is added from below, the fluid on the bottom of the tank warms and becomes less dense, so a light lower fluid underlies a denser upper fluid. Eventually, the density inversions increase to a magnitude sufficient for a slight lateral variation to occur spontaneously. Then a convection pattern starts. In plan view, the cells are two-dimensional cylinders which rotate about their horizontal axes. The hot material rises along one side of the cylinder, and the cold material sinks along the other side. As heating proceeds these two-dimensional cylinders become unstable, and a second set of cylindrical cells develops perpendicular to the first set. This rectangular planform is called *bimodal*. As the heating continues, this bimodal pattern changes into a spoke pattern. In plan view these convection cells are hexagons, with hot material rising in the centres and cold material descending around the edges. With heating, the fluid convects more and more vigorously, with the upgoing and downgoing limbs of a cell confined increasingly to the centre and edges of the cell, respectively. Finally, with extreme heating, the regular cell pattern breaks up, and hot material rises at random; the flow is then irregular.

7.8.2 Equations Governing Thermal Convection

The derivation and solution of the differential equations governing the flow of a heated viscous fluid are beyond the scope of this book, but it is of value to look at the differential equations governing two-dimensional thermal convection in an incompressible Newtonian viscous fluid.

The general equation of conservation of fluid is

$$\frac{\partial u_x}{\partial x} + \frac{\partial u_z}{\partial z} = 0 \tag{7.95}$$

The two-dimensional heat equation in a moving medium is

$$\frac{\partial T}{\partial t} = \frac{k}{\rho c_p}\left(\frac{\partial^2 T}{\partial x^2} + \frac{\partial^2 T}{\partial z^2}\right) - u_x \frac{\partial T}{\partial x} - u_z \frac{\partial T}{\partial z} \tag{7.96}$$

The horizontal equation of motion is

$$\frac{\partial P}{\partial x} = \eta\left(\frac{\partial^2 u_x}{\partial x^2} + \frac{\partial^2 u_x}{\partial z^2}\right) \tag{7.97}$$

The vertical equation of motion is

$$\frac{\partial P}{\partial z} = \eta\left(\frac{\partial^2 u_z}{\partial x^2} + \frac{\partial^2 u_z}{\partial z^2}\right) - g\rho' \tag{7.98}$$

where $\mathbf{u} = (u_x, u_z)$ is the velocity at which the fluid is flowing, P is the pressure generated by the fluid flow, η is the dynamic viscosity, g is gravity and ρ' is the density disturbance. When the density disturbance is of thermal origin,

$$\rho' = \rho - \rho_0 = -\rho_0 \alpha (T - T_0) \tag{7.99}$$

where ρ_0 is density at reference temperature T_0, and α is the volumetric coefficient of thermal expansion. To evaluate thermal convection occurring in a layer of thickness d, heated from below, the four differential equations 7.95–7.98 have to be solved with appropriate boundary conditions. Usually, these boundary conditions are a combination of

(i) $z = 0$ or $z = d$ is at a constant specified temperature (i.e., they are isotherms), or the heat flux is specified across $z = 0$ or $z = d$;
(ii) no flow of fluid occurs across $z = 0$ and $z = d$;
(iii) $z = 0$ or $z = d$ are solid surfaces, in which case there is no horizontal flow (no slip) along these boundaries, or $z = 0$ or $z = d$ are free surfaces, in which case the shear stress is zero at these boundaries.

Solution of the equations with appropriate boundary conditions indicates that convection does not occur until a dimensionless number called the *Rayleigh number* (Ra) exceeds some critical value $\mathrm{Ra_c}$. The

Rayleigh number is given by

$$\mathrm{Ra} = \frac{\alpha g d^4 (Q + Ad)}{k \kappa \nu} \tag{7.100}$$

where α is the volume coefficient of thermal expansion, g gravity, d the thickness of the layer, Q the heat flow through the lower boundary, A the internal heat generation, κ the thermal diffusivity, k the thermal conductivity and ν the kinematic viscosity (kinematic viscosity = dynamic viscosity/density). The critical value of the Rayleigh number further depends on the boundary conditions:

1. For no shear stress on the upper and lower boundaries, the upper boundary held at a constant temperature and all heating from below ($A = 0$), $\mathrm{Ra_c} = 27\pi^4/4 = 658$. At this Rayleigh number the horizontal dimension of a cell is $2.8d$.
2. For no slip on the boundaries, the upper boundary held at a constant temperature and all heating from below ($A = 0$), $\mathrm{Ra_c} = 1708$. At this Rayleigh number the horizontal dimension of a cell is $2.0d$.
3. For no shear stress on the boundaries, a constant heat flux across the upper boundary and all heating from within the fluid ($Q = 0$) $\mathrm{Ra_c}$ is 868. At this Rayleigh number the horizontal dimension of a cell is $3.5d$.
4. For no slip on the boundaries, a constant heat flux across the upper boundary and all heating from within the fluid ($Q = 0$), $\mathrm{Ra_c}$ is 2772. At this Rayleigh number the horizontal dimension of a cell is $2.4d$.

Thus, although the exact value of the critical Rayleigh number $\mathrm{Ra_c}$ depends on the shape of the fluid system, the boundary conditions and the details of heating, it is clear in all cases that $\mathrm{Ra_c}$ is of the order of 10^3 and that the horizontal cell dimension at the critical Rayleigh number is about two to three times the thickness of the convecting layer. For convection to be vigorous with little heat transported by conduction, the Rayleigh number must be about 10^5. If the Rayleigh number exceeds 10^6, then convection is likely to become more irregular.

A second dimensionless number used in fluid dynamics, the *Reynold's number* (Re), indicates whether fluid flow is laminar or turbulent.

$$\mathrm{Re} = \frac{ud}{\nu} \tag{7.101}$$

where u is the velocity of the flow, d the depth of the fluid layer and ν the kinematic viscosity. For $\mathrm{Re} \ll 1$, the flow is laminar. For $\mathrm{Re} \gg 1$, the flow is turbulent. A third dimensionless number, the *Nusselt number* (Nu), is the ratio of heat flow to conducted heat flow:

$$\mathrm{Nu} = \frac{Qd}{k \Delta T} \tag{7.102}$$

where Q is the heat flow, d the thickness of the layer, k the thermal conductivity and ΔT the difference in temperature between the top and bottom of the layer. The Nusselt number is approximately proportional to the fourth root of the Rayleigh number. In regions where upwelling occurs, such as beneath a midocean ridge axis, heat is carried upwards, or advected, by the rising material. The thermal *Peclet Number* ($\mathrm{Pe_t}$) is a measure of the relative importance of advective to conductive heat transport,

$$\mathrm{Pe_t} = \frac{vl}{\kappa} \tag{7.103}$$

where v is the velocity at which the material is moving, l a length scale and κ the thermal diffusivity. If $\mathrm{Pe_t}$ is much larger than unity, advection dominates; if $\mathrm{Pe_t}$ is much smaller than unity, conduction dominates. Under midocean ridges $\mathrm{Pe_t}$ is about 30, showing that the heat is transported mainly by advection.

Table 7.5 gives approximate values of Ra calculated for the mantle using Eq. 7.100 with $A = 0$ and assuming the following values: $\alpha = 2 \times 10^{-5}\,^{\circ}\mathrm{C}^{-1}$ and $\kappa = 10^{-6}\,\mathrm{m^2\,s^{-1}}$; for the upper mantle $\nu = 10^{17}\,\mathrm{m^2\,s^{-1}}$ and $Q/k = 10^{-3}\,^{\circ}\mathrm{C\,m^{-1}}$; for the lower mantle $\nu = 10^{16}\,\mathrm{m^2\,s^{-1}}$ and $Q/k = 10^{-4}\,^{\circ}\mathrm{C\,m^{-1}}$. Similar values for the Rayleigh number are obtained from Eq. 7.100 if one assumes Q to be zero and the internal heat generation of the mantle to be about $10^{-11}\,\mathrm{W\,kg^{-1}}$.

It is clear from Table 7.5 that the Rayleigh number is considerably larger than the critical Rayleigh number ($\sim 10^3$) whether flow is considered to occur throughout the whole mantle or to be confined to the upper and lower mantle separately. Although the exact value of the Rayleigh number changes, depending on the exact values chosen for the constants, its order of magnitude does not change. The most unrealistic assumption in these calculations is that of a constant viscosity mantle. It is probable that the viscosity of the mantle is highly temperature dependent and may change by as much as an order of magnitude for each 100°C temperature change. Estimates of the Reynolds number give values of around 10^{-19}, indicating that the flow is almost certainly laminar.

The possibility that the convective flows in the upper and lower mantle could be separate systems is suggested by several observations. There is a jump in the seismic P-wave velocity at 670 km (see Sect. 4.3.1) which may be due to a phase change (670 km is also the depth of a transition zone in density). Along the convergent plate boundaries this is observed to be the maximum depth at which most earthquakes occur, and it seems that the descending slab may sometimes break off at this level. These results are in agreement with some geochemical models which imply that the upper mantle is depleted and has been almost separate from the lower mantle throughout the earth's history; however, the exact variation of mantle viscosity with depth and the question of whole mantle convection or separate upper and lower mantle convection systems remain subjects of debate.

Table 7.5. *Possible Rayleigh numbers for the mantle*

	Thickness (km)	Rayleigh number, Ra
Upper mantle	700	10^6
Lower mantle	2000	3×10^7
Whole mantle	2700	10^8

7.8.3 Models of Convection in the Mantle

The actual pattern of the convection cells can be investigated in two ways. Numerical models can be simulated on a computer, or physical laboratory models can be made by choosing material of an appropriate viscosity to yield flow on a measurable time scale (Tate and Lyle's golden syrup, glycerine or silicone oil are frequent choices). The dynamic viscosity of water is 10^{-3} Pa s and that of thick syrup is perhaps 10 Pa s; compare these values with the values of 10^{21} Pa s for the mantle (Sect. 5.7.2).

Initial numerical models of flow at high Rayleigh numbers (10^4–10^6), appropriate for the mantle, indicated that only cells with an aspect ratio of about 1 were stable. Cells with a large aspect ratio were unstable. Figure 7.18a shows an example of the temperature and flow lines for a numerical model with heat supplied from below and the temperature fixed on the upper boundary. There is a cold thermal boundary layer at the surface, which could represent the lithospheric plates. This material sinks and descends almost to the base of the model. This sinking material could represent the descending plate at a convergent plate boundary. There is also a hot thermal boundary layer at the base of the box which rises as hot material at the 'ridges'. If the flows in the upper and lower mantle are indeed separate, then simple models such as this imply that the horizontal scale of the cells in the upper mantle is of the order of their depth (the aspect ratio of the cells is about one). Thus, it is not possible in these simple models for the convection in the upper mantle to be directly related to the motions of the plates, with the downgoing cold flow representing the descending plates along the convergent boundary and the upwelling hot flow representing the midocean ridge system, because the horizontal scale of these motions is of the order of 10,000 km. It has, however, been proposed that there could be a two-scale convective flow in the upper mantle. The large-scale flow would represent the plate motions, and for this flow the upper boundary layer is the strong, cold mechanical plate. The small-scale flow aligned in the direction of shear would exist beneath the plates; its upper boundary layer, not rigidly attached to the plate, would be the thermal boundary layer. Three-dimensional laboratory experiments with silicone oil and a moving, rigid upper boundary have indicated that such a two-scale flow can occur. Figure 7.19 shows such a convection system. Another laboratory experiment, which modelled the thermal effect of the subducted lithosphere by cooling one of the side walls, gave rise to a single, stable large-aspect-ratio convection cell. Again this illustrates that large-aspect-ratio cells can be stable; however, the exact form of instabilities and secondary flow depends on the particular experimental model and its boundary conditions.

It now seems clear that the initial numerical models of two-dimensional large-aspect-ratio convection cells in the upper mantle (e.g., Fig. 7.18a) were unstable because of the particular simple boundary conditions chosen. If, instead of a constant temperature on the upper boundary, a constant heat flow across the upper boundary is assumed, then large-aspect-ratio convection cells can be stable (Fig. 7.18b). This shows results from the same model as Figure 7.18a but with constant heat flow across both the upper and the lower boundaries. For these boundary conditions, large-aspect-

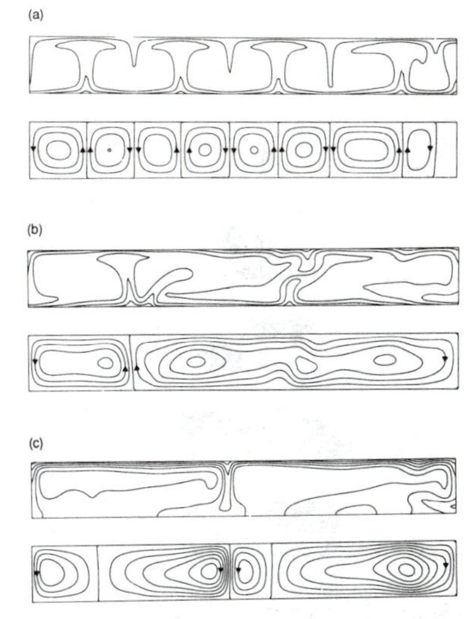

Figure 7.18. Temperature (upper) and fluid flow lines (lower) for computer models of convection in the upper mantle. Rayleigh number, 2.4×10^5. No vertical exaggeration. (a) Temperature constant on upper boundary, heat flow constant on lower boundary. (b) Heat flow constant across upper and lower boundaries. (c) All heat supplied from within, heat flow constant on upper boundary. (After Hewitt et al. 1980.)

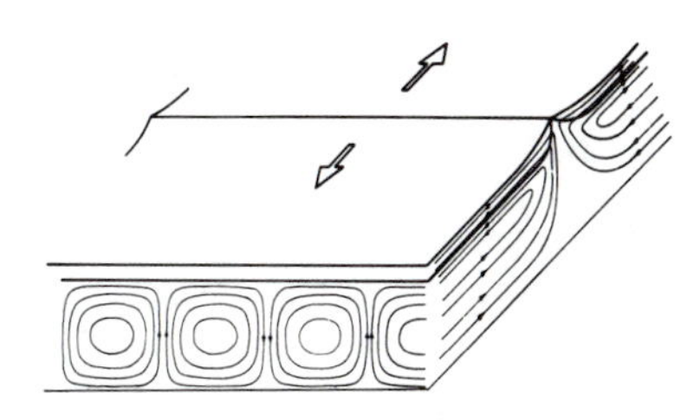

Figure 7.19. Laboratory experiments with a moving rigid upper boundary indicate that flow in the upper mantle could take this form. (After Richter and Parsons 1975.)

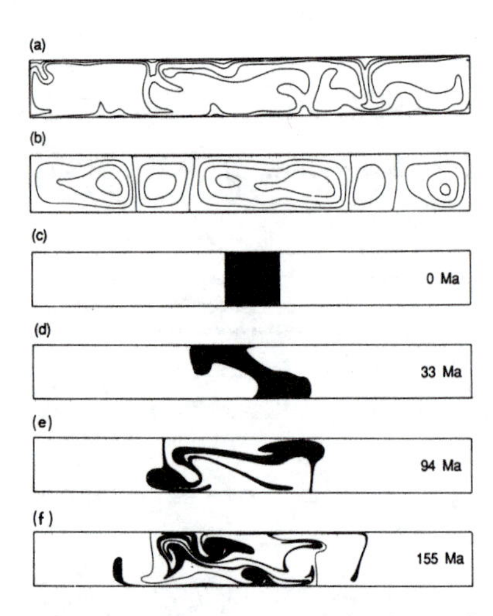

Figure 7.20. Computer model of convection in the upper mantle. Half the heat is supplied from below, and half is supplied internally. Rayleigh number is 1.4×10^6. The model has several adjacent cells, each with separate circulation, although over time the cell boundaries move and material is exchanged between adjacent cells. (a) Isotherms (temperature contours), (b) fluid flow lines and (c) location of marked fluid. (a), (b) and (c) are all at the starting time. Deformation of the marked fluid at subsequent times: (d) 33 Ma, (e) 94 Ma and (f) 155 Ma. (From Hoffman and McKenzie 1985.)

ratio convection cells are stable. The small-scale instabilities on both boundaries of this model do not break up the large-scale flow. Figure 7.18c shows the results of the same experiment but with all the heat supplied from within. Again, large-aspect-ratio cells are stable. The main difference between Figures 7.18b and c is that when all the heat is supplied from within, there are no rising sheets of hot material.

Isotopic ratios of oceanic basalts are very uniform and are quite different from those of the bulk earth. This means that the mantle must be very well mixed. Figure 7.20 shows a computer model of mantle convection. A square patch of mantle with physical properties identical to the rest of the model is marked, and its deformation and distribution throughout the mantle are traced at subsequent times. Within several hundred million years, the convective process is able to mix upper mantle material thoroughly. This time is short compared with the half-lives of the measured radioactive isotopes, indicating that upper mantle convection is well able to account for the general uniformity of isotopic ratios in oceanic basalts. For the upper mantle model illustrated in Figure 7.20, any body smaller than 1000 km is reduced to less than 1 cm thick within 825 Ma.

The isotopic ratios of oceanic island basalts (OIB) require a source for these magmas which is less depleted than the source of midocean ridge basalts (MORB). The efficient mixing of the upper mantle or the whole mantle by convection suggests that the source of oceanic island basalts was a recent addition to the mantle. If not, it would be mixed into the mantle too well to allow the characteristic isotopic signatures of oceanic island basalts to develop. The source of oceanic island basalts is a matter of considerable conjecture. It is possible that there is a contribution from the lower mantle. Another idea, which is in agreement with the convection and geochemical models, is that the subcontinental lithosphere provides a source for oceanic island basalts. Isotopic anomalies can easily form in the deep lithosphere beneath the continents. Deep continental material could become dense and fall into the upper mantle, perhaps triggered by a continent–continent collision. Such a cold body would fall to the base of the upper mantle where it would warm before rising to the surface as part of the convection system. It would remain a viable magma source for about 100–300 Ma. After 150 Ma a body that was originally 100 km thick would be mixed into 5 km thick sheets. Another proposal for the origin of oceanic island basalts is that they are the result of partial melting of material which has risen from a separately convecting primitive lower mantle (Fig. 7.21). Even though this model cannot explain why these basalts do not have the same isotopic composition as the bulk earth, it is attractive in its simplicity.

7.9 Thermal Structure of the Core

7.9.1 Temperatures in the Core

Attempts to calculate the temperature at the centre of the earth using conduction models (Sect. 7.4) fail because heat is primarily convected

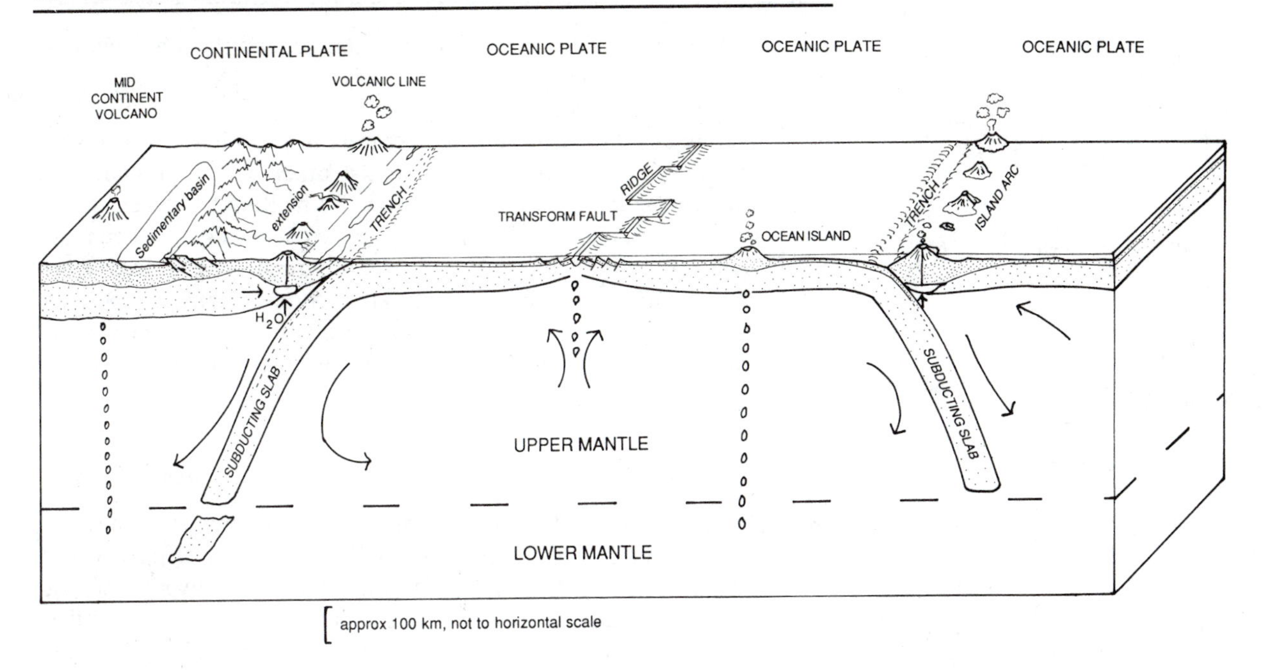

Figure 7.21. Schematic illustrating the formation of new oceanic lithosphere along the midocean ridges and its eventual subduction back into the mantle. Lithosphere is stippled. Crust is indicated by dense stipple. Oceanic island basalts may be derived from the lower mantle.

through much of the earth. The fine detail of the temperature structure of the mantle depends on its dynamic structure. Figures 7.16a, b showed two possible temperature models, one for a separately convecting upper and lower mantle and the other for the whole mantle convecting with no boundary at 700 km depth. The temperature structure of the core is another important constraint on the temperature structure of the mantle because it controls the amount of heat crossing the core–mantle boundary. Conversely, to calculate the temperatures in the core, it is necessary to start with a temperature for the base of the mantle. The other major unknowns are the physical properties, at very high temperatures and pressures, of the iron and iron alloys of which the core is composed (see Sect. 4.3.5). Nevertheless, despite these difficulties the core temperatures have been estimated.

Laboratory experiments which allow materials to be studied at the very high temperatures and pressures of the core have recently been developed (1986–7). These experiments differ from the shock-wave experiments (see Sect. 4.3.5) in that they allow samples under study to be maintained at core temperatures and pressures. In the shock-wave experiments, the samples are subjected to the core pressures only instantaneously, and core temperatures are not attained. Direct experiments on the behaviour of materials at core temperatures and pressures are now possible. The pressure at the core–mantle boundary is about 136 GPa, whereas the pressure at the centre of the earth is about 362 GPa (see Sect. 4.3.2). Thus, the pressure at the core–mantle boundary is 1.36 million times atmospheric pressure. The results from these high-pressure melting experiments on iron suggest that previously projected temperatures were too low. The new estimate of the temperature at the centre of the earth is $6600 \pm 1000°C$ (6900 ± 1000 K). Figure 7.16c shows the new estimates of the temperature structure within the earth. The upper curve for the mantle is similar to

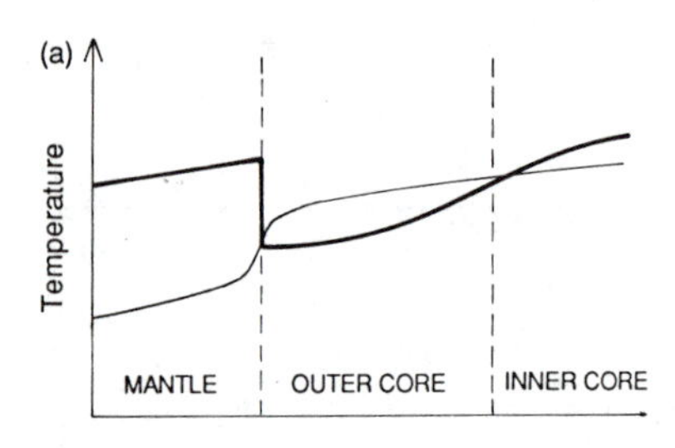

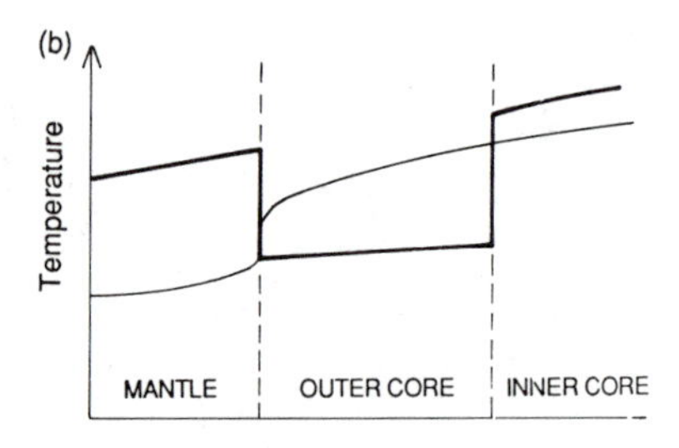

Figure 7.22. Schematic of possible melting temperatures for the mantle and core and the actual temperature profile. Heavy line, melting curve; lighter line, actual temperature profile. (a) Chemically homogeneous core. As the core cools, the inner core grows. (b) The inner and outer core have different chemical compositions and hence different melting temperatures. An outer core composed of an Fe–S or Fe–O alloy would have a much lower melting temperature than a pure iron inner core.

that shown in Fig. 7.16a, b for a two-layer mantle, and the lower curve is for a single-layer mantle.

That the outer core is liquid and the inner core solid is accounted for by the melting curve for iron. The temperature in the outer core is higher than iron's melting temperature and so the outer core is molten. The temperature in the inner core is lower than the melting temperature and so the inner core is solid (Fig. 7.22). If the core is chemically homogeneous and if it is slowly cooling, the inner core will progressively grow with time. If the inner core and outer core are of different compositions, it is possible that their boundary is below the melting temperature of iron because of the depression of the melting point caused by impurities in the liquid outer core.

7.9.2 Convection in the Outer Core and the Earth's Magnetic Field

The first suggestion that the earth's magnetic field is similar to that of a uniformly magnetized sphere came from William Gilbert in 1600 (see Sect. 3.1.2). Karl Friedrich Gauss (1777–1855) later formally showed that the magnetized material or the electric currents which produce the field are not external to the earth but are internal. Figure 7.23 shows four possible models for producing the earth's main dipole field: a magnetic dipole at the centre of the earth, a uniformly magnetized core, a uniformly magnetized earth, or an east–west electrical current system flowing around the core–mantle boundary. Because the mantle is composed of silicates (see Sect. 4.3.5), it is not a candidate for the origin of the magnetic field. The other reason why permanent magnetization of the mantle or core cannot produce the earth's magnetic field is that temperatures in the deep interior exceed the Curie temperatures for magnetic minerals (see Sect. 3.1.3). These two facts rule out the uniformly magnetized earth model; however, the core is predominantly composed of iron and could produce the magnetic field. The earth's magnetic field is not a constant in time but

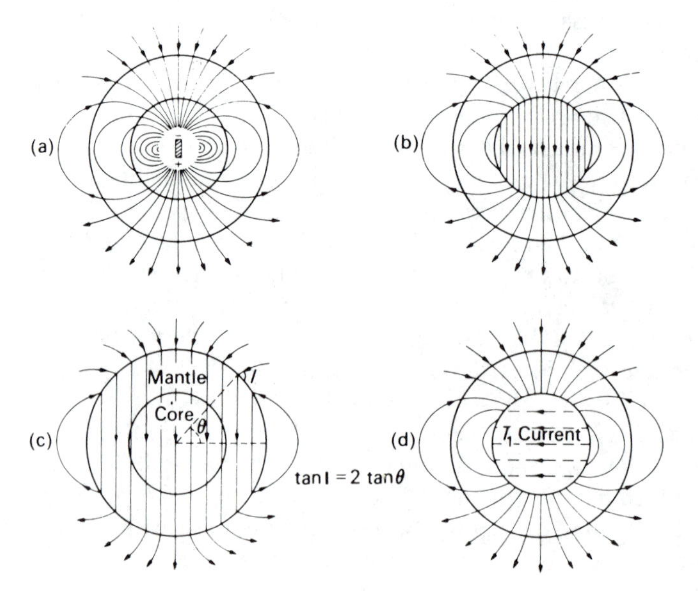

Figure 7.23. Four possible models for producing the earth's main dipole field: (a) a dipole at the centre of the earth, (b) a uniformly magnetized core, (c) a uniformly magnetized core and mantle and (d) a current system flowing east–west around the core–mantle boundary. (From Bott 1982.)

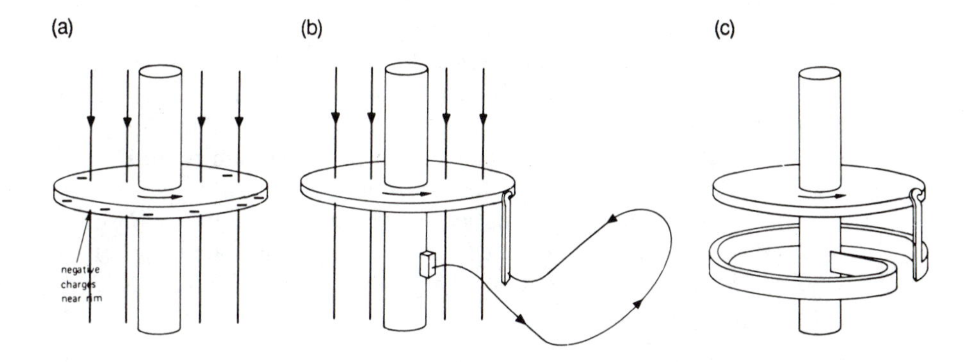

Figure 7.24. Development of a self-exciting dynamo. (a) Metal disc rotating on an axle in a magnetic field. Charge collects on the rim of the disc but cannot go anywhere. (b) A wire joining the rim of the disc to the axle enables current to flow. (c) The wire joining the rim to the axle is modified so that it is a coil looping around the axle. Now the current flowing reinforces the magnetic field, which will induce more current, thus sustaining the magnetic field. This is a self-exciting dynamo. (From Bullard 1972.)

at present is slowly decreasing in strength and drifting westwards. It undergoes irregular reversals as discussed and used in Chapter 3. This changeability indicates that it is unlikely that the core is uniformly magnetized or that there is a magnetic dipole at the centre of the earth. This leaves an electrical current system as the most plausible model for producing the magnetic field. The problem with such an electrical current system is that it must be constantly maintained. If it were not, it would die out in much less than a million years; yet we know from palaeomagnetic studies that the magnetic field has been in existence for at least 3500 Ma.

The model which best explains the magnetic field and what we know of the core is called the *geomagnetic dynamo* or *geodynamo*. A model of a self-exciting dynamo was developed in the 1940s by W. M. Elsasser and Sir Edward Bullard. Figure 7.24 shows how this model works. A simple dynamo is sketched in Figure 7.24a; this model consists of a metal disc on an axle rotating in a magnetic field. The disc is constantly cutting the magnetic field, and so a potential difference (voltage) is generated between the axle and the rim of the disc. However, because there is nowhere for current to flow, the charge can only build up around the rim. In Figure 7.24b, a wire is connected between the rim and the axle so that current is able to flow; but if the external magnetic field is removed, the current stops flowing. In Figure 7.24c, the wire connecting the rim to the axle is coiled around the axle; now the current flowing in the coil gives rise to a magnetic field which reinforces the original field. Thus, when the disc rotates fast enough, the system is self-sustaining, producing its own magnetic field. Unlike a bicycle dynamo, which has a permanent magnet, this dynamo does not need a large constant magnetic field to operate: A slight transient magnetic field can be amplified by the dynamo. All that is necessary is for the disc to be rotating. For this reason this model is often called a *self-exciting dynamo*. The input of energy to power the dynamo is that required to drive the disc.

An interesting feature of the dynamo shown in Figure 7.24c is that it works either with the current and field as illustrated or with both reversed. This means that, like the earth's dynamo, such a dynamo is capable of producing a reversed magnetic field. However, unlike the earth's dynamo, the dynamo in the figure cannot reverse itself unless the circuit includes a shunt. It is most unlikely that the earth's dynamo is like this self-exciting disc dynamo. To start with, because the disc dynamo has a hole in it

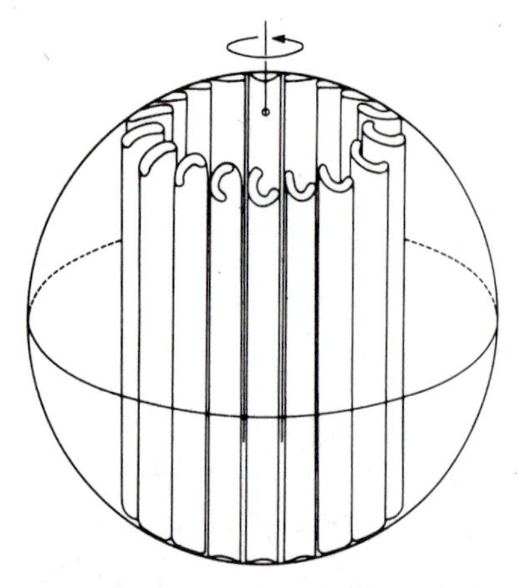

Figure 7.25. Convection currents in a laboratory model of the outer core: a rotating sphere containing a concentric liquid shell and interior sphere. Thermal convection in the fluid was produced by maintaining a temperature difference between the inner and outer spheres. The convection cells which resulted were slowly spinning rolls; those in the northern and southern hemispheres had opposite polarity. Such a convection system in an electrically conductive outer core would be capable of generating the earth's dipole field. These convection rolls drift in the same direction as the rotation (arrow). (From Gubbins 1984, modified from Busse 1970.)

and is antisymmetric, it is topologically different from the core. Also, it is hard to imagine that such an electrical current system could operate in the core without short-circuiting itself somewhere. Nevertheless, it has been demonstrated that there are motions of the liquid outer core which can generate a magnetic field which can undergo random reversals.

The whole subject of magnetic fields in fluids is known as *magnetohydrodynamics*. The mathematical equations governing fluid motion in the outer core and generation of a magnetic field are a very complicated interrelated set of partial differential equations. They can, however, be separated (after Jacobs 1987) into four groups:

(a) the electromagnetic equations relating the magnetic field to the velocity of the fluid in the outer core;
(b) the hydrodynamic equations, including conservation of mass and momentum and the equation of motion for the fluid in the outer core;
(c) the thermal equations governing the transfer of heat in a flowing fluid or the similar equations governing compositional convection; and
(d) the boundary and initial conditions.

Simultaneous solution of all these equations is exceedingly difficult, in part because the equations are nonlinear. However, in special situations solutions can be found for some of the equations. One such simpler approach is to assume a velocity field for the flow in the outer core and then to solve the electromagnetic equations of group (a) to see what type of magnetic field it would generate. Another line of work has been to investigate group (b), possible fluid motions in a fluid outer core sandwiched between a solid mantle and a solid inner core. Figure 7.25 shows the fluid motions observed in a scaled laboratory experiment using a rotating spherical model, with the fluid outer core subjected to a temperature gradient. The convection cells in this model core were cylindrical rolls, with the fluid spiralling in opposite directions in the northern and southern hemispheres. The problem with applying flow patterns such as these directly to dynamo models is that any flow pattern is markedly altered by the magnetic field it generates. Figure 7.26 shows a schematic illustration of the interaction between magnetic field and fluid flow for one dynamo model, the *Parker–Levy dynamo*. For this particular dynamo model to be self-sustaining, four conditions must be met:

1. The initial dipole field must be aligned along the Earth's spin axis.
2. The fluid outer core must be rotating.
3. Upwelling thermal convection currents must be present in the outer core.
4. A spiralling motion of the convection system caused by the Coriolis force is required. The spiralling motions have opposite polarity in the northern and southern hemispheres.

The rotation of the electrically conducting fluid in the outer core will stretch the original dipole magnetic field lines and wind them into a toroidal field. The interaction of this toroidal magnetic field with the convecting rolls then results in a magnetic field with loops that are aligned with the rotation axis. If the loops have the same sense as the original field, that dipole field can be regenerated; but if the loops have the opposite sense, the original dipole field can be reversed.

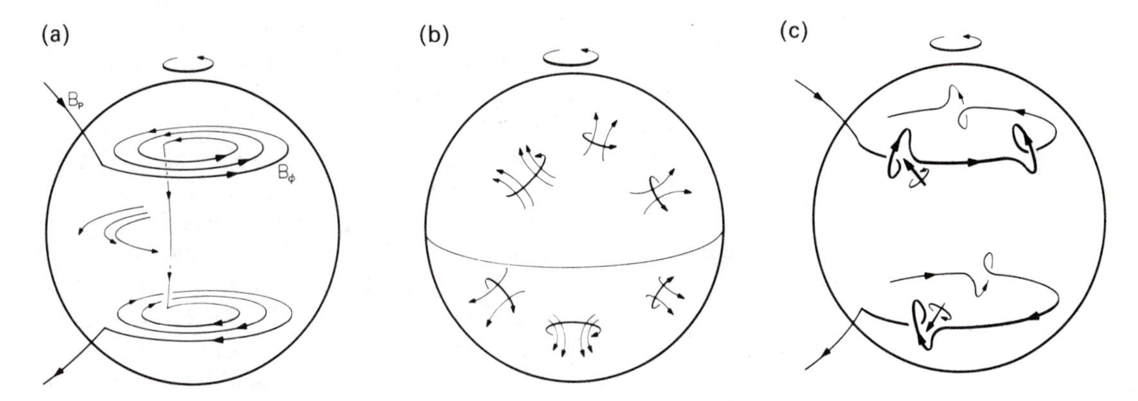

Figure 7.26. The Parker–Levy dynamo. (a) Rotation of an electrically conducting fluid outer core results in the stretching of the magnetic dipole field lines; they are wound into a toroidal field. Toroidal magnetic fields cannot be detected at the earth's surface because of the intervening insulating mantle. (b) The Coriolis force acting on the convecting fluid gives rise to spiralling motions (as in Fig. 7.25). The motion has opposite polarity in the northern and southern hemispheres. Such cyclonic motions are analogous to atmospheric cyclones and anticyclones. (c) The toroidal field lines shown in (a) are further deformed into loops by the spiralling motions shown in (b). These loops tend to rotate into longitudinal planes and so effectively regenerate the original dipole field. (From Levy 1976. Reproduced with permission from the *Annual Reviews of Earth and Planetary Sciences*, Vol. 5, © 1976 by Annual Reviews Inc.)

Although it can be shown that a convecting outer core can act as a dynamo which undergoes intermittent polarity reversals, exactly why these reversals occur is as yet unknown. They could be due to the random character of the irregular fluid convection and the nonlinear coupling of the fluid motion with the magnetic field or to changing boundary conditions. It seems from palaeomagnetic measurements that during a reversal the magnitude of the field diminished to about 10% of its normal value, and the path followed by the north magnetic pole was not a simple line of longitude from north to south but appears to have been a complex wandering from north to south. The length of time necessary to complete a reversal was short, approximately ten thousand years.

The convection in the outer core is not completely separate from the convection in the mantle. The two convection systems are coupled, even if only very weakly. This means that convection cells in the outer core tend to become aligned with convection cells in the mantle, with upwelling in the outer core occurring beneath hot regions in the mantle, and downwelling in the outer core occurring beneath cold regions in the mantle. A major driving force for convection in the mantle is the heat flux from the core. Estimates of this heat range from 10^{12} to 10^{13} W (compare this with the total heat loss from the earth in Table 7.3).

7.9.3 What Drives Convection in the Outer Core?

What drives or powers the dynamo? Convection in the outer core must occur because there is an inherent density instability there, with less dense material lying beneath denser material. Such a density instability could be due to heating (thermal buoyancy), or it could be due to chemical differences in the core. A number of heat sources could contribute to thermally driven convection in the outer core. One possible source is the radioactive isotopes ^{235}U and ^{40}K, which are present in the crust and the mantle, and possibly in the core. However, since there is not yet enough evidence to be certain, this heat source is speculative. If the inner core is cooling, solidifying and separating from the liquid outer core, then the latent heat of solidification (crystallization) could provide heat to help power the dynamo. Another possible source of heat is the 'primordial heat': that heat which resulted from the formation of the earth, which the core is probably slowly losing. It is probable that chemical differences could provide sufficient energy to power the dynamo; in this case, density instabilities arise if the outer core is crystallizing dense crystals. These crystals, being denser than the liquid iron alloy of the outer core, fall towards the inner core, and the less dense liquid rises.

Much of what has been suggested here about the core is speculative. In our quest to understand the workings of the core we are hampered by the very high temperatures and pressures which must be attained in experiments, by the thick insulating mantle which prevents complete measurement and understanding of the magnetic field and by not having any sample of core material. It has not been possible to construct a laboratory model in which the convective flow of a fluid generates a magnetic field. Materials available for laboratory experiments are not sufficiently good conductors for models to be of a reasonable size. Queen Elizabeth I, whose physician was William Gilbert, regarded Canada as Terra Meta Incognita. In this century the label should perhaps be transferred to the core.

7.10 Forces Acting on the Plates

7.10.1 Introduction

The cold upper thermal boundary layer which forms in models of thermal convection of the mantle is assumed to represent the lithosphere. The motion of these lithospheric plates relative to each other and the mantle is associated with a number of forces, some of which drive the motion and some of which resist the motion. Figure 7.27 shows the main driving and resistive forces. If the plates are moving at a constant velocity, then there must be a force balance: driving forces = resistive forces.

7.10.2 Driving Forces

The *ridge-push* force acts at the midocean ridges on the edges of the plates. It is made up of two parts: the pushing by the upwelling mantle material and the tendency of newly formed plate to slide down the sides of the ridge.

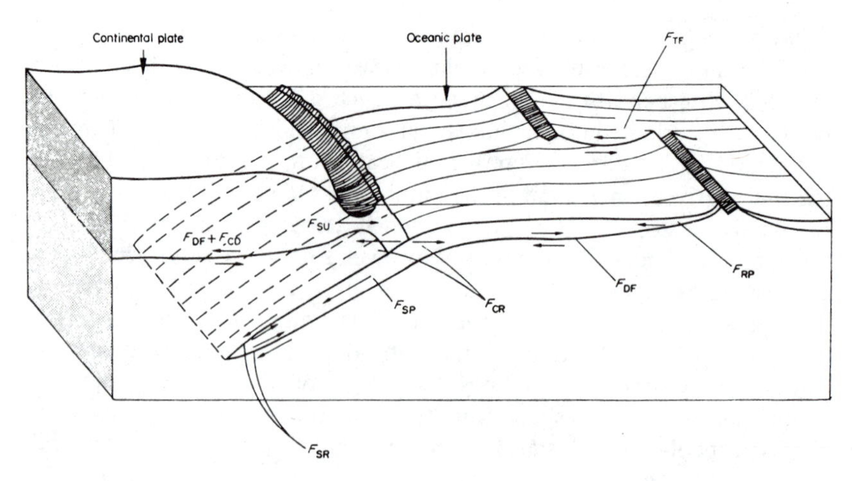

Figure 7.27. Possible forces acting on the lithospheric plates: F_{DF}, mantle-drag; F_{CD}, extra mantle-drag beneath continents; F_{RP}, ridge-push; F_{TF}, transform fault resistance; F_{SP}, slab-pull; F_{SR}, slab resistance on the descending slab as it penetrates the asthenosphere; F_{CR}, colliding resistance acting on the two plates in equal magnitude and opposite directions; F_{SU}, a suctional force that may pull the overriding plate towards the trench. (From Forsyth and Uyeda 1975.)

Of these two, the plate sliding contribution is approximately an order of magnitude smaller than the upwelling contribution.

An estimate of the total ridge-push per unit length of the ridge axis, F_{RP}, is

$$F_{RP} = ge(\rho_m - \rho_w)\left(\frac{L}{3} + \frac{e}{2}\right) \tag{7.104}$$

where e is the elevation of ridge axis above the cooled plate, ρ_m the density of the mantle at the base of the plate, ρ_w the density of sea water and L the plate thickness (Richter and McKenzie 1978). Equation 7.104 gives F_{RP} as $2 \times 10^{12}\,\mathrm{N\,m^{-1}}$ (N, newton) for the following values: L, 8.5×10^4 m; e, 3×10^3 m; ρ_w, $10^3\,\mathrm{kg\,m^{-3}}$; ρ_m, $3.3 \times 10^3\,\mathrm{kg\,m^{-3}}$; g, $9.8\,\mathrm{m\,s^{-2}}$.

The other main driving force is the negative buoyancy of the plate being subducted at a convergent plate boundary. This arises because the subducting plate is cooler and therefore more dense than the mantle into which it is descending. This force is frequently known as *slab-pull*. An estimate of the slab-pull force per unit length of subduction zone, $F_{SP}(z)$, acting at depth z and caused by the density contrast between the cool plate and the mantle is given by

$$F_{SP}(z) = \frac{8g\alpha\rho_m T_1 L^2\,\mathrm{Re}}{\pi^4}\left[\exp\left(-\frac{\pi^2 z}{2\mathrm{Re}\,L}\right) - \exp\left(-\frac{\pi^2 d}{2\mathrm{Re}\,L}\right)\right] \tag{7.105}$$

where z is the depth beneath the base of the plate, α the coefficient of thermal expansion, T_1 the temperature of the mantle, $d + L$ the thickness of the upper mantle, c_p the specific heat, v the rate at which the slab sinks and Re the *thermal Reynolds number*, given by

$$\mathrm{Re} = \frac{\rho_m c_p v L}{2k} \tag{7.106}$$

The total force available is F_{SP} evaluated at $z = 0$, $F_{SP}(0)$. $F_{SP}(z)$ decreases with depth into the mantle, until by $z = d$, it is zero, $F_{SP}(d) = 0$. Figure 7.28 shows the dependence of this total force on the consumption velocity v. An additional driving force in the sinking slab is caused by the elevation of the olivine–spinel phase change within the slab compared with the mantle, which occurs at depths of 300–400 km (Fig. 8.43). The magnitude of this force is about half that caused by the temperature difference between the slab and mantle. The total slab-pull force is estimated to be $10^{13}\,\mathrm{N\,m^{-1}}$ in magnitude, which is greater than the $10^{12}\,\mathrm{N\,m^{-1}}$ magnitude of the ridge-push force. Both slab-pull and ridge-push are caused by the density difference between hot and cold mantle; hot mantle can only rise because cold mantle sinks.

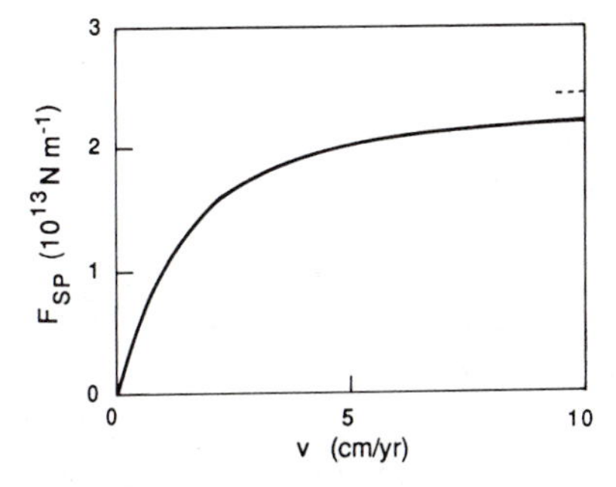

Figure 7.28. Total driving force available from the subducting slab: $F_{SP}(0)$ as a function of subduction velocity v. Horizontal dashed line shows the asymptotic limit as v increases to infinity. (After Richter and McKenzie 1978.)

7.10.3 Resistive Forces

Resistive forces occur locally at the ridge axis (occurrence of shallow earthquakes), along the bases of the plates as *mantle-drag* (assuming that the mantle flow is less than the plate velocity; if the reverse is true, then this would be a driving force), along transform faults (earthquake occurrence) and on the descending slab. Estimates of these forces suggest that the resistive force acting on the top of the sinking slab is greater than the shear force acting on its sides. The resistive force acting on the base of the plate is proportional to the area of the plate but is of the same magnitude as the resistive forces acting on the descending slab. These resistive forces cannot easily be estimated analytically and must be calculated numerically from the differential equations for flow in a fluid. The forces are proportional to the product of mantle viscosity η and plate velocity v and are about $10^{13}\,\mathrm{N\,m^{-1}}$ in magnitude (depending on the value of mantle viscosity assumed). For a 6000–10,000 km long plate, they would total $80\text{–}100\eta v$.

It is difficult to estimate the resistive forces acting on faults. However, the stress drop for large earthquakes is $10^{6}\,\mathrm{N\,m^{-2}}$ in magnitude. Earthquakes at ridge axes are shallow and small, and their contribution to the resistive forces can be ignored in comparison with the fluid dynamical drag forces. The resistive forces acting on transform faults are harder to evaluate. Earthquakes on transform faults are usually shallow, even though the plates can be perhaps 80 km in thickness (Sect. 8.5.3). It is probable that their total resistive contribution is of the same, or smaller, magnitude as the ridge-push driving force. Estimates of the resistive force acting on thrusts at the convergent plate boundaries, as indicated by earthquakes, give values of $10^{12}\,\mathrm{N\,m^{-1}}$. Again, this is less than the mantle-drag force.

To summarize, the main driving force is slab-pull, and the main resistive forces occur as drag along the base of the plate and on the descending slab.

7.10.4 What Drives the Plates?

The question of whether convection in the mantle drags the plates around or whether the forces acting at the edges of the plates drive the plates, which in turn drag the mantle, is complicated. From analysis of the driving and resistive forces, it is clear that the pull of the descending slab is a major factor in determining the mantle flow. If the only locations for ridges were above the rising limbs of convection cells, then in simple schemes (e.g., Fig. 7.18a) each plate should have one edge along a ridge and the other along a subduction zone. Clearly this is not the case; for example, the Antarctic and African plates are bounded almost entirely by ridges. Where could the return flow go? In these instances it seems reasonable to assume that the ridges form where the lithosphere is weakest and that mantle material rises from below to fill the gap. Plates which have a subducting portion move with higher velocities than those without (see Fig. 2.2), in agreement with the earlier estimate of the importance of the slab-pull driving force.

Thus, in conclusion, though there is still much that is not understood about flow in the mantle and the motion of lithospheric plates, the pull of the descending plate at convergent boundaries due to its decreased temperature seems to be a major factor in both the thermal modelling of the mantle flow and the mechanical models of the forces involved.

7.10.5 Did Plate Tectonics Operate in the Archaean?

A force-balancing model can be used to investigate the possibility of plate tectonics operating in the Archaean and to estimate probable plate velocities. The earth was probably much hotter then than now, with temperatures at the top of the asthenosphere of about 1700°C compared with 1300–1400°C today. The ridge-push force would then be about $4 \times 10^{11}\,\mathrm{N\,m^{-1}}$ and the slab-pull $8 \times 10^{12}\,\mathrm{N\,m^{-1}}$ (from Eqs. 7.104 and 7.105). Equating driving and resistive forces enables estimates of viscosity and plate velocity to be made. Plate tectonics could operate very effectively over an upper mantle with dynamic viscosity 10^{18} Pa s. Velocities could have been high, about $50\,\mathrm{cm\,yr^{-1}}$. High plate velocities may have been necessary in the Archaean to maintain a high heat loss through the oceans since, despite the higher temperatures and heat generation prevalent at that time, the thermal gradients determined from Archaean continental metamorphic rocks are relatively low. This topic is discussed further in Section 9.5.

7.11 Metamorphism: Geotherms in the Continental Crust

7.11.1 Introduction

Metamorphism is yet another process which is controlled by the transfer and generation of heat, and understanding of the thermal constraints on metamorphism is important in attempts to deduce past tectonic and thermal settings from the metamorphic evidence available to geologists today. Thus, in this section, considering heat to be transferred by conduction, we study the thermal evolution of some two-dimensional models of the crust.

Two-dimensional thermal models are conceptually easier to understand than one-dimensional models, but except for a few limited cases, simple analytical solutions to the differential equations are not possible. For the examples shown here, the two-dimensional heat conduction equation with erosion or sedimentation (Eq. 7.19) has been solved numerically by finite difference methods.

Three models are illustrated: a model of burial metamorphism, a model of intrusion and a model of overthrusting. These have been chosen to demonstrate a variety of possible metamorphic environments and by no means represent the possible range existing in the earth. No metamorphic rock is exposed at the surface without erosion or tectonic accident; but initially, we discuss hypothetical cases with no erosion or sedimentation.

7.11.2 Two-Dimensional Conductive Models

Burial Metamorphism A model of a typical burial terrain consists of a granitic country rock in which a rectangular trough of sediment has been deposited. Beneath both granite and sediment is a gneissic continental crust overlying the mantle (Fig. 7.29a). The initial temperature gradient in the country rock is the equilibrium gradient. Initially, we arbitrarily

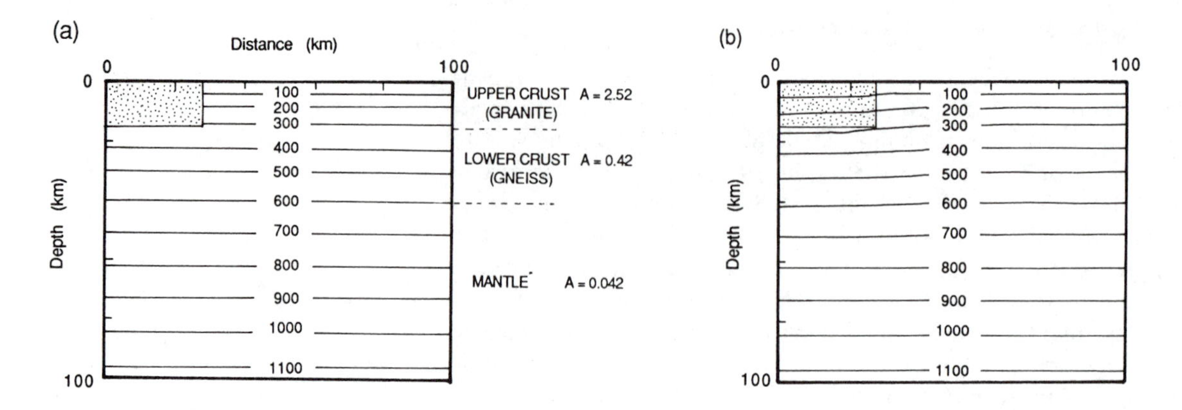

Figure 7.29. (a) Dimensions and physical parameters of the two-dimensional burial and intrusion models. Initial equilibrium temperature A in μW m^{-3}. Because the model is symmetrical about the left-hand edge, only half is shown. Stippled region denotes sediment or intrusion. (b) Burial model after 20 Ma. Sediment had an initial temperature of 100°C and has a radioactive heat generation of 0.84 μW m^{-3}. The sediment has very little effect on the crustal temperatures. (From Fowler and Nisbet 1982.)

assume the sediment to be at 100°C throughout and to have radioactive heat generation of 0.84 μW m^{-3}. A model such as this could be similar to a sedimentary trough formed on a continent above a subduction zone or to an Archaean greenstone belt filled with thick sediment and set in a granitic terrain. Figure 7.29b shows how the model evolves after 20 Ma. The sediment rapidly equilibrates towards the temperature of the surrounding country rock and is strongly influenced by the heat production in the surrounding granite. If the country rock had been mafic rather than granitic (see Table 7.1), the equilibrium gradient would have been lower and thus the final temperature at the base of the sediment would be some 50 to 100°C lower.

Intrusion Metamorphism Figure 7.29a also illustrates a family of models in which a large igneous body is intruded into the country rock. During the period immediately after the intrusion, hydrothermal convection cells occur around the hot body, especially if the intrusion is in a relatively wet country rock. These cells dominate the heat transfer process, so the simple conductive models considered here should be regarded only as rough guides to the real pattern of metamorphism. Convective heat transfer, which moves heat more quickly than conduction alone, tends to speed up the cooling of the intrusion. Furthermore, it tends to concentrate the metamorphic effects near the source of heat because that is where convection is most active. In granites, radioactive heat generation may prolong the action of convection cells. The presence of water also has profound effects on the mineralogical course of the metamorphism. Nevertheless, simple conductive models are useful for a general understanding of metamorphism around plutons.

Basic Intrusion The igneous body is assumed to intrude at 1100°C and to have radioactive heat generation of 0.42 μW m^{-3}; the latent heat of crystallization, 4.2×10^5 J kg^{-1}, is released over a 1 Ma cooling interval. Figure 7.30 shows the temperature field after 1 Ma and after 20 Ma. Contact metamorphism of the country rock is an important transient phenomenon, but if such basic intrusions are to be a major cause of *regional* as opposed to *local* metamorphism, the intrusions must form a large proportion of the total rock pile.

Granitic Intrusion The granite is assumed to intrude at 700°C and to have radioactive heat generation of $4.2\,\mu W\,m^{-3}$; the latent heat of crystallization, $4.2 \times 10^5\,J\,kg^{-1}$, is released over 2 Ma. Figure 7.31 shows the thermal evolution of this model. It is clear that there is less contact metamorphism than for the basic intrusion, but there is extensive deep-level or regional metamorphism. Indeed, massive lensoid granitic bodies are common in calc-alkaline mountain chains such as the Andes, and may be an important cause of regional metamorphism. Beneath and around the intrusion at depth, temperatures may be raised to such an extent that some local partial melting takes place in the country rock, which may cause further intrusion. If this occurs, then low $^{87}Sr/^{86}Sr$ initial ratios will be measured at the top of the original intrusion; younger, high $^{87}Sr/^{86}Sr$ initial ratios and young partial melting textures will be present at the base of the pluton, and below. Unravelling the history of the intrusion would be very difficult.

Overthrusts A wholly different type of metamorphism is produced by overthrusts. Figure 7.32a illustrates an example with a large overthrust slice of granite–gneiss material emplaced over mafic rock. Real parallels include subduction zones or an area such as the eastern Alps where a thick overthrust crystalline block has produced metamorphism below it. In this simple model, thrusting is assumed to be instantaneous. The most interesting feature of this model is that one thrusting event necessarily leads to two very distinct metamorphic events:

1. very early and rapid *retrogression* (cooling) in the upper block and *progression* (heating) beneath, followed by

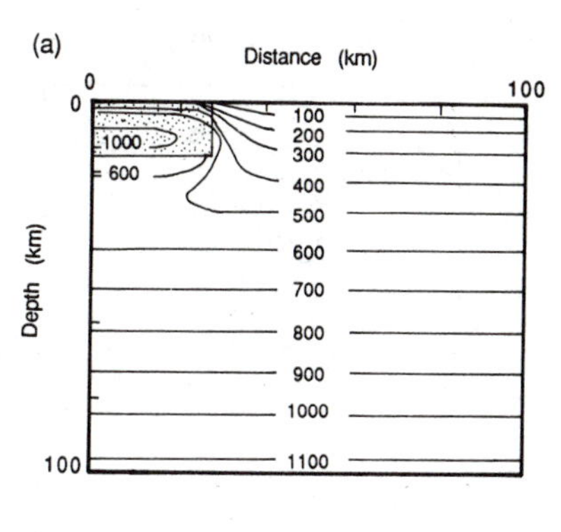

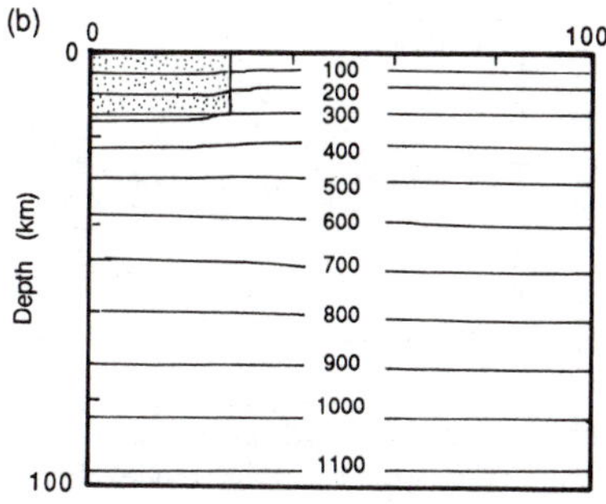

Figure 7.30. Stippled region denotes a large basalt body which intruded at 1100°C. (a) 1 Ma after intrusion the basalt body is cooling, and the country rock is being heated. (b) 20 Ma after intrusion the basalt body has solidified and cooled. (From Fowler and Nisbet 1982.)

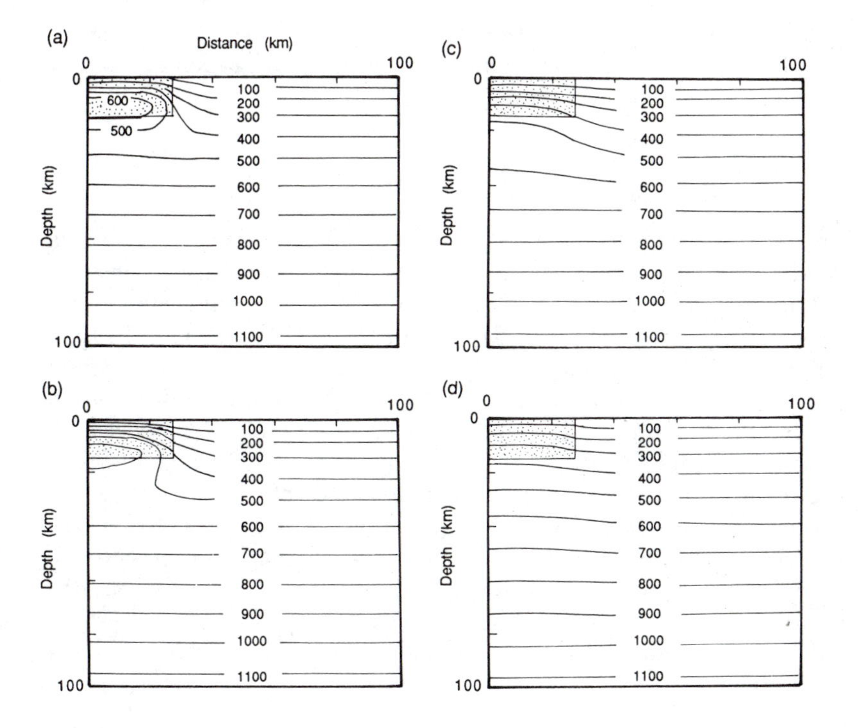

Figure 7.31. Stippled region denotes a large granite body which intruded at 700°C. (a) 1 Ma, (b) 2 Ma and (c) 5 Ma after intrusion, the granite is cooling and solidifying and heating the country rock. (d) 20 Ma after intrusion, the high heat generation of the granite means that temperatures in and around it are elevated. (From Fowler and Nisbet 1982.)

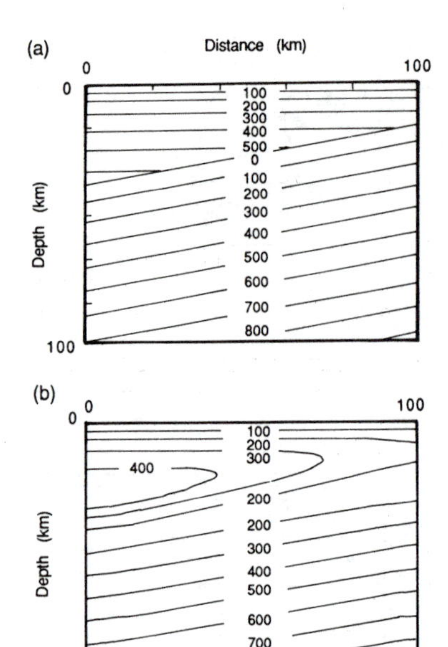

Figure 7.32. (a) Overthrust model initial temperatues. (b) Overthrust model after 1 Ma of cooling. (From Fowler and Nisbet 1982.)

2. slow progression (heating) throughout and finally partial melting and 'late' intrusion to high levels.

Immediately after thrusting (Fig. 7.32b), the hot base of the overthrust block thermally reequilibrates with the cool underthrust rocks beneath. This initial thermal reequilibration is very rapid, and inverted thermal gradients are probably very short-lived. The resulting geotherm (temperature–depth curve) in the thrust zone and below is of the order of a few degrees Celsius per kilometre. If the cool lower slab is rich in volatiles, rapid retrograde metamorphism takes place in the upper block. At the same time, equally rapid prograde high-pressure metamorphism occurs in the lower slab. At the deeper end of the overthrust block, local partial melting may take place if large amounts of volatiles move from the lower slab into the hot crust of the upper block. This can produce shallow granitic intrusions.

After this initial reequilibration comes a long period (perhaps 30–50 Ma, or more, depending on the size of the pile) in which a slow buildup of the geotherm, which in reality would be affected by uplift and erosion, takes place. This is a period of prograde metamorphism throughout the pile with the removal of water to higher levels during recrystallization. Finally, partial melting takes place at the base of the pile, and the radioactive heat production is redistributed until thermal stability is reached. This upwards redistribution of the radioactive elements is an episodic process. When a partial melt forms, it tends to be rich in potassium, thorium and uranium. Thus, over time this process of melting and intrusion effectively scours the deep crust of heat-producing elements and leads to their concentration in shallow-level intrusions (which would be recognized as 'late' or 'posttectonic') and in pegmatites. Eventually, after erosion, they tend to be concentrated in sediments and sea water. The net effect is a marked concentration of heat production in the upper crust; whatever the initial distribution of heat production, this leads to the stabilization of the rock pile to a nonmelting equilibrium.

7.11.3 Erosion and Deposition

Erosion and deposition are two processes which are able to change a geotherm rapidly. They are also interesting to geologists because no sedimentary rock can exist without deposition nor can any metamorphic rock become exposed at today's surface without erosion. Erosion represents the solar input to the geological machine, and the volcanism and deformation that provide the material to be eroded are driven from the interior. Figure 7.33a shows the effect of eroding the model rock column of Figure 7.3a at a rate of 1 km $\mathrm{Ma^{-1}}$ for 25 Ma. The shallow-level geotherm is raised to 50°C $\mathrm{km^{-1}}$ after the 25 Ma of erosion, after which it slowly relaxes towards the new equilibrium. If, instead of erosion, the model column is subject to sedimentation, then the shallow-level geotherm is depressed. Sedimentation at 0.5 km $\mathrm{Ma^{-1}}$ for 25 Ma depresses the shallow-level geotherm to about 23°C $\mathrm{km^{-1}}$ (Fig. 7.33b). After sedimentation ceases, the temperatures slowly relax towards the new equilibrium.

Alternatively, instead of considering the effects of erosion and deposition on the geotherm, we could trace the temperature history of a particular rock (e.g., the rock originally at 30 km depth). In the erosion example, the temperature of this rock decreases, dropping some 500 °C during erosion and a further 200 °C during the reequilibration. In the depositional example, the temperature of this rock is not affected much during the deposition, but it increases some 400°C during the subsequent slow reequilibration.

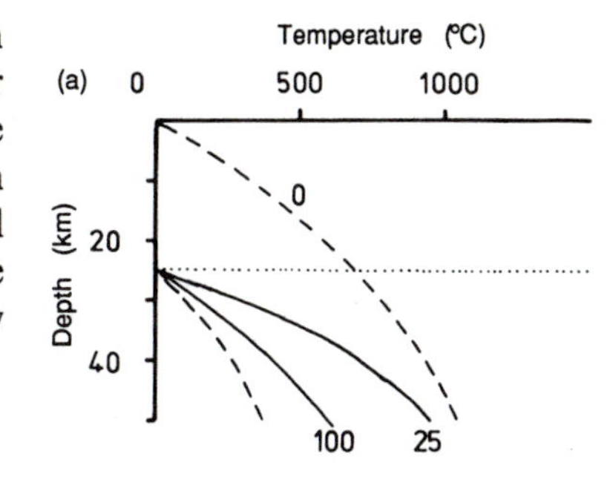

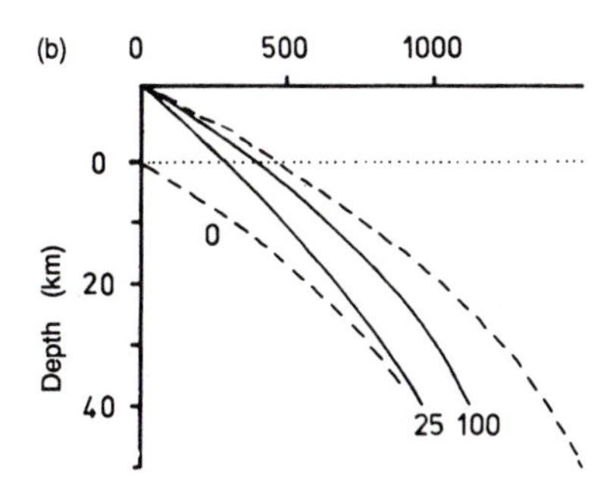

Figure 7.33. Geotherms for standard model shown in Figure 7.3a for two cases. (a) Erosion at 1 km/Ma for 25 Ma, and then no further erosion. Dotted horizontal line is the surface after erosion. (b) Deposition at 0.5 km/Ma for 25 Ma. Dotted horizontal line is the original surface before deposition. Curve 0, standard equilibrium geotherm of Figure 7.3a; curve 25, geotherm immediately after erosion/deposition for 25 Ma; curve 100, geotherm after 100 Ma; unlabelled dashed line, final equilibrium geotherm. (From Nisbet and Fowler 1982.)

7.11.4 Erosional Models: The Development of a Metamorphic Geotherm

Erosion is essential in the formation of a metamorphic belt since without it no metamorphic rocks would be exposed at the surface. However, as illustrated with the simple one-dimensional model, the process of erosion itself has a profound effect on the geotherm (Fig. 7.33) and on the pressure–temperature (P–T) path through which any metamorphic rock passes. The shape of the *metamorphic geotherm*, that is, the P–T trajectory inferred from the metamorphic rocks exposed at the surface, is also strongly influenced by erosion and, as is shown later, often does not at any time represent the actual geotherm.

The intrusion, burial and overthrust models of the previous sections are now subjected to erosion and deposition to illustrate the effects of these processes. Two erosion models are used for the burial and intrusion models: In space across the model, the first has strong erosion of the country rock and deposition on the trough, whereas the second has strong erosion of the trough and deposition on the country rock. For the overthrust, the erosion is taken to be constant across the model. All erosion rates decay with time.

Burial Model with Erosion In the first erosional model, deposition occurs in the centre of the trough at an initial rate of 1.1 km Ma^{-1} while erosion occurs at the edges of the model at an initial rate of 3.3 km Ma^{-1}. Figure 7.34 shows this burial model after 20 Ma when the sedimentary trough has been further covered by sediment and deep erosion has taken place in the country rock. Figure 7.34 also indicates the maximum temperatures attained during the 20 Ma period by the rocks finally exposed

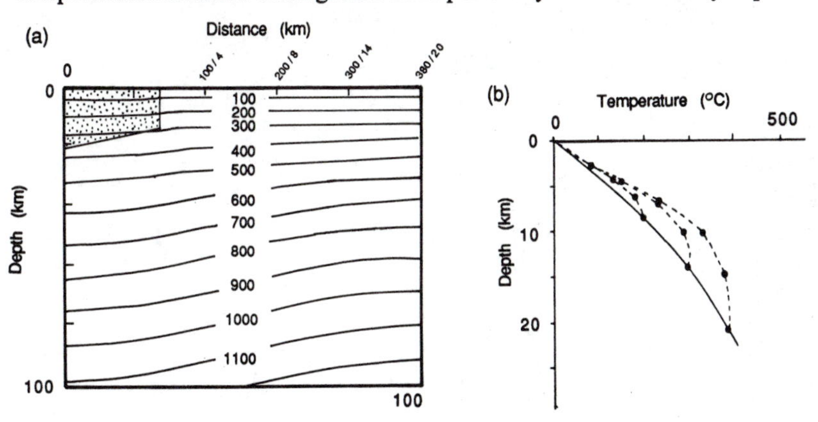

Figure 7.34. (a) Burial model after 20 Ma of erosion of the country rock and deposition on the sediment. Numbers along surface are the maximum temperature/depth (°C/km) attained by rocks exposed at the surface. (b) Temperature–depth paths followed by rocks originally at 8, 14 and 20 km depth and finally exposed at the surface. Solid line, initial equilibrium geotherm in the country rock; dots, temperature every 2 Ma from the start of erosion. (From Fowler and Nisbet 1982.)

at the surface. These are the rocks available to a field geologist. This maximum temperature is not necessarily preserved by the highest-grade minerals, but it is sufficient here to assume that the mineral assemblages exposed at the surface are those formed when the rock reaches its highest temperature. As it cools, the rock equilibrates to a lower temperature, and the mineral composition alters. However, the reaction kinetics become markedly slower as the rock cools, and there is a good chance of preserving some of the higher-grade minerals if erosion is fast enough.

In this simple model of burial metamorphism, the trough of buried sediment has no heating effect on the country rock. Therefore, the P–T curve (which can be plotted from the highest-grade minerals in the exposed rocks) is simply that of the initial, equilibrium thermal gradient in the rock. In this case, the metamorphic geotherm is identical to the equilibrium geotherm, no matter what the rate of erosion (provided that erosion is fast enough to 'quench' the mineral compositions at their highest temperatures). The *metamorphic facies series* (temperatures and pressures recorded in the rocks) produced by the event is that of a normal equilibrium geotherm in the country rock (facies series 1 in Fig. 7.35).

Intrusion Models with Erosion If the country rock is eroded, little metamorphic effect is seen even from these very large intrusions. With the exception of a localized contact zone (of the order of 5 km across) in both cases, the country rock gives a metamorphic facies series identical to the equilibrium facies series 1, and the net result is similar to that in Figure 7.34a. A real example of this could be the Great Dyke of Zimbabwe, which has only a restricted contact zone. Figure 7.34b again shows depth–temperature paths followed by individual points exposed on the surface after 20 Ma.

On the other hand, when the intrusion is eroded, and deposition takes place on the country rock, marked effects are seen because deep-seated rocks close to the intrusion are now being eroded. The resulting facies series is one of very low dP/dT (high temperatures at low pressures: facies

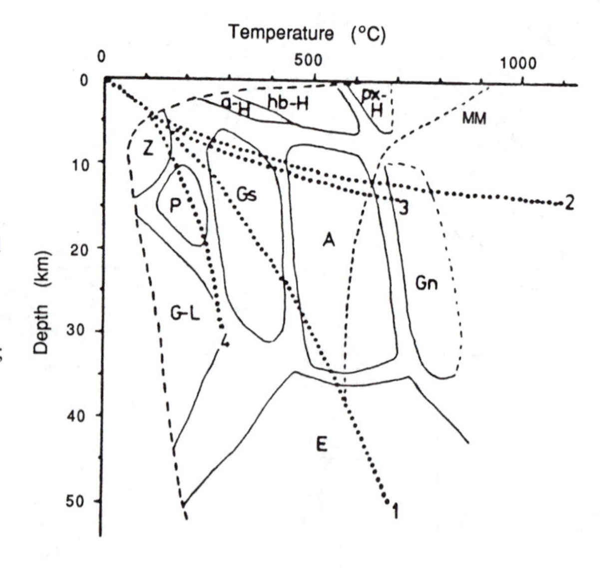

Figure 7.35. Pressure–temperature curves (metamorphic facies series) obtained from exposed rocks in the models after erosion. Dotted lines show facies series: curve 1, equilibrium series (Fig. 7.29), high grade; curve 2, basalt intrusion after erosion (Fig. 7.36); curve 3, granite intrusion after erosion (Fig. 7.36); and curve 4, overthrust model (Fig. 7.37), low grade. Facies fields: Z, zeolite; P, prehnite-pumpelleyite; G–L, glaucophane–lawsonite; Gs, greenschist; A, amphibolite; E, eclogite; Gn, granulite; a-H, hb-H and px-H, albite, hornblende and pyroxene hornfels. Dashed line MM, minimum melting for some metamorphic mineral assemblages. (From Fowler and Nisbet 1982.)

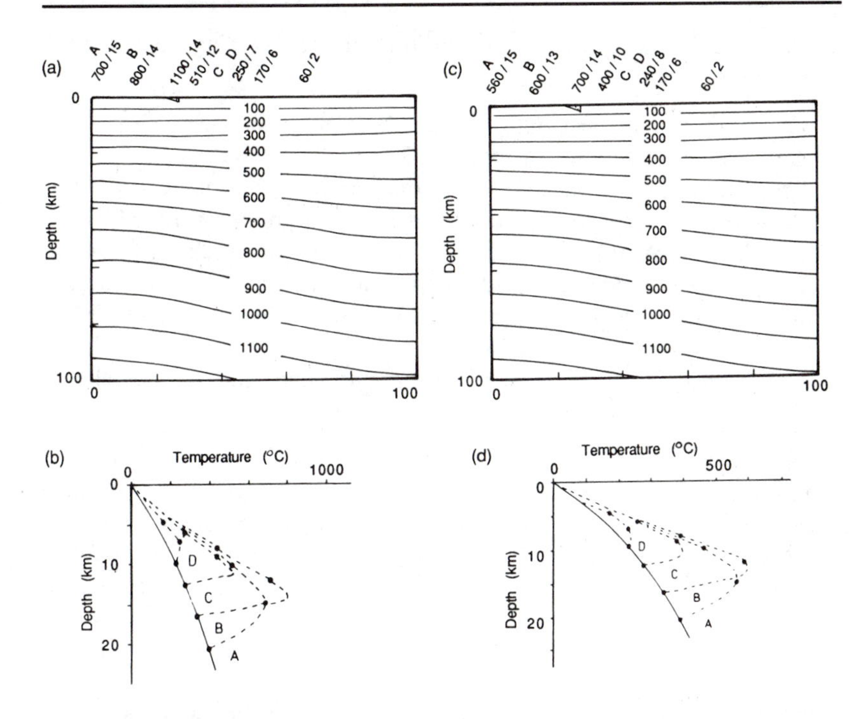

Figure 7.36. (a) Basalt intrusion model after 20 Ma of erosion of intrusion and country rock. (b) Temperature–depth paths for rocks finally exposed at the surface in model (a). (c) Granite intrusion after 20 Ma of erosion of intrusion and country rock. (d) Temperature–depth paths for rocks finally exposed at the surface in model (c). Notation as in Figure 7.34. (From Fowler and Nisbet 1982.)

series 2 for basalt and facies series 3 for granite, as shown in Fig. 7.35). It can be seen from Figure 7.36 that, although the original intrusions have been almost completely eroded, the metamorphic imprint of intrusion and erosion is widespread and lasting. The small body of granite present today (Fig. 7.36c) appears to have had a major metamorphic effect. Figures 7.36b,d show some depth–temperature paths for points exposed on the surface after 20 Ma.

Overthrust Model with Erosion Erosion of the overthrust model provides clues to the origin of paired metamorphic belts. For thinner thrust sheets, the results would be similar to those illustrated here but with lower temperatures and pressures.

Initially, only the overthrust unit is exposed at the surface. Since the thrusting event has a general cooling effect on the upper block, the highest temperature any part of the upper block experiences is its initial temperature prior to thrusting. Rocks from the upper block probably retain relict minerals from their previous high-grade environment. The extent of the mineral reequilibration is probably dependent on the availability of volatiles rising from the underthrust sheet into the hot base of the upper sheet. If local partial melting occurs (e.g., in the upper sheet on the extreme left of the model), prograde effects are seen.

Simultaneously with the metamorphism of the upper block, the underthrust block experiences prograde metamorphism. As erosion takes place, the deep level rocks are lifted towards the surface. This has a cooling effect which progressively halts and reverses the rise in temperature of the rocks in the underthrust block. The maximum temperatures attained in this underthrust block are controlled by the time of initiation and the rate of erosion, and also by any shear heating in the upper slab. Eventually the thrust surface is exposed by the erosion, and the following metamorphic events are seen:

1. An early (prior to thrusting) high-grade event in the overthrust block, shown as facies series 1 in Figure 7.35.
2. A post-thrusting retrograde event in the remaining part of the overthrust block which overprints (1). In real rocks this event is controlled by the introduction of large amounts of volatiles from the underthrust rocks. The overthrust block is rapidly cooled and hydrated, and the degree of overprinting depends on reaction kinetics (exponentially related to temperature) and the availability of volatiles.
3. A high-pressure low-temperature event in the underthrust rocks, shown as facies series 4 in Figure 7.35.

An overthrust of the dimensions modelled here produces twin metamorphic belts: one high grade (facies series 1) and one low grade (facies series 4). Much smaller overthrusts (e.g., at shallow levels in mountain belts) would be qualitatively similar, but the metamorphic effects might be restricted by kinetic factors or be small or difficult to distinguish.

7.11.5 Dating and Metamorphism

The age of a radiometric system such as a mineral generally depends on the way in which the daughter product became sealed into the system as it cooled. In a typical system a blocking temperature exists: This is the temperature below which the system can be thought to be closed (see Sect. 6.2). Different minerals, dating techniques and rates of cooling all produce different blocking temperatures; thus, we have a powerful tool for working out the thermal history of a mountain belt.

For moderate cooling rates, suggested closure temperatures are 70–110°C for apatite fission tracks, about 175–225°C for zircon fission tracks, 280 ± 40°C for biotite K/Ar, and 530 ± 40°C for hornblende K/Ar (see Table 6.4 and Fig. 6.7). In many cases, the Rb/Sr whole-rock ages probably reflect the original age of the rock. Consider first the basaltic and granitic intrusions illustrated in Figures 7.30 and 7.31. Underneath the intrusion, the radiometric clocks of minerals which close around 500°C would start recording about 1–2 Ma after the start of cooling. Under the basalt intrusion, which has little internal heat generation, minerals which close at around 300°C would not start recording until 20 Ma. In the case of the granite, a mineral with a 300°C blocking temperature would not close at all. The depth–temperature paths in Figure 7.36 show the great spread in radiometric ages that would be obtained from various mineral 'clocks'. Furthermore, the blocking temperatures are dependent on the rate of cooling, which is different in each case.

Figure 7.37 shows the effect of using these various dating techniques across the overthrust model's 30 Ma erosion surface; the dates shown are the dates that would be measured by a geologist working on this surface. It can be seen that these dates provide a useful tool for studying the cooling and erosional history of the pile. In many natural examples, there can be profound differences in closure ages of minerals which were initially produced by the same tectonic event.

MINERAL	MINERAL COOLING AGES (Ma)						
ap	9	10	10	13	13	12	11
zr	16	16	16	17	17	16	16
bi	20	20	21	23	-	-	-
hb	28	29	-	-	-	-	-
T_m (C)	562	562	562	272	256	248	245
D_t (km)	36	36	36	24	21	19	19

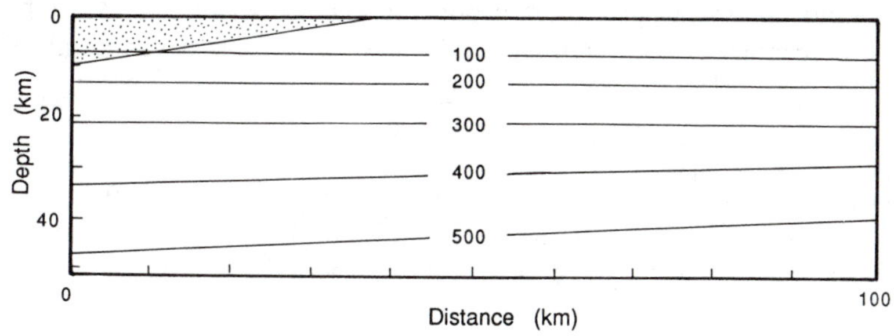

Figure 7.37. Overthrust model after 30 Ma of erosion. Stipple, remnants of overthrust block: T_m, maximum temperature attained by rock finally at the surface; D_t, depth at which T_m was reached. Figure also shows time of closure (Ma before present) of various radiometric systems; ap, apatite fission track: zr, zircon fission track (maximum age based on ~175°C closure); bi, biotite K/Ar; hb, hornblende K/Ar. (From Fowler and Nisbet 1982.)

Implications To interpret a metamorphic terrain fully, it is not sufficient to know the pressure–temperature conditions undergone by the rocks: A full interpretation of the thermal history of the rocks also involves studying the erosional and radiometric history of the terrain. These thermal models demonstrate that a general knowledge of the stratigraphy of an area is essential before its metamorphic history can be properly unravelled.

Convective movement of fluid has not been discussed, but it can be a major factor in heat transport around plutons and during dehydration of overthrust terrains. As a general rule, fluid movement speeds thermal reequilibration and reduces to some degree the extent of aureoles around intrusions; at the same time it promotes metamorphic reactions and has a major impact on the thermodynamics of reequilibration.

PROBLEMS

1. Estimate Table 7.2 for 4000 Ma and 4500 Ma. Discuss the implications of the relative heat-generation values in this table, particularly with respect to the Archaean earth.
2. Calculate the phase difference between the daily and annual surface temperature variation that would be measured at depths of 2 m and 5 m in a sandstone.
3. Taking T_0 as 40°C for the annual variation in surface temperature, calculate the depth at which the variation is 5°C. What is the phase difference (in weeks) between the surface and this depth?
4. Calculate an equilibrium geotherm for $0 \leqslant z \leqslant d$ from the one-dimensional heat flow equation, given these boundary conditions:

   (i) $T = 0$ at $z = 0$ and

   (ii) $T = T_d$ at $z = d$.

   Assume that there is no internal heat generation.
5. Calculate an equilibrium geotherm from the one-dimensional heat flow equation given these boundary conditions:

(i) $\partial T/\partial z = 30°\text{C km}^{-1}$ at $z = 0$ km and

(ii) $T = 700°\text{C}$ at $z = 35$ km.

Assume that the internal heat generation is $1\ \mu\text{W m}^{-3}$ and the thermal conductivity is $3\ \text{W m}^{-1}\ °\text{C}^{-1}$.

6. On missions to Venus the surface temperature was measured as 740 K, and at three sites heat-producing elements were measured (in percentage of total volume) as follows (ppm, parts per million):

	Venera 8	Venera 9	Venera 10
K	$0.47 \pm 0.08\%$	$0.30 \pm 0.16\%$	$4 \pm 1.2\%$
U	0.60 ± 0.16 ppm	0.46 ± 0.26 ppm	2.2 ± 0.2 ppm
Th	3.65 ± 0.42 ppm	0.70 ± 0.34 ppm	6.5 ± 2 ppm

The density of the Venusian crust can be taken from a measurement by Venera 9 as $2.8 \times 10^3\ \text{kg m}^{-3}$. Calculate heat generation in $\mu\text{W m}^{-3}$ at each site. (From Nisbet and Fowler 1982.)

7. Using the one-dimensional equilibrium heat conduction equation, calculate and plot the Venus geotherms (Aphroditotherms) of Problem 6 down to 50 km depth at each site. Assume that the conductivity is $2.5\ \text{W m}^{-1}\ °\text{C}^{-1}$ (a typical value for silicates) and that at a depth of 50 km the heat flow from the mantle and deep lithosphere of Venus is $21 \times 10^{-3}\ \text{W m}^{-2}$. What have you assumed in making this calculation? What do these Aphroditotherms suggest about the internal structure of the planet?
8. Calculate the geotherms for the models shown in Figure 7.14. Discuss the reason for the difference at depth between these geotherms and the geotherm shown as a solid line in Figure 7.14.
9. Calculate an equilibrium geotherm for the model Archaean crust shown in Figure 7.4. Discuss your estimates.
10. To what depth are temperatures in the earth affected by ice ages? (Use thermal conductivity $2.5\ \text{W m}^{-1}\ °\text{C}^{-1}$ and specific heat $10^3\ \text{J kg}^{-1}\ °\text{C}^{-1}$.)
11. Calculate the equilibrium geotherm for a two-layered crust. The upper layer, 10 km thick, has an internal heat generation of $2.5\ \mu\text{W m}^{-3}$, and the lower layer, 25 km thick, has no internal heat generation. Assume that the heat flow at the base of the crust is $20 \times 10^{-3}\ \text{W m}^{-2}$ and that the thermal conductivity is $2.5\ \text{W m}^{-1}\ °\text{C}^{-1}$.
12. Repeat the calculation of Problem 11 when the upper layer has no internal heat generation and the lower layer has internal heat generation of $1\ \mu\text{W m}^{-3}$. Comment on the effect that the distribution of heat-generating elements has on geotherms.
13. Calculate geotherms for a layered continental crust and comment on the significance of your results for the following cases:
    (a) A 10 km thick upper layer with heat generation of $2.5\ \mu\text{W m}^{-3}$ overlying a 30 km thick layer with heat generation of $0.4\ \mu\text{W m}^{-3}$.
    (b) A 30 km thick upper layer with heat generation of $0.4\ \mu\text{W m}^{-3}$ overlying a 10 km thick layer with heat generation of $2.5\ \mu\text{W m}^{-3}$.

For both cases, assume a surface temperature of zero, heat flow from the mantle of $20 \times 10^{-3}\,\mathrm{W\,m^{-2}}$ and thermal conductivity of $2.5\,\mathrm{W\,m^{-1}\,°C^{-1}}$.

14. A 1 m wide dyke with a temperature of 1050°C is intruded into country rock at a temperature of 50°C.
    (a) Calculate how long the dyke will take to solidify.
    (b) After two weeks what will the temperature of the dyke be?
    (Assume a diffusivity of $10^{-5}\,\mathrm{m^{-2}\,s^{-1}}$ and a solidus temperature of 800°C.)
15. Volcanic flood basalts can be several kilometres thick and extend over very large areas (the Karoo basalt in southern Africa is one example). A 2 km thick basalt is erupted at 1200°C. If the solidus temperature is 900°C, estimate the time required for the basalt to solidify. If this basalt is later eroded and the underlying rocks exposed, indicate how far you would expect the metamorphism to extend from the basalt. State all your assumptions in answering this question.
16. (a) Calculate the difference in depth of the sea bed at the intersection of a midocean ridge and a transform fault. Assume that the ridge is spreading at $5\,\mathrm{cm\,yr^{-1}}$ and that the ridge axis is offset 200 km by the transform fault.
    (b) Calculate the difference in depth on either side of the same fault at 1000 km from the ridge axis and at 3000 km from the ridge axis. (See Sect. 8.5 for information on transform faults.)
17. Calculate the 60 Ma geotherm in the oceanic lithosphere for the simple model of Section 7.5.2. What is the thickness of the 60 Ma old lithosphere? Use an asthenosphere temperature of 1300°C and assume a temperature of 1150°C for the base of the lithosphere
18. Assume that the earth is solid and that all heat transfer is by conduction. What value of internal heat generation distributed uniformly throughout the earth is necessary to account for the earth's mean surface heat flow of $82 \times 10^{-3}\,\mathrm{W\,m^{-2}}$? How does this value compare with the actual estimated values for the crust and mantle?
19. Calculate the rate at which heat is produced in (a) the crust and (b) the mantle. Assume that the crust is 10 km thick and that the volumetric heat generation rates are $1.5 \times 10^{-6}\,\mathrm{W\,m^{-3}}$ in the crust and $1.5 \times 10^{-8}\,\mathrm{W\,m^{-3}}$ in the mantle.
20. Calculate the steady-state surface heat flow for a model solid earth with constant thermal properties: k, $4\,\mathrm{W\,m^{-1}\,°C^{-1}}$; A, $2 \times 10^{-8}\,\mathrm{W\,m^{-3}}$.
21. It takes about 4 min to boil a 60 gm hen's egg to make it edible for most people. For how long would it be advisable to boil an ostrich egg which weighs about 1.4 kg? (From Thompson 1987.)
22. (a) Calculate the conductive characteristic time for the whole earth.
    (b) Calculate the thickness of the layer that has a characteristic time of 4500 Ma.
    (c) Comment on your answers to (a) and (b).
23. (a) A sphere has radius r and uniform density ρ. What is the gravitational energy released by bringing material from infinitely far away and adding a spherical shell, density thickness δr, to the original shell?
    (b) By integrating the expression for the gravitational energy over r

from 0 to R, calculate the gravitational energy released in assembling a sphere of density ρ and radius R.

(c) Use the result of (b) to estimate the gravitational energy released as a result of the accretion of the earth.

(d) Assume that all the energy calculated in (c) became heat and estimate the rise in temperature of the primeval earth. Comment on your answer.

BIBLIOGRAPHY

Anderson, D. L. 1986. Properties of iron at the earth's core conditions. *Geophys. J. R. Astr. Soc.*, *84*, 561–79.

Bloxham, J., and Gubbins, D. 1985. The secular variation of earth's magnetic field. *Nature*, *317*, 777–81.

1987. Thermal core–mantle interactions. *Nature*, *325*, 511–13.

Bott, M. H. P. 1982. *The interior of the earth: Its structure, composition and evolution.* Elsevier, Amsterdam.

Brett, R. 1976. The current status of speculations on the composition of the core of the earth. *Rev. Geophys. Space Phys.*, *14*, 375–83.

Bullard, E. C. 1949. Electromagnetic induction in a rotating sphere. *Proc. Roy. Soc. Lond. A*, *199*, 413.

1949. The magnetic field within the earth. *Proc. Roy. Soc. Lond. A*, *197*, 433–53.

1972. The earth's magnetic field and its origin. In I. G. Gass, P. J. Smith and R. C. L. Wilson eds., *Understanding the earth*, 2nd ed. Artemis Press, Sussex, 71–9.

Bullard, E. C., and Gellman, H. 1954. Homogeneous dynamos and terrestrial magnetism. *Phil. Trans. Roy. Soc. Lond. A*, *247*, 213–78.

Busse, F. H. 1975. A model of the geodynamo. *Geophys. J. R. Astr. Soc.*, *42*, 437–59.

1983. Recent developments in the dynamo theory of planetary magnetism. *Ann. Rev. Earth Planet. Sci.*, *11*, 241–68.

1970. Thermal instabilities in rapidly rotating systems. *J. Fluid Mech.*, *44*, 441–60.

Busse, F. H., and Carrigan, C. R. 1976. Laboratory simulation of thermal convection in rotating planets and stars. *Science*, *191*, 31–8.

Carrigan, C. R., and Busse, F. H. 1983. An experimental and theoretical investigation of the onset of convection in rotating spherical shells. *J. Fluid Mech.*, *126*, 287–305.

Carrigan, C. R., and Gubbins, D. 1979. The source of the earth's magnetic field. *Sci. Am.*, *240*, 2, 118–130.

Carslaw, H. S., and Jaeger, J. C. 1959. *Conduction of heat in solids*, 2nd ed. Oxford Univ. Press, New York.

Célérier, B. 1988. Paleobathymetry and Geodynamic Models for subsidence. *Palaios*, *3*, 454–63.

Clark, S. P. 1966. Thermal conductivity. *In* S. P. Clark, ed., *Handbook of physical constants*, Vol. 97 of Geol. Soc. Am. Mem., 459–82.

Clark, S. P., Peterman, Z. E., and Heir, K. S., 1966. Abundances of uranium, thorium and potassium. *In* S. P. Clark, ed., *Handbook of physical constants*, Vol. 97 of Geol. Soc. Am. Mem., 521–41.

Davis, E. E., and Lister, C. R. B. 1974. Fundamentals of ridge crest topography. *Earth Planet. Sci. Lett.*, *21*, 405–13.

Elsasser, W. M. 1946. Induction effects in terrestrial magnetism. Part I: Theory. *Phys. Rev.*, *69*, 106–16.

1946. Induction effects in terrestrial magnetism. Part II: The secular variation. *Phys. Rev.*, *70*, 202–12.

1947. Induction effects in terrestrial magnetism. Part III: Electric Modes. *Phys. Rev.*, *72*, 821–33.

1950. The earth's interior and geomagnetism. *Rev. Mod. Phys.*, *22*, 1–35.

England, P. C., and Richardson, S. W. 1977. The influence of erosion upon the mineral facies of rocks from different tectonic environments. *J. Geol. Soc. Lond.*, *134*, 201–13.

Forsyth, D. W., and Uyeda, S. 1975. On the relative importance of the driving forces of plate motion. *Geophys. J. R. Astr. Soc.*, *43*, 163–200.

Fowler, C. M. R., and Nisbet, E. G. 1982. The thermal background to metamorphism II. Simple two-dimensional conductive models. *Geoscience Canada*, *9*, 208–14.

Giardini, D., Xiang-Dong, Li, and Woodhouse, J. M. 1987. Three-dimensional structure of the earth from splitting in free-oscillation spectra. *Nature*, *325*, 405–11.

Gubbins, D. 1974. Theories of the geomagnetic and solar dynamos. *Rev. Geophys. Space Phys.*, *12*, 137–54.

1976. Observational constraints on the generation process of earth's magnetic field. *Geophys. J. R. Astr. Soc.*, *46*, 19–39.

1984. The earth's magnetic field. *Contemp. Phys.*, *25*, 269–90.

1987. Mapping the mantle and core. *Nature*, *325*, 392–3.

Gubbins, D., and Bloxham, J. 1986. Morphology of the geomagnetic field: Implications for the geodynamo and core–mantle coupling. *EOS Trans. Am. Geophys. Un.*, *67*, 916.

1987. Morphology of the geomagnetic field and its implications for the geodynamo. *Nature*, *325*, 509–11.

Gubbins, D., and Masters, T. G. 1979. Driving mechanisms for the earth's dynamo. *Adv. Geophys.*, *21*, 1–50.

Gubbins, D., Masters, T. G., and Jacobs, J. A. 1979. Thermal evolution of the earth's core. *Geophys. J. R. Astr. Soc.*, *59*, 57–99.

Hager, B. H., Clayton, R. W., Richards, M. A., Cormer, R. P., and Dziewonski, A. M. 1985. Lower mantle heterogeneity, dynamic topography and the geoid. *Nature*, *313*, 541–5.

Harrison, T. M. 1987. Comment on 'Kelvin and the age of the earth'. *J. Geology*, *95*, 725–7.

Hayes, D. E. 1988. Age–depth relationships and depth anomalies in the southeast Indian Ocean and South Atlantic Ocean. *J. Geophys. Res.*, *93*, 2937–54.

Hewitt, J. M., McKenzie, D. P., and Weiss, N. O. 1975. Dissipative heating in convective fluids. *J. Fluid Mech.*, *68*, 721–38.

1980. Large aspect ratio cells in two-dimensional thermal convection. *Earth Planet. Sci. Lett.*, *51*, 370–80.

Hoffman, N. R. A., and McKenzie, D. P. 1985. The destruction of geochemical heterogeneities by differential fluid motions during mantle convection. *Geophys. J. R. Astr. Soc.*, *82*, 163–206.

Houseman, G. 1983. Large aspect ratio convection cells in the upper mantle. *Geophys. J. R. Astr. Soc.*, *75*, 309–34.

Jacobs, J. A. 1984. *Reversals of the earth's magnetic field.* Adam Hilger, Bristol.

1987. *The earth's core*, 2nd ed. Academic, London.

Jeanloz, R. 1983. The earth's core. *Sci. Am.*, *249*, 3, 13–24.

1988. High-pressure experiments and the Earth's deep interior. *Physics Today*, *41*, 1, S44–5.

Jeanloz, R., and Richter, F. M. 1979. Convection, composition and the thermal state of the lower mantle. *J. Geophys. Res.*, *84*, 5497–504.

Jessop, A. M., and Lewis, T. 1978. Heat flow and heat generation in the Superior Province of the Canadian Shield. *Tectonophys.*, *50*, 55–77.

Knittle, E., and Jeanloz, R. 1986. High-pressure metallization of FeO and implications for the earth's core. *Geophys. Res. Lett.*, *13*, 1541–4.

Knittle, E., Jeanloz, R., Mitchell, A. C., and Nellis, W. J. 1986. Metallization of

$Fe_{0.94}O$ at elevated pressures and temperatures observed by shock-wave electrical resistivity measurements. *Solid State Commun., 59*, 513–15.

Lachenbruch, A. H. 1970. Crustal temperature and heat production; implications of the linear heat flow relation. *J. Geophys. Res., 75*, 3291–300.

Larmor, J. 1919. How could a rotating body such as the sun become a magnet? *Rep. Br. Assoc. Advmt. Sci.*, 159.

Levy, E. H. 1972. Kinematic reversal schemes for the geomagnetic dipole. *Astrophys. J., 171*, 635–42.

1972. On the state of the geomagnetic field and its reversals. *Astrophys. J., 175*, 573–81.

1976. Generation of planetary magnetic fields. *Ann. Rev. Earth Planet. Sci., 4*, 159–85.

1979. Dynamo magnetic field generation. *Rev. Geophys. Space Phys., 17*, 277–81.

Loper, D. E. 1984. Structure of the core and lower mantle. *Adv. Geophys., 26*, 1–34.

McKenzie, D. P. 1967. Some remarks on heat flow and gravity anomalies. *J. Geophys. Res., 72*, 6261–73.

1969. Speculations on the consequences and causes of plate motions. *Geophys. J. R. Astr. Soc., 18*, 1–32.

1983. The earth's mantle. *Sci. Am., 249*, 3, 66–113.

McKenzie, D., and Bickle, M. J. 1988. The volume and composition of melt generated by extension of the lithosphere. *J. Petrology, 29*, 625–79.

McKenzie, D. P., and O'Nions, R. K. 1982. Mantle reservoirs and ocean island basalts. *Nature, 301*, 229–31.

McKenzie, D. P., and Richter, F. 1976. Convection currents in the earth's mantle. *Sci. Am., 235*, 5, 72–89.

1981. Parametrized thermal convection in a layered region and the thermal history of the earth. *J. Geophys. Res., 86*, 11667–80.

McKenzie, D. P., Roberts, J. M., and Weiss, N. O. 1974. Convection in the earth's mantle: Towards a numerical simulation. *J. Fluid Mech., 62*, 465–538.

Melchior, P. 1986. *The physics of the earth's core: An introduction.* Pergamon, Oxford.

Moffatt, H. K. 1978. *Magnetic field generation in electrically conducting fluids.* Cambridge Univ. Press.

Nisbet, E. G., and Fowler, C. M. R. 1982. The thermal background to metamorphism. I. Simple one-dimensional conductive models. *Geoscience Canada, 9*, 161–4.

Oldenberg, D. W. 1975. A physical model for the creation of the lithosphere. *Geophys. J. R. Astr. Soc., 43*, 425–52.

O'Nions, R. K., and Carter, S. R. 1981. Upper mantle geochemistry. *In* C. Emiliani, ed., *The sea*, Vol. 7, 49–71.

Parker, E. N. 1955. Hydromagnetic dynamo models. *Astrophys. J., 122*, 293–314.

1969. The occasional reversal of the geomagnetic field. *Astrophys. J., 158*, 815–27.

1979. *Cosmical magnetic fields.* Oxford Univ. Press, New York.

Parker, R. L., and Oldenberg, D. W. 1973. Thermal models of mid-ocean ridges. *Nature Phys. Sci., 242*, 137–9.

Parsons, B., and McKenzie, D. P. 1978. Mantle convection and the thermal structure of the plates. *J. Geophys. Res., 83*, 4485–96.

Parsons, B., and Sclater, J. G. 1977. An analysis of the variation of ocean floor bathymetry and heat flow with age. *J. Geophys. Res., 32*, 803–27.

Richter, F. M. 1986. Kelvin and the age of the earth. *J. Geology, 94*, 395–401.

Richter, F. M., and McKenzie, D. P. 1978. Simple plate models of mantle convection. *J. Geophys., 44*, 441–71.

Richter, F. M., and Parsons, B. 1975. On the interaction of two scales of convection in the mantle. *J. Geophys. Res., 80*, 2529–41.

Ringwood, A. E. 1977. Composition of the earth and implications for origin of the earth. *Geochem. J., 11*, 111–35.

1979. Composition and origin of the earth. *In* M. W. McElhinny, ed., *The earth: Its origin, structure and evolution.* Academic, New York, 1–58.

Rochester, M. G., and Crossley, D. J. 1987. Earth's third ocean: The liquid core. *EOS Trans. Am. Geophys. Un., 67*, 481.

Roy, R. F., Decker, E. R., Blackwell, D. D., and Birch, F. 1968. Heat flow in the United States. *J. Geophys. Res., 73*, 5207–21.

Runcorn, S. K., Creer, K. M., and Jacobs, J. A., eds. 1982. The earth's core: Its structure, evolution and magnetic field. *Phil. Trans. Roy. Soc. Lond. A, 306*, 1–289.

Sclater, J. G., Jaupart, C., and Galson, D. 1981. The heat flow through oceanic and continental crust and the heat loss of the earth. *Rev. Geophys. Space Phys., 18*, 269–311.

Sclater, J. G., Parsons, B., and Jaupart, C. 1981. Oceans and continents: similarities and differences in the mechanisms of heat loss. *J. Geophys. Res., 86*, 11535–52.

Silver, P. G., Carlson, R. W., and Olson, P. 1988. Deep slabs, geochemical heterogeneity and the large-scale structure of mantle convection: Investigation of an enduring paradox. *Ann. Rev. Earth Planet. Sci., 16*, 477–541.

Stacey, F. D., and Loper, D. E. 1984. Thermal histories of the core and mantle. *Phys. Earth Planet. Int., 36*, 99–115.

Stephenson, D. J. 1981. Models of the earth's core. *Science, 214*, 611–19.

Thompson, N., ed. 1987. *Thinking like a physicist: Physics problems for undergraduates.* Adam Hilger, Bristol.

White, R. S. 1988. The earth's crust and lithosphere. *J. Pet.* in Menzies, M. A. and Cox, K. eds., *Oceanic and Continental Lithosphere: Similarities and differences, J. Pet.*, Special Lithosphere Issue, 1–10.

Williams, Q., Jeanloz, R., Bass, J., Svendsen, B., and Ahrens, T. J. 1987. The melting curve of iron to 250 GPa: constraint on the temperature at the earth's centre. *Science, 236*, 181–2.

8

The Oceanic Lithosphere

Ridges, Transforms, Trenches and Oceanic Islands

8.1 Introduction

Three-fifths of the surface of the solid earth is oceanic lithosphere, all of which has been formed during the last 160 Ma at the midocean ridges. Understanding the structure of the oceanic lithosphere and the midocean ridges is particularly important because it provides a key to understanding the mantle.

8.1.1 Beneath the Waves

Bathymetric profiles across the oceans reveal the rugged nature of some of the seabed and something of the scale of its topography (Figs. 8.1 and 8.2). The deepest point on the surface of the earth was discovered during the voyage of *H. M. S. Challenger* (1873–6). The bottom of the Challenger Deep in the Marianas Trench (western Pacific Ocean) is more than 11.5 km below sea level, and Mauna Kea on the island of Hawaii rises to 4.2 km above sea level from an ocean basin more than 5 km deep. Such features dwarf even Mount Everest (8.85 km above sea level). The average global land elevation is 0.84 km, whereas the average depth of the oceans in 3.8 km. Although the seabed is hidden from us by the oceans, the imprint of its shape is revealed by the sea surface and gravity (see Fig. 5.10).

The sea floor can be classified into four main divisions: midocean ridges, ocean basins, continental margins and oceanic trenches.

Midocean Ridges The *midocean ridges* are a chain of undersea mountains with a total length of over 60,000 km. The ridges typically rise to heights of more than 3 km above the ocean basins and are many hundreds of kilometres in width. The Mid-Atlantic Ridge was discovered while the first trans-Atlantic telegraph cable was being laid. The East Pacific Rise was discovered by *H. M. S. Challenger* in 1875. As was discussed in Chapter 2, the midocean ridges mark the constructive boundaries of the plates. Hot material rises from the asthenosphere along the axis of the midocean ridges and fills the space left by the separating plates; as the material cools, it becomes part of the plates. For this reason, midocean ridges are often called *spreading centres.* The examples in Chapter 2 show that present-day spreading rates of the midocean ridges vary between approximately 0.5 and 10 cm yr^{-1}. Spreading rates are generally quoted as half the plate separation rate. For example, the North American and Eurasian plates are

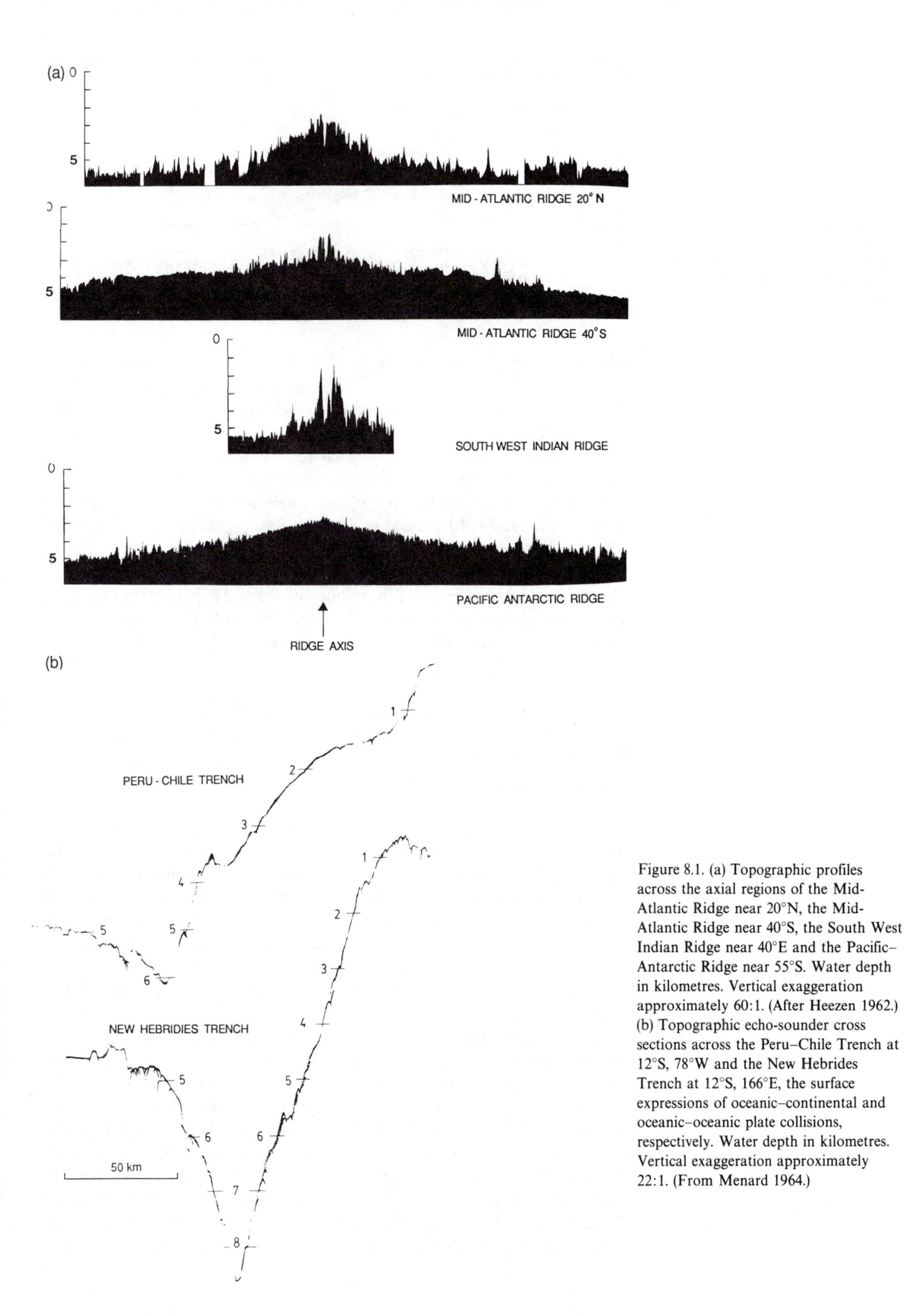

Figure 8.1. (a) Topographic profiles across the axial regions of the Mid-Atlantic Ridge near 20°N, the Mid-Atlantic Ridge near 40°S, the South West Indian Ridge near 40°E and the Pacific–Antarctic Ridge near 55°S. Water depth in kilometres. Vertical exaggeration approximately 60:1. (After Heezen 1962.) (b) Topographic echo-sounder cross sections across the Peru–Chile Trench at 12°S, 78°W and the New Hebrides Trench at 12°S, 166°E, the surface expressions of oceanic–continental and oceanic–oceanic plate collisions, respectively. Water depth in kilometres. Vertical exaggeration approximately 22:1. (From Menard 1964.)

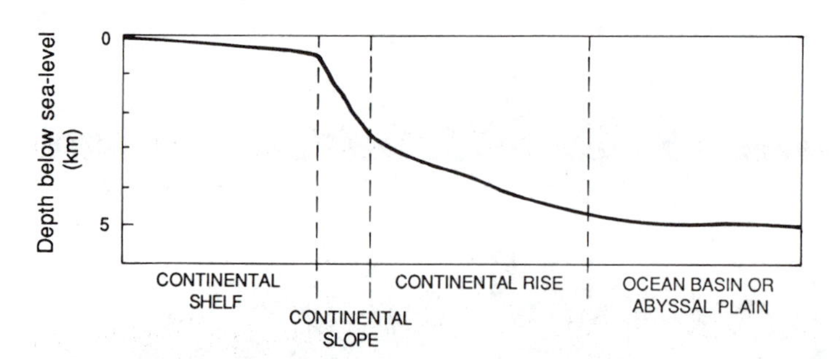

Figure 8.2. The bathymetry of the regions of a passive continental margin such as the margins of the Atlantic Ocean. Slopes are approximate; vertical exaggeration is 100–200:1.

separating at approximately 2 cm yr^{-1}, so the Mid-Atlantic Ridge is said to be spreading at a rate of 1 cm yr^{-1}. The fastest spreading ridge today is that portion of the East Pacific Rise between the Nazca and Pacific plates; its rate is almost 10 cm yr^{-1}.

Bathymetric profiles across the Mid-Atlantic Ridge, South West Indian Ridge and Pacific–Antarctic Ridge are shown in Figure 8.1. For sea floor younger than about 70 Ma, the relation between the mean bathymetric depth d and its age t (discussed in detail in Sect. 7.5) is given by Eq. 7.57a:

$$d = 2.5 + 0.35t^{1/2}$$

where d is in kilometres and t in millions of years. This holds for all ridges irrespective of their spreading rates. At ages greater than 70 Ma, the mean depth increases more slowly than $t^{1/2}$ and is given by Eq. 7.57b:

$$d = 6.4 - 3.2e^{-t/62.8}$$

The East Pacific Rise is wider than the more slowly spreading Mid-Atlantic Ridge because the width of any ridge is proportional to its spreading rate. The detailed cross sections of the axial region of these three ridges reveal more differences among them. These are discussed in Section 8.4.1.

Ocean Basins Sedimentation occurs throughout the oceans. As a result of this and the subsidence of the seabed (due to cooling and contraction of the plates), over time the rugged, faulted topography of the midocean ridges is buried under sediments. The almost flat regions of the seabed, thousands of kilometres in width and some 5–6 km below sea level, are often called *ocean basins* or *abyssal plains.* Isolated volcanic islands occur in most of the oceans, often in chains (e.g., the Emperor Seamount chain and the Hawaiian Islands in the northern Pacific Ocean, as shown in Fig. 2.19). Such island chains apparently form when a plate passes over a 'hot spot', a localized region where magma is rising from deep in the mantle. The depth from which the magma comes can be roughly estimated from the chemistry of the lava. In addition, the islands can be dated and the history of the passage of the plate over the magma source determined. This information was used in the determination of the plate motions in Chapter 2. The *aseismic ridges* (such as the Walvis Ridge and the Rio Grande Rise in the South Atlantic Ocean, and the Iceland–Faeroe Ridge in the North Atlantic Ocean) are submarine volcanic mountain chains, typically elevated some 2–4 km above the seabed. They are not tectonically active (hence their name) and seem to have formed 50–110 Ma ago. The origin of each is controversial, but one possibility is that they also were formed as the plate passed over a hot region in the mantle.

Continental Margins The *continental margins*, as their name suggests, mark the transition between the continent and the ocean floor. At a *passive margin* (Fig. 8.2), such as occurs off the east coast of the Americas or the matching west coast of Europe and Africa, the first structure seawards of the land is the gently sloping *continental shelf*, which is merely continent covered by shallow water. At the outer edge of the shelf, the gradient abruptly increases and the seabed deepens rapidly; this is the *continental slope*. At the base of the slope, the gradient is again much less; this region is the *continental rise*. The transition from continental crust to oceanic crust occurs beneath the continental slope.

An *active margin*, such as the west coast of North and South America, is so called because of the igneous and tectonic activity occurring at the plate boundary. There the continental shelf is often, but not always, very narrow. Where the plate boundary is a transform fault, the seabed characteristically drops rapidly from the shelf to oceanic depths. Where the plate boundary is a subduction zone, there is usually a trench, typically many kilometres deep.

Oceanic Trenches The *oceanic trench* marks the surface location of a subduction zone, at which one plate is overriding another; for example, the continental South American Plate is overriding the oceanic Nazca Plate along the west coast of South America. Here the Peru–Chile Trench extends to depths of 8 km and is considerably less than 200 km in width (see Fig. 8.1). Many of the trenches around the western margin of the Pacific Plate occur where an oceanic plate is overriding the oceanic Pacific Plate. At such destructive plate boundaries, an island arc runs parallel to the trench, and frequently sea-floor spreading occurs on the consuming plate behind the island arc, thus forming a *back-arc*, or *marginal, basin*. The trenches of the western Pacific are frequently 10 km or more deep.

Classification of igneous rocks

The classification of igneous rocks is a vexed business that has kept igneous petrologists amused, annoyed and employed for the best part of a century. The discussion that follows is not rigorous but merely attempts to outline the arguments for those who are not igneous petrologists.

The problem is that an igneous rock can be classified in many ways, each being useful in some circumstance. Classification began in the obvious way when geologists simply looked at igneous rocks and the minerals they contain. One obvious difference is between coarse-grained *plutonic* rocks and fine-grained or partly glassy *volcanic* rocks. Plutonic rocks are derived from magmas that cooled slowly at depth in the interior of the earth and therefore grew large crystals. Volcanic rocks erupted at the earth's surface cooled quickly and thus grew small crystals. Indeed, if a magma cools very rapidly, much or all of the rock may be glassy. Of course, to confuse the issue, a magma may carry to the surface some larger crystals that formed at depth (called *phenocrysts*).

In general, the colour of igneous rocks is a good guide to their chemical

Table 8.1. *Classifications of igneous rocks*

Colour (% dark minerals)	Type	Example
Less than 40	Felsic	Granite
40–70	Intermediate	Diorite
70–90	Mafic	Gabbro
over 90	Ultramafic	Peridotite

Silica (% SiO_2)	Type	Example
Over 66	Acid	Rhyolite
52–66	Intermediate	Andesite
47–52	Basic	Basalt
Less than 47	Ultrabasic	Komatiite

Average chemical composition (%)

	Granite	Grano-diorite	Diorite	Gabbro	Peri-dotite	Andesite	Basalt	Komatiite
SiO_2	70.4	66.9	51.9	48.4	43.5	54.9	50.8	44.9
TiO_2	0.4	0.6	1.5	1.3	0.8	1.0	2.0	0.3
Al_2O_3	14.4	15.7	16.4	16.8	4.0	17.7	14.1	3.1
Fe_2O_3	1.0	1.3	2.7	2.6	2.5	2.4	2.9	2.3
FeO	1.9	2.6	7.0	7.9	9.8	5.6	9.1	11.5
MnO	0.1	0.1	0.2	0.2	0.2	0.2	0.2	0.2
MgO	0.8	1.6	6.1	8.1	34.0	4.9	6.3	33.0
CaO	2.0	3.6	8.4	11.1	3.5	7.9	10.4	3.8
Na_2O	3.2	3.8	3.4	2.3	0.6	3.7	2.2	0.01
K_2O	5.0	3.1	1.3	0.6	0.3	1.1	0.8	0.01
P_2O_5	0.2	0.2	0.4	0.2	0.1	0.3	0.2	0.03

Note: These are families of rocks and show much variation around the typical; thus, an average is somewhat meaningless. The table presents only the rough chemical content to be expected.
Source: Chemical compositions from Nockolds et al. (1978) and Smith and Erlank (1982).

composition, and it usually reflects the mineralogy of the rock (Table 8.1). Light-coloured rocks, rich in silica and alumina, are *felsic* rocks. These include granite (a plutonic rock) and rhyolite (the volcanic equivalent of granite). The detailed nomenclature of the granitoid rocks is very complex: *Granites* and *rhyolites* are rich in quartz and alkali feldspar minerals; *granodiorites* (plutonic) and their volcanic equivalents, *dacites*, are rich in plagioclase feldspar minerals and quartz but have less alkali feldspar; *tonalites* are plutonic rocks that contain quartz and plagioclase feldspar but very little alkali feldspar.

Igneous rocks which contain moderate quantities of dark minerals, such as pyroxene or hornblende, are termed *intermediate*. An example is *diorite*, a plutonic rock that contains plagioclase feldspar, is moderately rich in calcium, and has some dark minerals such as hornblende, biotite or pyroxene. Its volcanic equivalent is *andesite*, a rock that is typical of many volcanoes above subduction zones.

Igneous rocks which are rich in dark minerals are called *mafic*, shorthand for magnesium and iron (Fe) rich. These rocks include *gabbro*, a plutonic rock containing calcic plagioclase, pyroxene and often olivine or hornblende; its volcanic equivalent is *basalt*.

The *ultramafic* igneous rocks (Fig. 4.34) are almost exclusively made of dark minerals. They include *peridotites*, which are plutonic rocks rich in olivine, with some pyroxene or hornblende. *Dunite* is a peridotite that is 90% or more olivine. *Harzburgite* is mostly olivine, plus some orthopyroxene and perhaps a little clinopyroxene; *lherzolite* contains olivine, orthopyroxene and clinopyroxene. The mantle of the earth is ultramafic. Ultramafic rocks have rarely erupted as lavas in the relatively recent geological past, but they were more common in Archaean strata. Ultramafic lavas are called *komatiites*; fresh komatiites have small olivine and pyroxene crystals set in a glassy matrix.

Another good way to classify rocks is by dividing them according to their chemical content. Rocks rich in SiO_2 such as rhyolite are called *acid; intermediate* rocks have rather less SiO_2; *basic* rocks have around 50% SiO_2 or less and include basalt; *ultrabasic* rocks are very poor in SiO_2.

Chemistry can be of further help. Not content with analysing fine-grained and glassy rocks, petrologists considered what a lava would have turned into if it had cooled slowly, crystallizing completely. Various schemes were devized to estimate the minerals that would have formed if a lava of given composition had crystallized. Of course, these schemes are artificial because so many variables are involved, but eventually standard ways of recalculating the chemical analysis were accepted. The standard, or *normative*, composition can then be used to classify the rock.

Another system of classification, which dates back to the 1930s but has introduced names which are deeply embedded in the literature, is to study a suite of rocks for their CaO and ($Na_2O + K_2O$) contents. Rock suites which are rich in the alkalis are called *alkaline*; rock suites which are rich in CaO are *calcic*. Andesite volcanoes typically have *calc-alkaline* trends.

For geophysicists, the difference between basalt and andesite is especially important. Basalts must have less than 52% SiO_2, whereas andesites have more. Basalts have over 40% by weight of dark minerals, typically pyroxene, whereas andesites have less than 40% of dark minerals and often (though not always) contain hornblende. The plagioclase in basalt is more calcium-rich than that in andesite. In thin section, basalt and andesite can be seen to have different microscopic textures. The rock types between basalt and andesite are called *andesitic basalts* and *basaltic andesite*.

Basalts are, in fact, a family of lavas, divided according to their content of normative minerals into *tholeiites* (slightly more silica relative to some other components) and *alkali–olivine basalts* (relatively more alkali), with a transitional group known as *olivine tholeiites*. Most ocean-floor basalts and most but not all basalts in Hawaii are olivine tholeiites. Alkali basalts occur in zones of continental rifting. For more detail, the reader should consult any textbook on igneous petrology.

8.2 Oceanic Lithosphere

8.2.1 Oceanic Crust

Worldwide gravity surveys indicate that the oceanic regions are in approximate isostatic equilibrium with the continents; that is, the pressure at an arbitrary depth in the mantle is the same beneath continents and oceans. This means that, at arbitrary depth in the mantle, a column of continental crust and underlying mantle and a column of oceanic crust and its underlying mantle have the same mass. This fact enables us to make a simple estimate of the thickness of the oceanic crust (see Sect. 5.5 for the method). If we assume Airy-type compensation, densities of sea water, crust and mantle of 1.03, 2.9 and $3.3 \times 10^3 \, kg \, m^{-3}$, respectively, and an average ocean basin depth of 5 km, then a typical 35 km thick continental crust would be in isostatic equilibrium with an oceanic crust 6.6 km in thickness. This rough calculation tells us the important fact that the oceanic crust is one-fifth the thickness of the continental crust.

The details of the seismic structure of the oceanic crust have been determined by using seismic refraction and reflection profiling and wide-angle reflection techniques. In the absence of direct sampling of the crust, its composition must be estimated from measurements of its physical properties (e.g., seismic velocity and density) which vary with lithology. These estimates are frequently ambiguous.

The most direct way to determine the composition of the oceanic crust is to collect rock samples from each of the oceanic plates. Dredging samples from the seabed is not particularly difficult or expensive, but it is often frustrating: Imagine trying to manoeuvre a large bucket, hanging on 5 km of wire over a scarp slope, which you can see only on the ship's echo sounder, and then attempting to collect rocks from the debris at the scarp base. Dredging is also not particularly representative: A sample may not be typical. A much more precise way of sampling the seabed is to use a submersible, a minisubmarine capable of descending to great depths; there the geologist can sample the exact outcrop and rock type wanted. Submersibles have enabled scientists to make detailed studies of small areas of the seabed, particularly the axial zones of the midocean ridges. Such operations are, however, extremely expensive and still sample only the surface. Sampling the rock outcropping at the seabed does not tell us what rocks make up the lower oceanic crust (even if the fault scarps are such that deeper rocks could be exposed at their bases). However, it does enable us to make informed guesses; for instance, since seabed samples include basalts, gabbros, serpentinites and recent sediments, one would not guess at a deep crust made up of granite (it took scientists a while to realize that granite samples dredged from the seabed were not representative of the ocean crust but had been dropped by icebergs, ice-rafted granite). Another way to obtain samples is to drill. Drilling into the oceanic crust is an expensive and difficult operation compared with drilling on land. Not only is the rock hard and frequently fractured, but there are many kilometres of sea water between the drilling ship and the top of the drillhole (Fig. 8.3). Drilling into the oceanic crust is an international cooperative venture overseen by the Joint Oceanographic Institutions for Deep Earth Sampling (JOIDES).

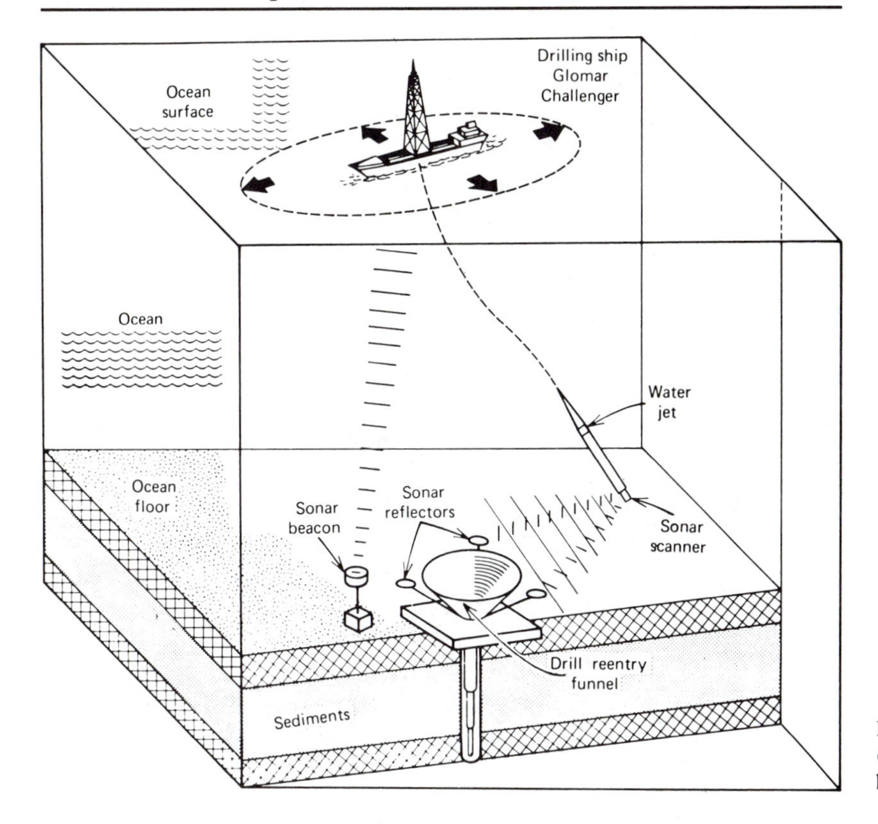

Figure 8.3. Drilling into the oceanic crust. Finding the hole for re-entry, a long way down. (From Wyllie 1976.)

Drilling, which started in 1968 as the Deep Sea Drilling Project (DSDP), and then in 1985 entered a new phase as the Ocean Drilling Program (ODP), has tremendously advanced our knowledge of the geological and geophysical structure of the uppermost crust. Unfortunately, however, drilling into the lower crust and upper mantle remains a project for the future.

Most of the seabed is sedimented. Figure 8.4 shows the sediment thickness over the western North Atlantic. Thicknesses gradually increase from zero at the active ridge axis to 0.5–1 km in the ocean basins. The greatest accumulation of sediments occurs beneath the continental margins; over 10 km of sediment is not uncommon, but thicknesses rarely reach 15 km. These sediment accumulations, a consequence of the rifting apart of the old continent, are discussed in Section 9.3.

Researchers began widespread oceanic seismic refraction experiments in the 1950s and by the 1960s had made enough determinations of crustal seismic velocities over the oceans for it to be apparent that, unlike the continents, the oceanic crustal structure varied little. Table 8.2 gives the standard seismic structure of mature oceanic crust beneath the ocean basins as determined from first P-wave arrival travel times. However, subsequent improvements in the techniques for the inversion of seismic wide-angle reflection and refraction data (including computation of synthetic seismograms; see Sect. 4.4) have provided a much improved resolution of seismic structure of the oceanic crust. The original four layers of constant velocity of Table 8.2 have given way to a discontinuous series of velocity

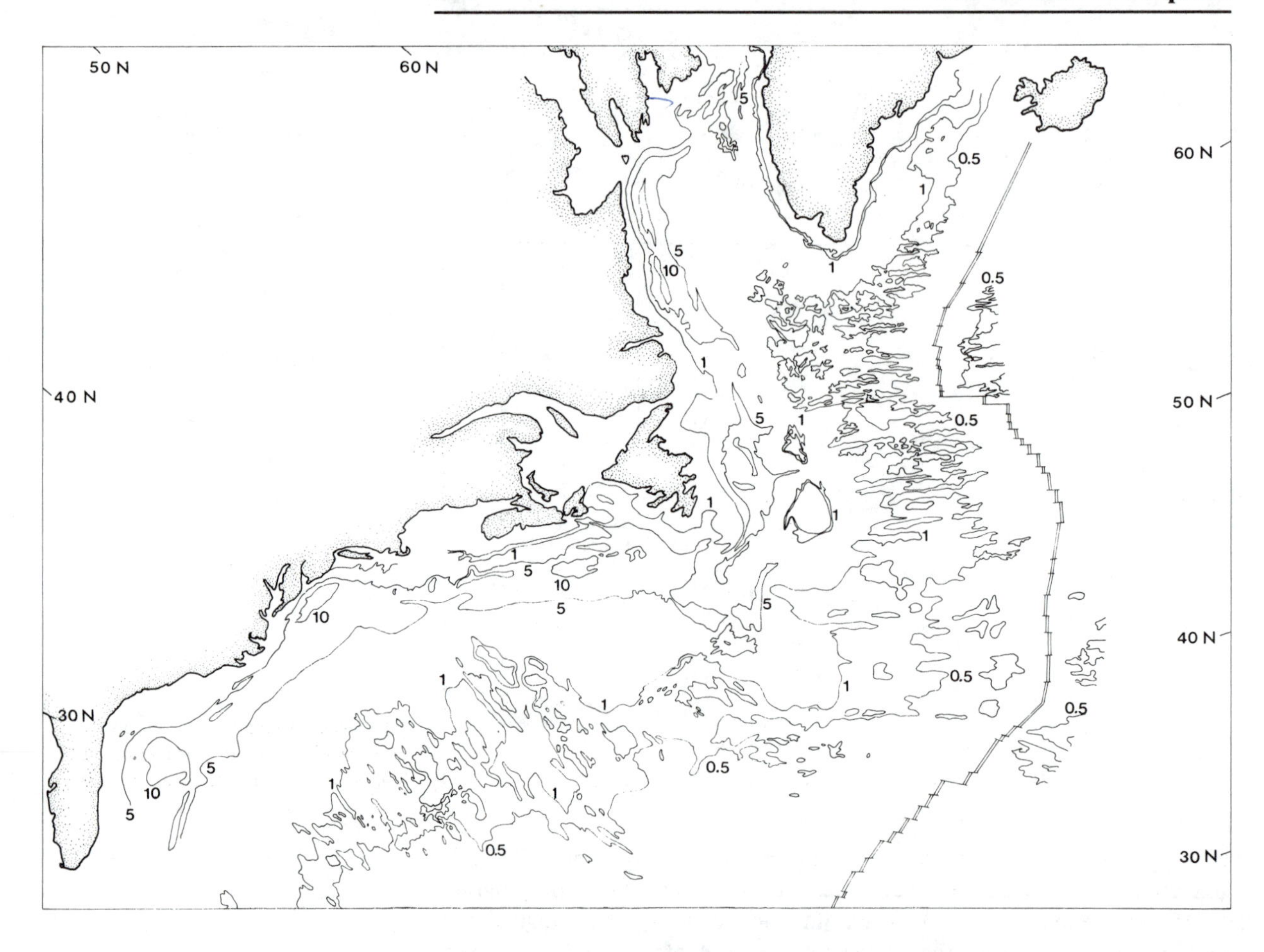

Figure 8.4. Sediment thickness over the western North Atlantic (contours 0.5, 1, 5 and 10 km). (After Tucholke 1986.)

gradients in which the velocity increases with depth. Figure 8.5a shows all (as of 1984) velocity–depth profiles for seismic refraction lines in the North Atlantic that were constrained by synthetic seismogram modelling. It is clear that although there are local differences, the general variation of seismic velocity with depth is fairly constant within each age zone. It should be remembered when looking at velocity–depth profiles such as these that seismic refraction experiments generally cannot resolve structure much less than one wavelength in thickness. (For layer 2, with velocity 5 km s^{-1} and a

Table 8.2. *Raitt–Hill layering of the oceanic crust beneath the ocean basins*

Layer	Thickness (km)	Seismic velocity (km s^{-1})
Sea water	4.5	1.5
Layer 1 (sediment, variable)	About 0.5	About 2
Layer 2 (volcanic layer)	1.71 ± 0.75	5.07 ± 0.63
Layer 3 (oceanic layer)	4.86 ± 1.42	6.69 ± 0.26
Layer 4 (upper mantle)		8.13 ± 0.24

Note: This layering is the 'average' seismic structure of 'normal' oceanic crust.
Source: After Raitt (1963) and Hill (1957).

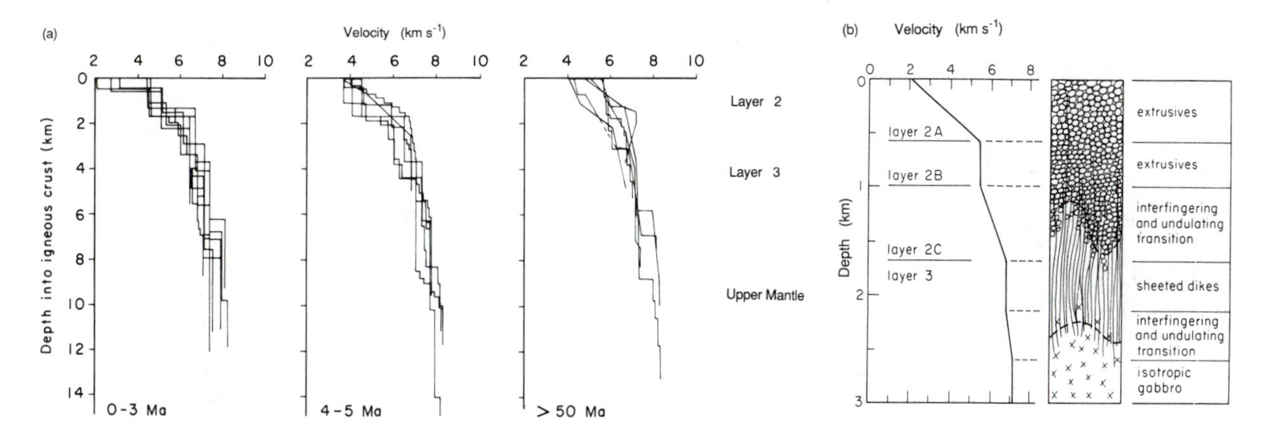

Figure 8.5. (a) Velocity–depth models from the North Atlantic, grouped according to the age of the lithosphere. All these models were determined from seismic refraction data by calculating synthetic seismograms and adjusting the model until the travel times and the amplitudes agreed well with the data. (From White 1984.) (b) Velocity–depth model obtained from the East Pacific Rise correlated with the DSDP hole 504B drilled into 6.2 Ma Pacific oceanic crust. (From Bratt and Purdy 1984.)

7 Hz signal, a wavelength is 0.7 km.) Therefore, much of the fine detail and staircase-like appearance of these profiles is a representation of velocity gradients.

Layer 1, the sedimentary layer, thickens with distance from the ridge axis. The P-wave velocity at the top of the sediments is generally close to 1.5 km s^{-1} (the velocity of sound in sea water) but increases with depth as the sediments consolidate. (This layer is not shown in Fig. 8.5.)

At the top of *layer 2*, the *volcanic layer*, there is a sudden increase in P-wave velocity to approximately 4 km s^{-1}. Reflection profiling has shown that the top of layer 2, the *oceanic basement*, is very rough for crust that was formed at a slow-spreading ridge. It was proposed in the 1970s that layer 2 should be subdivided into three layers, 2A, 2B and 2C, with P-wave velocities of 3.5, 4.8–5.5 and 5.8–6.2 km s^{-1}, respectively, and this terminology is often used in the literature. However, layer 2 is probably best described as a region of the oceanic crust in which seismic velocity increases rapidly with depth (gradients of 1–2 km s^{-1} per kilometre of depth). In very young crust, the P-wave velocity in the top few hundred metres of layer 2 may be less than 2.5 km s^{-1}. Drilling into the top of layer 2 has shown that it is made up of sediments, basaltic pillow lavas and lava debris in varying degrees of alteration and metamorphism. Deeper drilling has found more consolidated basalts. Towards the base of this basaltic layer are some dykes which cause the P-wave velocity to increase further. Together, this variability in composition and gradation from sediments to basalt to dykes appears to account for the rapid and variable increase in seismic velocity with depth (Fig. 8.5b).

Layer 3, sometimes called the *oceanic layer*, is thicker and much more uniform than layer 2. Typical P-wave velocities are 6.5–7.0 km s^{-1}, with gradients of around 0.1 km s^{-1} per kilometre depth. Layer 3 is generally presumed to be gabbroic in composition. Some seismic experiments have shown that the basal part of this layer (sometimes termed layer 3B) has a higher P-wave velocity (7.2–7.7 km s^{-1}), indicating perhaps a change to more cumulate-rich gabbros at the base of the crust. A few experimenters have suggested that there is a low-velocity P-wave zone at the base of layer 3, but this is not generally thought to be a universal feature of the oceanic layer. In 1962, Hess proposed that layer 3 is partially serpentinized peridotite, formed as a result of hydration of the upper mantle. His proposal

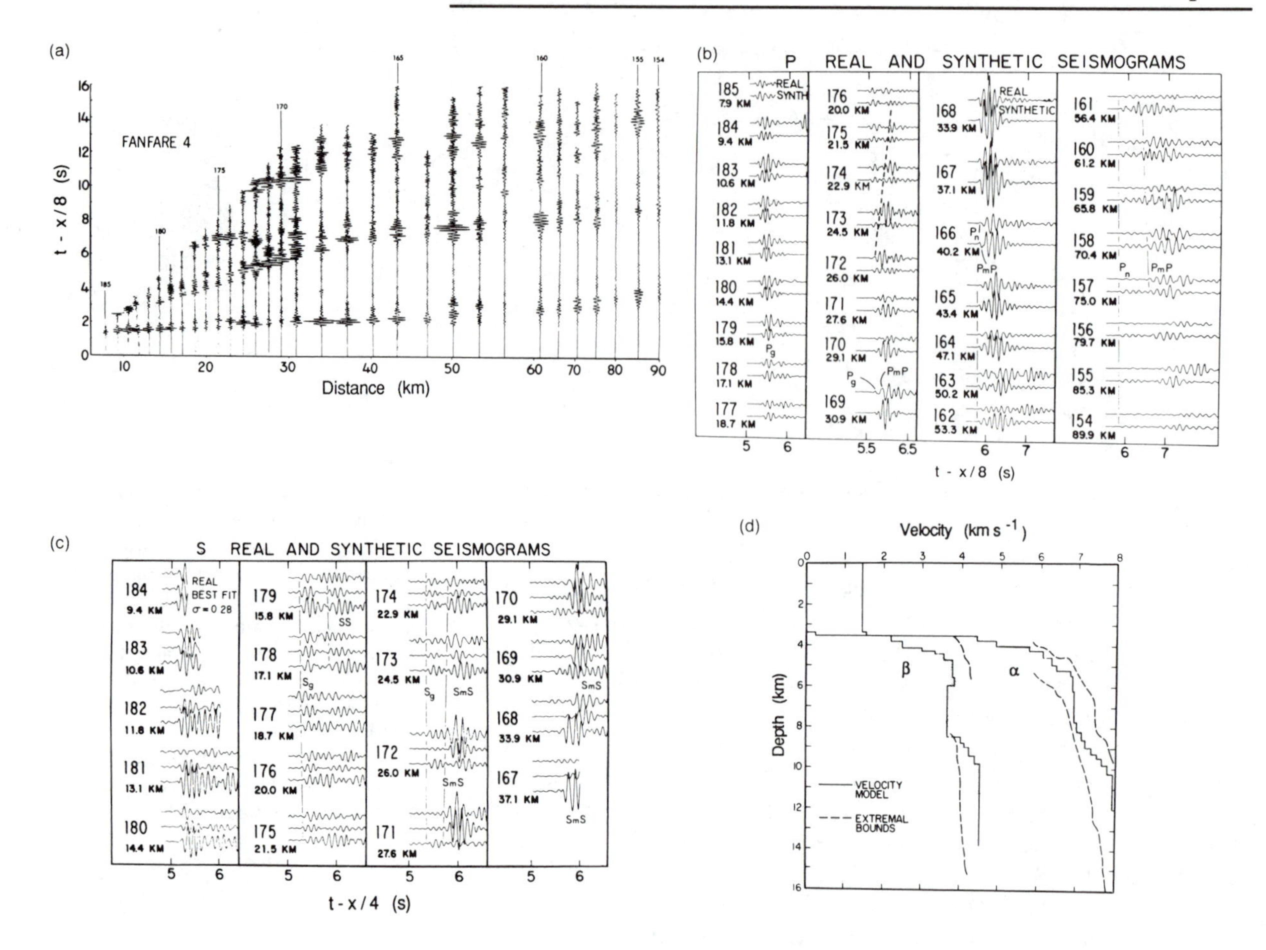

Figure 8.6. Real and synthetic seismogram for a refraction line shot between Guadalupe Island and Baja California on 15 Ma old oceanic crust. (a) Real data: reduced record section. (b) P-wave seismogram pairs. Synthetic seismograms computed from the model in (d). (c) Real and synthetic S-wave seismograms. Best-fitting synthetic seismograms were computed from the S-wave model in (d); synthetic seismograms marked $\sigma = 0.28$ (the third in each set) were computed from the P-wave velocity model in (d) with Poisson's ratio $\sigma = 0.28$. (d) P-wave and S-wave velocity models. All velocity–depth models satisfying the travel-time data lie within the extremal bounds as determined from the travel-time inversion. (From Spudich and Orcutt 1980.)

was largely discounted in the 1970s, but recent studies of the axial zones of the midocean ridges have renewed interest in it.

In the same way that P-waves are used to model the seismic structure of the crust, we can use S-waves to make an S-wave velocity model. Refraction data of sufficient quality for making detailed S-wave models are very rare, and so, unfortunately, crustal S-wave velocity models are uncommon. Figure 8.6 shows a detailed model of the P- and S-wave velocity structures of the oceanic crust east of Guadalupe Island off the west coast of Baja California. The sea floor at this location is about 15 Ma old and is part of what is left of the Farallon Plate (see Sect. 3.3.4). Of particular interest here are the low velocities in the shear-wave model for layer 3. The evidence for this low-velocity zone is very good. Synthetic S-wave seismograms, computed for the best P-wave velocity model assuming that Poisson's ratio σ is 0.28 throughout the crust, do not match the shear reflection from the mantle (S_mS): The critical distance for S_mS is too large, and the arrival time is too early. The critical distance could be decreased by elevating the S-wave Moho, but this would cause S_mS to arrive earlier still. A solution can be obtained, however, by introducing a low-velocity S-wave zone into the lower crust; this reduces the S_mS critical distance and delays the arrival times without affecting the P-wave structure at all (see Fig. 8.6). It is suggested that these well-defined low S-wave velocities could indicate the relative abundance of hornblende in comparison to augite and olivine in the gabbros which make up layer 3. This interpretation is not the only solution: The low velocities could also be produced by fluid-filled pores.

A major two-ship multichannel seismic experiment, the North Atlantic Transect, collected some 4000 km of reflection data between the east coast of the United States at 32°N and the Mid-Atlantic Ridge at 23°N. A sharp, apparently continuous reflection from the Moho was observed along most of the transect about 2.5 s beneath the top of the oceanic basement. Eleven expanding spread (wide-angle) profiles were also recorded along this transect; Figure 8.7a shows the record section for one of these profiles. Modelling of these travel times and amplitudes yielded a detailed structure

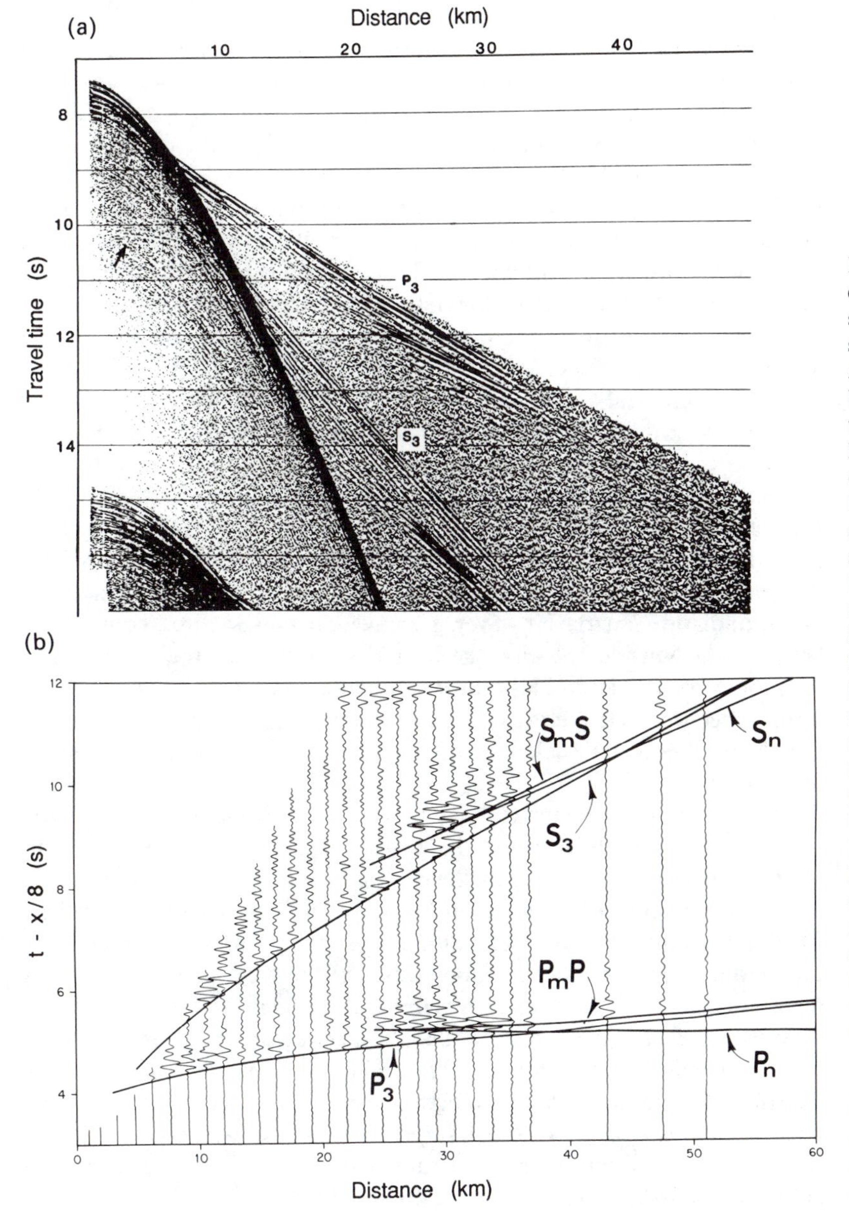

Figure 8.7. (a) Record section of an expanding spread (wide-angle) profile shot over 118 Ma old oceanic crust southwest of Bermuda on the North American Plate. The records have been band-pass filtered at 6–30 Hz and plotted with a gain (amplification) that increases linearly with distance and as the square of the travel time. The P-wave arrival, P_3, is a lower crustal refraction. The S_3 arrival is the lower crustal S-wave. The strong secondary P-wave arrival seen between 20 and 35 km is a refraction from the upper mantle. Arrow; near normal-incidence reflection from the Moho. The large-amplitude hyperbola, which is the first arrival at distances less than about 6 km, is the reflection from the seabed. (From NAT Study Group 1985.) (b) Reduced record section of a refraction profile shot over 140 Ma oceanic crust in the western central Atlantic Ocean on the North American Plate. The seismograms have been low-pass filtered at 15 Hz, corrected for varying explosive charge sizes and plotted with a gain that increases linearly with distance. P_3 and S_3 are the crustal refractions, P_mP and S_mS the reflections from the crust-mantle boundary, P_n and S_n the upper mantle refractions. The S phases are converted shear waves (converted from P to S and then back from S to P at the sediment–crust interface beneath the shot and receiver, respectively). (From Purdy 1983.)

of the oceanic crust at this location, which is some 7.5 km in thickness. Study of the amplitudes of those seismograms shows that there may be a low-velocity zone in the lower crust, but it is not necessarily needed to satisfy the date. A reduced record section from a seismic refraction line shot over 140 Ma old oceanic crust in the western central Atlantic Ocean is shown in Figure 8.7b. The main features of these typical seismograms are the large-amplitude mantle reflections P_mP, the weak mantle refraction P_n and the clear shear waves.

Thus, a number of lithologies have been proposed for layer 3, and we will not be able to determine definitively the composition of layer 3 and its subdivision until a hole (better yet, several holes) is drilled completely through the oceanic crust.

Ophiolites (a sequence of rocks characterized by basal ultramafics overlain by gabbro, dykes, lavas and deep-sea sediments) are often regarded as examples of oceanic crust. However, because they are now tectonically emplaced on land, they are atypical and may not represent normal oceanic crust. Nevertheless, ophiolites do have extensive thicknesses of basaltic and gabbroic material. Figure 8.8 shows the principal ophiolite belts of the world. They may be samples of oceanic crust produced in back-arc basins behind subduction zones (Sect. 8.6). One well-studied ophiolite is the Semail ophiolite in the Sultanate of Oman, which is believed to have been formed some 95 Ma ago at a spreading centre in the Tethyan Sea and later emplaced on the Arabian Plate as Tethys closed when Africa and Arabia moved northwards and collided with Asia (see Sect. 3.3.3). The results of the extensive geological and seismic velocity studies of this classical ophiolite are summarized in Figure 8.9. It is apparent that the seismic velocity structure is similar to that of the oceanic crust and upper mantle, with gradients throughout layer 2, a rapid increase in velocity at the layer 2–layer 3 boundary and a relatively high velocity layer 3B at the base of the crust. The transition from crust to mantle is sharp. Similar results have been determined for the Blow Me Down Massif of the Bay of Islands ophiolite in Newfoundland, Canada. However, it is unwise to make a direct comparison between the seismic velocities measured on ophiolite rock samples at ultrasonic frequencies in a laboratory and the seismic velocities determined from reflection and refraction experiments using very-low-frequency signals; the frequency scale and the length scale of the velocity determinations differ.

The oceanic crust seems to have more or less the same thickness and velocity structure throughout the world's oceans, on all plates. This observation can be explained by the fact that the depth and temperature of melting are likely to be similar in both fast- and slow-spreading ridges, which means that the extent of melting is probably similar. An alternative explanation which has been proposed is that the crust–mantle boundary represents the depth to which water circulates through (and therefore cools) the crust and is thus the depth at which the cracks close. The Pacific oceanic crust is similar to but much more uniform than the Atlantic crust, presumably due to a faster magma supply rate. However, to fully understand the structure and composition of the oceanic crust, it is necessary to determine the structure of the midocean ridges and the processes occurring there where the crust is formed.

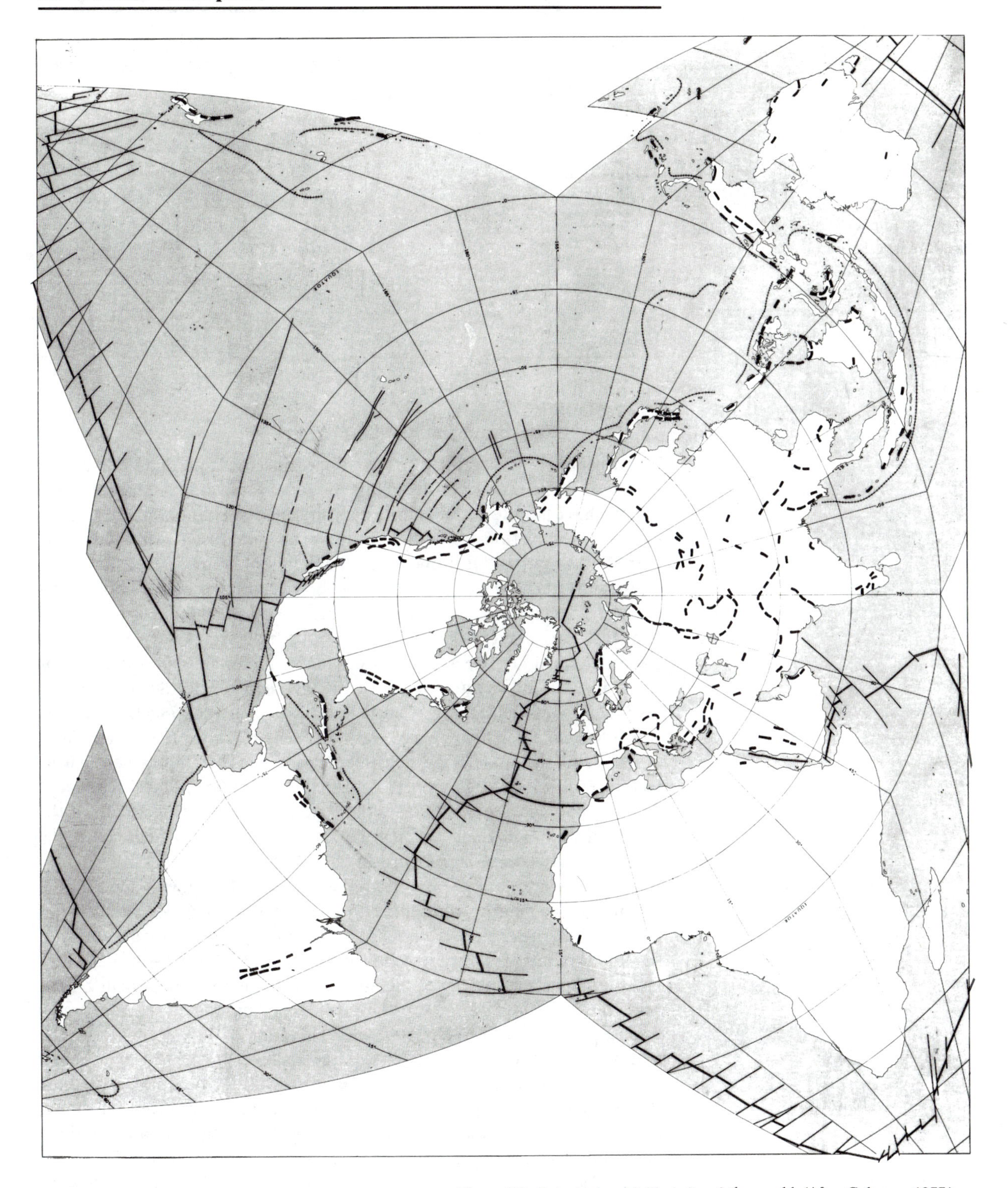

Figure 8.8. Principal ophiolite belts of the world. (After Coleman 1977.)

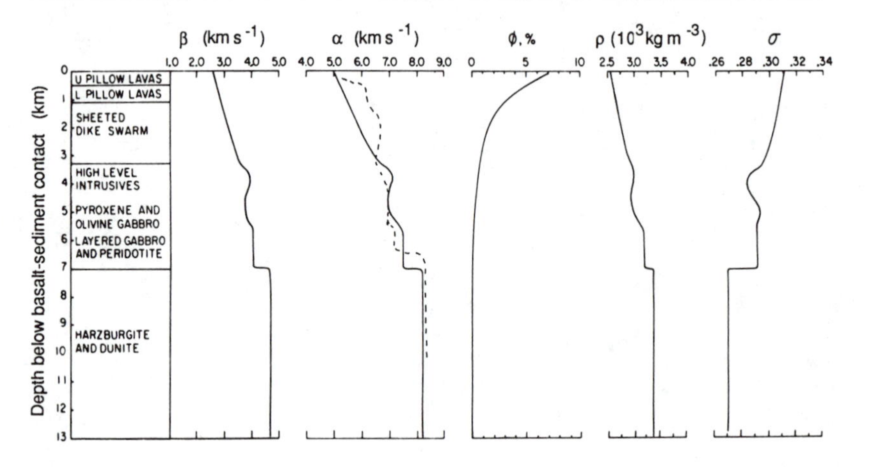

Figure 8.9. Shear wave velocity β, compressional wave velocity α, effective porosity φ, density ρ, and Poisson's ratio σ, versus depth and rock type as measured in the laboratory, on samples taken from the Semail ophiolite in Oman. Dashed line shows the compressional wave velocity measured on samples from the Blow Me Down Massif, an ophiolite in Newfoundland, Canada. (After Christensen and Smewing 1981.)

8.2.2 Oceanic Upper Mantle

The seismic P-wave velocity of the uppermost oceanic mantle averages 8.1 km s^{-1}. The oceanic crust–mantle boundary may be a sharp transition but can have a finite variable thickness and is often a thin gradient zone. It marks the transition from the basaltic-gabbroic crust to the peridotitic mantle. As with all seismic and geologic boundaries, there is some uncertainty about the exact correlation between the seismic boundary and the petrological boundary – hence expressions such as the 'seismic Moho' and the 'petrological Moho', which may not be exactly coincident. Correctly speaking, the Moho is a seismic boundary, and presumably it is the transition downwards into ultramafic rock. Petrologists, however, distinguish between zones of cumulate ultramafic rock (such as dunite, which is made of olivine; see Fig. 4.34), which have precipitated from a melt, and the underlying deformed residual upper mantle known as tectonite ultramafic. To a petrologist, this boundary is the significant dividing line because it separates the top of the residual mantle from the overlying material which is ultimately derived from the partial melt at the ridge axis. This boundary is termed the *petrological Moho.*

There is a gradual increase of seismic P-wave velocity with depth in the oceanic upper mantle, amounting to about 0.01 km s^{-1} per kilometre of depth. In addition, there is evidence of anisotropy: The P-wave velocity of the upper mantle is dependent on the azimuth of the ray path. Velocities measured perpendicular to the midocean ridge axis are greater than velocities measured parallel to the ridge axis. This velocity anisotropy is believed to be caused by the preferential aligning of olivine crystals in the mantle parallel to the flow direction (see Sect. 4.3.4).

Surface wave dispersion curves (see Sect. 4.1.3) for oceanic paths show that Rayleigh waves phase velocities are significantly reduced for young oceanic lithosphere and asthenosphere (Fig. 8.10). Interpretation of these dispersion curves suggests that the lithospheric thickness increases from some 30 km at 5 Ma to 100 km at 100 Ma, whereas lithospheric S-wave velocities increase from 4.3 km s^{-1} to 4.6 km s^{-1} with age. Similarly, asthenosphere S-wave velocities increase from 4.1 km s^{-1} to 4.3 km s^{-1} with age.

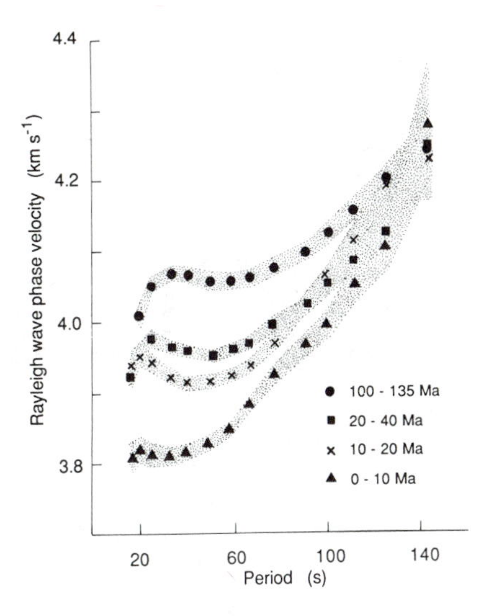

Figure 8.10. Rayleigh wave phase velocity dispersion curves (see Sect. 4.1.3) for oceanic lithosphere of various ages. Shading shows one standard deviation in the measurements. Notice that for periods less than 100 s the phase velocity increases with the age of the plate. For periods greater than 100 s the velocities are within error of each other. (After Forsyth 1977.)

8.3 Deep Structure of Midocean Ridges

8.3.1 Geophysical Evidence

The free-air gravity anomaly across the midocean ridges is not zero, indicating that the ridges are not in total isostatic equilibrium (see Sect. 5.5 for the definition and use of these terms). Partial compensation is attained by the presence of low-density material in the upper mantle beneath the ridge. Because gravity models are nonunique, they need to be constrained. Figure 8.11 shows four gravity models for the Mid-Atlantic Ridge. In the first model, the low-density zone is narrow and extends 200 km deep into the mantle; in the other three models, compensation is achieved by a broad and fairly shallow low-density zone. All adequately satisfy the gravity data. For reasons explained later, model c (see figure) is believed to be the best representation of the density structure.

The most direct evidence of the deep structure of the midocean ridges has come from earthquake seismology. Early studies of the Mid-Atlantic Ridge north of 50°N, using earthquakes recorded in Iceland and Greenland, indicated that the upper mantle there had a very low P-wave velocity and that this low-velocity zone extended to perhaps 250 km and was a few hundred kilometres wide. The detailed three-dimensional S-wave velocity structure of the upper mantle obtained by the inversion of long-period seismic recordings shows that beneath the midocean ridges in the depth interval 25–250 km the S-wave velocity is reduced by 2–8%. Seismic tomography (Sect. 4.3.4) confirms the depth extent of the low-velocity regions beneath the midocean ridges as about 250 km (Figs. 4.32 and 4.33). Seismologists working with teleseismic data from earthquakes have reported a 'gap in the lithosphere' beneath the midocean ridge system, across which S_n (a seismic shear wave which propagates in the uppermost mantle) is very poorly propagated. S_n propagates well across stable regions such as continental shields and deep ocean basins but only very inefficiently when its path crosses the midocean ridge system or the concave side of

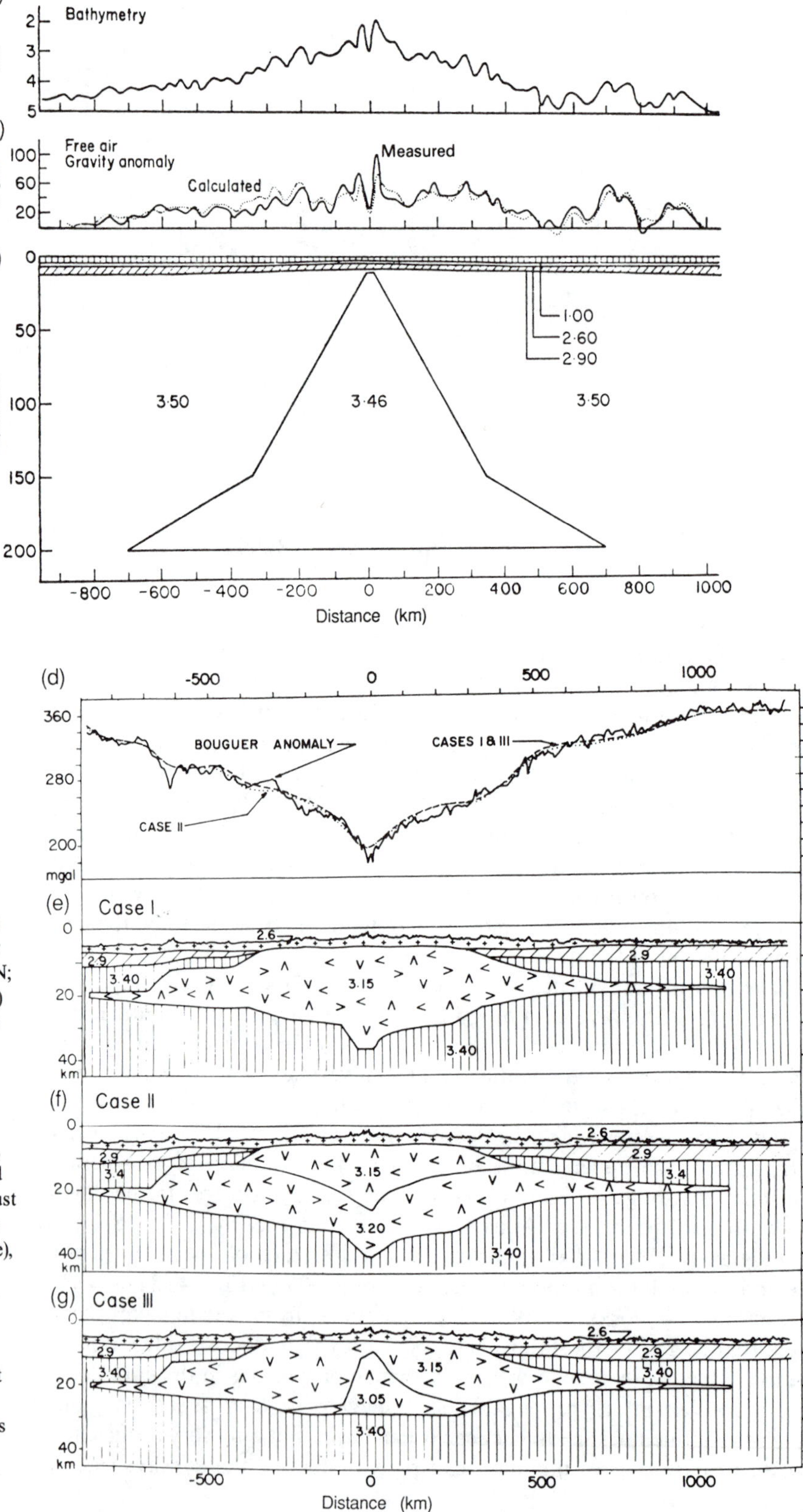

Figure 8.11. Gravity models for the Mid-Atlantic Ridge: (a) bathymetry; (b) free-air gravity anomaly and (c) density model for the Mid-Atlantic Ridge at 45°N; (d) Bouguer gravity anomaly and (e), (f) and (g) density models which all satisfy the anomaly shown in (d). These four density models – (c), (e), (f) and (g) – illustrate the nonuniqueness of models based on gravity data. A low-density zone lies beneath the ridge, but its dimensions need to be constrained by other methods also. The oceanic crust is assumed to be continuous across the ridge axis in model (c); but in models (e), (f) and (g) there is a zone some 800 km wide and centred on the ridge axis in which normal oceanic crust and uppermost mantle are absent. The density model in (c) is in better agreement with everything that is known about midocean ridge structure than the models in (e), (f) and (g). Densities are given in 10^3 kg m^{-3}. (From Talwani et al. 1965 and Keen and Tramontini 1970.)

island arcs. In addition, it is generally noted that magnitudes of ridge-crest earthquakes are often lower than those of earthquakes with comparable surface-wave magnitude but which are located away from the midocean ridge system. Frequently, surface waves are observed from midocean ridge earthquakes for which no body waves are detected. All of these observations can be explained by the presence of an absorptive zone in the upper mantle beneath the midocean ridges. Such a zone, which is also limited in extent, can most readily and reasonably be explained as being due to partial melting, occurring because there the upper mantle material is raised above its solidus. Seismic refraction experiments along and parallel to the crest of the Mid-Atlantic Ridge also suggest that in some places there is a zone of strong P- and S-wave absorption extending some 25 km on either side of the ridge axis and indicate that the top of this zone lies at least 7 km beneath the seabed.

Detailed studies of the source mechanism of large, ridge axis earthquakes on the Mid-Atlantic, South West Indian, American–Antarctic ridges and the Gorda Rise and Galapagos spreading centres indicate that all the foci were extremely shallow (1–6 km) and were located beneath the median valley. The mechanisms were nearly pure normal faulting on planes dipping at 45° with strike parallel to the local trend of the ridge axis. In addition, the focal depths decreased with increasing spreading rate, consistent with the theory that the maximum epicentral depth is representative of the depth at which the lithosphere ceases to deform in a brittle manner and ductile deformation takes over.

Figure 8.12 shows detailed P-wave and S-wave velocity models of a midocean ridge. These models were derived from a thermal model similar to the models discussed in Section 7.5.2 by estimating the dependence of seismic velocity on temperature and melting for a wet (0.1% water)

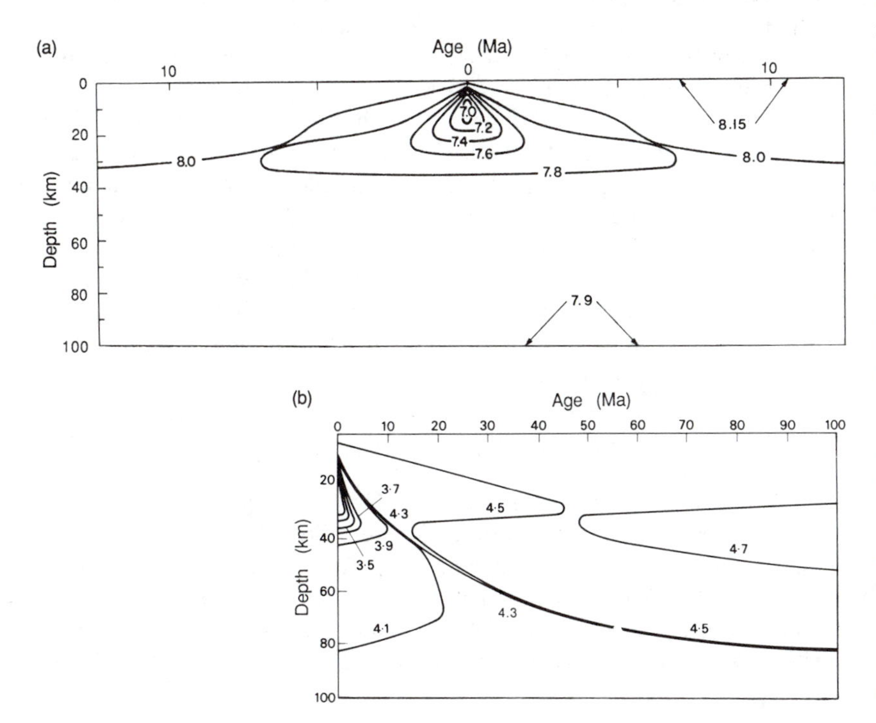

Figure 8.12. Seismic velocity models of midocean ridges calculated from a thermal model similar to those in Section 7.5.2 and theoretical estimates of the dependence of seismic velocity on temperature and melting of mantle material (assumed to be peridotite with 0.1% water). (a) Contours of constant P-wave velocity in the oceanic lithosphere and asthenosphere. The 5 Ma wide, 30 km deep low-velocity region centred on the ridge axis is modelled as a zone of extensive melting. (After Solomon and Julian 1974.) (b) Contours of constant S-wave velocity in the oceanic lithosphere and asthenosphere. The hydrous (wet) and anhydrous (dry) solidus for peridotite are approximately delineated by the 4.3 km s^{-1} and 3.7 km s^{-1} contours, respectively. The 4.3 km s^{-1} contour therefore represents the boundary between the lithosphere and asthenosphere if the base of the lithosphere is defined as the depth at which partial melting first starts. Note the different horizontal scales for (a) and (b). (From Duschenes and Solomon 1977.)

peridotitic mantle. The large velocity gradients in the low-velocity zone (1 km s^{-1} over distances of 15–25 km) are the result of partial melting of mantle material. The S-wave velocities predicted by this model are in broad agreement with all the surface-wave dispersion results. Were the mantle drier than was assumed for these models, the melting temperatures and hence the velocities would be somewhat higher.

None of these geophysical investigations can distinguish between in situ partial melting and an aggregation of melt that has separated from a region of lesser melt content. However, it has been suggested on chemical grounds that the melting beneath the ridge axis occurs in the top 60 km of a region more than 100 km wide. Partial melting of 18% of the mantle in this zone would be sufficient to produce the oceanic crust and to maintain the melt fraction in the zone at less than 2%.

The global variation in major element composition of the oceanic crust can be accounted for by 8–20% partial melting of the mantle at pressures of 0.5–1.6 GPa (depths of 15–50 km). It is possible that the proportion of some incompatible elements present in midocean ridge basalts correlates with the spreading rate of the ridge. Some correlation of major element chemistry with the depth of the ridge axis (depth of the seabed at the intrusion zone) is also observed. This presumably indicates that axial depth is dependent on the thermal structure of the underlying mantle. Where the mantle is hottest (and ridge axis shallowest), the rising mantle material intersects the solidus earlier, at greater depth, and so melts more (Fig. 7.17). Maximum temperature differences of some 200°C in the subsolidus mantle may be necessary to account for this observed correlation.

All of the seismic velocity models are dependent on the elastic properties of partially molten systems; unfortunately, few data exist on the properties of such systems. An example of the effect of partial melting on seismic velocity and attenuation is given by an NaCl–ice system: For experiments on a sample with 1% concentration of NaCl, reductions in the P- and S-wave velocities of 9.5 and 13.5% were measured at the onset of melting, and attenuation increased by 37 and 48%, respectively. These measurements were made for a 3.3% melt. Partial melting, therefore, can have a large effect on seismic velocity and attenuation even at very low melt percentages. Experiments on basaltic and peridotitic samples indicate that a large velocity decrease occurs over the liquidus–solidus temperature range (Fig. 8.13). Basalt P-wave velocities decreased from about 5.5 km s^{-1} to about 3.5 km s^{-1}, and peridotite P-wave velocities decreased from 7.5 km s^{-1} to 5.5 km s^{-1}. The premelting dependence of P-wave velocity on temperature for these rock samples was an almost linear decrease from 0°C to the onset of melting.

8.3.2 Melting under Midocean Ridges

Ridges are mostly passive structures. The interior of the mantle is everywhere so hot that if mantle rocks were brought up to the surface without temperature loss, decompressing from mantle pressures to 1 atmosphere, they would melt. Fertile mantle (that is, mantle which has not already been depleted of some of its chemical components by a partial melting event) will always melt when it is brought up adiabatically to 40 km

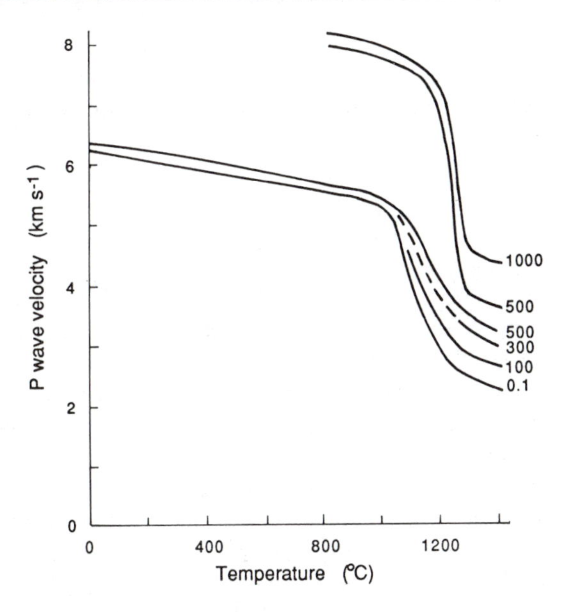

Figure 8.13. Effects of temperature and pressure on P-wave velocities. Ultrasonic P-wave velocity measurements made on dry basaltic (lower four curves) and peridotitic (upper two curves) samples in a laboratory. The numbers beside each curve denote the pressure (MPa) at which the measurements were made. Note the sudden decrease in velocity at the solidus temperature and the steady decrease over the solidus–liquidus interval (approximately 1000–1200°C for basaltic samples and 1200–1300°C for peridotitic samples). These measurements were made on dry samples. The presence of water would reduce the solidus and liquidus temperatures but would not greatly change the shape of curves: They would merely be shifted to the left and slightly downwards. (After Khitarov et al. 1983.)

or less below the surface. Upwelling must occur under any rift, to fill the vacated space: If the extension on rifting is infinite, the melting produces a midocean ridge.

McKenzie and Bickle (1988) successfully modelled this melting by finding expressions for the solidus and liquidus temperatures of upwelling mantle and for the degree of melting in typical mantle material raised to a given pressure and temperature. They assumed that fertile mantle is garnet peridotite and found for the solidus temperature

$$P = \frac{T_s - 1100}{136} + 4.968 \times 10^{-4} e^{1.2 \times 10^{-2}(T_s - 1100)} \tag{8.1}$$

where P is the pressure in GPa and T_s the solidus temperature in °C. For the liquidus temperature, T_l, in °C,

$$T_l = 1736.2 + 4.343P + 180 \tan^{-1}(P/2.2169)$$

They defined a dimensionless temperature, T',

$$T' = \frac{(T - (T_s + T_l)/2)}{T_l - T_s} \tag{8.2}$$

and found that the degree of melting, as a fraction by weight of the rock, x, was given by

$$x - 0.5 = T' + (T'^2 - 0.25)(0.4256 + 2.988T') \tag{8.3}$$

Surprisingly, there was no clear evidence for variation of $x(T')$ with pressure.

To generate the 7 km thick oceanic crust, 7 km of melt are needed from the upwelling mantle. To do this, the potential temperature (Eq. 7.94a) of the source region must be about 1280°C. Rising melt crosses the solidus at a depth of about 45 km, at 1300°C, reaching the surface at 1200°C. McKenzie and Bickle obtained an average melting depth of 15 km and a melt fraction of 10–15%. The magma is about 10% MgO, and the melt fraction does not exceed 24%. Where the oceanic crust is thicker, as under Iceland (27 km), higher potential temperatures are implied, up to 1480°C.

8.4 Shallow Structure of Midocean Ridges

8.4.1 Topography

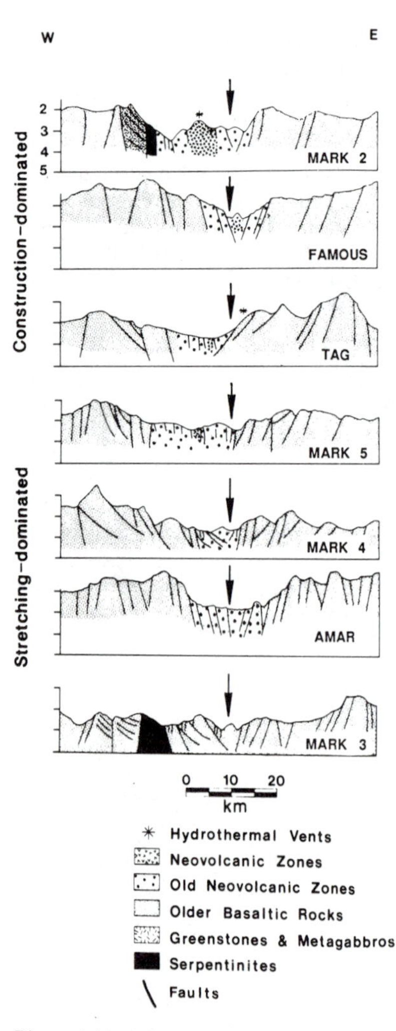

Figure 8.14. Schematic geological cross sections of the axis of the Mid-Atlantic Ridge. The MARK cross sections are at about 15 km intervals along the axis at 23°N, with MARK 2 being the most northerly and MARK 5 the most southerly. FAMOUS is at 37°N, AMAR at 36°N, TAG at 26°N. Arrow: ridge axis. Vertical exaggeration 3:1. Depth beneath sea surface in km. (From Karson et al. 1987.)

The axial topography of midocean ridges varies widely (Fig. 8.1). The crest of the Mid-Atlantic Ridge is very rugged with faulted blocks and a narrow axial *median valley*. In contrast, the axial regions of the Pacific–Antarctic Ridge and the East Pacific Rise are much smoother and generally lack the median valley. However, the presence of a median valley cannot be controlled purely by spreading rate, with slow-spreading ridges having median valleys and fast-spreading ridges lacking them, because there are exceptions. For example, the slow-spreading Reykjanes Ridge does not have a median valley, and in some locations the fast-spreading East Pacific Rise has an axial valley. These differences in axial topography can be modelled by the ductile (i.e., nonbrittle) extension of a viscous or plastic lithosphere which thickens with distance from the ridge axis, as well as the effect of magma accumulation beneath the plate. For slow-spreading ridges, the depth of the rift valley primarily depends on the strength of the model lithosphere and the rate at which it thickens; for fast-spreading ridges, topography is due mainly to the buoyancy of the magma. This theory is in agreement with observations on the East Pacific Rise: Where there is an axial high, a magma chamber is present; but at 23°N where no magma chamber is evident, there is an 18 km wide, 1 km deep axial rift valley.

Figure 8.14 illustrates the extent of variations in topography, tectonics and geology found on the Mid-Atlantic Ridge. The area which has undergone the most recent volcanic constructional event is shown at the top of the figure, and the area at the bottom has the most developed extensional tectonic features. The other areas are arranged in order between these two extremes. It is believed that this variability, which occurs over even short distances along the axis, is the result of a cyclic process (with a period of perhaps 10,000–50,000 years) with more or less continuous extension (moving apart of the two plates) but only periodic or intermittent magmatism. The magmatism in neighbouring ridge segments, or cells (a cell generally extends for a few tens of kilometres from one transform fault to the next), need not be synchronous with each other. The last major volcanic event would have occurred in the last few hundred years for the area at the top of the figure, whereas those areas at the bottom are in the last extensional phases of their cycle.

8.4.2 Crustal Magma Chambers

As has been shown, the region of partial melting in the upper mantle is probably relatively broad. This is in sharp contrast to the very narrow zone of intrusion at the ridge axis: Magnetic data (see Sect. 3.2) suggest that the standard deviation of the distribution of dyke injections must be only a few

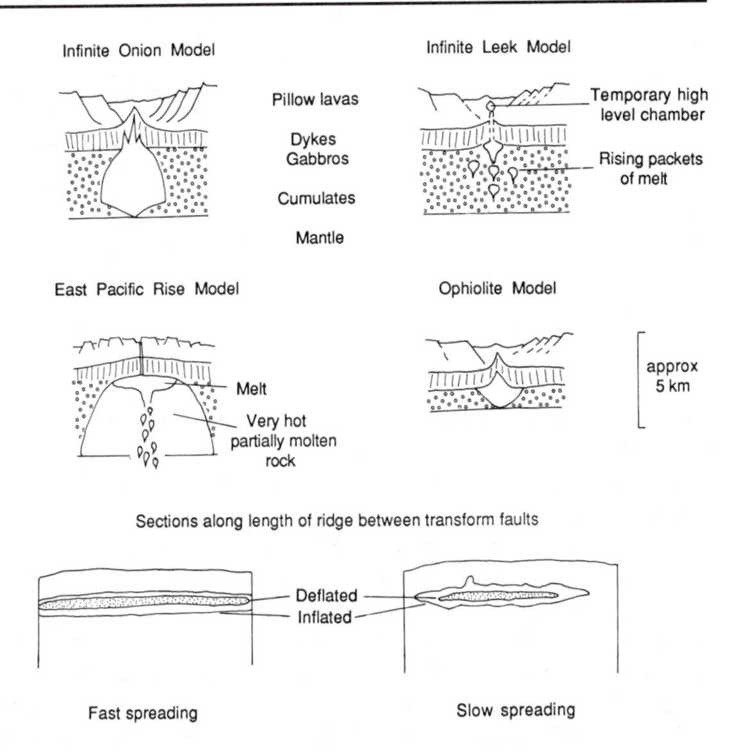

Figure 8.15. Four models of midocean ridge magma chambers (left-hand pair, fast spreading; right-hand pair, slow spreading). No vertical exaggeration. Since the 'infinite onion' model (Cann 1974, Bryan and Moore 1977) and the 'ophiolite' model (Pallister and Hopson 1981) have large steady-state magma chambers in the lower crust, they can easily produce the lavas and dykes (layer 2) and the gabbros and cumulates (layer 3) of the oceanic crust. Seismic experiments should be able to detect such chambers. The 'infinite leek' model (Nisbet and Fowler 1978) does not require a steady-state magma chamber; rather magma is envisaged as rising intermittently in packets from the mantle, nucleating a crack in the basaltic crust and rising to the surface. Each crack is likely to be in the same location as the previous crack. A small, shallow magma chamber could be a temporary feature of this model. The 'East Pacific Rise' model (Macdonald 1986, 1989) has a mushroom-shaped central magma chamber overlying a zone of very hot rock which is at most a few percent partially molten. The two vertical sections show the possible extent of the magma chambers for fast and slow spreading ridges with the magma supply low, 'deflated' (stippled) and plentiful, 'inflated'. The infinite onion and ophiolite models can be compared with inflated magma chambers whereas the infinite leek and East Pacific Rise models can be compared with deflated magma chambers. Infinite onion and infinite leek models can be regarded as end-members of a continuum of models whose character depends on spreading rate and degree of inflation.

kilometres or less though the actual extrusion zone of lavas onto the sea floor is probably only a few hundred metres wide.

The comparison of ophiolites with the oceanic crust initially resulted in a number of models (of varying degrees of complexity) of magma chambers in the crust at shallow depth beneath the ridge axis. These models were nicknamed 'infinite onions'; *infinite* because of their continuity in time and along the axis of the midocean ridge system, terminated only by transform faults, and *onion* because of their shape and because they peel off layers of oceanic crust at their sides. These magma chambers were thought to produce the basaltic lavas which are erupted onto the sea floor, as well as the underlying dykes and cumulate layers which are present in ophiolites and which are thought to make up much of seismic layers 2 and 3. A number of thermal models of crustal magma chambers have been made to determine limits on their size and dependence on spreading rate, crystal settling and thermal parameters. Clearly, an infinite onion magma chamber such as is illustrated here, extending as it does through much of the crust and being several kilometres in width, should be easily detectable seismically. Many detailed seismic reflection, refraction, gravity and deep-two surveys have been made to delineate these magma chambers on the Mid-Atlantic Ridge, East Pacific Rise and Juan de Fuca Ridge. Figure 8.15 shows some of the variations of magma chamber models and their probable geographic settings.

8.4.3 Thermal Models

It is easy to see how a large infinite onion magma chamber could produce the layered lava, dyke, and gabbro cumulate crust observed in ophiolites

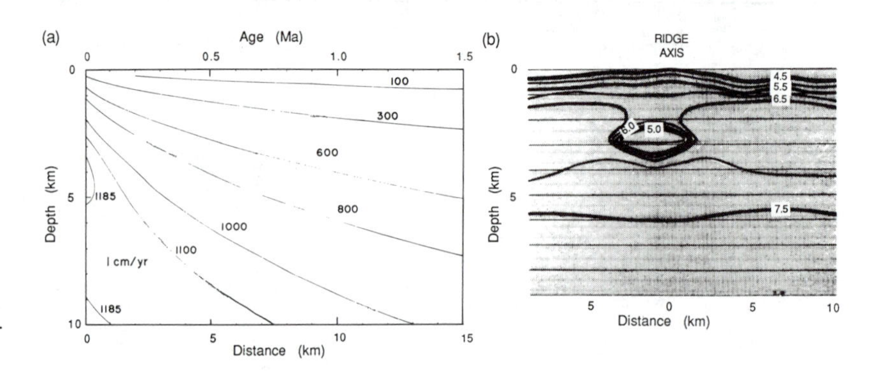

Figure 8.16. (a) Temperature in oceanic crust calculated for a midocean ridge with a half-spreading rate of 1 cm yr^{-1}. The edge of the magma chamber is indicated by the 1185°C isotherm (the solidus). The chamber is confined to depths appropriate for layer 3 and is only about 1 km wide. (From Sleep 1975.) (b) A seismic P-wave velocity model for the East Pacific Rise at 9°N (half-spreading rate 6.1 cm yr^{-1}) based on the first thermal model discussed in Section 8.4.3. (From Hale et al. 1982.)

and presumed to comprise the oceanic crust. Thermal models of the formation of the oceanic crust indicate, however, that such an extensive steady-state magma chamber cannot be present beneath slow-spreading ridges such as the Mid-Atlantic Ridge.

To calculate thermal models of midocean ridges, we need to know something about the eruption and solidification temperatures of likely crustal and upper mantle materials. Estimates of 1203–1246°C have been made for the eruption temperature of basalts collected from the FAMOUS area of the Mid-Atlantic Ridge. In melting experiments on a Mid-Atlantic Ridge basaltic glass, clinopyroxene disappeared at 1166°C, plagioclase at 1215°C and olivine at 1245°C. When melting a Mid-Atlantic Ridge olivine tholeiite (at 0.15 GPa), experiments found that clinopyroxene disappeared at 1160°C, plagioclase at 1180°C and olivine at 1210°C.

Thermal models of midocean ridges are computed numerically by using the heat conduction Eq. 7.59. The exact size and shape of any magma chamber depends on the assumptions made about the latent heat of solidification and the temperatures for the boundary conditions, as well as the spreading rate. Figure 8.16a shows one steady-state model made for a ridge spreading with a half-rate of 1 cm yr^{-1}. In this model, the temperature at the base of the lithosphere was assumed to be 1290°C, the edge of the magma chamber was defined by the 1185°C isotherm, and all the latent heat was released at the axis. The maximum half-width of this magma chamber is about 0.5 km. In another model, the temperature at the base of the lithosphere and along the intrusion axis was assumed to be 1125°C, the edge of the magma chamber was defined by the 1000°C isotherm and the latent heat was distributed over the liquidus–solidus interval. The maximum half-width of this chamber is about 4 km (or 3 km if the latent heat is released entirely at the solidus). Some of the apparent difference between these two models is due to the different temperature assumptions, and some is due to the different treatments of latent heat and upper boundary conditions. The temperature measurements discussed previously suggest that the temperature assumptions of the second model are too low. When this second model is recalculated using more appropriate liquidus temperatures of 1230–1250°C and solidus temperatures of 1160–1185°C, the maximum half-width of the magma chamber is about 2 km (or less if the latent heat is released entirely at the solidus). Latent heat of about 4.2×10^5 J kg^{-1} is a reasonable estimate for basalt.

The effect of accumulation of crystals settling to the bottom of a magma chamber can also be included into the thermal models. The width of any chamber is reduced by crystal settling: A 2:1 ratio of solidified material forming the chamber roof to that falling to the bottom reduces the maximum width of the magma chamber to about one-half its previous value, and a 1:1 ratio results in a one-third reduction of width. Crystal settling also has a significant effect on the shape of the magma chamber, changing it from triangular (apex at top) to a diamond shape.

Thermal models calculated for medium- and fast-spreading ridges are very similar to those for slow-spreading ridges, the only significant difference being that the magma chambers are much wider, as one would expect. Thermal considerations therefore imply that a *steady-state crustal* magma chamber can exist on fast-spreading ridges, though for half-spreading rates less than 1 cm yr^{-1} it is questionable whether any steady-state crustal magma chamber can exist. For the Mid-Atlantic Ridge, the maximum half-width is probably 0.5–1.0 km, and any magma chamber would be confined to depths equivalent to the lower part of layer 3 (i.e., 4–6 km below the sea floor). For the East Pacific Rise, a half-width of 3–5 km would be generally appropriate. Note that if much hydrothermal cooling of the crust occurs, then these conductive thermal models are not applicable because cooling would be very much faster. This means that the size of any magma chambers would be much reduced.

Figure 8.16b shows a possible velocity model for a fast-spreading ridge, the East Pacific Rise at 9°N (half-spreading rate 6.1 cm yr^{-1}). This model was derived using the first thermal model discussed, P-wave velocities measured on the Semail ophiolite samples, a rate of change of seismic velocity with temperature $d\alpha/dT$ of -8×10^{-4} km s^{-1} $°C^{-1}$ and an arbitrary 1.5 km s^{-1} reduction in seismic velocity in the magma chamber to model the partial melt (compare with Fig. 8.13). The maximum width of this magma chamber is about 5 km, and its top is some 2 km beneath the seabed.

8.4.4 Hydrothermal Circulation in Young Oceanic Crust

Hot water is less dense than cold water. Any hot water that has entered the porous, rubbly flank of a volcano heats and rises, to be replaced by cold water from rain or from the sea. At midocean ridges this process is intense, and water penetrates very deeply into the new oceanic crust. The system is so vigorous that huge, well-organized circulations are set up, with cold water entering, warming as it penetrates to depths of several kilometres and then rising. The rising water may pass through well-defined channels to debouch back into the ocean at temperatures that are nearly critical, often approaching 400°C at the ambient pressures of the ocean floor. Enormous volumes of water pass through these systems, so great that in total a volume equal to the entire volume of the oceans passes through the ridge hydrothermal systems in about 10 million years.

This hydrothermal circulation through young ocean floor is one of the most important geochemical and geophysical processes on earth. It plays a major role in controlling the chemistry of sea water, in the operation of subduction zones, in the growth of continents and in managing the earth's heat budget. The hydrothermal circulation is the cooling radiator of the front of the earth's engine, like the splendid radiator on a Rolls-Royce, but it is much more than just a radiator.

There are immediate chemical results of the hydrothermal circulation. As sea water passes through rock, it exchanges cations such as sodium and calcium, dissolving some, moving or precipitating others. This changes the chemistry and the mineralogy of the rock, altering and metamorphosing it, depending on the depth and extent of hydration. Slightly away from the axis, the new basalt is already a much rearranged metamorphic rock, typically with its fine-grained minerals and with any glassy groundmass partly or substantially changed. Water has been added; and in many places, CO_2 has also been added, typically as calcium carbonate precipitated in the pores and fissures through which the water flowed.

When the water leaves the rock, it is often very hot. In many places on the midocean ridges, these vents are well-defined pipes that build up into chimneys as minerals precipitate from the water when it is suddenly cooled on reentering the ocean, which is around 4°C, near freezing. Clouds of dark, precipitated chemicals vent from these chimneys, called *black smokers*. *White smokers* are similar chimneys formed by somewhat cooler ascending water. Exotic communities of life have been discovered around these hot-water vents: giant tubeworms up to 3 cm in diameter and 3 m long, giant clams, crabs and unusual bacteria. These communities of unusual organisms were discovered by geophysicists, much to the surprise of biologists. The lives of these organisms do not depend directly on the sun (as all other life does) but on chemical energy from the hydrothermal system. They are not wholly disconnected from the rest of the biosphere, however, since they depend on the oxidation state of the environment, which is set by photosynthetic life. The flux of water leaving the smokers is very great and is one of the most important controls on the chemistry of the sea. In the last century, geologists attempted to calculate the age of the earth by measuring the amount of salt in the sea and dividing it by the amount of salt annually brought down by rivers from the land. We now know that the hydrothermal circulation is one of the principal controls on sodium; so the early calculation was invalid.

The hydrothermal circulation cools the new oceanic plate very rapidly in geological terms. This cooling has an effect on plate thickness (see Sect. 7.5) because the top few kilometres of plate are 'instantly' cooled. In effect, the cooling reduces the mean age of the oceanic crust before subduction. Perhaps 25% of the total heat loss of the earth takes place via the hydrothermal systems near the ridges. As the oceanic plate moves away from the ridge and cools, the hydrothermal systems slowly lose their vigour.

At subduction zones, the influence of the hydrothermal circulation is again felt. On a totally dry planet, the subducting crust (if subduction took place at all) would be without water. On earth, the descending plate includes much water and CO_2 (even if virtually all the sediments lying on top of the oceanic crust are scraped off; see Sect. 9.2.2) because the slab is so heavily hydrated. The water and CO_2 in the descending plate are eventually driven off and rise into the overlying mantle wedge. There the water plays a critical role in promoting melting since wet rocks melt at much lower temperatures than dry rocks. This melting (Sect. 9.2.1) is the process that leads eventually to the magmas that have formed the continents.

Finally, a very small amount of water and CO_2 can descend with the plate into the deeper interior and can recirculate through the upper mantle over billions of years, eventually to rise again with the mantle under midocean ridges. Even the very small amount of water that exists in ridge lavas has important geochemical and tectonic effects, such as promoting the melting process.

8.4.5 Seismic Structure

The possibility of the existence of extensive, crustal infinite onion magma chambers (Fig. 8.15) has led to a number of seismic experiments over the axial regions of the midocean ridges in an attempt to delineate these chambers. Such magma chambers should be characterized by low seismic velocities and high attenuation.

In the early model of the seismic structure of the oceanic crust (Fig. 8.11e, f, g), it appeared as though there was a very wide zone centred on the ridge axis in which normal layer 3 and upper mantle velocities were not measured. However, subsequent experiments have shown that the 'normal' oceanic crustal structure is present except over a very narrow axial zone (about 20 km wide in some cases). In this zone, layer 3 often appears to be absent or has a reduced velocity, and normal upper mantle velocities are frequently not observed. Instead, the highest velocity measured is generally 7.1–7.6 km s^{-1}.

Slow-Spreading Ridges A seismic experiment shot on the crest of the Reykjanes Ridge at 60°N gave an upper mantle velocity of only 7.1 km s^{-1}. A number of experiments shot on the Mid-Atlantic Ridge have yielded upper mantle velocities at the ridge axis of 7.2–7.6 km s^{-1}. The final, well-constrained seismic velocity models for those experiments, for which synthetic seismograms were used as part of the modelling procedure, are shown in Figure 8.17. None of these experiments indicates the presence of a crustal low-velocity zone. However, an experiment on the Mid-Atlantic Ridge at 23°N, just south of the Kane Fracture Zone, has found near-normal oceanic crustal and upper mantle velocities beneath a portion of the median valley. These differences are thought to reflect the time since the beginning of the last volcanic cycle. At 23°N, the crust is older and cooler and so is closer to the velocity of normal oceanic crust.

A seismic reflection survey shot across the axis of the Mid-Atlantic Ridge at 37°N detected a small, shallow low-velocity zone in the crust. This appears to be a zone of extensive cracking at the injection centre associated with the extrusion of lavas at the seabed. Seismic experiments on the Juan de Fuca Ridge show no evidence of crustal magma chamber.

Perhaps the most conclusive test of the presence of a partially molten zone can be provided by shear waves, which should be greatly attenuated when they cross such a zone. Arrays of three-component seismographs, with which it is possible positively to identify shear waves, have monitored the microearthquake activity on the Mid-Atlantic Ridge (Fig. 8.18). There is no evidence for any special crustal attenuation of the shear waves since the microearthquakes show large-amplitude shear waves. The crustal shear

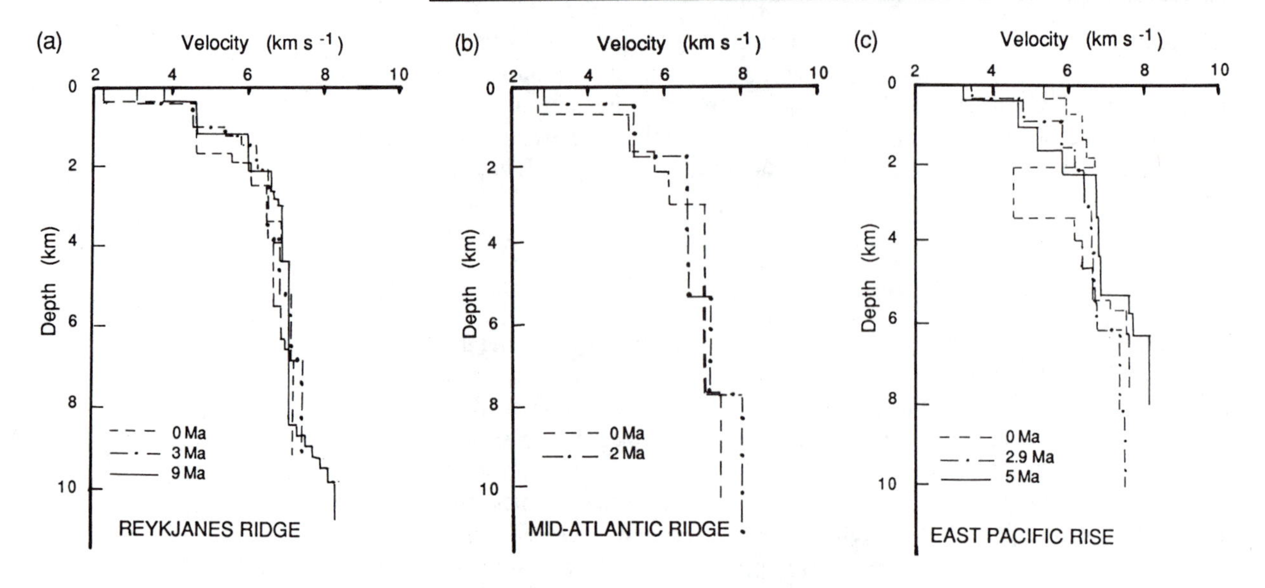

Figure 8.17. Seismic P-wave velocity structures for the axial regions of midocean ridges: (a) Reykjanes Ridge at 60°N, (b) Mid-Atlantic Ridge at 37°N and (c) East Pacific Rise at 9°N. All these velocity structures were determined using synthetic seismograms. (After Bunch and Kennett 1980, Fowler 1976 and Orcutt et al. 1976.)

waves have travel times consistent with their having crossed the axial zone at depths to about 5 km beneath the seabed. At these depths they would have had a wavelength of up to 1 km.

Examination of all the seismic and thermal evidence strongly suggests that no large, continuous crustal magma chamber (as in Fig. 8.15) can be present beneath the axis of slow-spreading ridges though the possible presence of small pockets of melt, up to perhaps 2 km in diameter, cannot be excluded.

Fast-Spreading Ridges In contrast to the evidence from the slow-spreading ridges, seismic refraction and reflection experiments shot on the East Pacific Rise have given strong support to the existence of a crustal magma chamber on this fast-spreading ridge. At 9°N, where the half-

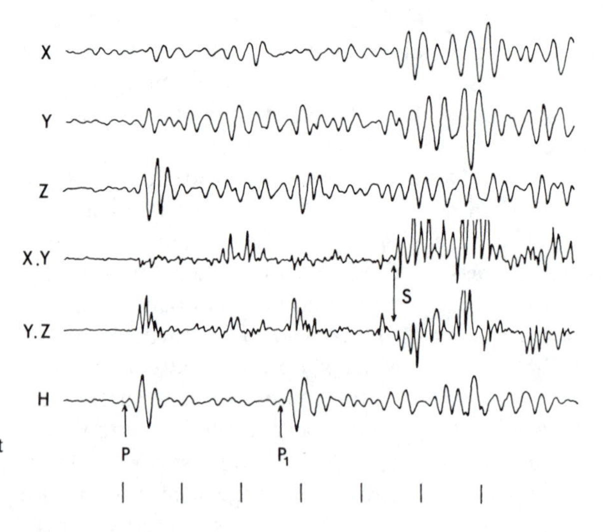

Figure 8.18. Record from an ocean-bottom seismometer (OBS) of an explosive charge 33 km away and on the opposite side of the axis of the Mid-Atlantic Ridge. X and Y (horizontal) and Z (vertical) seismometer components; H, hydrophone; time marks every second; X.Y (product of X and Y) and Y.Z (product of Y and Z) are displayed with a smaller scaling factor than X, Y and Z. P is the first-arriving P-wave; it has crossed the ridge axis in the lower part of layer 3; P_1 is the sea surface reflection (multiple) of P; S is the first-arriving shear wave, which has also crossed the ridge axis in the lower crust. The arrival time of the shear wave can be picked most accurately from the products X.Y and Y.Z. (From Fowler 1976.)

spreading rate is 6 cm yr^{-1}, a zone with a velocity of around 5 km s^{-1} was detected at the ridge axis by a seismic refraction experiment using ocean bottom seismometers (Fig. 8.17c). This low-velocity zone was not present in the models for 2.9 Ma and 5.0 Ma old crust. Along the ridge axis and at age 2.9 Ma, the mantle arrivals indicate low P-wave velocities of 7.6–7.7 km s^{-1}. However, by age 5 Ma, the mantle P-wave velocity is 8 km s^{-1}, and a more typical oceanic crustal layering appears to have developed. This model was determined from an analysis of the travel-time and distance data, as well as from waveform and amplitude studies using synthetic seismograms. Other refraction data from the same area confirm these results and indicate that the low-velocity zone thins with age and is probably confined to the 10 km wide axial block.

Another refraction experiment on the East Pacific Rise (at 11°20′N) using three-component ocean bottom seismographs indicated that both crustal P- and S-waves crossed the ridge crest with high amplitudes and no apparent time delay. If magma is present within the crust at this location, these results indicate that it must be confined to narrow dikes or isolated bodies less than about 1 km in vertical extent.

Evidence for the existence of a narrow magma chamber beneath the East Pacific Rise axis at 21°N was provided by the differing attenuation of P- and S-waves (Fig. 8.19). The P-waves were well transmitted to a seismometer located on the ridge axis, but the S-waves were severely attenuated. In contrast, both P-waves and S-waves were well transmitted to a seismometer located 10 km from the ridge axis. This suggests that a narrow magma chamber is present beneath the ridge axis at this location. A schematic of the crest of the East Pacific Rise at 21°N is shown in Fig. 8.20.

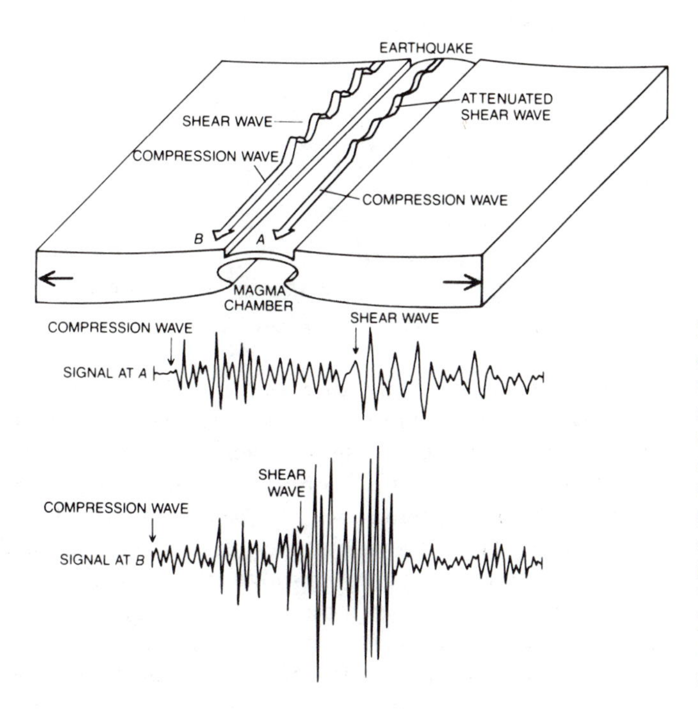

Figure 8.19. Compressional waves from a distant earthquake are well transmitted to an ocean-bottom seismometer A, situated on the East Pacific Rise axis, but the shear waves are attenuated. Both compressional and shear waves are well transmitted to a seismometer B, 10 km from the ridge axis. These observations suggest that a narrow magma chamber is present beneath the ridge axis. (From The crest of the East Pacific Rise, Macdonald and Luyendyk. Copyright © 1981 by Scientific American, Inc. All rights reserved.)

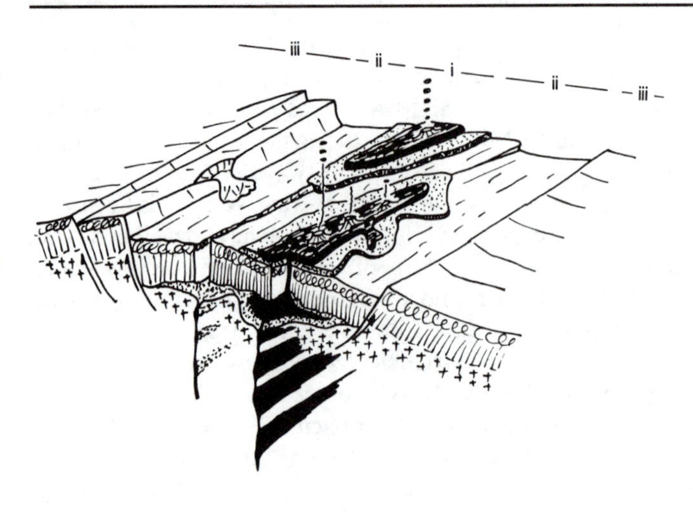

Figure 8.20. Schematic of the crest of the East Pacific Rise at 21°N. The ridge crest at this location consists of an axial block some 5 km wide bounded by 80 m high fault scarps (off the edges of this diagram). In this axial block are three domains: (i) Axial extrusive zone characterized by volcanism and active hydrothermal vents or smokers. This zone has a 1.7 km offset in the study area. (ii) Horst and graben zone with open fissures and some sediment covering basalt flows. (iii) Active tectonic zone consisting of tilted fault blocks. The dark material beneath the axial zone is olivine basalt in a magma chamber. This has recently erupted and is associated with the active smokers. Stippled pattern is older basalt which erupted through the same fissures and formed a residual, partially solidified boundary layer sheath in the magma chamber. There is some seismic evidence indicating the presence of a low-velocity zone, a few kilometres wide and centred on the ridge axis. (After Hekinian and Walker 1987.)

Example: seismic velocity and width of a low-velocity zone

The velocity of the lower crust in a model of the East Pacific Rise at 12°N is 7 km s^{-1}. Calculate the seismic P-wave velocity of a lower crustal magma chamber if ray paths crossing the axis are delayed by 0.1 s and the width of the magma chamber is assumed to be (a) 1 km, (b) 3 km and (c) 6 km. If w is the width of the chamber, the travel time in normal lower crustal material is $w/7.0$ s. Ray paths crossing the axis are 0.1 slower than this, so their travel time is $0.1 + w/7.0$ s. If the seismic velocity in the magma chamber is α km s^{-1}, then

$$\frac{w}{\alpha} = 0.1 + \frac{w}{7.0}$$

Rearranging this equation gives the seismic velocity in the chamber as

$$\alpha = \frac{7w}{w + 0.7}$$

Therefore, (a) for $w = 1$ km, $\alpha = 4.12$ km s^{-1}; (b) for $w = 3$ km, $\alpha = 5.68$ km s^{-1}; and (c) for $w = 6$ km, $\alpha = 6.27$ km s^{-1}.

Several thousands of kilometres of common-depth-point seismic reflection data have been obtained from the axial region of the East Pacific Rise in the search for the elusive crustal magma chamber. Data from a reflection line which crossed the ridge axis (one of a series shot in 1985) are shown in Fig. 8.21a. A large-amplitude reflector about 3 km wide and centred on the ridge axis can be seen at about 0.6 s beneath the seabed. This reflector, which is continuous beneath the ridge axis for many tens of kilometres (Fig. 8.21b), has negative polarity, indicating that it arises at an interface with a strong negative velocity contrast. All these data are consistent with this arrival being a reflection from the lid of a magma chamber. The magma chamber reflection was observed on about 60% of the 3500 km of reflection

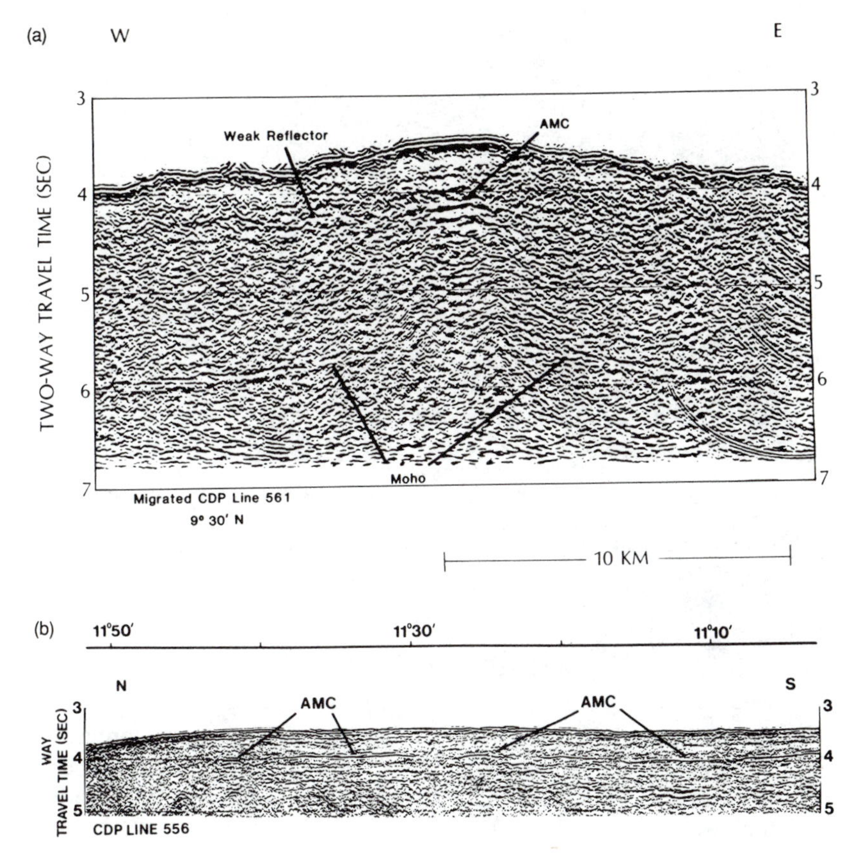

Figure 8.21. (a) Migrated CDP reflection record section across the axis of the East Pacific Rise at 9°30′N. The first reflector at 3.5–3.9 s is the seabed reflection (water depth 2.6–2.9 km). The large-amplitude event beneath the ridge axis is interpreted as the reflection from the lid of an axial magma chamber (AMC). Its extension as a very weak reflector out onto the flanks of the ridge could perhaps represent the top of a frozen chamber or the change from extrusives to dykes and gabbros. Reflections arriving at the correct time for Moho reflections extend to within 2–3 km of the ridge axis. (b) CDP reflection record section along 100 km of the axis of the East Pacific Rise (11–12°N). The axial magma chamber (AMC) can be traced as a continuous event for tens of kilometres. The AMC reflection was not recorded north of 11°50′N. (From Detrick et al. 1987.)

profile of this particular 1985 survey. The depth to the top of the magma chamber varies between about 1.2 and 2.4 km beneath the seabed and is correlatable with the depth of the ridge axis. Where the ridge axis is shallowest, the magma chamber lid is shallowest; where it is deepest, the magma chamber is discontinuous or nonexistent, which shows the effect of magma supply on the axial topography. The presence of a magma chamber is also indicated by three expanding spread reflection profiles which were shot parallel to the East Pacific Rise at 13°N (Fig. 8.22). The profile shot along the ridge axis (ESP9) shows a pronounced shadow zone for crustal arrivals beyond 11 km range, indicating the presence of a crustal low-velocity zone. The profile shot 1.5 km off the ridge axis, (ESP11) shows some evidence for a thin low-velocity zone (note the discontinuity in first arrivals between 8 and 10 km). The profile 3.6 km away from the axis (ESP13) shows no evidence for a low-velocity zone and has a typical oceanic velocity structure. Thus, at 13°N the low-velocity zone has a total width less than 5 km. The event at 6 s two-way time (Fig. 8.21a) is interpreted as being a reflection from the Moho. This reflection can be traced to within 2–3 km of the ridge axis, which again limits the maximum width of the magma chamber to about 5 km. There is no evidence as to whether the magma chamber is literally that – a chamber filled with liquid magma – or whether it is a chamber filled with partially solidified crystal mush. There were no reflections from beneath the chamber to enable interval velocities or

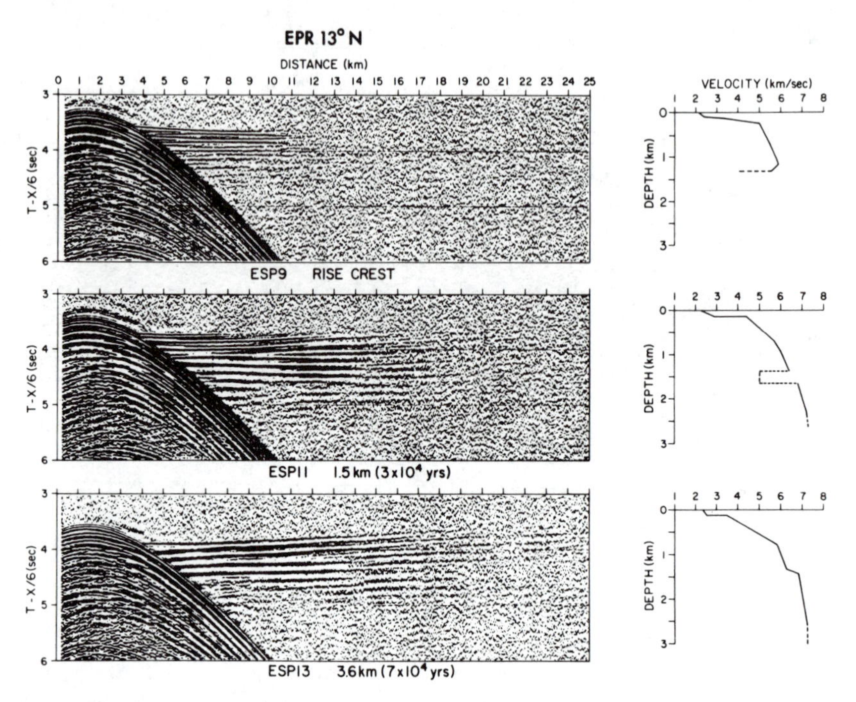

Figure 8.22. Reduced record sections for three expanding spread profiles (ESP) shot along and parallel to the East Pacific Rise axis at 13°N. ESP9 was shot along the ridge axis, ESP11 was shot 1.5 km away on 0.03 Ma crust and ESP13 was shot 3.6 km away on 0.07 Ma crust. At the right of each record section is the velocity–depth model determined from it by travel-time analysis. Note the low-velocity zone at the ridge axis which produces a shadow zone on ESP9. The slight discontinuity in first arrivals at 8–10 km on ESP11 may indicate that a thin low-velocity zone is still present there. Beneath ESP13, however, there is typical oceanic crust, and clearly no low-velocity zone is present. (From Detrick et al. 1987.)

estimates of attenuation to be made. A schematic of the magma chamber on the East Pacific Rise is shown in Figure 8.23. This model magma chamber is very much smaller than many that have been proposed as results of studies of ophiolites.

The preceding discussion of some of the seismic experiments carried out over the East Pacific Rise shows clearly that, just as is indicated for the Mid-Atlantic Ridge, the East Pacific Rise is not such a simple one-dimensional feature as was first thought. Although a magma chamber seems to be present in some locations, it apparently varies widely in size and extent along the axis and is not a continuous feature. In contrast, the Moho discontinuity appears to be continuous under all oceanic crust except for a very narrow zone some few kilometres in width along the ridge axis.

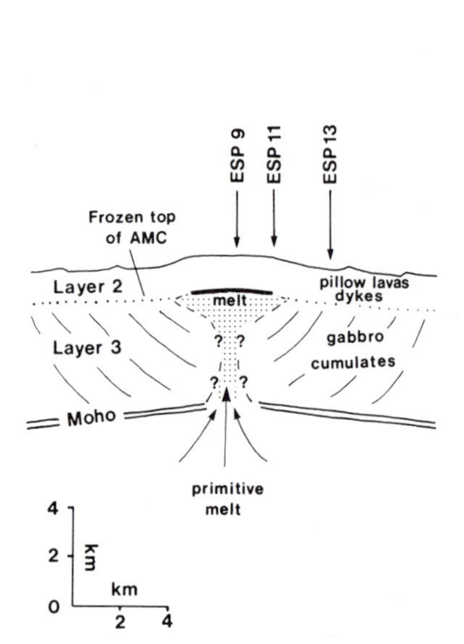

Figure 8.23. Schematic of the magma chamber beneath the East Pacific Rise, based on the results of the reflection profiling illustrated in Figs. 8.21 and 8.22. (From Detrick et al. 1987.)

8.5 Transform Faults

8.5.1 Geometry

Transform faults are plate boundaries at which material is neither created nor destroyed. The relative motion between the two plates is parallel to the strike of the fault. By referring to Sections 2.3 and 2.4, you will appreciate that since lines of constant velocity are small circles about the pole of rotation between two plates, transform faults are arcs of small circles. From this fact, the relative motions of the plates can be determined, as discussed in Chapter 2. If the relative motion between the two plates is not exactly parallel to the fault, there is a small component of convergence or divergence. A transform with a small component of divergence is called a *leaky transform fault*. An example of a leaky fault is the plate boundary

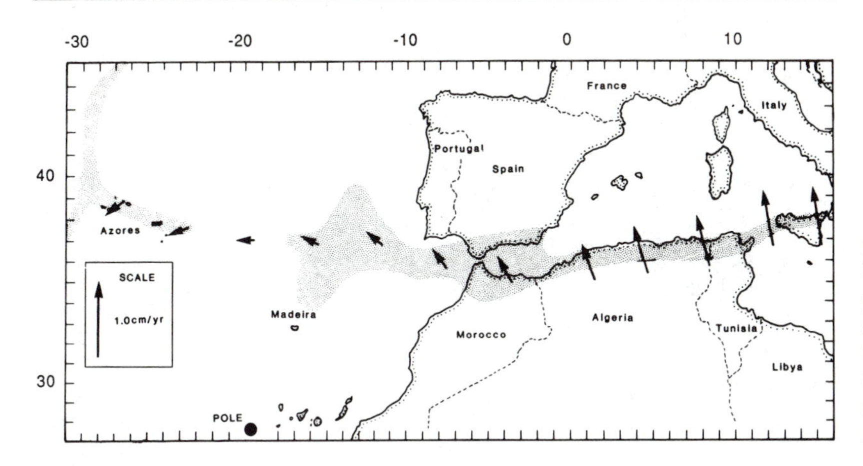

Figure 8.24. Slip vectors for the motion of Africa relative to Eurasia along their plate boundary. The length of each arrow indicates the magnitude of the velocity. Solid circle denotes the pole position. Shading indicates the regions of most intense seismicity. (From Anderson and Jackson 1987.)

between Eurasia and Africa, from the Azores Triple Junction on the Mid-Atlantic Ridge eastwards towards Gibraltar (Fig. 8.24). Earthquake fault-plane solutions for this boundary near the Azores illustrate the small amount of extension occurring there.

Table 8.3 lists some of the main transform faults on the midocean ridge system. Some transform faults, such as the San Andreas Fault in California, occur on land, but they usually occur between ridge segments at sea. Many Indian and Pacific transform faults are not listed because data are not yet available for them; the details of much of the ocean bed remain unknown. Due to financial and logistical considerations, the most easily accessible parts of the ocean bed are the most studied and best understood. The term *transform fault* is correctly applied only to the active region of a fault (i.e., the part between the offset ridge segments). Extending beyond the active region is a zone of fracturing called a *fracture zone*, or an 'inactive fossil transform fault'; examples are the Kurchatov Fracture Zone and the Clipperton Fracture Zone. There is no horizontal slip motion on the fracture zones, but slight vertical adjustments do occur.

Transform faults range from very long to very short. The equatorial transform faults in the Atlantic are many hundreds of kilometres long. However, the more detailed the bathymetric, magnetic and reflection surveys become, the more numerous small transform faults are seen to be. In the FAMOUS area at 37°N on the Mid-Atlantic Ridge, the ridge segments are 20–60 km in length and are offset by small transform faults about 20 km in length. Even on this small scale, the regular geometric pattern is maintained.

Figure 8.25 illustrates a transform fault between two ridge segments. In this example, the offset is 200 km. The fault juxtaposes materials that differ in age by 20 Ma. Note that although the spreading rate is 1 cm yr^{-1}, the slip rate along the fault (which is twice the spreading rate) is 2 cm yr^{-1}. This slip motion occurs on the fault only between the two ridge segments. This is demonstrated by the location of earthquake epicentres: Outside the active transform faults, seismic activity is negligible (Fig. 8.26).

Even though transform faults are active only between the two offset ridge segments, they remain major features on bathymetric, gravity (Fig. 5.10) and magnetic anomaly maps outside the active zone because of the

Table 8.3. *Details of some transform faults*

Name	Location[a]	Offset (km)	Approx. age contrast (Ma)	Approx. topo-graphic relief	Approx. truncated lithosphere thickness[b]	Present slip rate (cm yr^{-1})
Charlie Gibbs	MAR 52°N	260	20	2.5	50	2.5
Kurchatov	MAR 40°N	22	2	1.5	15	2.5
A }	FAMOUS	20	2	1.5	15	2
B }	area: MAR 37°N	20	2	1	15	2
Oceanographer	MAR 35°N	120	13	4	40	2
Kane	MAR 24°N	145	10	3	35	2.5
Vema	MAR 11°N	315	20	3	50	3
Romanche	Equatorial MAR	935	60	4	85	3
Blanco	JdFR	300	10	—	35	6
Mendocino	JdFR 40°N	1150	27	2	55	2–4
Tamayo	EPR 23°N	80	3	1	20	6
Clipperton	EPR 10°N	85	2	1	15	11
Gofar	EPR 5°S	180	2	2	15	15
Wilkes	EPR 9°S	55	1	1	11	16
Atlantis II	SWIR 32°S	210	22	—	50	1
Udintsev	PAR 58°S	300	10	1	35	6
Eltanin system	PAR 55°S	900	30–34	—	60	6

[a]MAR, Mid-Atlantic Ridge; JdFR, Juan de Fuca Ridge; EPR, East Pacific Rise; SWIR, South West Indian Ridge; PAR, Pacific–Antarctic Ridge.
[b]From Eq. 7.65.

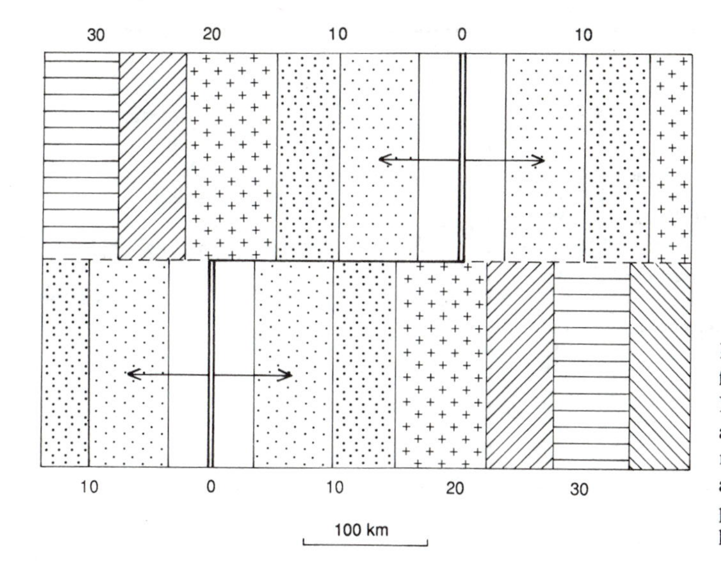

Figure 8.25. Idealized map of a transform fault offsetting two ridge segments. Numbers are age in Ma. Age provinces are progressively shaded. Note that the motion on this fault is sinistral not dextral as it would have been described in the pre-plate-tectonic, pre-magnetic-stripe literature.

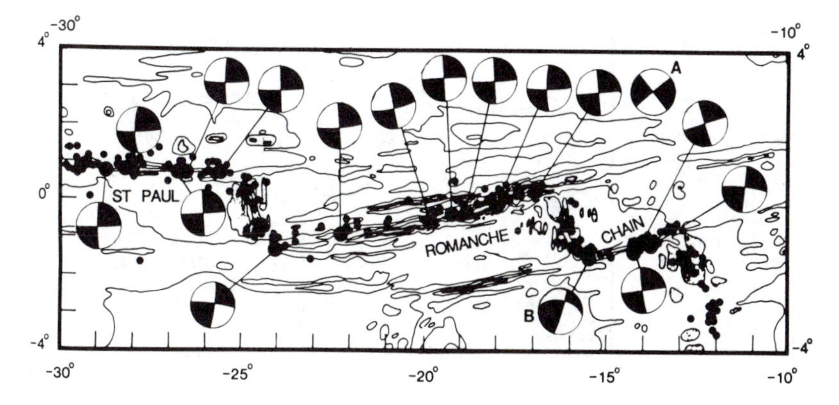

Figure 8.26. Bathymetric map of the St. Paul, Romanche and Chain transform faults in the equatorial Atlantic (Table 8.3). Dots indicate earthquake epicentres. The large dots are those for which the fault-plane solutions are shown. Notice how closely the epicentres follow the ridge axis and transform faults. Because the foci were all shallow (less than 5 km), and the depth range of faulting is about 0–10 km, the foci were above (shallower than) the 600°C isotherm. (From Engeln et al. 1986.)

constant age contrast across the fault. The majority Atlantic fracture zones are bathymetric features from America to Africa and start at the continental margin. This observation was puzzling to earth scientists before the development of plate tectonics, and the problem of *transcurrent faults* (as they were then called) was the subject of much debate. Before sea-floor spreading was known, the fault shown in Figure 8.25 would have been thought to be caused by a right lateral offset of the ridge axis. In this case, the entire fault would be active, not just the central portion, and the slip motion on the fault would be in the opposite direction to that shown in the figure. This gave rise to the question "What happens to the fault on the continent?" or "Where do transcurrent faults end?" Plate tectonics neatly solved the problem.

There are two main reasons why there are transform faults offsetting ridge axes. First, the major faults are determined by the geometry of the initial break between the two continents; this break occurs along lines of weakness which probably are associated with old geological structures. The break need not be a straight line and may be quite irregular. Transform

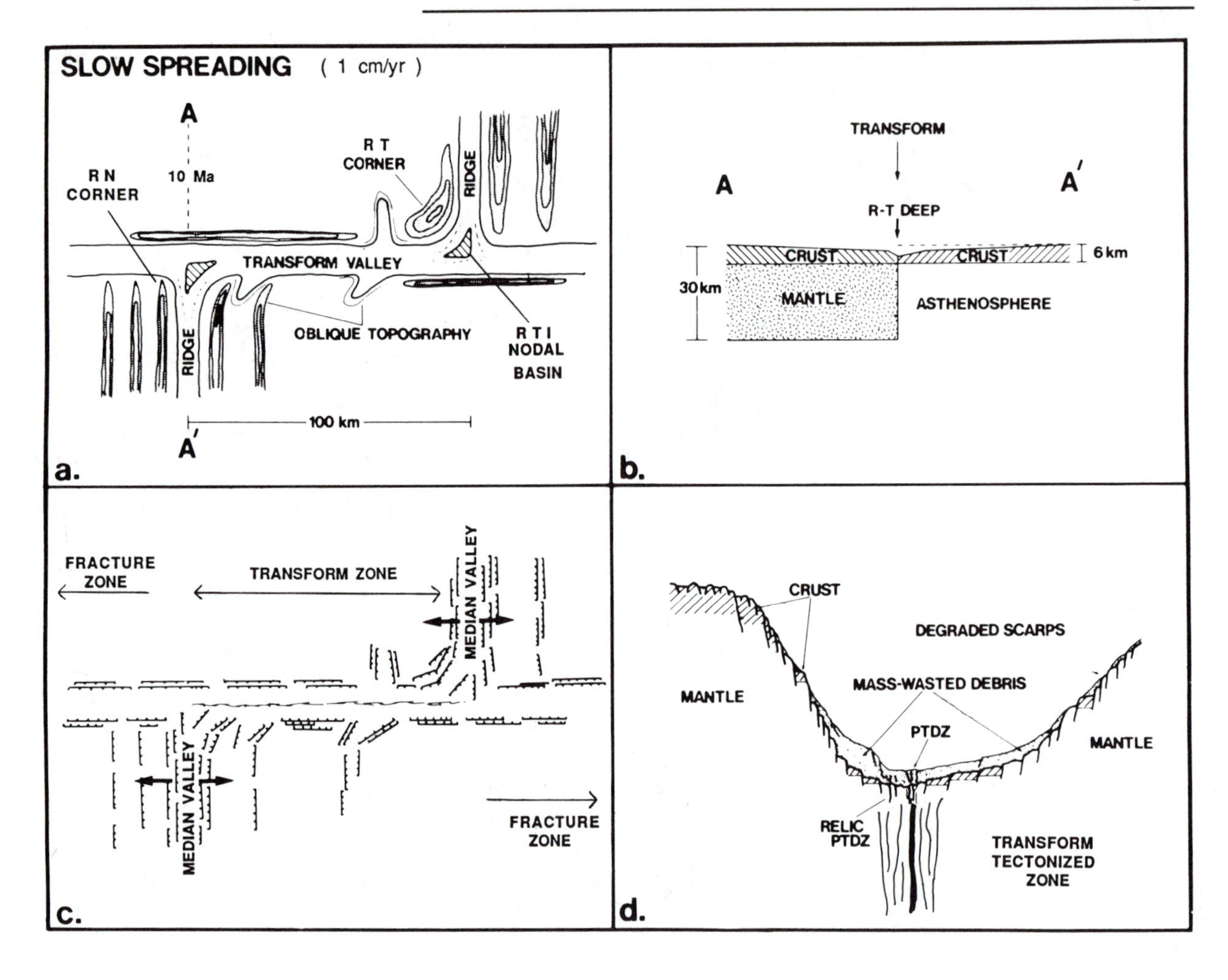

faults tend to develop in locations where the relative motion is approximately parallel to the break and are necessary to accommodate this geometry with the pole of rotation. Thus, the preferred spreading geometry configuration consists of straight ridge segments offset by orthogonal transform faults. Second, the small offset transform faults probably form to accommodate the change in relative plate motion with distance from the rotation pole. In this way, ridge segments with constant spreading rates are created.

8.5.2 Topography and Crustal Structure

Transform faults are major bathymetric features, visible on magnetic anomaly maps and marked by earthquake epicentres. Figure 8.25 and problem 9 suggest that transform faults might ideally be marked by a single vertical fault; but instead, active transforms are anomalous linear valleys which are usually less than 15 km wide, bounded by inward-facing scarps and depressed from 1 to 5 km. In the idealized single fault model, the oceanic crust is exposed along the fault wall and easily accessible to geologists for sampling by dredging and drilling. However, in practice,

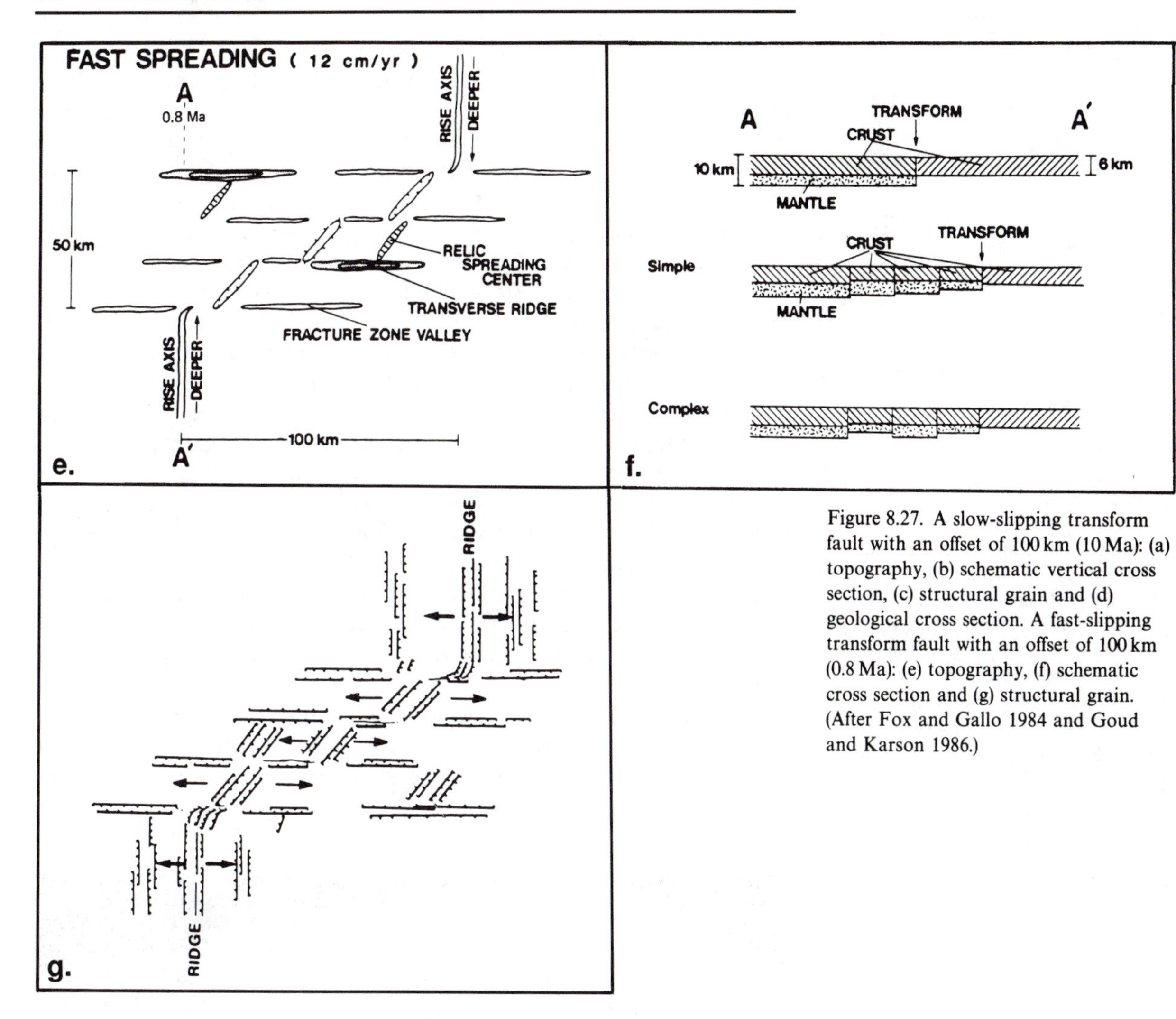

Figure 8.27. A slow-slipping transform fault with an offset of 100 km (10 Ma): (a) topography, (b) schematic vertical cross section, (c) structural grain and (d) geological cross section. A fast-slipping transform fault with an offset of 100 km (0.8 Ma): (e) topography, (f) schematic cross section and (g) structural grain. (After Fox and Gallo 1984 and Goud and Karson 1986.)

instead of a single fault scarp, there are usually a large number, each typically with a throw of only a few hundreds of metres or less.

Some of the jargon used in the literature to describe the various provinces along transform faults is illustrated in Figure 8.27, a schematic of the bathymetry and faulting associated with a slow-slipping transform. The valley along the line of a transform fault is the *transform valley*. The transform domain or *transform tectonized zone* (TTZ) is the region which has been affected by deformation associated with the ridge offset (i.e., the shear zone). The zone of interconnecting faults which are presently active (the active fault zone) is called the *principal transform displacement zone* (PTDZ). At the ridge–transform fault intersection (RTI), there is a deep, closed-contour depression called a *nodal basin*. The junctions of the median valley walls with the transform valley walls are called the ridge–transform (RT) and ridge–nontransform (RN) corners.

On fast-spreading ridges, transform faults may not occur as distinct faults at regular intervals; instead, they are less frequent and often extend over wide shear zones, perhaps up to 50 km wide, and have a series of ridge–

fault segments. It should not be surprising that fast-slipping transforms operate somewhat differently from slow-slipping transforms. There is a difference of up to a factor of ten in their slip rates, which means that their thermal structures are very different. Figure 8.27 shows the features which might reasonably be expected to result at a large offset, fast-slipping transform fault. It is clear that there is a wide shear zone made up of a series of ridge–fracture segments. Fast-slipping transform faults have been unstudied until very recently. The East Pacific Rise between the Nazca and Pacific plates, the fastest-spreading ridge segment in the world, is offset by a number of long transform faults. Results from various bathymetric and tectonic studies of these faults show that they can be best described as 20–150 km wide shear zones of complex topography.

Work on the East Pacific Rise has resulted in more new terminology. Although the terms *transform fault* and *fracture zone* are used, respectively, for the active and inactive portions of faults, the terms *deviations in axial linearity* (DEVAL) and *overlapping spreading centre* (OSC) have also been introduced. DEVAL refers to very small offsets or deviations which do not appear to be regular transform faults but which appear to mark a petrological segmentation of the ridge; OSC refers to a phenomenon which occurs on fast-spreading ridges when the transform fault does not mark the abrupt termination of both ridge segments and instead the segments overlap slightly (Fig. 8.28).

Detailed studies of the Oceanographer Transform Fault on the Mid-Atlantic Ridge have shown that its active region is centred on the deepest part of the valley and is confined to a zone a few hundred metres to a few

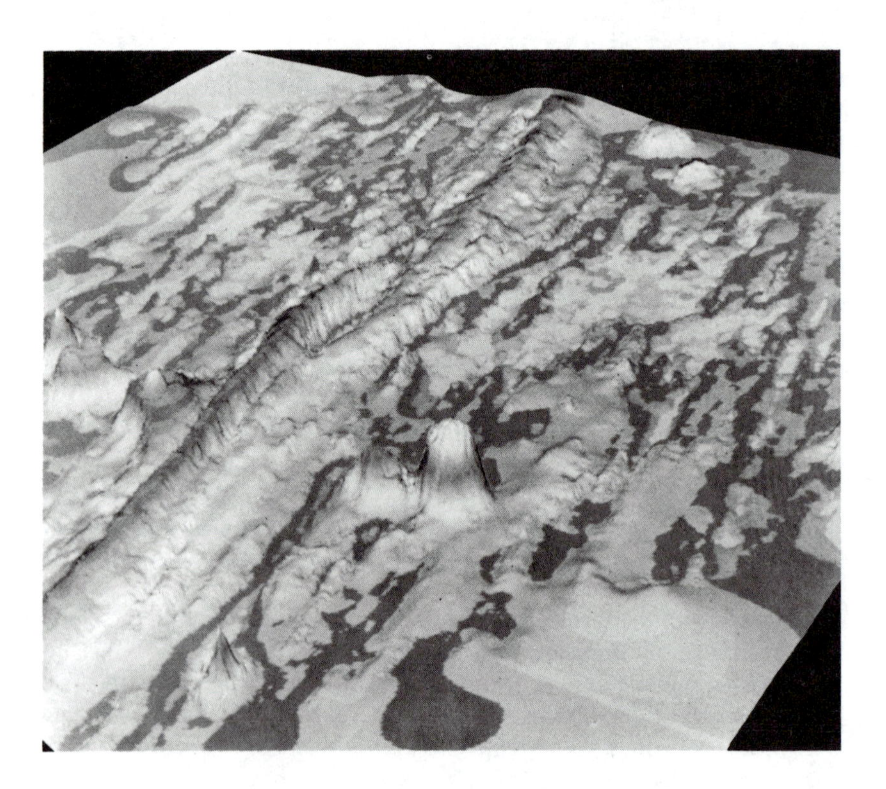

Figure 8.28. A sonar image of the overlapping spreading centres on the East Pacific Rise at 9°N. There is no transform fault between the two tips of the ridge segments which extend past each other. Such overlapping spreading centres are temporary phenomena caused by misalignment of tensional cracks in neighbouring ridge segments. As the magma rises episodically, the tips of the ridge segments move back, forth and laterally. (From Macdonald et al. 1988.)

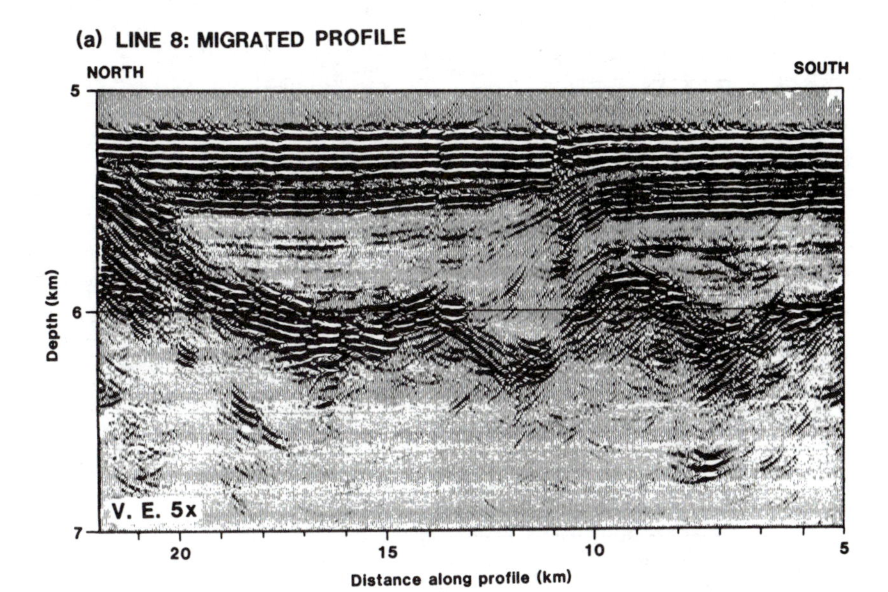

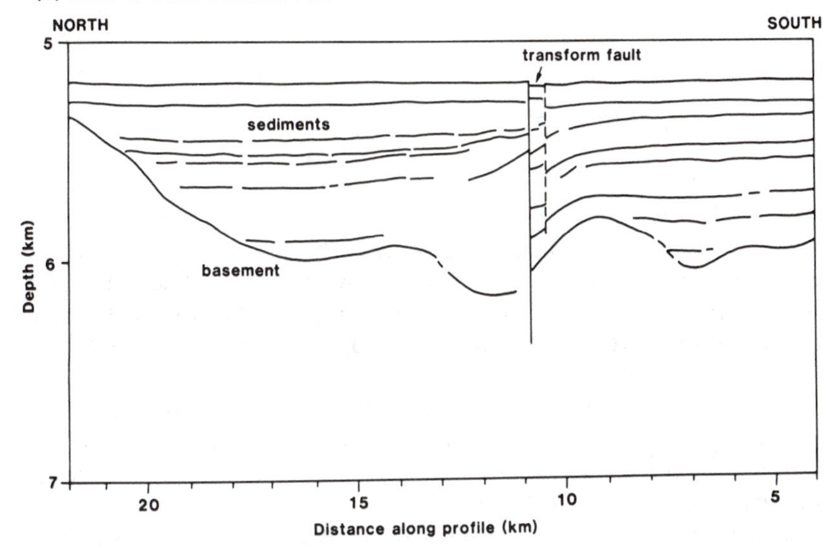

Figure 8.29. (a) Migrated seismic reflection profile across the Vema Transform Fault on the Mid-Atlantic Ridge at 6°N, approximately midway between the two ridge segments. (b) Interpretation of the reflection profile shown in (a). Note that the sediments are undisturbed except in a narrow zone about 0.5 km wide. On the seabed, the position of the active fault is marked by a narrow groove 30 m deep and about 500 m wide. (From Bowen and White 1986.)

kilometres wide. Figure 8.29a, a seismic reflection profile across the Vema transform fault at 6°N on the Mid-Atlantic Ridge, shows a narrow region of faulting which marks the boundary between the African and South American plates. The deformation in the sediments is confined to a narrow zone less than a kilometre across.

A number of detailed seismic refraction experiments have been shot along and across large and small offset Atlantic transform faults which indicate that their crustal structures are anomalous. Instead of normal 5–8 km thick oceanic crust, the fracture zone crust is generally thinner, in places only 2–3 km thick. This thin crust is confined to a region less than 10 km wide. In addition, lower than normal compressional crustal velocities are characteristic, and layer 3 velocities are frequently not observed.

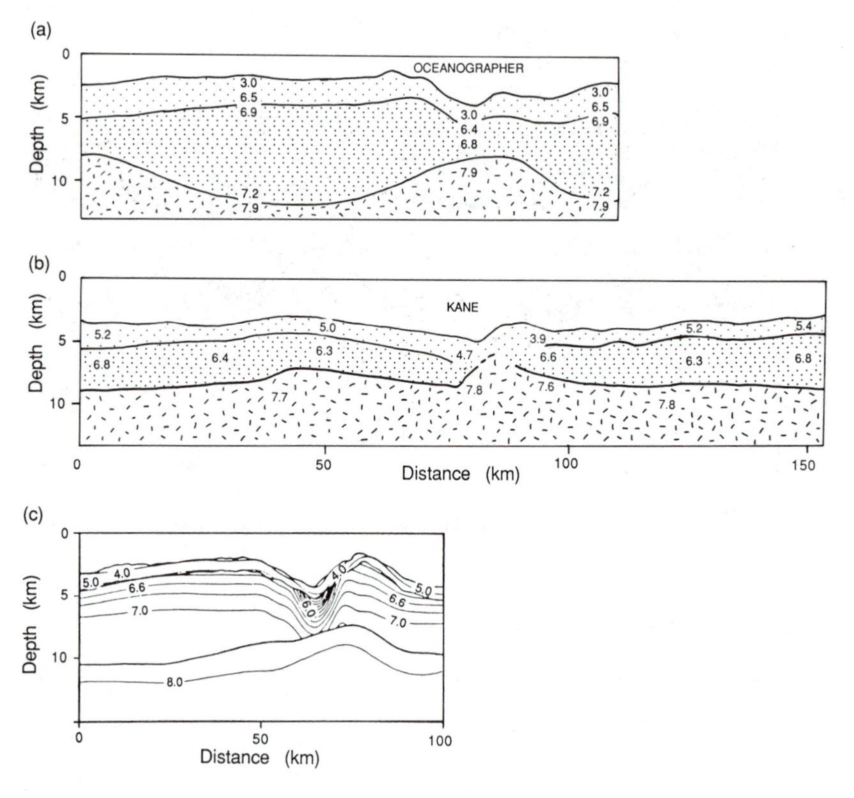

Figure 8.30. Seismic cross sections of three slow-slipping Atlantic transform faults: (a) Oceanographer, (b) Kane and (c) Charlie Gibbs. All sections are normal to the transform fault (parallel to the ridge) and cross the active fault between the two offset ridge segments. Note the low velocities and thin crust in the fault and the gradual thinning of the crust towards the fault. (After Sinha and Louden 1983, Detrick and Purdy 1980, Whitmarsh and Calvert 1986 and White et al. 1984.)

Figure 8.30 shows seismic models crossing the Oceanographer, Kane and Charlie Gibbs transform faults along the strike of the median valley. It is clear that over some tens of kilometres the crustal thickness decreases towards the transform fault, in addition to the very thin crust occurring in the transform fault itself. The Kane and Oceanographer transform faults have also been studied in some detail by submersibles. It is clearly documented that rocks usually associated with the lower crust and upper mantle (i.e., gabbroic and ultramafic rocks) outcrop on the escarpments. This is hard to explain if the crust is of normal thickness and again suggests that transform fault crust is thinner than normal.

Reflection lines have been shot across the Blake Spur Fracture Zone (a small western Atlantic fracture zone with about 2 Ma offset). The Moho deepens under one part of this fracture zone and then thins adjacent to it. However, it is important here to keep in mind that the Moho is strictly a seismic boundary. This seismic Moho and the petrologic Moho may not be locally coincident beneath fracture zones. Water penetrating through the fracture zone could have serpentinized the uppermost mantle. Serpentinization results in a reduction of the seismic velocity of peridotites such that the seismic Moho would occur *beneath* any serpentinized peridotite, whereas the petrologic crust–mantle boundary would be drawn *above* it. Localized deepening of the Moho beneath a fracture zone is therefore not irreconcilable with an overall thinning of the crust.

As yet no seismic models of the fastest-slipping transform faults have been made. However, beneath the Tamayo and Orozco transform faults, which both have slip rates of approximately 6 cm yr^{-1}, the crust is 1–3 km thinner than normal oceanic crust.

Gravity anomalies are usually associated with transform faults; the local topography is uncompensated. The interpretation of the excess mass is a matter of current debate (gravity models are not unique), but it could be partially due to anomalous material at a shallow depth. Until the details of the tectonics of transform fault are better understood, this issue probably will not be resolved.

The age contrast across transform faults varies greatly (Table 8.3). This means that the thickness of the lithosphere changes across the fault (see Sect. 7.5.2). A 37 Ma age contrast across the St. Paul's Transform Fault in the North Atlantic implies an increase in lithospheric thickness from perhaps 7 km at the ridge axis to over 60 km, almost an order of magnitude. It is this presence of young, hot lithosphere next to old, cold thick lithosphere which probably controls much of the topography, magmatism and morphology of the transform faults. The nodal basins are considered to be the result of a thermally induced loss of viscous head in the rising asthenosphere beneath the ridge axis, where the ridge is truncated by the older, cold thick lithosphere.

Figure 8.27 illustrates the two extremes in transform fault structures: large offset slow-slipping and large offset fast-slipping. It is important to realize that it is not the length of the offset but the age difference across the transform which controls the magnitude of the topography and the type of transform fault formed. In the slow-slipping example, the faulting takes place across a narrow zone because the thick cold lithosphere places severe mechanical constraints on any fault. In the fast-slipping example, the lithosphere is much thinner, and so it is much easier for a wide shear zone system to develop. Detailed studies of the Mid-Atlantic Ridge have been made in the FAMOUS area, which includes a number of short ($\sim$ 20 km) offset transform faults. Two of these transform faults, A and B, which are at opposite ends of a 50 km ridge segment, illustrate the inadvisability of generalization. They are similar in terms of their broad structure (Table 8.3), but detailed submersible studies of morphology, faulting, location of scarps and exposures of fresh volcanic material show that transform fault A looks very like the large offset slow-slipping model of Figure 8.27 with faulting confined to a narrow zone, whereas in transform fault B, shear deformation is occurring across a wide ($>$ 6 km) zone. Interpretations of the bathymetry and magnetics support these observations and suggest that at transform fault B the ridge may well have undergone propagation and recession in the past (instability in the location of the PTDZ), giving rise to the more complicated structure observed there. Figure 8.31 is a scheme of such an unstable PTDZ.

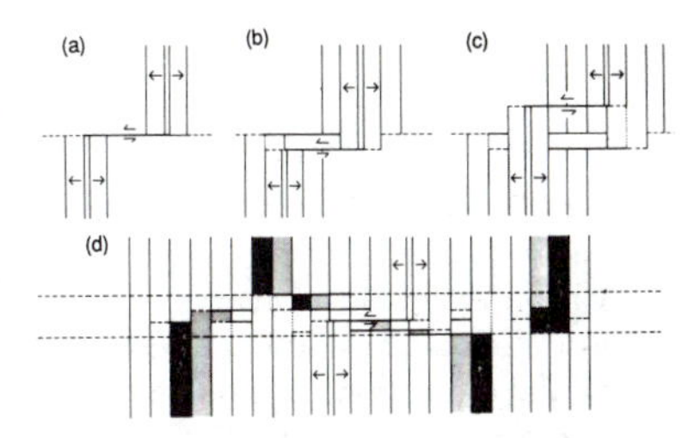

Figure 8.31. Diagram illustrating the evolution of topography and the lithospheric age distribution for a right lateral transform fault. (a) Simple model. (b), (c), (d) Progressive evolution when the ridge shown in (a) propagates and recedes. Dashed lines mark the extent of the fracture zone. At any one time the active transform fault would link the termination of the ridge segments. By time (d), it is clear that near the active transform the age of the lithosphere and the faulting are very complicated and that many old small lithospheric blocks are retained there. (From Goud and Karson 1985.)

The thinner than normal crust in the fracture zones and the gradual thinning of the crust over some tens of kilometres towards the fracture zones, as illustrated for the Atlantic in Figure 8.30, is thought to be the result of generalization of two effects. The old, cold lithosphere juxtaposed against the ridge axis has a significant thermal effect on the magma supply and tectonics. Thermal modelling (detailed in Chapter 7) shows that for a 10 Ma offset fracture zone, $\pm$ 100°C anomalous te·nperatures would be observed within about 10 km from the transform fault while on the ridge

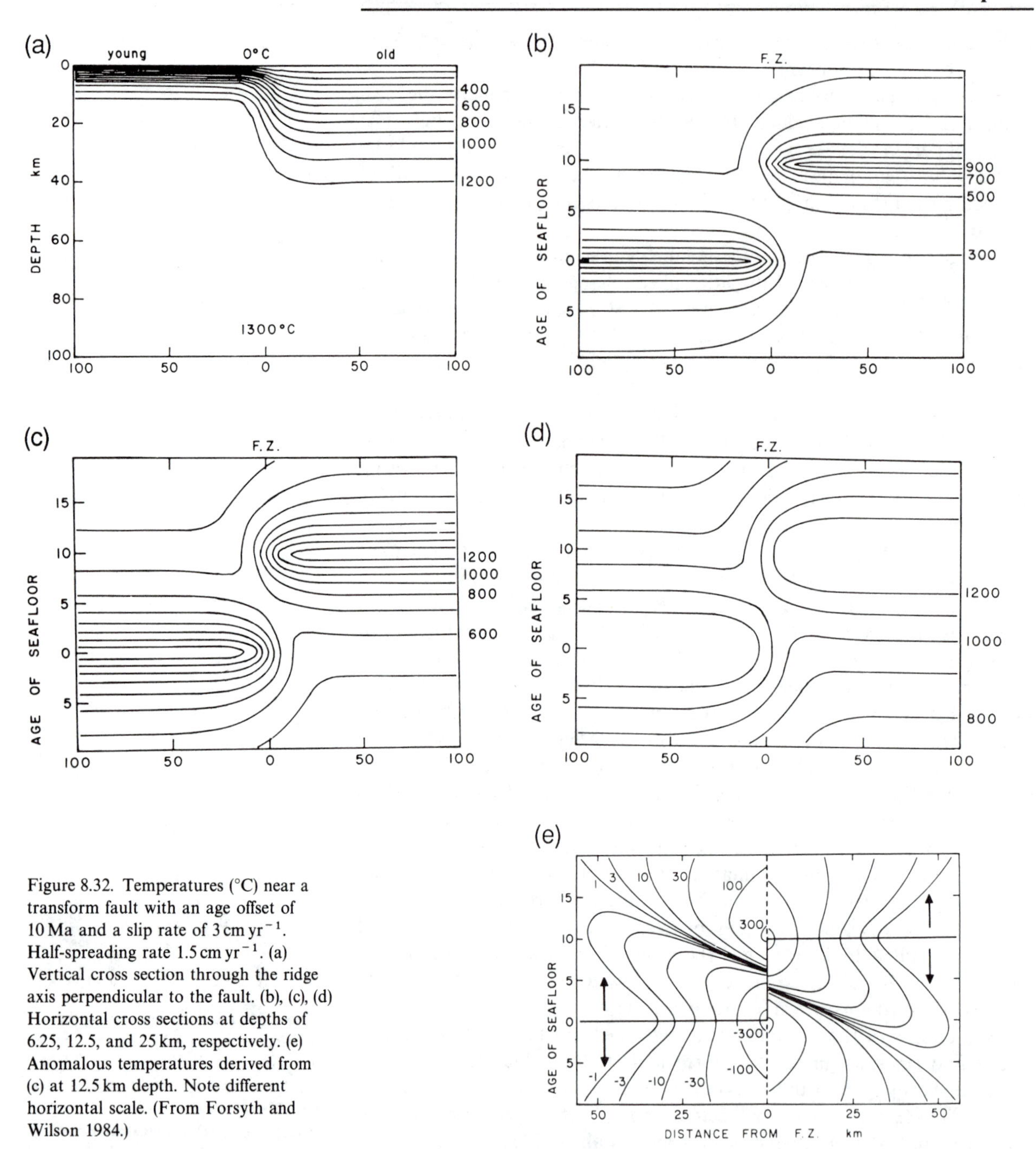

Figure 8.32. Temperatures (°C) near a transform fault with an age offset of 10 Ma and a slip rate of 3 cm yr^{-1}. Half-spreading rate 1.5 cm yr^{-1}. (a) Vertical cross section through the ridge axis perpendicular to the fault. (b), (c), (d) Horizontal cross sections at depths of 6.25, 12.5, and 25 km, respectively. (e) Anomalous temperatures derived from (c) at 12.5 km depth. Note different horizontal scale. (From Forsyth and Wilson 1984.)

axis close to the intersection with the transform fault, the temperature would be about 300°C too low (Fig. 8.32). From such a temperature difference, it is fairly simple to calculate the resulting contribution to the change in depth of the median valley as the transform fault is approached. These anomalous temperatures provide an explanation for thin crust in large offset transform faults, such as the Kane, Oceanographer or Charlie Gibbs, but cannot explain thin crust in small offset transform faults nor the gradual thinning of the crust over about 40 km adjacent to the transform fault and the deepening of the median valley towards the nodal basin. These effects are better explained by a magma 'plumbing' system, in which each ridge segment is fed by a single subcrustal, centrally located magma injection zone. In this way, the crust near large and small offset transform faults would be thinner because these faults are at the far ends of the magma supply systems. Additionally, the regular 40–80 km spacing between transform faults may indicate the horizontal distance over which magma from a single central supply point can feed a slow-spreading ridge segment. The more infrequent transform faults on fast-spreading ridges could in this way be an effect of a more plentiful supply of magma.

8.5.3 Seismic Activity at Transform Faults

The active part of a transform fault is that portion between the two ridge segments. Figure 8.26 shows how closely the epicentres follow the axis of the midocean ridge and transform faults. These epicentres are determined by observations of the earthquakes made by the WWSSN system, which, being on land, are necessarily far from many of the midocean ridge epicentres. This means that only fairly large-magnitude earthquakes (body wave magnitude greater than about 5 in the case of the Mid-Atlantic Ridge) are detected and located and that location is subject to some error (to determine the location one has to assume a velocity structure, which may not be particularly close to the exact structure at the epicentre). Nevertheless, it is clear that the earthquake activity is confined to a very narrow zone centred on the median valley at the ridge axis and to the transform valley on the transform faults. To detect smaller-magnitude earthquakes on the midocean ridges, it is necessary to deploy seismometers directly on the seabed in an array close to and within the median and transform valleys. These instruments, called ocean bottom seismographs (OBS), must withstand pressures of at least 5 km of water and record for 10 days or so; they are necessarily very expensive. Launching and recovery and the determination of their exact position on the seabed are difficult and important procedures. To use these instruments to locate earthquakes, at least three OBS must be recording simultaneously because four unknowns – origin time and epicentre location (x, y, z) – must be determined (see Sect. 4.2.1). The level of seismic activity along transform faults is quite variable; often 10–50 microearthquakes per day are recorded over regions 40–50 km long, though this has been observed to vary by up to two orders of magnitude during a few months on adjacent transform faults. Earthquake swarms are common. Figure 8.33 shows the type of earthquake which could occur along a transform fault and at its intersections with the midocean ridge. Notice in particular that anomalous fault-plane solutions could occur at the inner corner between the ridge and the transform fault. Figure 8.26 shows how fault-plane solutions along the equatorial Atlantic transform faults conform to this model.

Depths determined for microearthquakes and for WWSSN-recorded events are generally shallower than 7–9 km. This confirms that the faulting is occurring above the 400–600°C isotherms (half-space and cooling plate models, Section 7.5.2, are indistinguishable at these young ages). In 1925, a

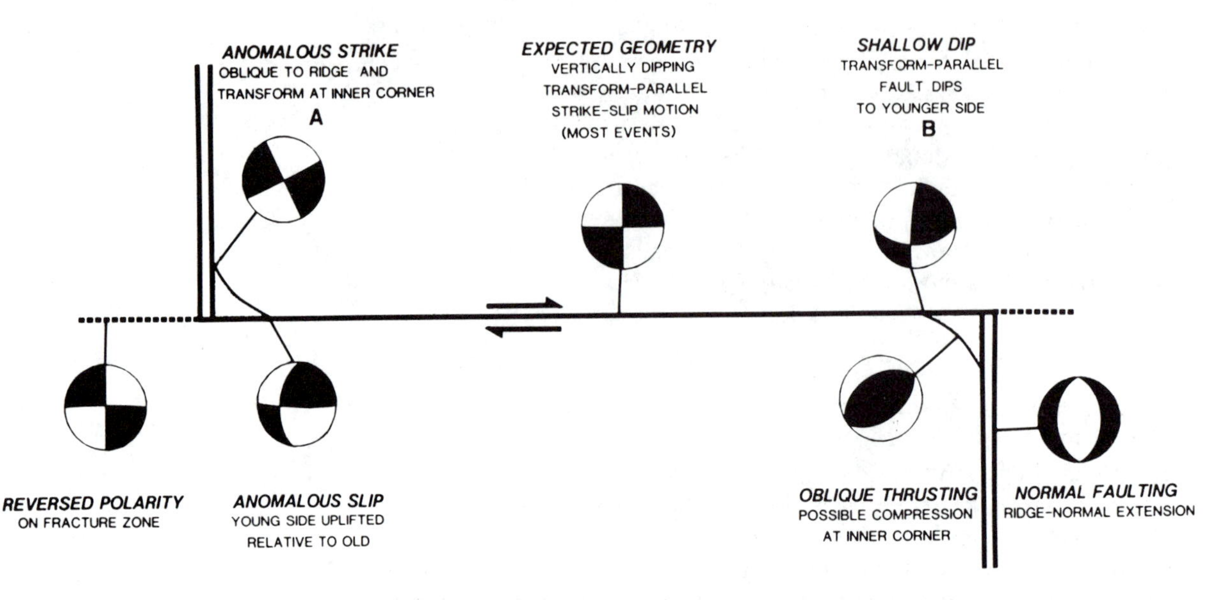

Figure 8.33. Schematic transform fault showing the types of earthquake which would occur in each location. Earthquakes A and B on the Romanche and Chain transform faults (Fig. 8.26) have the anomalous geometries illustrated here. (From Engeln et al. 1986.)

magnitude 7.5 earthquake occurred on the eastern half of the Vema Transform; that was the largest Atlantic transform earthquake recorded to date.

8.6 Subduction Zones

8.6.1 Introduction

A *subduction zone*, a convergent plate boundary, is the zone where old, cold lithospheric plate descends (is subducted) into the earth's mantle. The surface expression of a subduction zone is the deep *oceanic trench* on the oceanic plate and the line of volcanoes on the overriding plate (Fig. 8.34).

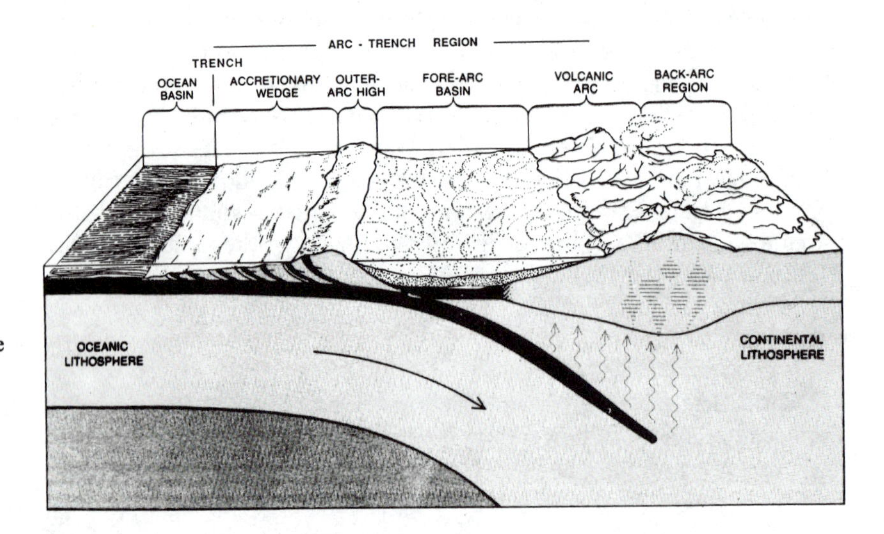

Figure 8.34. An active convergent plate boundary is characterized by a sequence of geological features. Extension in the back-arc region forms a back-arc, or marginal, basin. (From The continental crust, Burchfiel. Copyright © 1983 by Scientific American, Inc. All rights reserved.)

The volcanoes either form an *island arc*, such as those around the western Pacific, Indonesia and Caribbean, or are continental, as in the Andes. The horizontal distance between the trench and the volcanic arc is hundreds of kilometres – some 270 km for the Japan Subduction Zone.

Back-arc basins, otherwise known as *marginal basins*, occur behind the island arcs of the Western Pacific and Caribbean as well as the Scotia Arc. The crust in these basins is similar to oceanic crust, and magnetic lineations have been deciphered over some of them (e.g., the West Philippine Basin and the South Fiji Basin). The ophiolites which today are found obducted (thrust up) on land may be crust formed in back-arc basins.

The deep oceanic trench and the outer rise are a consequence of the bending of the lithospheric plate. Trenches often exceed 8 km in depth, the deepest being the Marianas Trench (see Sect. 8.1.1). The deformation of a thin elastic plate under an external load has been discussed in Section 5.7.1. This theoretical deformation has been used to explain the bending of the lithosphere at oceanic trenches and to yield estimates of its elastic parameters (see Figs. 5.14 and 5.15). For a subduction zone adjacent to a continent, the trench can be almost nonexistent because a high sedimentation rate may have filled it with sediments. In such cases, deformed sediments form a substantial *accretionary wedge* and *outer-arc high*. The *volcanic arc* itself is behind this sedimentary fore-arc basin. In the case of the Makran Subduction Zone (part of the boundary between the Arabian and Eurasian plates), which is an example of a shallow-dipping subduction zone with an extensive accretionary wedge resulting from a 7 km thickness of undisturbed sediment on the oceanic plate, the distance from the first deformation of sediment in the accretionary wedge (the *frontal fold*; see Fig. 9.9) to the volcanic arc is more than 400 km.

Oceanic trenches and island arcs are characteristically concave towards the overriding plate. This shape can be explained by imagining the lithosphere to be an inextensible spherical shell (like a ping-pong ball). If a portion of the shell is bent inwards, its edge is circular, and the indented portion has the same curvature as the original sphere (Fig. 8.35). The radius r of the indented circle is therefore $R\theta$, where R is the radius of the sphere and 2θ is the angle subtended by the circle at the centre of the sphere. The angle of dip of the indented circle, measured at the surface, is equal to 2θ. This simple model can be applied to the earth and tested against observed dip angles of trenches and subduction zones and their radii of curvature. An arc radius of about 2500 km gives a dip angle of 45°, which is in reasonable agreement with values for many subduction zones but certainly not for all (Table 8.4 and Fig. 2.2). Thus, this simple model provides a partial explanation for the concave shape of many oceanic trenches and island arcs even though it is an oversimplification of the problem.

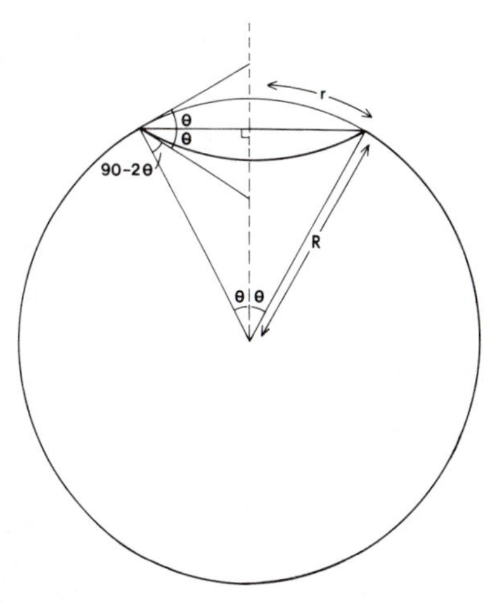

Figure 8.35. Geometry of an indented spherical cap on an inextensible spherical shell (e.g., a ping-pong ball). Radius of shell, R; radius of spherical cap, r; angle subtended by spherical cap at centre of sphere, 2θ.

8.6.2 Seismic Activity at Subduction Zones

Subduction zones are delineated by earthquake activity (see Figs. 2.1 and 2.2) extending from the surface down to depths of, in some cases, almost 700 km. The volcanic and seismic *ring of fire* around the margins of the Pacific Ocean is due to the subduction of oceanic plates (Fig. 2.2). It has been estimated that 80% of the total energy at present being released

Table 8.4. *Details of major subduction zones*

Subduction zone	Plates	Length of zone (km)	Approx. subduction rate (cm yr^{-1})	Approx. dip angle (deg)	Geometry
Kurile–Kamchatka–Honshu	Pacific under Eurasia	2800	6–13	40	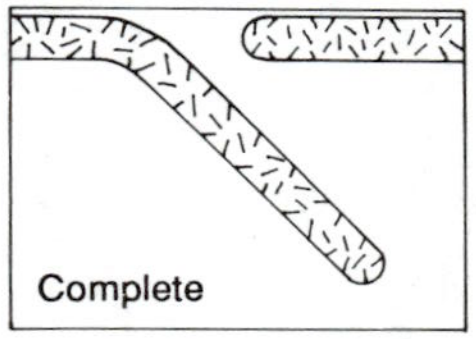 Complete
Tonga–Kermadec–New Zealand	Pacific under India	3000	8	60	
Middle America	Cocos under N. America	1900	9	70	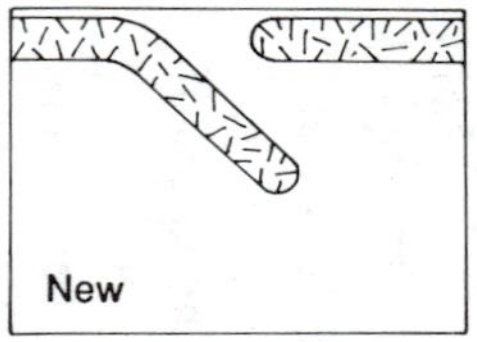 New
Aleutians	Pacific under N. America	3800	3	50	
Sundra–Java–Sumatra–Burma	India under Eurasia	5700	7	70	
South Sandwich–Scotia	S. America under Scotia	650	2	—	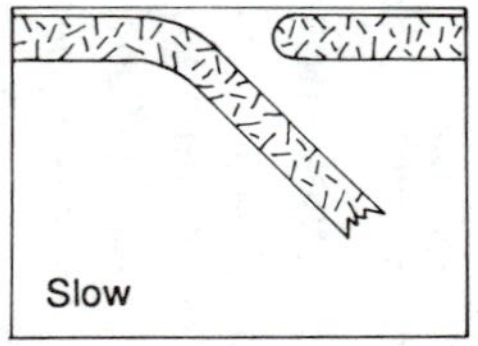 Slow
Caribbean	S. America under Caribbean	1350	0.5	50	
Aegean	Africa under Eurasia	1550	3	—	
Solomon–New Hebrides	India under Pacific	2750	10	70	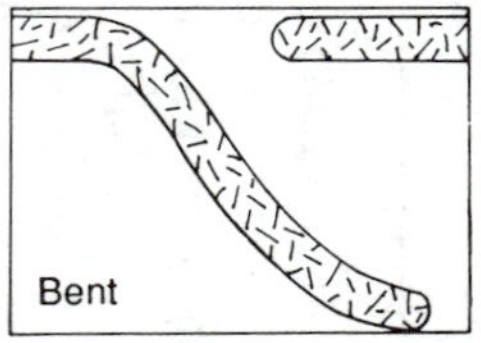 Bent
Izu–Bonin–Marianas	Pacific under Philippine	4450	10	60	
Iran	Arabian under Eurasia	2250	5	5	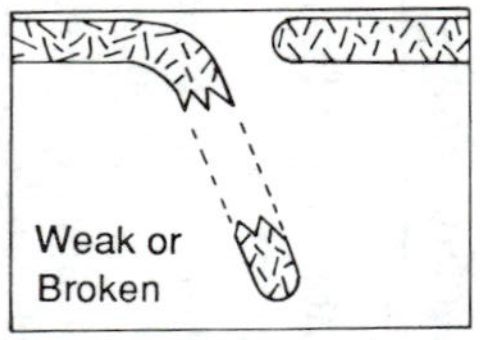 Weak or Broken
Himalayan	India under Eurasia	2400	5	—	
Ryukyu–Philippine	Philippine under Eurasia	4750	7	45	
Peru–Chile	Nazca under S. America	6700	9	30	

Note: The subducted plate is oceanic except for the Iran and Himalayan subduction zones, for which all or part of the subducted plate is continental.
Source: After Toksöz (1975) and Furlong and Chapman (1982).

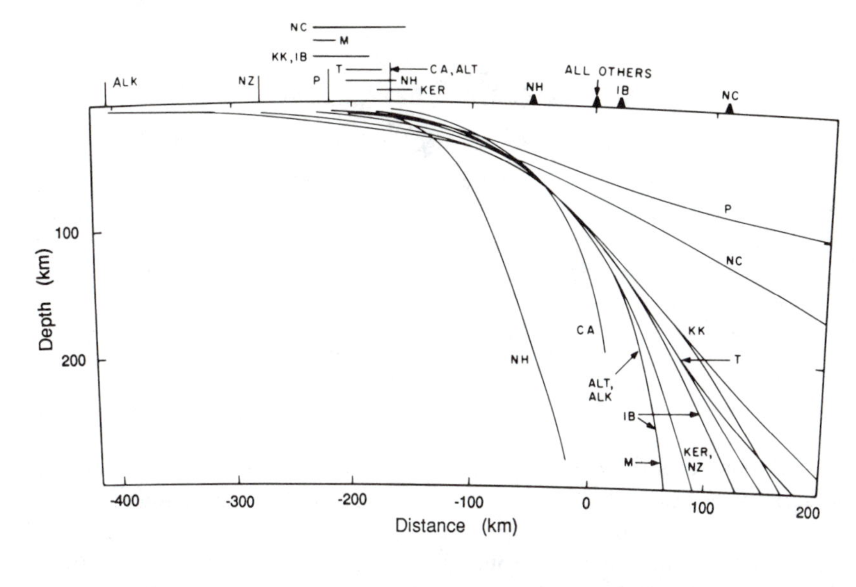

Figure 8.36. Shallow geometry of subduction zones as defined by earthquake foci. Abbreviation of subduction zone names: NH, New Hebrides; CA, Central America; ALT, Aleutian; ALK, Alaska; M, Mariana; IB, Izu–Bonin; KER, Kermadec; NZ, New Zealand; T, Tonga; KK, Kurile–Kamchatka; NC, North Chile; CC, Central Chile; SC, South Chile; P, Peru. Solid triangle marks the volcanic line; vertical or horizontal lines mark the location or extent of the oceanic trench. Some sections have, for clarity, been offset from the others. (After Isacks and Barazangi 1977.)

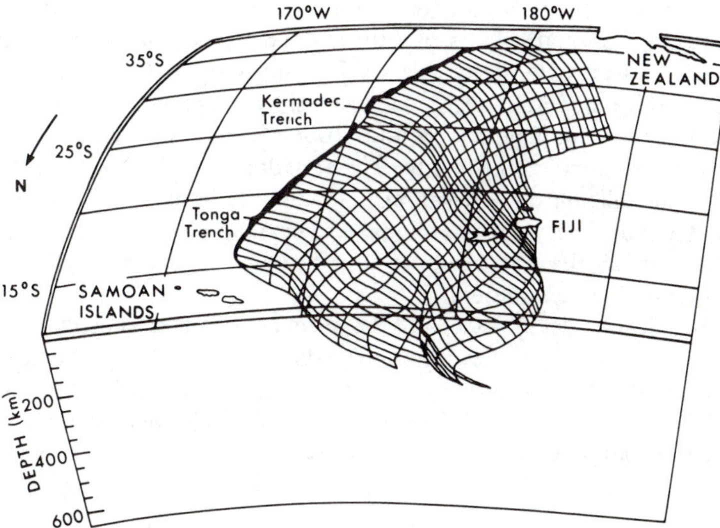

Figure 8.37. Shape of the Tonga–Kermadec subduction zone as defined by earthquake foci. (After Billington 1980.)

worldwide by earthquakes comes from earthquakes located in this ring and that another 15% is released by earthquakes in the broader seismic belt which extends eastwards from the Mediterranean and across Asia and includes the Alps, Turkey, Iran and the Himalayas.

Figure 8.36 shows the distribution of earthquake foci with depth for the major subduction zones. These deeply dipping seismic zones are sometimes termed Wadati–Benioff zones after the Japanese discoverer of deep earthquakes, Kiyoo Wadati, and his American successor Hugo Benioff. Figure 8.36 shows the shallow geometry of the major subduction zones, and Figure 8.37 shows the shape of the Tonga–Kermadec subduction zone as defined by earthquake foci. This subduction zone, which extends to more than 600 km in depth, is S-shaped in plan view. Figure 8.38 shows the focal distribution of microearthquakes along a cross section perpendicular to the

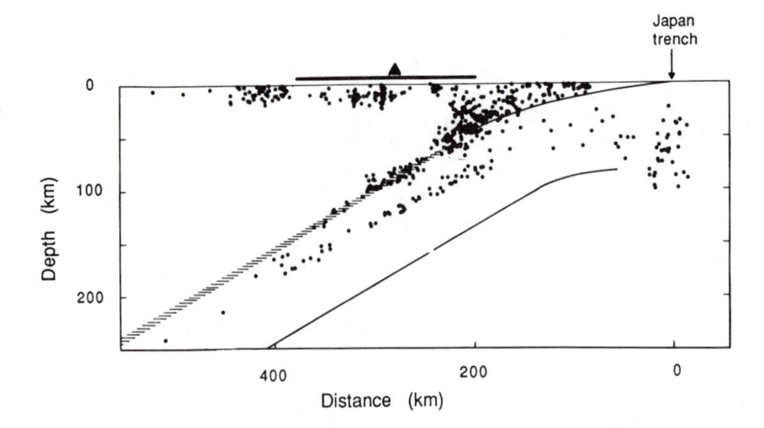

Figure 8.38. Focal depth distribution of microearthquakes recorded by a seismic network in northeast Japan marks the subducting Pacific Plate (earthquakes were located between 39 and 40°N and occurred between April and December 1975). Thick horizontal bar indicates the position of mainland Japan. Solid triangle marks the volcanic front. The 80 km thick subducting slab has P- and S-wave velocities 6% higher than the surrounding mantle. (After Hasegawa et al. 1978.)

trench axis beneath northeastern Japan. A large amount of very shallow seismicity is associated with volcanism and shallow deformation and thrusting. The deeper foci, however, clearly delineate the descending lithospheric plate. These foci apparently define two almost parallel planes: The upper plane is defined by earthquakes with reverse faulting or down-dip compressional stresses, and the lower plane is defined by earthquakes with down-dip extensional stresses. The upper and lower boundaries of an 80 km thick model subducting plate are shown superimposed onto the foci. This subducting plate has P- and S-wave velocities which are 6% higher than in the surrounding mantle. The boundaries of the plate were obtained from the difference in arrival time between the ScS and ScSp phases for a nearby, large, very deep earthquake. (The ScSp phase was converted to p from the ScS phase at the upper boundary of the subducting plate; refer to Section 4.2.7 for notation.)

Results of a detailed study of the propagation of the short-wave seismic phases P_n and S_n, in the region extending from the Tonga Trench to the Fijian islands, are shown in Figure 8.39. A zone of extremely high attenuation (Q as low as 50 for P-waves and less for S-waves) lies beneath the Lau marginal basin and above the subducted plate. This is presumably

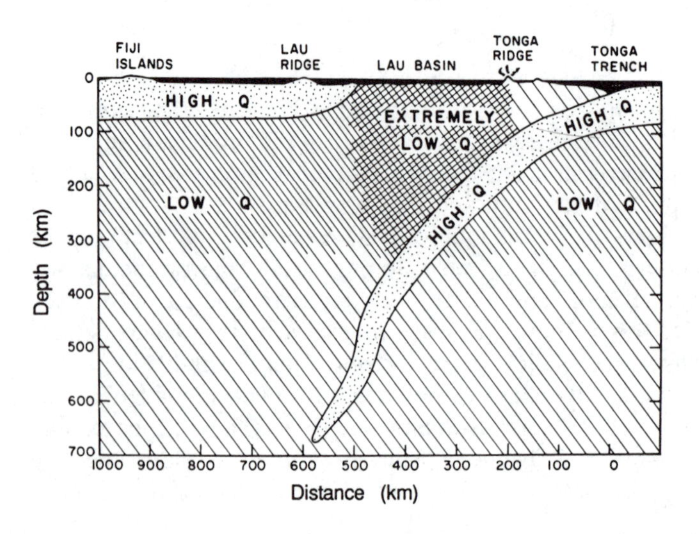

Figure 8.39. Cross section perpendicular to the Tonga Trench showing the location of the subducting plate relative to the zones of high and low seismic attenuation (Q). (After Barazangi and Isacks 1971.)

related to the thermal structure and extension occurring in this region. The subducted plate itself is characterized by low attenuation, with Q values of perhaps 1000 or more. The low-Q zones beneath the subducted plate and to the west of the marginal basin are in the depth range generally associated with the asthenosphere and low-velocity zone and are in agreement with other observations. It has been suggested that a similar highly attenuative zone is present at depth above the subducting Pacific Plate beneath Japan. In the same region above the slab, a low-velocity zone with velocities some 5–10% less than normal has been determined. The high attenuation and low velocities above these subducting plates are probably general features of subduction zones.

Seismic activity at subduction zones is not constant along the length of the zone: Many quiet regions, or *seismic gaps*, have been observed. Such gaps are receiving particular attention since it is likely that a gap will be the location of a future large earthquake. Variations in the level of seismicity along the length of a subduction zone may also be controlled by factors such as geological or bathymetric structures in either of the two plates.

Figure 8.40 shows the seismicity along the Tonga–Kermadec Arc. The Tonga–Kermadec subduction zone has very high seismicity; approximately 1 earthquake with m_b greater than 4.9 occurred every 2 kilometres along the arc during the 18 years of recording shown. There are clear

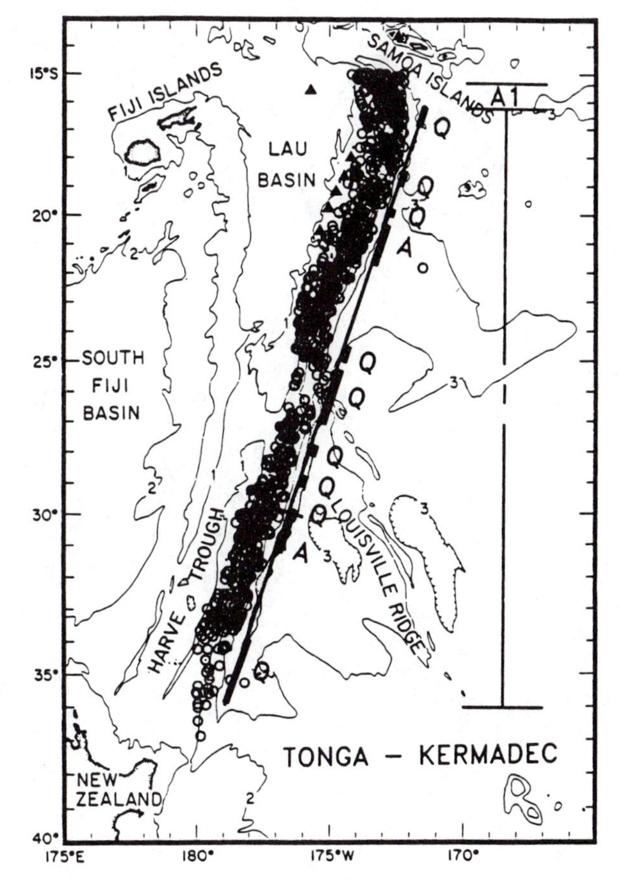

Figure 8.40. Map of the Tonga–Kermadec Arc showing the location of all the shallow, main shock epicentres (aftershocks are not shown) with m_b greater than 4.9 which occurred between 1963 and 1981. Regions with high and low seismicity are marked A (active) and Q (quiet), respectively. The very active zone marked A1 is associated with the change in strike of the plate boundary from northerly to westerly. Bathymetric contours in kilometres. (From Haberman et al. 1986.)

variations in the level of the seismicity on a length scale ranging from tens to hundreds of kilometres.

One particular 250 km long seismic gap in the Aleutian Arc is the Shumagin gap (Figs. 8.41 and 8.42). Analysis of background seismicity along the Aleutian Arc shows that activity there is low although very large earthquakes have occurred along this arc in the past. It has been proposed that the amount of stress built up along the Shumagin gap since the last

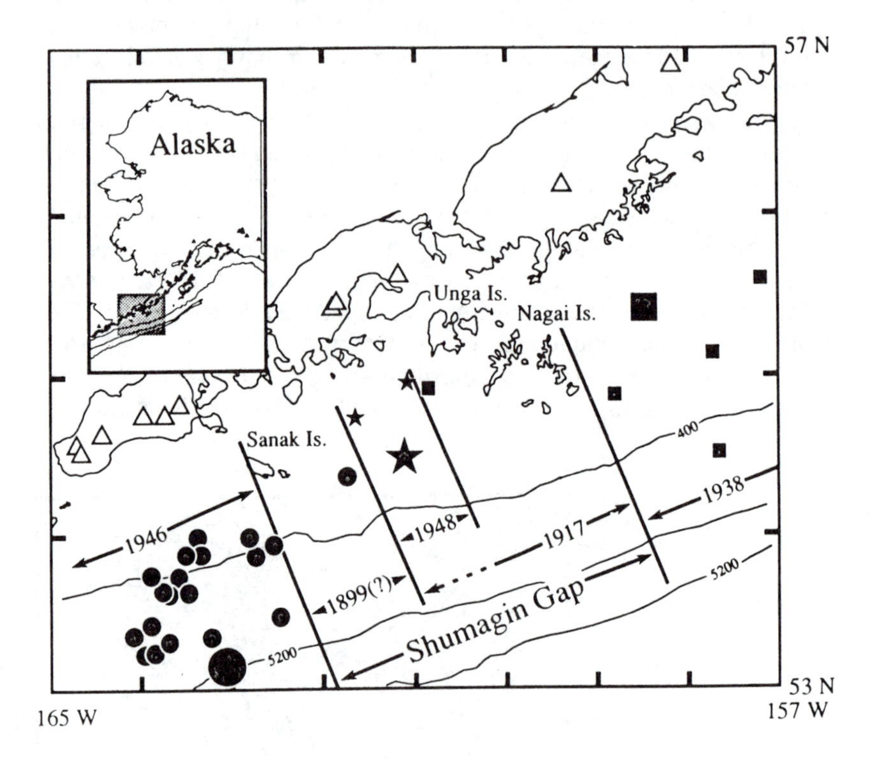

Figure 8.41. Part of the Aleutian Arc showing the location of the Shumagin gap in relation to the previous large earthquakes in the region. Main shocks and aftershock sequences are shown as well as estimates of the extent of rupture for each event. ■, 1938, $M_w = 8.2$; ●, 1946, $M_w = 7.3$; ★, 1948, $M_s = 7.5$. Probable rupture extent of the 1917, $M_s = 7.9$, and the 1899, $M_s = 7.2$, events are also shown. Δ, location of volcanoes. Bathymetric contours in metres. (From Boyd et al. 1988.)

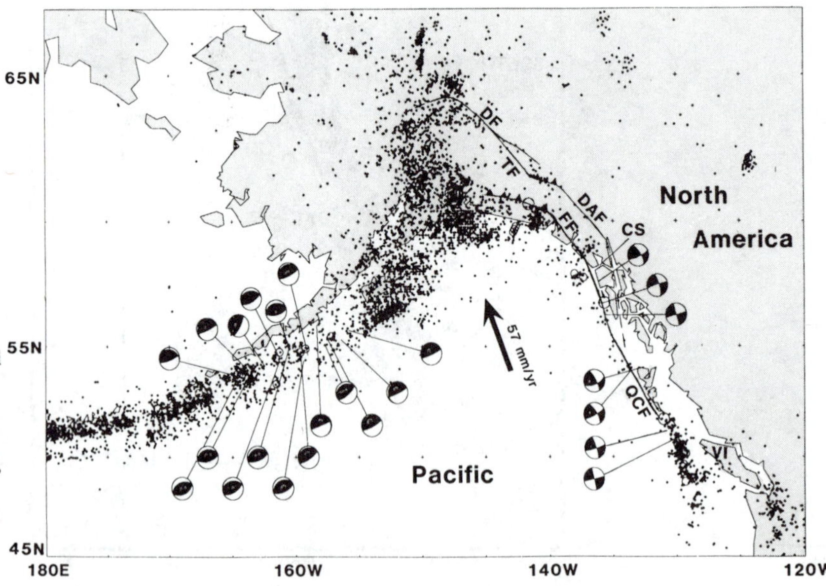

Figure 8.42. Focal mechanism for selected earthquakes along the Queen Charlotte Islands transform fault (QCF) and along the Alaskan end of the Aleutian subduction zone. VI, Vancouver Island; DF, TF, FF, DAF and CS, major faults. Black dots, epicentres of earthquakes shallower than 50 km. (Courtesy C. Demets and R.G. Gordon.)

major rupture there would be sufficient to result in a 8.3–8.5 magnitude earthquake in the near future, unless aseismic slip is occurring (i.e., the plates are moving smoothly past each other). The limited strain measurements available do not show any significant strain accumulation, indicating that aseismic slip could be occurring. Unfortunately for those who live in the vicinity, only time may provide the answer to the question about the origin of the Shumagin gap.

A detailed study of the focal mechanisms of large subduction zone earthquakes indicates that they are depth dependent. Shallow and medium-depth earthquakes usually have focal mechanisms with down-dip extension, whereas deeper earthquakes (depths greater than 300–400 km) have down-dip compression. Some subduction zones have no seismic activity at intermediate depths; the slab may be broken, although another explanation is that this is merely a stress-free zone. The shallow earthquakes occurring in the overriding plate have predominantly thrust faulting fault-plane solutions, whereas the shallow earthquakes in the subducting plate generally have normal faulting fault-plane solutions.

The observation that intermediate and deep earthquakes have down-dip compressive focal mechanisms is often taken to indicate that there is a physical barrier at 650–700 km depth which stops any subducting plate. Any portion of a deep detached plate would then be absorbed back into the mantle. Therefore, this is one piece of evidence in favour of an upper mantle convection system which is separate from the lower mantle. However, it can also be argued that below this depth (and hence at higher temperature and pressure) fracturing does not occur; instead, the subducting plate deforms plastically. If this were the case, there would be no deeper earthquakes. Nevertheless, the compressive deep fault-plane solutions do indicate that the phase changes from olivine to spinel to postspinel forms which are associated with the upper mantle–lower mantle boundary do provide significant resistance to the downwards motion of the plate. Evidence from seismic modelling of deep earthquakes has revealed that the subducting plate may indeed penetrate into the lower mantle, as an anomalous high-velocity body, reaching depths of at least 1000 km.

A viscosity increase of several orders of magnitude between the upper and lower mantle, which would mean that the slab was unable to penetrate into the lower mantle, is probably not warranted from viscosity studies (see Sect. 5.7). A smaller viscosity increase could result in the observed down-dip compression and still allow the slab to penetrate the lower mantle. Such an increase might also result in deformation of the slab in the lower mantle.

8.6.3 Thermal Structure

Measurements of heat flow on a cross section perpendicular to the Japan Trench and the Vancouver Island subduction zone are shown in Figures 8.46 and 8.48. The heat flow is low in the region between the trench and the volcanic arc. At the volcanic arc, there is a sudden jump in heat flow from about $40 \times 10^{-3}\,\mathrm{W\,m^{-2}}$ to $75\text{–}100 \times 10^{-3}\,\mathrm{W\,m^{-2}}$ within a very short horizontal distance. Heat flow then remains high across the volcanic arc and into the marginal basins. Measurements across other subduction zones

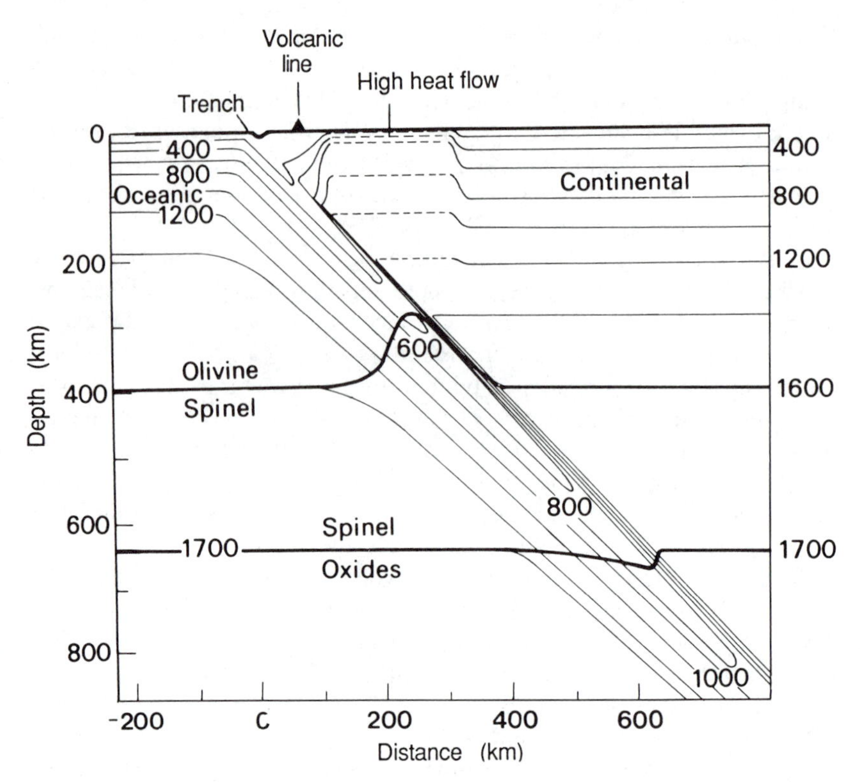

Figure 8.43. Thermal structure of a thick subducting lithospheric plate. Temperature, °C. Calculations include the effects of frictional heating (to a maximum value of 100 MPa) and latent heat for the olivine–spinel and spinel–postspinel structure phase changes. The temperature gradient for the oceanic lithosphere away from the subduction zone is taken to be that for the simple oceanic plate model (Sect. 7.5.2), and the overriding continental plate has a typical geotherm. (Refer to Fig. 7.14 to see that oceanic and continental geotherms are similar.) Mantle temperatures are adiabatic. Solid lines superimposed on geotherms mark the location of the olivine–spinel and the spinel–postspinel structure phase changes. Within the subducting plate, the depth of olivine–spinel structure phase change is elevated and that of the spinel–postspinel structure phase change is depressed. Dashed temperature contours in the back-arc region are estimates. (After Schubert et al. 1975.)

show the same pattern. The sudden increase in heat flow at the volcanic arc is due to shallow heat sources: namely, magma beneath the volcanoes.

Subduction zones have been modelled by many researchers, and although there are differences between their starting models and the resulting isotherms and heat flow, the results are broadly the same for all and could probably already be anticipated by a reader of Chapter 7. Figure 7.32 shows the thermal reequilibration of a thrust fault. A subduction zone is merely a type of thrust that keeps moving. Figure 8.43 shows one conductive thermal model of the temperature structure in and around a subducting oceanic plate. This rather thick plate is subducting at 8 cm yr^{-1} at a dip of 45°. The model includes an allowance for shear (frictional) heating along the upper boundary of the subducting plate and includes heat release (1.7×10^5 J kg^{-1}) and absorption (7.5×10^4 J kg^{-1}) for the olivine–spinel and spinel–postspinel phase changes, respectively. Note that the model subducting slab has low temperatures compared with adjacent mantle, even at depths of more than 700 km, which results in high seismic velocities and densities for the slab.

The fine details of subduction zone temperature structure remain unknown. Many presently rather poorly determined factors affect the temperature estimates. Estimates of the magnitude of frictional heating shear stresses range up to 100 MPa, though values around 10–40 MPa appear to be more reasonable based on the values of the heat flow in the trench–volcanic arc region. Of central importance to the shallow thermal structure is the role of water, which transports heat along the thrust fault and through the overriding plate. In addition, the melting behaviour of the subducting plate and the overriding mantle depends very strongly on the amount of water present. Heat is advected by rising magma beneath the volcanic arc and the back-arc basin. A small-scale convection system tends to operate beneath the back-arc basin, giving rise to the ocean-type crust and magnetic anomalies there (discussed further in Sect. 9.2.1).

The main changes which occur in the subducting plate are the shallow reaction of the oceanic crust to eclogite and the deeper mantle changes of olivine to a spinel structure and then to postspinel structures (see Sect. 4.3.4). These changes result in increases in the density of the subducting slab. Figure 8.44 shows equilibrium pressure–temperature curves for the olivine–spinel and spinel–postspinel structures. At lower temperatures, the change of olivine to spinel occurs at a lower pressure (shallower depth). Thus, in the subducting plate this phase change is elevated (i.e., occurs closer to the surface). For the model shown (Fig. 8.43), the phase change is elevated over 100 km in the subducting plate. The density increase associated with this phase change is about 300 kg m^{-3}. The lower temperature of the subducting plate means that the spinel–postspinel phase change occurs there at a higher pressure (greater depth) than in the mantle. However, the thermodynamics of this change is not yet well understood. The phase change may be endothermic (heat-absorbing), as assumed here in Figures 8.43 and 8.44, but could be exothermic (heat-releasing), in which case it would have the effect of increasing rather than decreasing the downwards force on the sinking plate. Thus, the downwards force on the subducting plate has three components: that due to the thermal contraction of the plate, that due to the elevation of the olivine–spinel phase boundary and that due to the depression of the spinel–postspinel boundary. The third contribution is of opposite sign to the first two, acting to impede the descent of the plate, but is smaller in magnitude. Thermal contraction provides the greatest contribution to the overall driving force. (See Sect. 7.10 for discussion of the driving forces of plate tectonics.)

8.6.4 Gravity across Subduction Zones

Very large gravity anomalies occur over subduction zones. The anomalies across the Aleutian Trench, the Japan Arc and the Andes are shown in

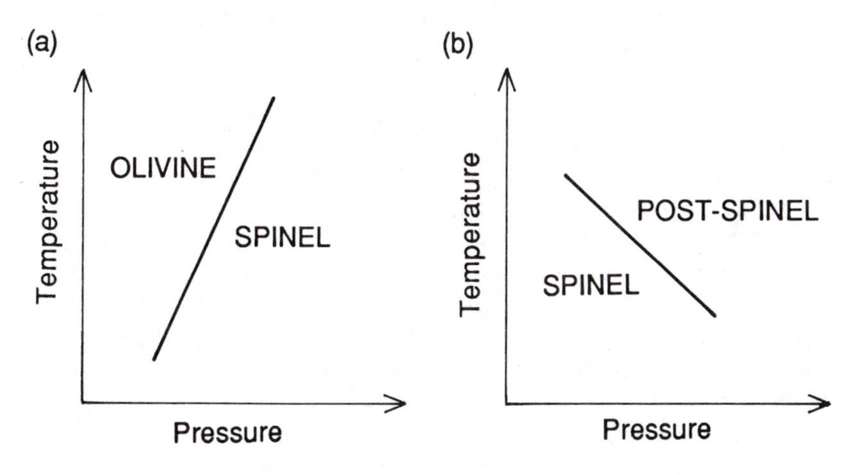

Figure 8.44. Equilibrium pressure–temperature curves (Clapeyron curves) for (a) the olivine–spinel and (b) the spinel–postspinel structure phase changes.

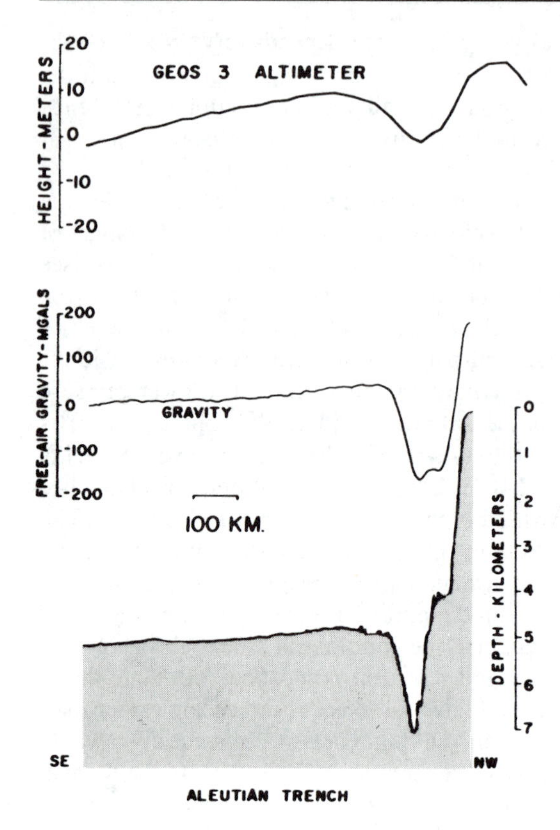

Figure 8.45. Geoid height anomaly as measured by the Geos3 satellite altimeter, free-air gravity anomaly and bathymetry along a profile perpendicular to the Aleutian Trench. (From Chapman and Talwani 1979.)

Figures 8.45–8.47. The general feature of gravity profiles over convergent plate boundaries is a parallel low–high pair of anomalies of total amplitude between 100 and 500 mgal and separated by about 100–150 km. The *low* is situated over the trench, and the *high* is near to and on the ocean side of the volcanic arc.

Density models which can account for these gravity anomalies include the dipping lithospheric plate and thick crust on the overriding plate. Details which are also included in the gravity modelling are the transformation of the basaltic oceanic crust to eclogite with a density increase of about 400 kg m^{-3} by about 30 km depth.

8.6.5 Seismic Structure of Subduction Zones

Earthquake data, seismic refraction and reflection profiling are all used to determine the seismic velocity structure around subduction zones. The large-scale and deep structures are determined from earthquake data. The subducting plate, being a cold, rigid, high-density slab, is a high-velocity zone with P- and S-wave velocities about 5–10% higher than normal mantle material at the same depth. The asthenosphere above the subducting plate, which is associated with convection and back-arc spreading in the marginal basin, is a region with low seismic velocities.

The crustal seismic structure across Japan and the Japan Sea is shown in Figure 8.46c. Japan has a 30 km thick continental type of crust whereas the

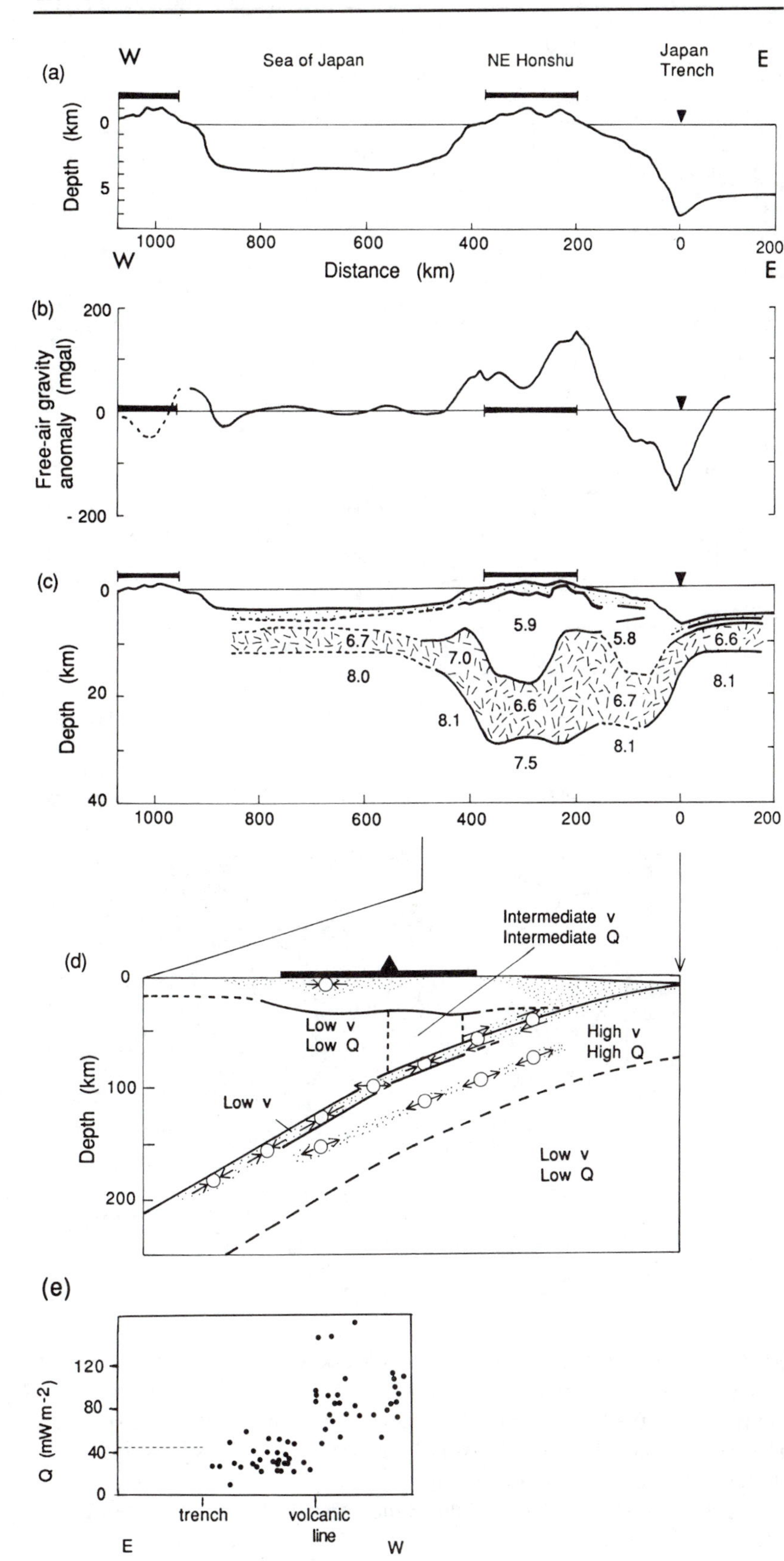

Figure 8.46. Cross sections across the Japan Trench and Arc. Solid horizontal bar denotes land; ▼, trench; ▲, volcanic front. (a) Topography, vertical exaggeration 25:1. (b) Free-air gravity anomaly. (c) Crustal seismic P-wave velocity (km s^{-1}) structure, vertical exaggeration 10:1. (d) Summary of the seismic structure, true scale. Stippled areas, seismically active regions. Typical earthquake focal mechanisms are shown. The low-Q, low-velocity regions are the asthenosphere beneath the 30 km thick overriding plate and beneath the 80 km thick subducting Pacific Plate. Focal depth distribution of microearthquakes along this cross section is shown in Figure 8.38. The thin low-velocity layer at the top of the subducting plate is considered to be the subducted oceanic crust. (e) Heat flow measurements. The dashed line is the theoretical heat flow for 120 Ma oceanic lithosphere, the age of the Pacific Plate beneath Japan. (After Yoshii 1979, Matsuzawa et al. 1986 and van den Beukel and Wortel 1986.)

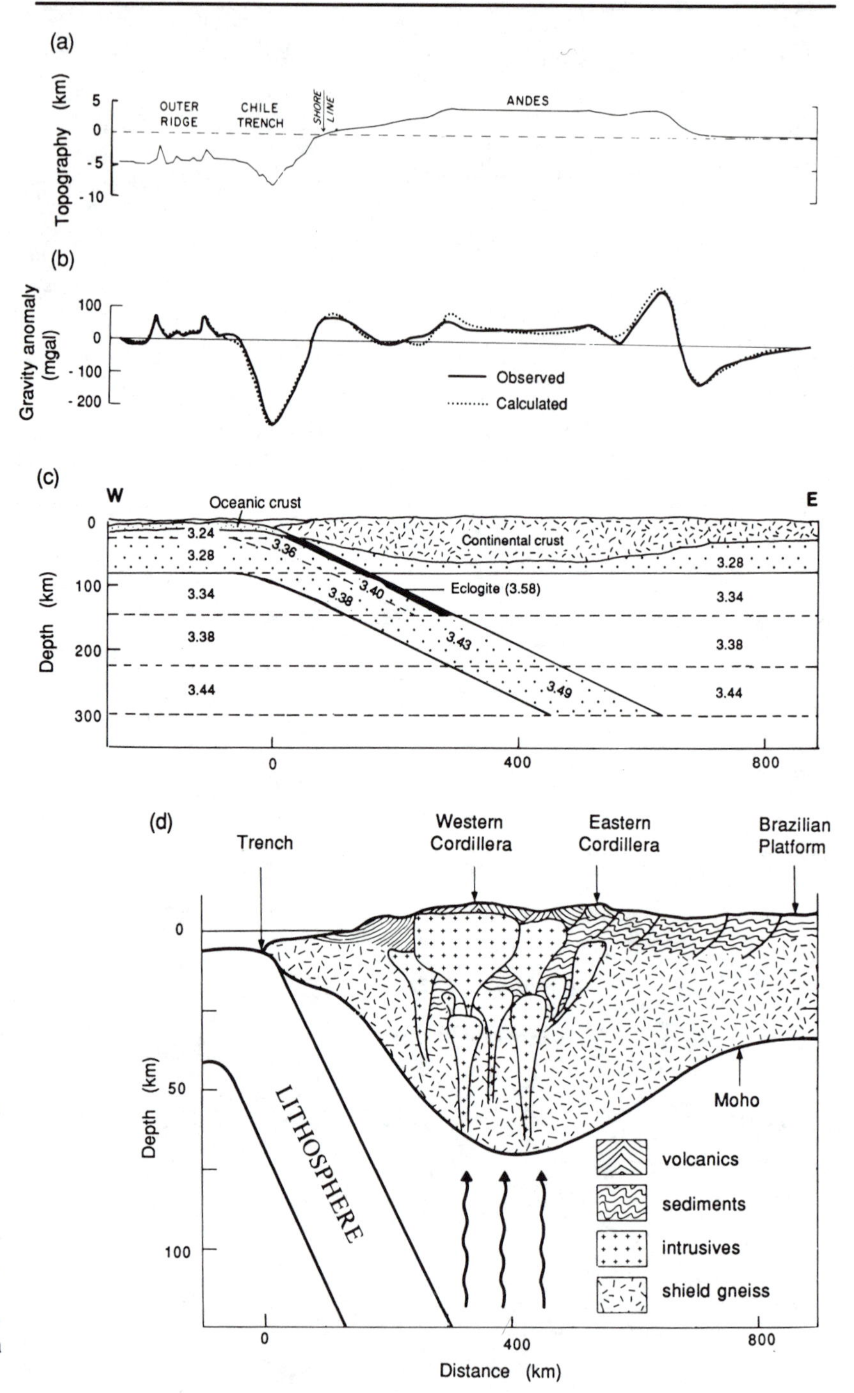

Figure 8.47. Cross sections across the Chile Trench and the Andes at 23°S: (a) topography, vertical exaggeration 10:1; (b) free-air gravity anomaly; (c) density model, true scale (densities in $10^3\,kg\,m^{-3}$). (After Grow and Bowin 1975.) (d) Schematic geological cross section through the central Andes. Arrows indicate rising magma. Most of the new material being added to the crust at this destructive plate boundary is in the form of huge diorite intrusions. The andesite volcanics provide only a small proportion of the total volume. Vertical exaggeration 5:1. (From Brown and Hennessy 1978.)

crust in the marginal basin is only 8–9 km thick with velocities near those of oceanic crust. Normal upper mantle velocities are present beneath the oceanic plate, the trench and the marginal basin, but not beneath Honshu. There the highest velocity measured was $7.5\,km\,s^{-1}$. This low velocity is characteristic of the asthenosphere.

The density model for the Chile Trench and Andes was constrained by refraction data, which indicate that the crust beneath the Andes is some 60 km thick, the upper 30 km having a seismic P-wave velocity of 6.0 km s^{-1} and the lower 30 km a velocity of 6.8 km s^{-1}. This thick crust appears to have grown from underneath by the addition of andesitic material from the subduction zone rather than by compression and deformation of sediments and preexisting crustal material.

The convergent plate boundary off the west coast of North America, where the North American Plate is overriding the young Juan de Fuca Plate (Fig. 2.2) at about 2 cm yr^{-1}, is an example of a subduction zone with no easily distinguishable bathymetric trench. The dip of the subducting plate here is very shallow at about 15°. The subducting oceanic plate is overlain by a complex of accreted terranes which are exposed on Vancouver Island and the mainland. This is, therefore, not a simple subduction zone but one where subduction has assembled a complex assortment of materials and pushed or welded them onto the North American continent. Figures 8.48 and 8.49b show the thermal structure and the seismic velocity structure across this subduction zone. The subducting oceanic plate and the 35 km thick continental crust are clearly visible in the seismic model. One unusual feature of the velocity model is the wedges of somewhat higher-velocity material and bands of low-velocity material immediately above the Juan de Fuca Plate. One interpretation is that the low-velocity material is associated with water lost from the subducting plate. Another interpretation is that the wedge structure is tectonically underplated oceanic lithosphere. Such underplating could have been a continuous process, scraping off slivers of oceanic crust, or it could have occurred rapidly if the subduction zone jumped westwards. Nevertheless, it should be stressed that however convincing any schematic model appears, it is only as good as the data on which it is based. Other interpretations of the seismic data may be possible, and this wedge may in reality not be exactly as shown in this figure.

The fine seismic structure of this convergent margin has been imaged by

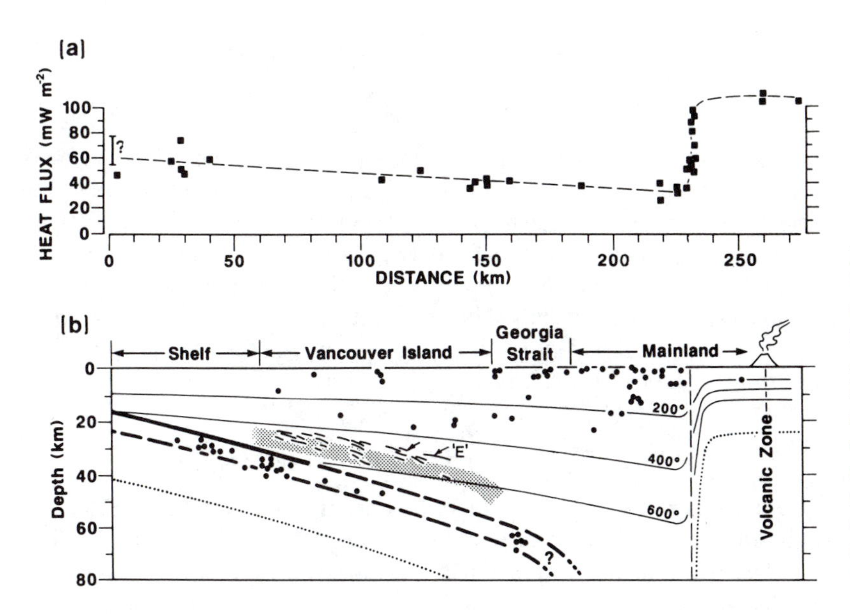

Figure 8.48. Structure across the Vancouver Island Subduction Zone of western Canada where the Juan de Fuca Plate is descending beneath the North American Plate. Land is shown by solid horizontal bar. (a) Surface heat flow measurements. (b) Estimated isotherms, earthquake foci (dots), strongest E reflectors (short lines) and zone of high electrical conductivity assumed to be associated with water (shaded zone). (From Lewis et al. 1988 and Hyndman 1988.)

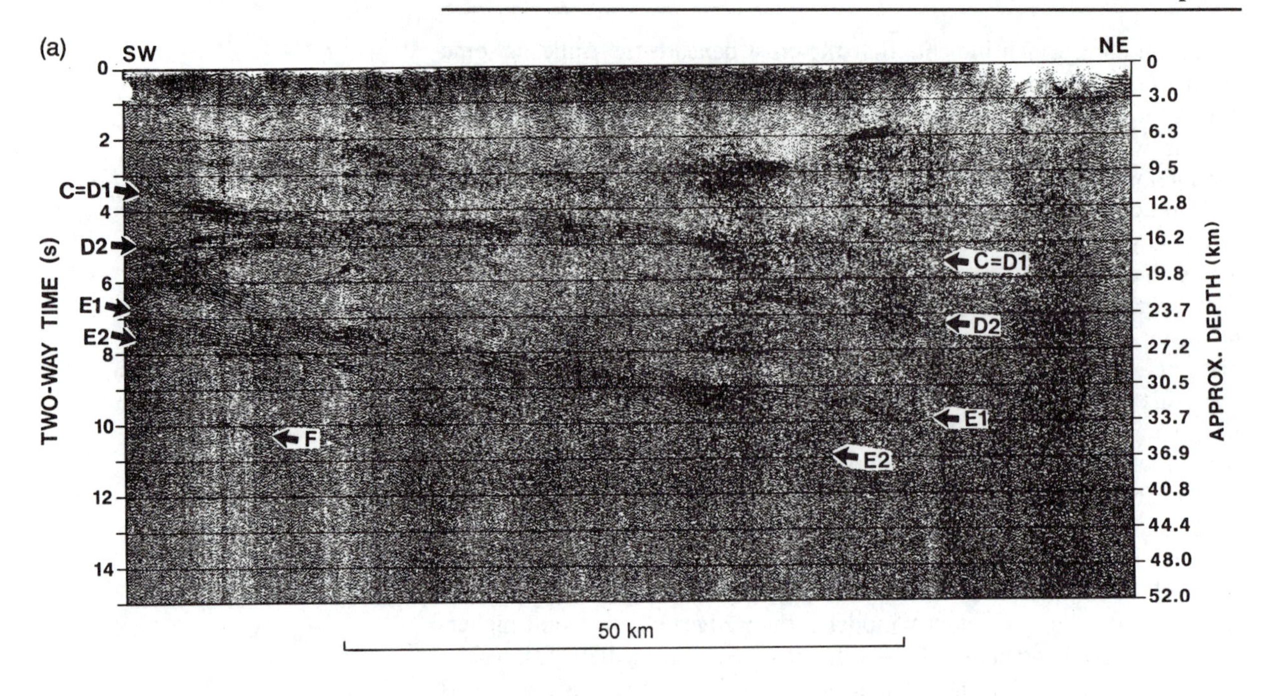

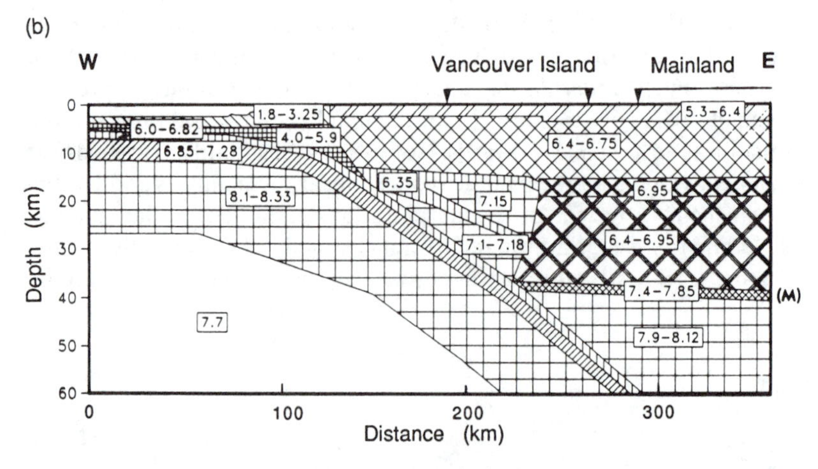

Figure 8.49. (a) Migrated seismic reflection section across Vancouver Island. Reflections C and D are associated with the base of a major accreted terrane; reflections E may be associated with water in the crust; reflection F is from the top of the subducting Juan de Fuca Plate. (Courtesy of R.M. Clowes.) (b) P-wave velocity model determined from refraction and reflection data. M, Moho. (From Drew and Clowes 1989.)

several deep reflection lines. Figure 8.49a shows data from a line shot across Vancouver Island coincident with the refraction line shown in Figure 8.49b. Two very clear laminated reflections, here marked C and E, were seen on all the Vancouver Island reflection lines. The C reflection, which dips at 5–8°, is believed to be from the decollement zone at the base of one of the accreted terranes (called Wrangellia). The E reflector, which dips at 9–13°, may represent a zone of porous sediments and volcanic rocks or may mark the location of trapped water in the crust.

8.6.6 Chemistry of Subduction Zone Magmas

The igneous rocks above subduction zones include granites, basalts and andesites as well as some ultramafic rocks (see Sect. 8.1.1, Table 8.1). The igneous rocks of the young Pacific island arcs such as the Tonga and Marianas arcs are primarily basalt and andesite. However, the older island arcs such as the Japan Arc are characterized by andesite volcanoes as well as diorite intrusions. Figure 8.47d shows a schematic geological cross section through the Andes. Although the andesite volcanoes provide the surface evidence of the active subduction zone beneath, the considerable thickening of the crust beneath the Andes is presumed to reflect the presence of large igneous intrusions.

The subducting plate produces partial melting in a number of ways. The basalts erupted above subduction zones result from partial melting of the mantle above the subducting plate. The loss even of small quantities of water from the subducting plate into the overriding mantle is sufficient to lower the melting temperature considerably (see Fig. 9.5). However, the magma which produces the andesite volcanics and the diorite intrusions forms either from partial melting of the subducted oceanic crust and sediments or, mostly, from melting in the overriding mantle wedge. The melt then collects beneath the overriding crust where it fractionates. The subducted oceanic mantle does not undergo partial melting because it is already depleted mantle material. Thus, the 'volcanic line' marks the depth at which material in the subducted plate or overlying mantle first reaches a high enough temperature for partial melting to occur. At shallow levels, partial melting is likely to produce basaltic magma; at greater depth, the degree of partial melting decreases because much of the water has been lost from the subducting plate. This means that the magmas produced are likely, and are observed to be, more alkaline (more andesitic). It is also possible that they will be altered on their ascent through the greater thickness of mantle and crust. Ultramafic rocks found above subduction zones are presumably tectonically emplaced pieces of the residue remaining after partial melting of the overriding mantle. This subject is discussed more fully in Section 9.2.1.

The chemical compositions of lavas in island arcs are spatially zoned with respect to the subduction zone. The strontium isotope ratio shows some correlation with the depth to the subduction zone. In the Indonesian Arc, this ratio appears to increase slightly with depth to the subduction zone. In the central Andes, the ratio increases with distance from the trench; but this is an age effect as well as a depth effect since the youngest rocks are inland and the oldest on the coast (Fig. 8.50). In this case, the isotopic data indicate that the magma source moved progressively eastwards with time. The smallest, and oldest, initial ratio of 0.7022 is in good agreement with the 0.702–0.704 which would be expected for the oceanic crustal basalts and mantle that presumably melted to form the sampled rocks. As time progressed and the magma source moved eastwards the rising magma would have to migrate through increasing thicknesses of crust; thus, the likelihood of contamination of the strontium isotope ratio is high. Subducted sea water, which would have a ratio of about 0.707, could also contaminate the magma. Thus, the greatest measured value of 0.7077 is still entirely consistent with a mantle or oceanic crust origin for these rocks.

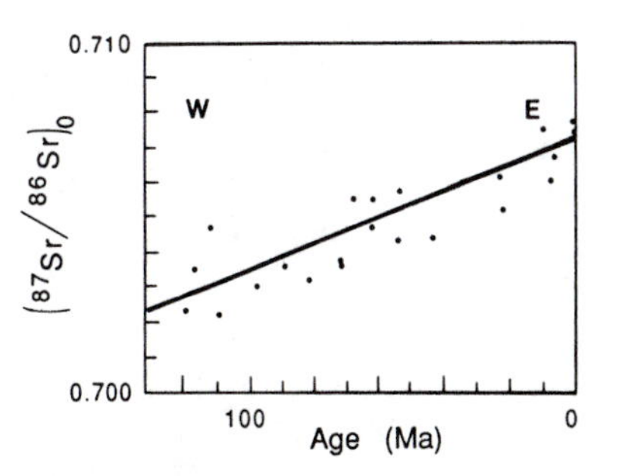

Figure 8.50. Initial $^{87}Sr/^{86}Sr$ ratio versus age for volcanic and plutonic rocks from the central Andes 26°–29°S. (Data from McNutt et al. 1975.)

8.7 Oceanic Islands

Oceanic island chains represent anomalies in the oceanic lithosphere, being locations away from the plate boundaries where considerable volcanic and

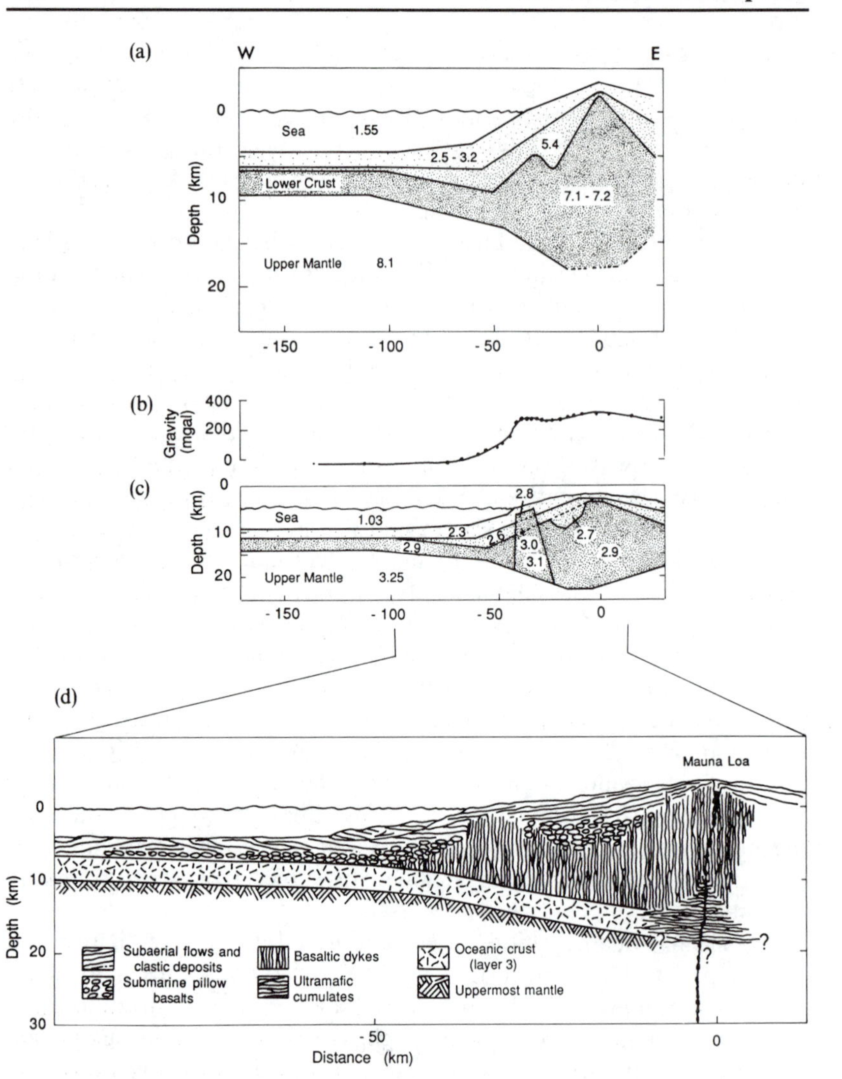

Figure 8.51. WSW–ENE profiles across the west coast of the island of Hawaii and the summit of Mauna Loa volcano. (a) P-Wave velocity structure determined from seismic refraction experiments. Vertical exaggeration 4:1. (b) Bouguer gravity anomaly onshore, free-air gravity anomaly offshore. Dots, measured anomalies; solid line, a matching anomaly computed from the density model in (c). (c) Density structure based on seismic velocity shown in (a). Vertical exaggeration 2:1. (d) Schematic geological structure based on the velocity and density models. The central magma conduit and the summit magma chamber are shown solid black. True scale. (From Hill and Zucca 1987.)

microearthquake activity is taking place. However, because of this they have been very useful in advancing an understanding of the physical properties of the plates and the underlying mantle.

Many of the details of seamount chains and oceanic islands have been discussed elsewhere in this book: dating of seamount chains and the hot-spot reference frame (Chapter 2), gravity and flexure of the oceanic lithosphere due to the loading of the Hawaiian Ridge (Sect. 5.7.) and possible origins for oceanic island basalts (Sect. 7.8.3). In this section some details of the seismicity beneath Hawaii and the seismic structure of the crust and upper mantle are presented.

Hawaii is the best-studied active oceanic island. Work by the Hawaii Volcano Observatory, located on the rim of the Kilauea crater, and others has built up a detailed picture of the processes taking place beneath the island. Figure 8.51 shows the P-wave velocity and density structure of the island of Hawaii on a profile crossing the Mauna Loa volcano which, together with the neighbouring volcano Kilauea and seamount Loihi, mark

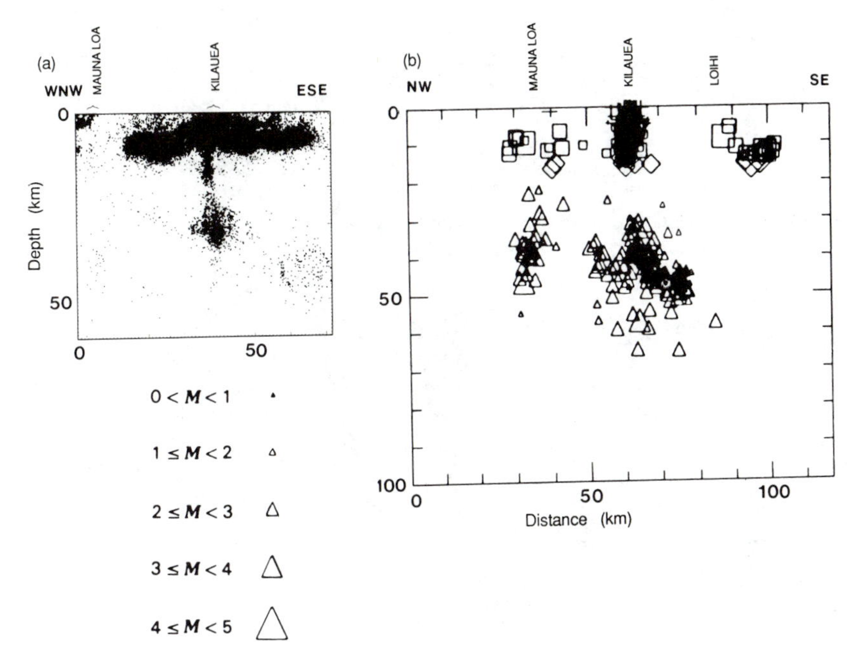

Figure 8.52. Distribution of earthquakes recorded by the Hawaiian Volcano Observatory along two cross sections through the island of Hawaii, approximately perpendicular to those shown in Figure 8.51. (a) All earthquakes, 1970–83. True scale. (b) Long-period earthquakes, 1972–84. The magnitudes of the events are indicated by the size of the symbol. □, Events shallower than 13 km; ◇, 13–20 km; △, > 20 km. True scale. (From Klein et al. 1987 and Koyanagi et al. 1987.)

the present location of the Hawaiian hot spot. The density values for the layers defined by the P-wave velocity were calculated by assuming a relation between velocity and density similar to that shown in Figure 4.2. Some small adjustments to the seismic layering were necessary in order to fit the gravity data well. The seismic velocity and density models, taken together, show that the crust beneath Mauna Loa thickens to some 18 km and that most of this material is of high density and velocity. The oceanic crust on which the volcano has formed is bent downwards by the load, in accordance with the flexural models. A schematic geological interpretation of these models is shown in Figure 8.51d. The high-velocity, high-density intrusive core of the volcano is interpreted as a sequence of densely packed dykes similar to the sheeted-dyke complex proposed for the upper portions of oceanic layer 3 (see Fig. 8.5b).

Figure 8.52 shows the seismic activity occurring down to depths of 60 km along two profiles across the island of Hawaii. In Figure 8.52a the epicentres are for all earthquakes from 1970 to 1983, and in Figure 8.52b the epicentres are for all long-period earthquakes from 1972 to 1984. Both sets of data show an extensive shallow zone of activity (less than about 13 km depth). Based on the seismic and density models, these shallow earthquakes, which often occurred in swarms, were all in the crust and are judged to be of volcanic origin. The deeper earthquakes are larger in magnitude and appear to be tectonic. The Kilauea magma conduit can be traced down to about 30 km. Deeper than 30 km the events merge into a broad zone. The magma transport systems for Mauna Loa, Kilauea and the seamount Loihi appear to connect to this deep zone.

The chemistry and dynamics of the Hawaiian magma supply have been studied in great detail. The top of the magma source zone beneath Hawaii is at about 60 km. From there it rises through the plumbing system to the

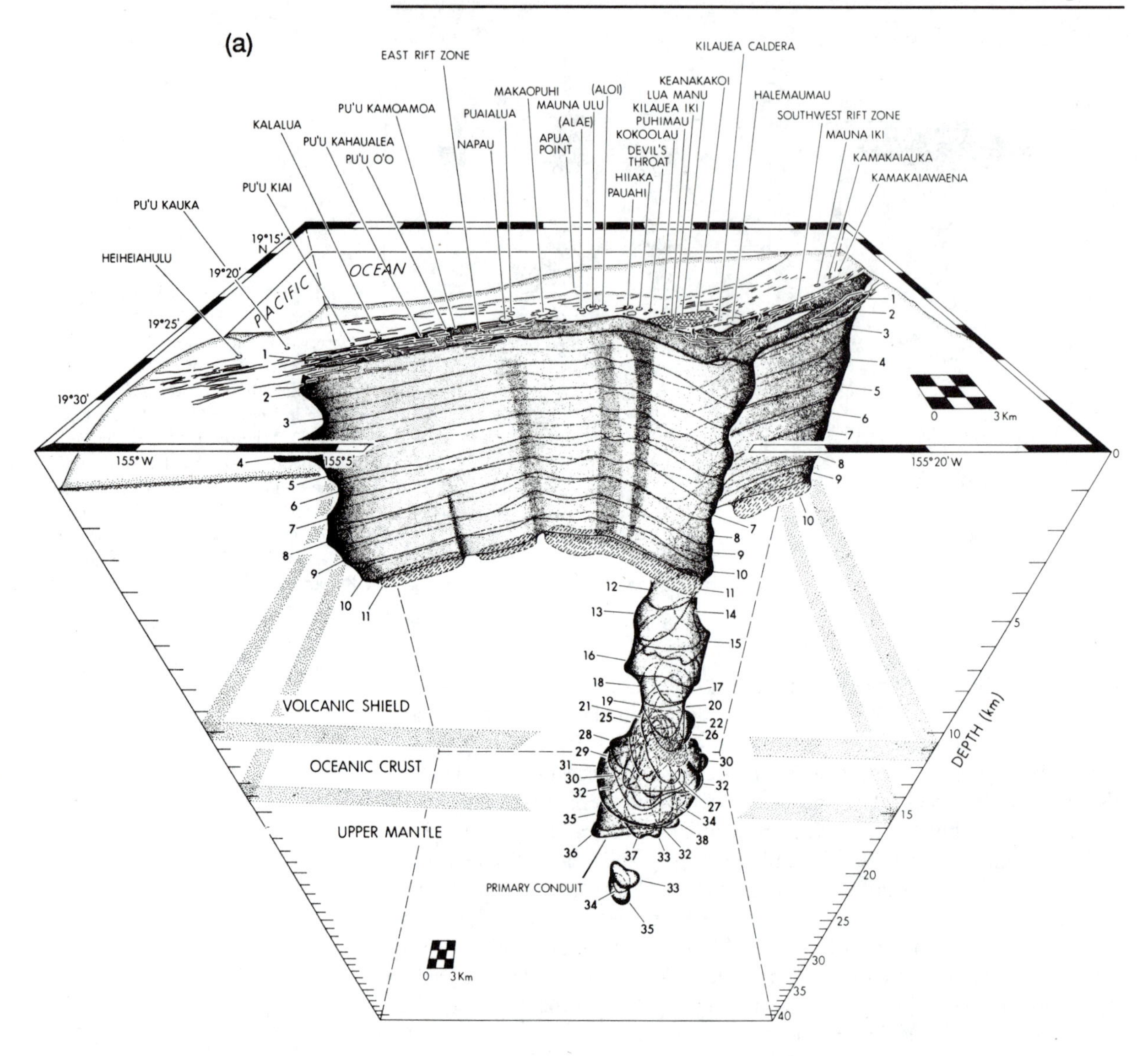

Figure 8.53. (a) View southward of the magma plumbing system beneath Kilauea volcano. The transition zones from the volcanic shield to oceanic crust and from oceanic crust to the upper mantle are stippled. Square 1-km scale grids are located at the surface and the base of the model (37 km). Individual conduit cross sections are labelled with their depths, beneath Kiluaea's caldera floor.

active vents. Figure 8.53 shows a detailed view of the magma conduit beneath Kilauea. This model was defined by well-located, magma-related earthquake foci; such definition is possible because fractures in the rocks around the magma chambers and conduits open as the pressure of the magma increases. Thus, the maximum lateral extent of the conduit at any depth corresponds to the region of hydraulically induced seismicity at that depth. The conduit system enables the primary conduit to supply magma to any of the structure vents or fissures. There are magma supply pathways from the summit magma reservoir beneath Kiluaea caldera to both the East and the Southwest rift zones.

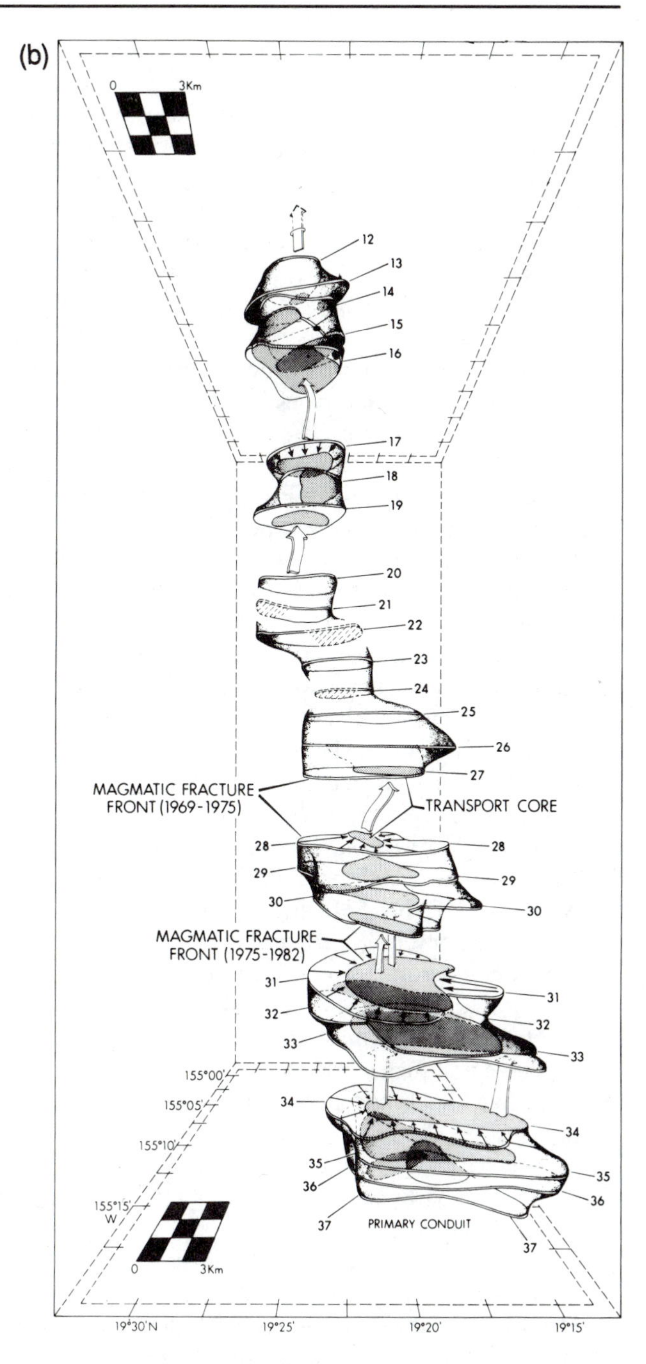

Figure 8.53 (*cont.*). (b) View eastward of the internal structure of the primary conduit, which is displayed as a series of segments to show its internal structure. The inner transport core of the conduit is shaded light grey. Where parts of two sections overlap, the core is shaded medium grey; and where three overlap, dark grey. Solid arrows show the retreat of the zone through which magma ascended from the 1969–75 interval to the 1975–82 interval when magma ascent was restricted to the core of the conduit. (From Ryan 1988.)

PROBLEMS

1. (a) Assuming isostatic equilibrium and Airy-type compensation, calculate the thickness of the oceanic crust if the continents averaged 50 km thick.
   (b) What is the minimum possible thickness of the continental crust? (Use densities of sea water, crust and mantle of $1.03 \times 10^3\,\mathrm{kg\,m^{-3}}$, $2.9 \times 10^3\,\mathrm{kg\,m^{-3}}$ and $3.3 \times 10^3\,\mathrm{kg\,m^{-3}}$; ocean basin depth, 5 km.)

2. Determine the oceanic crustal structure for the wide-angle reflection–refraction data shown in Figure 8.7a.
   (a) Use the normal incidence two-way travel times to estimate the depth of the seabed. (Use 1.5 km s^{-1} for the velocity of sound in sea water.)
   (b) Use the slope-intercept method to estimate the following: (i) an upper crustal P-wave velocity (use first arrivals at less than 10 km distance), (ii) the lower crustal P-wave velocity (use P_3), (iii) the upper mantle velocity, (iv) an upper crustal S-wave velocity (use S-wave arrivals at 10–15 km distance), (v) the lower crustal S-wave velocity (use S_3), (vi) the upper mantle S-wave velocity and (vii) the thickness of the oceanic crust at this location.
   (c) Compare the crustal thickness obtained in (vii) with the value obtained using normal-incidence two-way times.
   (d) Calculate the ratios of P-wave velocity to S-wave velocity. Do these values fall within the expected range?
3. Derive and plot the relationship between continental crustal thickness and ocean depth. Assume isostatic equilibrium and Airy-type compensation.
4. If the magma chamber on the East Pacific Rise at 12°N described in the example in Section 8.4 were filled with molten basaltic magma with seismic P-wave velocity of 3 km s^{-1}, how wide would it be? At the other extreme, if it contained only 10% partial melt, what would its width be? Comment on the likelihood of detection of such extreme magma chambers by seismic reflection and refraction experiments.
5. What would be the likelihood of delineating the magma chambers of Problem 4 by seismic methods if the dominant frequency of your signal were (a) 5 Hz, (b) 15 Hz and (c) 50 Hz?
6. What are the criteria for deciding whether the earth's crust at any location is oceanic or continental in origin?
7. The structure of the oceanic crust is very much the same whether it was created at a slow- or fast-spreading ridge. Discuss why. Should we also expect to find such worldwide lack of variation in the continental crust?
8. Calculate the lithospheric thicknesses at the ridge–transform fault intersection shown in Figure 8.25. Use the lithospheric thickness–age model $z(t) = 11t^{1/2}$ with z (in km) and t (in Ma).
9. (a) Draw a bathymetric profile along the north–south 10 Ma isochron of Figure 8.25.
   (b) Draw a similar profile along a north–south line crossing the fracture zone continuation 500 km to the east.
10. Draw fault-plane solutions for earthquakes occurring within the active zone of the transform fault shown in Figure 8.25. Discuss the relative frequency of earthquakes occurring between the two ridge segments and outside this zone. Draw a fault-plane solution for an earthquake occurring on the fault at the 50 Ma isochron on (a) the west and (b) the east side of the fault.
11. Calculate the topographic relief across a transform fault if the age offset is (a) 10 Ma, (b) 20 Ma and (c) 50 Ma. Plot the relief against distance (age) from the ridge axis for these three faults. What can you deduce about earthquakes on the inactive portion of these faults?

12. The boundary between the African and Eurasian plates between the Azores Triple Junction and Gibraltar runs approximately east–west. What is the nature of the boundary? Draw fault–plane solutions for earthquakes occurring along it.
13. What would happen to the plates shown in Figure 8.25 if the rotation pole were suddenly to move so that the spreading rate remained unchanged in magnitude but altered 20° in azimuth? Illustrate your answer with diagrams.
14. It is possible to make a first approximation to the temperatures along a transform fault by assuming that the temperature at any point along the transform is the average of the temperatures on either side. Using the cooling half-space model (Sect. 7.5.2), plot 400, 600 and 800°C isotherms beneath a 20 Ma offset transform fault. What assumptions have you made? If the base of the plate is assumed to be the 1000°C isotherm, how does this model agree with the depth to which faulting occurs on transform faults?

BIBLIOGRAPHY

Anderson, H., and Jackson, J. A. 1987. Active tectonics of the Adriatic region. *Geophys. J. R. Astr. Soc., 91*, 937–84.

Anderson, R. N., DeLong, S. E., and Schwarz, W. M. 1978. Thermal model for subduction with dehydration in the downgoing slab. *J. Geol., 86*, 731–9.

1980. Dehydration, asthenospheric convection and seismicity in subduction zones. *J. Geol., 88*, 445–51.

Ansell, J. H., and Gubbins, D. 1986. Anomalous high-frequency wave propagation from the Tonga–Kermadec seismic zone to New Zealand. *Geophys. J. R. Astr. Soc., 85*, 93–106.

Barazangi, M., and Isacks, B. L. 1971. Lateral variations of seismic-wave attenuation in the upper mantle above the inclined earthquake zone of the Tonga Island arc: Deep anomaly in the upper mantle. *J. Geophys. Res., 76*, 8493–516.

Benioff, H. 1949. Seismic evidence for the fault origin of oceanic deeps. *Geol. Soc. Am. Bull., 60*, 1837–56

Bergman, E. A., and Solomon, S.C. 1984. Source mechanisms of earthquakes near mid-ocean ridges from body waveform inversion, implications for the early evolution of oceanic lithosphere. *J. Geophys. Res., 89*, 11415–41.

Bevis, M., and Isacks, B. L. 1984. Hypocentral trend surface analysis: Probing the geometry of Benioff zones. *J. Geophys. Res., 89*, 6153–70.

Billington, S. 1980. The morphology and tectonics of the subducted lithosphere in the Tonga–Fiji–Kermadec region from seismicity and focal mechanism solutions. Ph.D. thesis, Cornell University, Ithaca.

Bird, P. 1978. Stress and temperature in subduction shear zones. *Geophys. J. R. Astr. Soc., 55*, 411–34.

Bowen, A. N., and White, R. S. 1986. Deep-tow seismic profiles from the Vema transform and ridge transform intersection. *J. Geol. Soc. Lond., 143*, 807–17.

Boyd, T. M., Taber, J. J., Lerner-Lam, A. L., and Beavan, J. 1988. Seismic rupture and arc segmentation within the Shumagin Islands Seismic Gap, Alaska. *Geophys. Res. Lett., 5*, 201–4.

Bratt, S. R., and Purdy, G. M. 1984. Structure and variability of oceanic crust on the flanks of the East Pacific Rise between 11° and 13°N. *J. Geophys. Res., 89*, 6111–25.

Bratt, S. R., and Solomon, S. C. 1984. Compressional and shear wave structure of the East Pacific Rise at 11°20′N, constraints from three-component ocean bottom seismometer data. *J. Geophys. Res., 89*, 6095–110.

Brown, G. C., and Hennessy, J. 1978. The initiation and thermal diversity of granite magmatism. *Phil. Trans. R. Soc. Lond., 288A*, 631–43.

Brownlow, A. H. 1979. *Geochemistry.* Prentice-Hall, Englewood Cliffs, N.J.

Bryan, W. B., and Moore, J. G. 1977. Compositional variation of young basalts in the mid-Atlantic Ridge rift valley near lat. 36°49′N. *Geol. Soc. Am. Bull., 88*, 556–70.

Bunch, A. W. H., and Kennett, B. L. N. 1980. The crustal structure of the Reykjanes Ridge at 59°30′N. *Geophys. J. R. Astr. Soc., 61*, 141–66.

Burchfiel, B. C. 1983. The continental crust. *Sci. Am. 249, 3*, 57–65.

Cann, J. R. 1974. A model for oceanic crustal structure developed. *Geophys. J. R. Astr. Soc., 39*, 169–87.

Cary, P. W., and Chapman, C. H. 1988. Waveform inversion of expanding spread profile 5 from the North Atlantic transect. *J. Geophys. Res., 93*, 13575–88.

Chapman, M., and Talwani, M. 1979. Comparison of gravimetric geoids with Geos3 altimeter. *J. Geophys. Res., 84*, 3803–16.

Chiang, C. S., and Detrick, R. S. 1985. The structure of the lower oceanic crust from synthetic seismogram modelling of near-vertical and wide-angle reflections and refractions near DSDP Site 417 in the western North Atlantic. *EOS Trans. Am. Geophys. Un., 66*, 956.

Christensen, N. I., and Smewing, J. D. 1981. Geology and seismic structure of the northern section of the Oman Ophiolite. *J. Geophys. Res., 86*, 2545–55.

Clague, D. A., and Dalrymple, G. B. 1987. The Hawaiian–Emperor volcanic chain. *In* R. W. Decker, T. L. Wright and P. H. Stauffer, eds., *Volcanism in Hawaii*, U.S. Geological Survey Professional Paper 1350, 5–54.

Clowes, Ron. 1987. LITHOPROBE: Exploring the subduction zone of western Canada. *Leading Edge, 6*, 6, 12–19.

Coleman, R. G. 1977. *Ophiolites, ancient oceanic lithosphere*? Springer-Verlag, Berlin.

Coulbourn, W. T. 1981. Tectonics of the Nazca plate and the continental margin of western South America, 18°S to 23°S. *In* L. D. Kulm, J. Dymond, E. J. Dasch and D. M. Hussong, eds., *Nazca plate crustal formation and Andean convergence*, Vol. 154 of Mem. Geol. Soc. Am., 587–618.

Cox, A. 1969. Geomagnetic Reversals. *Science, 163*, 237–45.

Creager, K. C., and Jordan, T. H. 1986. Slab penetration into the lower mantle beneath the Mariana and other island arcs of the northwest Pacific. *J. Geophys. Res.*, 91, 3573–89.

Decker, R. W. 1987. Dynamics of Hawaiian volcanoes: An overview. *In* R. W. Decker, T. L. Wright and P. H. Stauffer, eds., *Volcanism in Hawaii*, U. S. Geological Survey Professional Paper 1350, 997–1018.

DeMets, C., Gordon, R. G., Argus, D. F., and Stein, S. 1990 . Current plate motions. *Geophysical J. Int., 101*, 425–78.

Detrick, R. S., Buhl, P., Vera, E., Mutter, J., Orcutt, J., Madsen, J., and Brocher, T. 1987. Multichannel seismic imaging of a crustal magma chamber along the East Pacific Rise. *Nature, 326*, 35–41.

Detrick, R. S., and Purdy, G. M. 1980. The crustal structure of the Kane fracture zone from seismic refraction studies. *J. Geophys. Res., 85*, 3759–77.

Drew, J. J., and Clowes, R. M. 1989. A re-interpretation of the seismic structure across the active subduction zone of western Canada. *In Studies of laterally heterogeneous structures using seismic refraction and reflection data*, Proceedings of the 1987 Commission on Controlled Source Seismology Workshop, Geological Survey of Canada Paper 89–13.

Driscoll, M., and Parsons, B. 1988. Cooling of the oceanic lithosphere – evidence from geoid anomalies across the Udintsev and Eltanin fracture zones. *Earth Planet. Sci. Lett.*, *88*, 289–307.

Duschenes, J. D., and Solomon, S. C. 1977. Shear wave travel time residuals from oceanic earthquakes and the evolution of the oceanic lithosphere. *J. Geophys. Res.*, *82*, 1985–2000.

Edmond, J. H., and Von Damm, K. 1983. Hot springs on the ocean floor. *Sci. Am.*, *248*, 4, 78–93.

Eicher, D. L., McAlester, A. L., and Rottman, M. L. 1984. *The history of the earth's crust.* Prentice-Hall, Englewood Cliffs, N.J.

EMSLAB Group. 1988. The EMSLAB electromagnetic sounding experiment. *EOS Trans. Am. Geophys. Un.*, *69*, 89.

Engeln, J. F., Wiens, D. A., and Stein, S. 1986. Mechanisms and depths of Mid-Atlantic Ridge transform faults. *J. Geophys. Res.*, *91*, 548–78.

Ernst, W. G. 1976. *Petrologic phase equilibria.* Freeman, San Francisco.

Fischer, K. M., Jordan, T. H., and Creager, K. C. 1988. Seismic constraints on the morphology of deep slabs. *J. Geophys. Res.*, *93*, 4773–83.

Forsyth, D. W. 1977. The evolution of the upper mantle beneath mid-ocean ridges. *Tectonophys.*, *38*, 89–118.

Forsyth, D. W., and Wilson, B. 1984. Three-dimensional temperature structure of a ridge–transform–ridge system. *Earth Planet. Sci. Lett.*, *70*, 355–62.

Fowler, C. M. R. 1976. Crustal structure of the Mid-Atlantic Ridge crest at 37°N. *Geophys. J. R. Astr. Soc.*, *47*, 459–491.

1978. The Mid-Atlantic Ridge: Structure at 45°N. *Geophys. J. R. Astr. Soc.*, *54*, 167–82.

Fowler, S. R., White, R. S., and Louden, K. E. 1985. Sediment dewatering in the Makran accretionary prism. *Earth Planet. Sci. Lett.*, *75*, 427–38.

Fox, P. J., and Gallo, D. G. 1984. A tectonic model for ridge–transform–ridge plate boundaries: Implications for the structure of oceanic lithosphere. *Tectonophys.*, *104*, 205–42.

1989. Transforms of the Eastern Central Pacific, In E. L. Winterer, D. M. Hussong, and R. W. Decker, eds., *The Geology of North America, Vol. N, The eastern Pacific and Hawaii, Geol. Soc. Am.*, 111–24.

Francis, T. J. G., Porter, I. T., and Lilwall, R. C. 1978. Microearthquakes near the eastern end of St. Paul's fracture zone. *Geophys. J. R. Astr. Soc.*, *53*, 201–17.

Frank, F. C. 1968. Curvature of island arcs. *Nature*, *220*, 363.

Furlong, K. P., and Chapman, D. S. 1982. Thermal modelling of the geometry of the tectonics of the overriding plate. *J. Geophys. Res.*, *87*, 1786–802.

Goud, H. R., and Karson, J. A. 1986. Tectonics of short-offset, slow slipping transform zones in the FAMOUS area, Mid-Atlantic Ridge. *Mar. Geophys. Res.*, *7*, 489–514.

Goto, K., Suzuki, Z., and Hamaguchi, H. 1987. Stress distribution due to olivine–spinel phase transition in descending plate and deep focus earthquakes. *J. Geophys. Res.*, *92*, 13811–20.

Green, A. G., Clowes, R. M., Yorath, C. J., Spencer, C., Kanasewich, E. R., Brandon, M. T., and Sutherland-Brown, A. 1986. Seismic reflection imaging of the subducting Juan de Fuca Plate. *Nature*, *319*, 210–13.

Green, A. G., Milkereit, B., Mayrand, L., Spencer, C., Kurtz, R., and Clowes, R. M. 1987. Lithoprobe seismic reflection profiling across Vancouver Island: Results from reprocessing. *Geophys. J. R. Astr. Soc.*, *89*, 85–90.

Grimison, N. L., and Chen, W.-P. 1988. Source mechanisms of four recent earthquakes along the Azores–Gibraltar boundary. *Geophys. J. R. Astr. Soc.*, *92*, 391–402.

Grow, J. A., and Bowin, C. O. 1975. Evidence for high-density crust and mantle beneath the Chile Trench due to the descending lithosphere. *J. Geophys. Res.*, *80*, 1449–58.

Haberman, R. E., McCann, W. R., and Perin, B. 1986. Spatial seismicity variations along convergent plate boundaries. *Geophys. J. R. Astr. Soc., 85*, 43–68.

Hale, L. D., Morton, C., and Sleep, N. H. 1982. Reinterpretation of seismic reflection data over the East Pacific Rise. *J. Geophys. Res. 87*, 7707–17.

Hasegawa, A., Umino, N., and Takagi, A. 1978. Double-planed deep seismic zone and upper-mantle structure in northeastern Japan Arc. *Geophys. J. R. Astr. Soc., 54*, 281–96.

Hay, W. W., Sloan, J. L., II, and Wold, C. N. 1988. Mass/age distribution and composition of sediments on the ocean floor and the global rate of sediment subduction. *J. Geophys. Res., 93*, 14933–40.

Heezen, B. C. 1962. The Deep-Sea-Floor. *In* S. K. Runcorn, ed., *Continental drift*. Academic, New York, 235–68.

Heirtzler, J. R., Dickson, G. O., Herron, E. M., Pitman, W. C., III, and LePichon, X. 1968. Marine magnetic anomalies, geomagnetic field reversals and motions of the ocean floor and continents. *J. Geophys. Res., 73*, 2119–36.

Hekinian, R., and Walker, D. 1987. Diversity and spatial zonation of volcanic rocks from the East Pacific Rise near 21°N. *Contrib. Mineral. Petrol., 96*, 265–80.

Hill, D. P., and Zucca, J. J. 1987. Geophysical constraints on the structure of Kilauea and Mauna Loa volcanoes and some implications for seismomagmatic processes. *In* R. W. Decker, T. L. Wright and P. H. Stauffer, eds., *Volcanism in Hawaii*, U.S. Geological Survey Professional Paper 1350, 930–17.

Hill, M. N. 1957. Recent geophysical exploration of the ocean floor. *Phys. Chem. Earth, 2*, 129–63.

Honda, S. 1985. Thermal structure beneath Tohoku, northeast Japan: A case study for understanding the detailed thermal structure of the subduction zone. *Tectonophys., 112*, 69–102.

Huang, P. Y., and Solomon, S. C. 1987. Centroid depths and mechanisms of mid-ocean ridge earthquakes in the Indian Ocean, Gulf of Aden and Red Sea. *J. Geophys. Res., 92*, 1361–82.

1988. Centroid depths of mid-ocean ridge earthquakes: Dependence on spreading rate. *J. Geophys. Res., 93*, 13445–7.

Huang, P. Y., Solomon, S. C., Bergman, E. A., and Nabelek, J. L. 1986. Focal depths and mechanisms of Mid-Atlantic Ridge earthquakes from body waveform inversion. *J. Geophys. Res., 91*, 579–98.

Hyndman, R. D. 1988. Dipping seismic reflectors, electrically conductive zones and trapped water in the crust over a subducting plate. *J. Geophys. Res., 93*, 13391–405.

Isacks, B., and Molnar, P. 1969. Mantle earthquake mechanisms and the sinking of the lithosphere. *Nature, 223*, 1121–4.

Isacks, B., Oliver, J., and Sykes, L. 1968. Seismology and the new global tectonics. *J. Geophys. Res., 73*, 5855–99.

Isacks, B. L., and Barazangi, M. 1977. Geometry of Benioff zones: Lateral segmentation and downwards bending of the subducted lithosphere. *In* M. Talwani and W. C. Pitman III, eds., *Island arcs, deep-sea trenches and back-arc basins*, Am. Geophys. Un. Maurice Ewing Series 1, 99–114.

Jackson, J. A., and McKenzie, D. P. 1984. Active tectonics of the Alpine–Himalayas belt between western Turkey and Pakistan. *Geophys. J. R. Astr. Soc., 77*, 185–266.

Karson, J. A., and Dick, H. J. B. 1983. Tectonics of ridge–transform intersections at the Kane Fracture Zone. *Mar. Geophys. Res., 6*, 51–98.

Karson, J. A., Thompson, G., Humphris, S. E., Edmond, J. M., Bryan, W. B., Brown, J. R., Winters, A. T., Pockalny, R. A., Casey, J. F., Campbell, A. C., Klinkhammer, G., Palmer, M. R., Kinzler, R. J., and Sulanowska, M. M. 1987. Along axis variability in seafloor spreading in the MARK area. *Nature, 328*, 681–5.

Keen, C. E., and Tramontini, C. 1970. A seismic refraction survey on the Mid-Atlantic Ridge. *Geophys. J. R. Astr. Soc., 20*, 473–91.

Khitarov, N. I., Lebedev, E. B., Dorfman, A. M., and Bagdasarov, N. S. 1983. Study of process of melting of the Kirgurich basalt by the wave method. *Geochimica, 9*, 1239–46.

Klein, F. W., Koyanagi, R. Y., Nakata, J. S., and Tanigawa, W. R. 1987. The seismicity of Kilauea's magma system. *In* R. W. Decker, T. L. Wright and P. H. Stauffer, eds., *Volcanism in Hawaii*, U.S. Geological Survey Professional Paper, 1350, 1019–186.

Koyanagi, R. Y., Chouet, B., and Aki, K. 1987. Origin of volcanic tremor in Hawaii, Part I; Data from the Hawaiian Volcano Observatory, 1969–1985. *In* R. W. Decker, T. L. Wright and P. H. Stauffer, eds. *Volcanism in Hawaii*, U.S. Geological Survey Professional Paper 1350, 1221–58.

Kusznir, N. J., and Bott, M. H. P. 1976. A thermal study of the formation of oceanic crust. *Geophys. J. R. Astr. Soc., 47*, 83–95.

Lambert, I. B., and Wyllie, P. J. 1972. Melting of gabbro (quartz eclogite) with excess water to 35 kilobars, with geological applications. *J. Geol., 80*, 693–708.

Langmuir, C. H. 1987. A magma chamber observed? *Nature, 326*, 15–16.

Larson, R. L., and Hide, T. W. C. 1975. A revised timescale of magnetic reversals for the Early Cretaceous and Late Jurassic. *J. Geophys. Res., 80*, 2586–94.

Laughton, A. S., Matthews, D. H., Fisher, R. L. 1970. The structure of the Indian Ocean. *In* A. E. Maxwell, ed., *The sea*, Vol. 4, Part II. Wiley-Interscience, New York, 543–86.

Lewis, B. T. R., and Garmany, J. D. 1982. Constraints on the structure of the East Pacific Rise from seismic refraction data. *J. Geophys. Res., 87*, 8417–25.

Lewis, T. J., Bentkowski, W. H., Davies, E. E., Hyndman, R. D., Souther, J. G., and Wright, J. A. 1988. Subduction of the Juan de Fuca Plate: Thermal consequences. *J. Geophys. Res., 93*, 15207–25.

Lewis, T. J., Jessop, A. M., and Judge, A. S. 1985. Heat flux measurements in southwestern British Columbia: The thermal consequences of plate tectonics. *Candian J. Earth Sci., 22*, 1262–73.

Lilwall, R. C., Francis, T. J. G., and Porter, I. T. 1977. Ocean-bottom seismograph observations on the Mid-Atlantic Ridge near 45°N. *Geophys. J. R. Astr. Soc., 51*, 357–69.

Lindwall, D. 1988. A two-dimensional seismic investigation of crustal structure under the Hawaiian Islands near Oahu and Kauai. *J. Geophys. Res., 93*, 12107–22.

Louden, K. E., White, R. S., Potts, C. G., and Forsyth, D. W. 1986. Structure and seismotectonics of the Vema Fracture Zone, Atlantic Ocean. *J. Geol. Soc. Lond., 143*, 795–805.

Macdonald, K. C. 1986. The crest of the Mid-Atlantic Ridge: Models for crustal generation processes and tectonics. In P. R. Vogt and B. E. Tucholke, eds., *The geology of North America*, Vol. M, *The western North Atlantic region*, Geol. Soc. Am., 51–68.

1989. Anatony of a magma reservoir. *Nature, 339*, 178–9.

Macdonald, K. C., Castillo, D. A., Miller, S. P., Fox, P. J., Kastens, K. A., and Bonatti, E. 1986. Deep-tow studies of the Vema fracture zone: 1. Tectonics of a major slow slipping transform fault and its intersection with the Mid-Atlantic Ridge. *J. Geophys. Res., 91*, 3334–54.

Macdonald, K. C., and Fox, P. J. 1988. The axial summit graben and cross-sectional shape of the East Pacific Rise as indicators of axial magma chambers and recent volcanic eruptions. *Earth Planet. Sci. Lett., 88*, 119–31.

Macdonald, K. C., Fox, P. J., Perram, L. J., Eisen, M. F., Haymon, R. M., Miller, S. P., Carbotte, S. M., Cormier, M.-H., and Shor, A. N. 1988. A new view of the mid-ocean ridge from the behaviour of ridge-axis discontinuities. *Nature, 335*, 217–25.

Macdonald, K. C., and Luyendyk, B. P. 1981. The crest of the East Pacific Rise. *Sci. Am., 244*, 5, 100–17.

Macdonald, K. C., Sempere, J. C., Fox, P. J., and Tyce, R. 1987. Tectonic evolution of ridge axis discontinuities by the meeting, linking, or self-decapitation of neighbouring ridge segments. *Geology, 15*, 993–7.

Mahlberg Kay, S., and Kay, R. W. 1985. Role of crystal cumulates and the oceanic crust in the formation of the lower crust of the Aleutian arc. *Geology, 13*, 461–4.

Matsuzawa, T., Umino, N., Hasegawa, A., and Takagi, A. 1986. Upper mantle velocity structure estimated from PS-converted wave beneath the north-eastern Japan Arc. *Geophys. J. R. Astr. Soc., 86*, 767–87.

Maxwell, A. E., Von Herzen, R. P., Hsü, K. J., Andrews, J. E., Saito, T., Percival, S. F. Jr., Millow, E. D., and Boyce, R. E. 1970. Deep sea drilling in the South Atlantic. *Science, 168*, 1047–59.

McCarthy, J., Mutter, J. C., Morton, J. L., Sleep, N. H., and Thompson, G. A. 1988. Relic magma chamber structures preserved within the Mesozoic North Atlantic crust? *Geol. Soc. Am. Bull., 100*, 1423–36.

McKenzie, D. P. 1969. Speculations on the consequences and causes of plate motions. *Geophys. J. R. Astr. Soc., 18*, 1–32.

McKenzie, D. P., and Bickle, M. J. 1988. The volume and composition of melt generated by extension of the lithosphere. *J. Petrol., 29*, 625–79.

McKenzie, D. P., and Sclater, J. G. 1971. The evolution of the Indian Ocean since the Late Cretaceous. *Geophys. J. R. Astr. Soc. 25*, 437–528.

McNutt, R. H., Crocket, J. H., Clark, A. H., Caelles, J. C., Farrar, E., Haynes, S. J., and Zentilli, M. 1975. Initial $^{87}Sr/^{86}Sr$ ratios of plutonic and volcanic rocks of the central Andes between latitudes 26° and 29° South. *Earth Planet. Sci. Lett., 27*, 305–13.

Menard, H. W. 1964. *Marine geology of the Pacific.* McGraw-Hill, New York.

Monger, J. W. H., Clowes, R. M., Price, R. A., Simony, P. S., Riddihough, R. P., and Woodsworth, G. J. 1985. *Continent–ocean transect B2: Juan de Fuca Plate to Alberta plains.* Geological Society of America, Centennial Continent/Ocean Transect No. 7.

Morgan, J. P., and Forsyth, D. W. 1988. Three-dimensional flow and temperature perturbations due to a transform offset: Effects on oceanic crustal and upper mantle structure. *J. Geophys. Res., 93*, 2955–66.

Morton, J. L., Sleep, N. H., Normark, W. R., and Tompkins, D. H. 1987. Structure of the southern Juan de Fuca Ridge from seismic reflection records. *J. Geophys. Res., 92*, 11315–26.

NAT Study Group. 1985. North Atlantic Transect: A wide-aperture, two-ship multichannel seismic investigation of the oceanic crust. *J. Geophys. Res., 90*, 10321–41.

Ness, G., Levi, S., and Couch, R. 1980. Marine magnetic anomaly timescales for the Cenozoic and Late Cretaceous: A precis, critique and synthesis. *Rev. Geophys. Space Phys. 18*, 753–70.

Nisbet, E. G., and Fowler, C. M. R. 1978. The Mid-Atlantic Ridge at 37 and 45°N: Some geophysical and petrological constraints. *Geophys. J. R. Astr. Soc., 54*, 631–60.

Nockolds, S. R., Knox, R. W. O'B., and Chinner, G. A. 1978. *Petrology for students.* Cambridge Univ. Press.

Orcutt, J. A., Kennett, B. L. N., and Dorman, L. M. 1976. Structure of the East Pacific Rise from an ocean bottom seismometer survey. *Geophys. J. R. Astr. Soc., 45*, 305–20.

Pallister, J. S., and Hopson, C. A. 1981. Semail ophiolite plutonic suite: Field relations, phase variation, cryptic variation and layering and a model of a spreading ridge magma chamber. *J. Geophys. Res., 86*, 2593–644.

Phipps Morgan, J. E., Parmentier, M., and Lin, J. 1987. Mechanisms for the origin of mid-ocean ridge axial topography: implications for the thermal and mechanical structure of accreting plate boundaries. *J. Geophys. Res., 92*, 12823–36.

Pitman, W. C., III, and Heirtzler, J. R. 1966. Magnetic Anomalies over the Pacific-Antarctic Ridge. *Science, 154*, 1164–71.

Prothero, W. A., Reid, I., Reichle, M. S., and Brune, J. N. 1976. Ocean-bottom seismic measurements on the East Pacific Rise and Rivera Fracture Zone. *Nature, 262*, 121–4.

Purdy, G. M. 1983. The seismic structure of 140 Myr old crust in the western central Atlantic Ocean. *Geophys. J. R. Astr. Soc., 72*, 115–37.

Purdy, G. M., and Detrick, R. S. 1986. The crustal structure of the Mid-Atlantic Ridge at 23°N from seismic refraction studies. *J. Geophys. Res., 91*, 3739–62.

Purdy, G. M., and Ewing, J. 1986. Seismic structure of the ocean crust. *In* P. R. Vogt and B. E. Tucholke, eds., *The geology of North America, Vol. M, The western North Atlantic region*, Geol. Soc. Am., 313–30.

Raitt, R. W. 1963. The crustal rocks. *In* M. N. Hill, ed., *The sea*, Vol. 3, Interscience, New York, 85–102.

Reid, I., Orcutt, J. A., and Prothero, W. A. 1977. Seismic evidence for a narrow zone of partial melting underlying the East Pacific Rise at 21°N. *Geol. Soc. Am. Bull., 88*, 678–82.

Richter, F. M. 1979. Focal mechanisms and seismic energy release of deep and intermediate earthquakes in the Tonga–Kermadec region and their bearing on the depth extent of mantle flow. *J. Geophys. Res., 84*, 6783–95.

Rise Project Group. 1980. East Pacific Rise: Hot springs and geophysical experiments. *Science, 207*, 1421–33.

Rona, P. A. 1988. Hydrothermal mineralisation at oceanic ridges. *Canadian Mineralogist, 26*, 431–65.

Rowlett, H., and Forsyth, D. W. 1984. Recent faulting and microearthquakes at the intersection of the Vema fracture zone and the Mid-Atlantic Ridge. *J. Geophys. Res., 89*, 6079–94.

Ryan, M. P. 1988. A close look at parting plates. *Nature, 332*, 779–80.

1988. The mechanics and three-dimensional internal structure of active magmatic systems: Kiluaea Volcano, Hawaii, *J. Geophys. Res., 93*, 4213–48.

Ryan, M. P., Koyanagi, R. Y., and Fiske, R. S. 1981. Modelling the three-dimensional structure of macroscopic magma transport systems: Application to Kilauea Volcano, Hawaii. *J. Geophys. Res., 86*, 7111–29.

Sacks, I. S., and Okada, H. 1974. A comparison of the anelasticity structure beneath western South America and its tectonic significance. *Phys. Earth Planet. Interiors, 9*, 211–19.

Searle, R. C. 1983. Multiple, closely spaced transform faults in fast-slipping fracture zones. *Geology, 11*, 607–10.

Schubert, G., Yuen, D. A., and Turcotte, D. L. 1975. Role of phase transitions in a dynamic mantle. *Geophys. J. R. Astr. Soc., 42*, 705–35.

Schweller, W. J., Kuln, L. D., and Prince, R. A. 1981. Tectonics, structure and sedimentary framework of the Peru–Chile Trench. *In* L. D. Kuln, J. Dymond, E. J. Dasch and D. M. Hussong, eds., *Nazca plate, crustal formation and Andean convergence*, Vol. 154 of Geol. Soc. Amer. Memoirs, 323–50.

Sinha, M. C., and Louden, K. E. 1983. The Oceanographer fracture zone – I. Crustal structure from seismic refraction studies. *Geophys. J. R. Astr. Soc., 75*, 713–36.

Sleep, N. H. 1975. Formation of oceanic crust: Some thermal constraints. *J. Geophys. Res., 80*, 4037–42.

1983. Hydrothermal convection at ridge axes. *In* P. A. Rona, K. Bostrom, L. Laubier and K. L. Smith, eds., *Hydrothermal processes at seafloor spreading centres*, Plenum, New York, 71–82.

Smith, H. S., and Erlank, A. J. 1982. Geochemistry and petrogenesis of komatiites from the Barberton greenstone belt, South Africa. *In* N. T. Arndt and E. G. Nisbet, eds., *Komatiites*, George Allen and Unwin, London, 347–97.

Solomon, S. C., and Julian, B. R. 1974. Seismic constraints on oceanic-ridge mantle structure: Anomalous fault plane solutions from first motions. *Geophys. J. R. Astr. Soc., 38*, 265–85.

Spence, G. D., Clowes, R. M., and Ellis, R. M. 1985. Seismic structure across the active subduction zone of western Canada. *J. Geophys. Res., 90*, 6754–72.

Spudich, P., and Orcutt, J. 1980. Petrology and porosity of an oceanic crustal site: Results from waveform modelling of seismic refraction data. *J. Geophys. Res. 85*, 1409–33.

Stauder, W., and Maulchin, L. 1976. Fault motion in the larger earthquakes of Kuie-Kamchatka Arc and of the Kurile, Hokkaido Corner. *J. Geophys. Res., 81*, 297–308.

Stoffa, P. L., Buhl, P., Henon, T. J., Kan, T. K., and Ludwig, W. J. 1980. Mantle reflections beneath the crustal zone of the East Pacific Rise from multichannel seismic data. *Mar. Geology, 35*, 83–97.

Suyehiro, K., and Sacks, I. S. 1983. An anomalous low velocity region above the deep earthquakes in the Japan subduction zone. *J. Geophys. Res., 88*, 10429–38.

Talwani, M., LePichon, X., and Ewing, M. 1965. Crustal structure of the mid-ocean ridges 2: Computed model from gravity and seismic refraction data. *J. Geophys. Res., 70*, 341–52.

Toksöz, M. N. 1975. The subduction of the lithosphere. *In* J. T. Wilson, ed., *Continents adrift and continents aground*, Scientific American, 113–22.

Toomey, D. R., Solomon, S. C., Purdy, G. M., and Murray, M. H. 1985. Microearthquakes beneath the median valley of the Mid-Atlantic Ridge near 23°N: Hypocentres and focal mechanisms. *J. Geophys. Res., 90*, 5443–58.

Tucholke, B. E. 1986. Structure of the basement and distribution of sediments in the western North Atlantic. *In* P. R. Vogt and B. E. Tucholke, eds., *The geology of North America*, Vol. M, Geol. Soc. Am., 331–40.

Turcotte, D. L., and Schubert, G. 1973. Frictional heating of the descending lithosphere. *J. Geophys. Res., 78*, 5876–86.

van den Beukel, J., and Wortel, R. 1986. Thermal modelling of arc-trench regions. *Geologie en Mijnbouw, 65*, 133–43.

Vassiliou, M. S. 1984. The state of stress in subducting slabs as revealed by earthquakes analysed by moment tensor inversion. *Earth Planet. Sci. Lett., 69*, 195–202.

Vassiliou, M. S., Hager, B. H., and Raefsky, A. 1984. The distribution of earthquakes with depth and stress in subducting slabs. *J. Geodyn., 1*, 11–28.

Vine, F. J. 1966. Spreading of the ocean floor: New evidence. *Science, 154*, 1405–15.

Vine, F. J., and Matthews, D. H. 1963. Magnetic anomalies over oceanic ridges. *Nature, 199*, 947–9.

Wadati, K. 1928. Shallow and deep earthquakes. *Geophys. Mag., 1*, 162–202.

1935. On the activity of deep-focus earthquakes in the Japan Islands and neighbourhoods. *Geophys. Mag., 8*, 305–25.

Watanabe, T., Langseth, M. G., and Anderson, R. N. 1977. Heat flow in back-arc basins of the Western Pacific. *In* M. Talwani and W. C. Pitman, III, eds., *Island arcs, deep-sea trenches and back-arc basins*, Am. Geophys. Un. Maurice Ewing Ser. 1, 137–62.

Watts, A. B., ten Brink, U. S., Buhl, P., and Brocher, T. M. 1985. A multi-channel seismic study of lithospheric flexure across the Hawaii–Emperor seamount chain. *Nature, 315*, 105–11.

Watts, A. B., Weissel, J. K., and Larson, R. L. 1977. Sea-floor spreading in marginal basins of the western Pacific. *Tectonophys., 37*, 167–81.
White, R. S. 1984. Atlantic oceanic crust: Seismic structure of a slow spreading ridge. *In* I. G. Gass, S. J. Lippard, and A. W. Shelton, eds., Ophiolites and oceanic lithosphere, Geol. Soc. Lond. 34–44.
White, R. S., Detrick, R. S., Mutter, J. C., Buhl, P., Minshull, T. A., and Morris, E. 1990. New seismic images of oceanic crustal structure. *Geology.18*, 462–5.
White, R. S., Detrick, R. S., Sinha, M. C., and Cormier, M. H. 1984. Anomalous seismic crustal structure of oceanic fracture zones. *Geophys. J. R. Astr. Soc., 79*, 779–98.
White, R. S., and Louden, K. E. 1982. The Makran continental margin: Structure of a thickly sedimented convergent plate boundary. *In* J. S. Watkins and C. L. Drake, eds., *Studies in continental margin geology*, Vol. 34 of Am. Assoc. Petrol. Geol. Memoirs, 499–518.
Whitmarsh, R. B., and Calvert, A. J. 1986. Crustal structure of Atlantic fracture zones – I. The Charlie-Gibbs Fracture Zone. *J. R. Astr. Soc., 85*, 107–38.
Wiens, D. A., and Stein, S. 1983. Age dependence of oceanic intraplate seismicity and implications for lithospheric evolution. *J. Geophys. Res., 88*, 6455–68.
Wyllie, P. J. 1976. *The way the earth works.* Wiley, New York.
1981. Experimental petrology of subduction, andesites and basalts. *Trans. Geol. Soc. S. Afr., 84*, 281–91.
Yoshii, T. 1972. Features of the upper mantle around Japan as inferred from gravity anomalies. *J. Phys. Earth, 20*, 23–34.
1977. Crust and upper-mantle structure beneath northeastern Japan. *Kagaku, 47*, 170–76 (in Japanese).

9

The Continental Lithosphere

9.1 Introduction

9.1.1 Complex Continents

We have seen something of the general simplicity of the earth's internal structure and the detailed complexity of the motions of tectonic plates and convective systems. The clues to this simplicity and complexity come from the oceans, the study of whose structures has led to an understanding of the plates, of the mantle beneath and to some extent of the core, via its magnetic properties.

Although complex details must be sorted out and theories may change slightly, we can now be reasonably confident that the oceans are understood in their broad structure. In contrast, the continents are not understood at all well. Yet we need to understand the continents because in their geological record lies most of the history of the earth and its tectonic plates, from the time that continental material first formed about 4000 Ma ago (see Sect. 6.10). The oldest oceanic crust is only about 160 Ma old, so the oceanic regions can yield no earlier information.

In the broadest terms, the continents often have their oldest material at their centres, flanked by younger material which represents many events of mountain building, collision, rifting and plate convergence and subsidence. Figure 3.30 illustrates the motions of the plates over the last 280 Ma, and it is clear that, even during that brief time, continents have collided and been torn asunder.

A major problem in the geological and geophysical study of continents is that we can only observe what is exposed at or near the surface. To extend that knowledge to tens of kilometres deep, let alone to hundreds of kilometres, demands conjecture which cannot be tested directly. Oil and mineral exploration companies have developed sophisticated techniques for surveying the upper few kilometres of the crust in search of deposits and have significantly advanced our knowledge of sedimentation, oil maturation and ore genesis. The proof of the pudding is in the eating – oil companies are accountable, and they finally have to drill to verify their interpretations and conclusions. If they are wrong too often, they become bankrupt. Their methods must be good since the results are tested.

In our study of the earth we are at present unable to sample directly

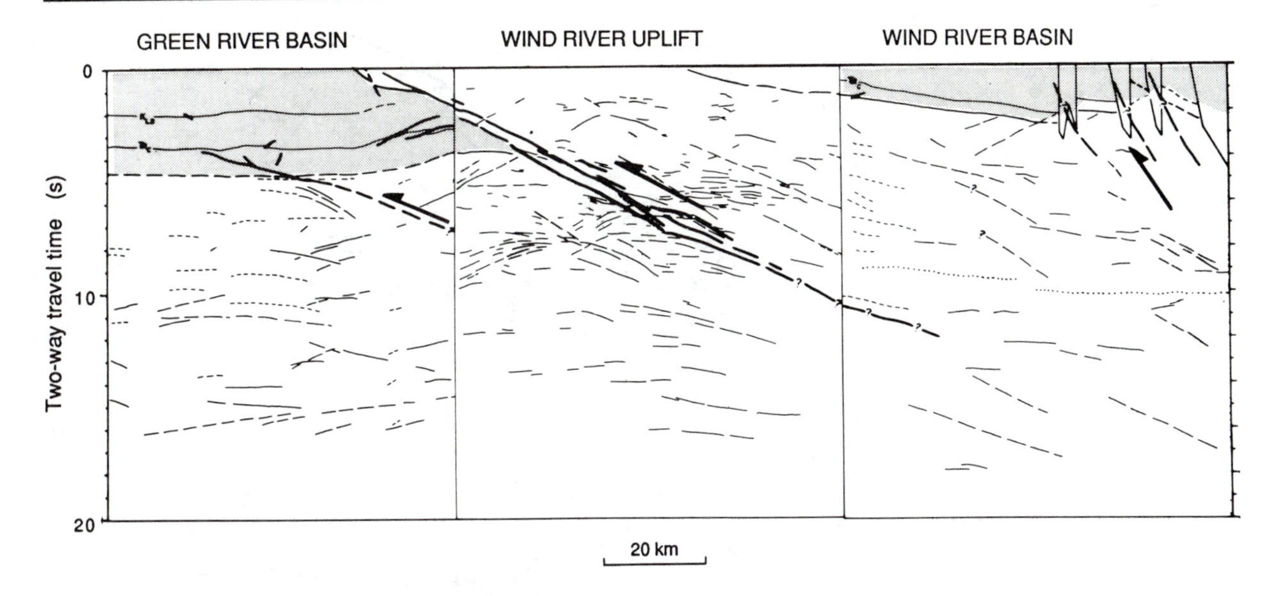

Figure 9.1. Deep reflection line shot by COCORP across the Wind River Mountains in Wyoming, U.S.A. Heavy lines, thrusts and faults. Shading, sediments. Short dashed lines, possible multiples. Dotted line at 10 s, an enigmatic low-frequency event. Event at ~15 s beneath Green River Basin, a possible Moho reflector. Dashed line at 4.0–4.5 s, the reflection from the base of the Green River Basin sediments. This is cut off by the Wind River Thrust. (After Brewer et al. 1980.)

the deep interior and so are at the disadvantage of being unable to test our models; however, we do have the questionable advantage that no one can prove us wrong. Academics need not go bankrupt through their mistakes. The deepest hole drilled into the crust is in the Kola peninsula of northwestern USSR. Drilling there began in the early 1970s and has penetrated, so far, to some 13 km. Technically, drilling such a hole is an exceedingly difficult enterprise. One of the problems with deep seismic reflection profiling (see Sect. 4.5.5) is that differing interpretations of the various reflectors can sometimes be made. Nevertheless, deep seismic profiling is very successful and has given an immediate solution to some geological puzzles; for instance, it enabled COCORP to trace the Wind River Thrust as a 30–35° dipping reflector in Wyoming (United States), from its surface exposure to some 25 km deep (Fig. 9.1), and to learn without doubt that compressional and not vertical forces were the cause of the uplifted basement blocks.

Geochemistry is probably less hindered than structural geology by our inability to obtain deep, fresh samples – although, of course, the exact composition of the lower crust and the nature of the Moho and upper mantle are matters of current debate, which fresh samples could resolve.

In this chapter some of the major geophysical and geological features of the continents are described and discussed in terms of their relation to the internal processes of the earth.

9.1.2 Geophysical Characteristics of Continents

The continental crust averages 35 km thick. We have already used this value in calculations of isostasy in Section 5.5 and in the calculation of the thickness of the oceanic crust (Sect. 8.2.1). Although a 'normal' or 'standard' oceanic crustal structure can be defined, it is much more difficult to give a standard continental crustal structure. The variability of the structure of the continental crust is, like all the other properties of the

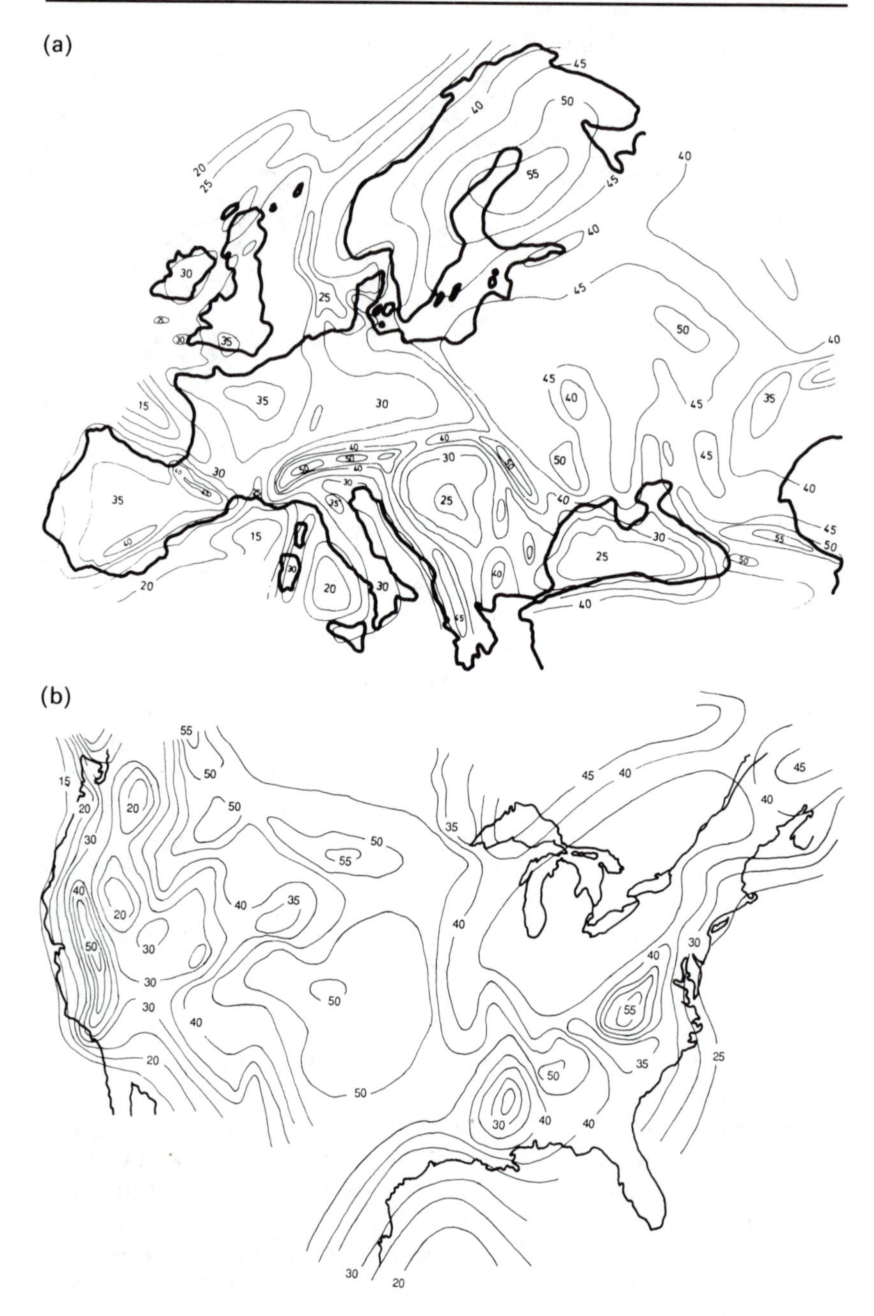

Figure 9.2. Depth of the Moho discontinuity below sea level. Contour interval, 5 km. (a) Europe and (b) United States. Not to same scale. (From Allenby and Schnetzler 1983; Meissner 1986.)

continents, a direct result of the diverse processes and the long time over which they have formed. Figure 9.2 shows maps of the crustal thickness beneath Europe and the United States of America. The large variations in crustal thickness are very clear. Generally, the crust is thick beneath young mountain ranges such as the Alps and Carpathians in Europe and the Sierra Nevada in the United States, moderately thick beneath the ancient shield regions, and thin beneath young basins and rifts such as the North Sea and Rhine Graben in Europe and the Basin and Range Province in the United States.

The seismic velocity structure of the crust is determined from long seismic refraction lines. The advent of deep reflection lines has delineated the fine structure of the crust very well, but such data usually cannot yield accurate velocity estimates (see discussion of stacking velocities in Sect. 4.5.3). The offset between the source and the further receiver must be increased considerably to obtain better deep velocity information from reflection profiling.

The direct wave which travels in the crystalline, continental basement, beneath surface soil and sedimentary cover, is termed P_g. This wave normally travels with a velocity of about 5.9–6.2 km s^{-1}. The velocity in the upper crust is usually in the range 6.0–6.3 km s^{-1} for the top 10 km or so; beneath that, the velocity is usually greater than 6.5 km s^{-1}. In some locations there is another, lower crustal layer with velocity greater than 7 km s^{-1}. Low-velocity zones, which have been described for various locations at all depths in the continental crust, may represent the complexity of the history of the crust. For many years, the *Conrad discontinuity* between the upper and lower crust was thought to be a universal feature of continental crust. This is no longer thought to be the case. Some localities show a well-developed discontinuity between upper and lower crust, but not others. The continental crust does not have a standard structure.

Figure 9.3 shows the range of laboratory measurements of the P-wave velocity for various rock types. For example, not every basalt has a velocity of 6.0 km s^{-1}, but velocities in the range 5.1–6.4 km s^{-1} are reasonable for basalts. Laboratory measurements show that the P-wave velocity increases with pressure. However, this does not necessarily mean that for a given rock unit the P-wave velocity should increase with depth in the crust. The increase of temperature with depth can either counteract or enhance the effect of increasing pressure, depending on the physical properties (e.g., pores and fissures) of a particular rock unit.

The Moho is in some places observed to be a velocity gradient, in other places it is a sharp boundary and in still others it is a thin laminated zone. The thickness of this transition from crust to mantle can be estimated from the wavelength of the seismic signals. Two kilometres is probably a maximum estimate of the transition. Geologically, however, the Moho probably represents the boundary between the lower crustal granulites and the ultrabasic upper mantle, which is predominantly olivine and pyroxene.

The gross structure of the continental crust as determined from surface waves was discussed in Section 4.1.3 and illustrated in Figure 4.6; we found the continental lithosphere to be thicker than the oceanic lithosphere. The thermal structure of the continents was discussed in Section 7.6, and we concluded that the oceanic geotherms for lithosphere older than 70 Ma can equally well be applied to the continental lithosphere (see Figs. 7.14 and 7.15).

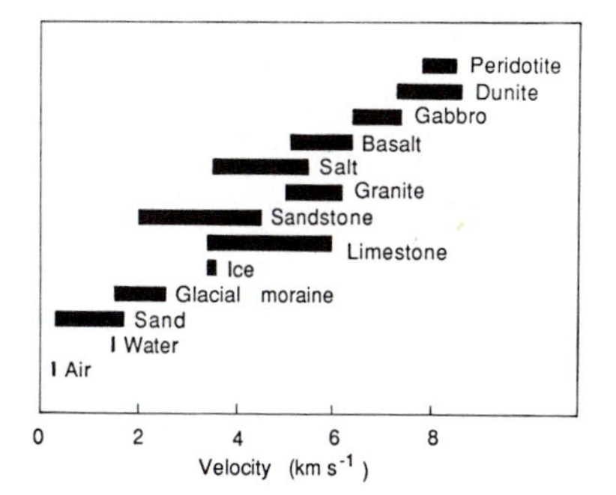

Figure 9.3. Range of laboratory measurements of the P-wave velocity in various rock types. (From data in Press 1966.)

9.1.3 Composition of the Continental Crust

The continental crust has been formed from mantle material over the lifetime of the earth by a series of melting, crystallization, metamorphic,

Table 9.1. *Estimate of the composition of the bulk continental crust and of oceanic crust*

Compound	Continental (%)	Oceanic (%)
SiO_2	57.3	49.5
TiO_2	0.9	1.5
Al_2O_3	15.9	16.0
FeO	9.1	10.5
MgO	5.3	7.7
CaO	7.4	11.3
Na_2O	3.1	2.8
K_2O	1.1	0.15

Source: Taylor and McLennan (1985).

erosional, depositional, subduction and endless reworking events. We saw in Chapter 6 how radiometric isotope methods unravel some of the complexity for us by dating samples, indicating whether they came from a crustal or mantle source, whether they were contaminated and where contamination may have occurred. The other tools used to decipher the history of rocks are, of course, all the methods of geology and geophysics.

The continental crust, despite its complexity and variation, has a fairly standard 'average' composition (Table 9.1). This composition is more silica-rich than that of oceanic basalts. In general terms, the composition of the continental upper crust is similar to granodiorite, and the lower crust is probably granulite. However, this is a gross oversimplification. The crust is far from being homogeneous and still retains the marks of its origins. Thus, sedimentary material buried during a thrusting event can be found deep in the crust, and oceanic-type rocks or even ultramafic rocks have been thrust up to the surface during mountain building.

Table 9.2 gives some idea of the worldwide extent of continental crust

Table 9.2. *Area of continental basement*

Age (Ma)	Area (10^6 km^2)	Percentage of total area
0–450	38.2	29.5
450–900	41.1	31.8
900–1350	14.6	11.3
1350–1800	8.7	6.7
1800–2250	19.4	15.0
2250–2700	6.2	4.8
2700–3150	1.1	0.9
Total	129.3	

Source: Hurley and Rand (1969).

of various ages. Only 30% of current basement rocks are younger than 450 Ma; the remaining 70% are older. Continental growth rates are discussed in Section 9.2.4.

It is immediately apparent from a map of the ages of the continents that the oldest material tends to concentrate towards the centre of a continent with younger material around it. These old continent interiors are termed *cratons* (Greek *cratos*, meaning strength, power or dominion). On the North American continent these cratons are the stable, flat interior regions (Fig. 9.4). To the east of the Archaean cratons are the Grenville and Appalachian rocks, which accreted much later as a result of continental collision. In the west, the Cordilleran rocks (western mountains) comprise a series of *accreted terranes*, which have been added, or accreted, to the continent during the last 200 Ma. These terranes, which comprise material of continental, oceanic and island-arc origin, have been added to the North American Plate as a result of plate tectonics and subduction. Terranes which are suspected of having originated far from the present location and of being transported and then accreted are descriptively referred to as *suspect terranes*! Such terranes were first identified in the early 1970s in the eastern Mediterranean in Greece and Turkey, when it was realized that the region is composed of small continental fragments with very different histories. The present-day subduction and volcanism along the western edge of the North American Plate is thus continuing a history that has been occurring there episodically for several hundred million years (see also Sect. 3.3.4).

This chapter does not proceed in chronological order from the beginning of the continents to the present but instead works from the present back into the past, or in the case of North America, from the edges to the centre. Our starting point is to continue the discussion of subduction zones from Chapter 8.

9.2 The Growth of Continents

9.2.1 Volcanism at Subduction Zones

The geophysical setting of subduction zones has been discussed in Section 8.6. The initial dip of the subducting plate is shallow, typically about 20° for the first 100 km, as seen horizontally from the trench, or on average about 25–30° in the region from the surface to the point at which the slab is 100 km deep. Volcanic arcs are characteristically located more than 150 km inland from their trench. This distance is variable, but it is clear from Figure 8.36 that, with the exception of the New Hebrides, which is a very steeply dipping subduction zone in a very complex region, the volcanic arcs are located above the place where the top of the subducting plate reaches a depth of about 100 km. The crust under volcanic arcs is usually fairly thick, in the range 25–50 km. The volcanic arcs are regions of high heat flow and high gravity anomalies. Despite the broad similarities, the settings of arc volcanism vary tremendously, and the styles of volcanism and the chemistry of the lavas vary in sympathy. The settings range from extensional to compressional and from oceanic to continental. In each case the product is different.

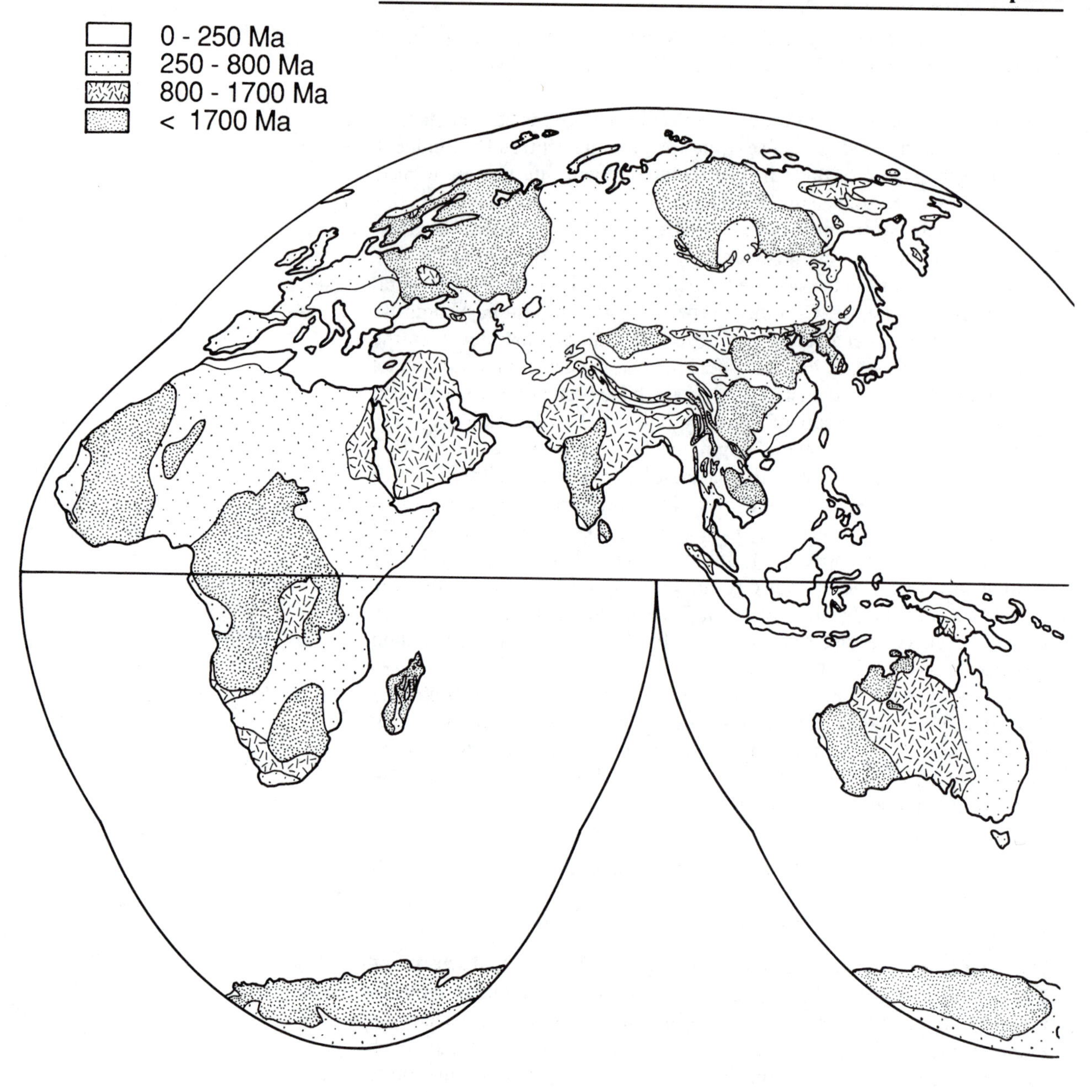

The Descending Slab: Dehydration of the Crust The subducting plate or descending slab (both terms are used) is cooler than the mantle. Thus, as it descends it is heated and undergoes a series of chemical reactions as the pressure and temperature increase. The oceanic crust is heavily faulted and cracked and water has usually circulated through it in hydrothermal systems that became active soon after the crust formed at the ridge. The crust of the descending slab is therefore strongly hydrated (up to several percent H_2O). All the chemical reactions which take place in the descending

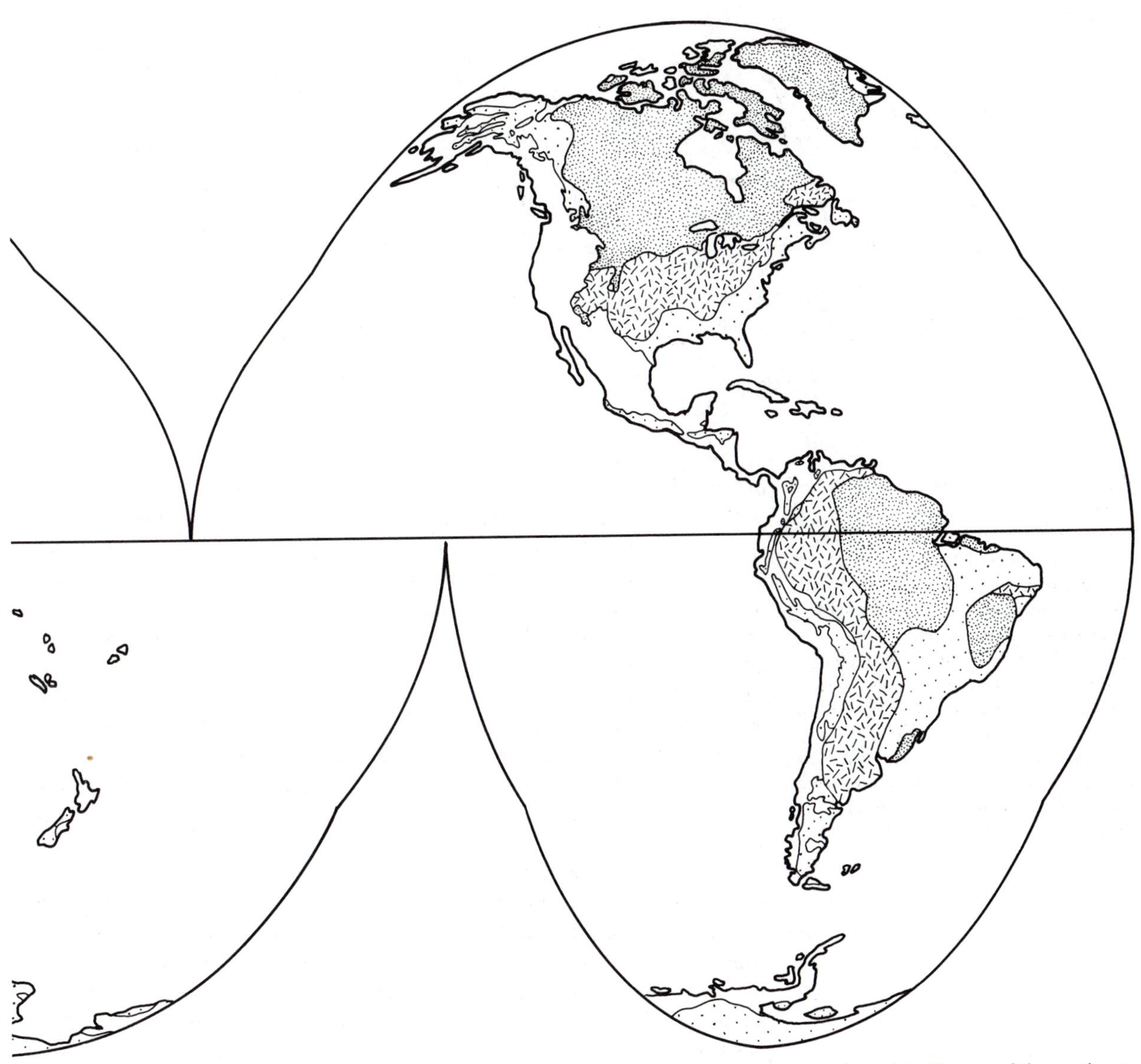

Figure 9.4. The age of the continents. (After Sclater et al. 1981.)

oceanic crust are *dehydration reactions*; that is, they involve a loss of water, usually in an endothermic process with a reduction in volume of the residue.

Prior to subduction, much of the oceanic crust is altered to low-grade brownstone, or at greater depth is in the greenschist metamorphic facies (Fig. 9.5a). This alteration was produced by the hydration and metamorphism of basalt in the near-ridge hydrothermal processes. In the initial stage of subduction, at shallow depths and low temperatures, the

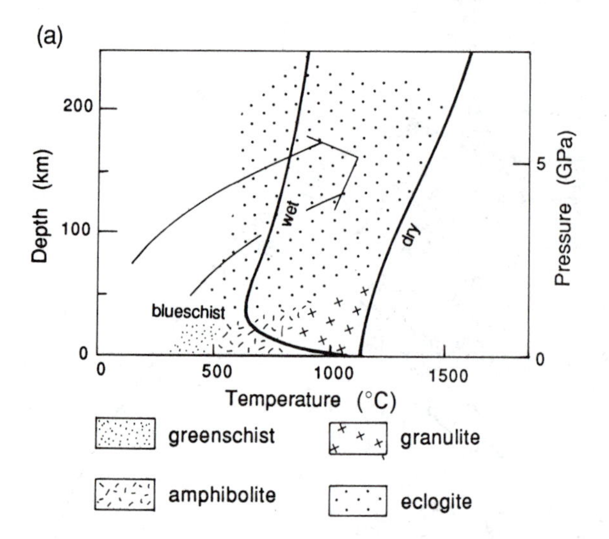

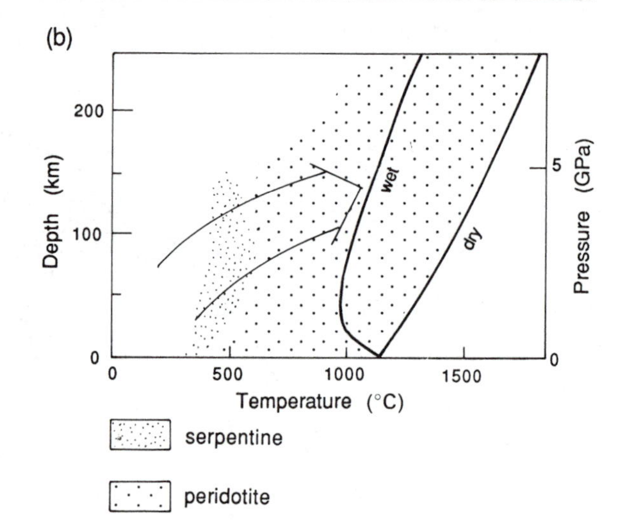

Figure 9.5. (a) Schematic of equilibrium metamorphic facies and melting of basaltic oceanic crust. Note that all the boundaries are gradational, not sharp, and that many of the reactions involved have not been particularly well defined. (b) Schematic of the breakdown of serpentinite and melting of peridotite. The two heavy lines are the solidus in the presence of excess water (wet melting) and the solidus for the dry rock (dry melting). Large arrow indicates the probable range of the temperature–depth profile of the subducting oceanic crust. (After Wyllie 1981 and Turner 1981.)

oceanic basalt passes through the pressure–temperature fields of the prehnite–pumpellyite and blueschist facies. At this stage, extensive dehydration and decarbonation take place as the basalt is metamorphosed, beginning with expulsion of unbound water and followed by significant metamorphic dehydration that commences at a depth of about 10–15 km. During the descent of the slab, the pressure increases and the slab slowly heats. The heat which warms the slab is transferred from the overlying wedge of mantle and is also produced by friction. The basalt then undergoes further dehydration reactions, transforming from blueschist to eclogite. During this compression, the released water, being light, moves upwards. Any entrained ocean-floor sediment also undergoes progressive dehydration and decarbonation. These processes are illustrated in Figure 9.5a.

The temperatures and pressures at which oceanic basalt, in the presence of excess water, produces significant amounts of melt are markedly different from the temperatures and pressures at which dry basalt produces copious melts (Fig. 9.5a). At 10 km depth, wet basalt begins to produce significant melt at about 850°C and dry basalt at 1200°C. For increasing amounts of water, the point at which copious melting begins lies at positions intermediate between the dry and wet extremes: Voluminous melting can start anywhere between the two curves, depending on the amount of water present. However, except in limited regions where heating is especially rapid, melting of the subducted oceanic crust is unlikely to occur in the early stages of descent.

The subducted oceanic crust carries wet oceanic sediment. At or near the trench, much of this sediment is scraped off and becomes part of the accretionary wedge (see Sect. 9.2.2). However, a small part of the sediment may be subducted. This subducted sediment melts at comparatively low temperatures and provides components (such as CO_2, K and Rb) to the stream of volatiles rising upwards.

The lower, plutonic (gabbro) portion of the subducted crust and the uppermost subducted mantle (peridotite) may also have been partially hydrated by sub-sea-floor hydrothermal processes. Figure 9.5b shows the controls on dehydration of hydrated ultramafic rock in the lowest crust and topmost mantle as water is driven off any serpentine in the rock. The subducted oceanic mantle, which is depleted and refractory, does not usually melt. Being cooler than normal upper mantle at these depths, the subducted mantle simply heats slowly towards the temperature of the surrounding mantle.

The Descending Slab: Heating The descending slab heats by conduction from the overlying hotter mantle, but other factors are also operating which combine to speed the heating process.

Some contribution to the heating of the slab may come from friction on its upper surface. This frictional or shear-stress heating is, however, not well quantified. Researchers have used a range of rather arbitrary estimates of shear stress (0–200 MPa) in their modelling. Detailed studies of the heat flow measured in the fore-arc region, where there is no thermal contribution from the volcanic arc, indicate that the shear stress is low and probably lies between 10 and 60 MPa. Figure 9.6 shows a conductive model of the thermal structure of the region between the trench and the volcanic line for a subduction zone dipping at 20° and with shear-stress heating increasing from zero at the surface to 70 MPa at 40 km depth and then decreasing to zero when the top of the slab reaches 100 km depth. The temperatures of the subducted oceanic crust are not high enough for it to melt in this region; the wet solidus for basalt (shown in Fig. 9.5a) is not attained in the upper part of the slab. Greatly increased values of the shear stress and/or much higher initial temperature gradients for the subducting and overriding plates would be necessary for melting to take place in this region. This is a good check on the validity of models and the physical conditions chosen: Since any models that imply that melting should take place between the trench and the volcanic line must fail, melting should first take place only beneath the volcanic line. The oceanic crust probably starts transforming to eclogite by the time it has subducted

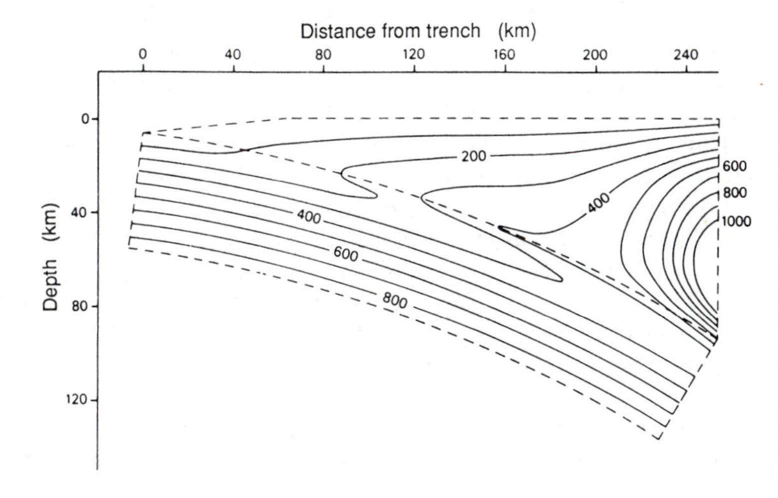

Figure 9.6. Thermal structure of the region between the trench and the volcanic line for a subduction zone dipping at approximately 20°, 50 Ma after subduction started. The temperatures were calculated using the two-dimensional heat conduction equation, a subduction rate of 8 cm yr^{-1}, a shear stress as described in the text and an initial boundary layer temperature structure for the subducting plate, with 1325°C being the temperature at the base of the lithosphere. Deep mantle temperatures are adiabatic. (From van den Beukel and Wortel 1988.)

to a depth of about 30 km, but not all of the reactions may be complete until it is much deeper than that because transformation takes time (though it is hastened by the presence of abundant fluid).

Many dehydration reactions of the oceanic crust are endothermic (the reactions require heat), and lack of heat may constrain the transformation. This means that temperatures in the real slab are lower than those obtained from computer models which do not include such heat requirements. Some estimates of the heat needed are $5.8 \times 10^4 \, J \, kg^{-1}$ for the mineral reactions involved in the greenschist–amphibolite change and $2.5 \times 10^5 \, J \, kg^{-1}$ for the serpentinite–peridotite reactions. The water released in these reactions rises into the overlying mantle, a process that further slows the heating of the slab and cools the overlying wedge. A total value of about $2.5 \times 10^5 \, J \, kg^{-1}$ for all the dehydration reactions is probably not unrealistic. None of these reactions takes place instantaneously (discussed further in Sect. 9.3.5). At low temperatures and pressures, the reactions proceed more slowly than at higher temperatures and pressure; furthermore, in very dry conditions, at great depth, transformation is also slow.

The released water moves out of the oceanic crust into the overlying upper mantle. Dehydration probably begins with the onset of subduction but at shallow depths has little effect on the overlying wedge of the overriding plate except to stream fluid through it and metamorphose it. Dehydration continues as the plate is subducted: A plate being subducted at an angle of 20° at $8 \, cm \, yr^{-1}$ for 1 Ma descends 27 km, with attendant heating, compression and metamorphic dehydration. Figure 9.5b indicates that dehydration of serpentine starts when the top of the slab reaches a depth of roughly 70 km. At such depths, but shallower than 100 km, water released from the slab is probably fixed as amphibole in the overlying mantle wedge, or streams upwards if the wedge is fully hydrated; but at a depth of 100 km, the melting of wet overlying mantle begins. In the descending slab, the wet solidus for basalt probably can be reached only at a depth of about 100–150 km, and so no melting of the subducted oceanic crust can take place much shallower than this. The precise depth depends on such factors as the age (and hence temperature and degree of hydration) of the slab and the angle and rate of subduction. In reality, probably only the melting of the overlying wedge occurs at 100 km; melting of the slab may not take place until it is much deeper because by this stage the slab must be highly dehydrated.

Isotopic work has indicated that the subducting slab provides only a small component of erupted lavas. Data from Chile have shown that the overriding mantle provides a significant proportion of all elements: Only volatile species such as H_2O and the *large-ion lithophile* elements such as Rb, K, Ba, Th and Sr were supplied in any great quantity by the subducting slab. These elements may have been transported upwards to the overlying mantle in the volatiles ascending from the subducted slab to become incorporated into the rising melt.

The Overriding Mantle Wedge Many thermal models of subduction zones assume that heat is transferred by conduction alone, an extreme simplification; more realistic models incorporate a viscous, convecting mantle. The inflow of overriding mantle ensures a supply of fertile asthenosphere, with a potential temperature (Eq. 7.94a) of about 1280°C, to the region above the descending slab.

Two conductive thermal models have been shown (Figs. 8.43 and 9.6). To illustrate the importance of convection, Figure 9.7 shows two thermal models with the same physical parameters except that one has a conductive and the other a convective upper mantle. For the convective model, the temperature of the descending slab is higher than for the conductive model, which affects the depth of the start of the dehydration reactions. However, the major difference between the two models is in the temperature structure of the overriding mantle; in the convective model, much higher temperatures are attained in the corner immediately above the subducting plate. This higher temperature is important because it means that the mantle temperatures are then in the range where addition of water results in copious partial melting (Fig. 9.5b). Most melt is generated in the upper mantle beneath the volcanic arc (the mantle wedge) as a result of the addition of water and other volatiles from the subducting slab. As discussed earlier, water is lost by the slab at all depths down to about 100 km, initially due to compaction (the closing of the pores) and then to the dehydration reactions. However, it is only at depth where the overlying mantle temperature is more than 1000°C that partial melting can take place in the mantle wedge; this condition seems to be met at about 100 km depth in most subduction zones. The thermal models shown in Figure 9.7 may have temperatures which are too low, since many thermal models require temperatures of about 1300°C or more at the base of the plate. Thus, Figure 9.7b should be regarded as a cool model; actual temperatures may be higher.

The effects of subduction on the overriding mantle wedge can be summarized as follows: (1) an influx from the descending slab of upwards-moving volatiles; (2) some rising melt from deeper parts of the slab; and (3) the driving of convective flow in the wedge, the flow lines of which show movement of mantle material from the distant part of the wedge (on the right in Fig. 9.7) and pulled downwards (counterclockwise in Fig. 9.7) by the slab.

The upwards-moving volatiles from the descending slab consist of H_2O and CO_2, probably accompanied by a substantial flux of mobile elements such as Rb, K, Ba, Th and Sr. Melting of any subducted sediment may be important. The contribution of melt (as opposed to volatiles, etc.) from the slab itself is probably relatively small. The deeper parts of the slab may leak some melt upwards, leaving residual quartz-eclogite behind. Compositionally, any melt rising from the slab is probably hydrous, siliceous magma, which is roughly similar to calc-alkaline magmas erupted in island arcs but probably with a higher CaO/(FeO + MgO) ratio.

The addition of streams of volatiles, plus perhaps some melt at deeper levels, probably causes partial melting in the warm peridotite overlying the subducted slab (Fig. 9.8). There is still much uncertainty about the exact location of melt generation, but it is possible that much of the melting occurs in the warm overlying wedge immediately above the locus where

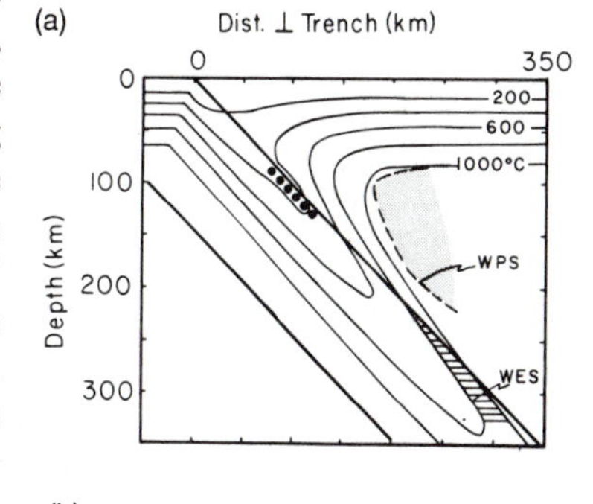

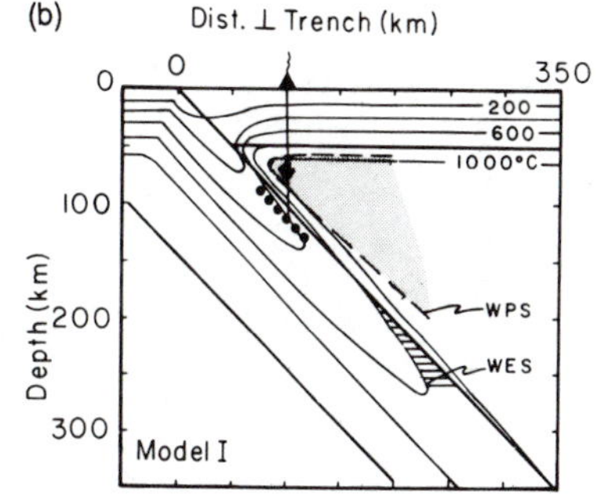

Figure 9.7. Thermal model of a subduction zone dipping at 45° and subducting at 6 cm yr^{-1}. Temperature at base of subducting plate is 1200°C. Contour interval is 200°C. (a) Heat is transferred through the overriding mantle wedge by conduction. (b) The mantle in the overriding wedge is allowed to convect. WPS, wet peridotite solidus; WES, wet eclogite solidus; dots, dehydration; diamond, possible melting. (From Anderson et al. 1980.)

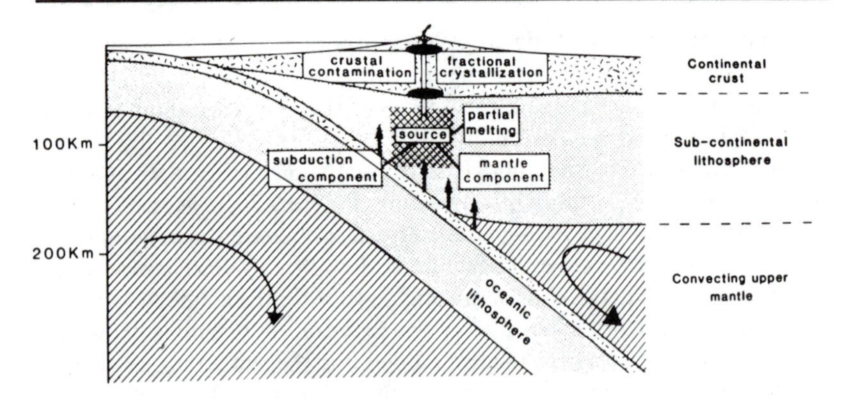

Figure 9.8. Cross section of a subduction zone showing the loss of water and volatiles from the downgoing oceanic plate and the onset of melting when that plate reaches 100 km depth. (From Pearce 1983.)

the slab reaches 100 km depth. This melting in the wedge, at depths of 60–90 km, would produce liquids which may be parental to basaltic andesite. Andesite, which is the commonest magma erupted from arc volcanoes, cannot be directly produced by melting of the mantle wedge except by melting of very wet (around 15% H_2O) mantle peridotite at depths around 40 km, or in other very restricted circumstances. Most probably, the fluids from the subducted slab promote melting of the mantle wedge, producing basic melts which rise because they are lighter than the surrounding residual peridotite and eventually reach the base of the continental crust. The system can be thought of as a conveyor belt which emits volumes of melt that rise upwards: Warm mantle convects into place above the subduction zone; fluid enters from the slab below and initiates melting; melt escapes upwards; the residual mantle is carried away by convection; and new warm material takes its place.

The Base of the Continental Crust Partial melts generated in the mantle wedge rise either as diapirs of melt and crystals, which increasingly melt and become more basic as they rise and decompress, or as rapidly moving segregated melt. The rising melt probably follows a nearly adiabatic pressure–temperature path in a mature subduction system, cooling by up to 1°C km^{-1} (this is the adiabatic gradient in a magmatic liquid, in contrast to 0.5°C km^{-1} in solid mantle). Much depends on the release of latent heat. At the base of the continental crust, lighter liquids can pass straight upwards to the surface, but most liquids are probably trapped by their density, the continental crust being less dense than the magma.

Rising magma carries heat with it. Hot magma collecting at the base of the crust loses heat to the overlying continent and cools and fractionates, precipitating minerals such as clinopyroxene together with garnet, olivine or orthopyroxene (depending on the depth of the melt and on its temperature, composition and percentage of water). After fractionation, the lighter liquids rise to the surface, most probably as basalt and basaltic andesite magmas. However, the heat transfer into the base of the continental crust also produces melting in the crust. High-temperature partial melting of deep continental crust at temperatures around 1100°C produces tonalite liquids, which are silica-rich melts that can rise to the surface and erupt as andesites. At somewhat lower temperatures (around 1000°C), in the presence of more water, the product of melting is granodiorite. Many of the large granitic intrusions of the continents above subduction zones are granodiorite. In the aftermath of large-scale continental collisions (such as in the Himalayas), there is often overthrusting of continental crust with partial melting of the underlying slab. After partial melting, the residual material left behind in the deep continental crust is granulite which is depleted of all its low-temperature melting fractions. This depletion includes the removal of the heat-producing elements (U, Th and K) which are carried upwards with the rising granitoid liquids. Because these heat-producing elements are carried upwards, the continents are self-stabilizing: Heat production is concentrated at the top, not the bottom, of a continent. This process, which has moved heat production to shallow levels, has had the effect of reducing the continental temperature gradient, making it more difficult to melt the crust (see Chapter 7, Problem 13).

It is chemically unlikely that granitic liquids are produced directly from partial melting of mantle peridotite or subducted oceanic crust; otherwise we should find granitoids in the oceanic lithosphere. The main geographic location of granitoids is above subduction zones and in continental collision zones, modern or ancient, which implies that granitoid generation is strongly linked to the processes of plate tectonics. In the modern earth, it is probable that most granitoids are generated in the presence of water. In the Archaean, when the mantle may have been hotter, tonalites appear to have been more common, generated at 1300°C from dry crust. Today, under cooler and wetter conditions, melts are granodioritic.

The upper continental crust above a subduction zone is characterized by large granodioritic (granitoid) intrusions. Above these are andesite volcanoes which erupt melt that originated in the mantle above the subduction zone, but which has fractionated on ascent and perhaps been contaminated by material derived from the continental crust. However, the broad similarity of volcanism in island arcs, where in some cases little continental material is present, to volcanism on continents above subduction zones indicates that magma must be produced within the mantle, irrespective of the type of overriding plate. The complication introduced by the continental crust is that it allows for the possibility of further chemical complexity and variation in the erupted and intruded melts.

9.2.2 Sediments at Subduction Zones

A schematic cross section of a subduction zone was shown in Figure 8.34, which illustrates the characteristic geological features. In reality, a subduction zone does not necessarily have all these features. The accretionary wedge is the region of folding, and then of faulting and thrusting of the sediments on the subducting oceanic plate. Then comes the outer-arc high. The fore-arc basin is an active sedimentary basin being filled largely by material (detritus) eroded from the adjacent arc. It may be underlain by oceanic crust which marks the position of the old passive continental margin before subduction began. The volcanic arc and the back-arc region were discussed in Sections 9.2.1 and 8.6.

One subduction zone with a well-developed accretionary wedge is the Makran Subduction Zone in the Gulf of Oman off Iran and Pakistan.

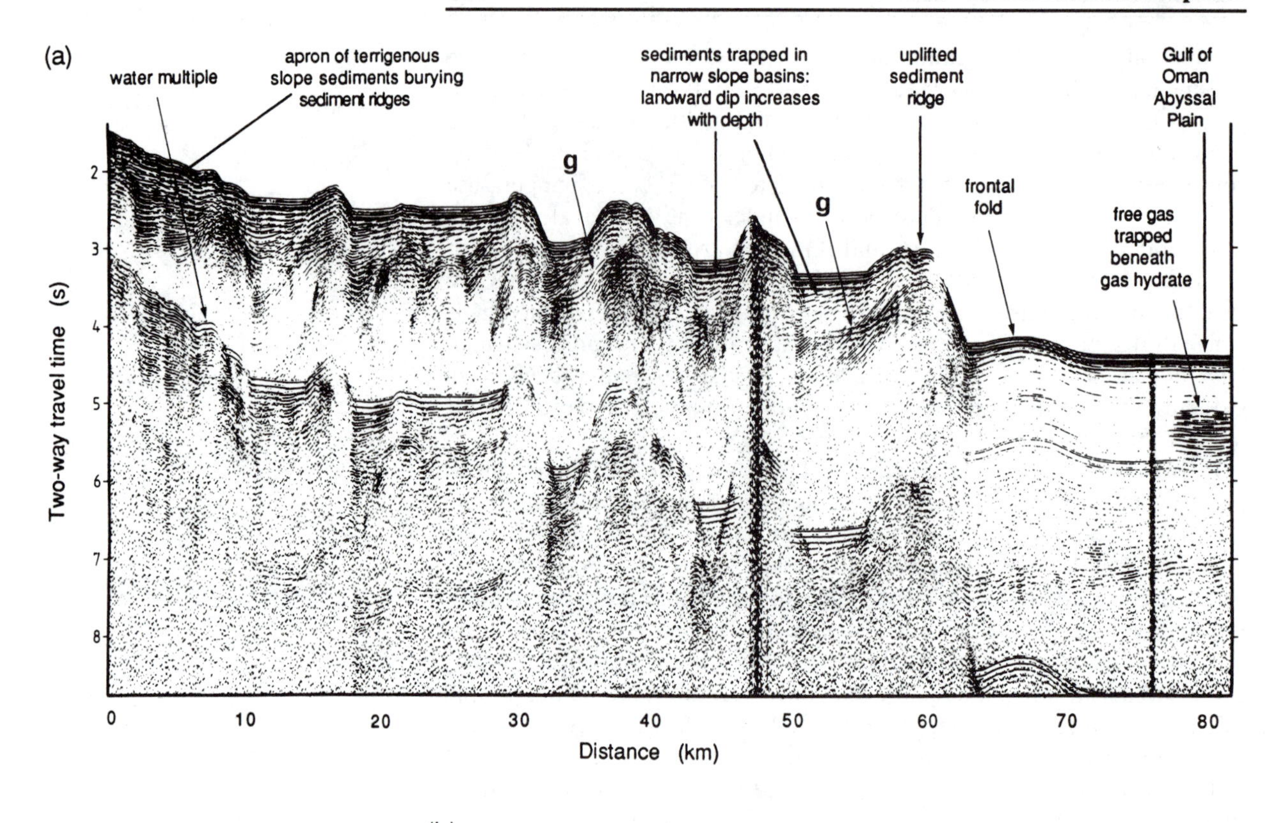

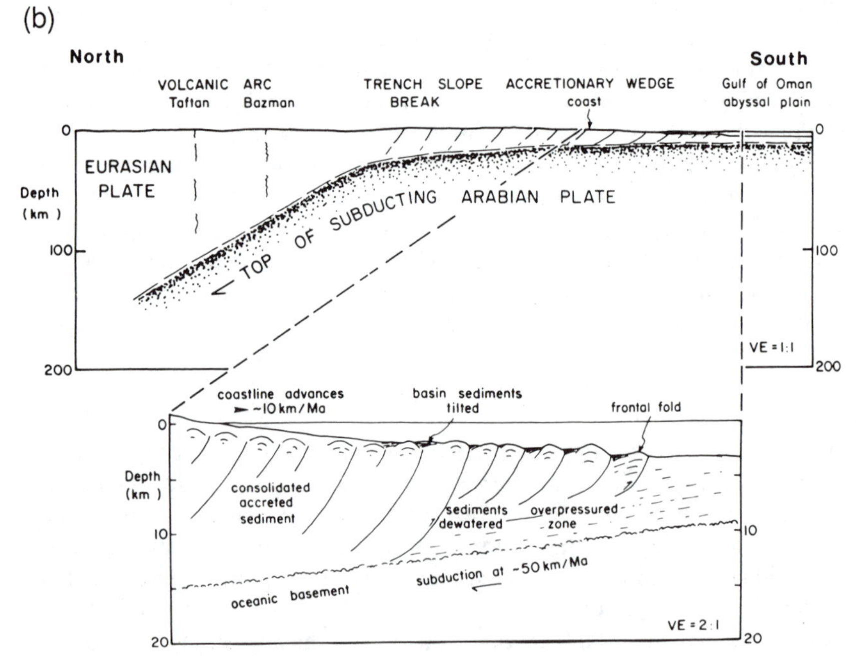

Figure 9.9. (a) Seismic reflection profile across the accretionary prism of the Makran subduction zone. Bottom simulating reflector of base of gas hydrates marked g. (b) Cross sections through the Makran subduction zone. (From White 1984.)

There the oceanic part of the Arabian Plate is being subducted beneath the Eurasian Plate. This 900 km long subduction zone is unusual in a number of ways:

1. the dip of the subducting Arabian Plate is very low;
2. there is no clear expression of an oceanic trench;
3. the background seismicity is very low; and
4. there is a prominent accretionary wedge, much of which is presently exposed on land in Iran and Pakistan.

Figure 9.9 shows a seismic reflection profile across part of the offshore portion of this thick accretionary wedge. The undeformed abyssal plain sediments on the Arabian Plate are 6–7 km thick. The deformation of these sediments as they are pushed against the accretionary wedge is clearly visible. A gentle frontal fold develops first, followed by a major thrust fault which raises the fold some 1200 m above the abyssal plain. Reflectors can be traced from the abyssal plain into the frontal fold but not beyond because deformation and faulting are too extreme farther into the wedge. Detailed wide-angle seismic velocity measurements show that dewatering of the sediments, and hence compaction, occurs in the frontal fold. The sediments are then sufficiently strong to support the major thrust fault. The continuous process of forming this accretionary wedge results in the southward advance of the coastline by 1 cm yr^{-1}. By this process, a considerable volume of material is being added to the Eurasian Plate every year. Figure 9.10 illustrates the possible stages in the development of a thick accretionary wedge. Sometime in the future, the situation shown in Figure 9.10 may be appropriate for the Makran: The subducted Arabian Plate may fall into the asthenosphere, resulting in extension and a new volcanic region in the present accretionary wedge.

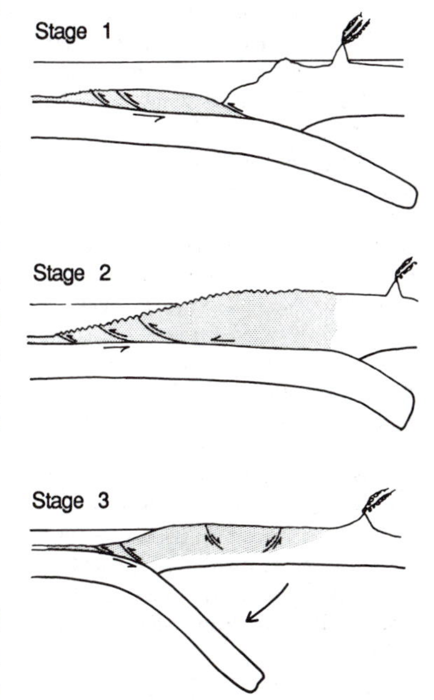

Figure 9.10. Three stages in the development of a subduction zone when there is a great thickness of sediment on the oceanic plate. Stage 1: A trench is visible. Stage 2: The large amount of sediment scraped off the subducting oceanic plate has choked the trench. The accreted sediments all belong to the overriding plate, and thrusting takes place beneath them. The Makran subduction zone is now at this stage. Stage 3: The subducted plate sinks into the asthenosphere, resulting in the gradual extinction of the volcanoes and, in the accretionary wedge, extension and (eventually) new volcanoes. (From Jackson and McKenzie 1984.)

9.2.3 Continent–Continent Collisions

Because continental lithosphere is not dense enough to be subducted as a whole into the mantle, the collision of two continents results in a complex process of thrusting and deformation and a reduction or cessation of relative motion. Other plates reorganize to take up the motion elsewhere. Two classic examples of young mountain ranges formed from such continental collisions are the Himalayas, which are the result of the collision of the Indian Plate with Eurasia, and the Alps, which are a result of the northward motion of the African Plate towards Eurasia.

The Himalayas Body and surface wave studies of earthquake data indicate that the crust beneath the Himalayas and the Tibetan Plateau is about 70 km thick. This is in contrast to the 40 km thick crust of the Indian shield. India has a typical shield S-wave velocity structure with a thick, high-velocity lithosphere overlying the asthenosphere. However, the lithosphere beneath Tibet appears to be thinner than that beneath India, indicating that at present the Indian Plate is not underthrusting the whole

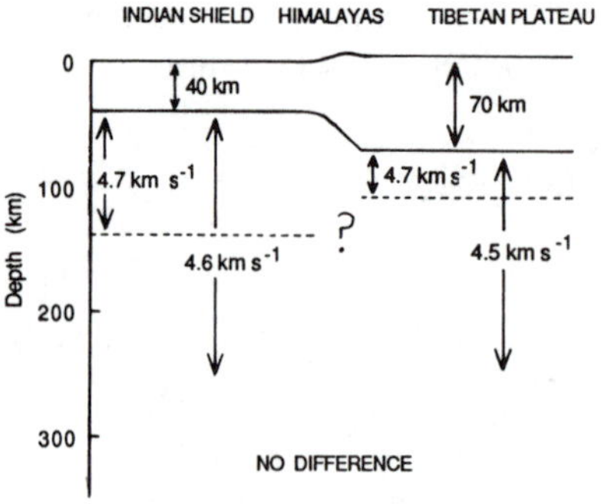

Figure 9.11. Summary of the S-wave velocity structure of the upper mantle beneath India and Tibet. Base of the crust, solid line. Approximate base of lithosphere, dashed line. Below 250–300 km there is no difference between the velocity structure of India and that of the Tibetan Plateau. (After Lyon-Caen 1986.)

of the Tibetan Plateau and that Tibet is not a typical shield region (Fig. 9.11).

The major tectonic blocks of the Himalayas and the Tibetan Plateau and their sutures are shown in Figure 9.12a. Figure 9.12b shows one attempt to explain the overall evolution of the region. This evolution has been much more complex than just a simple collision of India with Eurasia. The reconstruction starts at 140 Ma with the Kunlun and Qiangtang blocks already sutured to Eurasia and the subduction of an oceanic plate moving the Lhasa block northwards. By 100 Ma, the Lhasa block was attached to Eurasia and may have undergone internal thrusting and intrusion while subduction moved to its southern margin. Continuing subduction beneath the southern margin of the Lhasa block moved the Indian continent

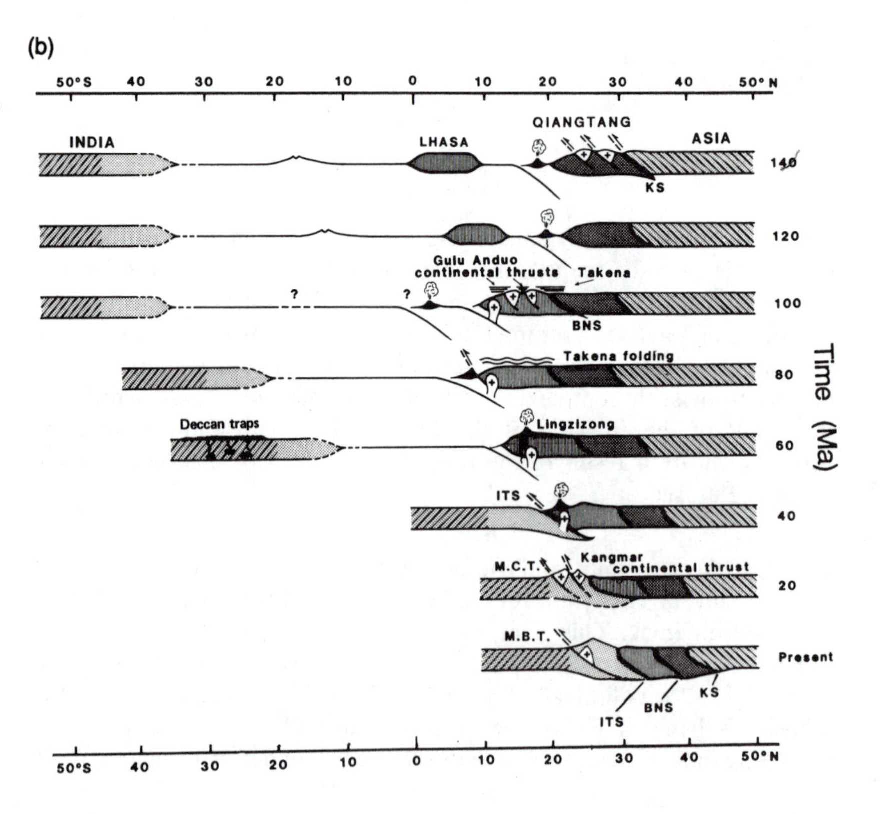

Figure 9.12. (a) The major suture zones and thrusts of the Himalayan region. CS, Chilien suture; KS, Kokoxili suture; BNS, Bangong Nujiang suture; ITS, Indus Tsangpo suture; MCT, main central thrust; MBT fault, main boundary thrust (or fault). The tectonic blocks are ASI, Asian plate; KUN, Kunlun block; QIA, Qiangtang block; LHA, Lhasa block; IND, Indian plate. (b) A reconstruction of Tibet and the Himalayas at 20 Ma intervals from 140 Ma to the present. The Qiangtang block is assumed to have sutured to Asia at about 200 Ma. 140–120 Ma: Small ocean basin between Asia and the Lhasa block closes. 100 Ma: Lhasa block is sutured to Asia along the BNS. 80–60 Ma: Subduction takes place beneath Lhasa continental margin (including the possible subduction of a volcanic arc). 40 Ma: Subduction ceases. Continental obduction or shortening takes place as Indian and Asian plates collide. 20 Ma: MCT is the main thrust. Present: MBT is the main thrust. (From Allègre et al. 1984.)

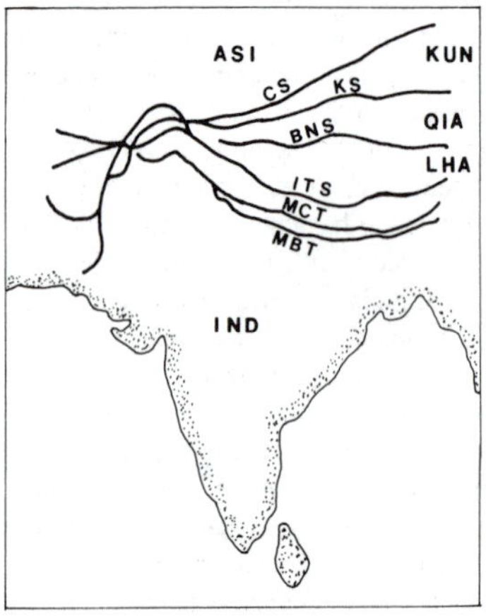

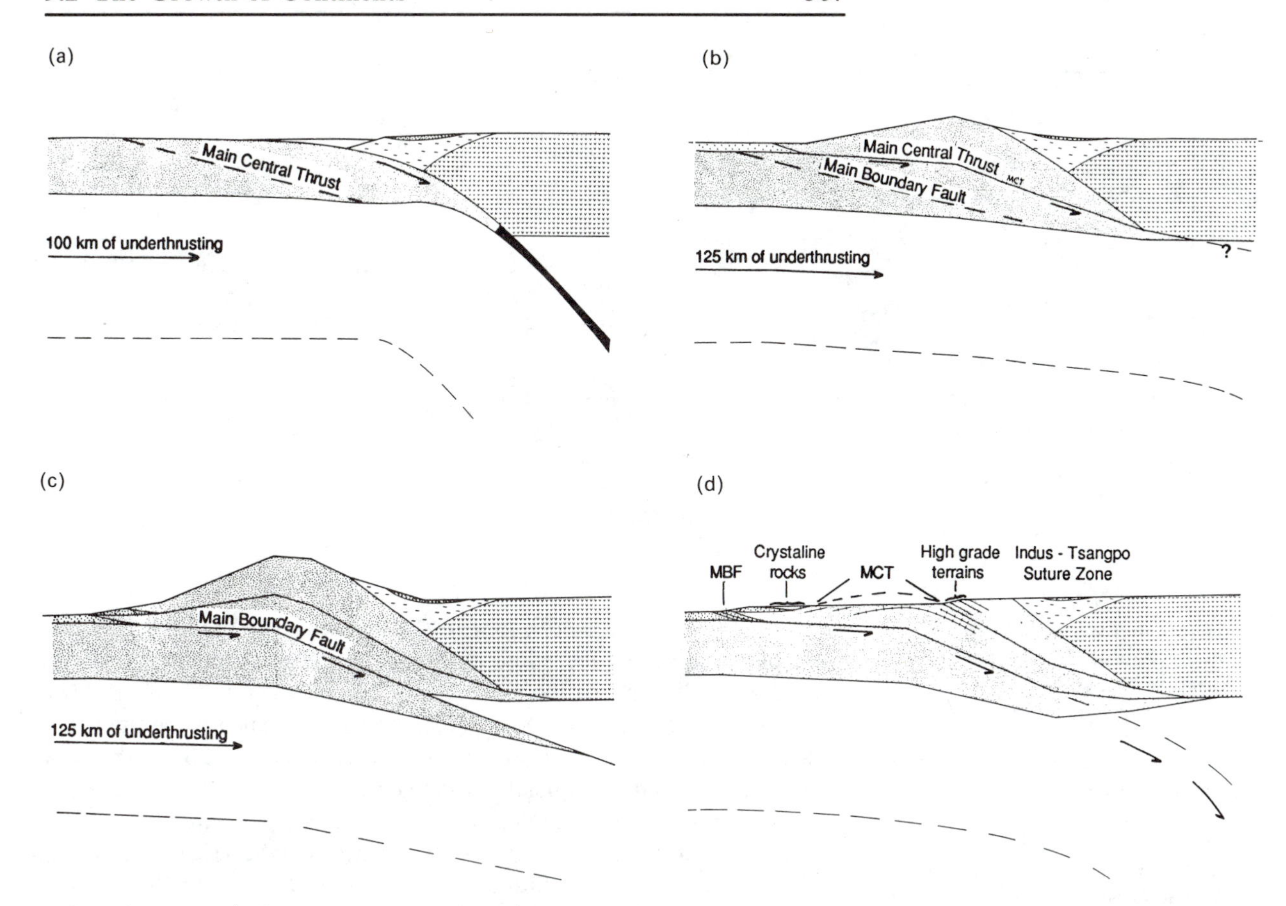

Figure 9.13. Cross sections illustrating the probable sequence of thrusting events which gave rise to the Himalayas as observed today. (a) The collision of India with Eurasia has taken place along the Indus–Tsangpo suture, and an arbitrarily assumed 100 km of subduction, the main central thrust (MCT) has formed. (b) Along the MCT, 125 km of underthrusting of India beneath Eurasia has occurred, and the main boundary fault (MBF) has formed. (c) At present, 125 km of underthrusting has taken place along the MBF. (d) Same as (c), but the uplifted material has been eroded to the level of the present-day topography, exposing the MCT in two places and the high-grade metamorphic (lower crustal) rocks above the MCT. (From Lyon-Caen and Molnar 1983.)

northwards (at about $10\,\mathrm{cm\,yr^{-1}}$; see Sect. 3.3.3) until by 40 Ma the continental collision had occurred at which time the convergence rate between India and Eurasia suddenly dropped to $5\,\mathrm{cm\,yr^{-1}}$. The continental collision did not result in the overriding of India by Eurasia; instead, it is thought that, as underthrusting proceeded, the main thrust repeatedly jumped southwards, each time leaving a thick slice of Indian crust attached to the Eurasian Plate (Fig. 9.13). Initially, thrusting took place along the Indus–Tsangpo suture zone. Thus, all the present Himalayan rocks were once part of India and not Eurasia. Then, after some 100 km of underthrusting of the Indian continent, the main central thrust (MCT) developed. Again, after perhaps 125 km of underthrusting of India along the MCT, another thrust, the main boundary fault (MBF or MBT) developed to the south, and the MCT become inactive. This is thought to represent the present situation, with the main thrust plane between India and Eurasia being the main boundary fault and the underthrust Indian Plate not extending beneath the entire Tibetan Plateau. The fate of the lower Indian lithosphere as the thrust zone jumps southwards is not clear, and more seismic work is required to delineate it. The main problem and the reason for uncertainty in the geology and in the formation and structure of the Himalayan region is the extreme size and ruggedness of the terrain, which makes access difficult.

The Himalayas are seismically active: Magnitude 8 earthquakes are not

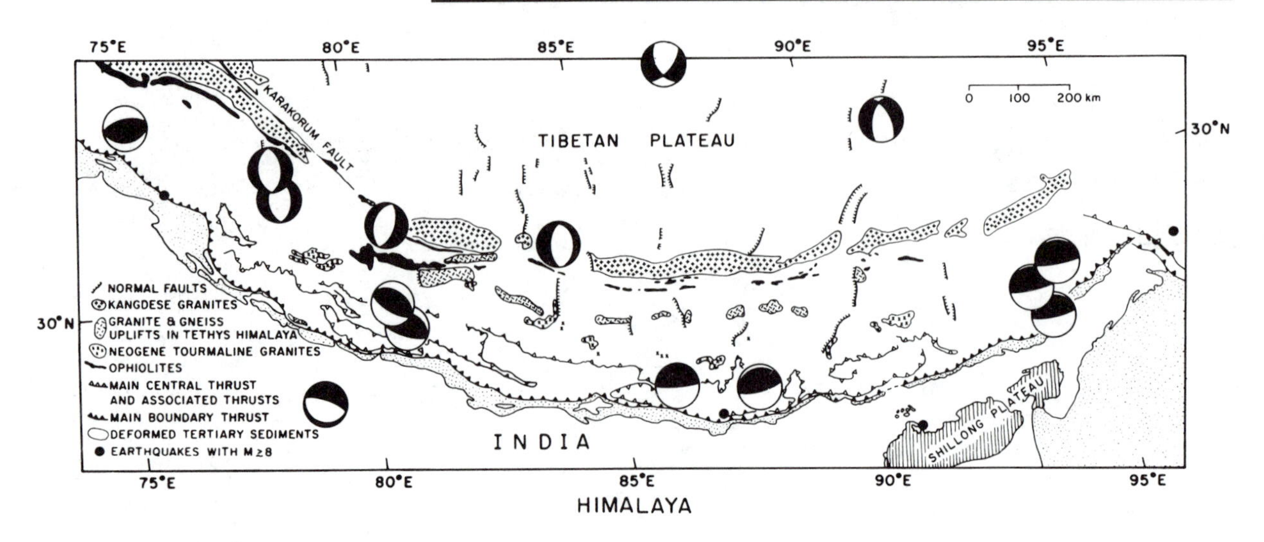

Figure 9.14. Tectonic map and earthquake fault-plane solutions for the Himalayas. The southernmost thrust, the main boundary fault, is the present location of the plate boundary between India and Eurasia. The string of ophiolites delineates the Indus–Tsangpo suture, the original collision zone between India and Eurasia. Black dots show the epicentres of the $M \geqslant 8$ earthquakes which occurred in 1905, 1934, 1897 and 1950 from west to east and for which fault-plane solutions are not available. (From Molnar and Chen 1983.)

uncommon (there have been four in the last century). Figure 9.14 shows fault-plane solutions for some earthquakes in the Himalayas. All the fault-plane solutions for the earthquakes immediately north of the MBF show thrust faulting. The nodal plane, which is assumed to be the thrust plane, dips at less than 10° in the east but increases to 20° in the west. This indicates some complexity in the geometry of the thrusting of India beneath Eurasia. The focal depths of these earthquakes are about 15 km which, since the epicentres are some 100 km north of the MBF, is consistent with the occurrence of earthquakes at the top of the Indian Plate as it is subducted beneath Eurasia. North of these thrust faulting earthquakes, the deformation style changes: The focal mechanisms indicate that normal faulting and east–west extension is taking place. The earthquake that occurred at 78°E in the Indian Plate south of the Himalayas shows normal faulting and therefore is presumably indicative of the extension taking place in the top of the Indian Plate as it bends prior to subducting beneath the Himalayas.

The present-day rate of convergence in the Himalayas is estimated at 1–2 cm yr^{-1}, which is less than half the estimated convergence between India and Eurasia. The remaining convergence is believed to be taking place over a very large area north of the Himalayas, on the basis of the extensive tectonic, seismic and local volcanic activity in that region. Figure 9.15b shows results from a plasticine model of Southeast Asia and the deformation that resulted when a rigid block (India) was pushed northwards into it. The large-scale internal deformation and eastward squeezing of regions appropriate for Tibet and China are very clear.

A number of major seismic refraction lines have been shot across the Himalayas to determine the crustal and uppermost mantle structure. These refraction lines include those of the Indo-Soviet Geodynamics Project in the western Himalayas (74°E), and in the eastern Himalayas (86–90°E) in the vicinity of Mt. Everest where French and Chinese expeditions have been working. The two main features of the crustal structure in both the eastern and western Himalayas are the very thick crust (on average 70 km) and the major faults which offset the Moho as it thickens from 35–40 km

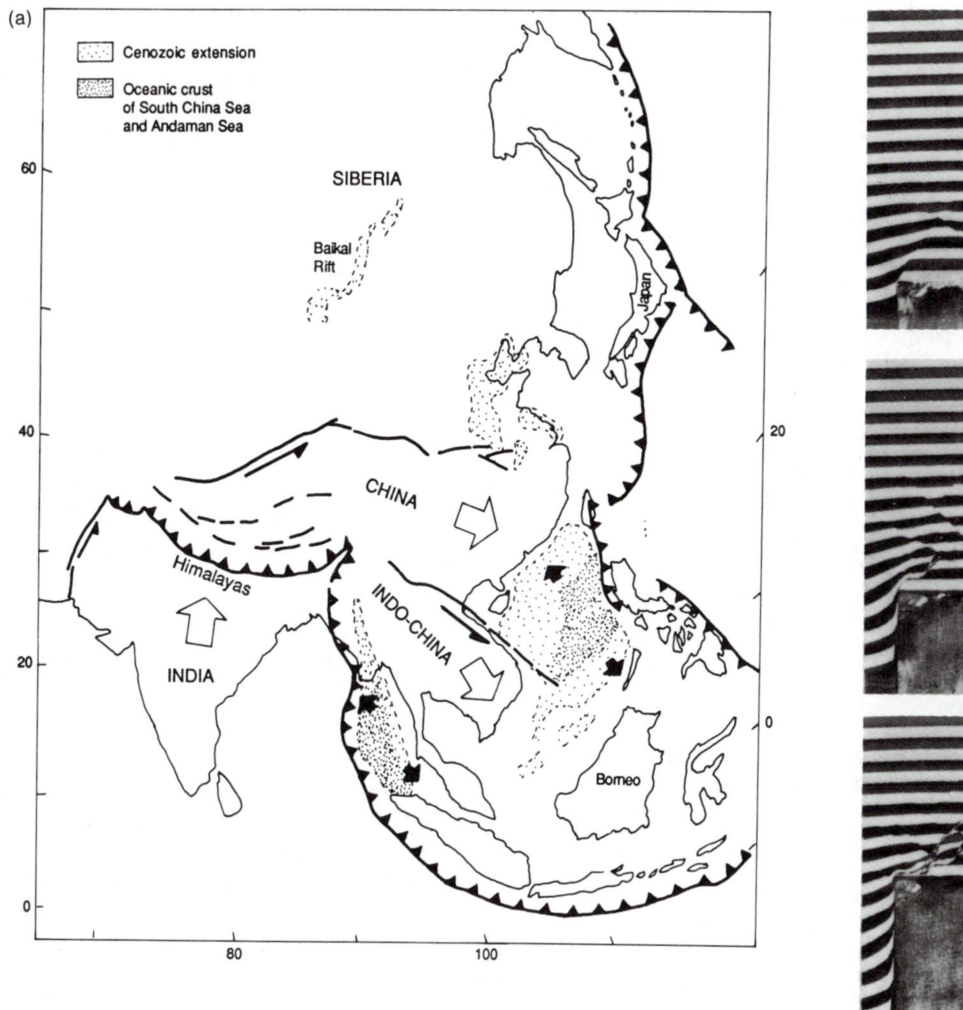

Figure 9.15. (a) Scheme of subduction zones, thrusts, large faults and Cenozoic extension in eastern Asia. Heavy lines, major plate boundaries or faults. White arrows, motion of India and the two major extruded blocks (China and Indochina) with respect to the Siberian block. Black arrows, direction of extension. (b), (c), (d) Plan view of three successive stages in the indentation by a rigid indenter (India) into a striped block of plasticine (Asia). The plasticine was confined on only the left-hand side, leaving the right-hand side representing China and Indochina to deform freely. The resulting extrusion (d) of two large blocks to the right and the faulting and rifting ahead of the indenter are similar to the large-scale deformation, as shown in (a). (After Tapponier et al. 1982.)

beneath India to depths of more than 70 km beneath Tibet. The main difference between the two regions is that a thick lower crust with a seismic velocity of 7 km s^{-1} is apparently present beneath the western Himalayas, whereas in the east the velocity in the lower crust is much less, the average being only 6.25 km s^{-1}. Figure 9.16 shows details of some of the refraction data recorded in the eastern Himalayas. The large and sudden offsets in the wide-angle reflection from the Moho indicate that major lateral changes in crustal and upper mantle structure are a feature of the Himalayas and Tibet. It appears that the segments of Moho may dip towards the south rather than towards the north. If substantiated by further study, this will require modification of the thrusting process as illustrated in the model in Figure 9.13.

The entire Himalayan mountain chain is a region of large negative Bouguer gravity anomalies. Figure 9.17 shows the Bouguer gravity anomaly along a profile perpendicular to the Himalayas at 78°E. Also shown in the figure is the anomaly that has been calculated by assuming

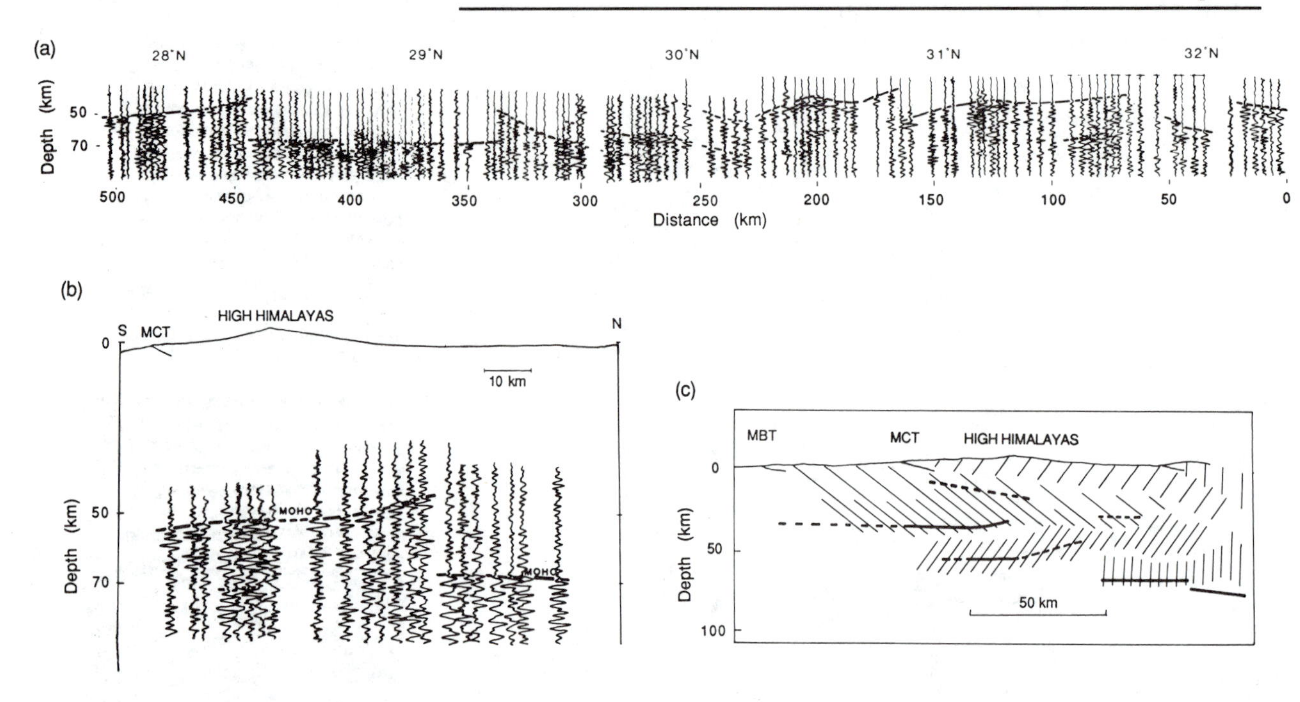

Figure 9.16. (a) A time-section of wide-angle reflections from the Moho recorded on a fan profile from 28°N to 32°N at about 90°E. The seismograms have all been corrected to a constant shot-to-receiver distance of 250 km, assuming that the average crustal velocity is constant (original ranges were 200–250 km). The wide-angle reflection from the Moho, which has large amplitude, is observed to have large offsets. (From Hirn et al. 1984b.)
(b) Details of the wide-angle reflections from the Moho, displayed as in (a), beneath the high Himalayas immediately north of the MCT at 87°E. (From Hirn et al. 1984a.) (c) Possible crustal structure between the MBT and the Indus–Tsangpo suture in the eastern Himalayas. Dashed line, interface between upper and lower crust. Solid line, Moho discontinuity. Wide hatching, upper crust. Tight hatching, lower crust. Possible correspondence of upper and lower crust is shown by hatching in the same direction. (After Hirn et al. 1984a.)

that the surface topography is locally isostatically compensated by crustal thickening (i.e., Airy's hypothesis; see Sect. 5.5.2). These calculated anomalies are different from those actually observed, being too small over the sedimentary Ganga (Ganges) basin to the south of the mountains and too large over the mountains. This means that the Ganga Basin is over-compensated (there is a mass deficiency relative to the isostatic model) and that the Himalayas are undercompensated (there is a mass excess relative to the isostatic model) by as much as 100 mgal. A model in which the Indian Plate underthrusts the mountains and is flexed downwards by the load of part of the mountains accounts for these differences (Fig. 9.17b). The Ganga Basin, which forms in front of the Himalayas because the Indian Plate is flexed downwards there, is filled with sediment eroded from the mountains. The small jog in the gravity data between 0 and 50 km is modelled by a thin layer of sediment which has been underthrust beneath the Himalayas. The observed gravity anomalies south of the Ganga Basin (at −200 to −400 km in Fig. 9.17a) are some 25 mgal less than the model anomalies. This is not thought to be related to the basin but is a general feature of the Indian Shield. The model shown in Figure 9.17b has the Indian Plate underthrust as an intact unit for just over 100 km beneath the Lesser Himalayas, and then continuing in a thinner and weaker form for about another 100 km beneath the Greater Himalayas. The flexural rigidity of the two parts of the plate are 7×10^{24} N m and 3×10^{22} N m. Additional bending moments ($M_0 \simeq 4 \times 10^{17}$ N) and vertical shear forces per unit length ($F_0 \simeq 3 \times 10^{12}$ N m^{-1}) are included to bend the plate down sharply beneath the Greater Himalayas and to help support the load of the mountains while maintaining the necessary flexure beneath the Ganga Basin to the south (see Sect. 5.7 for discussion of flexure).

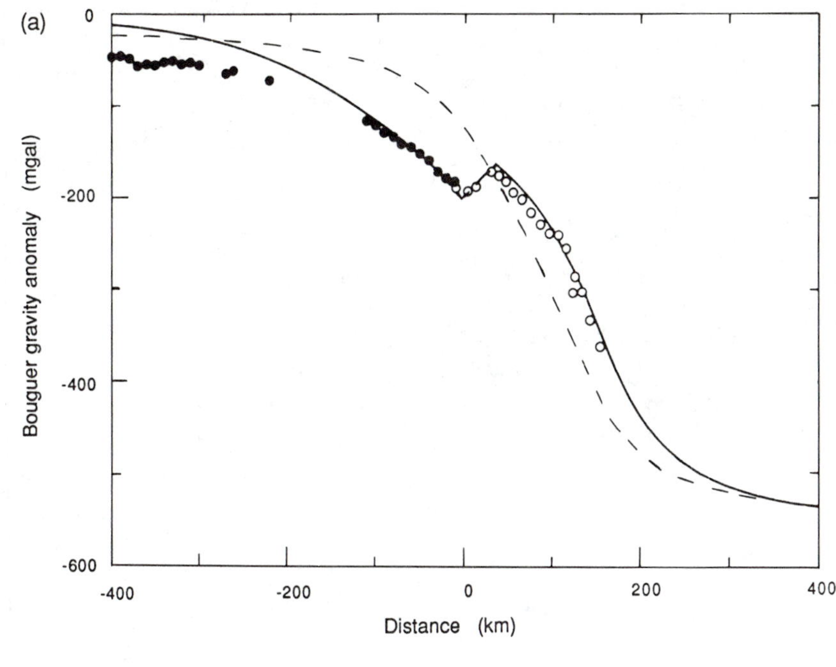

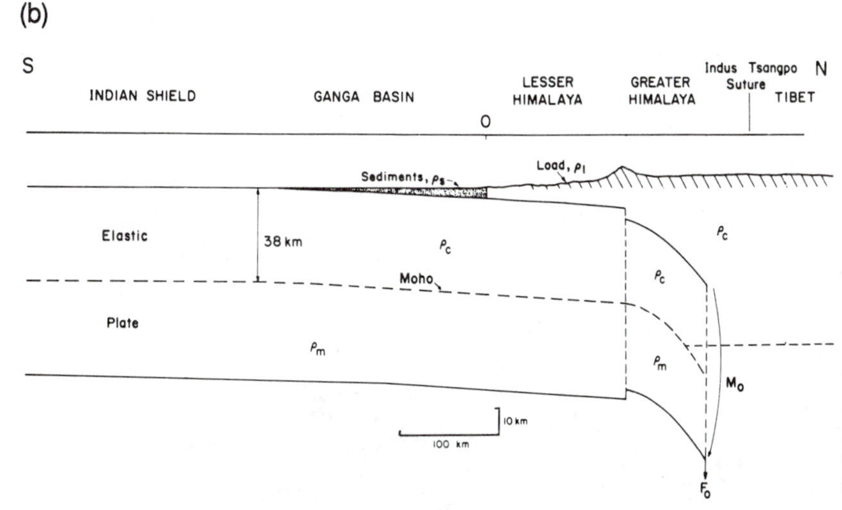

Figure 9.17. Profiles perpendicular to the Himalayas at approximately 79°E. (a) Bouguer gravity anomalies. Circles, observations; dashed line, computed anomaly assuming local isostatic compensation; solid line, anomaly computed using flexural model shown in (b). (b) Cross section of the flexural model of the Indian Plate and the Himalayas, which accounts for the gravity anomalies in (a). The assigned densities for the sediment, crust, mantle and load, ρ_s, ρ_c, ρ_m and ρ_l are 2.4, 2.8, 3.35 and 2.7×10^3 $kg\,m^{-3}$, respectively. The underthrusting Indian Plate has a reduced flexural rigidity beneath the Greater Himalayas and does not extend beneath Tibet. An additional bending moment, M_0, and shear force per unit length, F_0, cause the plate to bend sharply beneath the Greater Himalayas. (From Lyon-Caen and Molnar 1985.)

The Alps Although our understanding of the Himalayas is still at an early stage, and many years of painstaking fieldwork in difficult conditions will be required to fill in the details and gain a more complete picture of the formation of that mountain belt, the Alps have had much more attention. The Alps formed when the Adriatic promontory on the African Plate collided with the southern margins of the Eurasian Plate. The Alps were not the only mountains formed as a result of the convergence of Africa and Eurasia. Figure 9.18a shows the extensive Alpine fold system of the Mediterranean region. The complex present-day tectonics of the Mediterranean involve a number of microplates. The main rigid regions are Africa, Eurasia, Arabia, the Adriatic Sea, central Turkey and central

(a)

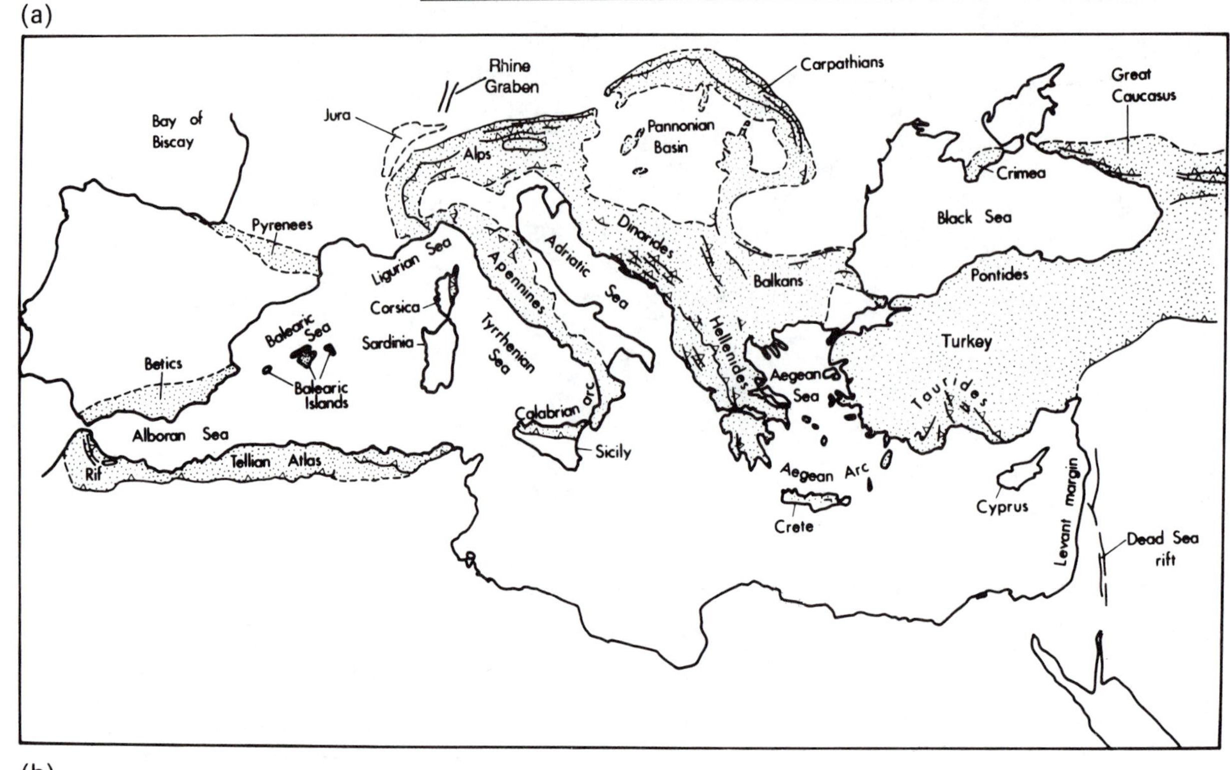

(b)

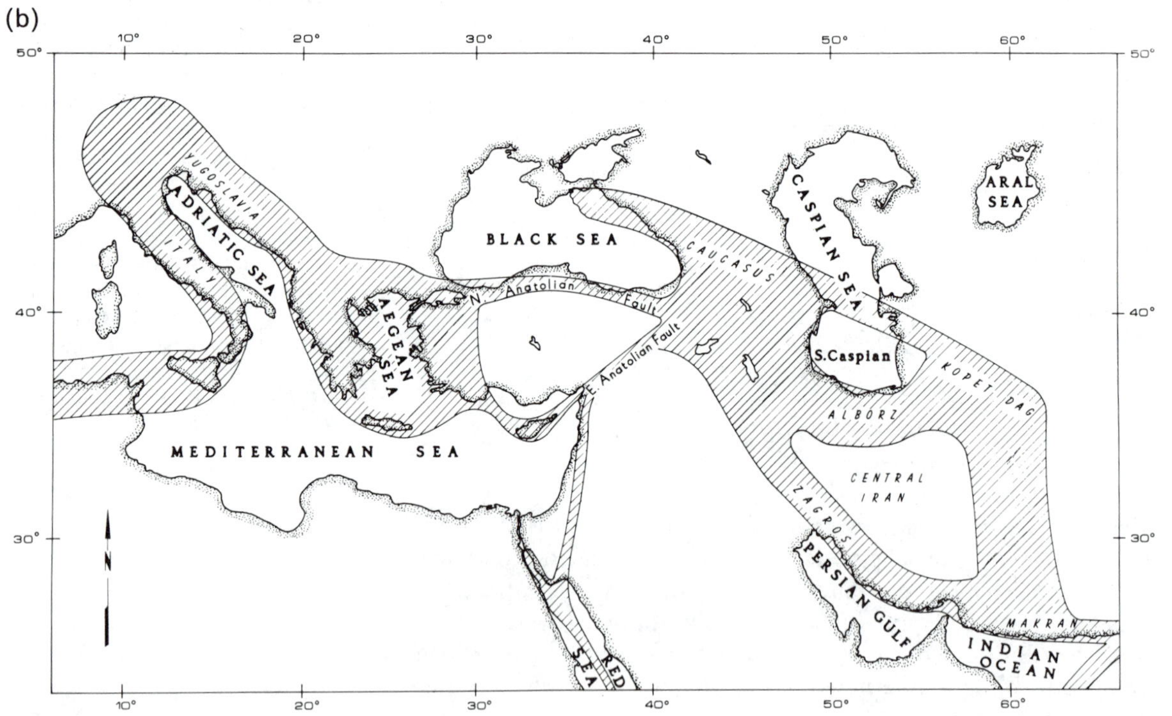

Figure 9.18. (a) The Alpine system of Europe (shaded) formed as a result of the convergence of Africa and Eurasia. (From Smith and Woodcock 1982.) (b) The principal seismic belts in the Mediterranean and Middle East. (From Jackson and McKenzie 1988.)

Iran (Fig. 9.18b). Palaeomagnetic data indicate that the Adriatic block, which was a northern promontory of the African Plate, has probably been separate from the African Plate since the early Tertiary and is now rotating separately. The African Plate is being subducted beneath Crete, and as a result the back-arc Aegean region is undergoing intense deformation.

Figure 9.19 shows the shear-wave structure of the upper mantle on a cross section through Switzerland and northern Italy. The velocities were obtained by the simultaneous inversion of all available surface-wave data. The lithosphere on the northern side of the Alps is somewhat thicker than that beneath Italy. The high shear-wave velocities, extending to 200 km depth beneath the Alps, are typical of the lithosphere and not the asthenosphere and indicate that lithosphere has probably been subducted to about that depth.

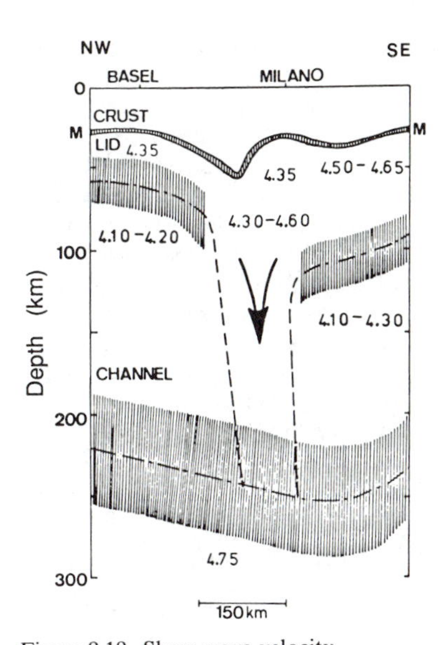

Figure 9.19. Shear-wave velocity structure of the upper mantle beneath the western Alps obtained from the simultaneous inversion of surface-wave dispersion data. S-wave velocities, in km s^{-1}. M, Moho discontinuity. 'Lid', lower lithosphere. 'Channel', asthenosphere. Hatching, uncertainty in boundaries. High velocity beneath the Alps indicates that material may have been subducted to about 200 km depth. (From Panza and Mueller 1979.)

Many seismic refraction lines have been shot in the Alps. The results have been used in the preparation of the Moho depth map in Figure 9.2: The crust thickens from 25 km beneath the Rhinegraben to 50 km in the Central Alps and then thins to 35 km on the southern side of the Alps. Figure 9.20a shows a seismic refraction record section from the Jura region, north of the Alps. There the crust is 27 km thick and has a complex structure with two distinct low-velocity zones, one in the upper crust and the second immediately above the Moho. These low-velocity zones have been detected because of the offset in the travel times between the wave which travelled in the overlying high-velocity material and the wave reflected from the base of the low-velocity zone (see Sect. 4.4.3). One prominent feature of this record section is the large amplitude of the P_n phase (the Moho headwave), which indicates that there is a strong positive velocity gradient in the upper mantle. Figure 9.20b, from the southern Swiss/Italian Alps, shows evidence of a low-velocity zone in the upper crust but not in the lower crust. The large amplitude of the wide-angle reflection from the Moho, P_mP, indicates that there is a large velocity contrast at the Moho beneath the southern Alps.

Figure 9.21 shows a series of cross sections along a northwest–southeast line across the Swiss Alps. Negative Bouguer gravity anomalies characterize the Alps. In the western Alps, a large positive gravity anomaly is caused by the Ivrea body, a slice of lower crustal–upper mantle material which was obducted from the southern plate and thrust to a shallow level. The high-velocity, 7.4 km s^{-1}, zone in the upper crust just south of the Insubric Line (one of the major Alpine faults) in Figure 9.21c is thought to be this Ivrea body. In the western Alps, the Ivrea body outcrops at the surface. The geology is very complex in detail. The northern ranges are molasse (sediments) from the foreland sedimentary basin. To the south, the rocks progressively age until finally the highly metamorphosed crystalline core is reached: These rocks were at deep levels in the crust until thrusting and erosion brought them to the surface. (See Sect. 7.11.4 on metamorphic belts.) The southern Alps, south of the Insubric Line, are thought to have originated on the continental shelf of the Adriatic promontory of the African continent. It is estimated that some 100 km of shortening has occurred (by folding and thrusting) across the Alps in the last 40 Ma.

The Alps are not particularly active seismically. Earthquakes do occur

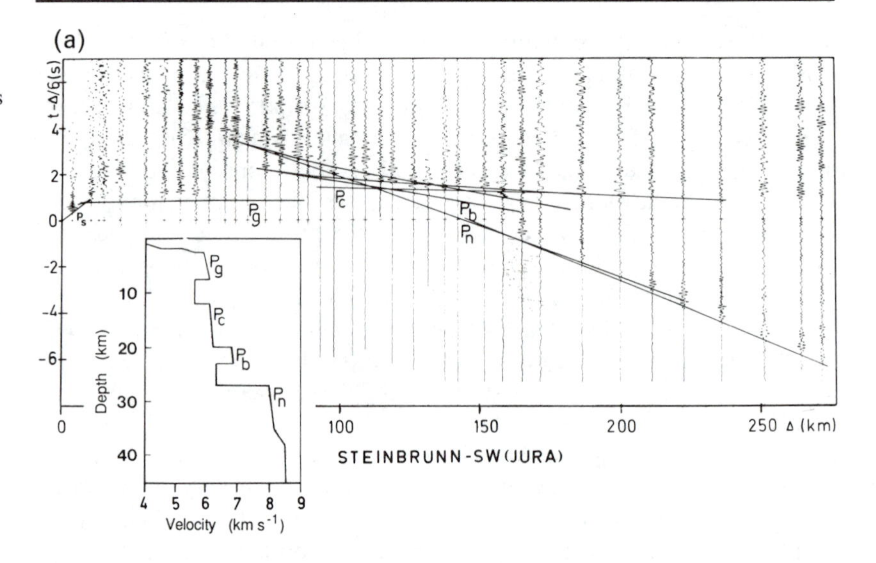

Figure 9.20. (a) Record section, reduced to $6\,km\,s^{-1}$, for a refraction line shot in the Jura in the northern part of the cross section shown in Figure 9.21b. The time offset between the crustal phases P_g and P_c indicates the presence of a low-velocity zone. Likewise, the low-velocity zone at the base of the crust is indicated by the time offset between phase P_b and the Moho reflection. (b) Record section, reduced to $6\,km\,s^{-1}$, for a refraction line shot in the southern Alps perpendicular to the cross section shown in Figure 9.21b. The postcritical Moho reflection P_mP is very strong on these records. (From Mueller et al. 1980.)

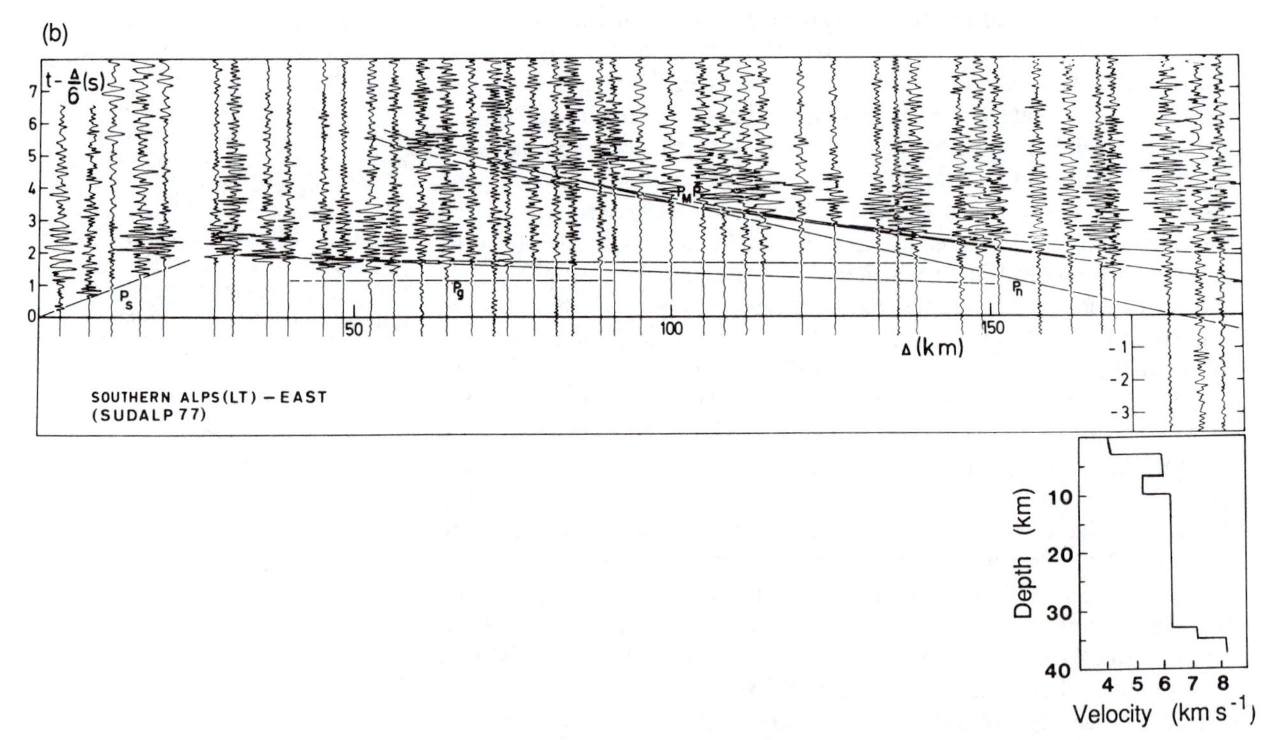

but not frequently, and although sometimes damaging, they are usually of smaller magnitude than the Himalayan events. The maximum uplift in the Alps is $0.15\,cm\,yr^{-1}$, almost an order of magnitude less than Himalayan values.

An interpretation of the stages of the evolution of the Alps is shown in Figure 9.22. In the late Permian to Triassic (250 Ma), two extensional rift systems were operating in the initial crust and mantle of Pangea (Fig. 9.22a). By the middle Jurassic (170 Ma) major extension was taking place, forming the so-called Neo-Tethys or Penninic Ocean. A minimum

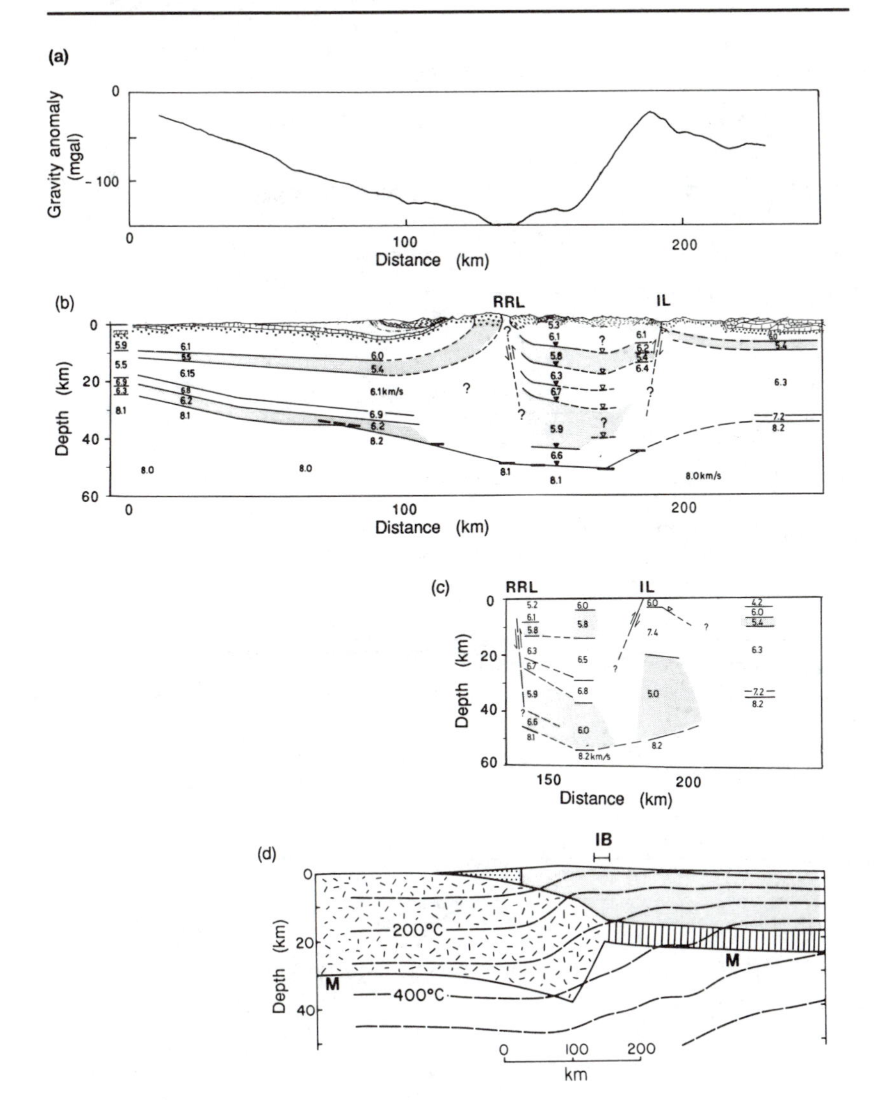

Figure 9.21. Cross sections along a NW–SE line across the Swiss Alps. (a) Bouguer gravity anomaly. (b) P-Wave velocity structure. Low-velocity zones are shaded. Major faults: RRL, Rhine–Rhone Line; IL, Insubric Line. Low-velocity zones shaded. (c) P-Wave velocity structure along a line 40 km to the east of (b) with RRL and IL as in (b). (d) Present-day thermal state of the Alps. IB, Ivrea body. Stippled pattern, sediments deposited in the fore-arc region. Random dashed pattern, continental crust. Vertical ruled pattern, oceanic crust. Gray, overthrust material. White, mantle. M, Moho. Note different horizontal scale. (After Klingelé and Oliver 1980, Mueller et al. 1980 and Stockmal and Beaumont 1987.)

of 250 km of oceanic-type crust was formed in the period up to about 110 Ma ago, and some of this material may now be represented in the ophiolite sequences which occur along the Alpine chain. By the middle to late Cretaceous (100 Ma), the Penninic Ocean was subducting beneath the northern margins of the African Plate (Fig. 9.22d); this continued until the Tertiary (about 40 Ma ago) when the main episode of continental collision started. This collision resulted in the major deformation, uplift and subsequent erosion which formed the Alps as we observe them today. The exact method of loading and deformation and the peeling off and stacking of slices of material from the colliding plates remain a matter of debate which will engage the geological community for some time.

An Ancient Continental Collision Interpreting ancient continental collision zones is a complex geological problem. The Caledonian orogeny occurred some 400 Ma ago when the ancient Iapetus Ocean between North

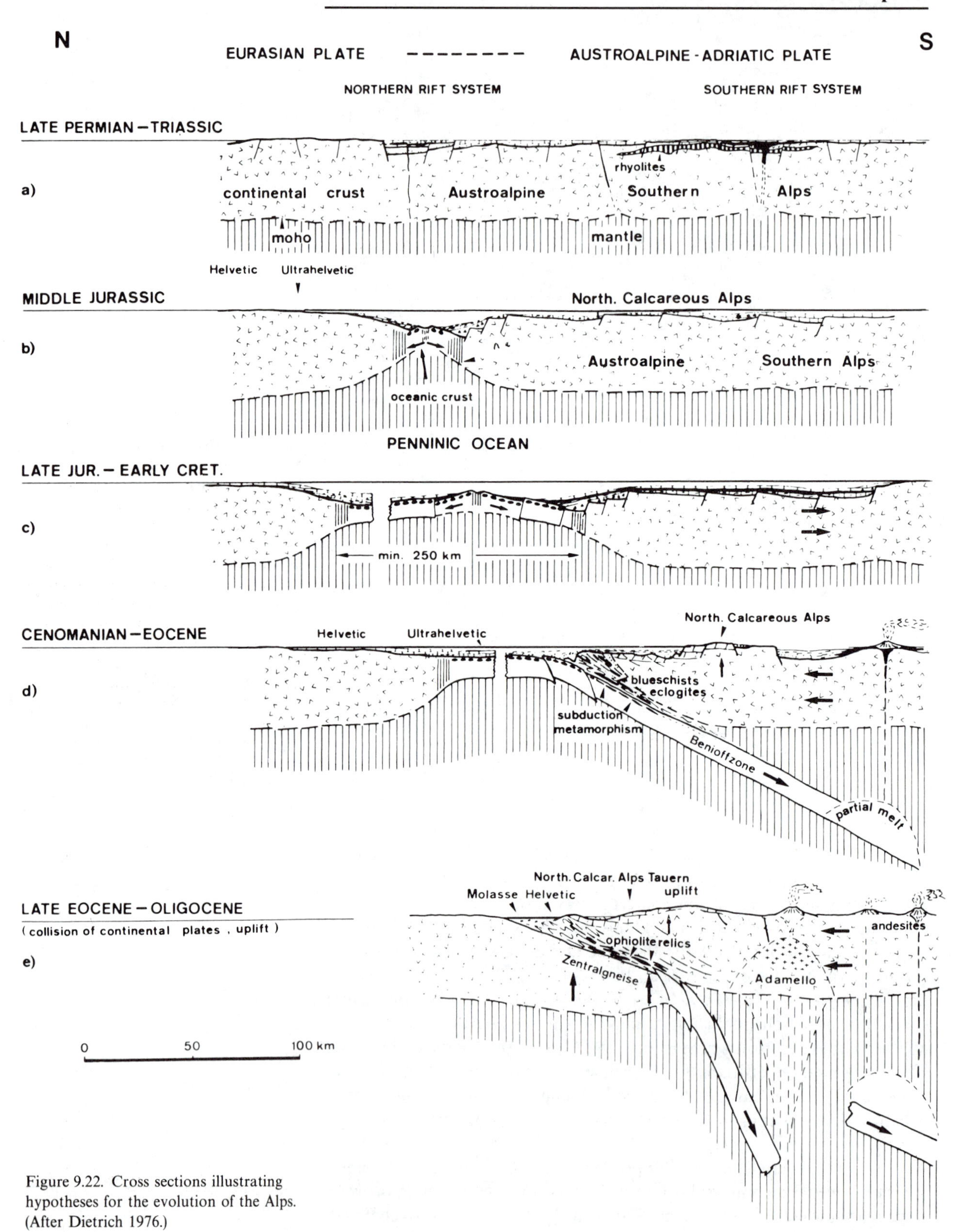

Figure 9.22. Cross sections illustrating hypotheses for the evolution of the Alps. (After Dietrich 1976.)

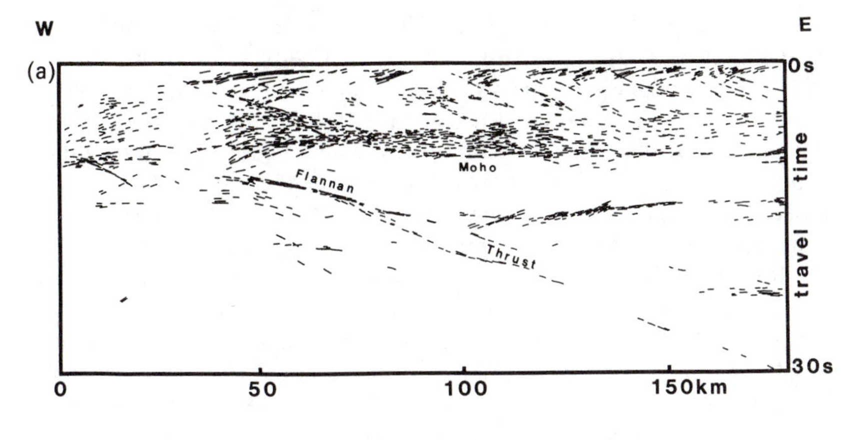

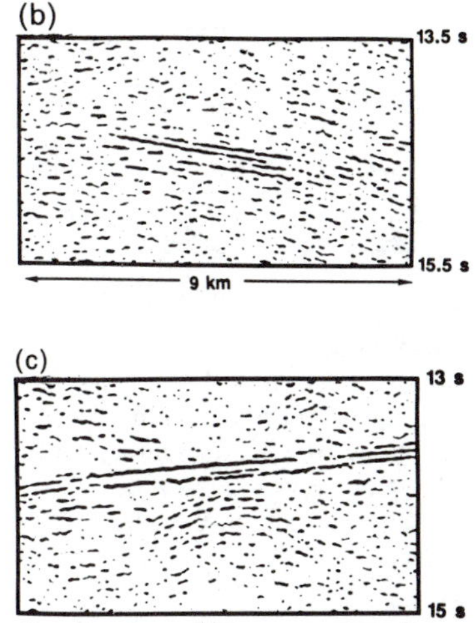

Figure 9.23. (a) Line drawing of the 15-fold unmigrated deep seismic reflection profile DRUM shot off the northern coast of Scotland (see also Fig. 4.57). (b) Details of the reflections from the Flannan Thrust. (c) Details of the reflections from the subhorizontal deep mantle reflector. (From McGeary and Warner 1985, and Warner and McGeary 1987.)

America and Europe closed during the formation of the supercontinent Pangea. The remnants of this collision are now in Scotland and eastern North America.

The Flannan Thrust off the northern coast of Scotland has been spectacularly imaged by deep-seismic reflection profiling (Fig. 9.23). This thrust originates in the lower crust, cuts (and may offset) the Moho and extends to a depth of 75–85 km. The upper crust is characterized by rotated half-grabens filled with sediment, which formed during a later period of Mesozoic extension. The lower crust is highly reflective with the Moho clearly visible as a bright reflector at its base. The dipping crustal reflector, which is visible between 30 and 80 km, is the Outer Isles Fault (see also Fig. 4.57). Two sets of clear, strong reflections originate within the mantle: the first, a 100 km long subhorizontal reflector at 13–15 s two-way time, and the second, the dipping Flannan Thrust reflector which extends from 7 s down to at least 27 and possibly 30 s two-way time (recording time of the survey was 30 s). Figure 9.24 shows the effect of depth migration (see Sect. 4.5.4) on these deep reflectors: The dipping reflectors steepen and migrate up-dip. The most plausible explanation for the Flannan Thrust reflector is that it is a fossil subduction zone dating from the Caledonian orogeny when the Iapetus Ocean closed. If so, it is a 400 Ma old thrust, though it could have been reactivated by the later Mesozoic extension in the region. Nevertheless, such strong reflections which clearly originate from within the lower part of the lithosphere show that the lower lithosphere can be structurally complex and can support localized strains over a long time.

Figure 9.24. The effect of depth migration on the deep reflections of Figure 9.23a. The dipping reflectors steepen and migrate up-dip. OIF, Outer Isles Fault; FT, Flannan Thrust. (From McGeary and Warner 1985.)

9.2.4 Continental Growth Rates

The growth of continents and the depletion of mantle (see Chapter 6 for mantle growth curves, etc.) are very much linked; the variation of initial isotope ratios with time illustrates this fact. So, in trying to estimate the growth rate of continental crust, we are also trying to estimate the depletion rate of the mantle.

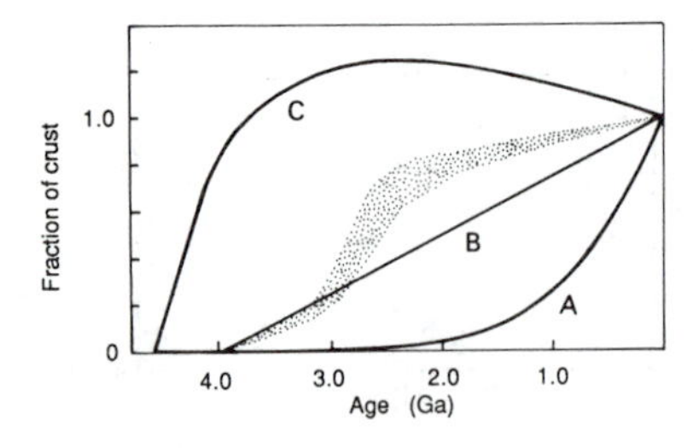

Figure 9.25. Growth curves for the continental crust. Model A has an increasing growth rate with time. In model B the growth rate is linear. Model C has an initial rapid growth rate. Generally accepted range of models is stippled.

Numerous models of crustal growth rates exist, but there is as yet no clear consensus as to which, if any, are correct. Figure 9.25 gives a summary of the basic models. At one extreme is model A, in which the rate of formation of continental crust increases with time; in this model, loss of sediments into the mantle at subduction zones is of minor importance. Model B has a linear rate of growth of continental crust with time; this model also assumes that all sediments are returned to the continents at subduction zones. Model C represents a rapid growth of crust early in the history of the earth, with the volume of crust staying more or less constant since then. It is possible to have a constant crustal volume if the sediments which are eroded from the continents and returned to the mantle at subduction zones have the same volume as the material which melts and is added to the continents above subduction zones. When the volume of volcanic material is less than that of the sediments, then the crustal volume decreases (as in model C).

Table 9.2 shows that some 70% of the present surficial area of the continents formed during the period which ended 450 Ma ago. Thus, if we bear in mind that in the past there must have been much more crust than this, the increasing rate of crustal growth depicted in model A is rather unlikely.

Most popular models for crustal growth lie between model B and model C, with the current isotopic data favouring models that are closer to B than C. The stippled region in Figure 9.25 shows a range of models which assume an increased growth rate from about 3.0 to 2.5 Ga to account for the large volume of Archaean volcanic rocks formed during that period (see Sect. 9.5). Such models probably provide a reasonable estimate of the growth of continental crust.

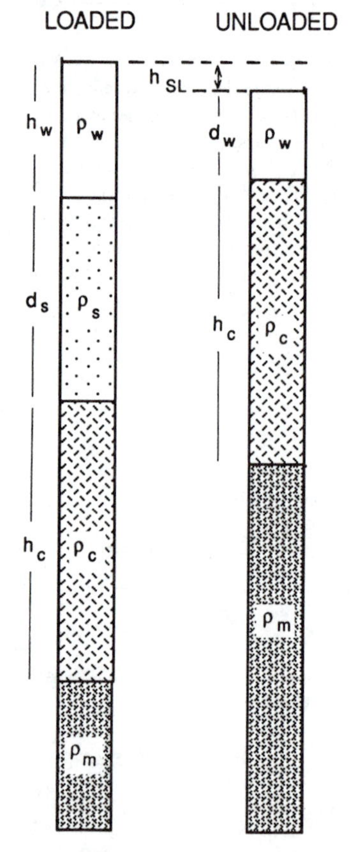

Figure 9.26. The loading effect of sediments. ρ is the density of w, water; s, sediment; c, crust; m, mantle. h_w is the depth of water in which sediments were deposited; d_s, the thickness of sediments; h_c, the thickness of crust; d_w, the backstripped depth of basement; h_{SL}, the change in sea level from the time at which sediments were deposited to the present day.

9.3 Sedimentary Basins and Continental Margins

9.3.1 Introduction

Sedimentary basins are of great economic importance because fossil fuels were formed there. Knowledge of the origin of a basin can provide information about the location and the type of thermal maturation of hydrocarbon deposits. A number of tectonic mechanisms can cause the formation of sedimentary basins. These are discussed in the following sections.

9.3.2 Loading Effect of Sediments

For a sedimentary basin to form, some factor must cause the basement to subside. However, let us initially suppose that a subaqueous depression of depth d_w in the basement exists (Fig. 9.26). Because holes simply tend to become filled, erosion of surrounding terrain automatically fills this depression with sediments. Assuming that it is completely filled with sediments and that isostatic equilibrium is maintained, we can use the methods of Section 5.5.2 to calculate the total thickness of these sediments. This thickness d_s is calculated by equating the mass of the water in the depression plus the mass of displaced mantle to the mass of the sediments:

$$d_w\rho_w + (d_s - d_w)\rho_m = d_s\rho_s$$

which on rearranging terms gives

$$d_s = d_w\left(\frac{\rho_m - \rho_w}{\rho_m - \rho_s}\right) \tag{9.1}$$

where ρ_m is the density of the mantle, ρ_w the density of water and ρ_s the density of the sediments. If we take ρ_m as $3.3 \times 10^3\,\mathrm{kg\,m^{-3}}$, ρ_s as $2.5 \times 10^3\,\mathrm{kg\,m^{-3}}$ and ρ_w as $1.0 \times 10^3\,\mathrm{kg\,m^{-3}}$, the presence of the sediments amplifies the original subaqueous depression by a factor d_s/d_w of 2.9. Thus, a basin filled with 5 km of sediments represents a subaqueous depression of the crust of about 1.5–2.0 km, the exact value depending on the density of the infilling sediments. To make comparisons between one sedimentary basin and another, we must first calculate the loading effect of the sediments so that this amplification can be removed. This process is called *backstripping*. The depth at which the basement would have been if there had been no infilling sediments, d_w, is given by Eq. 9.1 as

$$d_w = d_s\left(\frac{\rho_m - \rho_s}{\rho_m - \rho_w}\right) \tag{9.2}$$

However, suppose that the sediments were deposited in water of depth h_w instead of at sea level (Fig. 9.26). In this case, 9.2 should be written as

$$d_w = h_w + d_s\left(\frac{\rho_m - \rho_s}{\rho_m - \rho_w}\right) \tag{9.3}$$

Sea level has not remained constant throughout geological history but has risen and fallen many times. Figure 9.27 illustrates one model of the variation of Phanerozoic sea level. To take account of this variation when we backstrip subsidence data, we must further modify Eq. 9.3 to

$$d_w = h_w + d_s\left(\frac{\rho_m - \rho_s}{\rho_m - \rho_w}\right) - h_{SL}\left(\frac{\rho_m}{\rho_m - \rho_w}\right) \tag{9.4}$$

where h_{SL} is the change in sea level as shown in Figure 9.27. In any detailed study of a region, another factor that should be taken into account is the postdepositional compaction of sediments. As more sediments are deposited, pore water is expelled and the thickness of each sedimentary layer reduced. In Eqs. 9.1–9.4, d_s represents the thickness of sediments when they were laid down and ρ_s their saturated density, not the thickness and density measured in situ today. *Decompaction* is the process of calculating the original sediment thickness.

Equations 9.1–9.4 assume Airy-type local compensation. However, if compensation is achieved not locally but regionally, the loading effect of the sediments should be calculated by considering the flexural loading of a thin elastic or viscoelastic plate. This is much more complex mathematically. In such models, because the load of the sediments is spread over a broad area, the net result is that the amplification factor is somewhat reduced.

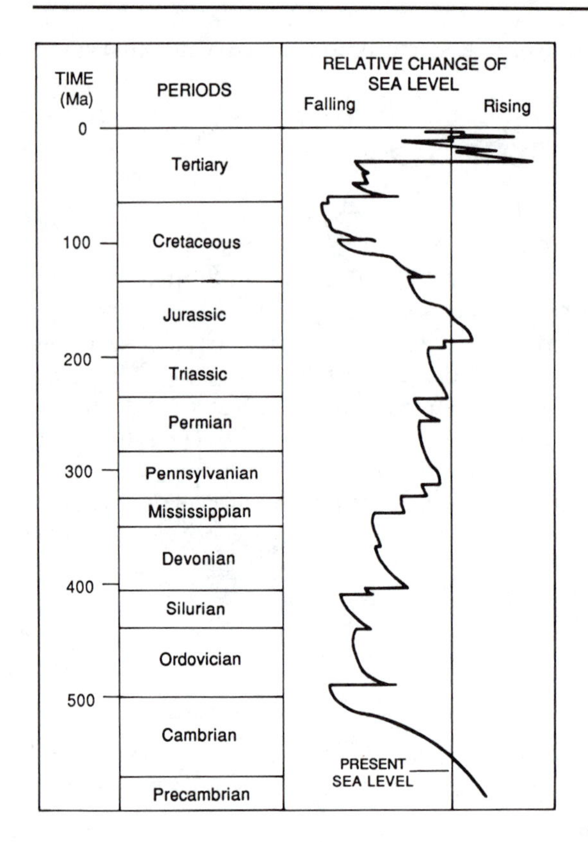

Figure 9.27. Changes in sea level through the Phanerozoic. (After Vail and Mitchum 1978, reprinted by permission.)

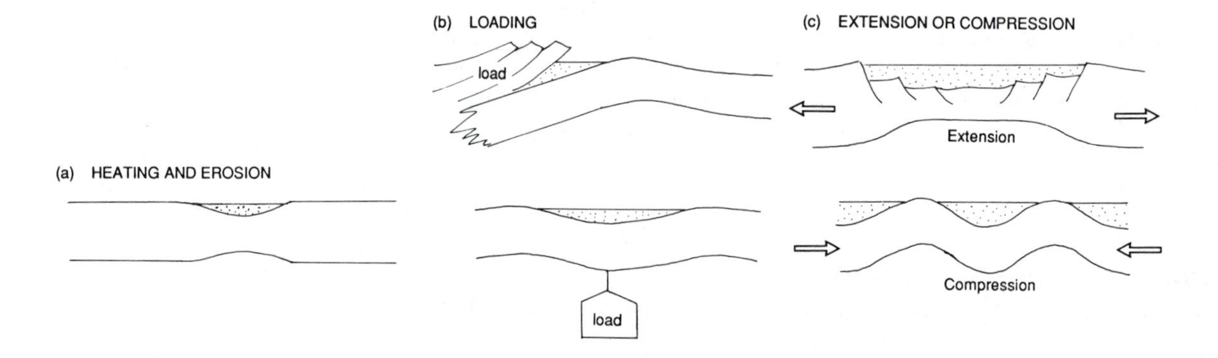

Figure 9.28. Various mechanisms that can form sedimentary basins: (a) Heating of the lithosphere, which results in expansion and uplift. If erosion takes place, a basin forms when the lithosphere cools and contracts. (b) Loading of the lithosphere from above (thrust piles) or below (intrusions, phase changes). (c) Extension, compression and faulting of the lithosphere.

9.3.3 Mechanisms for Producing Subsidence

A variety of geological processes can lead to the formation of thick sedimentary deposits on continental crust. They can, however, be divided into three main classes: (1) those formed by thermal events, (2) those formed as the result of flexure of the lithosphere by an imposed load and (3) those formed as the result of extension, compression or faulting of the basement (Fig. 9.28). The first class of basins subsides as a result of a heating event which causes uplift and erosion. Then, as cooling and contraction begin, sedimentation occurs. Examples are the Illinois and

Michigan basins in the United States. An example of the second class, a basin formed as a result of flexure of the lithosphere under an imposed load is the Canadian Alberta Basin. This foreland basin formed as a result of the loading provided by the Rocky Mountain fold thrust belt. Examples of the third class include the thick deposits laid down on the continental margins as the result of continental rifting (e.g., the Atlantic continental margins), basins formed after stretching of the continental lithosphere (e.g., the Tertiary North Sea Basin), basins formed in major fault grabens (e.g., the Basin and Range of the western United States) and the fault-controlled basins that apparently formed as a result of compression of the lithosphere (e.g., central Australian basins).

These classes of basins are by no means mutually exclusive or comprehensive. For example, the subsidence of the North Sea Basin is explained by extension and thinning of the lithosphere by a factor of about 1.25, which gave rise to an initial rapid subsidence. This was followed by cooling, which resulted in a subsequent, slower, exponential thermal subsidence.

9.3.4 Basins of Thermal Origin

A thermal event probably provides the simplest illustration of the formation of a sedimentary basin. If a region of crust and lithosphere is heated, there is expansion, the density changes and then uplift takes place. Erosion follows, reducing the crustal thickness. Subsequent cooling and contraction of the lithosphere forms a basin. (If no erosion takes place, the surface simply contracts back to its preheating level, and no basin forms; nothing changes.) Isostatic calculations based on reasonable values of the increase in temperature during heating give a maximum value for the initial uplift of about 2 km. This would result in a maximum erosion of 4 km, allowing for the isostatic amplification as material is removed (assuming that the erosional time constant is of similar magnitude to the thermal time constant), and therefore would result in a maximum of perhaps 3 km of sediment. This is significantly less than the thickness of sediments on the continental margins, where 10 km of sediment is often present (see Fig. 8.4), and somewhat less than the thickness of sediment in many of the intracontinental sedimentary basins. To deposit 10 km of sediment as a result of heating the lithosphere, a temperature increase of more than 1000 °C is required.

The subsidence due to thermal contraction is easily modelled mathematically. It is similar to the plate model for formation of the oceanic lithosphere (see Sect. 7.5.2). To first order, the depth to a particular bed (of age t) in a hole drilled into sediments can be written as

$$d = d_0[e^{(t/t_0)} - 1] \tag{9.5}$$

where d_0 is a slowly varying function of position of the well in the basin. The thermal time constant of the lithosphere t_0 is given by

$$t_0 = \frac{L^2}{\pi^2 \kappa} \tag{9.6}$$

where L is the thickness of the lithosphere and κ is the thermal diffusivity.

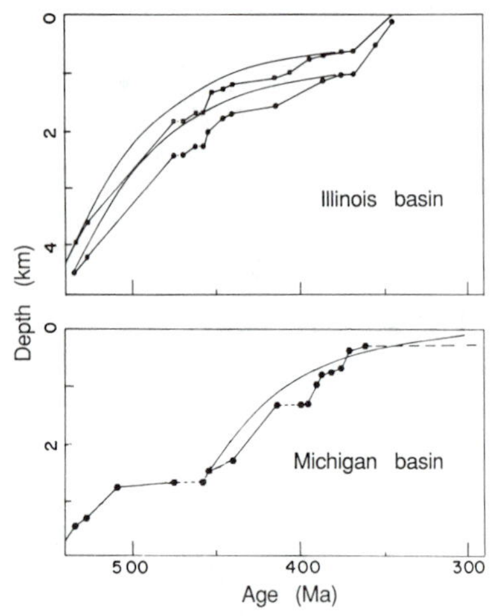

Figure 9.29. Depth to particular sedimentary horizons versus age of the sediment. Data are from two wells drilled in the Illinois basin and one in the Michigan basin. Solid curves are 50 Ma exponentials. If the subsidence was due to thermal contraction of the lithosphere, the depths should plot approximately on these exponential curves. Horizontal dashed lines indicate major unconformities in the sedimentary sequence. The Cambrian–Carboniferous (530–360 Ma) subsidence of the Illinois Basin and the Ordovician–Carboniferous (460–360 Ma) subsidence of the Michigan Basin could be due to thermal contraction of the lithosphere. (After Sleep et al. 1980. Reproduced with permission from the *Annual Reviews of Earth and Planetary Sciences*, vol. 8, © 1980 by Annual Reviews Inc.)

The thermal time constant for a 125-km-thick lithosphere with thermal diffusivity of $10^{-6}\,\mathrm{m^2\,s^{-1}}$ is 50 Ma. The Michigan and Illinois basins have been modelled on the basis of thermal contraction of the lithosphere (Fig. 9.29). The subsidence of the Illinois Basin between 530 and 360 Ma can be explained by one thermal event. However, a second thermal event is required to account for the rapid Carboniferous subsidence (360 Ma onwards). The neighbouring Michigan Basin has a somewhat different subsidence history. The Cambrian to Middle Ordovician (530–470 Ma) subsidence appears to be separate from the Middle Ordovician to Carboniferous (460–360 Ma) subsidence, which approximates a thermal subsidence curve. Some evidence exists for igneous activity at the time of the proposed Cambrian (530 Ma) heating event but not for the others. Invoking a heating event for which there is at present no direct external evidence other than the observed subsidence (i.e., no volcanic activity or known uplift) is not altogether satisfactory, and further work on the basement geology and subsidence must be done before these basins can be proven to be of thermal origin. Other explanations are possible.

9.3.5 Flexural Basins

Flexure of the lithosphere by a load at the surface produces a foreland basin, of which the Canadian Alberta Basin and the Appalachian Basin are excellent examples. If a load or thrust sheet is placed on the lithosphere, the lithosphere subsides. To study problems of this type, it is not enough to assume local isostatic equilibrium; rather, the lithosphere must be modelled as a thin, elastic or viscoelastic plate, and regional deformation must be calculated as discussed in Section 5.7. The width of a flexural basin is a function of the lithospheric thickness. For a given load, a thick lithosphere supports a wider basin than a thin lithosphere, as one would intuitively expect. Since isostatic balance requires the mass of mantle material displaced from beneath the lithosphere to equal the mass of the load, a wide basin on a thick lithosphere is shallower than a narrow basin on a thin lithosphere (Fig. 9.30a).

In the case of a viscoelastic lithosphere, the initial flexure is the same as for an elastic lithosphere, but the lithosphere relaxes with time. A viscoelastic material behaves elastically on a short time scale and viscously on a long time scale. The viscoelastic relaxation time τ, the time taken for any stress to relax to $1/e$ of its original value, is defined by $\tau = 2\eta/E$, where E is Young's modulus and η the dynamic viscosity. This means that the basin becomes deeper and narrower with time, evolving towards local isostatic equilibrium (Fig. 9.30b). The rate at which this relaxation occurs is characterized by the viscoelastic relaxation time τ.

Figure 9.31 shows how loading can produce a foreland basin. Figure 9.32 shows a cross section of the best-fitting flexural model for the Alberta Basin. This Canadian basin was formed on the North American Plate east of the Rocky Mountains as a consequence of thrusting and emplacement from the west of the Rocky Mountains. Thrusting in the Rocky Mountains began in the late Jurassic (approximately 140 Ma) and lasted about 100 Ma. From 35 Ma to the present, massive erosion (and hence isostatic uplift) has taken place across the region represented by the model. Almost 10 km of material has been eroded from the centre of the thrust

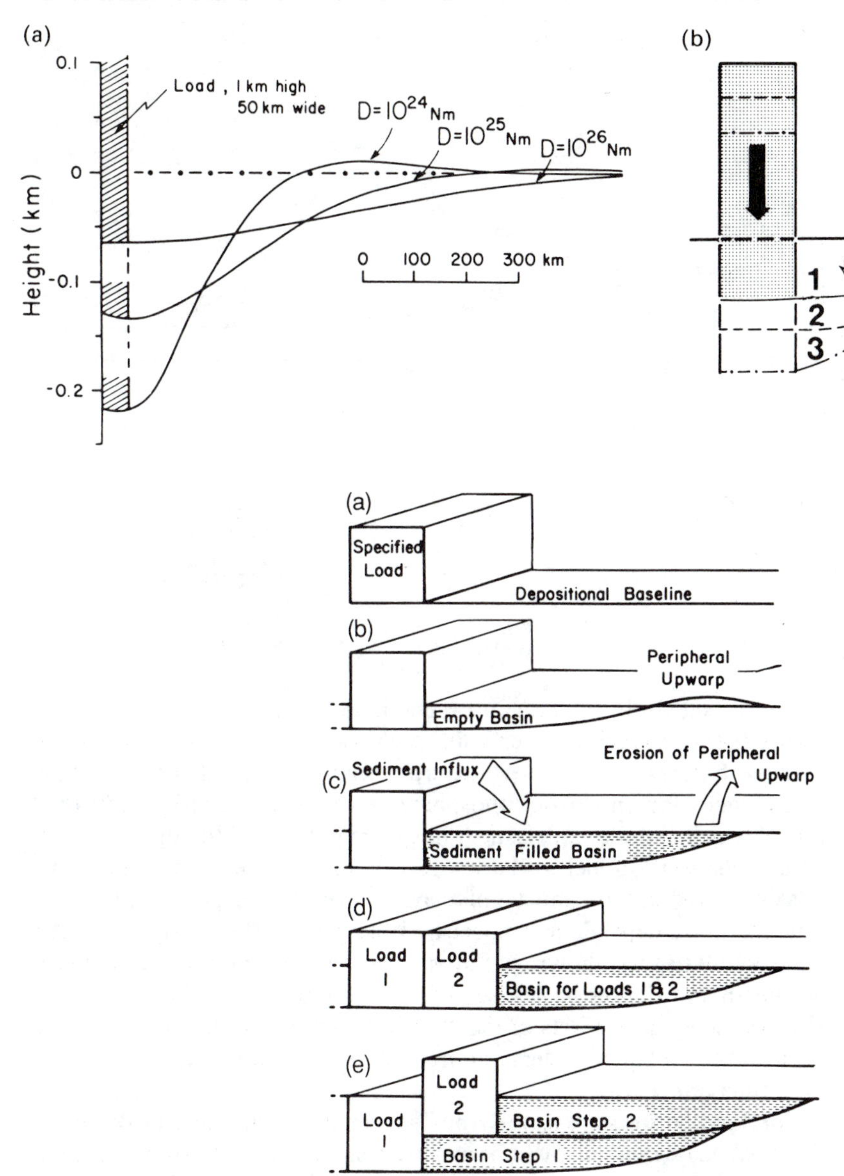

Figure 9.30. (a) Cross sections of sedimentary basins formed on an elastic lithosphere due to a 1 km high, 50 km wide two-dimensional load. Load density and sediment density are 2.4×10^3 kg m^{-3}. The three curves show the difference in basin cross sections for a lithosphere with flexural rigidity D, of 10^{24}, 10^{25} and 10^{26} N m. (From Beaumont 1981.) (b) Deformation of the surface of a viscoelastic lithosphere by a surface load. The initial deformation is the same as for an elastic lithosphere (a). However, with time, the viscoelastic lithosphere allows the stress to relax; so the deformation evolves to curves 2 and 3. The final stage if the load is left in place would be local isostatic equilibrium. (From Quinlan and Beaumont 1984.)

Figure 9.31. How a load on the surface of the lithosphere forms a depression which, when filled with sediment, becomes a foreland basin. (a) Load is emplaced. (b) Depression forms. (c) Sediment fills the depression, and erosion eliminates the forebulge. (d) Two loads are emplaced simultaneously. (e) Two loads are emplaced sequentially, resulting in a different basin from (d). (After Beaumont 1981.)

pile and about 3 km from its western edge. Most of the Paskapoo formation (dense stipple) has been eroded and many of the older units outcrop away from the Rocky Mountains. The model shown in Figure 9.32 has been calculated for a thin viscoelastic plate with flexural rigidity 10^{25} N m and a viscous relaxation time of 27.5 Ma overlying the asthenosphere.

The Appalachian Basin has been modelled similarly, but in this case, the best fit is obtained with a temperature-dependent viscoelastic model which has an approximately elastic thickness for the lithosphere of 80–90 km. Sections across arches between the Appalachian and Illinois basins and between the Illinois and Michigan basins are shown in Figure 9.33. Thrusting and emplacement of loads on the Appalachian mountain system took place from about 470 until 200 Ma. The modellers used the observed subsidence of the neighbouring Michigan and Illinois

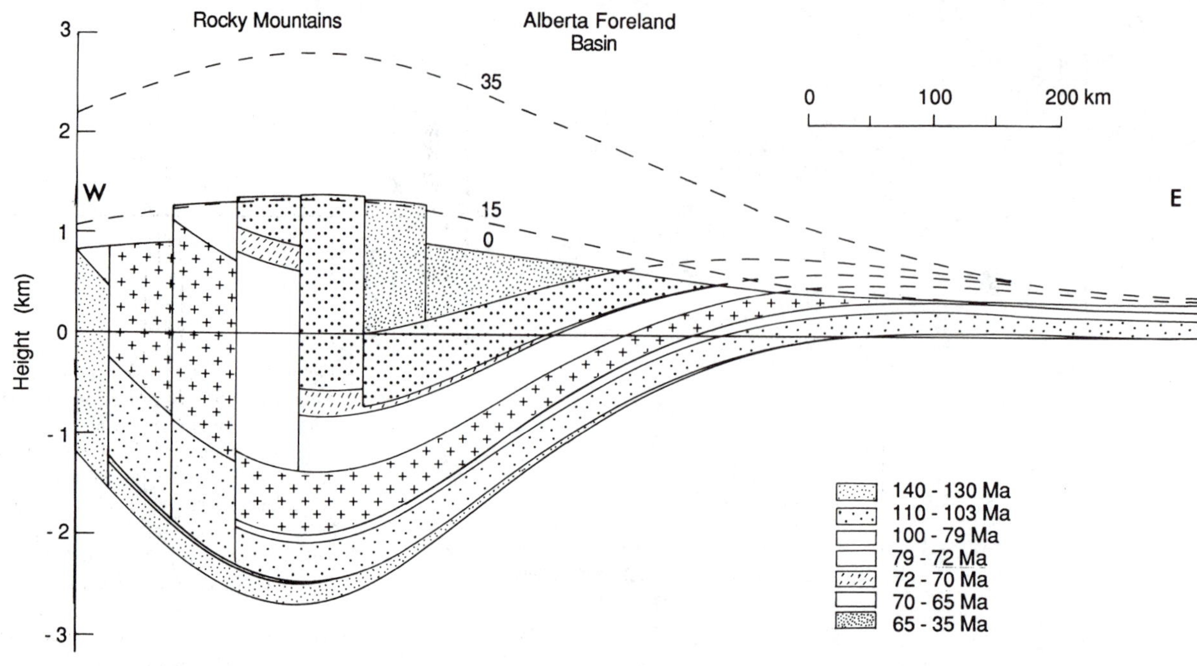

Figure 9.32. Cross section of a loading model for the Alberta Foreland Basin. The vertical column loads emplaced during each time interval are shaded like the sediments which filled the resulting depression. Dashed lines denote sediment surfaces which have since been eroded. Since loading stopped 35 Ma ago, massive erosion of the load (thrust pile) and the youngest sediments has taken place. Thus, 35 Ma ago the Paskapoo formation was some 1.5–2.0 km thicker than it is today, and almost 10 km of material has been removed from the centre of the thrust pile. The flexural rigidity of the lithosphere for this model was 10^{25} N m, and the viscoelastic relaxation time was 27.5 Ma. (After Beaumont 1981.)

basins, as shown in the previous section, and investigated the effect that their presence has on the intervening arches as a result of the Appalachian loading. Figures 9.33c, d show the cross sections calculated for the same model loads for an elastic lithosphere and a viscoelastic lithosphere. A model which assumes an elastic lithosphere gives too little uplift over the arches whereas a model with a viscoelastic lithosphere gives far too much relaxation and thus too much uplift and erosion. The best-fitting model has a temperature-dependent viscoelastic lithosphere. This study of the Appalachian Basin has shown that the subsidence–time plots for the adjacent Michigan and Illinois basins (see Fig. 9.29) should ideally be corrected to remove the loading effects of the Appalachians. When this correction has been applied, a better understanding of the subsidence of these two basins may emerge.

Foreland basins such as the Appalachian Basin, the Alberta Basin and the northern Alpine molasse basins are generally linear two-dimensional features, unlike many intercratonic basins, which are circular. It is clear that since subsidence–time plots for foreland basins are controlled by factors external to the basin and lithosphere, they may not be useful except in so far as they can indicate possible timing and magnitude of the imposed loads.

The lithosphere can be loaded from below as well as from above. An example of an intercratonic basin which may have formed as a result of loading in or beneath the lithosphere is the Williston Basin, which straddles the U.S.–Canadian border just east of the Alberta Basin (Fig. 9.34). Prior to the initiation of subsidence in the Cambrian, there had been no tectonic activity in the region since a probable continental collision at about 1800 Ma in the Hudsonian event and some late Proterozoic events further west. The basin appears to have subsided continuously at a slow, fairly constant rate for most of the Phanerozoic (over 400 Ma). A remarkable

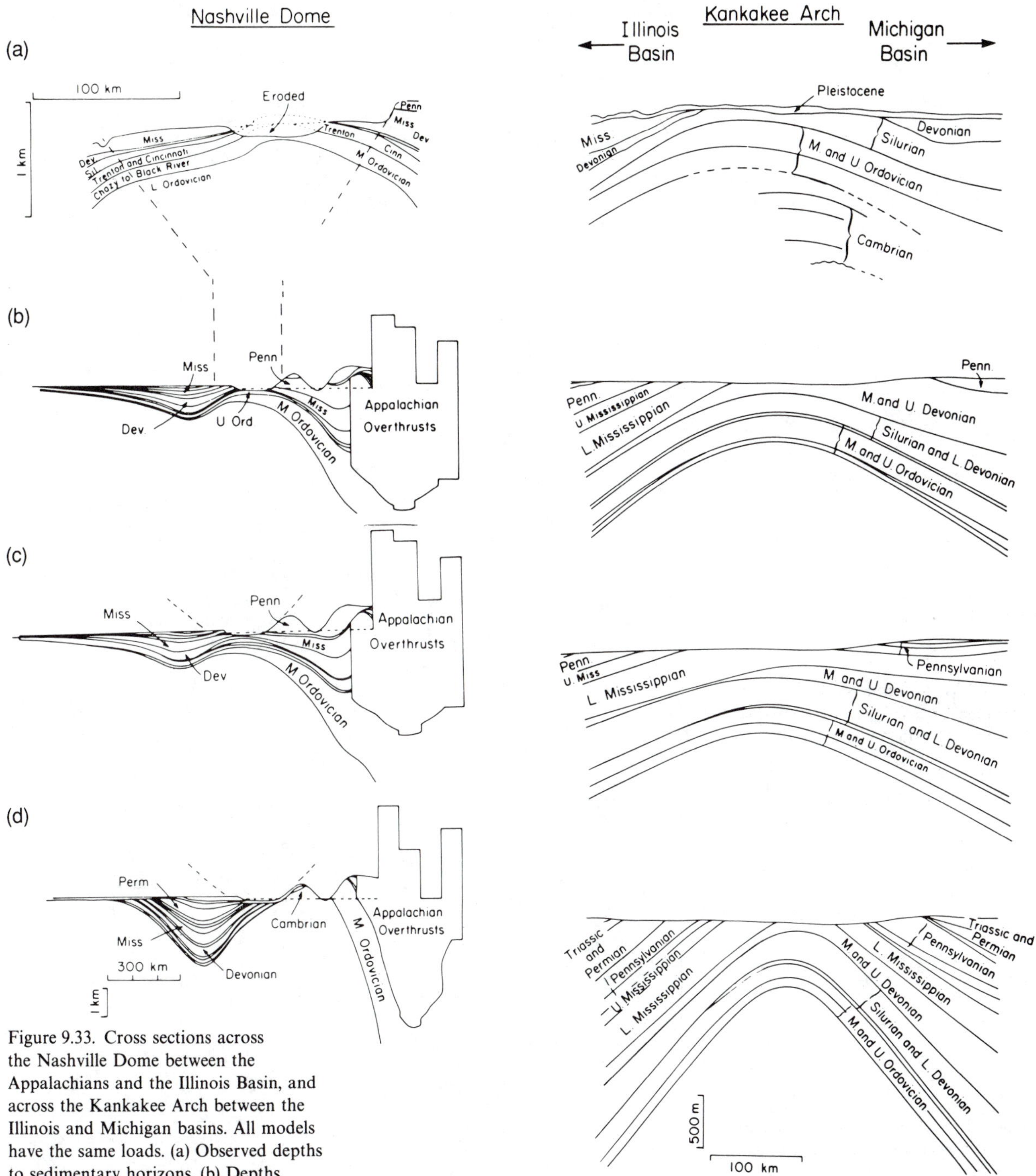

Figure 9.33. Cross sections across the Nashville Dome between the Appalachians and the Illinois Basin, and across the Kankakee Arch between the Illinois and Michigan basins. All models have the same loads. (a) Observed depths to sedimentary horizons. (b) Depths calculated for the best-fitting model which has a temperature dependent viscoelastic lithosphere. (c) Depths calculated for a model *elastic* lithosphere ($D = 10^{25}$ N m). (d) Depths calculated for a model *viscoelastic* lithosphere ($D = 10^{25}$ N m, $\tau = 27.5$ Ma). (After Quinlan and Beaumont 1984.)

feature of the subsidence is that the centre of depression of the basin remained almost in the same place throughout that time (Fig. 9.35). One simple model for the subsidence of the basin involves a steadily increasing load hung under the centre of the basin. This raises the obvious question: What is the load? One possibility is that, by some means, a region deep in the lithosphere was heated and then slowly cooled and contracted,

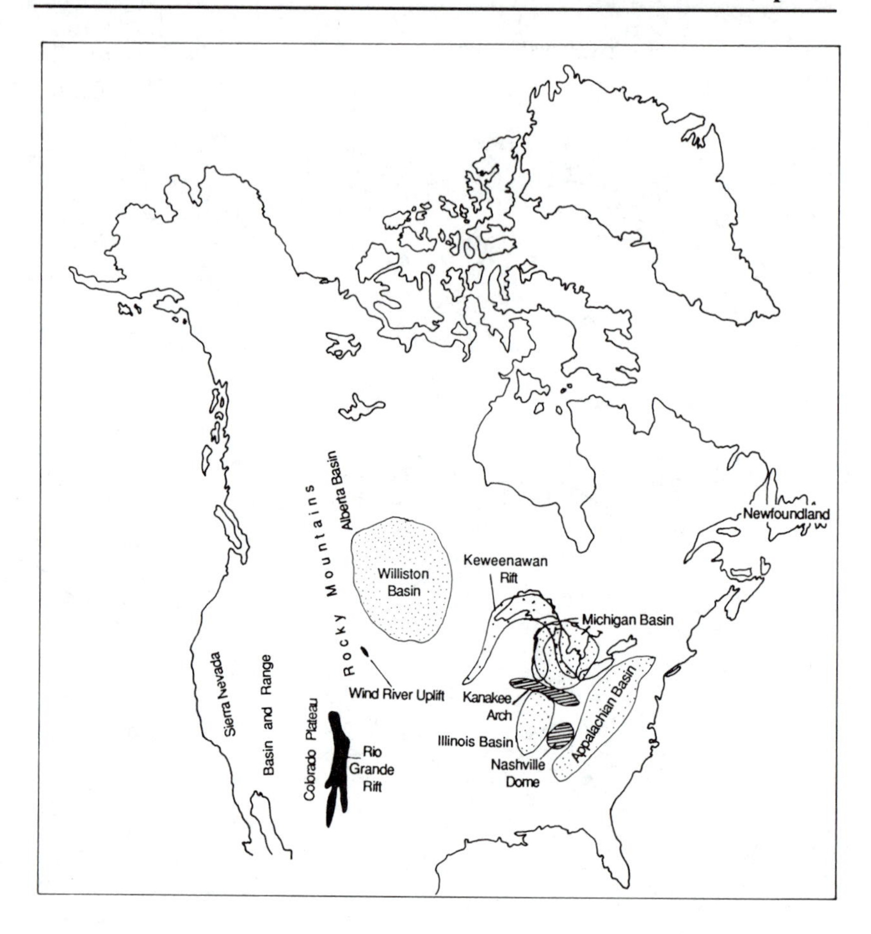

Figure 9.34. Location of some geographical, geological and geophysical features in North America.

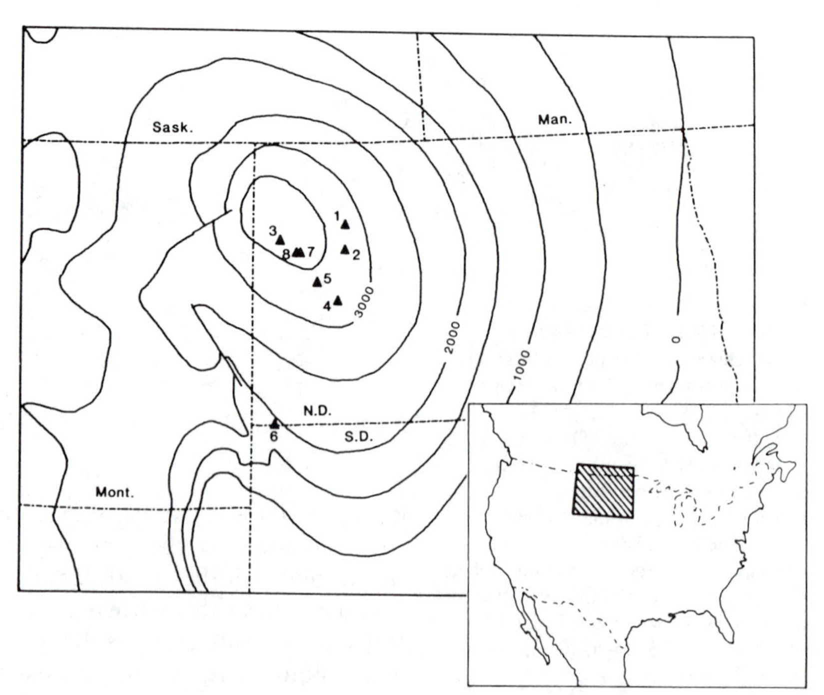

Figure 9.35. Depth to the Precambrian basement beneath the Williston Basin. Contour interval is 500 m. Triangles denote location of the centre of the basin as determined from the best-fitting flexural model of the plate during the following time intervals: 1, 450–352 Ma; 2, 352–328 Ma; 3, 328–324 Ma; 4, 324–239 Ma; 5, 328–239 Ma; 6, 239–168 Ma; 7, 168–0 Ma; 8, 114–0 Ma. (From Ahern and Ditmars 1985.)

becoming denser. The problem with this model is that this intrusive body must account for the subsidence without producing surface volcanism. One possibility is a very large, cool intrusion. An alternative explanation is that some part of the lithosphere is undergoing phase changes or metamorphic reaction. Geologically, the most probable reactions are those which involve the growth of high-pressure assemblages in the deep crust or upper mantle. As an example, the complete transformation from gabbro to eclogite at the base of the crust would increase the density from about 3.0 to $3.4 \times 10^3\,\mathrm{kg\,m^{-3}}$. This density change, about 30 times greater in magnitude than that caused by the cooling of gabbro through 150°C, means that the region undergoing the phase changes can be 30 times smaller than needed in an intrusion model and still give rise to the same load. Many other similar metamorphic changes can occur, each with particular rates of change. Such metamorphic changes occur slowly at rates controlled by the extent to which volatiles are present and by the temperature–time history of the rock. The rate of nucleation of a new phase can be expressed as

$$\text{rate} = K e^{(-\Delta G^*/RT)} e^{(-H_a/RT)} e^{(-P\Delta V/RT)} \tag{9.7}$$

where ΔG^*, the free energy of activation, is proportional to

$$\frac{1}{(T_{\text{equilibrium}} - T)^2}$$

H_a is the enthalpy of activation, P the pressure, ΔV the volume change, R the gas constant, T the temperature and K a constant. At temperatures close to equilibrium, ΔG^* is large, and so the transformation rate is slow. At intermediate temperatures or at much greater temperatures than the equilibrium temperature, the transformation rate is faster. At low temperatures, however, T is small, so (Eq. 9.7) the transformation rate is slow. The reaction rates are not known with any accuracy; but, for example, a gabbro layer at 50 km – such as would be produced by underplating in a major flood basalt event (comparable to the Karoo event in southern Africa) or by emplacing a slab of mafic material under the Williston area during the last stages of the Hudsonian plate collision – would cool and transform to eclogite over a time of the order of 10^9 or fewer years. Initially, massive uplift and erosion would occur, but then cooling, contraction and transformation to eclogite would take over and progressively load the lithosphere. This would produce nearly steady subsidence, as may have occurred in the Williston Basin. Either of the models discussed could allow for a mid-Proterozoic event to produce Phanerozoic subsidence. Such models are supported by independent seismological evidence: A COCORP deep reflection profile over the basin shows that the lowermost crust is characterized by relatively high-amplitude reflections. Detailed seismic refraction surveys have shown that beneath the basin, where the crust is some 45 km thick, there is a high-velocity lower crustal layer and some indication of a high P-wave velocity of $8.4\,\mathrm{km\,s^{-1}}$ for the upper mantle. This high-velocity reflective material can most simply be explained as a layer of eclogite.

These are simply models, and other models have also been proposed.

The origin of the basin is as yet unknown. The inversion of thermal and subsidence data is extremely nonunique.

9.3.6 Extensional Basins

Sediments deposited on the continental margins record the sedimentary history of the rifting apart of old continents. Figure 8.4 shows that sediment thickness increases rapidly with distance from the coast locally, beneath the continental shelf, reaching thicknesses of more than 10 km. This is considerably greater than the thickness of sediment on old oceanic crust. The subsidence rates in these *syn-rift basins*, as recorded in the sedimentary record, decrease with time since rifting. The crust beneath these basins is extensively cut by normal faults, which have been clearly imaged by reflection profiling. These faults are often observed to be *listric faults*, faults for which the angle of dip decreases with depth.

To explain these continental margin sedimentary basins, we must consider the effects of stretching the lithosphere and so allowing hotter asthenospheric material to rise. The first effect of such stretching is isostatic: Lithospheric material is replaced by asthenospheric material, and depending on the thicknesses and densities, subsidence or uplift ensures that isostatic equilibrium is reattained. The second effect is thermal: The stretched lithosphere and asthenosphere are not in thermal equilibrium. Cooling, and hence slow contraction and further subsidence, takes place until a new thermal equilibrium is eventually reached.

Stretching models can be used to explain continental margin sedimentary deposits formed during continental rifting and some extensional basins that formed when continental lithosphere was stretched but did not split. Figure 9.36 outlines the situations which occur when the lithosphere is uniformly stretched by a factor of β. The initial subaqueous subsidence S_i can be determined by assuming the *before* and *after* columns to be in isostatic equilibrium (see Sect. 5.5.2). In this case, we can write

$$h_c\rho_c + (h_l - h_c)\rho_l = \frac{h_c}{\beta}\rho_c + \frac{(h_l - h_c)}{\beta}\rho_l + \left(h_l - \frac{h_l}{\beta} - S_i\right)\rho_a + S_i\rho_w \quad (9.8)$$

where ρ_c is the average density of the crust, ρ_l the average density of the subcrustal lithosphere, ρ_a the density of the asthenosphere, ρ_w the density of water, h_c the thickness of the crust, h_l the thickness of the lithosphere and β the stretching factor. Rearranging Eq. 9.8 gives an expression for the initial subsidence:

$$S_i = \left[\frac{h_l(\rho_a - \rho_l) + h_c(\rho_l - \rho_c)}{\rho_a - \rho_w}\right]\left(1 - \frac{1}{\beta}\right) \quad (9.9)$$

If the temperature gradient in the lithosphere is assumed to be linear and the temperature of the asthenosphere T_a is assumed to be constant, the densities ρ_c, ρ_l and ρ_a are given by

$$\rho_c = \rho_{co}\left(1 - \frac{\alpha T_a h_c}{2h_l}\right)$$

$$\rho_l = \rho_{mo}\left(1 - \frac{\alpha T_a}{2} - \frac{\alpha T_a h_c}{2h_l}\right)$$

and

$$\rho_a = \rho_{mo}(1 - \alpha T_a)$$

where α is the coefficient of thermal expansion and ρ_{co} and ρ_{mo} are the densities of crust and mantle, respectively, at 0°C.

After this initial subsidence, the lithosphere gradually cools (Fig. 9.36), which results in a period of slower thermal subsidence. This can be modelled mathematically by using the following conditions:

$$\left.\begin{aligned} T &= \frac{T_a \beta z}{h_l} \quad &&\text{for} \quad 0 \leqslant z \leqslant \frac{h_l}{\beta} \\ T &= T_a \quad &&\text{for} \quad \frac{h_l}{\beta} \leqslant z \leqslant h_l \end{aligned}\right\} \quad \text{at } t = 0 \qquad (9.10)$$

for the one-dimensional heat-flow equation 7.13,

$$\frac{\partial T}{\partial t} = \frac{k}{\rho c_p} \frac{\partial^2 T}{\partial z^2}$$

where z is measured downwards from the surface. Problems such as this

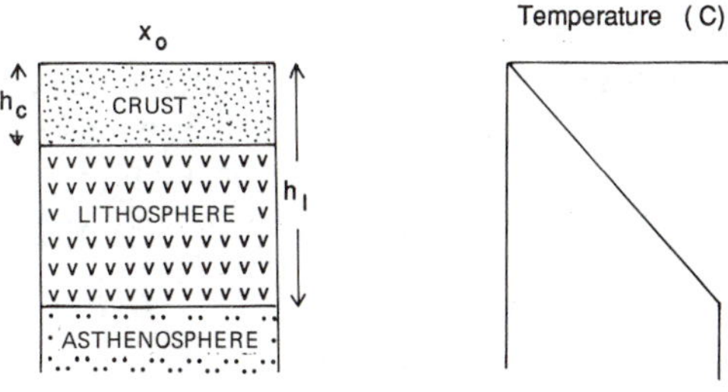

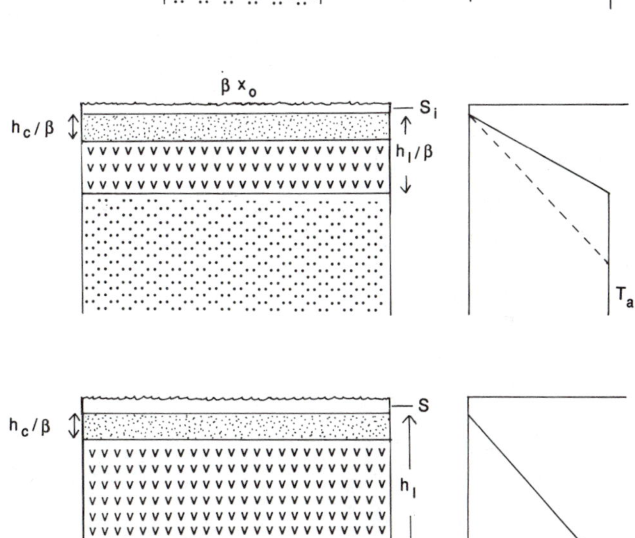

Figure 9.36. At time $t = 0$ the continental lithosphere has initial length x_0 and thickness h_l (top). It is instantaneously stretched by a factor β (middle). By isostatic compensation, hot asthenospheric material rises to replace the thinned lithosphere. The temperature of the lithosphere is assumed to be unaffected by this stretching, but the stretched lithosphere is out of thermal equilibrium (dashed line). The stretched lithosphere slowly cools and thickens until it finally reattains its original thickness (bottom). An initial subaqueous subsidence S_i occurs as a result of the replacement of light crust by denser mantle material. Further subsidence occurs as the lithosphere slowly cools. The final subsidence is S. (After McKenzie 1978.)

are best solved by Fourier expansion; readers are referred to Carslaw and Jaeger 1959 for details of these methods. To a first approximation, the thermal subsidence is an exponential with a time constant equal to the time constant of the oceanic lithosphere (e.g., Eqs. 9.5 and 9.6).

The total amount of subaqueous subsidence occurring after an infinite time S can be most simply expressed by assuming Airy-type isostasy:

$$S = \frac{h_l(\rho_l' - \rho_l) + h_c\left(\rho_l - \frac{\rho_l'}{\beta} + \frac{\rho_c'}{\beta} - \rho_c\right)}{\rho_l' - \rho_w} \tag{9.11}$$

where the average densities of the lithosphere and crust at infinite time are

$$\rho_l' = \rho_{mo}\left(1 - \frac{\alpha T_a}{2} - \frac{\alpha T_a h_c}{2\beta h_l}\right)$$

and

$$\rho_c' = \rho_{co}\left(1 - \frac{\alpha T_a h_c}{2\beta h_l}\right)$$

The total amount of thermal subsidence S_t is then the difference between the total subsidence S (Eq. 9.11) and the initial subsidence S_i (Eq. 9.9):

$$S_t = S - S_i \tag{9.12}$$

Obviously, models such as this, though complicated, are grossly oversimplified approximations to the real earth, neglecting lateral conduction of heat and the time over which extension takes place. However, it can be shown that, provided stretching takes place in less than 20 Ma, an instantaneous stretching model such as this is adequate.

Figure 9.37a shows the linear relationships between initial and thermal

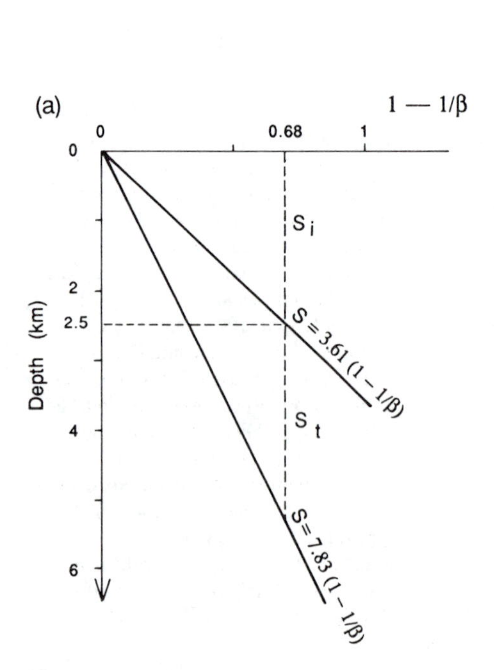

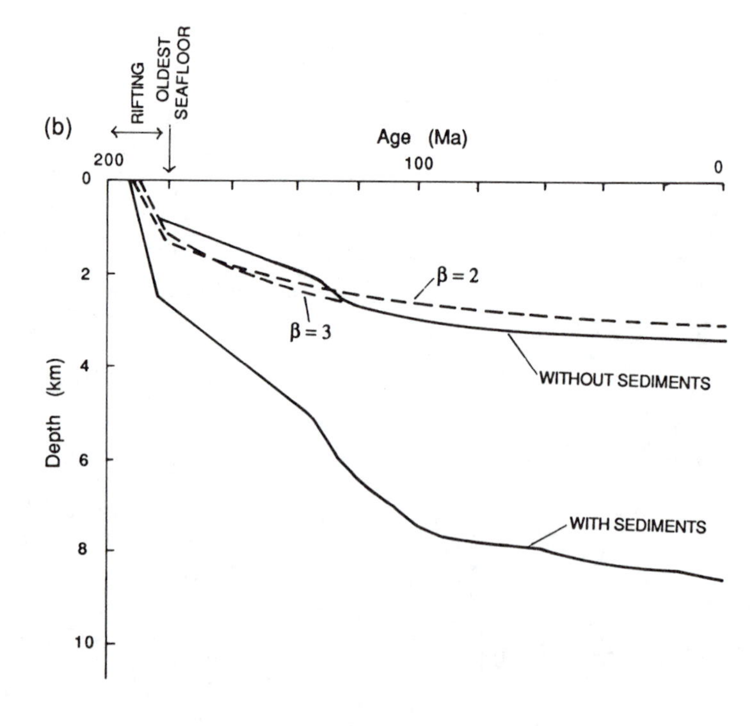

Figure 9.37. (a) Plot of relative thinning $(1-1/\beta)$ against subsidence for the simple stretching model of the lithosphere. Upper line represents the initial (immediately poststretching) subsidence S_i as given by Eq. 9.9. Lower line represents the total subsidence, $S = S_i + S_t$ at infinite time as given by Eq. 9.11, where S_t is the thermal subsidence. Values of parameters are h_l, 125 km; h_c, 30 km; ρ_0, 3350 kg m^{-3}; ρ_c, 2780 kg m^{-3}; ρ_w, 1030 kg m^{-3}; α, 3.28×10^{-5}°C^{-1}; T_a, 1333°C. (After Le Pichon and Sibuet 1981.) (b) Observed and computed subsidence curves for a well on the Nova Scotia continental margin off eastern Canada. Rifting period indicates the time during which extension was taking place and thus shows the initial subsidence S_i. Subsidence occurring since that time is due to thermal reequilibration. (After Keen and Cordsen 1981.)

subsidence and relative thinning $(1 - 1/\beta)$ for particular values of the parameters appropriate for the Bay of Biscay. The usual depth of the midocean ridges is 2.5 km below sea level; thus, this is the isostatic equilibrium level of the asthenosphere or, in other words, the surface of the asthenosphere's geoid. If crust is submerged to depths of more than 2.5 km, extensive volcanism and rupture of the lithosphere occur. Thus, the boundary between oceanic and continental crust should occur at the locus of points offshore at a depth of 2.5 km. For the Bay of Biscay, an initial subsidence of 2.5 km is attained for $(1 - 1/\beta) = 0.68$, for $\beta = 3.24$; this means that asthenospheric material could break through to the surface for β greater than 3.24. However, since the time of breakthrough of asthenospheric material (when the oceanic crust was first formed), additional thermal subsidence will have occurred. It is necessary to know the approximate age of the continental margin to calculate this thermal subsidence. Once this is known, the present depth at which the boundary between stretched continental crust and oceanic crust occurs can be calculated. This is useful because to make accurate continental reconstructions (see Chapter 3), it is desirable to calculate the original (prestretching) width of the continental margin instead of choosing some arbitrary present-day bathymetric contour.

Faulting is the surface expression of the initial subsidence caused by extension. Evidence from multichannel seismic reflection profiles and from fault-plane solutions for earthquakes indicates that although the normal faults outcropping at the surface have steep dips, the dips decrease with depth. Such faults are called listric, or rotational, faults. Estimates of β made from supposed normal faulting of the basement surface are, therefore, likely to be lower than the actual value. Measurement of the crustal thickness by seismic refraction experiments provides another method of estimating β, provided an adjacent unextended crustal thickness is known.

Figure 9.37b shows observed and computed subsidence curves for a well on the continental margin off eastern Canada. Figure 9.38 shows a cross section of the passive continental margin of the eastern United States.

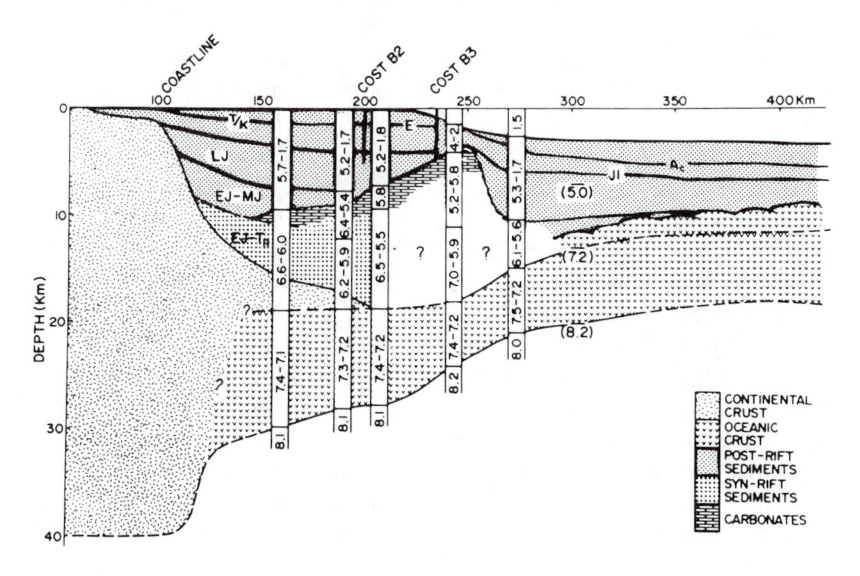

Figure 9.38. Cross section of the passive continental margin off the east coast (New Jersey) of the United States. The numbers are the seismic P-wave velocities in kilometres per second as determined from expanding spread (wide-angle) profiles. Vertical solid lines labelled COST give the locations of two deep commercial wells. The deep crustal layer having P-wave velocity 7.1–7.5 km s^{-1} is believed to be igneous material intruded at the time of rifting; thus, it is very similar to the oceanic layer 3 into which it merges. (From LASE Study Group 1986.)

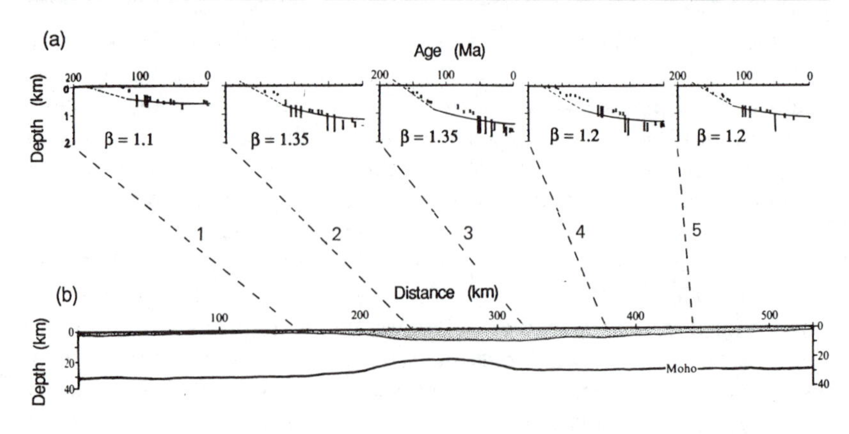

Figure 9.39. (a) Subsidence of the North Sea Basin when filled with water only (corrections have been made for sediment compaction, loading and palaeowater depth) for five wells that straddle the Central Graben of the basin. Bars denote range of basement level; dashed line, calculated fault controlled subsidence for a model with steady extension occurring over 60 Ma; solid line, calculated subsequent thermal subsidence. The model had an initial crustal thickness of 31 km for wells 1, 4 and 5, 25 km for well 2 and 28 km for well 3. (b) Crustal thickness beneath the North Sea as determined by seismic refraction experiments. (From Wood and Barton 1983.)

The transition between stretched and unstretched crust, termed the *hinge zone*, is narrow, being only 50 km across. Figure 9.39 shows observed and computed subsidence curves for five wells situated across the Central Graben of the North Sea (on a line between Scotland and Norway) and the variation in crustal thickness across the graben. To explain this subsidence satisfactorily, the original crustal thickness must be thinner beneath the Central Graben than on either side. This suggests that there may have been an earlier stretching event in the graben.

More complex modifications of this continental extension model involve depth-dependent extension – that is, more extension in the lower, more ductile part of the lithosphere than in the upper crustal part (Fig. 9.40b) – dyke intrusion or melt segregation and lateral variation of stretching (Fig. 9.40c). These complex models have been developed to explain why some continental margins and rift systems apparently show no initial subsidence but some uplift or doming, and to explain why thermal contraction is insufficient to account for the maximum depth of the ocean basins. However, the simple one-dimensional model described here is a reasonable approximation to the formation of many continental margins and basins.

Some rifted continental margins are characterized by a short period of intense volcanism. For example, it has been estimated that more than 2×10^6 km^3 of basalt were extruded onto the rifting margins of the northern North Atlantic over a period of about 2 Ma. Perhaps twice as much igneous rock was added to the lower crust. The extruded volcanic rocks show up on reflection profiles as a set of seaward-dipping reflectors. A thick lower crust which has an abnormally high P-wave velocity (7.1–7.6 km s^{-1}) is interpreted as the igneous rocks which were added to the lower crust. Estimates of the stretching factor β obtained from changes in crustal thickness are therefore too low. In contrast, further south, in the Bay of Biscay, there is no evidence for such an intense volcanic event. The volcanism which took place along the northern North Atlantic margins is believed to have been caused by elevated temperatures in the asthenosphere. If the temperature in the asthenosphere were 100–150°C higher than normal, rapid rifting would generate large quantities of melt (McKenzie and Bickle 1988). Proximity to the hotspot currently located

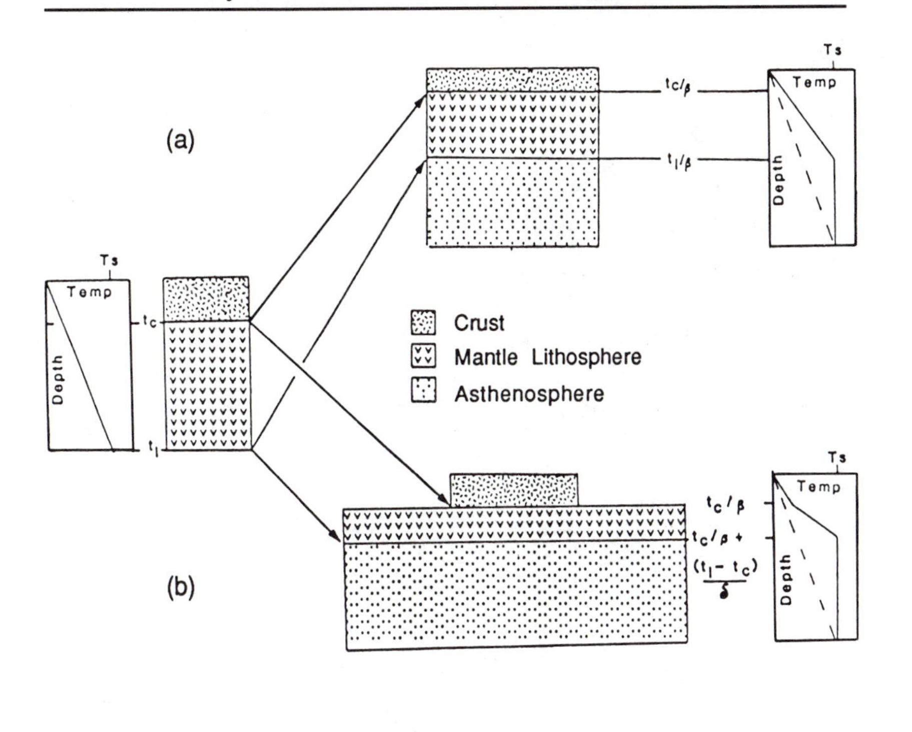

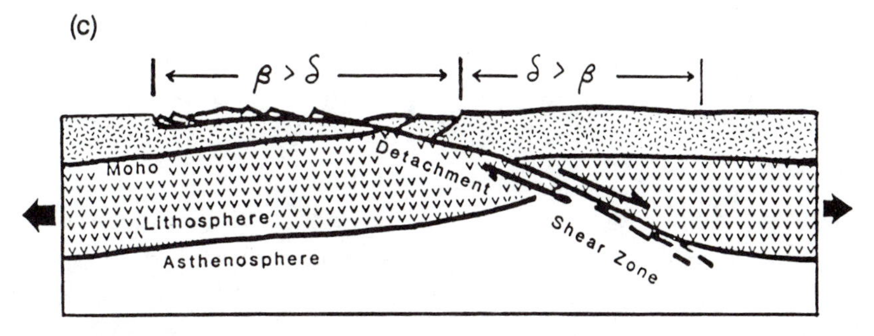

Figure 9.40. Various models of lithospheric extension: (a) Pure shear extension, as in Figure 9.36. t_c and t_l are the h_c and h_l of this text. They appear rather commonly in the literature, but have nothing to do with time. (b) Depth-dependent pure shear extension. The crust is extended by factor β, the mantle by a factor δ ($\delta > \beta$). Solid line denotes temperature profiles immediately after extension; dashed line, the final equilibrium temperature profile. (c) Simple shear extension. A detachment surface, or fault, extends right through the crust and mantle. For this model, the extension factors β and δ vary continually across the structure, in contrast to pure shear extension when they are constant across the structure. (After Quinlan 1988 and Wernicke 1985.)

beneath Iceland was probably the reason for the elevated temperatures beneath the northern North Atlantic at the time of rifting. This also explains the origin of continental flood basalts.

9.3.7 Compressional Basins

A striking feature of the gravity field of central Australia is the 600 km sequence of east–west anomalies with a north–south wavelength of about 200 km. These Bouguer anomalies range from −150 to +20 mgal (Fig. 9.41). This part of central Australia comprises a series of east–west trending intracontinental basins and arches, with the gravity highs corresponding to the arches and the lows to the sedimentary basins. The crust beneath this region apparently has an average thickness of about 40 km, but the Moho is depressed by up to 10 km beneath the basins and is similarly elevated beneath the arches. However, there is no indication of faulting at the bases of the basins. There has been no plate-boundary activity in the region since the late Proterozoic. Subsidence of the basins started

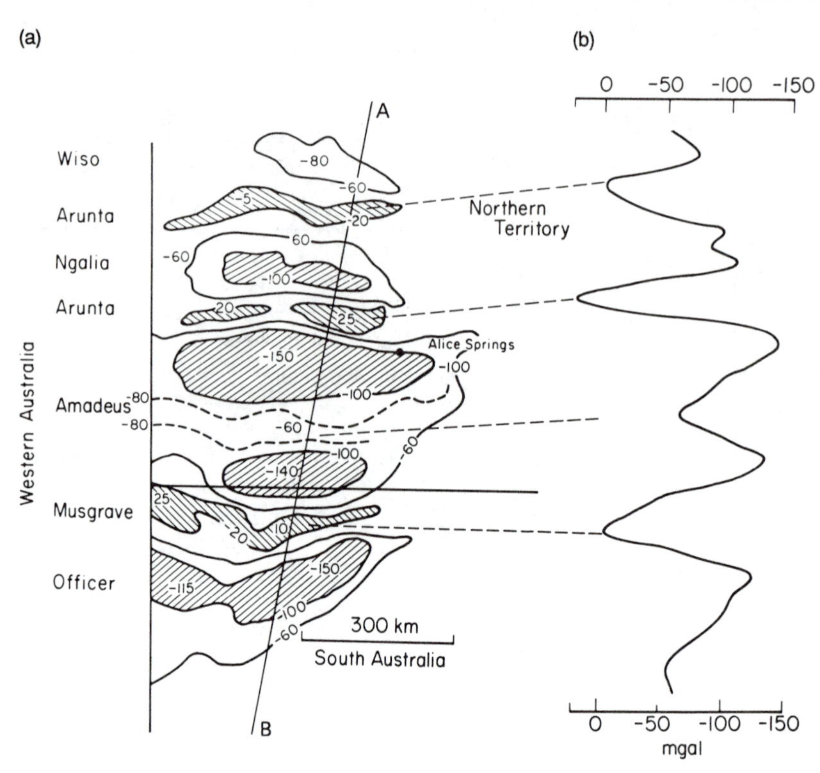

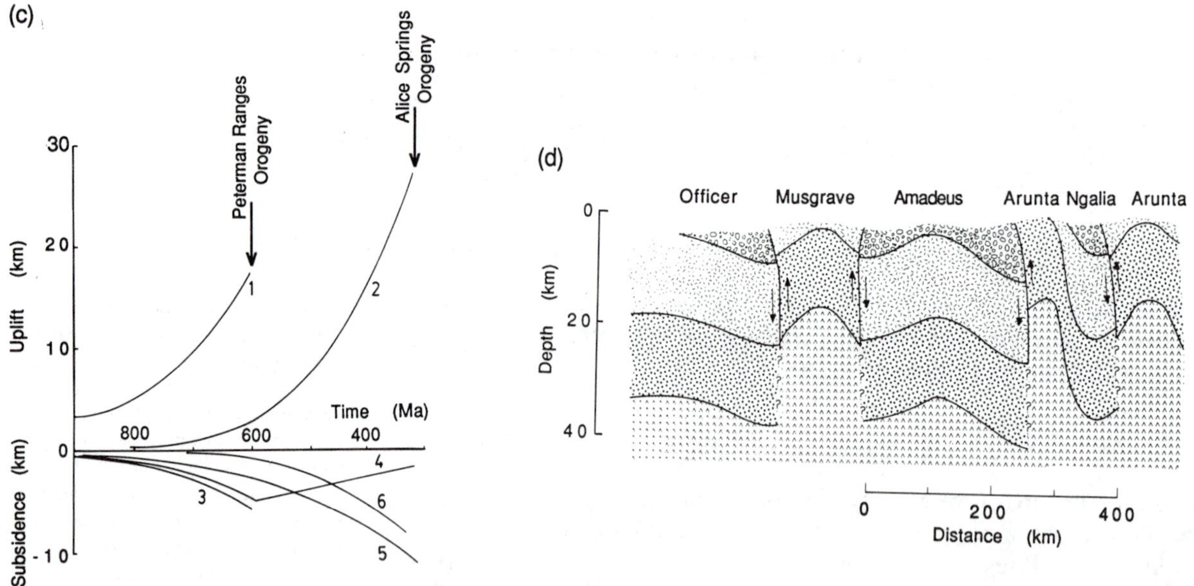

Figure 9.41. (a) Map of the regional Bouguer gravity field in central Australia. \\\\ anomalies less negative than −20 mgal; //// anomalies more negative than −100 mgal. (b) Bouguer gravity anomaly along line AB. (c) Predicted uplift of (1) the central part of the Musgrave block and (2) the southern Arunta block. Subsidence of (3) the southern Amadeus Basin, (4) central Amadeus Basin, (5) northern Amadeus Basin and (6) the central Ngalia Basin. Note that except for the central Amadeus Basin, all subsidence and uplift rates have increased with time. (d) Predicted cross section of the crust along line AB after the Alice Springs orogeny (about 320 Ma). The original crust was assumed to be 30 km thick and divided into 15 km thick upper and lower crustal layers. (After Lambeck 1983.)

about 1000 Ma ago and continued for some 700 Ma, and the subsidence rate has increased with time (Fig. 9.41c). These facts, taken together, indicate that neither an extensional nor a thermal model is appropriate for these basins and arches. It has been proposed that they formed instead as a result of compression of the lithosphere.

Let us initially consider a simple problem: an elastic plate, of flexural rigidity D, subjected only to a constant horizontal force H per unit width.

The deformation of this plate w satisfies Eq. 5.56:

$$D\frac{d^4w}{dx^4}+H\frac{d^2w}{dx^2}=0 \tag{9.13}$$

The solution to this equation is obtained by integrating twice, giving

$$D\frac{d^2w}{dx^2}+Hw=c_1x+c_2 \tag{9.14}$$

where c_1 and c_2 are constants of integration.

If we assume the plate to be of a finite length l with $d^2w/dx^2=0$ and $w=0$ at $x=0$ and $x=l$ (i.e., the plate is fixed at 0 and l), then both c_1 and c_2 must be zero. The solution to Eq. 9.14 is then sinusoidal:

$$w=c_3\sin\left(\sqrt{\frac{H}{D}}\,x\right)+c_4\cos\left(\sqrt{\frac{H}{D}}\,x\right) \tag{9.15}$$

where c_3 and c_4 are constants. Because the plate is fixed at $x=0$, c_4 must equal zero. The condition that w must also equal zero at $x=l$ is then only possible when c_3 is zero, in which case there is no deformation at all, or when

$$\sqrt{\frac{H}{D}}\,l=n\pi \qquad \text{for } n=1,2,\ldots \tag{9.16}$$

The smallest value of H for which deformation occurs is therefore given by $n=1$. This critical value of the horizontal force is π^2D/l^2. For horizontal forces less than this value, there is no deformation. At this critical value, the plate deforms into a sine curve given by

$$w=c_3\sin\left(\frac{\pi x}{l}\right) \tag{9.17}$$

However, this simple calculation is not directly applicable to the lithospheric plates (or to layers of rock) because the lithosphere is hydrostatically supported by the underlying mantle. A hydrostatic restoring force (see Eqs. 5.58 and 5.59) must be included in Eq. 9.13 to apply it to the lithosphere. In this case, Eq. 9.13 becomes

$$D\frac{d^4w}{dx^4}+H\frac{d^2w}{dx^2}=-(\rho_m-\rho_w)gw \tag{9.18}$$

The sine function $w=w_0\sin(2\pi x/\lambda)$ is a solution to this equation for values of λ given by

$$D\left(\frac{2\pi}{\lambda}\right)^4+H\left(\frac{2\pi}{\lambda}\right)^2=-(\rho_m-\rho_w)g \tag{9.19}$$

(To check this, differentiate the expression for w and substitute into Eq. 9.18.) Since Eq. 9.19 is a quadratic equation in $(2\pi/\lambda)^2$, the solution is

$$\left(\frac{2\pi}{\lambda}\right)^2=\frac{-H\pm\sqrt{H^2-4D(\rho_m-\rho_w)g}}{2D} \tag{9.20}$$

Because $(2\pi/\lambda)^2$ must be real, and not imaginary, the term under the square-root sign must be positive:

$$H^2 \geqslant 4D(\rho_m - \rho_w)g \tag{9.21}$$

The smallest value of H for which there is a real solution is given by

$$H = \sqrt{4D(\rho_m - \rho_w)g} \tag{9.22}$$

For values of the horizontal force less than this value, there is no deformation; but at this critical value, the plate deforms into a sine curve. The wavelength of the deformation for this critical force is then obtained by substituting the critical value for H from Eq. 9.22 into Eq. 9.20, which gives

$$\left(\frac{2\pi}{\lambda}\right)^2 = \frac{\sqrt{4D(\rho_m - \rho_w)g}}{2D} \tag{9.23}$$

Upon reorganization Eq. 9.23 yields

$$\lambda = 2\pi\left[\frac{D}{(\rho_m - \rho_w)g}\right]^{1/4} \tag{9.24}$$

For an elastic plate with flexural rigidity of 10^{25} N m, a value which may be appropriate for the lithosphere, the critical compressive force as given by Eq. 9.22 is therefore 10^{15} N m^{-1}, which corresponds to a critical horizontal compressive stress of 10^{10} N m^{-2} (10 GPa). Even a flexural rigidity of 10^{24} N m corresponds to a critical compressive stress of more than 6 GPa. Such values of the compressive stress are much greater than reasonable failure limits of the lithosphere. Buckling would not occur in reality: Failure would take place first by the formation of faults.

However, if the lithosphere is modelled as a viscoelastic plate which is subjected to some irregular normal load, it can be shown that under constant compression the initial deflections due to this load are magnified and increase with time. Such deformations occur for compressive forces which are an order of magnitude less than the critical buckling forces. With time, failure of the crust by thrust faulting presumably also occurs. Figure 9.41d shows the cross section of the crust predicted by such a model. All the main geological and geophysical features of the region are correctly predicted.

Oil as a metamorphic product

The main reason for the great commercial interest in sedimentary basins is, of course, the deposits of oil, gas and coal which they may contain. Organic deposits must undergo metamorphism – elevated temperatures and pressures for considerable times – before they become *hydrocarbons.* The subsidence history of a sedimentary basin combined with oil maturation history will show where in the basin the oil is likely to be found. In view of the immense economic importance of hydrocarbons, a short discussion of their formation is included here.

Organic remains deposited in a sedimentary basin are gradually heated and compacted as the basin subsides. These organic deposits are called *kerogens* (Greek *Keri*, "wax" or "oil"). There are three types: *Inert kerogen*, which is contained in all organic material, transforms into graphite; *labile kerogen*, which is derived from algae and bacteria, transforms into oil though a small proportion transforms directly into gas; *refractory kerogen* is derived from plants and transforms into gas. At elevated temperatures, oil also transforms to gas by a process called *oil to gas cracking*. These complex organic chemical reactions are not well understood, but we know that time and temperature are controlling factors. Of course, none of the reactions could take place if the organic material were not buried in sediment and protected from oxidation. In the Guaymas Basin in the middle of the Gulf of California, the planktonic carbon-rich silts have been heated to such a degree by the hydrothermal systems (Sect. 8.4.4) that the kerogens have been transformed into hydrocarbons. The sediments there smell like diesel fuel.

The rates at which these chemical reactions proceed are described mathematically by

$$\frac{dC}{dt} = -kC \tag{9.25}$$

where C is the concentration of the reactant (i.e., the kerogen) and k is the reaction rate coefficient. The Arrhenius equation (see also Eq. 6.24) defines the temperature–dependence of k as

$$k = Ae^{(-E/RT)} \tag{9.26}$$

where A is a constant (sometimes called the Arrhenius constant), E the activation energy, R the gas constant and T the temperature.

Data on the laboratory and geological transformation of kerogens to oil and gas are shown in Figure 9.42, which demonstrates the time-dependence of Eq. 9.25, showing the difference between heating labile and refractory kerogens at geological (natural) rates and heating at a fast rate in the laboratory. The calculations were performed using $A = 1.58 \times 10^{13}\,\mathrm{s}^{-1}$ and $E = 208\,\mathrm{kJ\,mol}^{-1}$ for labile kerogens and $A = 1.83 \times 10^{18}\,\mathrm{s}^{-1}$ and $E = 279\,\mathrm{kJ\,mol}^{-1}$ for refractory kerogens. For the kerogen-to-hydrocarbon reactions to take place in a reasonable time, very much higher temperatures have to be attained in the laboratory than are necessary in the earth. (As an illustration, contemplate cooking a turkey in an oven at 50°C,. 100°C, 150°C, 200°C or 250°C.) The temperature differences between the range of plausible geological heating rates are fairly small. Figure 9.43a shows an estimate of the effect of temperature on the time taken for oil to be transformed into gas. The *oil half-life* is the time necessary for half the oil to transform to gas. At a temperature of 160°C, the predicted half-life is less than 10 Ma, whereas at 200°C the half-life is less than 0.1 Ma. These times are short on the geological scale.

In summary, mathematical predictions based on geological and laboratory data indicate that temperatures of 100–150°C are necessary for labile kerogens to transform to oil, temperatures of 150–190°C are necessary for the cracking of oil to gas, and temperatures of 150–220°C are necessary for refractory kerogens to transform to gas.

A standard empirical relationship between temperature and time and the *hydrocarbon maturity* is called the *time–temperature index* (TTI). This

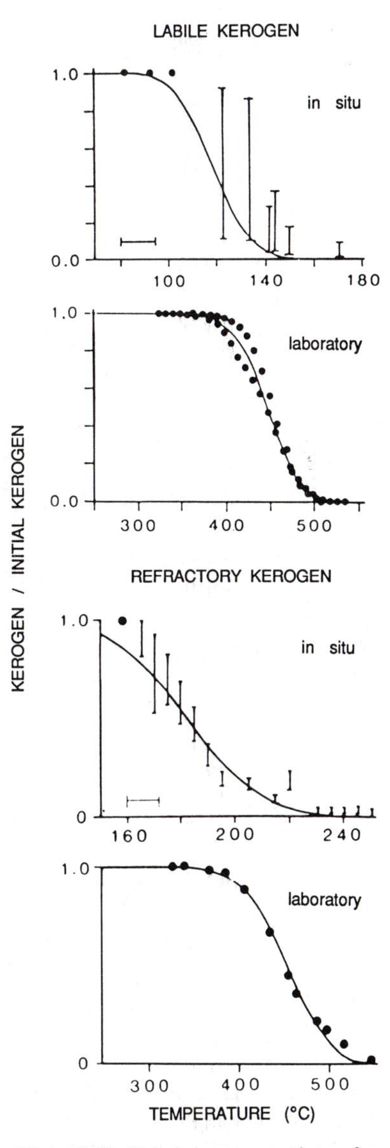

Figure 9.42. Relative concentrations of kerogen for labile kerogen (upper pair) and refractory kerogen (lower pair) as a function of maximum temperature attained: measured (dots and bars) and calculated (solid line). Upper graphs are geological measurements and represent the actual thermal history of the samples; estimates of average heating of these in situ kerogens are $1°\mathrm{C\,Ma}^{-1}$ for the labile kerogen and $6°\mathrm{C\,Ma}^{-1}$ for the refractory kerogen. Lower graphs are for laboratory heating measurements carried out at $25°\mathrm{C\,min}^{-1}$. The fit between measured and calculated relative concentrations is good. (From Quigley and McKenzie 1988.)

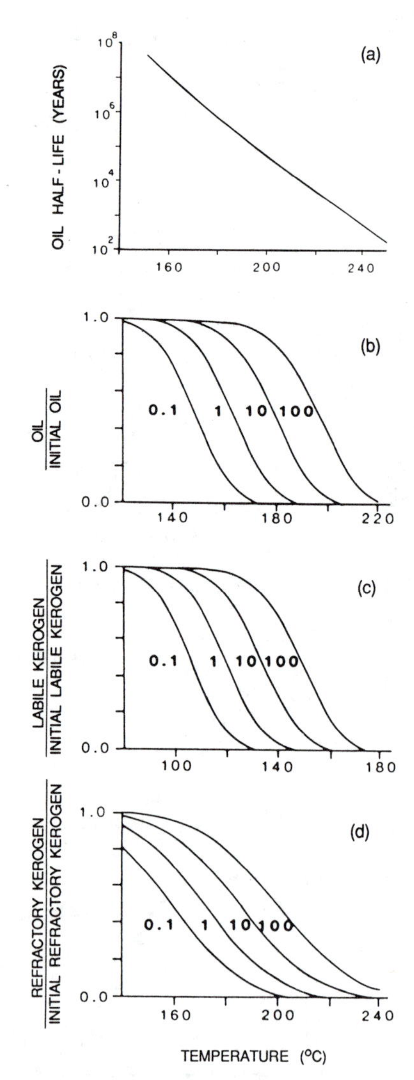

Figure 9.43. Calculated time and temperature dependence of kerogen reactions. (a) Time taken to convert half a given mass of oil to gas at a given temperature. (b) Relative concentration of oil as function of maximum temperature for heating rates of 0.1, 1, 10 and 100°C Ma^{-1}. (c) As in (b) but for labile kerogen. (d) As in (b) but for refractory kerogen. (From Quigley and McKenzie 1988.)

relationship states that the reaction rate doubles for each 10°C temperature rise. The total maturity of a hydrocarbon, or its TTI, is defined as

$$\mathrm{TTI} = \sum_{j=n_{\min}}^{n_{\max}} \Delta t_j 2^j \tag{9.27}$$

where Δt_j is the time in millions of years that it takes for the temperature of the material to increase from $100 + 10j$ to $100 + 10(j + 1)$°C, and $n_{\min}$ and $n_{\max}$ are the values of j for the lowest and highest temperatures to which the organic material was exposed. This empirical approach is generally appropriate for chemical reactions on laboratory time scales but has been extended to geological time scales. However, the data and predictions of transformation rates summarized here suggest that this widely used TTI approach may overestimate the importance of time and underestimate the importance of temperature on hydrocarbon maturation.

9.4 Continental Rift Zones

9.4.1 Introduction

A number of continental rift zones which are active today have not yet, and perhaps may never, become active midocean ridges (refer to Sect. 9.3.6). However, some features are common to all continental rift zones:

1. A rift or graben structure with a rift valley flanked by normal faults
2. Negative Bouguer gravity anomalies
3. Higher than normal surface heat flow
4. Shallow, tensional seismicity
5. Thinning of the crust beneath the rift valley

These features are in agreement with those expected for the early stages of extensional rifting.

Two of the best known rift zones are the East African Rift and the Rio Grande Rift, though there are others such as the Rhine Graben in Europe and the Baikal Rift in Asia. The Keweenawan Rift is a North American example of an ancient extinct continental rift.

9.4.2 The East African Rift

This long rift system stretches some 6000 km from the Gulf of Aden in the north towards Zimbabwe in the south (Fig. 9.44). In the Gulf of Aden it joins, at a triple junction, the Sheba Ridge and the Red Sea. Along this rift system, uplifting and splitting of the African continent may be in progress.

The crustal and upper mantle structure of the rift system have been studied using seismic and gravity data (Fig. 9.45). The region is in approximate isostatic equilibrium. The long-wavelength Bouguer gravity anomaly and the earthquake data can be explained by anomalous low-velocity, low-density material in the upper mantle. This anomalous zone is generally assumed to be partially molten peridotite. Earthquake and seismic refraction data show that the crust on either side of the rift valley itself has a simple, typical shield structure: two crustal layers with velocities of 5.9 and 6.5 km s^{-1} and a total thickness of about 43 km overlying a normal 8.0 km s^{-1} upper mantle. A positive Bouguer gravity anomaly lies immediately over the eastern rift in Kenya, which is interpreted as a zone of denser, molten upper mantle material penetrating the crust. Such a model is also consistent with teleseismic delay times for earthquakes.

The largest body-wave magnitudes for earthquakes occurring along the rift system are about 6.0–6.2. Source mechanisms for some earthquakes are shown in Figure 9.46. Earthquakes along the rift system are normal faulting events, with approximately east–west tension. The focal depths range between 7 and 33 km, with most events being shallower than 20 km. Thus, the majority of earthquakes are located in the upper crust with only a few occurring in the lower crust.

The East African Rift system is an active volcanic region as well as a seismic region. The first volcanism probably started about 25 Ma ago and, though not continuing at a constant rate, is still taking place. The lavas are rather alkaline, which is normal for continental volcanism in relatively undisturbed lithosphere. One volcano, on occasion, produces almost pure washing soda (Oldoinyo Lengai, in Tanzania). The geological estimate of

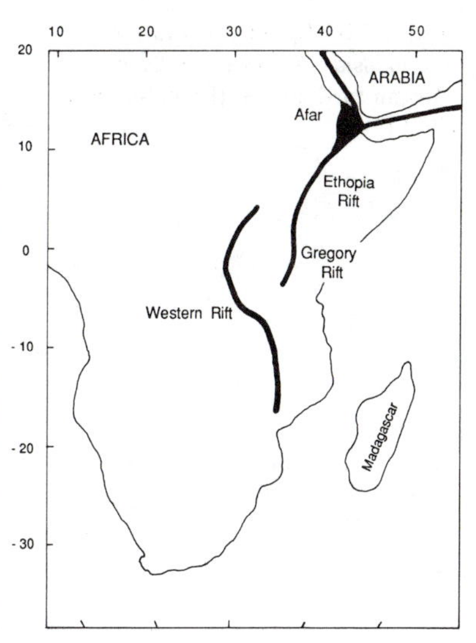

Figure 9.44. East African Rift system.

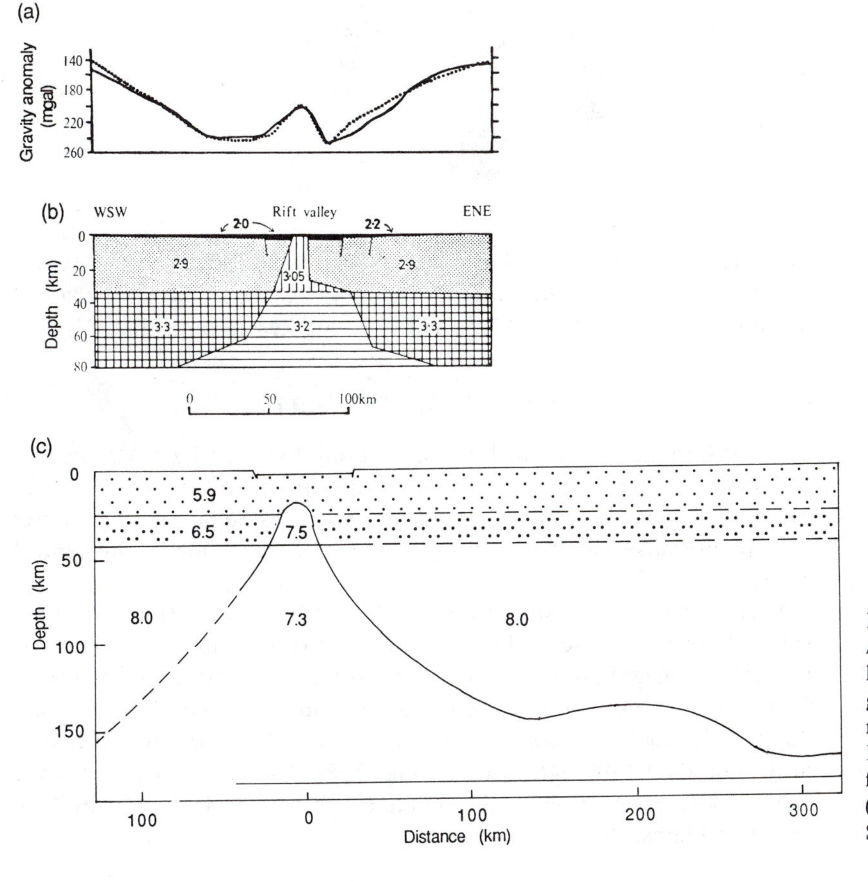

Figure 9.45. Cross sections over the East African Rift in Kenya. (a) Observed (solid line) and model (dotted line) Bouguer gravity anomalies. (b) Density model for model anomalies shown in (a). (c) Seismic P-wave velocity structure as determined from earthquake and refraction data. (After Baker and Wohlenberg 1971, and Savage and Long 1985.)

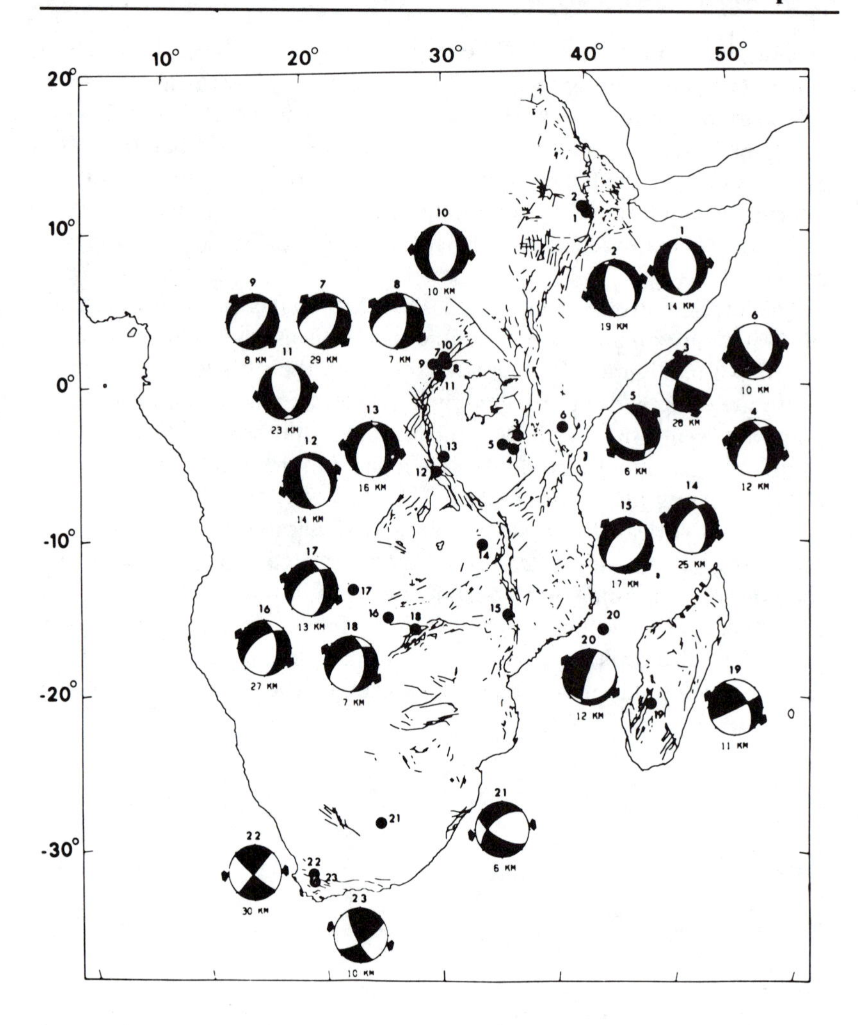

Figure 9.46. Epicentres, focal mechanisms and focal depths for east African earthquakes. (From Shudofsky 1985.)

the maximum extension which has taken place is 30 km. This gives a maximum spreading half-rate of 0.1 cm yr^{-1}, which is an order of magnitude less than values for the neighbouring Red Sea and Sheba Ridge.

9.4.3 The Rio Grande Rift

The Rio Grande Rift, a much smaller feature than the East African Rift system, has been studied in detail. Volcanism in this rift began 27–32 Ma ago in a Precambrian Shield. Visually, the two rift systems are very similar with platformlike rift blocks rising in steps on each side of the central graben.

Seismic refraction data indicate that the crust in the central part of the rift is about 30 km thick, which is some 10–15 km thinner than the crust beneath the neighbouring Great Plains to the east. The crustal structure for the rift is simple, with two layers, but the upper mantle velocity is only 7.7 km s^{-1}. Such a structure is also indicated by the surface-wave dispersion data from earthquakes (Fig. 9.47). The upper mantle S-wave value is also lower beneath the rift than beneath the Colorado Plateau or Great Plains.

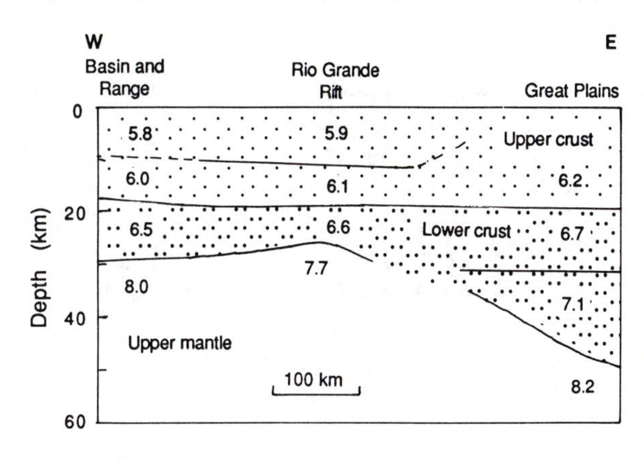

Figure 9.47. Cross section of crust and upper mantle structure across Rio Grande Rift. Velocities are based on refraction and surface wave results. (From Sinno et al. 1986.)

The discontinuity between the upper and lower crustal layers gives rise to a strong reflection on the seismic refraction data. The amplitude of this reflection has been modelled with synthetic seismogram programs. If the first few kilometres of the lower crustal layer have a low S-wave velocity then the amplitudes of the synthetic seismograms are in agreement with the data.

Figure 9.48 shows the gravity anomaly along a profile crossing the rift. A small positive Bouguer gravity anomaly is superimposed on a wide, low (-200 mgal) anomaly. The interpretation of this gravity data is ambiguous, but like the East African rift, the broad, low anomaly appears to be caused by thinning of the lithosphere beneath the rift, and the small positive anomaly beneath the rift is caused by shallow, dense intrusions at the rift itself.

Heat flow along the rift is high, about $120–130 \times 10^{-3}\,\mathrm{W\,m^{-2}}$ with

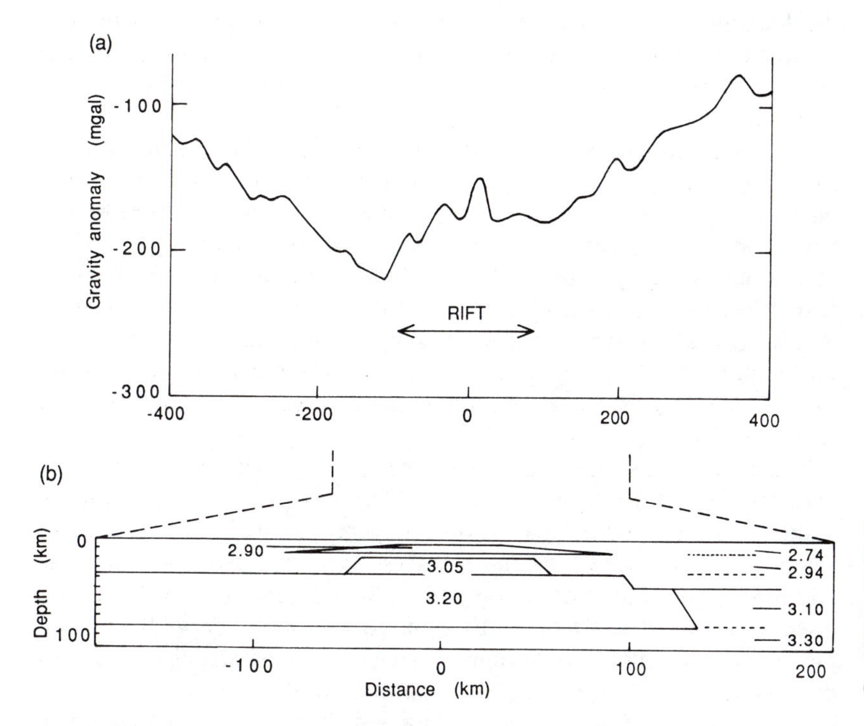

Figure 9.48. Rio Grande Rift. (a) Bouguer gravity anomaly on a profile at approximately 33°N. (After Cordell 1978.) (b) Interpretation of the Bouguer gravity anomaly after corrections have been made for the shallow structure. (After Ramberg 1978.)

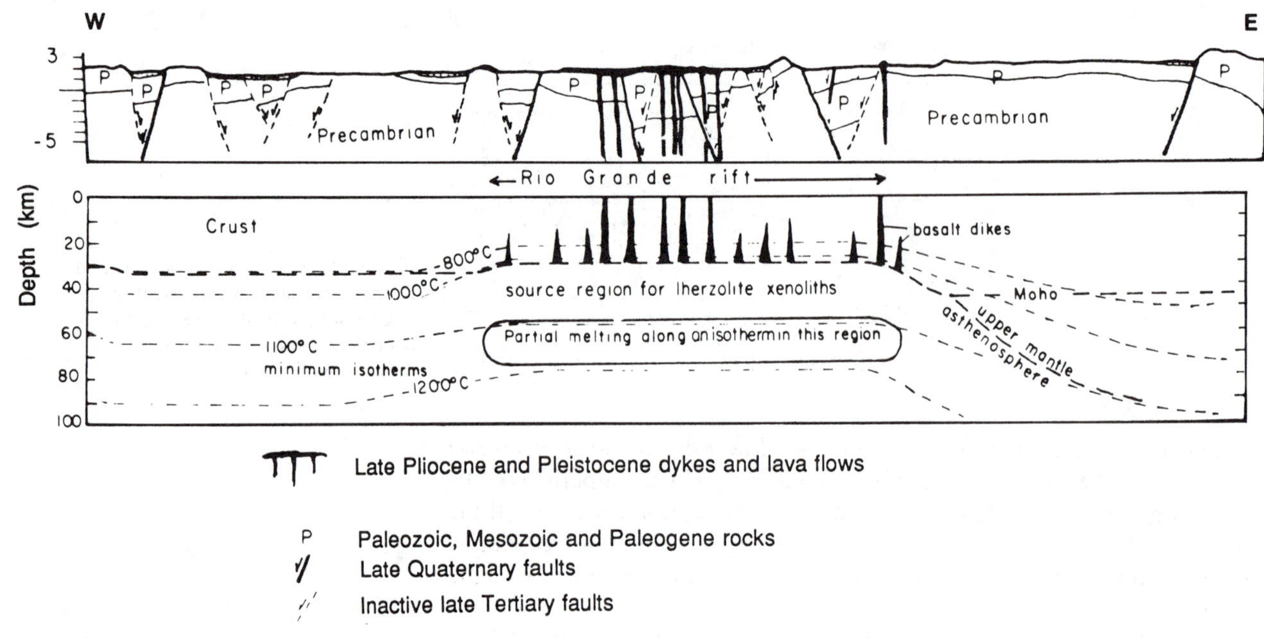

Figure 9.49. Cross sections of the (a) geological and (b) thermal structure beneath the Rio Grande Rift. (From Seager and Morgan 1979.)

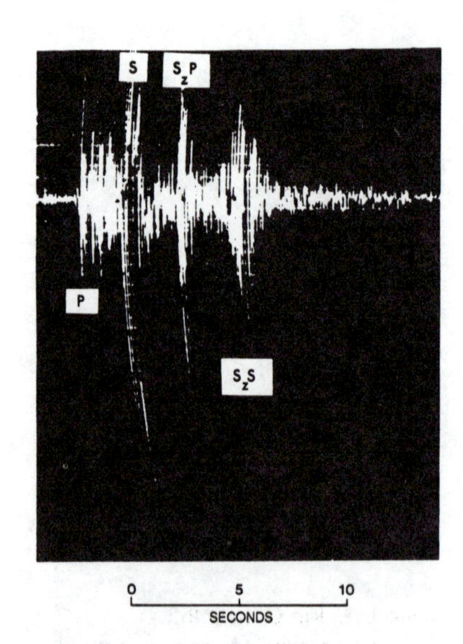

Figure 9.50. Microearthquake seismogram from the Rio Grande Rift. P and S denote first P and S wave phases; S_zS and S_zP are S-to-S and S-to-P reflections from the upper surface of an extensive 20 km deep magma body. (From Rinehart et al. 1979.)

local values up to $400 \times 10^{-3}\,\mathrm{W\,m^{-2}}$. Studies of xenoliths (fragments of rock brought up from depth) erupted from volcanoes indicate that their source is at a temperature greater than 1000 °C. Figure 9.49 illustrates the temperature field beneath the rift. The uppermost mantle is also the location of significant electrical conductivity anomalies with high conductivity (low resistivity) beneath the rift. This is yet another piece of evidence that the uppermost mantle beneath the rift is a region of very high temperatures.

Since detailed instrumental studies began in 1962, the seismicity in the rift has not been high by western American standards; on average there have been only two earthquakes per year with magnitude greater than 2. Several areas of concentrated microseismic activity exist in the rift and are associated with magma bodies in the middle and upper crust. An aseismic region has particularly high heat-flow values (up to $200 \times 10^{-3}\,\mathrm{W\,m^{-2}}$), indicating that the lack of seismic activity is due to high temperatures in the crust. Fault-plane solutions for the microearthquakes in the rift show that the focal mechanisms are predominantly normal faulting with some strike-slip faulting.

The microearthquake data have also been used to delineate the top of the midcrustal magma body beneath the rift. The seismograms for earthquakes in the region of Socorro, New Mexico, show pronounced secondary energy arriving after the direct P- and S-waves (Fig. 9.50). These arrivals, S_zP and S_zS, are an S-to-P reflection and an S-to-S reflection from a seismic discontinuity at 20 km depth. Ratios among the amplitudes of the various phases have been used to determine the nature of the material immediately beneath the reflecting horizon. A solid–liquid interface can account for the observations. However, this magma body cannot be more than about

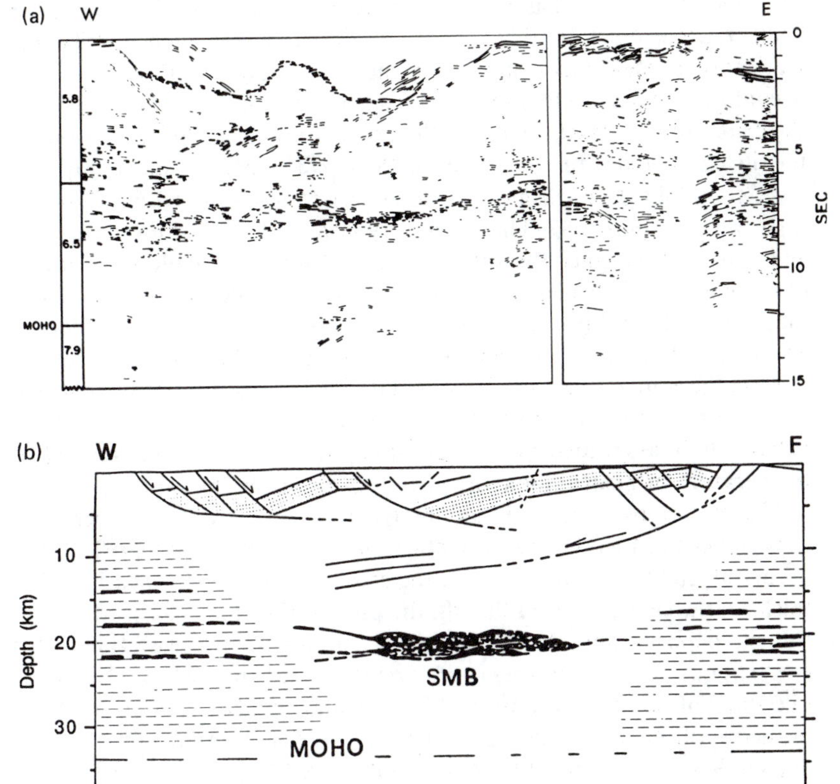

Figure 9.51. (a) Line drawing of the COCORP deep seismic reflection sections which cross the Rio Grande Rift near Socorro, New Mexico, at 34°N. Velocities from one of the refraction models are shown at the left-hand side. (From Brown et al. 1979.) (b) Interpretation of the COCORP deep reflection lines across the Rio Grande Rift. Dotted blocks represent prerift sedimentary strata. Note the deep fault which penetrates to midcrustal depths. Horizontal shading represents the horizontal compositional/deformational fabric of the lower part of the crust. SMB is the Socorro magma body. True scale. (From de Voogt et al. 1988.)

1 km thick (if completely molten) because it does not cause observable delays in P-wave teleseismic or refraction arrivals. The body lies beneath $1700\,km^2$ of the central part of the rift.

A series of 24-fold, deep seismic reflection lines were shot by COCORP across the Rio Grande Rift near Socorro in the region where this magma sill is located (Figs. 4.56 and 9.51). The shallow reflections show normal faults, some having an offset of more than 4 km. The extensional origin of the rift shows very clearly. A clear, rather complicated P-wave reflector at 7–8 s two-way time corresponds to the magma body. The time at which the reflections from the Moho arrive is consistent with the predictions from the surface wave and refraction data.

Ideally, it is possible to determine the presence of a solid–liquid interface by studying the polarity of its reflections. Unfortunately, it was not possible to determine unequivocally the polarity of reflections recorded on these COCORP lines, but they are consistent with a thin layer of magma in solid material. The complexity of the reflections, however, indicates that the reflector is not a simple continuous sill but may be layered in some way and/or discontinuous.

9.4.4 The Keweenawan Rift System

The Keweenawan or Mid-continent rift system is a 100 km wide, 2000 km long, extinct ($\sim$1100 Ma old) rift system extending from Kansas to Michigan in the United States (see Fig. 9.34). Beneath Lake Superior, the rift bends by 120°; it has been suggested that this is the location of an ancient triple junction. The rift is delineated by the high gravity and magnetic anomalies associated with the thick sequence of basaltic lavas it contains. Seismic refraction data from the rift indicate that the basalt deposits are very thick and that the crust is about 50 km thick beneath the rift, compared with a more typical 35–45 km for neighbouring regions.

A 24–30-fold, deep seismic reflection line across this rift in Lake Superior is shown in Figure 9.52. The rift is very clear indeed. On both northern and southern margins, the rift is bounded by normal faults. The major basin reflectors are believed to be lavas with some interlayered sediments; they extend downwards to almost 10 s two-way time (about 30 km depth). These reflections may originate either from the sediment–lava contacts or from the contacts between lavas of differing composition. Similar strong reflections have been observed on Atlantic continental margins and are believed to be associated with basaltic lavas which were erupted, at elevated temperature or near a hot spot, when the continent split apart.

The reflections labelled Ba are interpreted as the prerift basement and those labelled M as the crust–mantle boundary. These mantle reflections, occurring at 13–15 s, indicate that the crust in this region is nearly 50 km thick. In the central part of the rift, the present thickness of crust between the rift deposits and the Moho is only about 4 s two-way time. This corresponds to a thickness of about 12–14 km, which is about one-third of the normal crustal thickness and therefore about one-third of the prerifting crustal thickness. Thus, if the simple stretching model in Section 9.3.6 is assumed, the crust was extended by a factor $\beta = 3$ during the rifting. Such a value implies that complete separation of the crust may

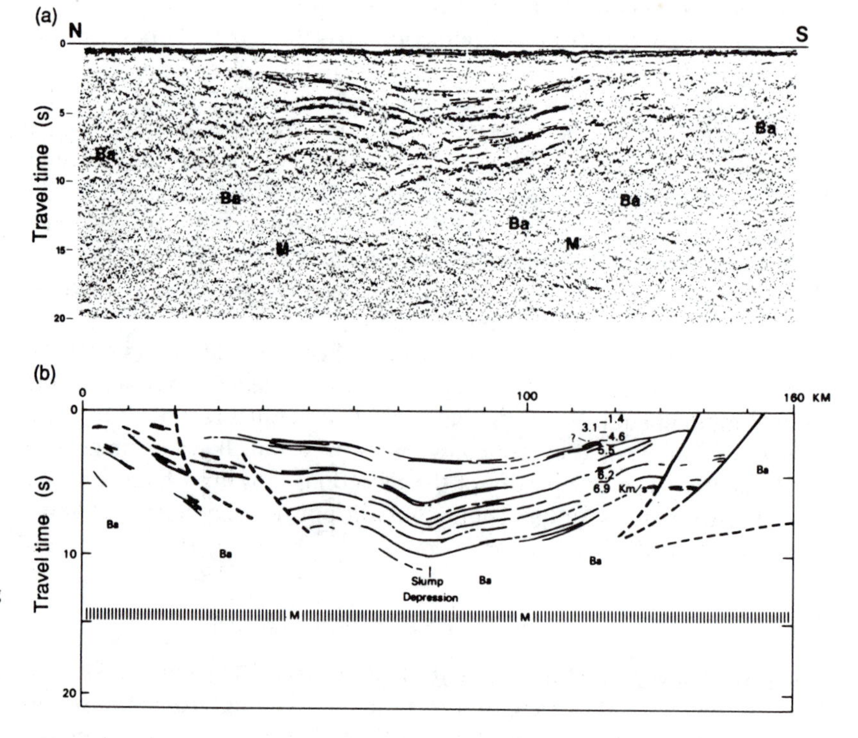

Figure 9.52. (a) Unmigrated seismic reflection record section from Lake Superior, crossing the Keweenawan Rift. Approximately true scale for 6 km s^{-1} material. Ba denotes reflection from prerift basement; M, reflection from crust–mantle boundary. (b) Line drawing of the migrated version of the reflection record section shown in (a). Vertical lines indicate the crust–mantle boundary. (From Behrendt et al. 1988.)

have occurred. The assumption that the present M reflector was the ancient as well as the present Moho is, of course, open to debate. The lowermost crust beneath the rift could easily be intrusive material, and the M reflection could be a new postrifting Moho. In that case, the value of β would be considerably greater than 3; so we can take 3 as a minimum value.

Whatever the final interpretation of the details of the Keweenawan structure, it is a major, old intercontinental rift filled with an incredibly thick sequence of lavas and sediments.

9.5 The Archaean

The Archaean, roughly speaking, is that time from the end of the earth's accretion until 2.5 Ga ago, comprising about 40% of the earth's lifetime. One of the problems facing geologists and geophysicists who are studying the Archaean is that the *uniformitarian* assumption, loosely stated as 'the present holds the key to the past', may be only partly correct. *Aktualism*, 'the present is the same as the past', is certainly not true. The earth may have behaved very differently in the beginning, with a different tectonic style, so our interpretations of structures, rocks and chemistry may be ambiguous.

When did the continents form? As was mentioned in Section 6.10, the oldest rocks are about 3.8–3.9 Ga old. The oldest known terrestrial material is some zircon crystals from Western Australia which have been dated at about 4.3 Ga old. These crystals are held in a younger (but still early Archaean) metasedimentary rock. A handful of zircon does not make a continent, but this material and the 3.8 Ga rocks suggest that some sort of continent was in existence at that time. The cratons which form the cores of the present continents are, for the most part, rafts of Archaean granitoids and gneisses, formed in a complex assortment of events from 3.5–2.7 Ga. Infolded into the granitoid gneiss cratons are belts of supracrustal lavas and sediment, including komatiitic lavas. These are highly magnesian lavas, perhaps originally up to 30% MgO. Experimental melting has shown that such lavas must have erupted at higher temperatures than modern basalts. Indeed, the temperatures are so high that some eruptions may have been blue hot. Young (less than 100 Ma old) komatiite does occur, with MgO around 20%, but it is very rare. To produce such hot lavas in abundance, the mantle may have been hotter in the Archaean than today. It is possible that some of the lavas arrived at the surface at temperatures as high as 1600–1650°C, implying temperatures of 1800–1900°C or more at their source.

Various questions can be asked about the Archaean earth: What was the continental crust like? Could plate tectonics have operated in the Archaean? Was there oceanic crust, and if so what was it like?

9.5.1 Archaean Continental Crust

Two tectonic accidents have resulted in exposures of Archaean crust. The Vredefort Dome in South Africa is a structure some 50 km in diameter in which a section of the Archaean crust aged 3.0–3.8 Ga is exposed. The

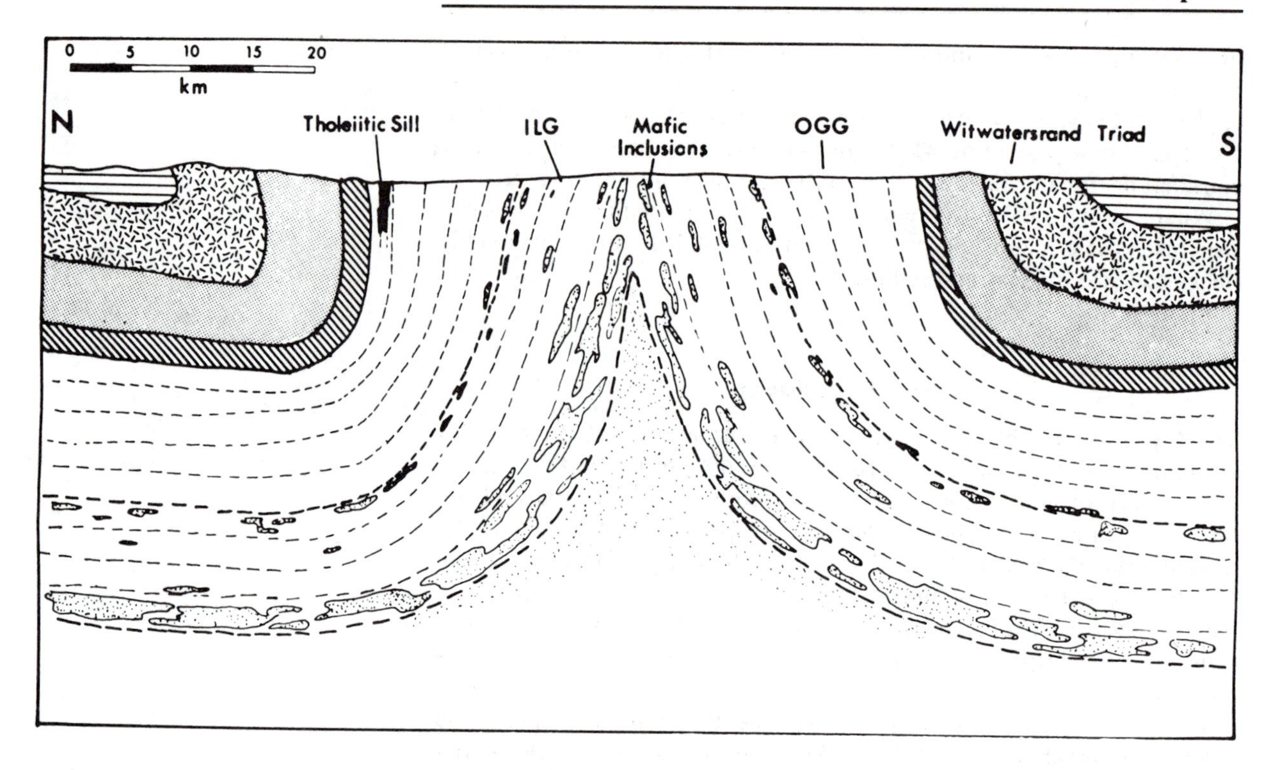

Figure 9.53. Geological cross section across the Vredefort dome structure in South Africa. OGG, outer granite gneiss; ILG, Inlandsee Leeucogranofels felsic rocks. (From Nicolaysen et al. 1981.)

Dome is thought to have formed at about 2.0 Ga as the result of deformation from within the earth, perhaps an explosive intrusion, though some authors have suggested it may be a meteorite impact structure. A cross section (Fig. 9.53) through the Dome shows that the sedimentary layers were underlain by a granite-gneiss upper crust and granulites in the middle crust. Approximately the upper 20 km of the crust are exposed here.

Another exposure of Archaean crust is in the Superior geological province of Canada. Figure 9.54 shows a generalized west–east cross section through the Kapuskasing zone. It appears that in this case a major thrust resulted in the uplifting of the deep crustal rocks. The upper

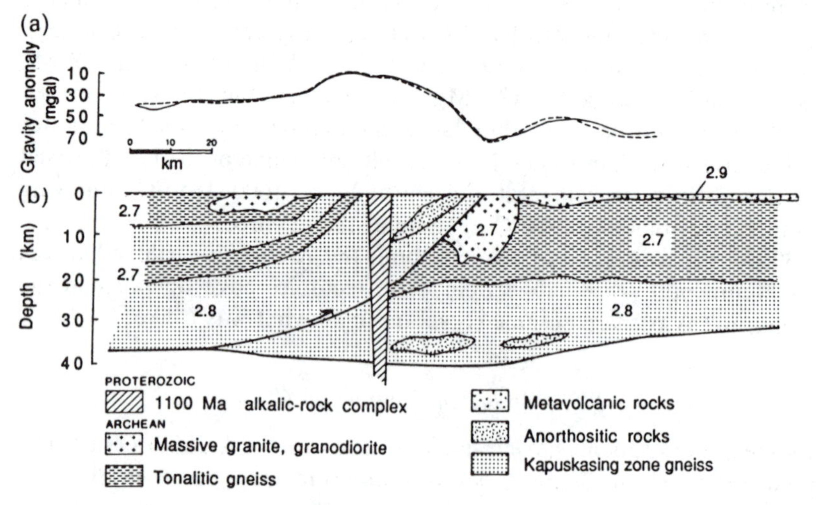

Figure 9.54. (a) Bouguer gravity across the Kapuskasing zone: observed (solid line) and calculated (dashed line).
(b) Crustal model based on geology and gravity (densities in $10^3\ kg\,m^{-3}$). (From Percival et al. 1983.)

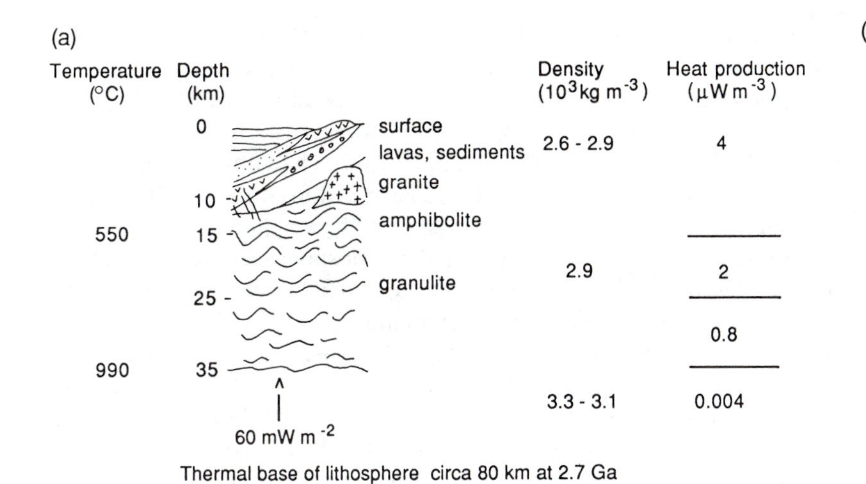

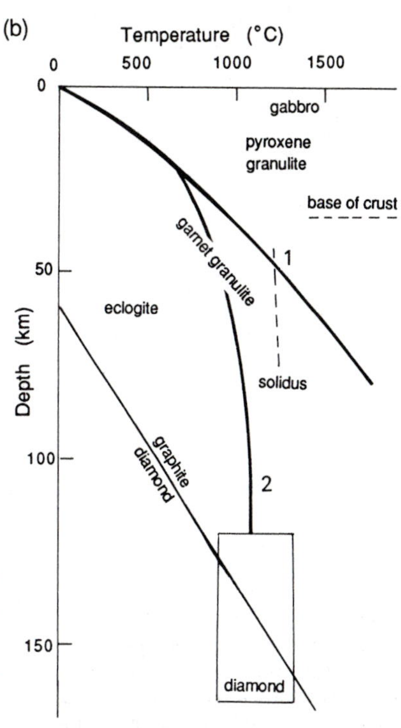

Figure 9.55. (a) Model of 2.8 Ga Archaean continental crust. (b) Model equilibrium Archaean continental geotherms based on (a). 1. Heat flow into the base of the crust is 63×10^{-3} W m^{-2}, and conductivity is 3.3 W m^{-1} °C^{-1}; this geotherm has $T = 550$°C at 15 km, as determined from the metamorphic assemblages. 2. Possible geotherm in an old cold continent as implied by the existence of Archaean diamonds. Box shows pressure–temperature field inferred from minerals included in diamonds. Dashed line, position of crustal solidus. (After Nisbet 1984 and 1987.)

crust is granitoid and the lower crust gneiss. The total thickness of exposed crust is about 25 km. This interpretation has been supported by deep seismic reflection studies.

The total thickness of the continental crust towards the end of the Archaean was probably about 35 km or more, similar to today's value. Problem 8.3 offers an insight into the consequences of this conclusion. The radioactive heat generation in the crust can be estimated by extrapolating backwards in time from the modern content of radiogenic elements. From these estimates, together with a knowledge of the metamorphic facies (and hence temperature and pressure) attained by Archaean rocks, Archaean equilibrium geotherms can be calculated. Some such models are shown in Figure 9.55. Geotherm 1 implies the thermal base of the lithosphere (1600–1700°C) was at about 80 km. However, in North America there is evidence that Archaean diamonds existed and they have even been mined from Archaean conglomerates in South Africa. The stability of diamonds is a major constraint on the thermal structure beneath the continents because they crystallize at about 150 km and 1150°C. These comparatively low temperatures indicate that the assumptions in the calculation of geotherm 1 may not have been valid everywhere. Some regions may have been relatively hot, others relatively old and cool (geotherm 2). These temperatures can be interpreted as differences between young (hot) and old (cold) continents. The hot, newly formed continents may have had a lithosphere 80 km or less in thickness, while coexisting colder continental regions may have had a 150–200 km thick lithosphere.

9.5.2 Archaean Tectonics and Ocean Crust

Thermal models are important in attempts to model Archaean tectonics. For the modern earth (see Sect. 7.4), about 65% of the heat loss results from the creation and destruction of plates and about another 17% is from radioactive heat produced in the crust. The heat flow from the mantle into the crust is about $29 \times 10^{-3}\,\mathrm{W\,m^{-2}}$. Most of the heat that is lost comes from the mantle as the earth cools. The Archaean earth had much higher radioactive heat generation rates than the modern earth. At 3 Ga, the internal heat production was 2.5–3.0 times its present value (see Table 7.2). It has been shown that if plate tectonics had not been operating in the Archaean, and if all this heat had been lost from the asthenosphere and had flowed through the lithosphere by conduction, then the equilibrium heat flow at the base of the lithosphere would be roughly $140 \times 10^{-3}\,\mathrm{W\,m^{-2}}$. If this value is used to calculate temperatures in the Archaean continents, geothermal gradients of about $50°\mathrm{C\,km^{-1}}$ are obtained. The temperature at the base of the crust would have been high enough to melt it. Indeed, if this model is correct, at 3.5 Ga the heat flow into the base of the lithosphere would have been about $190 \times 10^{-3}\,\mathrm{W\,m^{-2}}$, with a temperature of 800°C at 10 km depth. These results are contrary to the metamorphic record of deep crustal rocks preserved from the Archaean (Fig. 9.55b). The continental crust is clearly self-stabilizing: Heat production is moved to the surface by geochemical processes such as partial melting. The problems remain however: what to do with the heat, and how was it dissipated?

Massive volcanism on a large scale – in other words, spreading centres or midocean ridges – could provide a solution. However, the heat problem is not neatly resolved. To dissipate such large amounts of heat, spreading rates need to be very high, which in turn means that, assuming the earth did not expand, the destruction or subduction rates must also have been very high. The dilemma is that young, hot oceanic lithosphere does not subduct easily. What would drive the system? Would it not heat until a different tectonic pattern was attained? However, if (as seems probable) the Archaean midocean ridges did not create a basaltic crust but instead created a komatiitic crust (Fig. 9.56a), then subduction might have taken

Figure 9.56. (a) Komatiitic Archaean oceanic crust, 15 km thick, assuming an asthenospheric temperature of 1700°C. Upper crust would be komatiitic basalt with dykes and pillow lavas, and lower crust would be cumulates. (b) Cooling model of the Archaean oceanic lithosphere. Density of Archaean mantle at 1700°C was assumed to be $3.15 \times 10^3\,\mathrm{kg\,m^{-3}}$. The mechanical base of the plate was arbitrarily chosen as 900°C for refractory mantle. (After Nisbet and Fowler 1983.)

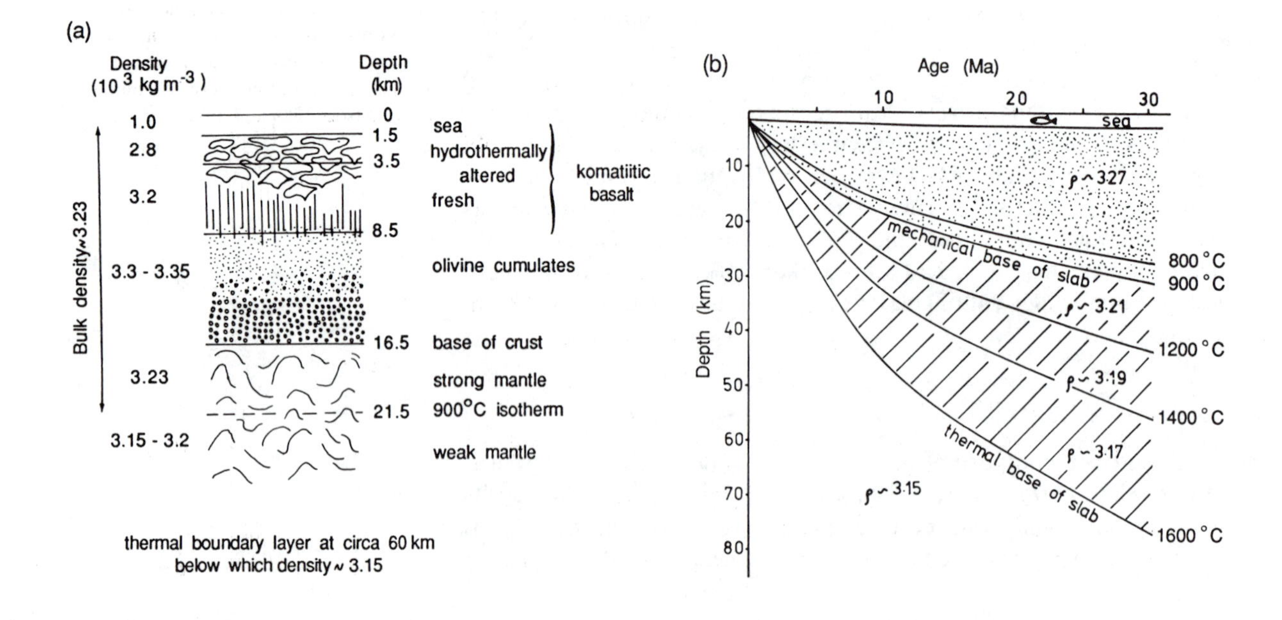

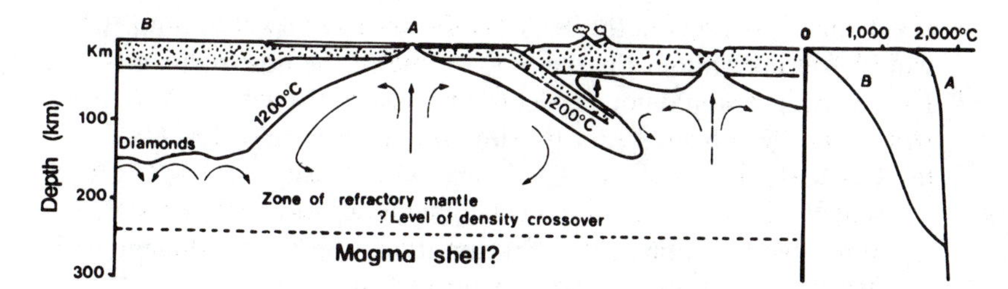

Figure 9.57. Diagram of some speculations about the Archaean upper mantle. Geotherm A at the right-hand side is for midocean ridges; geotherm B is for old cool continents. (From Nisbet 1985.)

place because a komatiitic crust would be denser than a basaltic crust. Such a crust would have been considerably thicker and denser (approximately 15 km and 3.23×10^3 kg m^{-3}) than modern oceanic crust. However, like the modern oceanic crust, the Archaean oceanic crust would probably have had a layered structure with lavas overlying dykes overlying cumulates. A thermal model for such Archaean oceanic lithosphere, based on the assumption that lithosphere is cooled mantle, is shown in Figure 9.56b. A schematic of Archaean plate tectonics is shown in Figure 9.57.

One unresolved controversy about the structure of the Archaean mantle is particularly interesting because it is in strong contrast to today's mantle. At depths greater than 250 km, it is possible that olivine was less dense than melt. This has produced the fascinating speculation that olivine may have floated above a buried magma ocean of melt (on the modern earth, melts, less dense everywhere than crystal residue, rise). Such a gravitationally stable deep magma 'ocean' has been nicknamed the LLLAMA (large laterally linked Archaean magma anomaly). If such a density difference existed in the Archaean, a magma shell could have surrounded the earth, as in Figure 9.57. It would have been overlain by a layer of olivine. All of this is the subject of debate, but it illustrates how very different the internal structure of the Archaean earth may have been. Of course, there is an analogous structure in the modern earth: the liquid outer core.

Finally, in this discussion of the diversity of continents and their history, it should be noted that the other planets have different tectonics. Even Venus, which is so similar to the earth, seems to have evolved in quite a different way. But that subject is planetology, and each planet deserves a book for itself, matching geophysics with aphroditophysics, aresophysics and even plutophysics, puzzles for the next generation of geophysicists.

PROBLEMS

1. Calculate the elastic thickness of the subducting Indian Plate beneath (a) the Lesser Himalayas and (b) the Greater Himalayas.
2. Fault-plane solutions similar to those shown in Figure 9.58 were obtained for earthquakes on the North Anatolian Fault in Turkey. Can you give a simple explanation for them?
3. Calculate how long it would take for three-fourths of a given mass of oil to convert to gas at the following temperatures: (a) 160°C (b) 180°C (c) 220°C.

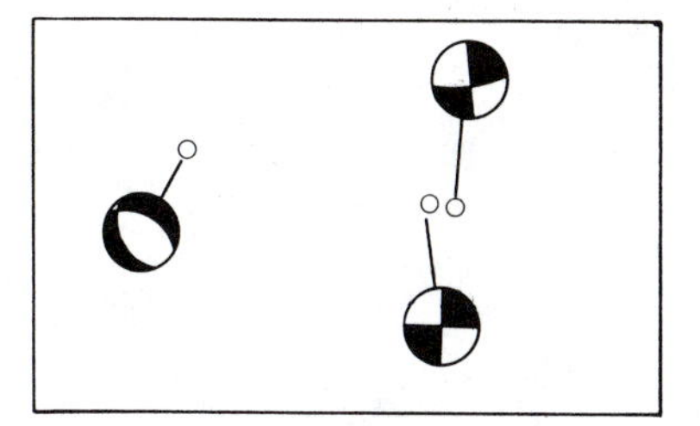

Figure 9.58. Earthquake focal mechanisms from the North Anatolian Fault in Turkey. (After Jackson and McKenzie 1984.)

4. If oil that had apparently been heated to 160°C were found, what would you infer about the tectonic setting of the host sediments?
5. How much gas would you expect to find associated with an oil deposit which had been heated to a maximum temperature of 150°C?
6. (a) Calculate the thickness of sediment with density $2.1 \times 10^3\,\mathrm{kg\,m^{-3}}$ that would be deposited in a subaqueous depression 0.5 km deep.
   (b) What would happen if a Precambrian ironstone sediment with density $4 \times 10^3\,\mathrm{kg\,m^{-3}}$ were deposited in the basin?
7. Assume that the asthenosphere behaves as a viscoelastic material. What is its viscous relaxation time? (Use 70 GPa for Young's modulus.) Does this value seem reasonable to you?
8. An elastic plate 1000 km long, with a flexural rigidity of $10^{25}\,\mathrm{N\,m}$, is fixed at each end.
   (a) Calculate the critical value of the horizontal compressive force for this plate.
   (b) Calculate the critical stress associated with the compressive force (stress = force per unit area).
   (c) Comment on the magnitude of your answers. What do they indicate about the behaviour of the lithospheric plates?
9. Calculate the critical value of the compressive stress for a 0.5 km thick rock layer which is isostatically supported by the underlying lithosphere. What is the wavelength of the initial deformation?
10. Calculate the effective elastic thicknesses for the three model lithospheres shown in Figure 9.30a. Which one would you intuitively expect to be appropriate for the North American Plate?
11. Derive an equation for the initial subsidence of an instantaneously stretched lithosphere, when no water fills the surface depression.
12. Assume that the earthquake shown in Figure 9.50 took place in the upper crust ($\alpha = 6\,\mathrm{km\,s^{-1}}, \beta = 3.5\,\mathrm{km\,s^{-1}}$).
    (a) Use the P and S arrival times to calculate the distance from the focus to the seismometer.
    (b) What estimates can you make about the height of the focus above the reflecting horizon and the depth of that horizon beneath the surface?
13. What can deep seismic profiling reveal about the structure of the continental lithosphere? (Cambridge University Natural Sciences Tripos II, 1986.)
14. A 35 km thick continental crust is heated, resulting in an instantaneous temperature increase of 500°C.
    (a) Calculate the resulting elevation of the surface.
    (b) Calculate the thickness of sediments which could finally be deposited if 500 m of crust was eroded while the surface was elevated. (Let ρ_s, ρ_c and ρ_m be 2.3, 2.8 and $3.3 \times 10^3\,\mathrm{kg\,m^{-3}}$, respectively; α, $3 \times 10^{-5}\,^\circ\mathrm{C}^{-1}$.)
15. Assume that continental lithosphere, thickness 125 km, undergoes instantaneous extension. What is the minimum value of the stretching factor β necessary for asthenospheric material to break through to the surface? For this value of β, what is the total amount of subsidence that would eventually occur? (Let crustal thickness be 35 km; water, crustal and mantle densities, 1.03, 2.8 and $3.35 \times 10^3\,\mathrm{kg\,m^{-3}}$,

respectively; coefficient of thermal expansion, 3×10^{-5} °C^{-1}; asthenosphere temperature, 1350°C.)

16. Calculate the amount of initial subsidence that would result from instantaneous extension of the continental lithosphere by factors of 2 and 10. Make reasonable assumptions for the thickness, density and temperature of the crust and mantle.
17. What evidence has been used to confirm the importance of a stretching and cooling mechanism in the formation of some sedimentary basins? (Cambridge University Natural Sciences Tripos IB, 1983.)
18. (a) Using the information available to you in this chapter, estimate the value of the stretching factor β for (i) the East African Rift and (ii) the Rio Grande Rift. Assume that both have formed as a result of uniform stretching of the lithosphere.
    (b) Based on the values of β estimated in (a), calculate initial and final subsidences for these two rifts. Do these values appear reasonable? (Remember these rifts are *not* subaqueous; assume $\rho_w = 0$.)
    (c) Now assume that the sea breaks through in the Gulf of Aden and floods the small portion of the rift valley that is below sea level there. What might happen?
    (d) Now assume that the drainage systems change and the entire East African Rift fills with water. What would happen in this eventuality?

BIBLIOGRAPHY

Ahern, J. L., and Ditmars, R. C. 1985. Rejuvenation of continental lithosphere beneath an intercratonic basin. *Tectonophys., 120*, 21–35.

Allenby, R. J., and Schnetzler, C. C. 1983. United States crustal thickness. *Tectonophys., 93*, 13–31.

Allègre, C. J. 1985. The evolving earth system. *Terra Cognita, 5*, 5–14.

Allègre, C. J., et al. 1984' Structure and evolution of the Himalaya–Tibet orogenic belt. *Nature, 307*, 17–22.

Anderson, R. N., DeLong, S. E., and Schwarz, W. M. 1978. Thermal model for subduction with dehydration in the downgoing slab. *J. Geol., 86*, 731–9.

1980. Dehydration asthenospheric convection and seismicity in subduction zones. *J. Geol., 88*, 445–51.

Baker, B. H., and Wohlenberg, J. 1971. Structure and evolution of the Kenya Rift Valley. *Nature, 229*, 538–42.

Barton, P. J., and Wood, R. 1984. Tectonic evolution of the North Sea basin: Crustal stretching and subsidence. *Geophys. J. R. Astr. Soc., 79*, 987–1022.

Beaumont, C. 1978. The evolution of sedimentary basins on a visco-elastic lithosphere: Theory and examples. *Geophys. J. R. Astr. Soc., 55*, 471–97.

1981. Foreland basins. *Geophys. J. R. Astr. Soc., 65*, 291–329.

Beaumont, C., Quinlan, G. M., and Hamilton, J. 1987. The Alleghanian orogeny and its relationship to the evolution of the eastern interior, North America. *In* C. Beaumont and A. J. Tankard, eds., *Sedimentary basins and basin-forming mechanisms*, Vol. 12 of Can. Soc. Petrol. Geol. Mem., 425–46.

Behrendt, J. C., Green, A. G., Cannon, W. F., Hutchinson, D. R., Lee, M. W., Milkereit, B., Agena, W. F., and Spencer, C. 1988. Crustal structure of the mid-continent rift system: Results from GLIMPCE deep seismic reflection profiles. *Geology, 16*, 81–5.

Beloussov, V. V., Belyaevsky, N. A., Borisov, A. A., Volvovsky, B. S., Volvovsky,

I. S., Resvoy, D. P., Tal-Virsky, B. B., Khamrabaev, I. K. H., Kaila, K. L., Narian, H., Marussi, A., and Finetti, J. 1980. Structure of the lithosphere along the deep seismic sounding profile: Tien Shan–Pamirs–Karakorum–Himalayas. *Tectonophys., 70,* 193–221.

Best, M. G. 1982. *Igneous and metamorphic petrology.* Freeman, San Francisco.

Bickle, M. J. 1978. Heat loss from the Earth: A constraint on Archaean tectonics from the relation between geothermal gradients and the rate of plate production. *Earth Planet. Sci. Lett., 40,* 301–15.

Bingham, D. K., and Klootwijk, C. J. 1980. Paleomagnetic constraints on Greater India's underthrusting of the Tibetan plateau. *Nature, 284,* 336–8.

Black, P. R., and Braille, L. W. 1982. P_n velocity and cooling of the continental lithosphere. *J. Geophys. Res., 87,* 10557–69.

Bott, M. H. P. 1982. *The interior of the earth,* 2nd ed. Edward Arnold, London.

Braille, L. W., and Smith, R. B. 1975. Guide to the interpretation of crustal refraction profiles. *Geophys. J. R. Astr. Soc., 40,* 145–76.

Brandon, C., and Romanowicz, B. 1986. A 'no lid' zone in the central Chang-Thang Platform of Tibet: Evidence from pure path phase velocity measurements of long-period Rayleigh waves. *J. Geophys. Res., 91,* 6547–64.

Brewer, J., Smithson, S., Oliver, J., Kaufman, S., and Brown, L. 1980. The Laramide orogeny: Evidence from COCORP deep crustal seismic profiles in the Wind River Mountains, Wyoming. *Tectonophys., 62,* 165–89.

Brown, C., and Girdler, R. W. 1980. Interpretation of African gravity and its implication for the breakup of the continents. *J. Geophys. Res., 65,* 287–304.

Brown, L. D., Krumhanst, P. A., Chapin C. E., Sanford, A. R., Cook, F. A., Kaufman, S., Oliver, J. F., and Schit, F. S. 1979. COCORP seismic reflection studies of the Rio Grande Rift. *In* R. E. Riecker, ed., *Rio Grande rift: Tectonics and magmatism,* Spec. Pub. 23, Am. Geophys. Un., 169–84.

Burchfiel, B. C. 1983. The continental crust. *Sci. Am. 249,* 3, 55–67.

Burchfiel, B. C., Oliver, J. E., and Silver, L. T., eds. 1980. *Studies in geophysics: Continental tectonics.* National Academy of Sciences.

Chen, W.-P., and Molnar, P. 1981. Constraints on the seismic wave velocity beneath the Tibetan plateau and their tectonic implications. *J. Geophys. Res., 86,* 5937–62.

1983. Focal depths of intracontinental and intraplate earthquakes and their implications for the thermal and mechanical properties of the lithosphere. *J. Geophys. Res., 88,* 4183–214.

Chapin, C. E. 1979. Evolution of the Rio Grande Rift – a summary. *In* R. E. Riecker, ed., Rio Grande rift: Tectonics and magmatism, Spec. Pub. 23, Am. Geophys. Un., 1–6.

Cheadle, M. J., McGeary, S., Warner, M. R., and Matthews, D. H. 1987. Extensional structures on the western U.K. continental shelf: A review of evidence from deep seismic profiling. *In* M. P. Coward et al., eds., *Continental extension tectonics,* Geol. Soc. Lond. Spec. Pub. 28, 445–65.

Cordell, L. 1978. Regional geophysical setting of the Rio Grande Rift. *Geol. Soc. Am. Bull., 89,* 1073–90.

Decker, E. R., and Smithson, S. B. 1975. Heat flow and gravity interpretation in southern New Mexico and West Texas. *J. Geophys. Res., 80,* 2542–52.

de Voogt, B., Serpa, L., and Brown, L. 1988. Crustal extension and magmatic processes: COCORP profiles from Death Valley and the Rio Grande Rift. *Geol. Soc. Am. Bull., 100,* 1550–67.

Dietrich, V. J. 1976. Evolution of the eastern Alps: A plate tectonics working hypothesis. *Geology, 4,* 147–52.

Edmond, J. M., and Von Damm, K. 1983. Hot springs on the ocean floor. *Sci. Am., 248,* 4, 78–93.

Egloff, R. 1979. Sprengseismische Untersuchungen der Erdkruste in der Schweiz, Diss, ETH (Swiss Federal Institute of Technology), Zürich, Switzerland.

England, P. C. 1983. Constraints on the extension of continental lithosphere. *J. Geophys. Res.*, *88*, 1145–52.

Fowler, C. M. R., and Nisbet, E. G. 1985. The subsidence of the Williston Basin. *Can. J. Earth Sci.*, *22*, 408–15.

Fowler, S. R., White, R. S., and Louden, K. E. 1985. Sediment dewatering in the Makran accretionary prism. *Earth Planet. Sci. Lett.*, *75*, 427–38.

Giese, P., Nicolich, R., and Reutter, K. J. 1982. Explosion seismic crustal studies in the Alpine–Mediterranean region and their implications to tectonic processes. *In* H. Berkehemer and K. Hsü, eds. *Alpine-Mediterranean geodynamics*, Geodynamics Series Vol. 7, Am. Geophys. Un. and Geol. Soc. Am., 39–74.

Giese, P., Prodehl, C., and Stein, A., eds. 1976. *Explosion seismology in central Europe*, Springer-Verlag, Berlin.

Goto, K., Suzuki, Z., and Hamaguchi, H. 1987. Stress distribution due to olivine–spinel phase transition in descending plate and deep focus earthquakes, *J. Geophys. Res.*, *92*, 13811–20.

Griffiths, D. H. 1972. Some comments on the results of a seismic refraction experiment in the Kenya Rift. *Tectonophys.*, *15*, 151–6.

Haq, B. U., Hardenbol, J., and Vail, P. R. 1987. Chronology of fluctuating sea levels since the Triassic, *Science*, *235*, 1156–67.

Hildreth, W., and Moorbath, S. 1988. Crustal contributions to arc magmatism in the Andes. *Contrib. Mineral. Petrol.*, *98*, 455–89.

Hirn, A., Jobert, G., Wittlinger, G., Zhong-Xin, X., and En-Yuan, G. 1984. Main features of the upper lithosphere in the unit between the High Himalayas and the Yarlung Zangbo Jiang suture. *Ann. Geophys.*, *2*, 113–18.

Hirn, A., Lepine, J.-C., Jobert, G., Sapin, M., Wittlinger, G., Zhong Xin, X., En Yuan, G., Xiang Jing, W., Ji Wen, T., Shao Bai, X., Pandey, M. R., and Tater, J. M. 1984. Crustal structure and variability of the Himalayan border of Tibet. *Nature*, *307*, 23–5.

Hirn, A., Nercessian, A., Sapin, M., Jobert, G., Zhong Xin, X., En Yuan, G., De Yuan, L., and Ji Wen, T. 1984. Lhasa block and bordering sutures – a continuation of a 500 km Moho traverse through Tibet. *Nature*, *307*, 25–7.

Hirn, A., and Sapin, M. 1984. The Himalayan zone of crustal interaction: Suggestions from explosion seismology. *Ann. Geophys.*, *2*, 123–30.

Houseman, G. A., McKenzie, D. P., and Molnar, P. 1981. Convective instability of a thickened boundary layer and its relevance for the thermal evolution of continental convergent belts. *J. Geophys. Res.*, *86*, 6115–32.

Hsui, A. T., and Toksöz, M. N. 1979. The evolution of thermal structures beneath a subduction zone. *Tectonophys.*, *60*, 43–60.

Hurley, P. M., and Rand, J. R. 1969. Pre-drift continental nuclei. *Science*, *164*, 1229–42.

Hyndman, R. D. 1988. Dipping seismic reflectors, electrically conductive zones and trapped water in the crust over a conducting plate. *J. Geophys. Res.*, *93*, 13391–405.

Jackson, J., and McKenzie, D. 1984. Active tectonics of the Alpine–Himalayan belt between western Turkey and Pakistan. *Geophys. J. R. Astr. Soc.*, *77*, 185–264.

1988. The relationship between plate motions and seismic moment tensors, and the rates of active deformation in the Mediterranean and Middle East. *Geophys. J.*, *93*, 45–73.

Jarvis, G. T., and McKenzie, D. P. 1980. Sedimentary basin evolution with finite extension rates. *Earth Planet. Sci. Lett.*, *48*, 42–52.

Jobert, N., Journet, B., Jobert, G., Hirn, A., and Sun, K. Z. 1985. Deep structure of southern Tibet inferred from the dispersion of Rayleigh waves through a long period seismic network. *Nature, 313*, 386–8.

Kaila, K. L., Roy Choudhury, K., Reddy, P. R., Krishna, V. G., Narain, H., Subbotin, S. I., Sollogub, V. B., Chekunov, A. V., Kharetchko, G. E., Lazarenko, M. A., and Ilchenko, T. V. 1979. Crustal structure along Kavali–Udipi profile in Indian Peninsula shield from deep seismic sounding. *J. Geol. Soc. India, 20*, 307–33.

Kaila, K. L., Roy Choudhury, K., Krishna, V. G. Dixit, M. M., and Narian, H. 1982. Crustal structure of Kashmir Himalaya and inferences about the asthenosphere layer from DSS studies along the international profile, Qarrakol–Zorkol–Nanga Parbat–Srinagar–Pamir. Himalaya Monograph. *Bull. Geofis. Teor. Appl., 25*, 221–34.

Kaila, K. L., Tripathi, K. M., and Dixit, M. M. 1984. Crustal structure along Wulan Lake–Gulmarg–Naoshera Profile across Pir Panjal range of the Himalayas from deep seismic soundings. *J. Geol. Soc. India, 25*, 706–19.

Karner, G. D., and Watts, A. B. 1983. Gravity anomalies and flexure of the lithosphere at mountain ranges. *J. Geophys. Res., 88*, 10449–77.

Kastens, K., and 20 others. 1988. ODP Leg 107 in the Tyrrhenian Sea: Insights into passive margin and back-arc basin evolution. *Geol. Soc. Am. Bull., 100*, 1140–56.

Keen, C. E., and Cordsen, A. 1981. Crustal structure, seismic stratigraphy and rift processes of the continental margin off eastern Canada: Ocean bottom seismic refraction results off Nova Scotia. *Can. J. Earth Sci., 18*, 1523–38.

Keller, G. R., Braile, L. W., and Schlue, J. W. 1979. Regional crustal structure of the Rio Grande rift from surface wave dispersion measurements. *In* R. E. Riecker, ed., *Rio Grande Rift: Tectonics and magmatism*, Spec. Pub. 23, Am. Geophys. Un., 115–26.

Klingelé, E., and Oliver, R. 1980. La nouvelle carte gravimetrique de la Suisse. *Beitr. Geol. Karte Schweiz, Ser. Geophys., 20.*

Kurtz, R. D., DeLaurier, J. M., and Gupta, J. C. 1986. A magnetotelluric sounding across Vancouver Island detects the subducting Juan de Fuca plate. *Nature, 321*, 596–9.

Lambeck, K. 1983. Structure and evolution of the intercratonic basins of central Australia. *Geophys. J. R. Astr. Soc., 74*, 843–86.

Lambeck, K., Burgess, G., and Shaw, R. D. 1988. Teleseismic travel-time anomalies and deep crustal structure in central Australia. *Geophys. J., 94*, 105–24.

LASE Study Group: Keen, C., Reid, I. Woodside, J., Nichols, B., Ewing, J. I., Purdy, G. M., Schouten, H., Diebold, J. B., Buhl, P., Mutter, J. C., Mithal, R., Alsop, J., Stoffa, P. L., Phillips, J. D., Stark, T., and O'Brien, T., 1986. Deep structure of the U.S. east coast passive margin from large aperture seismic experiments (LASE). *Mar. Petroleum Geol., 3*, 234–42.

Latham, T. S., Best, J., Chaimov, T., Oliver, J., Brown, L., and Kaufman, S. 1988. COCORP profiles from the Montana plains: The Archean cratonic crust and a lower crustal anomaly beneath the Williston Basin. *Geology, 16*, 1073–6.

Le Pichon, X., and Sibuet, J. C. 1981. Passive margins, a model of formation. *J. Geophys. Res., 86*, 3708–20.

Lépine, J.-C., Hirn, A., Pandey, M. R., and Tater, J. M. 1984. Features of the P-waves propagated in the crust of the Himalayas. *Ann. Geophys., 2*, 119–22.

Lewis, T. J., Bentkowski, W. H., Davis, E. E., Hyndman, R. D., Souther, J. G., and Wright, J. A. 1988. Subduction of the Juan de Fuca plate: Thermal consequences. *J. Geophys. Res., 93*, 15207–25.

Long, R. E., and Backhouse, R. W. 1976. The structure of the western flank of the Gregory Rift. Part II. The mantle. *Geophys. J. R. Astr. Soc., 44*, 677–88.

Luetgert, J. H., and Meyer, R. P. 1982. Structure of the western basin of Lake Superior from cross structure refracting profiles. *In* J. R. Wold and W. J. Hinze, eds., *Geology and tectonics of the Lake Superior Basin*, Vol. 156 of Geol. Soc. Am. Mem., 245–56.

Lyon-Caen, H. 1986. Comparison of the upper mantle shear wave velocity structure of the Indian Shield and the Tibetan Plateau and tectonic implications. *Geophys. J. R. Astr. Soc., 86*, 727–49.

Lyon-Caen, H., and Molnar, P. 1983. Constraints on the structure of the Himalayas from an analysis of gravity anomalies and a flexural model of the lithosphere. *J. Geophys. Res., 88*, 8171–91.

1985. Gravity anomalies, flexure of the Indian plate and the structure support and evolution of the Himalaya and Ganga Basin. *Tectonics, 4*, 513–38.

McGeary, S., and Warner, M. 1985. Seismic profiling the continental lithosphere. *Nature, 317*, 795–7.

McKenzie, A. S., and McKenzie, D. P. 1983. Isomerization and aromatization of hydrocarbons in sedimentary basins formed by extension. *Geol. Mag., 120*, 417–528.

McKenzie, D. P. 1969. Speculations on the consequences and causes of plate motions. *Geophys. J. R. Astr. Soc., 18*, 1–32.

1978. Some remarks on the development of sedimentary basins. *Earth Planet. Sci. Lett., 40*, 25–32.

McKenzie, D. P., and Bickle, M. J. 1988. The volume and composition of melt generated by extension of the lithosphere. *J. Petrology, 29*, 625–79.

Meissner, R. 1986. *The continental crust: A geophysical approach.* Academic, Orlando.

Molnar, P. 1984. Structure and tectonics of the Himalaya: Constraints and implications of geophysical data. *Ann. Rev. Earth Planet. Sci., 12*, 489–518.

Molnar, P., and Chen, W.-P. 1983. Seismicity and mountain building. *In* K. J. Hsü, ed., *Mountain building processes*, Academic, London, 41–57.

Molnar, P., and Taponnier, P. 1975. Caenozoic tectonics of Asia: The effects of a continental collision. *Science, 189*, 419–26.

Mueller, S. 1983. Deep structure and recent dynamics in the Alps. *In* K. J. Hsü ed., *Mountain building processes*, Academic, London, 181–200.

Mueller, S., Ansorge, J., Egloff, R., and Kissling, E. 1980. A crustal cross-section along the Swiss Geotraverse from the Rhinegraben to the Po Plain. *Eclogae Geol. Helv., 73*, 463–83.

Mueller, S. 1977. A new model of the continental crust. In J. G. Heacock, ed., *The earth's crust*, Vol. 20 of Geophys. Monogr., Am. Geophys. Un., 289–317.

Ni, J., and Barazangi, M. 1984. Seismotectonics of the Himalayan collision zone: Geometry of the underthrusting Indian plate beneath the Himalayas. *J. Geophys. Res., 89*, 1147–63.

Nicolaysen, L. O., Hart, R. J., and Gale, N. H. 1981. The Vredefort radioelement profile extended to supracrustal strata at Carltonville, with implications for continental heat flow. *J. Geophys. Res., 86*, 10653–61.

Nisbet, E. G. 1984. The continental and oceanic crust and lithosphere in the Archaean: Isostatic, thermal and tectonic models. *Can. J. Earth Sci., 21*, 1426–41.

1985. Putting the squeeze on rocks. *Nature, 315*, 541.

1986. Archaean mantle models. *Nature, 320*, 306–7.

1987. *The young earth.* Allen and Unwin, London.

Nisbet, E. G., and Fowler, C. M. R. 1983. Model for Archaean plate tectonics. *Geology, 11*, 376–9.

Nisbet, E. G., and Walker, D. W. 1982. Komatiites and the structure of the Archaean mantle. *Earth Planet. Sci. Lett, 60*, 105–13.

Nowroozi, A. A., and Mohajer-Ashjai, A. 1985. Fault movements and tectonics of eastern Iran: Boundaries of the Lut plate. *Geophys. J. R. Astr. Soc., 83*, 215–37.

Olsen, K. H., Keller, G. R., and Stewart, J. N. 1979. Crustal structure along the Rio Grande Rift from seismic refraction profiles. *In* R. E. Riecker, ed., Rio Grande Rift: Tectonics and magmatism, Spec. Publ. 23, Am. Geophys, Un., 127–43.

Panza, G. F., and Mueller, St. 1979. The plate boundary between Eurasia and Africa in the Alpine area. *Memorie de Scienze Geologiche, 33*, 43–50.

Pavoni, N. 1979. Investigation of recent crustal movements in Switzerland. *Schweiz. Mineral. Petrol. Mitt., 59*, 117–26.

Pearce, J. A. 1983. Role of the sub-continental lithosphere in magma genesis of active continental margins. *In* C. J. Hawkesworth and M. J. Norry, eds., *Continental basalts and mantle xenoliths*, Shiva, Nantwich, 230–49.

Percival, J. A., Card, K. D., Sage, R. P., Jensen, L. S., and Luhta, L. E. 1983. The Archaean crust in the Wawa–Chapleau–Timmins region. *In* L. D. Ashwal and K. D. Card, eds., *Workshop on cross-section of Archaean crust*, L.P.I. Tech. Rpt., 83–03, Lunar and Planetary Institute, Houston, 99–169.

Press, F. 1966. Seismic Velocities. *In* S. P. Clark, ed., *Handbook of physical constants*, Vol. 97 of Geol. Soc. Am. Mem., 195–218.

Quigley, T. M., and McKenzie, A. S. 1988. The temperatures of oil and gas formation in the sub-surface. *Nature, 333*, 549–52.

Quigley, T. M., McKenzie, A. S., and Gray, J. R. 1987. Kinetic theory of petroleum generation. *In* B. Doligez, ed., *Migration of hydrocarbons in sedimentary basins*, Technip, Paris, 649–65.

Quinlan, G. 1987. Models of subsidence mechanisms in intracratonic basins and their applicability to North American examples. *In* C. Beaumont, and A. J. Tankard, eds., *Sedimentary basins and basin-forming mechanisms*, Vol. 12 of Can. Soc. Petrol. Geol. Mem., 463–81.

1988. The thermal signatures of tectonic environments. *In* E. G. Nisbet and C. M. R. Fowler, eds., *Heat, metamorphism and tectonics*, Short Course Vol. 14, Min. Assoc. Can., 213–57.

Quinlan, G. H., and Beaumont, C. 1984. Appalachian thrusting, lithospheric flexure, and the Paleozoic stratigraphy of the Eastern Interior of North America. *Can. J. Earth Sci., 21*, 973–96.

Ramberg, I. B., Cook, F. A., and Smithson, S. B. 1978. Structure of the Rio Grande Rift in southern New Mexico and west Texas based on gravity interpretation. *Geol. Soc. Am. Bull., 89*, 107–23.

Rinehart, E. J., Sanford, A. R., and Ward, R. M. 1979. Geographic extent and shape of an extensive magma body at mid-crustal depths in the Rio Grande Rift near Socorro, New Mexico. *In* R. E. Riecker, ed., *Rio Grande Rift: Tectonics and magmatism*, Spec. Pub. 23, Am. Geophys. Un., 237–51.

Rosendahl, G. R. 1987. Architecture of continental rifts with special reference to East Africa. *Ann. Rev. Earth Planet Sci., 15*, 445–503.

Rybach, L., Mueller, St., Milnes, A. G., Ansorge, J., Bernoulli, D., and Frey, M. 1980. The Swiss Geotraverse Basel-Chiasso – a review. *Eclogae Geol. Helv., 73*, 437–62.

Sanford, A. R. 1983. Magma bodies in the Rio Grande Rift in central New Mexico. *In New Mexico Geol. Soc., 34th Field Conference Guidebook*, 123–5.

Sanford, A. R., Mott, R. P., Jr., Shuleski, P. J., Rinehart, E. J., Caravella, F. J., Ward, R. M., and Wallace, T. C. 1977. Geophysical evidence for a magma body in the crust in the vicinity of Socorro, N. M. *In* J. G. Heacock, ed., *The earth's crust*, Vol. 20 of Am. Geophys. Un. Monog., 385–403.

Savage, J. E. G., and Long, R. E. 1985. Lithospheric Structure beneath the Kenya Dome. *Geophys. J. R. Astr. Soc.*, *82*, 461–77.

Sclater, J. G., Parson, B., and Jaupart, C. 1981. Oceans and continents: Similarities and differences in the mechanisms of heat loss. *J. Geophys. Res.*, *86*, 11535–52.

Seager, W. R., and Morgan, P. 1979. Rio Grande Rift in southern New Mexico, West Texas and northern Chihuahua. *In* R. E. Riecker, ed., *Rio Grande Rift: Tectonics and magmatism*, Spec. Pub. 23, Am. Geophys. Un., 87–106.

Searle, R. C. 1970. Evidence from gravity anomalies for thinning of the lithosphere beneath the Rift Valley in Kenya. *Geophys. J. R. Astr. Soc.*, *21*, 13–21.

Searle, R. C., and Gouin, P. 1972. A gravity survey of the central part of the Ethopian Rift Valley. *Tectonophys.*, *15*, 41–52.

Shudofsky, G. N. 1985. Source mechanisms and focal depths of East African earthquakes using Rayleigh-wave inversion and body-wave modelling. *Geophys. J. R. Astr. Soc.*, *83*, 563–614.

Sinno, Y. A., Daggett, P. H., Keller, G. R., Morgan, P., and Harder, S. H. 1986. Crustal structure of the southern Rio Grande Rift determined from seismic refraction profiling. *J. Geophys. Res.*, *91*, 6143–56.

Sleep, N. H. 1971. Thermal effects of the formation of Atlantic continental margins by continental break-up. *Geophys. J. R. Astr. Soc.*, *24*, 325–50.

1976. Platform Subsidence Mechanisms and 'eustatic' sea-level changes. Tectonophys., *36*, 45–56.

Sleep, N. H., Nunn, J. A., and Chou, L. 1980. Platform basins. *Ann. Rev. Earth Planet. Sci.*, *8*, 17–34.

Sleep, N. H., and Sloss, L. L. 1978. A deep borehole in the Michigan Basin. *J. Geophys. Res.*, *83*, 5815–19.

Sleep, N. H., and Snell, N. S. 1976. Thermal contraction and flexure of mid-continent and Atlantic marginal basins. *Geophys. J. R. Astr. Soc.*, *45*, 125–54.

Smith, A. G., and Woodcock, N. H. 1982. Tectonic synthesis of the Alpine–Mediterranean region: A review. *In* H. Berkehemer and K. Hsü, eds., *Alpine-Mediterranean geodynamics*, Vol. 7 of Geodynamics Series, Am. Geophys. Un. and Geol. Soc. Am., 15–38.

Steckler, M. S., and Watts, A. B. 1978. Subsidence of the Atlantic type margin off New York. *Earth Planet. Sci. Lett.*, *41*, 1–13.

Stockmal, G. S. 1983. Modelling of large-scale accretionary wedge deformation. *J. Geophys. Res.*, *88*, 8271–87.

Stockmal, G. S., and Beaumont, C. 1987. Geodynamic models of convergent margin tectonics: The southern Canadian Cordillera and the Swiss Alps. *In* C. Beaumont and A. J. Tankard, eds., *Sedimentary basins and basin-forming mechanisms*, Vol. 12 of Can. Soc. Petrol. Geol. Mem., 393–412.

Tapponier, P., Peltzer, G., Le Dain, A. Y., Armijo, R., and Cobbold, P. 1982. Propagating extrusion tectonics in Asia: New insights from simple experiments with plasticine. *Geology*, *10*, 611–16.

Tapponier, P., et al. 1981. The Tibetan side of the India–Eurasia collision. *Nature*, *294*, 404–10.

Tatsumi, Y., Sakuyama, M. Fukuyama, H., and Kushiro, I. 1983. Generation of arc basalt magmas and thermal structure of the mantle wedge in subduction zones. *J. Geophys. Res.*, *88*, 5815–25.

Taylor, S. R., and McLennan, S. M. 1985. *The continental crust: Its composition and evolution.* Blackwell Scientific, Oxford.

Turner, F. J. 1981. *Metamorphic petrology: Mineralogical, field and tectonics aspects.* McGraw-Hill, New York.

Vail, P. R., and Mitchum, R. M., Jr. 1979. Global cycles of relative changes of sea-level from seismic stratigraphy. *In* J. S. Watkins, L. Montadert, and

P. W. Dickerson, eds., *Geological and geophysical investigations of continental margins*, Am. Assoc. Petrol. Geol. Memoir 29.

Van den Beukel, J., and Wortel, R. 1986. Thermal modelling of arc–trench regions. *Geologie en Mijnbouw*, *65*, 133–43.

1987. Temperatures and shear stresses on the upper part of a subduction zone. *Geophys. Res. Lett.*, *14*, 1057–60.

1988. Thermomechanical modelling of arc–trench regions. *Tectonophys.*, *154*, 177–93.

Van Schmus, W. R., and Hinze, W. J. 1985. The mid-continent rift system. *Ann. Rev. Earth Planet. Sci.*, *13*, 345–83.

Walpes, D. W. 1980. Time and temperature in petroleum formation: Application of Lopatin's method to petroleum exploration. *Bull. Am. Assoc. Petrol. Geol.*, *64*, 916–26.

Warner, M., and McGeary, S. 1987. Seismic reflection coefficients from mantle fault zones. *Geophys. J. R. Astr. Soc.*, *89*, 223–230.

Watts, A. B., and Ryan, W. B. F. 1976. Flexure of the lithosphere and continental margin basins. *Tectonophys.*, *36*, 25–44.

Wernicke, B. 1985. Uniform-sense simple shear of the continental lithosphere. *Can. J. Earth Sci.*, *22*, 108–25.

Wernicke, B., and Axen, G. J. 1988. On the role of isostasy in the evolution of normal fault systems. *Geology*, *16*, 848–51.

White, R. S. 1984. Active and passive plate boundaries around the Gulf of Oman, north-west Indian Ocean. *Deep Sea Res.*, *31*, 731–45.

White, R. S., and Louden, K. E. 1982. The Makran continental margin: Structure of a thickly sedimented convergent plate boundary. *In* J. S. Watkins and C. L. Drake, eds., *Studies in continental margin geology*, Vol. 34 of Am. Assoc. Petrol. Geol. Mem., 499–518.

White, R. S., Spence, G. D., Fowler, S. R., McKenzie, D. P., Westbrook, G. K., and Bowen, A. N. 1987. Magmatism at rifted continental margins. *Nature*, *330*, 439–44.

White, R. S., Westbrook, G. K., Fowler, S. R., Spence, G. D., Barton, P. J., Joppen, M., Morgan, J., Bowen, A. N., Prestcott, C., and Bott, M. H. P. 1987. Hatton Bank (Northwest U.K.) continental margin structure. *Geophys. J. R. Astr. Soc.*, *89*, 265–72.

Williams, H., Hoffman, P., Monger, J., Lewry, J., Rivers, T., Muehlberger, W., and Trettin, H. Compilation, time of accretion. Postcards from Dept. Earth Sciences and Centre for Earth Resources Research, Memorial University of Newfoundland, St. Johns, Newfoundland, Canada.

Windley, B. F. 1984. *The evolving continents*, 2nd ed. Wiley, New York.

Wood, R., and Barton, P. J. 1983. Crustal thinning and subsidence in the North Sea. *Nature*, *302*, 134–6.

Wyllie, P. J. 1979. Magmas and volatile components. *American Mineralogist*, *64*, 469–500.

1981. Experimental petrology of subduction, andesites and batholiths. *Trans. Geol. Soc. S. Afr.*, *84*, 281–91.

Zucca, J. 1984. The crustal structure of the southern Rhine graben from reinterpretation of seismic refraction data. *J. Geophys.*, *55*, 13–22.

APPENDIX 1

Scalars, Vectors and Differential Operators

Scalars and Vectors

A *scalar* is a quantity which just has a magnitude. For example, the temperature outside today could be $+10°C$.

A *vector* is a quantity which has a magnitude and a direction. For example, the wind velocity in your city today could be 20 km hr^{-1} and due east. The speed (magnitude of the velocity) would be measured by an anemometer and the direction by a wind vane.

A vector is indicated in print by a boldface character such as **x**. Its magnitude is indicated by the same character in italic type, x, as is a scalar. A vector or a scalar can be either a constant or a function of some variable, which can itself be either a scalar or a vector. When the scalar (or vector) is a function of a variable, it is called a scalar (or vector) field. For example, the temperature at midday across the province of British Columbia, Canada, is a scalar field. That is, the temperature at each place depends on its position in the province and thus is written $T(x, y, z)$, where x, y and z are geographical and height coordinates within British Columbia. The wind velocity $\mathbf{V}$ across British Columbia at midday depends on geographical position and so is a vector field, written $\mathbf{V}(x, y, z)$. Note that if the coordinate system is changed, the scalar is unaffected, but the components of the vector must be recalculated.

Products of Scalars and Vectors

Many physical relationships are best expressed by using the products of scalars and vectors. The product of two scalars is another scalar, and everyone is well accustomed to the process called multiplication, learned laboriously in elementary school. The product of a scalar s and a vector $\mathbf{V} = (V_x, V_y, V_z)$ is another vector $s\mathbf{V}$. In cartesian coordinates (x, y, z), the product is simply

$$s\mathbf{V} = s(V_x, V_y, V_z) = (sV_x, sV_y, sV_z) \tag{A1.1}$$

Thus, the scalar multiplies each component of the vector.

When two vectors are involved, multiplication becomes more complicated. There are two products of vectors: One, called the *scalar product*, is a scalar; the other, called the *vector product*, is a vector.

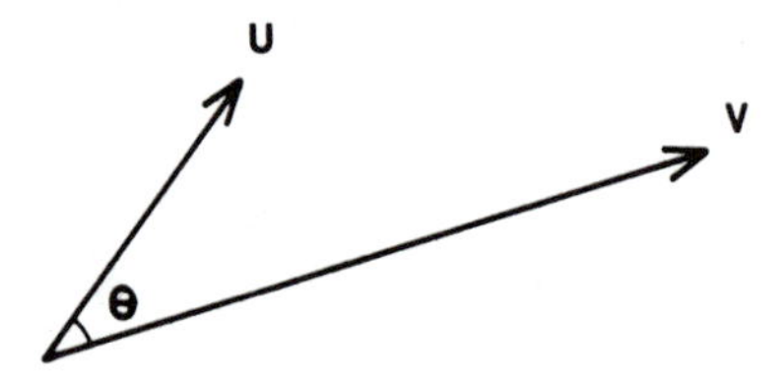

Figure A1.1. Two vectors **U** and **V**. The angle between them is θ.

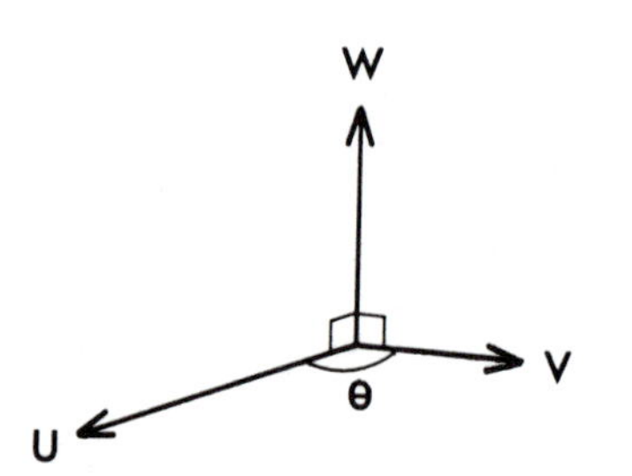

Figure A1.2. **W** is the vector product of **U** and **V**, $\mathbf{U} \wedge \mathbf{V}$.

The scalar product of two vectors **U** and **V** is written $\mathbf{U} \cdot \mathbf{V}$ and defined as

$$\mathbf{U} \cdot \mathbf{V} = UV \cos \theta \tag{A1.2}$$

where θ is the angle between the two vectors and U and V are the magnitudes of the vectors (Fig. A1.1). The scalar product is also known as the *dot product*. If **U** and **V** are parallel, then $\theta = 0$, $\cos = 1$, and so $\mathbf{U} \cdot \mathbf{V} = UV$. However, if **U** and **V** are perpendicular, then $\theta = 90°$ and $\mathbf{U} \cdot \mathbf{V} = 0$. Thus, the scalar product of two perpendicular vectors is zero. In cartesian coordinates (x, y, z), the scalar product is

$$\mathbf{U} \cdot \mathbf{V} = U_x V_x + U_y V_y + U_z V_z \tag{A1.3}$$

As an example of a scalar product, consider a force **F** acting on a mass m and moving that mass a distance **d**. The work done is then $\mathbf{F} \cdot \mathbf{d}$: Work is a scalar.

The vector product of two vectors **U** and **V** is written $\mathbf{U} \wedge \mathbf{V}$ or $\mathbf{U} \times \mathbf{V}$ and defined as

$$\mathbf{U} \wedge \mathbf{V} = \mathbf{W} \tag{A1.4}$$

where **W** is a vector perpendicular to both **U** and **V** (Fig. A1.2), with magnitude

$$W = UV \sin \theta \tag{A1.5}$$

The vector product is also known as the *cross product*. The vector product of two parallel vectors is zero since $\sin \theta = 0$ when $\theta = 0$. In cartesian coordinates (x, y, z), the vector product is expressed as

$$\mathbf{U} \wedge \mathbf{V} = (U_y V_z - U_z V_y, U_z V_x - U_x V_z, U_x V_y - U_y V_x) \tag{A1.6}$$

As an example of a vector product, consider a rigid body rotating about an axis with angular velocity $\boldsymbol{\omega}$ (the earth spinning about its north–south axis if you like). The velocity **V** of any particle at a radial position **r** is then given by

$$\mathbf{V} = \boldsymbol{\omega} \wedge \mathbf{r} \tag{A1.7}$$

Compare this with Eq. 2.3; rotation of the plates also involves the vector product.

Gradient

The gradient of a scalar T is a vector that describes the rate of change of T. The component of Grad T in any direction is the rate of change of T in that direction. Thus, the x component is $\partial T/\partial x$, the y component is $\partial T/\partial y$, and the z component is $\partial T/\partial z$. Grad T is an abbreviation for 'the gradient of T'.

$$\text{Grad } T \equiv \nabla T \tag{A1.8}$$

defines the vector operator ∇. The notations Grad T and ∇T are equivalent and are used interchangeably. In cartesian coordinates (x, y, z), ∇T is given by

$$\nabla T = \left(\frac{\partial T}{\partial x}, \frac{\partial T}{\partial y}, \frac{\partial T}{\partial z} \right) \tag{A1.9}$$

Grad T is normal (perpendicular) to surfaces of constant T. To show this, consider the temperature T at point (x, y, z). A small distance $\boldsymbol{\delta r} = (\delta x, \delta y, \delta z)$ away, the temperature is $T + \delta T$, where

$$\begin{aligned} \delta T &= \frac{\partial T}{\partial x}\delta x + \frac{\partial T}{\partial y}\delta y + \frac{\partial T}{\partial z}\delta z \\ &= (\nabla T)\cdot\boldsymbol{\delta r} \end{aligned} \tag{A1.10}$$

On the surface $T = \text{constant}$, $\delta T = 0$. This means that the scalar product $(\nabla T)\cdot\boldsymbol{\delta r}$ is zero on a surface of constant T and hence that ∇T and $\boldsymbol{\delta r}$ are perpendicular. Since $\boldsymbol{\delta r}$ is parallel to the surface and $T = \text{constant}$, ∇T must be perpendicular, or normal, to that surface.

Divergence

The divergence of a vector field $\mathbf{V}$, Div $\mathbf{V}$, is a scalar field. It is written $\nabla\cdot\mathbf{V}$ and is defined as

$$\text{Div}\,\mathbf{V} \equiv \nabla\cdot\mathbf{V} = \frac{\partial V_x}{\partial x} + \frac{\partial V_y}{\partial y} + \frac{\partial V_z}{\partial z} \tag{A1.11}$$

where the components of $\mathbf{V}$ in cartesian coordinates are (V_x, V_y, V_z).

The divergence represents a net flux, or rate of transfer, per unit of volume. If the wind velocity is $\mathbf{V}$ and the air has a constant density ρ, then

$$\nabla\cdot(\rho\mathbf{V}) = \rho\nabla\cdot\mathbf{V} \tag{A1.12}$$

represents the net mass flux of air per unit volume. If no air is created or destroyed, then the total mass flux entering each unit volume is balanced by that leaving it, and the net mass flux is zero:

$$\rho\nabla\cdot\mathbf{V} = 0 \tag{A1.13}$$

A vector field for which $\nabla\cdot\mathbf{V} = 0$ is called *solenoidal.*

Curl

The curl of a vector field $\mathbf{V}$, Curl $\mathbf{V}$, is a vector function of position. It is written $\nabla \wedge \mathbf{V}$, or $\nabla \times \mathbf{V}$, and is defined in cartesian coordinates as

$$\text{Curl}\,\mathbf{V} \equiv \nabla \wedge \mathbf{V} = \left(\frac{\partial V_z}{\partial y} - \frac{\partial V_y}{\partial z}, \frac{\partial V_x}{\partial z} - \frac{\partial V_z}{\partial x}, \frac{\partial V_y}{\partial x} - \frac{\partial V_x}{\partial y}\right) \tag{A1.14}$$

It is related to rotation and is sometimes called *rotation*, or Rot. For example, the differential expression of *Ampère's law* for the magnetic field $\mathbf{H}$ due to a current J, is $\nabla \wedge \mathbf{H} = \mathbf{J}$. Alternatively, consider a body rotating with constant angular velocity $\boldsymbol{\omega}$. Equation A1.7 expresses the velocity at $\mathbf{r}$ in terms of the angular velocity:

$$\mathbf{V} = \boldsymbol{\omega} \wedge \mathbf{r}$$

Now, take the curl of $\mathbf{V}$:

$$\nabla \wedge \mathbf{V} = \nabla \wedge (\boldsymbol{\omega} \wedge \mathbf{r}) \tag{A1.15}$$

Since $\boldsymbol{\omega}$ is a constant, this equation can be simplified to

$$\begin{aligned}\nabla \wedge \mathbf{V} &= \boldsymbol{\omega}(\nabla \cdot \mathbf{r}) - (\boldsymbol{\omega} \cdot \nabla)\mathbf{r} \\ &= 3\boldsymbol{\omega} - \boldsymbol{\omega} \\ &= 2\boldsymbol{\omega} \end{aligned} \tag{A1.16}$$

Thus, the curl of the velocity is twice the angular velocity. A vector field for which $\nabla \wedge \mathbf{V} = 0$ is called *irrotational.*

Laplacian Operator

In cartesian coordinates, the Laplacian operator ∇^2 is defined by

$$\nabla^2 = \nabla \cdot \nabla = \frac{\partial^2}{\partial x^2} + \frac{\partial^2}{\partial y^2} + \frac{\partial^2}{\partial z^2} \tag{A1.17}$$

which is the divergence of the gradient. It is a scalar operator:

$$\nabla^2 T = \nabla \cdot \nabla T = \frac{\partial^2 T}{\partial x^2} + \frac{\partial^2 T}{\partial y^2} + \frac{\partial^2 T}{\partial z^2} \tag{A1.18}$$

To define a Laplacian operator ∇^2 for a vector, it is necessary to use the identity

$$\nabla \cdot (\nabla \mathbf{V}) = \nabla(\nabla \cdot \mathbf{V}) - \nabla \wedge (\nabla \wedge \mathbf{V}) \tag{A1.19}$$

In cartesian coordinates, this is the same as applying the Laplacian operator to each component of the vector in turn:

$$\nabla^2 \mathbf{V} = (\nabla^2 V_x, \nabla^2 V_y, \nabla^2 V_z) \tag{A1.20}$$

However, in curvilinear coordinate systems this is not true because unlike the unit vectors $(1, 0, 0)$, $(0, 1, 0)$ and $(0, 0, 1)$ in the cartesian coordinate system, those in curvilinear coordinate systems are not constants with respect to their coordinate system. The calculation of the Laplacian operator applied to a vector in cylindrical and spherical polar coordinates is long and is left to the reader as an extracurricular midnight activity. (Hint: Use Eqs. A1.19, A1.22, A1.23, A1.24, A1.28, A1.29, A1.30.)

Curvilinear Coordinates

In geophysics it is frequently advantageous to work in curvilinear instead of cartesian coordinates. The curvilinear coordinates which exploit the symmetry of the earth, and are thus the most often used, are cylindrical polar coordinates and spherical polar coordinates. Although not every gradient, divergence, curl and Laplacian operator is used in this book in each of these coordinate systems, all are included here for completeness.

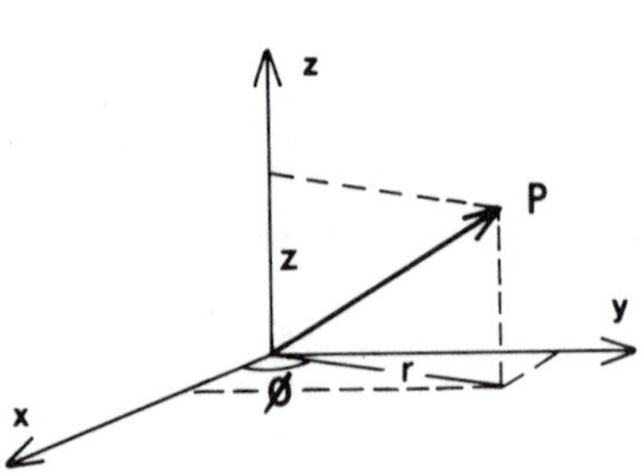

Figure A1.3. The cylindrical polar coordinates of point P are (r, ϕ, z). Any point P is defined by radius r, longitude ϕ and height z.

Cylindrical Polar Coordinates (r, ϕ, z)

In cylindrical polar coordinates (Fig. A1.3), a point P is located by specifying r the radius of the cylinder on which it lies, ϕ the longitude or azimuth in the xy plane and the z the distance from the xy plane to the point P, where $r \geqslant 0$, $0 \leqslant \phi \leqslant 2\pi$ and $-\infty < z < \infty$. From Figure A1.3 it can be seen that

$$\begin{aligned} x &= r\cos\phi \\ y &= r\sin\phi \\ z &= z \end{aligned} \qquad \text{(A1.21)}$$

In these cylindrical polar coordinates (r, ϕ, z), the gradient, divergence, curl and Laplacian operators are

$$\nabla T = \left(\frac{\partial T}{\partial r}, \frac{1}{r}\frac{\partial T}{\partial \phi}, \frac{\partial T}{\partial z}\right) \qquad \text{(A1.22)}$$

$$\nabla \cdot \mathbf{V} = \frac{1}{r}\frac{\partial}{\partial r}(rV_r) + \frac{1}{r}\frac{\partial V_\phi}{\partial \phi} + \frac{\partial V_z}{\partial z} \qquad \text{(A1.23)}$$

$$\nabla \wedge \mathbf{V} = \left(\frac{1}{r}\frac{\partial V_z}{\partial \phi} - \frac{\partial V_\phi}{\partial z}, \frac{\partial V_r}{\partial z} - \frac{\partial V_z}{\partial r}, \frac{1}{r}\frac{\partial}{\partial r}(rV_\phi) - \frac{1}{r}\frac{\partial V_r}{\partial \phi}\right) \qquad \text{(A1.24)}$$

$$\nabla^2 T = \frac{1}{r}\frac{\partial}{\partial r}\left(r\frac{\partial T}{\partial r}\right) + \frac{1}{r^2}\frac{\partial^2 T}{\partial \phi^2} + \frac{\partial^2 T}{\partial z^2} \qquad \text{(A1.25)}$$

$$\nabla^2 \mathbf{V} = \left(\nabla^2 V_r - \frac{V_r}{r^2} - \frac{2}{r^2}\frac{\partial V_\phi}{\partial \phi}, \nabla^2 V_\phi + \frac{2}{r^2}\frac{\partial V_r}{\partial \phi} - \frac{V_\phi}{r^2}, \nabla^2 V_z\right) \qquad \text{(A1.26)}$$

Spherical Polar Coordinates (r, θ, ϕ)

In spherical polar coordinates (Fig. A1.4), a point P is located by specifying r the radius of the sphere on which it lies, θ the colatitude, and ϕ the longitude or azimuth where $r \geqslant 0$, $0 \leqslant \phi \leqslant 2\pi$, $0 \leqslant \theta \leqslant \pi$. From Fig. A1.4 it can be seen that

$$\begin{aligned} x &= r\sin\theta\cos\phi \\ y &= r\sin\theta\sin\phi \\ z &= r\cos\theta \end{aligned} \qquad \text{(A1.27)}$$

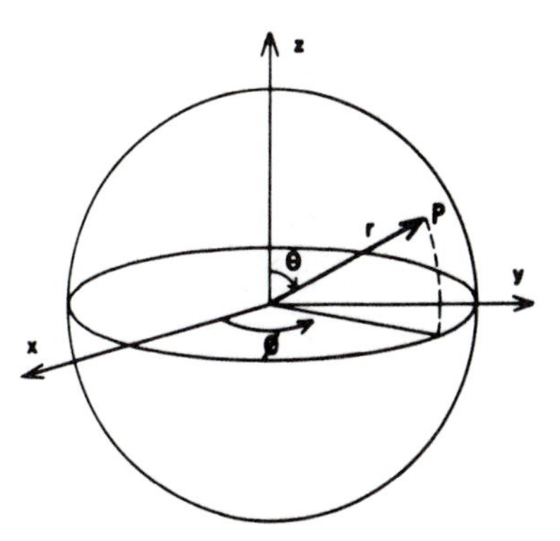

Figure A1.4. The spherical polar coordinates of point P are (r, θ, ϕ). Any point P is defined by radius r, colatitude ($90° -$ latitude) θ and longitude ϕ. For the earth $\theta = 0°$ is the North Pole and $\phi = 0°$ the Greenwich meridian.

In spherical polar coordinates (r, θ, ϕ), the gradient, divergence, curl and Laplacian operators are

$$\nabla T = \left(\frac{\partial T}{\partial r}, \frac{1}{r}\frac{\partial T}{\partial \theta}, \frac{1}{r\sin\theta}\frac{\partial T}{\partial \phi}\right) \qquad \text{(A1.28)}$$

$$\nabla \cdot \mathbf{V} = \frac{1}{r^2}\frac{\partial}{\partial r}(r^2 V_r) + \frac{1}{r\sin\theta}\frac{\partial}{\partial \theta}(\sin\theta\, V_\theta) + \frac{1}{r\sin\theta}\frac{\partial V_\phi}{\partial \phi} \qquad \text{(A1.29)}$$

$$\nabla \wedge \mathbf{V} = \left(\frac{1}{r\sin\theta}\frac{\partial}{\partial \theta}(\sin\theta\, V_\phi) - \frac{1}{r\sin\theta}\frac{\partial V_\theta}{\partial \phi}, \frac{1}{r\sin\theta}\frac{\partial V_r}{\partial \phi} - \frac{1}{r}\frac{\partial}{\partial r}(rV_\phi), \frac{1}{r}\frac{\partial}{\partial r}(rV_\theta) - \frac{1}{r}\frac{\partial V_r}{\partial \theta}\right) \qquad \text{(A1.30)}$$

$$\nabla^2 T = \frac{1}{r^2}\frac{\partial}{\partial r}\left(r^2 \frac{\partial T}{\partial r}\right) + \frac{1}{r^2 \sin\theta}\frac{\partial}{\partial \theta}\left(\sin\theta \frac{\partial T}{\partial \theta}\right) + \frac{1}{r^2 \sin^2\theta}\frac{\partial^2 T}{\partial \phi^2} \quad \text{(A1.31)}$$

$$\begin{aligned}\nabla^2 \mathbf{V} = \Bigg(& \nabla^2 V_r - \frac{2}{r^2} V_r - \frac{2}{r^2 \sin\theta}\frac{\partial}{\partial\theta}(\sin\theta\, V_\theta) - \frac{2}{r^2 \sin\theta}\frac{\partial V_\phi}{\partial \phi}, \\ & \nabla^2 V_\theta + \frac{2}{r^2}\frac{\partial V_r}{\partial \theta} - \frac{V_\theta}{r^2 \sin^2\theta} - \frac{2\cos\theta}{r^2 \sin^2\theta}\frac{\partial V_\phi}{\partial \phi}, \\ & \nabla^2 V_\phi + \frac{2}{r^2 \sin\theta}\frac{\partial V_r}{\partial \phi} + \frac{2\cos\theta}{r^2 \sin^2\theta}\frac{\partial V_\theta}{\partial \phi} - \frac{V_\phi}{r^2 \sin^2\theta}\Bigg)\end{aligned} \quad \text{(A1.32)}$$

APPENDIX 2

Theory of Elasticity and Elastic Waves

When a fixed solid body is subjected to an external force, it changes in size and shape. An *elastic solid* is a solid which returns to its original size and shape after the external deforming force is removed. For small deformations and on a short time scale (minutes not millions of years), rocks can be considered to be elastic.

Stress

Stress is defined as a force per unit area. When a deforming force is applied to a body, the stress is the ratio of the force to the area over which it is applied. If a force of 1 Newton (N) is applied uniformly to an area of 1 square metre, the stress is $1\,\mathrm{N\,m^{-2}} \equiv 1$ Pascal (Pa). If the force is normal (perpendicular) to the surface, then the stress is termed a *normal stress*; if tangential to the surface, the stress is termed a *shearing stress*. Usually, the force is neither entirely normal nor tangential but is at some arbitrary intermediate angle, in which case it can be resolved into components which are normal and tangential to the surface; so the stress is composed of both normal and shearing components. The sign convention is that tensional stresses are positive and compressional stresses negative.

Now consider a small parallelipiped with sides δx, δy, δz (Fig. A2.1) and imagine that it is being stressed by some external force. On each face, the stresses can be resolved into components in the x, y and z directions. The stresses acting on the shaded face are $-\sigma_{xx}$, $-\sigma_{xy}$ and $-\sigma_{xz}$. The notation is that σ_{xy} refers to the stress σ acting in the y direction on the face which is perpendicular to the x axis. The *normal stress* is thus $-\sigma_{xx}$ and the *shearing stresses* are $-\sigma_{xy}$ and $-\sigma_{xz}$.

If the parallelipiped is to be in *static equilibrium* (not moving), then the stresses on opposite faces must balance, and there must be no net couple which would rotate the parallelipiped. This requires that the stresses on opposite faces be equal in magnitude and opposite in direction. The shearing stresses on opposite faces of the parallelipiped (e.g., $-\sigma_{xy}$ and σ_{xy} on the back and front faces as shown in Fig. A2.1) provide a couple that will rotate the parallelipiped. Since the parallelipiped must not rotate, this couple must be balanced by the couple provided by the shearing stresses $-\sigma_{yx}$ and σ_{yx} acting on the two side faces. This means

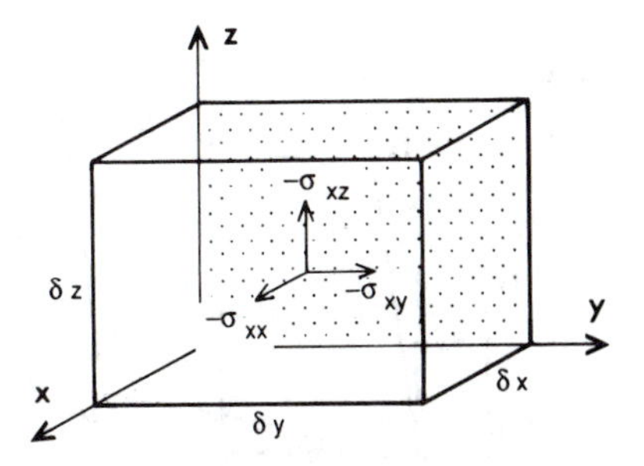

Figure A2.1. Parallelipiped with sides $\delta x, \delta y, \delta z$ in length. The parallelipiped is in static equilibrium. Stresses acting on the shaded back face are $(-\sigma_{xx}, -\sigma_{xy}, -\sigma_{xz})$. Stress acting on the front face are $(\sigma_{xx}, \sigma_{xy}, \sigma_{xz})$.

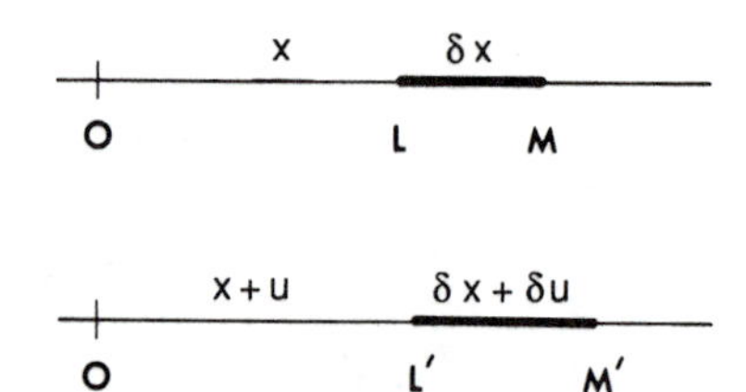

Figure A2.2. An elastic string is stretched: Point O is fixed, L moves to L', M moves to M'.

that σ_{xy} must equal σ_{yx}. The same conditions apply to the other shearing stresses: $\sigma_{xy}=\sigma_{yx}$, $\sigma_{xz}=\sigma_{zx}$, $\sigma_{yz}=\sigma_{zy}$ (i.e., the stress tensor must be symmetric).

Strain

When a body is subjected to stresses, the resulting deformations are called *strains*. Strain is defined as the relative change (i.e., the fractional change) in the shape of the body. First, consider a stress which acts in the x direction only on an elastic string (Fig. A2.2). The point L on the string moves a distance u to point L' after stretching, and point M moves a distance $u+\delta u$ to point M'. The strain in the x direction, termed e_{xx}, is then given by

$$e_{xx}=\frac{\text{change in length of } LM}{\text{original length of } LM}$$

$$=\frac{L'M'-LM}{LM}$$

$$=\frac{\delta x+\delta u-\delta x}{\delta x}$$

$$=\frac{\delta u}{\delta x} \qquad \text{(A2.1)}$$

In the limit when $\delta x\to 0$, the strain at L is

$$e_{xx}=\frac{\partial u}{\partial x} \qquad \text{(A2.2)}$$

To extend the analysis to two dimensions x and y, we must consider the deformation undergone by a rectangle in the x–y plane (Fig. A2.3).

Points L, M, N move to L', M', N' with coordinates

$$L=(x,y), \qquad L'=(x+u,y+v),$$

$$M=(x+\delta x,y), \quad M'=\left(x+\delta x+u+\frac{\partial u}{\partial x}\delta x,\, y+v+\frac{\partial v}{\partial x}\delta x\right)$$

$$N=(x,y+\delta y) \quad N'=\left(x+u+\frac{\partial u}{\partial y}\delta y,\, y+\delta y+v+\frac{\partial v}{\partial y}\delta y\right)$$

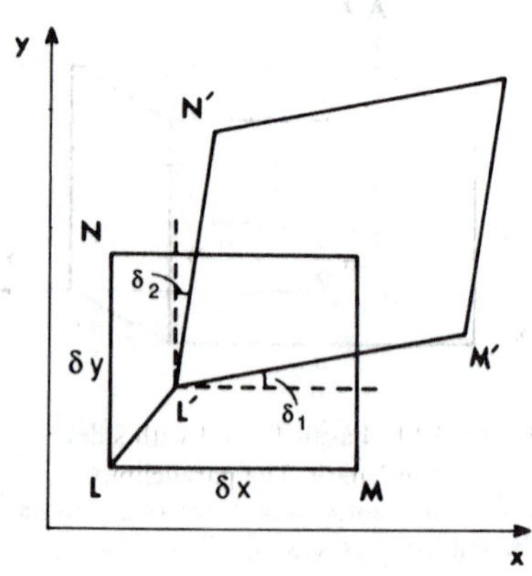

Figure A2.3. Deformation of a rectangle: Point L moves to L', M moves to M' and N moves to N'.

The strain in the x direction e_{xx} is given by

$$e_{xx}=\frac{\text{change in length of } LM}{\text{original length of } LM}$$

$$=\frac{\delta x+\dfrac{\partial u}{\partial x}\delta x-\delta x}{\delta x}$$

$$=\frac{\partial u}{\partial x} \qquad \text{(A2.3)}$$

Likewise, the strain in the y direction is

$$e_{yy} = \frac{\text{change in length of } LN}{\text{original length of } LN}$$

$$= \frac{\partial v}{\partial y} \tag{A2.4}$$

These are called the *normal strains*, the fractional changes in length along the x and y axes. For three dimensions, $e_{zz} = \partial w/\partial z$ is the third normal strain.

As well as changing size, the rectangle undergoes a change in shape (Fig. A2.3). The right angle NLM is reduced by an amount $\delta_1 + \delta_2$ called the *angle of shear*, where

$$\delta_1 + \delta_2 = \frac{\partial v}{\partial x} + \frac{\partial u}{\partial y} \tag{A2.5}$$

(We assume that products of $\partial u/\partial x$, $\partial v/\partial x$ and so on are small enough to be ignored, which is the basis of the *theory of infinitesimal strain.*) The quantity which measures the change in shape undergone by the rectangle is called the *shear component of strain* and is written e_{xy}. In three dimensions, there are six shear components of strain:

$$e_{xy} = e_{yx} = \frac{1}{2}\left(\frac{\partial u}{\partial y} + \frac{\partial v}{\partial x}\right) \tag{A2.6}$$

$$e_{xz} = e_{zx} = \frac{1}{2}\left(\frac{\partial u}{\partial z} + \frac{\partial w}{\partial x}\right) \tag{A2.7}$$

$$e_{zy} = e_{yz} = \frac{1}{2}\left(\frac{\partial w}{\partial y} + \frac{\partial v}{\partial z}\right) \tag{A2.8}$$

Note that the angle of shear is equal to twice the shear component of strain. As well as undergoing a change in shape, the whole rectangle is also rotated anticlockwise by an angle $\frac{1}{2}(\delta_1 - \delta_2)$, termed θ_z, where

$$\theta_z = \frac{1}{2}(\delta_1 - \delta_2)$$

$$= \frac{1}{2}\left(\frac{\partial v}{\partial x} - \frac{\partial u}{\partial y}\right) \tag{A2.9}$$

θ_z is an anticlockwise rotation about the z axis.

Extending the theory to three dimensions, the deformation $(\delta u, \delta v, \delta w)$ of any point $(\delta x, \delta y, \delta z)$ can be expressed as a power series, where to first order

$$\delta u = \frac{\partial u}{\partial x}\delta x + \frac{\partial u}{\partial y}\delta y + \frac{\partial u}{\partial z}\delta z$$

$$\delta v = \frac{\partial v}{\partial x}\delta x + \frac{\partial v}{\partial y}\delta y + \frac{\partial v}{\partial z}\delta z \tag{A2.10}$$

$$\delta w = \frac{\partial w}{\partial x}\delta x + \frac{\partial w}{\partial y}\delta y + \frac{\partial w}{\partial z}\delta z$$

Alternatively, Eqs. A2.10 can be split into symmetric and antisymmetric parts:

$$\begin{aligned}\delta u &= e_{xx}\delta x + e_{xy}\delta y + e_{xz}\delta z - \theta_z\delta y + \theta_y\delta z\\ \delta v &= e_{xy}\delta x + e_{yy}\delta y + e_{yz}\delta z + \theta_z\delta x - \theta_x\delta z\\ \delta w &= e_{xz}\delta x + e_{yz}\delta y + e_{zz}\delta z - \theta_y\delta x + \theta_x\delta y\end{aligned} \tag{A2.11}$$

where*

$$\begin{aligned}\theta_x &= \frac{1}{2}\left(\frac{\partial w}{\partial y} - \frac{\partial v}{\partial z}\right)\\ \theta_y &= \frac{1}{2}\left(\frac{\partial u}{\partial z} - \frac{\partial w}{\partial x}\right)\\ \theta_z &= \frac{1}{2}\left(\frac{\partial v}{\partial x} - \frac{\partial u}{\partial y}\right)\end{aligned} \tag{A2.12}$$

In more compact matrix form, Eq. A2.11 is

$$(\delta u, \delta v, \delta w) = \begin{pmatrix} e_{xx} & e_{xy} & e_{xz}\\ e_{xy} & e_{yy} & e_{yz}\\ e_{xz} & e_{yz} & e_{zz}\end{pmatrix}\begin{pmatrix}\delta x\\ \delta y\\ \delta z\end{pmatrix} + \begin{pmatrix} 0 & -\theta_z & \theta_y\\ \theta_z & 0 & -\theta_x\\ -\theta_y & \theta_x & 0\end{pmatrix}\begin{pmatrix}\delta x\\ \delta y\\ \delta z\end{pmatrix} \tag{A2.13}$$

Strain is a dimensionless quantity. Generally in seismology, the strain caused by the passage of a seismic wave is about 10^{-6} in magnitude.

The fractional increase in volume caused by a deformation is called *cubical dilatation* and is written Δ. The volume of the original rectangular parallelipiped is V, where

$$V = \delta x\,\delta y\,\delta z \tag{A2.14}$$

The volume of the deformed parallelipiped $V + \delta V$ is approximately

$$V + \delta V = (1 + e_{xx})\delta x(1 + e_{yy})\delta y(1 + e_{zz})\delta z \tag{A2.15}$$

The cubical dilatation Δ is then given by

$$\begin{aligned}\Delta &= \frac{\text{change in volume}}{\text{original volume}}\\ &= \frac{V + \delta V - V}{V}\\ &= \frac{(1 + e_{xx})(1 + e_{yy})(1 + e_{zz})\delta x\,\delta y\,\delta z - \delta x\,\delta y\,\delta z}{\delta x\,\delta y\,\delta z}\end{aligned} \tag{A2.16}$$

Therefore, to first order (recall that assumption of infinitesimal strain means that products of strains can be neglected), the cubical dilatation is

*The curl of the vector (u, v, w), $\nabla \wedge \mathbf{u}$, is equal to twice the roation $(\theta_x, \theta_y, \theta_z)$, as discussed in Appendix 1.

given by

$$\Delta = e_{xx} + e_{yy} + e_{zz}$$

or

$$\Delta = \frac{\partial u}{\partial x} + \frac{\partial v}{\partial y} + \frac{\partial w}{\partial z}$$

or

$$\Delta = \nabla \cdot \mathbf{u} \tag{A2.17}$$

Relationship between Stress and Strain

In practice, in a given situation, we want to calculate the strains when the stress is known. In 1676, the English physicist Robert Hooke proposed that, for small strains, any strain is proportional to the stress that produces it. This is known as *Hooke's law* and forms the basis of the theory of perfect elasticity. In one dimension x, Hooke's Law means that

$$\sigma_{xx} = c e_{xx}$$

where c is a constant. Extending the theory to three dimensions gives us 36 different constants:

$$\begin{aligned} \sigma_{xx} &= c_1 e_{xx} + c_2 e_{xy} + c_3 e_{xz} + c_4 e_{yy} + c_5 e_{yz} + c_6 e_{zz} \\ &\vdots \\ \sigma_{zz} &= c_{31} e_{xx} + c_{32} e_{xy} + c_{33} e_{xz} + c_{34} e_{yy} + c_{35} e_{yz} + c_{36} e_{zz} \end{aligned} \tag{A2.18}$$

If we assume that we are considering only isotropic materials (materials with no directional variation), the number of constants is reduced to two:

$$\begin{aligned} \sigma_{xx} &= (\lambda + 2\mu) e_{xx} + \lambda e_{yy} + \lambda e_{zz} \\ &= \lambda\Delta + 2\mu e_{xx} \\ \sigma_{yy} &= \lambda\Delta + 2\mu e_{yy} \\ \sigma_{zz} &= \lambda\Delta + 2\mu e_{zz} \\ \sigma_{xy} &= \sigma_{yx} = 2\mu e_{xy} \\ \sigma_{xz} &= \sigma_{zx} = 2\mu e_{xz} \\ \sigma_{yz} &= \sigma_{zy} = 2\mu e_{yz} \end{aligned} \tag{A2.19}$$

The constants λ and μ are known as the Lamé elastic constants (named after the nineteenth century French mathematician G. Lamé). In suffix notation, Eqs. A2.19 are written as

$$\sigma_{ij} = \lambda\Delta\delta_{ij} + 2\mu e_{ij} \qquad \text{for} \quad i, j = x, y, z \tag{A2.20}$$

where the *Kronecker delta*

$$\begin{aligned} \delta_{ij} &= 1 \qquad \text{where } i = j \\ &= 0 \qquad \text{where } i \neq j \end{aligned}$$

The Lamé elastic constant μ ($\mu = \sigma_{xy}/2e_{xy}$ from Eq. A2.19) is a measure of the resistance of a body to shearing strain and is often termed the *shear modulus* or the *rigidity modulus*. The shear modulus of a liquid or gas is zero.

Besides the Lamé elastic constants, other elastic constants are also used: Young's modulus E, Poisson's ratio σ (no subscripts) and the bulk modulus K.

Young's modulus E is the ratio of tensional stress to the resultant longitudinal strain for a small cylinder under tension at both ends. Let the tensional stress act in the x direction on the end face of the small cylinder, and let all the other stresses be zero. Equations A2.19 then give

$$\begin{aligned} \sigma_{xx} &= \lambda\Delta + 2\mu e_{xx} \\ 0 &= \lambda\Delta + 2\mu e_{yy} \\ 0 &= \lambda\Delta + 2\mu e_{zz} \end{aligned} \tag{A2.21}$$

and

$$0 = e_{xy} = e_{xz} = e_{yz} \tag{A2.22}$$

Adding Eqs. A2.21 gives

$$\sigma_{xx} = 3\lambda\Delta + 2\mu\Delta \tag{A2.23}$$

Substituting Eq. A2.23 into Eq. A2.21 gives

$$e_{xx} = (\lambda + \mu)\frac{\Delta}{\mu} \tag{A2.24}$$

Hence, Young's modulus is

$$E = \frac{\sigma_{xx}}{e_{xx}} = \frac{(3\lambda + 2\mu)\Delta\mu}{(\lambda + \mu)\Delta} = \frac{(3\lambda + 2\mu)\mu}{(\lambda + \mu)} \tag{A2.25}$$

Poisson's ratio σ (named after the nineteenth century French mathematician Siméon Denis Poisson) is defined as the negative of the ratio of the fractional lateral contraction to the fractional longitudinal extension for the same small cylinder under tension at both ends. Using Eqs. A2.23 and A2.21, Poisson's ratio is given by

$$\sigma = -\frac{e_{zz}}{e_{xx}} = \frac{\lambda\Delta}{2\mu}\,\frac{\mu}{(\lambda + \mu)\Delta} = \frac{\lambda}{2(\lambda + \mu)} \tag{A2.26}$$

Consider a small body subjected to a hydrostatic pressure (i.e., the body is immersed in a liquid). This pressure causes compression of the body. The ratio of the pressure to the resulting compression is called the *bulk modulus* or *incompressibility* K of the body. For hydrostatic pressure p, the stresses are

$$\begin{aligned} \sigma_{xx} &= \sigma_{yy} = \sigma_{zz} = -p \\ \sigma_{xy} &= \sigma_{xz} = \sigma_{yz} = 0 \end{aligned} \tag{A2.27}$$

Equations A2.19 then give

$$\begin{aligned} -p &= \lambda\Delta + 2\mu e_{xx} \\ -p &= \lambda\Delta + 2\mu e_{yy} \\ -p &= \lambda\Delta + 2\mu e_{zz} \end{aligned} \tag{A2.28}$$

and

$$0 = e_{xy} = e_{xz} = e_{yz} \tag{A2.29}$$

Adding Eqs. A2.28 gives

$$-3p = 3\lambda\Delta + 2\mu\Delta \tag{A2.30}$$

Finally, the bulk modulus is given by

$$\begin{aligned} K &= \frac{\text{pressure}}{\text{compression}} = \frac{\text{pressure}}{-\text{dilatation}} \\ &= \frac{p}{-\Delta} \\ &= \lambda + \tfrac{2}{3}\mu \end{aligned} \tag{A2.31}$$

Using these relations (Eqs. A2.25, A2.26 and A2.31) between the five elastic constants, we can write Eq. A2.19 or A2.20 in terms of any pair of the constants.

Poisson's ratio is dimensionless, positive and less than 0.5 (0.5 for a liquid since then $\mu = 0$). Young's modulus, the Lamé constants and the bulk modulus are all positive and (along with stress and pressure) are all quoted in units of $\mathrm{N\,m^{-2}}$ ($1\,\mathrm{Pa} \equiv 1\,\mathrm{N\,m^{-2}}$). (For rocks, E, K, λ and μ are generally 20–120 GPa). The two Lamé constants have almost the same value for rocks, so the approximation $\lambda = \mu$ is sometimes made. This approximation is called *Poisson's relation.*

Equations of Motion

Let us assume that the stresses on opposite faces of the small parallelipiped illustrated in Figure A2.1 do not exactly balance, so the parallelipiped is not in equilibrium: Motion is possible. In this case, although the stresses on the rear face are $(-\sigma_{xx}, -\sigma_{xy}, -\sigma_{xz})$, the stresses on the front shaded face can be written as $(\sigma_{xx} + \delta\sigma_{xx}, \sigma_{xy} + \delta\sigma_{xy}, \sigma_{xz} + \delta\sigma_{xz})$ (Fig. A2.4). The additional stress $(\delta\sigma_{xx}, \delta\sigma_{xy}, \delta\sigma_{xz})$ can be written as $[(\partial\sigma_{xx}/\partial x)\delta x, (\partial\sigma_{xy}/\partial x)\delta x, (\partial\sigma_{xz}/\partial x)\delta x]$.

Thus, the net force (stress multiplied by area) acting on the two faces perpendicular to the x axis is

$$\begin{aligned} &\left(-\sigma_{xx} + \sigma_{xx} + \frac{\partial\sigma_{xx}}{\delta x}\delta x, \; -\sigma_{xy} + \sigma_{xy} + \frac{\partial\sigma_{xy}}{\partial x}\delta x, \right. \\ &\qquad \left. -\sigma_{xz} + \sigma_{xz} + \frac{\partial\sigma_{xz}}{\partial x}\delta x\right)\delta y\,\delta z \\ &= \left(\frac{\partial\sigma_{xx}}{\partial x}, \frac{\partial\sigma_{xy}}{\partial x}, \frac{\partial\sigma_{xz}}{\partial x}\right)\delta x\,\delta y\,\delta z \end{aligned} \tag{A2.32}$$

and similarly for the other two pairs of faces. The total force acting on the

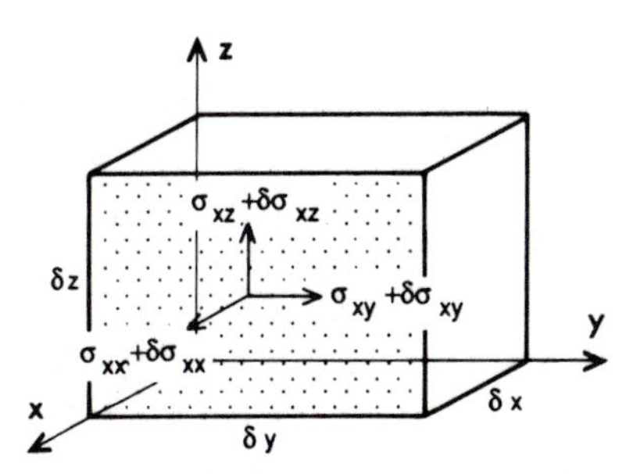

Figure A2.4. Small parallelipiped shown in Figure A2.1. Stresses acting on the front face (shaded) now do not balance those acting on the back face: The parallelipiped is not in equilibrium.

parallelipiped in the x direction thus is

$$\left(\frac{\partial\sigma_{xx}}{\partial x}+\frac{\partial\sigma_{xy}}{\partial y}+\frac{\partial\sigma_{xz}}{\partial z}\right)\delta x\,\delta y\,\delta z$$

and similarly for the force in the y and z directions. Using Newton's second law of motion (force = mass × acceleration), we can write

$$\left(\frac{\partial\sigma_{xx}}{\partial x}+\frac{\partial\sigma_{xy}}{\partial y}+\frac{\partial\sigma_{xz}}{\partial z}\right)\delta x\,\delta y\,\delta z=\rho\,\delta x\,\delta y\,\delta z\frac{\partial^2 u}{\partial t^2}\tag{A2.33}$$

where ρ is the density of the parallelipiped and u is the x component of the displacement. (We assume that all other body forces are zero; that is, gravity does not vary significantly across the parallelipiped.) This equation of motion, which relates the second differential of the displacement to the stress, can be simplified by expressing stress in terms of strain from Eqs. A2.19 and strain in terms of displacement from Eqs. A2.3, A2.6 and A2.7. Substituting for the stress from Eqs. A2.19 into Eqs. A2.33 gives

$$\rho\frac{\partial^2 u}{\partial t^2}=\frac{\partial}{\partial x}\left(\lambda\Delta+2\mu e_{xx}\right)+\frac{\partial}{\partial y}(2\mu e_{xy})+\frac{\partial}{\partial z}(2\mu e_{xz})\tag{A2.34}$$

Substituting for the strains from Eqs. A2.3, A2.6 and A2.7 into Eq. A2.34 gives

$$\rho\frac{\partial^2 u}{\partial t^2}=\frac{\partial}{\partial x}\left(\lambda\Delta+2\mu\frac{\partial u}{\partial x}\right)+\frac{\partial}{\partial y}\left(\mu\left(\frac{\partial v}{\partial x}+\frac{\partial u}{\partial y}\right)\right)+\frac{\partial}{\partial z}\left(\mu\left(\frac{\partial w}{\partial x}+\frac{\partial u}{\partial z}\right)\right)\tag{A2.35}$$

Assuming λ and μ to be constants, we can write

$$\begin{aligned}\rho\frac{\partial^2 u}{\partial t^2}&=\lambda\frac{\partial\Delta}{\partial x}+2\mu\frac{\partial^2 u}{\partial x^2}+\mu\frac{\partial^2 u}{\partial y^2}+\mu\frac{\partial^2 v}{\partial x\,\partial y}+\mu\frac{\partial^2 w}{\partial x\,\partial z}+\mu\frac{\partial^2 u}{\partial z^2}\\&=\lambda\frac{\partial\Delta}{\partial x}+\mu\frac{\partial}{\partial x}\left(\frac{\partial u}{\partial x}+\frac{\partial v}{\partial y}+\frac{\partial w}{\partial z}\right)+\mu\left(\frac{\partial^2 u}{\partial x^2}+\frac{\partial^2 u}{\partial y^2}+\frac{\partial^2 u}{\partial z^2}\right)\\&=\lambda\frac{\partial\Delta}{\partial x}+\mu\frac{\partial\Delta}{\partial x}+\mu\nabla^2 u\end{aligned}\tag{A2.36a}$$

where ∇^2 is the Laplacian operator $\equiv\partial^2/\partial x^2+\partial^2/\partial y^2+\partial^2/\partial z^2$ (see Appendix 1). Likewise, the y and z components of the forces are used to yield equations for v and w:

$$\rho\frac{\partial^2 v}{\partial t^2}=(\lambda+\mu)\frac{\partial\Delta}{\partial y}+\mu\nabla^2 v\tag{A2.36b}$$

$$\rho\frac{\partial^2 w}{\partial t^2}=(\lambda+\mu)\frac{\partial\Delta}{\partial z}+\mu\nabla^2 w\tag{A2.36c}$$

These three equations are the equations of motion for a general disturbance transmitted through a homogeneous, isotropic, perfectly elastic medium, assuming infinitesimal strain and no body forces. We now manipulate these equations to put them into a more useful form.

First, if we differentiate the u, v and w equations with respect to x, y and z, respectively, and add the results, we obtain

$$\rho \frac{\partial^2}{\partial t^2}\left(\frac{\partial u}{\partial x}+\frac{\partial v}{\partial y}+\frac{\partial w}{\partial z}\right)=(\lambda+\mu)\frac{\partial^2 \Delta}{\partial x^2}+\mu \nabla^2\left(\frac{\partial u}{\partial x}\right)$$

$$+(\lambda+\mu)\frac{\partial^2 \nabla}{\partial y^2}+\mu \nabla^2\left(\frac{\partial v}{\partial y}\right)$$

$$+(\lambda+\mu)\frac{\partial^2 \Delta}{\partial z^2}+\mu \nabla^2\left(\frac{\partial w}{\partial z}\right) \tag{A2.37}$$

or

$$\rho \frac{\partial^2 \Delta}{\partial t^2}=(\lambda+\mu)\nabla^2\Delta+\mu \nabla^2\Delta$$

$$=(\lambda+2\mu)\nabla^2\Delta \tag{A2.38}$$

This is a *wave equation* for a dilatational disturbance transmitted through the material with a speed

$$\alpha=\sqrt{\frac{\lambda+2\mu}{\rho}} \tag{A2.39}$$

In seismology, as discussed in Chapter 4, this type of wave involves only dilatation and no rotation and is termed the *primary* or *P-wave.*

Second, we can differentiate Eq. A2.36a with respect to y and Eq. A2.36b with respect to x:

$$\rho \frac{\partial^2}{\partial t^2}\left(\frac{\partial u}{\partial y}\right)=(\lambda+\mu)\frac{\partial^2 \Delta}{\partial x\,\partial y}+\mu \nabla^2\left(\frac{\partial u}{\partial y}\right) \tag{A.2.40}$$

and

$$\rho \frac{\partial^2}{\partial t^2}\left(\frac{\partial v}{\partial x}\right)=(\lambda+\mu)\frac{\partial^2 \Delta}{\partial x\,\partial y}+\mu \nabla^2\left(\frac{\partial v}{\partial x}\right) \tag{A2.41}$$

Subtracting Eq. A2.41 from A2.40 gives

$$\rho \frac{\partial^2}{\partial t^2}\left(\frac{\partial u}{\partial y}-\frac{\partial v}{\partial x}\right)=\mu \nabla^2\left(\frac{\partial u}{\partial y}-\frac{\partial v}{\partial x}\right) \tag{A2.42a}$$

By differentiating and subtracting derivatives, we obtain the other two equations:

$$\rho \frac{\partial^2}{\partial t^2}\left(\frac{\partial u}{\partial z}-\frac{\partial w}{\partial x}\right)=\mu \nabla^2\left(\frac{\partial u}{\partial z}-\frac{\partial w}{\partial x}\right) \tag{A2.42b}$$

$$\rho \frac{\partial^2}{\partial t^2}\left(\frac{\partial v}{\partial z}-\frac{\partial w}{\partial y}\right)=\mu \nabla^2\left(\frac{\partial v}{\partial z}-\frac{\partial w}{\partial y}\right) \tag{A2.42c}$$

However, since $\partial u/\partial y-\partial v/\partial x$ and so on are the components of Curl $\mathbf{u}$ (or $\nabla \wedge \mathbf{u}$; see Appendix 1), these three equations can be written

$$\rho \frac{\partial^2}{\partial t^2}(\mathrm{Curl}\,\mathbf{u})=\mu \nabla^2(\mathrm{Curl}\,\mathbf{u}) \tag{A2.43}$$

This is a vector wave equation for a rotational disturbance transmitted through the material with a speed

$$\beta = \sqrt{\frac{\mu}{\rho}} \tag{A2.44}$$

In seismology, as discussed in Chapter 4, this type of wave involves only rotation and no change in volume and is called the *secondary* or *S-wave*.

Displacement Potentials

We can use the method of Helmholtz to express the displacement **u** as the sum of the gradient of a scalar potential ϕ and the Curl of a vector potential $\boldsymbol{\psi}$. The divergence of the vector potential must be zero: $\nabla \cdot \boldsymbol{\psi} = 0$. The displacement is then expressed as

$$\mathbf{u} = \nabla \phi + \nabla \wedge \boldsymbol{\psi} \tag{A2.45}$$

The two potentials ϕ and $\boldsymbol{\psi}$ are called the *displacement potentials.* Substituting Eq. A2.45 into Eqs. A2.38 and A2.43 and using the vector identities $\nabla \cdot (\nabla \wedge \mathbf{v}) = 0$, $\nabla \wedge (\nabla S) = 0$ and $\nabla \wedge (\nabla \wedge \mathbf{V}) = \nabla(\nabla \cdot \mathbf{V}) - \nabla^2 \mathbf{V}$, where S is a scalar, gives

$$\frac{\partial^2}{\partial t^2}(\nabla^2 \phi) = \left(\frac{\lambda + 2\mu}{\rho}\right) \nabla^2 (\nabla^2 \phi) \tag{A2.46}$$

and

$$\frac{\partial^2}{\partial t^2}(\nabla^2 \boldsymbol{\psi}) = \frac{\mu}{\rho} \nabla^4 \boldsymbol{\psi} \tag{A2.47}$$

The potentials therefore satisfy the wave equations

$$\frac{\partial^2 \phi}{\partial t^2} = \left(\frac{\lambda + 2\mu}{\rho}\right) \nabla^2 \phi \tag{A2.48}$$

and

$$\frac{\partial^2 \boldsymbol{\psi}}{\partial t^2} = \frac{\mu}{\rho} \nabla^2 \boldsymbol{\psi} \tag{A2.49}$$

Equation A2.48 is thus an alternative expression of Eq. A2.38, the wave equation for P-waves, and Eq. A2.49 is an alternative expression of Eq. A2.43, the wave equation for S-waves.

Plane Waves

Consider the case in which ϕ is a function of x and t only. Then Eq. A2.48 simplifies to

$$\frac{\partial^2 \phi}{\partial t^2} = \frac{\lambda + 2\mu}{\rho} \frac{\partial^2 \phi}{\partial x^2} = \alpha^2 \frac{\partial^2 \phi}{\partial x^2} \tag{A2.50}$$

Any function of $x \pm \alpha t$, $\phi = f(x \pm \alpha t)$ is a solution to Eq. A2.50, provided that $\partial \phi / \partial x$, $\partial^2 \phi / \partial x^2$, $\partial \phi / \partial t$ and $\partial^2 \phi / \partial t^2$ are continuous. The simplest harmonic solution to Eq. A2.50 is

$$\phi = \cos \kappa(x - \alpha t) \tag{A2.51}$$

where κ is a constant called the *wave number*. Equation A2.51 describes a plane wave travelling in the x direction with velocity α. The displacement of the medium due to the passage of this wave is given by Eq. A2.45:

$$\begin{aligned}\mathbf{u} &= \nabla \phi \\ &= \left(\frac{\partial \phi}{\partial x}, \frac{\partial \phi}{\partial y}, \frac{\partial \phi}{\partial z} \right) \\ &= (-\kappa \sin \kappa(x - \alpha t), 0, 0)\end{aligned} \tag{A2.52}$$

The velocity at any point $\partial \mathbf{u}/\partial t$ is then given by

$$\frac{\partial \mathbf{u}}{\partial t} = (\alpha \kappa^2 \cos \kappa(x - \alpha t), 0, 0) \tag{A2.53}$$

The wavelength λ, angular frequency ω, frequency f and period T of this wave are given by

$$\begin{aligned}\lambda &= \frac{2\pi}{\kappa} \\ \omega &= \kappa \alpha \\ f &= \frac{\omega}{2\pi} \\ T &= \frac{\lambda}{\alpha} = \frac{2\pi}{\omega} = \frac{1}{f}\end{aligned} \tag{A2.54}$$

APPENDIX 3

Geometry of Ray Paths and Inversion of Earthquake Body Wave Time–Distance Curves

To be able to use the travel-time–distance curves for teleseismic earthquakes (Fig. 4.13) to determine the internal structure of the earth, it is necessary to devise equations relating seismic velocity and depth to travel time and distance.

Initially, consider an earth assumed to consist of spherically symmetric shells, each shell having constant seismic velocity. Consider part of the particular seismic ray (Fig. A3.1) which traverses three of these layers. Applying Snell's law (Sect. 4.4.2) to interface 1 gives

$$\frac{\sin i_1}{v_1} = \frac{\sin j_1}{v_2} \tag{A3.1}$$

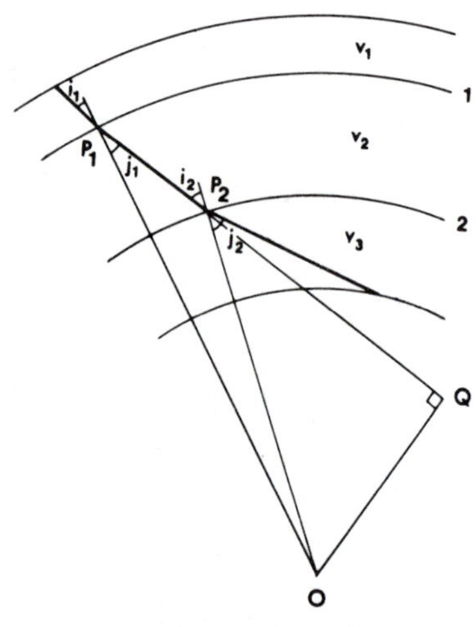

Figure A3.1. Cross section through a symmetric model earth. Geometry of a ray (heavy line) passing through three spherical shells. Each shell has a constant seismic velocity (v_1, v_2 or v_3). P_1 and P_2 are the points where the ray intersects interfaces 1 and 2, respectively. Radius OP_1 is r_1, and radius OP_2 is r_2. O is the centre of the earth. Line P_2Q is the extension of line P_1P_2. Angles of incidence on interfaces 1 and 2 are i_1 and i_2, and angles of refraction are j_1 and j_2, respectively.

and applying it to interface 2 gives

$$\frac{\sin i_2}{v_2} = \frac{\sin j_2}{v_3} \tag{A3.2}$$

However, from the right-angled triangles OP_1Q and OP_2Q, we can write

$$OQ = OP_1 \sin j_1 = r_1 \sin j_1 \tag{A3.3}$$

and

$$OQ = OP_2 \sin i_2 = r_2 \sin i_2 \tag{A3.4}$$

where $OP_1 = r_1$ and $OP_2 = r_2$. Thus, combining Eqs. A3.3 and A3.4, we have

$$r_1 \sin j_1 = r_2 \sin i_2 \tag{A3.5}$$

Multiplying Eq. A3.1 by r_1 and Eq. A3.2 by r_2 and using Eq. A3.5 means that

$$\frac{r_1 \sin i_1}{v_1} = \frac{r_1 \sin j_1}{v_2} = \frac{r_2 \sin i_2}{v_2} = \frac{r_2 \sin j_2}{v_3} \tag{A3.6}$$

At this point we define a parameter p as the *ray parameter*:

$$p = \frac{r \sin i}{v} \tag{A3.7}$$

where r is the distance from the centre of the earth O to any point P, v is the seismic velocity at P and i the angle of incidence at P. Equation A3.6 shows that p is a constant along the ray. At the deepest point to which the ray penetrates (the *turning point*), i is $\pi/2$, so Eq. A3.7 becomes

$$p = \frac{r_{min}}{v} \tag{A3.8}$$

where r_{min} is the radius of the turning point and v the velocity at the point. The value of the ray parameter p is different for each ray.

Now consider two adjacent rays (Fig. A3.2). The shorter ray A_1B_1 subtends an angle Δ at the centre of the earth, and the longer ray A_2B_2 subtends $\Delta + \delta\Delta$. The travel time for ray A_1B_1 is t, and the travel time for ray A_2B_2 is $t + \delta t$. In the infinitesimal right triangle A_1NA_2, the angle A_2A_1N is i_0 and

$$\sin i_0 = \frac{A_2N}{A_2A_1} \tag{A3.9}$$

Assuming that the surface seismic velocity is v_0,

$$A_2N = \tfrac{1}{2}v_0\,\delta t \tag{A3.10}$$

and

$$A_2A_1 = \tfrac{1}{2}r_0\,\delta\Delta \tag{A3.11}$$

Substituting Eqs. A3.10 and A3.11 into Eq. A3.9 gives

$$\sin i_0 = \frac{v_0\,\delta t}{r_0\,\delta\Delta} \tag{A3.12}$$

Comparison with Eq. A3.7 means that in the limit, when $\delta t, \delta\Delta \to 0$,

$$p = \frac{dt}{d\Delta} \tag{A3.13}$$

The ray parameter p is therefore the slope of the travel-time versus epicentral-angle curve (Fig. 4.13) and so, for any particular phase, is an observed function of the epicentral angle.

Let ds be the length of a short segment of a ray, as shown in Figure A3.3. Then, using Pythagoras' theorem on the infinitesimal triangle, we obtain

$$(ds)^2 = (dr)^2 + (r\,d\theta)^2 \tag{A3.14}$$

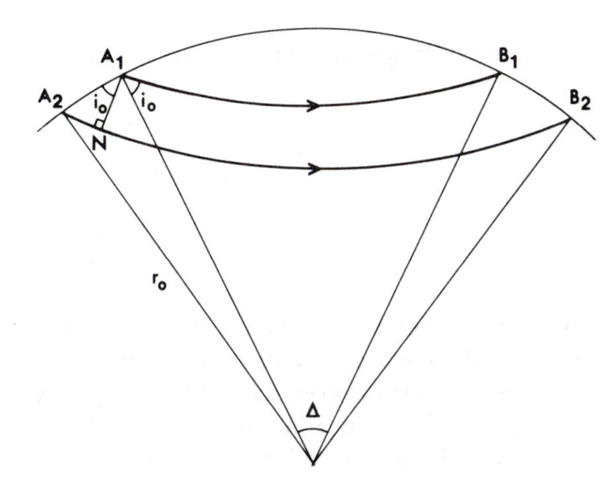

Figure A3.2. Schematic cross section through the earth showing two neighbouring rays A_1B_1 and A_2B_2. O is the centre of the earth, r_0 the radius. Epicentral angle of A_1B_1 is Δ; that of A_2B_2 is $\Delta + \delta\Delta$.

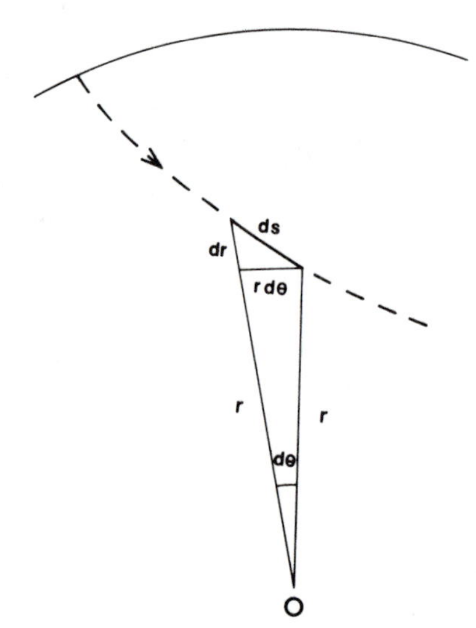

Figure A3.3. Geometry of a small portion, length *ds*, of one ray; O is the centre of the earth.

However, from Eq. A3.7 we have

$$p = \frac{r \sin i}{v} = \frac{r}{v} r \frac{d\theta}{ds} \tag{A3.15}$$

Eliminating *ds* from Eqs. A3.14 and A3.15 gives

$$\frac{r^4 (d\theta)^2}{p^2 v^2} = (dr)^2 + r^2 (d\theta)^2 \tag{A3.16}$$

which, upon rearranging, becomes

$$d\theta = \frac{p\, dr}{r\left(\dfrac{r^2}{v^2} - p^2\right)^{1/2}} \tag{A3.17}$$

Integrating this equation between the surface ($r = r_0$) and the deepest point ($r = r_{min}$) gives an expression for Δ:

$$\Delta = 2p \int_{r=r_{min}}^{r_0} \frac{dr}{r\left(\dfrac{r^2}{v^2} - p^2\right)^{1/2}} \tag{A3.18}$$

Eliminating $d\theta$ from Eqs. A3.14 and A3.15 yields

$$ds = \frac{r\, dr}{v\left(\dfrac{r^2}{v^2} - p^2\right)^{1/2}} \tag{A3.19}$$

The travel time dt along this short ray segment ds is ds/v. Integrating this along the ray between the surface ($r = r_0$) and the deepest point ($r = r_{min}$) gives an expression for t, the total travel time for the ray path:

$$t = 2 \int_{r=r_{min}}^{r_0} \frac{ds}{v} = 2 \int_{r=r_{min}}^{r_0} \frac{r\, dr}{v^2\left(\dfrac{r^2}{v^2} - p^2\right)^{1/2}} \tag{A3.20}$$

Sometimes for convenience another variable η, defined as

$$\eta = \frac{r}{v} \tag{A3.21}$$

is introduced. When this substitution is made, Eqs. A3.18 and A3.20 are written as

$$\Delta = 2p \int_{r=r_{min}}^{r_0} \frac{dr}{r(\eta^2 - p^2)^{1/2}} \tag{A3.22}$$

$$t = 2 \int_{r=r_{min}}^{r_0} \frac{\eta^2\, dr}{r(\eta^2 - p^2)^{1/2}} \tag{A3.23}$$

These two integrals can always be calculated: The travel times and epicentral distances can be calculated even for complex velocity–depth structures involving low velocity or hidden layers.

However, to use the t–Δ curves to determine seismic velocities, it is necessary to change the variable in Eq. A3.22 from r to η, which is possible only when η decreases monotonically with decreasing r:

$$\Delta = 2p \int_{\eta=\eta_m}^{\eta_0} \frac{1}{r(\eta^2 - p^2)^{1/2}} \frac{dr}{d\eta} \, d\eta \qquad \text{(A3.24)}$$

The limits of integration are $\eta_0 = r_0/v_0$ and $\eta_m = r_{min}/v(r = r_{min})$. However, since by Eq. A3.8, $p = r_{min}/v(r = r_{min})$, the lower limit of integration η_m is in fact equal to the ray parameter p for the ray emerging at Δ.

Now at $r = r_1$, where r_1 is any radius for which $r_0 \geqslant r_1 > r_{min}$, let η and v have values η_1 and v_1, respectively. Assume that there is a series of turning rays sampling only the region between r_0 and r_1 with values of p between η_0 ($\eta_0 = r_0/v_0$), which is the ray travelling at a tangent to the earth's surface and hence having $\Delta = 0$, and η_1 ($\eta_1 = r_1/v_1$), which is the ray whose turning point is r_1. Multiplying both sides of Eq. A3.24 by $1/(p^2 - \eta_1^2)^{1/2}$ gives

$$\frac{\Delta}{(p^2 - \eta_1^2)^{1/2}} = \frac{2p}{(p^2 - \eta_1^2)^{1/2}} \int_{\eta=p}^{\eta_0} \frac{1}{r(\eta^2 - p^2)^{1/2}} \frac{dr}{d\eta} \, d\eta \qquad \text{(A3.25)}$$

Now integrate Eq. A3.25 with respect to p between the limits η_1 and η_0:

$$\int_{p=\eta_1}^{\eta_0} \frac{\Delta}{(p^2 - \eta_1^2)^{1/2}} \, dp = \int_{p=\eta_1}^{\eta_0} \frac{2p}{(p^2 - \eta_1^2)^{1/2}} \left[\int_{\eta=p}^{\eta_0} \frac{1}{r(\eta^2 - p^2)^{1/2}} \frac{dr}{d\eta} \, d\eta \right] dp \qquad \text{(A3.26)}$$

It is mathematically permissible to change the order of integration on the right-hand side of Eq. A3.26 from η first and p second to p first and η second:

$$\int_{p=\eta_1}^{\eta_0} \frac{\Delta}{(p^2 - \eta_1^2)^{1/2}} \, dp = \int_{\eta=\eta_1}^{\eta_0} \left[\int_{p=\eta_1}^{\eta} \frac{2p}{r(p^2 - \eta_1^2)^{1/2}(\eta^2 - p^2)^{1/2}} \frac{dr}{d\eta} \, dp \right] d\eta \qquad \text{(A3.27)}$$

Integrating the left-hand side of Eq. A3.27 by parts gives

$$\int_{p=\eta_1}^{\eta_0} \frac{\Delta \, dp}{(p^2 - \eta_1^2)^{1/2}} = \left[\Delta \cosh^{-1}\left(\frac{p}{\eta_1}\right) \right]_{p=\eta_1}^{\eta_0} - \int_{p=\eta_1}^{\eta_0} \frac{d\Delta}{dp} \cosh^{-1}\left(\frac{p}{\eta_1}\right) dp \qquad \text{(A3.28)}$$

since

$$\int \frac{dx}{(x^2 - 1)^{1/2}} = \cosh^{-1}(x)$$

The first term on the right-hand side of Eq. A3.28 is zero because the epicentral angle Δ is zero when $p = \eta_0$, and when $p = \eta_1$, $\cosh^{-1}(p/\eta_1)$ is zero. The second term simplifies to

$$-\int_{\Delta=\Delta_1}^{0} \cosh^{-1}\left(\frac{p}{\eta_1}\right) d\Delta$$

or

$$\int_{\Delta=0}^{\Delta_1} \cosh^{-1}\left(\frac{p}{\eta_1}\right) d\Delta$$

where Δ_1 is the value of Δ for the ray with parameter η_1 (which has its deepest point at $r = r_1$) and $\Delta = 0$ for the ray with parameter η_0.

The right-hand side of Eq. A3.27 is handled by first performing the p integration:

$$\int_{p=\eta_1}^{\eta} \frac{p}{(p^2-\eta_1^2)^{1/2}(\eta^2-p^2)^{1/2}}\,dp$$

Making the substitution $x = p^2$, we obtain

$$\frac{1}{2}\int_{x=\eta_1^2}^{\eta^2} \frac{dx}{(x-\eta_1^2)^{1/2}(\eta^2-x)^{1/2}} = \left[\tan^{-1}\left(\frac{x-\eta_1^2}{\eta^2-x}\right)^{1/2}\right]_{x=\eta_1^2}^{\eta^2}$$

$$= \tan^{-1}(\infty) - \tan^{-1}(0) = \frac{\pi}{2}$$

(Reference works such as the 'Standard Mathematical Tables', ed. by S. M. Selby, Chemical Rubber Company, are invaluable in solving integrals such as these. Alternatively, if we make the substitution $p^2 = \eta_1^2 \sin^2\theta + \eta^2\cos^2\theta$ the integral simplifies to $\int_{\theta=0}^{\pi/2} d\theta$).

We now have the solution to Eq. A3.27:

$$\int_{\Delta=0}^{\Delta_1} \cosh^{-1}\left(\frac{p}{\eta_1}\right) d\Delta = \int_{\eta=\eta_1}^{\eta_0} \frac{2}{r}\frac{dr}{d\eta}\frac{\pi}{2}\,d\eta$$

$$= \pi\int_{r=r_1}^{r_0} \frac{dr}{r} = \pi[\log_e r]_{r=r_1}^{r_0} = \pi \log_e\left(\frac{r_0}{r_1}\right) \qquad \text{(A3.29)}$$

This equation now allows the velocity at any depth to be evaluated from the t–Δ curves provided certain conditions are met. As was shown in Eq. A3.13, p is the slope of the t–Δ curve, and $dt/d\Delta$ is a function of Δ. For chosen values of Δ_1 and η_1 (the value of $dt/d\Delta$ at Δ_1), the integral on the left-hand side of Eq. A3.29 can be evaluated and r_1 determined. Repeating the calculations for all possible values of η_1 means that r_1 is determined as a function of η_1. Recalling from Eq. A3.21 that $\eta = r/v$, this determination means that the seismic velocity has been determined as a function of radius.

Such an inversion (due to Herglotz, Wiechert, Rasch and others and dating from 1907) has been invaluable in enabling us to evaluate the seismic structure of the interior of the earth. It is generally called the *Herglotz–Wiechert inversion.* The main limitations of the method stem from the mathematical restriction that $\eta = r/v$ must decrease with depth (i.e., increase with increasing radius). Thus Eq. A3.29 cannot be used in situations in which r/v increases with depth, which in practice means in low-velocity regions. Within the earth, therefore, the method fails for the upper mantle where there are low-velocity zones and at the core–mantle boundary. Other difficulties occur because of the lack of exact spherical symmetry in the earth and the fact that the time–distance curves are not completely error free (this means that the S-wave structure is less well determined than the P-wave structure since S-wave arrival times are more difficult to pick).

APPENDIX 4

The Least-Squares Method

In geophysics it is often useful to be able to fit straight lines or curves to data (e.g., in radioactive dating and seismology). Although the eye is a good judge of what is and is not a good fit, it is unable to give any numerical estimates of errors. The method of least squares fills this need.

Suppose that $t_1, \ldots, t_n$ are the measured values of t (e.g., for travel times in seismology) corresponding to values $x_1, \ldots, x_n$ of quantity x (e.g., distance). Assume that the x values are accurate but the t values are subject to error. Further assume that we want to find the particular straight line

$$t = mx + c \tag{A4.1}$$

which fits the data best. If we substitute the value $x = x_i$ into Eq. A4.1, the resulting value of t does not equal t_i. There is some error e_i:

$$e_i = mx_i + c - t_i \tag{A4.2}$$

In the least-squares method, the values of m and c are chosen so that the sum of the squares of the errors e_i is least. In other words, $\sum_{i=1}^{n} e_i^2$ is minimized where

$$\sum_{i=1}^{n} e_i^2 = \sum_{i=1}^{n} (mx_i + c - t_i)^2 \tag{A4.3}$$

To minimize this sum, it must be partially differentiated with respect to m, the result equated to zero and the process repeated for c. The two equations are then solved for m and c:

$$\begin{aligned} 0 &= \frac{\partial}{\partial m}\left(\sum_{i=1}^{n} e_i^2\right) \\ &= \frac{\partial}{\partial m}\left[\sum_{i=1}^{n} (mx_i + c - t_i)^2\right] \\ &= \sum_{i=1}^{n} 2x_i(mx_i + c - t_i) \\ &= 2m\sum_{i=1}^{n} x_i^2 + 2c\sum_{i=1}^{n} x_i - 2\sum_{i=1}^{n} x_i t_i \end{aligned}$$

and

$$0 = \frac{\partial}{\partial c}\left(\sum_{i=1}^{n} e_i^2\right) \tag{A4.4}$$

$$= \frac{\partial}{\partial c}\left[\sum_{i=1}^{n}(mx_i + c - t_i)^2\right]$$

$$= \sum_{i=1}^{n} 2(mx_i + c - t_i)$$

$$= 2m\sum_{i=1}^{n} x_i + 2nc - 2\sum_{i=1}^{n} t_i \qquad \text{(A4.5)}$$

Rearranging Eqs. A4.4 and A4.5 gives

$$\sum_{i=1}^{n} x_i t_i = m\sum_{i=1}^{n} x_i^2 + c\sum_{i=1}^{n} x_i \qquad \text{(A4.6)}$$

and

$$\sum_{i=1}^{n} t_i = m\sum_{i=1}^{n} x_i + nc \qquad \text{(A4.7)}$$

Equations A4.6 and A4.7 are simultaneous equations which are solved to give m and c:

$$m = \frac{n(\sum_{i=1}^{n} x_i t_i) - (\sum_{i=1}^{n} x_i)(\sum_{i=1}^{n} t_i)}{n(\sum_{i=1}^{n} x_i^2) - (\sum_{i=1}^{n} x_i)^2} \qquad \text{(A4.8)}$$

$$c = \frac{(\sum_{i=1}^{n} t_i)(\sum_{i=1}^{n} x_i^2) - (\sum_{i=1}^{n} x_i)(\sum_{i=1}^{n} x_i t_i)}{n(\sum_{i=1}^{n} x_i^2) - (\sum_{i=1}^{n} x_i)^2} \qquad \text{(A4.9)}$$

The standard errors in these values of m and c, δm and δc (these are one standard deviation, 1σ, errors), are given by

$$(\delta m)^2 = \frac{n(\sum_{i=1}^{n} e_i^2)}{(n-2)[n(\sum_{i=1}^{n} x_i^2) - (\sum_{i=1}^{n} x_i)^2]} \qquad \text{(A4.10)}$$

and

$$(\delta c)^2 = \frac{(\sum_{i=1}^{n} x_i^2)(\sum_{i=1}^{n} e_i^2)}{(n-2)[n(\sum_{i=1}^{n} x_i^2) - (\sum_{i=1}^{n} x_i)^2]} \qquad \text{(A4.11)}$$

Equations A4.8–A4.11 can easily be programmed. Two standard deviation, 2σ, errors are generally quoted in geochronology.

The least-squares method can be applied also to curve fitting in exactly the same way as is shown here for straight lines. However, it becomes more difficult to solve the simultaneous equations when more than two coefficients are to be determined.

APPENDIX 5

The Error Function

The error function is defined as

$$\operatorname{erf}(x) = \frac{2}{\sqrt{\pi}} \int_{y=0}^{x} e^{-y^2}\, dy \tag{A5.1}$$

It is apparent that

$$\operatorname{erf}(-x) = -\operatorname{erf}(x) \tag{A5.2}$$

and

$$\operatorname{erf}(0) = 0 \tag{A5.3}$$

and

$$\operatorname{erf}(\infty) = 1 \tag{A5.4}$$

The complementary error function $\operatorname{erfc}(x)$ is defined as

$$\begin{aligned}\operatorname{erfc}(x) &= 1 - \operatorname{erf}(x)\\ &= \frac{2}{\sqrt{\pi}} \int_{x}^{\infty} e^{-y^2}\, dy\end{aligned} \tag{A5.5}$$

The error function is tabulated in Table A5.1. An easily programmable approximation to the error function is

$$\operatorname{erf}(x) = 1 - (a_1 t + a_2 t^2 + a_3 t^3)e^{-x^2} + \varepsilon(x) \tag{A5.6}$$

where $t = 1/(1 + 0.47047x)$, $a_1 = 0.3480242$, $a_2 = -0.0958798$ and $a_3 = 0.7478556$. The error in this approximation is $\varepsilon(x) \leqslant 2.5 \times 10^{-5}$. (C. Hastings, *Approximations for digital computers*, Princeton Univ. Press. Princeton, 1955.)

In this text, the error function appears in solutions of the heat conduction equation (see Sect. 7.3.6). In more detailed thermal problems, the solutions may include repeated integrations or derivatives of the error function. For example,

$$\int_{x}^{\infty} \operatorname{erfc}(y)\, dy = \frac{1}{\sqrt{\pi}} e^{-x^2} - x\operatorname{erfc}(x)$$

and

$$\frac{d}{dx}(\operatorname{erf}(x)) = \frac{2}{\sqrt{\pi}} e^{-x^2}$$

Table A5.1. *The error function*

x	erf(x)
0	0
0.05	0.056372
0.10	0.112463
0.15	0.167996
0.20	0.222703
0.25	0.276326
0.30	0.328627
0.35	0.379382
0.40	0.428392
0.45	0.475482
0.50	0.520500
0.55	0.563323
0.60	0.603856
0.65	0.642029
0.70	0.677801
0.75	0.711156
0.80	0.742101
0.85	0.770668
0.90	0.796908
0.95	0.820891
1.00	0.842701
1.1	0.880205
1.2	0.910314
1.3	0.934008
1.4	0.952285
1.5	0.966105
1.6	0.976348
1.7	0.983790
1.8	0.989091
1.9	0.992790
2.0	0.995322
2.5	0.999593
3.0	0.999978

(For more examples see H. S. Carslaw and J. C. Jaeger, *Conduction of heat in solids*, 2nd ed., Oxford Univ. Press, 1959.)

APPENDIX 6

Units and Symbols

Conversion Factors

Time

1 day = 1.44×10^3 minutes (min) = 8.64×10^4 seconds (s)
1 year (a) = 8.76×10^3 hours (hr) = 5.26×10^6 minutes = 3.16×10^7 s
1 Ma = 3.16×10^{13} s
1 Ga = 10^3 Ma = 10^9 yr (a)

Length

1 metre (m) = 100 cm = 10^3 millimetres (mm) = 10^6 micrometres (μm)
= 10^8 angstrom (Å)
1 kilometre (km) = 10^3 m
1 fathom = 6 ft = 1.8288 m
1 nautical mile = 1.852 km

Area

1 m^2 = 10^4 cm^2
1 km^2 = 10^6 m^2

Volume

1 m^3 = 10^3 litres
1 km^3 = 10^{12} litres

Velocity

1 m s^{-1} = 3.6 km hr^{-1}
1 km s^{-1} = 10^3 m s^{-1} = 3.6×10^3 km hr^{-1}

Angle

1 radian (rad) = $57.30°$ = $57°18'$
$1°$ = 0.01745 rad

Mass

1 kilogram (kg) = 1000 grams (g)

Force

1 newton (N) = $1\ kg\,m\,s^{-2} = 10^5$ dynes $= 10^5\ g\,cm\,s^{-2}$

Pressure

1 pascal (Pa) = $1\ N\,m^{-2} = 1\ kg\,m^{-1}\,s^{-2} = 10^{-5}$ bar (b)
$= 10^{-8}$ kilobars (kb)
1 MPa = 10^6 Pa = $10^6\ N\,m^{-2}$
1 GPa = 10^9 Pa = approximate pressure at the base of a 30 km high column of rock
1 atmosphere (atm) = pressure at the base of a 76 cm high column of mercury = 1.013×10^5 Pa

Energy, work, heat

1 joule (J) = $1\ kg\,m^2\,s^{-2} = 10^7$ ergs = 0.2389 calories (cal) = 2.389×10^{-4} kcal
1 kcal = 4185 J

Power

1 watt (W) = 1 joule/second ($J\,s^{-1}$) = $0.2389\ cal\,s^{-1}$
$= 2.389 \times 10^{-4}\ kcal\,s^{-1}$
1 kilowatt (kW) = 1000 W

Heat-flow rate across a surface

$1\ W\,m^{-2} = 2.389 \times 10^{-5}\ cal\,cm^{-2}\,s^{-1}$
$1\ cal\,cm^{-2}\,s^{-1} = 4.18 \times 10^4\ W\,m^{-2}$
1 heat flow unit (hfu) = $10^{-6}\ cal\,cm^{-2}\,s^{-1} = 4.18 \times 10^{-2}\ W\,m^{-2}$

Heat generation rate

$1\ W\,kg^{-1} = 7.54 \times 10^3\ cal\,g^{-1}\,a^{-1}$
$1\ W\,m^{-3} = 2.389 \times 10^{-7}\ cal\,cm^{-3}\,s^{-1}$
$1\ cal\,cm^{-3}\,s^{-1} = 4.18 \times 10^6\ W\,m^{-3}$
1 heat generation unit (hgu) = $10^{-13}\ cal\,cm^{-3}\,s^{-1}$
$= 4.18 \times 10^{-7}\ W\,m^{-3} = 0.418\ \mu W\,m^{-3}$

Thermal conductivity

$1\ W\,m^{-1}\,°C^{-1} = 2.389 \times 10^{-3}\ cal\,cm^{-1}\,s^{-1}\,°C^{-1}$
$1\ cal\,cm^{-1}\,s^{-1}\,°C^{-1} = 4.18 \times 10^2\ W\,m^{-1}\,°C^{-1}$

Specific heat

$1\ J\,kg^{-1}\,°C^{-1} = 2.389 \times 10^{-4}\ cal\,g^{-1}\,°C^{-1}$
$1\ cal\,g^{-1}\,°C^{-1} = 4.18 \times 10^3\ J\,kg^{-1}\,°C^{-1}$

Latent heat

$1\ J\,kg^{-1} = 2.389 \times 10^{-4}\ cal\,g^{-1}$
$1\ cal\,g^{-1} = 4.18 \times 10^3\ J\,kg^{-1}$

Diffusivity

$1\,m^2\,s^{-1} = 10^4\,cm^2\,s^{-1}$
$1\,cm^2\,s^{-1} = 10^{-4}\,m^2\,s^{-1}$

Temperature

degrees Kelvin (K) = degrees Celsius (°C) + 273.16

Density

$1\,kg\,m^{-3} = 10^{-3}\,g\,cm^{-3}$
$1\,g\,cm^{-3} = 10^3\,kg\,m^{-3}$

Dynamic viscosity

1 pascal second (Pa s) = $1\,N\,m^{-2}\,s$
1 Pa s = 10 poise = $10\,g\,cm^{-1}\,s^{-1}$

Kinematic viscosity

$1\,m^2\,s^{-1} = 10^4\,cm^2\,s^{-1}$

Frequency

1 hertz (Hz) = 1 cycle s^{-1}

Magnetic induction

1 tesla (T) = $1\,kg\,amp^{-1}\,s^{-2} = 10^4$ gauss = 10^9 gamma (γ)

Symbols

Symbol	Name	Units	First Eq.
A	Activity	s^{-1}	6.6
A	Radioactive heat generation rate per unit volume	$W\,m^{-3}$	7.7
A	Arrhenius constant		9.26
A	Amplitude		4.13
A	Area		4.22
a	Gravitational acceleration	$m\,s^{-2}$	5.3
a	Area	m^2	7.6
$\mathbf{B}$	Magnetic field	T	3.2
B	Amplitude		4.81
b	Radius	m	5.7
C	Concentration of reactant		9.25
c_p	Specific heat at constant pressure	$J\,kg^{-1}\,°C^{-1}$	7.9
D	Declination	degrees	3.18
D	Distance	m	5.7
D	Compensation depth	m	5.28
D	Flexural rigidity	N m	5.56

Symbol	Name	Unit	First Eq.
D	Number of daughter atoms		6.9
D	Diffusion coefficient	$m^2 s^{-1}$	6.24
d	Depth to sediment bed	m	9.5
d	Depth	m	4.125
d	Ocean depth	km	7.57
d_s	Sediment thickness	m	9.1
d_w	Water depth	m	9.1
D_I	Number of decays by induced fission	km	6.58
D_R	Number of radioactive decays		6.56
D_S	Number of decays by spontaneous fission		6.57
E	Energy	$kg\,m^2 s^{-2}$	4.27
E	Young's modulus	Pa	5.57
E	Activation energy	$J\,mol^{-1}$	6.24
e	Angle		4.80
e	Elevation	m	7.104
e	Strain		A2.1
F	Force	N	5.1
F_{RP}	Ridge-push force per unit length	$N\,m^{-1}$	7.104
F_{SP}	Slab-pull force per unit length	$N\,m^{-1}$	7.105
f	Frequency	s^{-1}	4.7
f	Angle		4.80
f	Ellipticity		5.17
G	Gravitational constant	$m^3 kg^{-1} s^{-2}$	4.32
g	Gravitational acceleration	$m\,s^{-2}$	4.30
ΔG^*	Free energy of activation	$J\,mol^{-1}$	9.7
g_e	Gravitational acceleration at the equator	$m\,s^{-2}$	5.19
g_{rot}	Gravitational acceleration of a rotating sphere	$m\,s^{-2}$	5.18
H	Horizontal component of the earth's magnetic field	T	3.15
H	Horizontal force per unit length	$N\,m^{-1}$	5.56
h	Focal depth		4.12
h	Height	m	5.23
Δh	Geoid height anomaly	m	5.48
H_a	Enthalpy of activation	$J\,mol^{-1}$	9.7
I	Angle of inclination	degrees	3.16
i	Angle		4.58
j	Angle		A3.1
K	Bulk modulus or incompressibility	$Pa \equiv N\,m^{-2}$	4.5
k	Thermal conductivity	$W\,m^{-1}\,°C^{-1}$	7.1
k	Reaction rate coefficient		9.25
L	Skin depth	m	7.41
L	Thickness of the lithosphere	m	7.63
l	Length	m	9.16

Symbol	Name	Unit	First Eq.
$\mathbf{M}$	Induced magnetization	T	3.21
M	Earthquake magnitude		4.13
M	Mass of a sphere	kg	5.15
M	Horizontal bending moment per unit length	N	5.64
$\mathbf{m}$	Dipole moment	$\mathrm{A\,m^2}$	3.1
m	Mass	kg	4.37
m_b	Body wave magnitude		4.17
M_E	Mass of the earth	kg	4.44
M_0	Seismic moment	N m	4.24
M_r	Mass of the earth within a sphere of radius r	kg	4.32
$\mathrm{M_s}$	Surface wave magnitude		4.14
M_W	Moment magnitude		4.23
N	Number		4.24
N	Number of parent atoms		6.1
n	Neutron dose	$\mathrm{cm^{-2}}$	6.58
N_I	Number of induced fission tracks		6.59
N_S	Number of spontaneous fission tracks		6.59
Nu	Nusselt number		7.102
P	Pressure	$\mathrm{Pa \equiv N\,m^{-2}}$	4.30
p	Seismic ray parameter	$\mathrm{s\,degree^{-1}}$	A3.7
$\mathrm{Pe_t}$	Peclet number		7.103
Q	Königsberger ratio		
Q	Quality factor		4.48
Q	Rate of flow of heat per unit area	$\mathrm{W\,m^{-2}}$	7.2
R	Radius of the earth	m	2.3
R	Gas constant	$\mathrm{J\,mol^{-1}\,°C^{-1}}$	6.24
r	Radius	m	3.1
r	Depth of root	m	5.23
Ra	Rayleigh number		7.100
Re	Reynold's number		7.101
R_e	Equatorial radius of the earth	m	5.17
S	Entropy	$\mathrm{J\,kg^{-1}\,°C^{-1}}$	7.86
T	Temperature	°C	6.24
T	Age of the earth		6.61
T	Period	s	4.13
t	Time	s	4.1
t	Thickness	m	5.31
T_p	Potential temperature	°C	7.94a
$T_{1/2}$	Half-life		6.7
TTI	Time–temperature index		9.27
U	Group velocity	$\mathrm{km\,s^{-1}}$	4.8
$\mathbf{u}$	Displacement	m	4.3
$\mathbf{u}$	Velocity	$\mathrm{m\,s^{-1}}$	7.18
V	Magnetic potential	A (amp)	3.1

Symbol	Name	Unit	First Eq.
V	Phase velocity	$km\,s^{-1}$	4.7
V	Volume	m^3	4.36
V	Gravitational potential	$m^2\,s^{-2}$	5.2
V	Vertical force per unit length	$N\,m^{-1}$	5.56
$\mathbf{v}$	Relative velocity	$cm\,yr^{-1}$	2.1
v	Seismic velocity	$km\,s^{-1}$	4.6
v_p	P-Wave velocity	$km\,s^{-1}$	
v_s	S-Wave velocity	$km\,s^{-1}$	
w	Width	m	4.125
$\mathbf{w}$	Vertical deflection	m	5.56
x	Horizontal distance	m	2.21
y	Horizontal distance	m	2.22
Z	Inward radial component of the earth's magnetic field	T	3.14
z	Depth	m	2.23
α	P-wave velocity	$km\,s^{-1}$	4.1
α	Flexural parameter	m	5.61
α	Coefficient of thermal expansion	$°C^{-1}$	4.45
β	Angle	degrees	2.8
β	S-wave velocity	$km\,s^{-1}$	4.2
β	Stretching factor		9.8
Δ	Angular distance	degrees	4.13
Δ	Cubical dilatation		A2.16
δ	Dip		4.67
η	Dynamic viscosity	Pa s	7.97
θ	Angle	degrees	2.3
θ	Colatitude	degrees	
κ	Thermal diffusivity	$m^2\,s^{-1}$	7.43
κ	Wave number		4.25
λ	Latitude	degrees	2.4
λ	Wavelength		
λ	Radioactive decay constant		6.1
λ	Lamé elastic constant	Pa	A2.19
μ	Shear modulus or Lamé elastic constant	Pa	4.4
μ_0	Magnetic permeability of free space	$T\,m\,A^{-1}$	3.2
ν	Kinematic viscosity	$m^2\,s^{-1}$	7.100
ρ	Density	$kg\,m^{-3}$	4.4
$\Delta\sigma$	Stress drop	Pa	4.24
σ	Poisson's ratio		5.57
σ	Neutron capture cross section		6.58
σ	Stress	Pa	A2.18
τ	Temperature difference	°C	4.45
ϕ	Longitude	degrees	2.6
Φ	Seismic parameter		
ϕ	Phase angle		7.42

Symbol	Name	Unit	First Eq.
ϕ	Seismic scalar displacement potential	m^2	4.1
χ	Magnetic susceptibility		3.21
$\boldsymbol{\psi}$	Seismic vector displacement potential	m^2	4.2
$\boldsymbol{\omega}$	Angular velocity	10^{-7} deg yr^{-1}	2.3
ω	Angular frequency		4.26

Multipliers for Powers of Ten

n	nano-	10^{-9}	k	kilo-	10^3
μ	micro-	10^{-6}	M	mega-	10^6
m	milli-	10^{-3}	G	giga-	10^9

Greek Alphabet

Alpha	A	α	Nu	N	ν
Beta	B	β	Xi	Ξ	ξ
Gamma	Γ	γ	Omicron	O	o
Delta	Δ	δ	Pi	Π	π
Epsilon	E	ε	Rho	P	ρ
Zeta	Z	ζ	Sigma	Σ	σ
Eta	H	η	Tau	T	τ
Theta	Θ	θ	Upsilon	Υ	υ
Iota	I	ι	Phi	Φ	ϕ
Kappa	K	κ	Chi	X	χ
Lambda	Λ	λ	Psi	Ψ	ψ
Mu	M	μ	Omega	Ω	ω

APPENDIX 7

Numerical Data

Physical Constraints

Gravitational constant, G	$6.673 \times 10^{-11}\, m^3\, kg^{-1}\, s^{-2}$
Gas constant, R	$8.3145\, J\, mol^{-1}\, {}^{\circ}C^{-1}$
Permeability of free space (vacuum) μ_0	$4\pi \times 10^{-7}\, kg\, m\, amp^{-2}\, s^{-2}$

The Earth

Age of the earth, T	4550 Ma
Angular velocity of the earth	$7.292 \times 10^{-5}\, rad\, s^{-1}$
Mean distance to the sun	1.5×10^{11} km
Average velocity around the sun	$29.77\, km\, s^{-1}$
Length of solar day	8.64×10^4 s
Length of year	3.1558×10^7 s
Equatorial radius, R_{eq}	6378.14 km
Polar radius, R_p	6356.75 km
Polar flattening, f	1/298.247
Radius of outer core	3480 km
Radius of inner core	1221 km
Volume	$1.083 \times 10^{21}\, m^3$
Volume of crust	approx $10^{19}\, m^3$
Volume of mantle	$9.0 \times 10^{20}\, m^3$
Volume of core	$1.77 \times 10^{20}\, m^3$
Mass of the sun	1.99×10^{30} kg
Mass of the moon	7.35×10^{22} kg
Mass of the earth, M_E	5.97×10^{24} kg
Mass of the oceans	1.4×10^{21} kg
Mass of the crust	2.8×10^{22} kg
Mass of the mantle	4.00×10^{24} kg
Mass of the core	1.94×10^{24} kg
Mean density	$5.52 \times 10^3\, kg\, m^{-3}$
Mean density of the mantle	$4.5 \times 10^3\, kg\, m^{-3}$
Mean density of the core	$1.1 \times 10^4\, kg\, m^{-3}$
Equatorial gravity at sea level, g_e	$9.7803185\, m\, s^{-2}$
Polar gravity at sea level, g_p	$9.8321773\, m\, s^{-2}$
Surface area	$5.10 \times 10^{14}\, m^2$

Area of continents and continental shelves	$2.01 \times 10^{14}\,m^2$
Area of oceans and ocean basins	$3.09 \times 10^{14}\,m^2$
Mean depth of the oceans	3.8 km
Mean height of land	0.84 km

Glossary

This is a compilation of some of the technical terms used in this book. For more formal definitions, refer to R. L. Bates and J. A. Jackson, eds., *Glossary of Geology*, American Geological Institute, Falls Church, 1987.

a abbreviation for year

abyssal plain deep, old ocean floor; well sedimented

accreted terrain(ane) terrane which has been accreted to continent

active margin continental margin characterized by volcanic activity and earthquakes (i.e., location of transform fault or subduction zone)

adiabat pressure–temperature path of a body which expands or contracts without giving or receiving heat

aeon (eon) longest division of geological time; also sometimes used for 10^9 years

alpha decay radioactive decay by emission of an alpha particle

alpha particle nucleus of a helium atom (two protons and two neutrons)

altered rocks rocks which have undergone changes in their chemical or mineral structure since they were formed

amphibolite intermediate-grade metamorphic rock; temperature attained above 400–450°C; characterized by amphibole minerals such as hornblende

andesite extrusive igneous rock usually containing plagioclase and mafic phase(s); about 55% SiO_2

anticline a fold, convex upwards, whose core contains stratigraphically older rocks

Archaean (Archean) division of geological time prior to $\sim$2500 Ma ago

aseismic region region with very infrequent earthquakes

asthenosphere region beneath the lithosphere where deformation is dominantly plastic and heat is transferred mainly by convection; now sometimes means the entire upper mantle beneath the lithosphere; Literally, the 'sick' or 'weak' sphere

atomic number number of protons in the nucleus of an atom

backarc basin basin behind the volcanic arc of a subduction zone

band-pass filter filters a signal to retain only those frequencies within the required range, eg., 5–40 Hz

basalt mafic igneous rock

basement rock continental crust which provides the substrate for later deposition

basin depression in which sediments collect

batholith large body of igneous rock, several kilometres thick and extending over areas up to thousands of square kilometres

bathymetry depth of the seabed

beta decay radioactive decay by emission of an electron

beta particle electron

blueschist low-grade metamorphic rock; formed at lower temperatures and higher pressures than greenschist; characterized by blue minerals

body wave seismic wave that travels through the interior of the earth; P-waves are longitudinal body waves; S-waves are transverse body waves

body-wave magnitude magnitude of an earthquake as estimated from the amplitude of body waves

bulk modulus (*K*) bulk property of a material; equal to the pressure acting on a sample divided by the resultant fractional decrease in volume of that sample

cation positively charged ion

centripetal acceleration acceleration, of a body with a circular motion, towards the centre of the circle; proportional to the square of the body's velocity and inversely proportional to the radius of the circle; depending on one's point of view, also called centrifugal acceleration

CHUR chondritic uniform reservoir (see Sect. 6.8)

compensation depth depth at which the overlying rocks are assumed to exert a constant pressure; below this depth, there are no large lateral variations in density

conduction transfer of heat by molecular collisions

continental rise part of the continental margin between the continental slope and the abyssal plain; slopes generally 1:40 to 1:2000

continental shelf part of the continental margin between the coast and the continental slope; slopes about 0.1°

continental slope part of the continental margin between the continental rise and the continental shelf; slopes about 3–6°

convection transfer of heat by the physical movement of molecules from one place to another; hot, less-dense fluid rises and cool denser fluid sinks

cordillera mountains of western North America from the Rocky mountains to the west coast

core iron-rich centre of the earth, 2885–6371 km below the surface

cosmic rays atomic nuclei, largely protons, travelling at or near relativistic speeds

craton (1) large stable part of a continent which has not been subject to deformation for a very long time (e.g., since Precambrian); (2) distinct, tectonically coherent, large region of granitoid crust

cumulate term applied to rocks formed by the accumulation (e.g., by precipitation) of crystals

Curie point or temperature temperature above which a mineral cannot be permanently magnetized

declination horizontal angle between geographic north and magnetic north

decollement zone detachment (unsticking) zone between strata, due to deformation such as folding or thrusting

dehydration loss of water

density ratio of the mass of a material to its volume; usual symbol, ρ

depleted mantle mantle which has been depleted by processes such as partial melting; residue after extraction of crust

dextral fault *see* right lateral fault

diapir body of light material (e.g., salt, magma) which pierces upwards into overlying strata

differentiate mathematical term; rock formed by magmatic differentiation (e.g., precipitation of crystals)

dip-slip fault fault on which the movement is parallel to the dip of the fault; thrust fault or normal fault

dunite rock dominantly composed of olivine

dyke small igneous body which has intruded into fissures which cut across the existing rock strata

earthquake sudden violent movement within the crust or upper mantle

earthquake epicentre point on the earth's surface immediately above the earthquake focus

earthquake focus location of the earthquake within the earth

eclogite type of dense rock formed by metamorphism of basalt

elastic limit maximum stress a body can withstand without being permanently deformed

electron capture absorption of an electron by a nucleus

endothermic reaction chemical reaction that requires heat in order to take place

erosion process by which rock is worn away and the material removed

extrusion eruption of magma

exothermic reaction chemical reaction that releases heat

facies character of a rock; can be applied to the appearance, composition or physical environment in which the rock originated

fault fracture in a rock body, along which motion has occurred or is still occurring

feldspar mineral family $(XAl)(AlSi)_3O_8$, where $X = K$, Na, Ca; plagioclase feldspars have $X = Na$, Ca; alkali feldspars have $X = Na$, K

forearc basin sedimentary basin on the trench side of the volcanic arc of a subduction zone

foreland basin sedimentary basin on the trench side of a continent–continent collision zone

fractionation separation of components in a magma (e.g., by precipitation of crystals)

free oscillation vibration of the whole earth after a major earthquake

frequency number of oscillations per unit time; unit is Hertz (Hz), which equals 1 cycle per second

Ga 10^9 years, giga-annum, (ago)

gabbro coarse-grained intrusive equivalent of basalt; composed of calcic plagioclase, clinopyroxene, with or without olivine, and/or orthopyroxene

gamma decay radioactive decay by emission of a gamma ray or photon; a short-wavelength electromagnetic wave

geotherm temperature–depth curve in the earth

gneiss foliated metamorphic rock rich in quartz and feldspar; can be derived from igneous or sedimentary rocks

Gondwanaland former continent which comprised present-day South America, Africa, India, Australia and Antarctica; started to break up about 150 Ma ago

granite type of plutonic rock with quartz together with plagioclase-feldspar, the plagioclase making up between 10 and 65% of total feldspar

granitoid family of granite-like rocks

granodiorite type of plutonic rock with quartz together with plagioclase-feldspar, the plagioclase making up between 65 and 90% of total feldspar

granulite high-grade (anhydrous) metamorphic rock; formed at temperatures of 700°C or more and depths of 10 km or more, sometimes much more

gravimeter instrument used to measure variations in the gravitational field

gravity anomaly difference between the observed (measured) value of gravity and the theoretical value (e.g., the Bouguer anomaly, free-air anomaly, isostatic anomaly)

greenschist low-grade metamorphic rock; temperature attained 300°C or more, characterized by presence of green minerals

greenstone belt belt of rocks usually including volcanic rocks and sediments at low to moderate metamorphic grade, often surrounded by granite or gneiss

Gutenberg discontinuity discontinuity in seismic velocity that marks the boundary between the core and the mantle; named after seismologist Beno Gutenberg

half-life time required for half of a given number of atoms of a radioactive element to decay

harzburgite peridotite composed of olivine and orthopyroxene

heat-producing element U, K, Th in the modern earth, also Al and Pu in the early earth

high-pass filter filters a signal to remove frequencies below a given frequency, passes only the high frequencies

hot spot localized region characterized by volcanism, high heat flow and uplift; believed to result from hot material rising from depth in the mantle

hydration incorporation of water into the mineral structure

hydrothermal circulation of hot water in rock (e.g., around a hot igneous body or a volcano)

igneous rocks rocks that were once molten but have cooled and solidified

inclination angle at which magnetic field lines dip

incompatible element element that enters the melt during partial melting (e.g., Ti, Zr, Y)

inner core central solid region of the earth's core, probably mostly iron; radius about 1221 km, discovered by Inge Lehmann in 1936

instantaneous velocity velocity of a body at a specific instant

interplate earthquake earthquake with its focus on a plate boundary

intracontinental sedimentary basin sedimentary basin which formed within a continent

intraplate earthquake earthquake with its focus within a plate

intrusion movement of magma into country rock

intrusive rock igneous rock which solidified in the earth

ion atom which is not electrically neutral; an atom with fewer than the normal number of electrons is a positive ion; an atom with more than the normal number of electrons is a negative ion

island arc chain of islands above a subduction zone (e.g., Japan, Aleutians)

isostasy the way in which the lithosphere 'floats' on the asthenosphere

isotherm line or surface of constant temperature

isotopes forms of an element with differing numbers of neutrons in the nucleus. For hydrogen, for example, the nucleus of hydrogen, ^{1}H, consists of one proton; the nucleus of the hydrogen isotope deuterium, ^{2}H, consists of one proton and one neutron; the nucleus of tritium, ^{3}H, has one proton and two neutrons.

kb, kbar kilobar, 0.1 GPa of pressure

kerogen organic material which can be transformed into oil and gas

kimberlite type of peridotite which erupts from great depth in a so-called pipe and which occasionally contains diamonds

kinematic viscosity dynamic viscosity of a fluid divided by its density; unit of kinematic viscosity, $m^2 s^{-1}$; symbol, ν.

kinetic energy energy which something has by virtue of its motion

komatiite highly magnesian lava

large ion lithophile (LIL) elements elements such as Rb, K, Ba, Th, Sr which have large ionic radii

latent heat or heat of fusion the amount of heat which must be supplied to a unit quantity of a given solid material at its melting point to change it into a liquid; conversely, the (same) amount of heat released when a unit quantity of the given liquid material at melting temperature is changed into a solid

latitude angular distance north or south of the equator of a point on the earth's surface

Laurasia former continent which comprised present-day North America, Greenland Europe and part of Asia; broke up during the Cretaceous, about 90 Ma ago.

lava molten rock when it is erupted at the earth's surface

left lateral fault, sinistral fault fault on which the displacement is such that, to an observer on the ground, the opposite side is displaced to the left

lherzolite peridotite composed of olivine, clinopyroxene and orthopyroxene

liquidus surface in temperature–pressure space above which the system is totally liquid

lithology physical character of rocks

lithophile element element usually found in silicate minerals

lithosphere the rigid outer skin of the earth, which includes the crust and the uppermost mantle, its base is defined either mechanically by strength or thermally as the level above which heat is dominantly transferred by conduction

LLLAMA large laterally linked Archaean magma anomaly; see Sect. 9.5

longitude angular distance east or west of the prime meridian (which passes through Greenwich, U.K.) of a point on the earth's surface

low-pass filter filters the signal to remove frequencies above a given frequency, passes only the low frequencies

Ma million years, mega-annum (million years ago)

mafic rock rock rich in magnesium (*ma*) and iron (*fic*) minerals

magma molten rock when it is within the earth

magmatism development and movement of magma

magnetic anomaly difference between the observed (measured) and the theoretical values of the magnetic field

magnetic poles ends of a permanent bar magnet; by convention, lines of force leave the north pole of the magnet and enter its south pole

magnetic variation *see* declination

magnetometer instrument that measures the earth's magnetic field

major elements the major components of the earth's crust: O, Si, Al, Fe, Ca, Na, K and Mg; in geochemical work, also Ti and P

mantle solid shell of the earth extending from the crust to the core; divided into the upper mantle (from the Moho down to 670 km depth) and the lower mantle (from 670 km depth to 2891 km depth of the core–mantle boundary)

metamorphic grade recorded temperature of metamorphism; sometimes, but less correctly, used to imply pressure

metamorphic rocks rocks which have been changed from their original state, usually by temperature or pressure

metamorphism changes in mineralogy and texture of a rock caused by temperature and pressure

minerals crystals which make up rocks

Moho seismic boundary between the crust and the mantle; named after A. Mohorovičić

molasse sedimentary rocks which are deposited in a forearc basin in front of a thrust-up mountain belt; often includes thick sandstones, conglomerates and shales

moment magnitude magnitude of an earthquake as estimated by using the seismic moment

momentum linear momentum of a body, the product of its mass and velocity; angular momentum of a body, the product of its moment of inertia and angular velocity

neutrino particle with no mass and no charge, but which can possess both energy and momentum; emitted in the decay of some elementary particles

nuclear fission splitting of a nucleus into fragments, with release of energy

obduction process which involves breaking off a piece of a subducting plate and thrusting it upwards onto the overriding plate; ophiolites were obducted into their present position

offset horizontal distance

olivine mineral $(Mg, Fe)_2SiO_4$

ophiolite fragment of crust and upper mantle rocks, now exposed on land, analogous to oceanic crust and mantle

orogen region or belt of rocks which have been deformed together at a particular time; literally, mountain creation.

orthopyroxene mineral $(Mg,Fe)_2Si_2O_6$

outer core outer liquid shell of the earth's core, probably iron with some oxygen; inner radius, 1221 km; outer radius, 3480 km

Pangaea the single supercontinent of the Permian and Triassic which broke up during the Jurassic

partial melting melting of part of a rock; because a rock is composed of different minerals, each with its own melting behaviour, melting does not take place at one temperature (as for ice at 0°C) but takes place over a range of temperatures; melting starts at the solidus temperature and continues, nonlinearly, as the temperature increases to the liquidus temperature when the rock is totally molten

passive margin continental margin formed during initial rifting apart of continents to form an ocean; frequently has thick sedimentary deposits

peridotite rock with over 90% mafic minerals, usually dominated by olivine

period the time taken for one complete wave to pass any point; the inverse of the frequency

phase change change of minerals from one crystallographic structure to another more compact form at the increased temperatures and pressures within the earth, or reverse

pillow lava lava showing pillow-like features, erupted underwater

plate tectonics the 'carpentry' or 'architecture' of the earth's surface; the system of large lithospheric plates which move across the earth's surface as spherical caps; most igneous and tectonic activity occurs along the boundaries between, rather than within, the plates

pluton body of magma which has solidified within the crust

potential energy energy of an entity by virtue of its position

quartz crystals of silicon dioxide, SiO_2

radioactive decay spontaneous decay of nucleus of an atom

range horizontal distance between a shot point and receiver

REE rare earth element; the fifteen metallic elements with atomic numbers 57–71: Ce, Pr, Nd, Pm, Sm, Eu, Gd, Tb, Dy, Ho, Er, Tm, Yb and Lu

refraction bending of rays (light or sound) when they pass from one medium to another

refractory mineral or element that remains solid during partial melting

rift region where the crust has split apart, usually marked by a rift valley (e.g., East African Rift, Rhine Graben)

right lateral fault, dextral fault fault on which the displacement is such that, to an observer on the ground, the opposite side is displaced to the right

scalar quantity which has a magnitude but not a direction

sedimentary rocks rock made up of fragments of other rocks usually resulting from the action of wind, water or ice

seismic discontinuity surface within the earth at which the seismic P-wave and or S-wave velocities change

seismicity distribution of earthquakes in space and time

seismograph, seismometer instrument which measures motion of the ground

seismology study of earthquakes and the passage of seismic waves through the earth

serpentine group of minerals in the family $(Mg, Fe)_3\ Si_2O_5\ (OH)_4$, formed by the hydration of ferromagnesium-rich silicates (e.g., olivine and pyroxene)

serpentinization alteration of olivine to serpentine by hydration

shale fine-grained sedimentary rock

silicate large class of minerals containing silicon and oxygen as SiO_4; make up most of earth's crust

sill small igneous body that has been intruded into and parallel with the existing rock strata

sinistral fault *see* left lateral fault

solidus surface in pressure–temperature space below which the system is wholly solid

specific heat amount of heat necessary to raise the temperature of 1 kg of a substance by 1°C; measured in $J\,kg^{-1}\,°C^{-1}$; symbol, c_p.

strain deformation of a body expressed as change in dimension (e.g., length) divided by original dimension (e.g., length)

stratigraphy classification of strata in sedimentary rocks

streamline line marking the position of a particle in a flowing fluid at successive times

stress force acting on a body, expressed as force per unit area

strike-slip fault fault on which the movement is horizontal and parallel to the strike of the fault

subduction process by which a lithospheric plate sinks into the mantle

subduction zone region where one lithospheric plate descends beneath another into the mantle

subsidence sinking of the surface of the earth; possible causes include loading and thermal contraction during cooling

surface wave seismic wave which travels in the surface layers of the earth; Love waves and Rayleigh waves are surface waves

surface-wave magnitude magnitude of an earthquake as estimated from measurements of the amplitude of surface waves

suspect terrain(ane) terrane which is suspected of having originated far from its present location

syncline a fold, concave upwards, whose core contains stratigraphically younger rocks

tectonics geological process that involves the movement of solid rock; the carpentry of the earth

terrain(ane) geological unit bounded by faults and differing markedly in structure or stratigraphy from its neighbours

tesla the unit of magnetic field; 1 tesla (T) is equal to 1 $kg\,A^{-1}\,s^{-2}$

thermal expansion change in volume of a sample of material divided by its original volume for a 1°C increase in temperature; units are $°C^{-1}$, symbol, α

tholeiitic family name of silica-saturated basalts

tonalite granitoid rock with quartz together with plagioclase-feldspar, the plagioclase making up more than 90% of total feldspar

trace element element that occurs in a concentration of less than 1%

trench long, narrow arcuate depression in the seabed which results from the bending of the lithospheric plate as it descends into the mantle at a subduction zone

triple junction point where three plates meet

ultramafic rock rock ultrarich in magnesium (*ma*) and iron (*fic*) minerals such as olivine and pyroxenes

unconformity rock strata above an unconformity which were deposited or extruded at a much later time, or after erosion or deformation, upon the strata beneath; marks a break or gap in the geologic record

velocity vector quantity which indicates both the speed and the direction a body is moving

vector quantity having both magnitude and direction

viscoelastic material a material which can behave as an elastic solid on a short time scale and as a viscous fluid on a long time scale

viscosity resistance to flow within a fluid; a measure of the internal friction of the fluid, also known as dynamic viscosity; unit of viscosity is $\mathrm{N\,s\,m^{-2} \equiv Pa\,s}$; symbol, η; *see also* kinematic viscosity

volatiles chemical components which go into a vapour phase during igneous and metamorphic processes (e.g., water, CO_2)

volcanic rock igneous rock which solidified at the earth's surface

volcanism geological process which involves the eruption of molten rock

wavefront imaginary surface or line that joins points at which the waves from a source are in phase (e.g., all at a maximum or all at a minimum)

wavelength distance between one crest of a wave and the next; usual symbol, λ

Index